2015 IFC®

INTERNATIONAL Fire Code®

2015 International Fire Code®

First Printing: May 2014

ISBN: 978-1-60983-474-6 (soft-cover edition)
ISBN: 978-1-60983-473-9 (loose-leaf edition)

Date of First Publication: May 30, 2014

PRINTED IN THE U.S.A.

PREFACE

Introduction

Internationally, code officials recognize the need for a modern, up-to-date fire code addressing conditions hazardous to life and property from fire, explosion, handling or use of hazardous materials and the use and occupancy of buildings and premises. The *International Fire Code*®, in this 2015 edition, is designed to meet these needs through model code regulations that safeguard the public health and safety in all communities, large and small.

This comprehensive fire code establishes minimum regulations for fire prevention and fire protection systems using prescriptive and performance-related provisions. It is founded on broad-based principles that make possible the use of new materials and new system designs. This 2015 edition is fully compatible with all of the *International Codes*® (I-Codes®) published by the International Code Council (ICC)®, including the *International Building Code*®, *International Energy Conservation Code*®, *International Existing Building Code*®, *International Fuel Gas Code*®, *International Green Construction Code*®, *International Mechanical Code*®, ICC *Performance Code*®, *International Plumbing Code*®, *International Private Sewage Disposal Code*®, *International Property Maintenance Code*®, *International Residential Code*®, *International Swimming Pool and Spa Code*™, *International Wildland-Urban Interface Code*® and *International Zoning Code*®.

The *International Fire Code* provisions provide many benefits, among which is the model code development process that offers an international forum for fire safety professionals to discuss performance and prescriptive code requirements. This forum provides an excellent arena to debate proposed revisions. This model code also encourages international consistency in the application of provisions.

Development

The first edition of the *International Fire Code* (2000) was the culmination of an effort initiated in 1997 by a development committee appointed by ICC and consisting of representatives of the three statutory members of the International Code Council: Building Officials and Code Administrators International, Inc. (BOCA), International Conference of Building Officials (ICBO) and Southern Building Code Congress International (SBCCI). The intent was to draft a comprehensive set of fire safety regulations consistent with and inclusive of the scope of the existing model codes. Technical content of the latest model codes promulgated by BOCA, ICBO and SBCCI was utilized as the basis for the development, followed by public hearings in 1998 and 1999 to consider proposed changes. This 2015 edition presents the code as originally issued, with changes reflected in the 2003, 2006, 2009 and 2012 editions and further changes approved through the ICC Code Development Process through 2014. A new edition such as this is promulgated every 3 years.

This code is founded on principles intended to establish provisions consistent with the scope of a fire code that adequately protects public health, safety and welfare; provisions that do not unnecessarily increase construction costs; provisions that do not restrict the use of new materials, products or methods of construction; and provisions that do not give preferential treatment to particular types or classes of materials, products or methods of construction.

Adoption

The International Code Council maintains a copyright in all of its codes and standards. Maintaining copyright allows ICC to fund its mission through sales of books, in both print and electronic formats. The *International Fire Code* is designed for adoption and use by jurisdictions that recognize and acknowledge the ICC's copyright in the code, and further acknowledge the substantial shared value of the public/private partnership for code development between jurisdictions and the ICC.

The ICC also recognizes the need for jurisdictions to make laws available to the public. All ICC codes and ICC standards, along with the laws of many jurisdictions, are available for free in a non-downloadable form on the ICC's website. Jurisdictions should contact the ICC at adoptions@iccsafe.org to learn how to adopt and distribute laws based on the *International Fire Code* in a manner that provides necessary access, while maintaining the ICC's copyright.

Maintenance

The *International Fire Code* is kept up to date through the review of proposed changes submitted by code enforcing officials, industry representatives, design professionals and other interested parties. Proposed changes are carefully considered through an open code development process in which all interested and affected parties may participate.

The contents of this work are subject to change through both the code development cycles and the governmental body that enacts the code into law. For more information regarding the code development process, contact the Codes and Standards Development Department of the International Code Council.

While the development procedure of the *International Fire Code* ensures the highest degree of care, the ICC, its members and those participating in the development of this code do not accept any liability resulting from compliance or noncompliance with the provisions because the ICC does not have the power or authority to police or enforce compliance with the contents of this code. Only the governmental body that enacts the code into law has such authority.

Code Development Committee Responsibilities (Letter Designations in Front of Section Numbers)

In each code development cycle, proposed changes to the code are considered at the Committee Action Hearings by the International Fire Code Development Committee, whose action constitutes a recommendation to the voting membership for final action on the proposed change. Proposed changes to a code section that has a number beginning with a letter(s) in brackets are considered by a different code development committee. For example, proposed changes to code sections that have [BE] in front of them (e.g., [BE] 607.3) are considered by the appropriate International Building Code Development Committee (IBC – Means of Egress) at the code development hearings.

The content of sections in this code that begin with a letter designation is maintained by another code development committee in accordance with the following:

[A] = Administrative Code Development Committee;

[BE] = IBC – Means of Egress Code Development Committee;

[BF] = IBC – Fire Safety Code Development Committee;

[BG] = IBC – General Code Development Committee;

[BS] = IBC – Structural Code Development Committee;

[EB] = International Existing Building Code Development Committee;

[FG] = International Fuel Gas Code Development Committee;

[M] = International Mechanical Code Development Committee; and

[P] = International Plumbing Code Development Committee.

For the development of the 2018 edition of the I-Codes, there will be three groups of code development committees and they will meet in separate years. Note that these are tentative groupings.

Group A Codes (Heard in 2015, Code Change Proposals Deadline: January 12, 2015)	**Group B Codes (Heard in 2016, Code Change Proposals Deadline: January 11, 2016)**	**Group C Codes (Heard in 2017, Code Change Proposals Deadline: January 11, 2017)**
International Building Code – Fire Safety (Chapters 7, 8, 9, 14, 26) – Means of Egress (Chapters 10, 11, Appendix E) – General (Chapters 2-6, 12, 27-33, Appendices A, B, C, D, K)	Administrative Provisions (Chapter 1 of all codes except IRC and IECC, administrative updates to currently referenced standards, and designated definitions)	International Green Construction Code
International Fuel Gas Code	International Building Code –Structural (Chapters 15-25, Appendices F, G, H, I, J, L, M)	
International Existing Building Code	International Energy Conservation Code	
International Mechanical Code	**International Fire Code**	
International Plumbing Code	International Residential Code – IRC-Building (Chapters 1-10, Appendices E, F, H, J, K, L, M, O, R, S, T, U)	
International Private Sewage Disposal Code	International Wildland-Urban Interface Code	
International Property Maintenance Code		
International Residential Code – IRC-Mechanical (Chapters 12-24) – IRC-Plumbing (Chapters 25-33, Appendices G, I, N, P)		
International Swimming Pool and Spa Code		
International Zoning Code		

Note: Proposed changes to the ICC *Performance Code* will be heard by the code development committee noted in brackets [ ] in the text of the code.

Code change proposals submitted for code sections that have a letter designation in front of them will be heard by the respective committee responsible for such code sections. Because different committees hold code development hearings in different years, proposals for this code will be heard by committees in both the 2015 (Group A) and the 2016 (Group B) code development cycles.

For example, Section 907.2.13.1.2 of this code (and the IBC) is designated as the responsibility of the International Mechanical Code Development Committee. This committee will conduct its code development hearings in 2015 to consider code change proposals in its purview, which includes any proposals to Section 907.2.13.1.2.

Note also that the majority of the sections of Chapter 1 of this code are designated as the responsibility of the Administrative Code Development Committee, and that committee is part of the Group B portion of the hearings. This committee will conduct its code development hearings in 2016 to consider most code change proposals for Chapter 1 of this code and proposals for Chapter 1 of all I-Codes except the *International Energy Conservation Code*, the ICC *Performance Code* and the *International Residential Code*. Therefore, any proposals received for the sections of Chapter 1 precluded by the designation [A] will be deferred for consideration in 2016 by the Administrative Code Development Committee.

It is very important that anyone submitting code change proposals understand which code development committee is responsible for the section of the code that is the subject of the code change proposal. For further information on the code development committee responsibilities, please visit the ICC website at www.iccsafe.org/scoping.

Marginal Markings

Solid vertical lines in the margins within the body of the code indicate a technical change from the requirements of the 2012 edition. Deletion indicators in the form of an arrow (➡) are provided in the margin where an entire section, paragraph, exception or table has been deleted or an item in a list of items or a table has been deleted.

A single asterisk [*] placed in the margin indicates that text or a table has been relocated within the code. A double asterisk [**] placed in the margin indicates that the text or table immediately following it has been relocated there from elsewhere in the code. The following table indicates such relocations in the 2015 edition of the *International Fire Code*.

2012 LOCATION	2015 LOCATION
408.11.3	311.6
408	403
903.3.5.2	914.3.2
908.7	915
1014.3, 1015, 1021	1006
1015.2, 1021.3	1007
1009.3	1019
2311.8	2309.6

Coordination between the International Building and Fire Codes

Because the coordination of technical provisions is one of the benefits of adopting the ICC family of model codes, users will find the ICC codes to be a very flexible set of model documents. To accomplish this flexibility some technical provisions are duplicated in some of the model code documents. While the *International Codes* are provided as a comprehensive set of model codes for the built environment, documents are occasionally adopted as a stand-alone regulation. When one of the model documents is adopted as the basis of a stand-alone code, that code should provide a complete package of requirements with enforcement assigned to the entity for which the adoption is being made.

The model codes can also be adopted as a family of complementary codes. When adopted together there should be no conflict of any of the technical provisions. When multiple model codes are adopted in a jurisdiction it is important for the adopting authority to evaluate the provisions in each code document and determine how and by which agency(ies) they will be enforced. It is important, therefore, to understand that where technical provisions are duplicated in multiple model documents that enforcement duties must be clearly assigned by the local adopting jurisdiction. ICC remains committed to providing state-of-the-art model code documents that, when adopted locally, will reduce the cost to government of code adoption and enforcement and protect the public health, safety and welfare.

Italicized Terms

Selected terms set forth in Chapter 2, Definitions, are italicized where they appear in code text. Such terms are not italicized where the definition set forth in Chapter 2 does not impart the intended meaning in the use of the term. The terms selected have definitions that the user should read carefully to better understand the code.

EFFECTIVE USE OF THE INTERNATIONAL FIRE CODE

The *International Fire Code*® (IFC®) is a model code that regulates minimum fire safety requirements for new and existing buildings, facilities, storage and processes. The IFC addresses fire prevention, fire protection, life safety and safe storage and use of hazardous materials in new and existing buildings, facilities and processes. The IFC provides a total approach of controlling hazards in all buildings and sites, regardless of the hazard being indoors or outdoors.

The IFC is a design document. For example, before one constructs a building, the site must be provided with an adequate water supply for fire-fighting operations and a means of building access for emergency responders in the event of a medical emergency, fire or natural or technological disaster. Depending on the building's occupancy and uses, the IFC regulates the various hazards that may be housed within the building, including refrigeration systems, application of flammable finishes, fueling of motor vehicles, high-piled combustible storage and the storage and use of hazardous materials. The IFC sets forth minimum requirements for these and other hazards and contains requirements for maintaining the life safety of building occupants, the protection of emergency responders, and to limit the damage to a building and its contents as the result of a fire, explosion or unauthorized hazardous material discharge.

Arrangement and Format of the 2015 IFC

Before applying the requirements of the IFC it is beneficial to understand its arrangement and format. The IFC, like other codes published by the International Code Council, is arranged and organized to follow sequential steps that generally occur during a plan review or inspection. In the 2012 edition, the IFC was reorganized into seven parts as illustrated in the tables below. Each part represents a broad subject matter and includes the chapters that logically fit under the subject matter of each part. It is also foreseeable that additional chapters will need to be added in the future as regulations for new processes or operations are developed. Accordingly, the reorganization was designed to accommodate such future chapters by providing reserved (unused) chapters in several of the parts. This will allow the subject matter parts to be conveniently and logically expanded without requiring a major renumbering of the IFC chapters.

ORGANIZATION OF THE IFC	
Parts and Chapters	**Subject Matter**
Part I – Chapters 1 and 2	Administrative and definitions
Part II – Chapters 3 and 4	General safety provisions
Part III – Chapters 5 through 11	Building and equipment design features
Part III – Chapters 12 through 19	Reserved for future use
Part IV – Chapters 20 through 37	Special occupancies and operations
Part IV – Chapters 38 through 49; 52	Reserved for future use
Part V – Chapters 50, 51 and 53 through 67	Hazardous materials
Part V – Chapters 68 through 79	Reserved for future use
Part VI – Chapter 80	Referenced standards
Part VII – Appendices A through M	Adoptable and informational appendices

The IFC requirements for fire-resistive construction, interior finish, fire protection systems, means of egress and construction safeguards are directly correlated to the chapters containing parallel requirements in the IBC, as follows:

IFC Chapter	Subject
7	Fire and smoke protection features
8	Interior finish, decorative materials and furnishings
9	Fire protection systems
10	Means of egress
33	Fire safety during construction and demolition

The following is a chapter-by-chapter synopsis of the scope and intent of the provisions of the *International Fire Code*:

PART I—ADMINISTRATIVE

Chapter 1 Scope and Administration. This chapter contains provisions for the application, enforcement and administration of subsequent requirements of the code. In addition to establishing the scope of the code, Chapter 1 identifies which buildings and structures come under its purview. Chapter 1 is largely concerned with maintaining "due process of law" in enforcing the regulations contained in the body of the code. Only through careful observation of the administrative provisions can the code official reasonably expect to demonstrate that "equal protection under the law" has been provided.

Chapter 2 Definitions. All terms that are defined in the code are listed alphabetically in Chapter 2. While a defined term may be used in one chapter or another, the meaning provided in Chapter 2 is applicable throughout the code.

Where understanding of a term's definition is especially key to or necessary for understanding of a particular code provision, the term is shown in *italics* wherever it appears in the code. This is true only for those terms that have a meaning that is unique to the code. In other words, the generally understood meaning of a term or phrase might not be sufficient or consistent with the meaning prescribed by the code; therefore, it is essential that the code-defined meaning be known.

Guidance regarding tense, gender and plurality of defined terms as well as guidance regarding terms not defined in this code are also provided.

PART II—GENERAL SAFETY PROVISIONS

Chapter 3 General Requirements. The open burning, ignition source, vacant building, miscellaneous storage, roof gardens and landscaped roofs, and hazards to fire fighters requirements and precautions, among other general regulations contained in this chapter, are intended to improve premises safety for everyone, including construction workers, tenants, operations and maintenance personnel, and emergency response personnel. As with other chapters of the *International Fire Code*, Section 302 contains a list of terms that are defined in Chapter 2 and are applicable to the chapter contents.

Chapter 4 Emergency Planning and Preparedness. This chapter addresses the human contribution to life safety in buildings when a fire or other emergency occurs. The requirements for continuous training and scheduled fire, evacuation and lockdown drills can be as important as the required periodic inspections and maintenance of built-in fire protection features. The level of preparation by the occupants also improves the emergency responders' abilities during an emergency. The *International Building Code* (IBC) focuses on built-in fire protection features, such as automatic sprinkler systems, fire-resistance-rated construction and properly designed egress systems, whereas this chapter fully addresses the human element. As with other chapters of the *International Fire Code*, Section 402 contains a list of terms that are defined in Chapter 2 and are applicable to the chapter contents.

PART III—BUILDING AND EQUIPMENT DESIGN FEATURES

Chapter 5 Fire Service Features. The requirements of this chapter apply to all buildings and occupancies and pertain to access roads; access to building openings and roofs; premises identification; key boxes; fire protection water supplies; fire command centers; fire department access to equipment and emergency responder radio coverage in buildings. As with other chapters of the *International Fire Code*, Section 502 contains a list of terms that are defined in Chapter 2 and are applicable to the chapter contents.

Chapter 6 Building Services and Systems. This chapter focuses on building systems and services as they relate to potential safety hazards and when and how they should be installed. This chapter brings together all building system- and service-related issues for convenience and provides a more systematic view of buildings. The following building services and systems are addressed: fuel-fired appliances (Section 603), emergency and standby power systems (Section 604), electrical equipment, wiring and hazards (Section 605), mechanical refrigeration (Section 606), elevator recall and maintenance (Section 607), stationary storage battery systems (Section 608), commercial kitchen hoods (Section 609), commercial kitchen cooking oil storage (610) and hyperbaric facilities (611). As with other chapters of the *International Fire Code*, Section 602 contains a list of terms that are defined in Chapter 2 and are applicable to the chapter contents.

Chapter 7 Fire and Smoke Protection Features. The maintenance of assemblies required to be fire-resistance rated is a key component in a passive fire protection philosophy. Chapter 7 sets forth requirements to maintain required fire-resistance ratings of building elements and limit fire spread. The required maintenance of fire-resistance-rated assemblies and opening protectives is described in Section 703 while Section 704 covers the enclosure requirements for shafts in existing buildings. As with other chapters of the *International Fire Code*, Section 702 contains a list of terms that are defined in Chapter 2 and are applicable to the chapter contents.

Chapter 8 Interior Finish, Decorative Materials and Furnishings. The overall purpose of Chapter 8 is to regulate interior finishes, decorative materials and furnishings in new and existing buildings so that they do not significantly add to or create fire hazards within buildings. The provisions tend to focus on occupancies with specific risk characteristics, such as vulnerability of occupants, density of occupants, lack of familiarity with the building and societal expectations of importance. This chapter is consistent with Chapter 8 of the *International Building Code* (IBC), which regulates the interior finishes of new buildings. As with other chapters of the *International Fire Code*, Section 802 contains a list of terms that are defined in Chapter 2 and are applicable to the chapter contents.

Chapter 9 Fire Protection Systems. Chapter 9 prescribes the minimum requirements for active systems of fire protection equipment to perform the functions of detecting a fire, alerting the occupants or fire department of a fire emergency, controlling smoke and controlling or extinguishing the fire. Generally, the requirements are based on the occupancy, the height and the area of the building, because these are the factors that most affect fire-fighting capabilities and the relative hazard of a specific building or portion thereof. This chapter parallels and is substantially duplicated in Chapter 9 of the *International Building Code;* however, this chapter also contains periodic testing criteria that are not contained in the IBC. In addition, the special fire protection system requirements based on use and occupancy found in Chapter 4 of the IBC are duplicated in Chapter 9 of the IFC as a user convenience. As with other chapters of the *International Fire Code*, Section 902 contains a list of terms that are defined in Chapter 2 and are applicable to the chapter contents.

Chapter 10 Means of Egress. The general criteria set forth in Chapter 10 regulating the design of the means of egress are established as the primary method for protection of people in buildings by allowing timely relocation or evacuation of building occupants. Both prescriptive and performance language is utilized in this chapter to provide for a basic approach in the determination of a safe exiting system for all occupancies. It addresses all portions of the egress system (i.e., exit access, exits and exit discharge) and includes design requirements as well as provisions regulating individual components. The requirements detail the size, arrangement, number and protection of means of egress components. Functional and operational characteristics also are specified for the components that will permit their safe use without special knowledge or effort. The means of egress protection requirements work in coordination with other sections of the code, such as protection of vertical openings (see Chapter 7), interior finish (see Chapter 8), fire suppression and detection systems (see Chapter 9) and numerous others, all having an impact on life safety. Sections 1002 through 1030 are duplicated text from Chapter 10 of the IBC; however, the IFC contains an additional Section 1031 on maintenance of the means of egress system in existing buildings. Retroactive minimum means of egress requirements for existing buildings are found in Chapter 11. As with other chapters of the *International Fire Code*, Section 1002 contains a list of terms that are defined in Chapter 2 and are applicable to the chapter contents.

Chapter 11 Construction Requirements for Existing Buildings. Chapter 11 applies to existing buildings constructed prior to the adoption of the code and intends to provide a minimum degree of fire and life safety to persons occupying existing buildings by providing for alterations to such buildings that do not comply with the minimum requirements of the *International Building*

Code. Prior to the 2009 edition, its content existed in the IFC but in a random manner that was neither efficient nor user-friendly. In the 2007/2008 code development cycle, a code change (F294-07/08) was approved that consolidated the retroactive elements of IFC/2006 Sections 607, 701, 704, 903, 905, 907 and 3406 (then 2506) and all of then-Section 1027 (Means of Egress for Existing Buildings) into a single chapter for easier and more efficient reference and application to existing buildings. As with other chapters of the *International Fire Code*, Section 1102 contains a list of terms that are defined in Chapter 2 and are applicable to the chapter contents.

Chapters 12 through 19. Reserved for future use.

PART IV–SPECIAL OCCUPANCIES AND OPERATIONS

Chapter 20 Aviation Facilities. Chapter 20 specifies minimum requirements for the fire-safe operation of airports, heliports and helistops. The principal nonflight operational hazards associated with aviation involve fuel, facilities and operations. Therefore, safe use of flammable and combustible liquids during fueling and maintenance operations is emphasized. Availability of portable Class B:C-rated fire extinguishers for prompt control or suppression of incipient fires is required. As with other chapters of the *International Fire Code*, Section 2002 contains a list of terms that are defined in Chapter 2 and are applicable to the chapter contents.

Chapter 21 Dry Cleaning. The provisions of Chapter 21 are intended to reduce hazards associated with use of flammable and combustible dry cleaning solvents. These materials, like all volatile organic chemicals, generate significant quantities of static electricity and are thus readily ignitable. Many flammable and nonflammable dry cleaning solvents also possess health hazards when involved in a fire. As with other chapters of the *International Fire Code*, Section 2102 contains a list of terms that are defined in Chapter 2 and are applicable to the chapter contents.

Chapter 22 Combustible Dust-producing Operations. The requirements of Chapter 22 seek to reduce the likelihood of dust explosions by managing the hazards of ignitable suspensions of combustible dusts associated with a variety of operations including woodworking, mining, food processing, agricultural commodity storage and handling and pharmaceutical manufacturing, among others. Ignition source control and good housekeeping practices in occupancies containing dust-producing operations are emphasized. As with other chapters of the *International Fire Code*, Section 2202 contains a list of terms that are defined in Chapter 2 and are applicable to the chapter contents.

Chapter 23 Motor Fuel-dispensing Facilities and Repair Garages. This chapter provides provisions that regulate the storage and dispensing of both liquid and gaseous motor fuels at public and private automotive, marine and aircraft motor fuel-dispensing facilities, fleet vehicle motor fuel-dispensing facilities and repair garages. As with other chapters of the *International Fire Code*, Section 2302 contains a list of terms that are defined in Chapter 2 and are applicable to the chapter contents.

Chapter 24 Flammable Finishes. Chapter 24 requirements govern operations where flammable or combustible finishes are applied by spraying, dipping, powder coating or flow-coating processes. As with all operations involving flammable or combustible liquids and combustible dusts or vapors, controlling ignition sources and methods of reducing or controlling flammable vapors or combustible dusts at or near these operations are emphasized. As with other chapters of the *International Fire Code*, Section 2402 contains a list of terms that are defined in Chapter 2 and are applicable to the chapter contents.

Chapter 25 Fruit and Crop Ripening. Chapter 25 provides guidance that is intended to reduce the likelihood of explosions resulting from improper use or handling of ethylene gas used for crop-ripening and coloring processes. This is accomplished by regulating ethylene gas generation; storage and distribution systems and controlling ignition sources. Design and construction of facilities for this use are regulated by the *International Building Code* to reduce the impact of potential accidents on people and buildings.

Chapter 26 Fumigation and Insecticidal Fogging. This chapter regulates fumigation and insecticidal fogging operations which use toxic pesticide chemicals to kill insects, rodents and other vermin. Fumigants and insecticidal fogging agents pose little hazard if properly applied; however, the inherent toxicity of all these agents and the potential flammability of some makes special precautions necessary when they are used. Requirements of this chapter are intended to protect both the public and fire fighters from hazards associated with these products. As with other chapters of the *International Fire Code*, Section 2602 contains a list of terms that are defined in Chapter 2 and are applicable to the chapter contents.

Chapter 27 Semiconductor Fabrication Facilities. The requirements of this chapter are intended to control hazards associated with the manufacture of electrical circuit boards or microchips, commonly called semiconductors. Though the finished product possesses no unusual hazards, materials commonly associated with semiconductor manufacturing are often quite hazardous and include flammable liquids, pyrophoric and flammable gases, toxic substances and corrosives. The requirements of this chapter are concerned with both life safety and property protection. However, the fire code official should recognize that the risk of extraordinary property damages is far more common than the risk of personal injuries from fire. As with other chapters of the *International Fire Code*, Section 2702 contains a list of terms that are defined in Chapter 2 and are applicable to the chapter contents.

Chapter 28 Lumber Yards and Agro-industrial, Solid Biomass and Woodworking Facilities. Provisions of this chapter are intended to prevent fires and explosions, facilitate fire control and reduce exposures to and from facilities storing, selling or processing wood and forest products, including sawdust, wood chips, shavings, bark mulch, shorts, finished planks, sheets, posts, poles, timber and raw logs and the hazard they represent once ignited. Also included are solid biomass feedstock and raw products associated with agro-industrial facilities. This chapter requires active and passive fire protection features to reduce on- and off-site exposures, limit fire size and development and facilitate fire fighting by employees and the fire service. As with other chapters of the *International Fire Code*, Section 2802 contains a list of terms that are defined in Chapter 2 and are applicable to the chapter contents.

Chapter 29 Manufacture of Organic Coatings. This chapter regulates materials and processes associated with the manufacture of paints as well as bituminous, asphaltic and other diverse compounds formulated to protect buildings, machines and objects from the effects of weather, corrosion and hostile environmental exposures. Paint for decorative, architectural and industrial uses comprises the bulk of organic coating production. Painting and processes related to the manufacture of nonflammable and noncombustible or water-based products are exempt from the provisions of this chapter. The application of organic coatings is covered by Chapter 24. Elimination of ignition sources, maintenance of fire protection equipment and isolation or segregation of hazardous operations are emphasized. As with other chapters of the *International Fire Code*, Section 2902 contains a term that is defined in Chapter 2 and is applicable to the chapter contents.

Chapter 30 Industrial Ovens. This chapter addresses the fuel supply, ventilation, emergency shutdown equipment, fire protection and the operation and maintenance of industrial ovens, which are sometimes referred to as industrial heat enclosures or industrial furnaces. Compliance with this chapter is intended to reduce the likelihood of fires involving industrial ovens which are usually the result of the fuel in use or volatile vapors given off by the materials being heated or to manage the impact if a fire should occur. As with other chapters of the *International Fire Code*, Section 3002 contains a list of terms that are defined in Chapter 2 and are applicable to the chapter contents.

Chapter 31 Tents and Other Membrane Structures. The requirements in this chapter are intended to protect temporary as well as permanent tents and air-supported and other membrane structures and temporary stage canopies from fire and similar hazards by regulating structure location and access, anchorage, egress, heat-producing equipment, hazardous materials and operations, combustible vegetation, ignition sources, waste accumulation and requiring regular inspections and certifying continued compliance with fire safety regulations. As with other chapters of the *International Fire Code*, Section 3102 contains a list of terms that are defined in Chapter 2 and are applicable to the chapter contents.

Chapter 32 High-piled Combustible Storage. This chapter provides guidance for reasonable protection of life from hazards associated with the storage of combustible materials in closely packed piles or on pallets, in racks or on shelves where the top of storage is greater than 12 feet in height. It provides requirements for identifying various classes of commodities; general fire and life safety features including storage arrangements, smoke and heat venting, fire department access and housekeeping and maintenance requirements. The chapter attempts to define the potential fire severity and, in turn, determine fire and life safety protection measures needed to control, and in some cases suppress, a potential fire. This chapter does not cover miscellaneous combustible materials storage regulated in Section 315. As with other chapters of the *International Fire Code*, Section 3202 contains a list of terms that are defined in Chapter 2 and are applicable to the chapter contents.

Chapter 33 Fire Safety during Construction and Demolition. Chapter 33 outlines general fire safety precautions for all structures and all occupancies during construction and demolition operations. In general, these requirements seek to maintain required levels of fire protection, limit fire spread, establish the appropriate operation of equipment and promote prompt response to fire emergencies. Features regulated include fire protection systems, fire fighter access to the site and building, means of egress, hazardous materials storage and use and temporary heating equipment and other ignition sources. With the 2012 reorganization, this chapter now correlates with Chapter 33 of the IBC.

Chapter 34 Tire Rebuilding and Tire Storage. The requirements of Chapter 34 are intended to prevent or control fires and explosions associated with the remanufacture and storage of tires and tire byproducts. Additionally, the requirements are intended to minimize the impact of indoor and outdoor tire storage fires by regulating pile volume and location, segregating the various operations, providing for fire department access and a water supply and controlling ignition sources.

Chapter 35 Welding and Other Hot Work. This chapter covers requirements for safety in welding and other types of hot work by reducing the potential for fire ignitions that usually result in large losses. Several different types of hot work would fall under the requirements found in Chapter 35, including both gas and electric arc methods and any open-torch operations. Many of the activities of this chapter focus on the actions of the occupants. As with other chapters of the *International Fire Code*, Section 3502 contains a list of terms that are defined in Chapter 2 and are applicable to the chapter contents.

Chapter 36 Marinas. Chapter 36 addresses the fire protection and prevention requirements for marinas. It was developed in response to the complications encountered by a number of fire departments responsible for the protection of marinas as well as fire loss history in marinas that lacked fire protection. Compliance with this chapter intends to establish safe practices in marina areas, provide an identification method for mooring spaces in the marina, provide fire fighters with safe operational areas and fire protection methods to extend hose lines in a safe manner. As with other chapters of the *International Fire Code*, Section 3602 contains a list of terms that are defined in Chapter 2 and are applicable to the chapter contents.

Chapter 37 Combustible Fibers. Chapter 37 (which was formerly Chapter 52) establishes the requirements for storage and handling of combustible fibers, including animal, vegetable and synthetic fibers, whether woven into textiles, baled, packaged or loose. Operations involving combustible fibers are typically associated with salvage, paper milling, recycling, cloth manufacturing, carpet and textile mills and agricultural operations, among others. The primary hazard associated with these operations is the abundance of materials and their ready ignitability. As with other chapters of the *International Fire Code*, Section 3702 contains a list of terms that are defined in Chapter 2 and are applicable to the chapter contents.

Chapters 38 through 49. Reserved for future use.

PART V—HAZARDOUS MATERIALS

Chapter 50 Hazardous Materials—General Provisions. This chapter contains the general requirements for all hazardous chemicals in all occupancies. Hazardous chemicals are defined as those that pose an unreasonable risk to the health and safety of operating or emergency personnel, the public and the environment if not properly controlled during handling, storage, manufacture, processing, packaging, use, disposal or transportation. The general provisions of this chapter are intended to be companion provisions with the specific requirements of Chapters 51 through 67 regarding a given hazardous material. As with other chapters of the *International Fire Code*, Section 5002 contains a list of terms that are defined in Chapter 2 and are applicable to the chapter contents.

Chapter 51 Aerosols. Chapter 51 addresses the prevention, control and extinguishment of fires and explosions in facilities where retail aerosol products are displayed or stored. It is concerned with both life safety and property protection from a fire; however, historically, aerosol product fires have caused property loss more frequently than loss of life. Requirements for storing aerosol products are dependent on the level of aerosol product, level of sprinkler protection, type of storage condition and quantity of aerosol products. As with other chapters of the *International Fire Code*, Section 5102 contains a list of terms that are defined in Chapter 2 and are applicable to the chapter contents.

Chapter 52. Reserved for future use.

Chapter 53 Compressed Gases. This chapter regulates the storage, use and handling of all flammable and nonflammable compressed gases, such as those that are used in medical facilities, air separation plants, industrial plants, agricultural equipment and similar occupancies. Standards for the design, construction and marking of compressed gas cylinders and pressure vessels are referenced. Compressed gases used in welding and cutting, cryogenic liquids and liquefied petroleum gases are also regulated under Chapters 35, 55 and 61, respectively. Compressed gases that are classified as hazardous materials are also regulated in Chapter 50, which includes general requirements. As with other chapters of the *International Fire Code*, Section 5302 contains a list of terms that are defined in Chapter 2 and are applicable to the chapter contents.

Chapter 54 Corrosive Materials. Chapter 54 addresses the hazards of corrosive materials that have a destructive effect on living tissues. Though corrosive gases exist, most corrosive materials are solid and classified as either acids or bases (alkalis). These materials may pose a wide range of hazards other than corrosivity, such as combustibility, reactivity or oxidizing hazards, and must conform to the requirements of this code with respect to all their known hazards. The focus of this chapter is on materials whose primary hazard is corrosivity; that is, the ability to destroy or irreparably damage living tissue on contact. As with other chapters of the *International Fire Code*, Section 5402 contains a list of terms that are defined in Chapter 2 and are applicable to the chapter contents.

Chapter 55 Cryogenic Fluids. This chapter regulates the hazards associated with the storage, use and handling of cryogenic fluids through regulation of such things as pressure relief mechanisms and proper container storage. These hazards are in addition to the code requirements that address the other hazards of cryogenic fluids such as flammability and toxicity. These other characteristics are dealt with in Chapter 50 and other chapters, such as Chapter 58 dealing with flammable gases. Cryogens are hazardous because they are held at extremely low temperatures and high pressures. Many cryogenic fluids, however, are actually inert gases and would not be regulated elsewhere in this code. Cryogens are used for many applications but specifically have had widespread use in the biomedical field and in space programs. As with other chapters of the *International Fire Code*, Section 5502 contains a list of terms that are defined in Chapter 2 and are applicable to the chapter contents.

Chapter 56 Explosives and Fireworks. This chapter prescribes minimum requirements for the safe manufacture, storage, handling and use of explosives, ammunition and blasting agents for commercial and industrial occupancies. These provisions are intended to protect the general public, emergency responders and individuals who handle explosives. Chapter 56 also regulates the manufacturing, retail sale, display and wholesale distribution of fireworks, establishing the requirements for obtaining approval to manufacture, store, sell, discharge or conduct a public display, and references national standards for regulations governing manufacture, storage and public displays. As with other chapters of the *International Fire Code*, Section 5602 contains a list of terms that are defined in Chapter 2 and are applicable to the chapter contents.

Chapter 57 Flammable and Combustible Liquids. The requirements of this chapter are intended to reduce the likelihood of fires involving the storage, handling, use or transportation of flammable and combustible liquids. Adherence to these practices may also limit damage in the event of an accidental fire involving these materials. These liquids are used for fuel, lubricants, cleaners, solvents, medicine and even drinking. The danger associated with flammable and combustible liquids is that the vapors from these liquids, when combined with air in their flammable range, will burn or explode at temperatures near normal living and working environment. The protection provided by this code is to prevent the flammable and combustible liquids from being ignited. As with other chapters of the *International Fire Code*, Section 5702 contains a list of terms that are defined in Chapter 2 and are applicable to the chapter contents.

Chapter 58 Flammable Gases and Flammable Cryogenic Fluids. Chapter 58 sets requirements for the storage and use of flammable gases. For safety purposes, there is a limit on the quantities of flammable gas allowed per control area. Exceeding these limitations increases the possibility of damage to both property and individuals. The principal hazard posed by flammable gas is its ready ignitability, or even explosivity, when mixed with air in the proper proportions. Consequently, occupancies storing or handling large quantities of flammable gas are classified as Group H-2 (high hazard) by the *International Building Code*. As with other chapters of the *International Fire Code*, Section 5802 contains a list of terms that are defined in Chapter 2 and are applicable to the chapter contents.

Chapter 59 Flammable Solids. This chapter addresses general requirements for storage and handling of flammable solids, especially magnesium; however, it is important to note that several other solid materials, primarily metals including, but not limited to, such metals as titanium, zirconium, hafnium, calcium, zinc, sodium, lithium, potassium, sodium/potassium alloys, uranium, thorium and plutonium which, under the right conditions, can be explosion hazards. Some of these metals are almost exclusively laboratory materials but because of where they are used, fire service personnel must be trained to handle emergency situations. Because uranium, thorium and plutonium are also radioactive materials, they present still more specialized problems for fire service personnel. As with other chapters of the *International Fire Code*, Section 5902 contains a list of terms that are defined in Chapter 2 and are applicable to the chapter contents.

Chapter 60 Highly Toxic and Toxic Materials. The main purpose of this chapter is to protect occupants, emergency responders and those in the immediate area of the building and facility from short-term, acute hazards associated with a release or general exposure to toxic and highly toxic materials. This chapter deals with all three states of toxic and highly toxic materials: solids, liquids and gases. This code does not address long-term exposure effects of these materials, which are addressed by agencies such as the Environmental Protection Agency (EPA) and Occupational Safety and Health Administration (OSHA). As with other chapters of the *International Fire Code*, Section 6002 contains a list of terms that are defined in Chapter 2 and are applicable to the chapter contents.

Chapter 61 Liquefied Petroleum Gases. Chapter 61 establishes requirements for the safe handling, storing and use of LP-gas to reduce the possibility of damage to containers, accidental releases of LP-gas and exposure of flammable concentrations of LP-gas to ignition sources. LP-gas (notably propane) is well known as a camping fuel for cooking, lighting, heating and refrigerating and also remains a popular standby fuel supply for auxiliary generators as well as being widely used as an alternative motor vehicle fuel. Its characteristic as a clean-burning fuel having resulted in the addition of propane dispensers to service stations throughout the country. As with other chapters of the *International Fire Code*, Section 6102 contains a list of terms that are defined in Chapter 2 and are applicable to the chapter contents.

Chapter 62 Organic Peroxides. This chapter addresses the hazards associated with the storage, handling and use of organic peroxides and intends to manage the fire and oxidation hazards of organic peroxides by preventing their uncontrolled release. These chemicals possess the characteristics of flammable or combustible liquids and are also strong oxidizers. This unusual combination of properties requires special storage and handling precautions to prevent uncontrolled release, contamination, hazardous chemical reactions, fires or explosions. The requirements of this chapter pertain to industrial applications in which significant quantities of organic peroxides are stored or used; however, smaller quantities of organic peroxides still pose a significant hazard and, therefore, must be stored and used in accordance with the applicable provisions of this chapter and Chapter 50. As with other chapters of the *International Fire Code*, Section 6202 contains a list of terms that are defined in Chapter 2 and are applicable to the chapter contents.

Chapter 63 Oxidizers, Oxidizing Gases and Oxidizing Cryogenic Fluids. Chapter 63 addresses the hazards associated with solid, liquid, gaseous and cryogenic fluid oxidizing materials, including oxygen in home use, and establishes criteria for their safe storage and protection in indoor and outdoor storage facilities, minimizing the potential for uncontrolled releases and contact with fuel sources. Although oxidizers themselves do not burn, they pose unique fire hazards because of their ability to support combustion by breaking down and giving off oxygen. As with other chapters of the *International Fire Code*, Section 6302 contains a list of terms that are defined in Chapter 2 and are applicable to the chapter contents.

Chapter 64 Pyrophoric Materials. This chapter regulates the hazards associated with pyrophoric materials, which are capable of spontaneously igniting in the air at or below a temperature of 130°F (54°C). Many pyrophoric materials also pose severe flammability or reactivity hazards. This chapter addresses only the hazards associated with pyrophoric materials. Materials that pose multiple hazards must conform to the requirements of the code with respect to all hazards. As with other chapters of the *International Fire Code*, Section 6402 contains a list of terms that are defined in Chapter 2 and are applicable to the chapter contents.

Chapter 65 Pyroxylin (Cellulose Nitrate) Plastics. This chapter addresses the significant hazards associated with pyroxylin (cellulose nitrate) plastics, which are the most dangerous and unstable of all plastic compounds. The chemically bound oxygen in their structure permits them to burn vigorously in the absence of atmospheric oxygen at a rate 15 times greater than comparable common combustibles. Strict compliance with the provisions of this chapter, along with proper housekeeping and storage arrangements, helps to reduce the hazards associated with pyroxylin (cellulose nitrate) plastics in a fire or other emergencies.

Chapter 66 Unstable (Reactive) Materials. This chapter addresses the hazards of unstable (reactive) liquid and solid materials as well as unstable (reactive) compressed gases. In addition to their unstable reactivity, these materials may pose other hazards, such as toxicity, corrosivity, explosivity, flammability or oxidizing potential. This chapter, however, intends to address those materials whose primary hazard is unstable reactivity. Materials that pose multiple hazards must conform to the requirements of the code with respect to all hazards. Strict compliance with the provisions of this chapter, along with proper housekeeping and storage arrangements, helps to reduce the exposure hazards associated with unstable (reactive) materials in a fire or other emergency. As with other chapters of the *International Fire Code*, Section 6602 contains a list of terms that are defined in Chapter 2 and are applicable to the chapter contents.

Chapter 67 Water-reactive Solids and Liquids. This chapter addresses the hazards associated with water-reactive materials that are solid or liquid at normal temperatures and pressures. In addition to their water reactivity, these materials may pose a wide range of other hazards, such as toxicity, flammability, corrosiveness or oxidizing potential. This chapter addresses only those materials whose primary hazard is water reactivity. Materials that pose multiple hazards must conform to the requirements of the code with respect to all hazards. Strict compliance with the requirements of this chapter, along with proper housekeeping and storage arrangements, helps to reduce the exposure hazards associated with water-reactive materials in a fire or other emergency. As with other chapters of the *International Fire Code*, Section 6702 contains a list of terms that are defined in Chapter 2 and are applicable to the chapter contents.

Chapters 68 through 79. Reserved for future use.

PART VI—REFERENCED STANDARDS

Chapter 80 Referenced Standards. This code contains several references to standards that are used to regulate materials and methods of construction. Chapter 80 contains a comprehensive list of all standards that are referenced in this code. The standards are part of the code to the extent of the reference to the standard (see Section 102.7). Compliance with the referenced standard is necessary for compliance with this code. By providing specifically adopted standards, the construction and installation requirements necessary for compliance with this code can be readily determined. The basis for code compliance is, therefore, established and available on an equal basis to the code official, contractor, designer and owner.

Chapter 80 is organized in a manner that makes it easy to locate specific standards. It lists all of the referenced standards alphabetically by acronym of the promulgating agency of the standard. Each agency's standards are then listed in either alphabetical or numeric order based upon the standard identification. The list also contains the title of the standard; the edition (date) of the standard referenced; any addenda included as part of the ICC adoption; and the section or sections of this code that reference the standard.

PART VII—APPENDICES

Appendix A Board of Appeals. This appendix contains optional criteria that, when adopted, provide jurisdictions with detailed appeals, board member qualifications and administrative procedures to supplement the basic requirements found in Section 108 of this code. Note that the provisions contained in this appendix are not mandatory unless specifically referenced in the adopting ordinance (see sample ordinance on page xxi).

Appendix B Fire-flow Requirements for Buildings. This appendix provides a tool for the use of jurisdictions in establishing a policy for determining fire-flow requirements in accordance with Section 507.3. The determination of required fire flow is not an exact science, but having some level of information provides a consistent way of choosing the appropriate fire flow for buildings throughout a jurisdiction. The primary tool used in this appendix is a table that presents fire flow based on construction type and building area based on the correlation of the Insurance Services Office (ISO) method and the construction types used in the *International Building Code*. Note that the provisions contained in this appendix are not mandatory unless specifically referenced in the adopting ordinance (see sample ordinance on page xxi).

Appendix C Fire Hydrant Locations and Distribution. This appendix focuses on the location and spacing of fire hydrants, which is important to the success of fire-fighting operations. The difficulty with determining the spacing of fire hydrants is that every situation is unique and has unique challenges. Finding one methodology for determining hydrant spacing is difficult. This particular appendix gives one methodology based on the required fire flow that fire departments can work with to set a policy for hydrant distribution around new buildings and facilities in conjunction with Section 507.5. Note that the provisions contained in this appendix are not mandatory unless specifically referenced in the adopting ordinance (see sample ordinance on page xxi).

Appendix D Fire Apparatus Access Roads. This appendix contains more detailed elements for use with the basic access requirements found in Section 503, which gives some minimum criteria, such as a maximum length of 150 feet and a minimum width of 20 feet, but in many cases does not state specific criteria. This appendix, like Appendices B and C, is a tool for jurisdictions looking for guidance in establishing access requirements and includes criteria for multiple-family residential developments, large one- and two-family subdivisions, specific examples for various types of turnarounds for fire department apparatus and parking regulatory signage. Note that the provisions contained in this appendix are not mandatory unless specifically referenced in the adopting ordinance (see sample ordinance on page xxi).

Appendix E Hazard Categories. This appendix contains guidance for designers, engineers, architects, code officials, plans reviewers and inspectors in the classifying of hazardous materials so that proposed designs can be evaluated intelligently and accurately. The descriptive materials and explanations of hazardous materials and how to report and evaluate them on a Material Safety Data

Sheet (MSDS) are intended to be instructional as well as informative. Note that this appendix is for information purposes and is not intended for adoption.

Appendix F Hazard Ranking. The information in this appendix is intended to be a companion to the specific requirements of Chapters 51 through 67, which regulate the storage, handling and use of all hazardous materials classified as either physical or health hazards. These materials pose diverse hazards, including instability, reactivity, flammability, oxidizing potential or toxicity; therefore, identifying them by hazard ranking is essential. This appendix lists the various hazardous materials categories that are defined in this code, along with the NFPA 704 hazard ranking for each. Note that the provisions contained in this appendix are not mandatory unless specifically referenced in the adopting ordinance (see sample ordinance on page xxi).

Appendix G Cryogenic Fluids—Weight and Volume Equivalents. This appendix gives the fire code official and design professional a ready reference tool for the conversion of the liquid weight and volume of cryogenic fluid to their corresponding volume of gas and vice versa and is a companion to the provisions of Chapter 55 of this code. Note that this appendix is for information purposes and is not intended for adoption.

Appendix H Hazardous Materials Management Plan (HMMP) and Hazardous Materials Inventory Statement (HMIS) Instructions. This appendix is intended to assist businesses in establishing a Hazardous Materials Management Plan (HMMP) and Hazardous Materials Inventory Statement (HMIS) based on the classification and quantities of materials that would be found on site in storage and/or use. The sample forms and available Material Safety Data Sheets (MSDS) provide the basis for the evaluations. It is also a companion to IFC Sections 407.5 and 407.6, which provide the requirement that the HMIS and HMMP be submitted when required by the fire code official. Note that the provisions contained in this appendix are not mandatory unless specifically referenced in the adopting ordinance (see sample ordinance on page xxi).

Appendix I Fire Protection Systems—Noncompliant Conditions. The purpose of this IFC appendix, which was developed by the ICC Hazard Abatement in Existing Buildings Committee, is to provide the fire code official with a list of conditions that are readily identifiable by the inspector during the course of an inspection utilizing the *International Fire Code*. The specific conditions identified in this appendix are primarily derived from applicable NFPA standards and pose a hazard to the proper operation of the respective systems. While these do not represent all of the conditions that pose a hazard or otherwise may impair the proper operation of fire protection systems, their identification in this adoptable appendix will provide a more direct path for enforcement by the fire code official. Note that the provisions contained in this appendix are not mandatory unless specifically referenced in the adopting ordinance (see sample ordinance on page xxi).

Appendix J Building Information Sign. This appendix provides design, installation and maintenance requirements for a Building Information Sign (BIS), a fire service tool to be utilized in the crucial, initial response of fire fighters to a structure fire. The BIS placard is designed to be utilized within the initial response time frame of an incident to assist fire fighters in their tactical size-up of a situation as soon as possible after arrival on the scene of a fire emergency. The BIS design is in the shape of a fire service Maltese Cross and includes five spaces (the four wings plus the centerpiece of the cross symbol) in which information is placed about the tactical considerations of construction type and hourly rating, fire protection systems, occupancy type, content hazards and special features that could affect tactical decisions and operations. Note that the provisions contained in this appendix are not mandatory unless specifically referenced in the adopting ordinance (see sample ordinance on page xxi).

Appendix K Construction Requirements for Existing Ambulatory Care Facilities. This new adoptable appendix was created by the ICC Ad Hoc Committee on Healthcare (AHC) and its intent is to provide jurisdictions with an option for assessing minimum fire and life safety requirements for buildings containing ambulatory care facilities. While this appendix is written with the intent to apply retroactive minimum standards, the AHC recognized that the ambulatory care requirements are relatively recent additions to the *International Building Code*. For that reason, these requirements are presented as an appendix so that the adopting authority can exercise judgment in the adoption and application of this section. This appendix would also be useful for those local and state jurisdictions that are specifically focused on ensuring the safety for existing ambulatory care facilities by providing minimum criteria that could be used to bring older facilities into compliance with the current standards at the discretion of the adopting jurisdiction. The technical requirements are based on the current IBC language, which is consistent with the overall concept of the current federal requirements. Note that the provisions contained in this appendix are not mandatory unless specifically referenced in the adopting ordinance (see sample ordinance on page xxi).

Appendix L Requirements for Fire Fighter Air Replenishment Systems. This new adoptable appendix provides for the design, installation and maintenance of permanently installed fire fighter breathing air systems in buildings designated by the jurisdiction. Breathing air is critical for fire-fighting operations. Historically, fire departments have supplied air bottles by means of a "bottle brigade," whereby fire fighters manually transport air bottles up stairways, which is an extraordinarily fire fighter-intensive process and takes fire fighters away from their primary mission of rescue and fire fighting. Technology now exists to address the issue using in-building air supply systems. Fire fighter breathing air systems were introduced in the late 1980s and are now required in a number of communities throughout the United States. The system has been called a "standpipe for air" and consists of stainless steel, high-pressure piping that is supplied by on-site air storage or fire department air supply units. Air filling stations are then strategically located throughout the building allowing fire fighters to refill breathing air cylinders inside the fire building, negating the required "bottle brigade," and making more fire fighters available for search, rescue and fire suppression operations. Note that the provisions contained in this appendix are not mandatory unless specifically referenced in the adopting ordinance (see sample ordinance on page xxi).

Appendix M High-rise Buildings—Retroactive Automatic Sprinkler Requirement. This new adoptable appendix was created by the ICC Fire Code Action Committee (FCAC) and its intent is to provide an option for adoption by jurisdictions that choose to require existing high-rise buildings to be retrofitted with automatic sprinklers. Modern fire and building codes require complete automatic fire sprinkler protection and a variety of other safety features in new high-rise construction. Many older high-rise buildings lack automatic sprinkler protection and other basic fire protection features necessary to protect the occupants, emergency responders and the structure itself. Without complete automatic sprinkler protection, fire departments cannot provide the level of protection that high-rise buildings demand. Existing high-rise buildings that are not protected with automatic sprinklers represent a significant hazard to occupants and fire fighters, and can significantly impact a community's infrastructure and economic viability in the event of a fire loss. The FCAC recognized that not all jurisdictions may choose to or may not have legal authority to enact a retroactive construction requirement of this nature, so the proposal has been included as an adoptable appendix. Note that the provisions contained in this appendix are not mandatory unless specifically referenced in the adopting ordinance (see sample ordinance on page xxi).

LEGISLATION

Jurisdictions wishing to adopt the 2015 *International Fire Code* as an enforceable set of regulations for the safeguarding of life and property from fire and explosion hazards arising from the storage, handling and use of hazardous substances, materials and devices, and from conditions hazardous to life or property in the occupancy of buildings and premises should ensure that certain factual information is included in the adopting legislation at the time adoption is being considered by the appropriate governmental body. The following sample adoption legislation addresses several key elements, including the information required for insertion into the code text.

SAMPLE LEGISLATION FOR ADOPTION OF THE *INTERNATIONAL FIRE CODE* ORDINANCE NO.________

A[N] **[ORDINANCE/STATUTE/REGULATION]** of the **[NAME OF JURISDICTION]** adopting the 2015 edition of the *International Fire Code*, regulating and governing the safeguarding of life and property from fire and explosion hazards arising from the storage, handling and use of hazardous substances, materials and devices, and from conditions hazardous to life or property in the occupancy of buildings and premises in the **[NAME OF JURISDICTION]**; providing for the issuance of permits and collection of fees therefor; repealing **[ORDINANCE/STATUTE/REGULATION]** No. ______ of the **[NAME OF JURISDICTION]** and all other ordinances or parts of laws in conflict therewith.

The **[GOVERNING BODY]** of the **[NAME OF JURISDICTION]** does ordain as follows:

Section 1. That a certain document, three (3) copies of which are on file in the office of the **[TITLE OF JURISDICTION'S KEEPER OF RECORDS]** of **[NAME OF JURISDICTION]**, being marked and designated as the *International Fire Code*, 2012 edition, including Appendix Chapters **[FILL IN THE APPENDIX CHAPTERS BEING ADOPTED]** (see *International Fire Code* Section 101.2.1, 2015 edition), as published by the International Code Council, be and is hereby adopted as the Fire Code of the **[NAME OF JURISDICTION]**, in the State of **[STATE NAME]** regulating and governing the safeguarding of life and property from fire and explosion hazards arising from the storage, handling and use of hazardous substances, materials and devices, and from conditions hazardous to life or property in the occupancy of buildings and premises as herein provided; providing for the issuance of permits and collection of fees therefor; and each and all of the regulations, provisions, penalties, conditions and terms of said Fire Code on file in the office of the **[NAME OF JURISDICTION]** are hereby referred to, adopted, and made a part hereof, as if fully set out in this legislation, with the additions, insertions, deletions and changes, if any, prescribed in Section 2 of this ordinance.

Section 2. That the following sections are hereby revised:

Section 101.1. Insert: **[NAME OF JURISDICTION]**

Section 109.4. Insert: **[OFFENSE, DOLLAR AMOUNT, NUMBER OF DAYS]**

Section 111.4. Insert: **[DOLLAR AMOUNT IN TWO LOCATIONS]**

Section 1103.5.3. Insert: **[DATE BY WHICH SPRINKLER SYSTEM MUST BE INSTALLED]**

Section 3. That the geographic limits referred to in certain sections of the 2015 *International Fire Code* are hereby established as follows:

Section 5704.2.9.6.1 (geographic limits in which the storage of Class I and Class II liquids in above-ground tanks outside of buildings is prohibited): **[JURISDICTION TO SPECIFY]**

Section 5706.2.4.4 (geographic limits in which the storage of Class I and Class II liquids in above-ground tanks is prohibited): **[JURISDICTION TO SPECIFY]**

Section 5806.2 (geographic limits in which the storage of flammable cryogenic fluids in stationary containers is prohibited): **[JURISDICTION TO SPECIFY]**

Section 6104.2 (geographic limits in which the storage of liquefied petroleum gas is restricted for the protection of heavily populated or congested areas): **[JURISDICTION TO SPECIFY]**

Section 4. That **[ORDINANCE/STATUTE/REGULATION]** No. ______ of **[NAME OF JURISDICTION]** entitled **[FILL IN HERE THE COMPLETE TITLE OF THE LEGISLATION OR LAWS IN EFFECT AT THE PRESENT TIME SO THAT THEY WILL BE REPEALED BY SPECIFIC REFERENCE]** and all other ordinances or parts of laws in conflict herewith are hereby repealed.

Section 5. That if any section, subsection, sentence, clause or phrase of this legislation is, for any reason, held to be unconstitutional, such decision shall not affect the validity of the remaining portions of this ordinance. The **[GOVERNING BODY]** hereby declares that it would have passed this law, and each section, subsection, clause or phrase thereof, irrespective of the fact that any one or more sections, subsections, sentences, clauses and phrases be declared unconstitutional.

Section 6. That nothing in this legislation or in the Fire Code hereby adopted shall be construed to affect any suit or proceeding impending in any court, or any rights acquired, or liability incurred, or any cause or causes of action acquired or existing, under any act or ordinance hereby repealed as cited in Section 4 of this law; nor shall any just or legal right or remedy of any character be lost, impaired or affected by this legislation.

Section 7. That the **[JURISDICTION'S KEEPER OF RECORDS]** is hereby ordered and directed to cause this legislation to be published. (An additional provision may be required to direct the number of times the legislation is to be published and to specify that it is to be in a newspaper in general circulation. Posting may also be required.)

Section 8. That this law and the rules, regulations, provisions, requirements, orders and matters established and adopted hereby shall take effect and be in full force and effect **[TIME PERIOD]** from and after the date of its final passage and adoption.

TABLE OF CONTENTS

Part I—Administrative

CHAPTER 1

SCOPE AND ADMINISTRATION

PART 1—GENERAL PROVISIONS

SECTION 101
SCOPE AND GENERAL REQUIREMENTS

[A] 101.1 Title. These regulations shall be known as the *Fire Code* of **[NAME OF JURISDICTION]**, hereinafter referred to as "this code."

[A] 101.2 Scope. This code establishes regulations affecting or relating to structures, processes, premises and safeguards regarding all of the following:

1. The hazard of fire and explosion arising from the storage, handling or use of structures, materials or devices.
2. Conditions hazardous to life, property or public welfare in the occupancy of structures or premises.
3. Fire hazards in the structure or on the premises from occupancy or operation.
4. Matters related to the construction, extension, repair, alteration or removal of fire suppression or alarm systems.
5. Conditions affecting the safety of fire fighters and emergency responders during emergency operations.

[A] 101.2.1 Appendices. Provisions in the appendices shall not apply unless specifically adopted.

[A] 101.3 Intent. The purpose of this code is to establish the minimum requirements consistent with nationally recognized good practice for providing a reasonable level of life safety and property protection from the hazards of fire, explosion or dangerous conditions in new and existing buildings, structures and premises, and to provide a reasonable level of safety to fire fighters and emergency responders during emergency operations.

[A] 101.4 Severability. If a section, subsection, sentence, clause or phrase of this code is, for any reason, held to be unconstitutional, such decision shall not affect the validity of the remaining portions of this code.

[A] 101.5 Validity. In the event any part or provision of this code is held to be illegal or void, this shall not have the effect of making void or illegal any of the other parts or provisions hereof, which are determined to be legal; and it shall be presumed that this code would have been adopted without such illegal or invalid parts or provisions.

SECTION 102
APPLICABILITY

[A] 102.1 Construction and design provisions. The construction and design provisions of this code shall apply to:

1. Structures, facilities and conditions arising after the adoption of this code.
2. Existing structures, facilities and conditions not legally in existence at the time of adoption of this code.
3. Existing structures, facilities and conditions where required in Chapter 11.
4. Existing structures, facilities and conditions that, in the opinion of the *fire code official*, constitute a distinct hazard to life or property.

[A] 102.2 Administrative, operational and maintenance provisions. The administrative, operational and maintenance provisions of this code shall apply to:

1. Conditions and operations arising after the adoption of this code.
2. Existing conditions and operations.

[A] 102.3 Change of use or occupancy. Changes shall not be made in the use or occupancy of any structure that would place the structure in a different division of the same group or occupancy or in a different group of occupancies, unless such structure is made to comply with the requirements of this code and the *International Building Code*. Subject to the approval of the *fire code official*, the use or occupancy of an existing structure shall be allowed to be changed and the structure is allowed to be occupied for purposes in other groups without conforming to all of the requirements of this code and the *International Building Code* for those groups, provided the new or proposed use is less hazardous, based on life and fire risk, than the existing use.

[A] 102.4 Application of building code. The design and construction of new structures shall comply with the *International Building Code*, and any *alterations*, additions, changes in use or changes in structures required by this code, which are within the scope of the *International Building Code*, shall be made in accordance therewith.

[A] 102.5 Application of residential code. Where structures are designed and constructed in accordance with the *International Residential Code*, the provisions of this code shall apply as follows:

1. Construction and design provisions of this code pertaining to the exterior of the structure shall apply including, but not limited to, premises identification, fire apparatus access and water supplies. Where interior or exterior systems or devices are installed, construction permits required by Section 105.7 of this code shall apply.
2. Administrative, operational and maintenance provisions of this code shall apply.

[A] 102.6 Historic buildings. The provisions of this code relating to the construction, *alteration*, repair, enlargement, restoration, relocation or moving of buildings or structures shall not be mandatory for existing buildings or structures identified and classified by the state or local jurisdiction as historic buildings where such buildings or structures do not constitute a distinct hazard to life or property. Fire protection in designated historic buildings shall be provided with an *approved* fire protection plan as required in Section 1103.1.1.

[A] 102.7 Referenced codes and standards. The codes and standards referenced in this code shall be those that are listed in Chapter 80, and such codes and standards shall be considered to be part of the requirements of this code to the prescribed extent of each such reference and as further regulated in Sections 102.7.1 and 102.7.2.

[A] 102.7.1 Conflicts. Where conflicts occur between provisions of this code and referenced codes and standards, the provisions of this code shall apply.

[A] 102.7.2 Provisions in referenced codes and standards. Where the extent of the reference to a referenced code or standard includes subject matter that is within the scope of this code, the provisions of this code, as applicable, shall take precedence over the provisions in the referenced code or standard.

[A] 102.8 Subjects not regulated by this code. Where applicable standards or requirements are not set forth in this code, or are contained within other laws, codes, regulations, ordinances or bylaws adopted by the jurisdiction, compliance with applicable standards of the National Fire Protection Association or other nationally recognized fire safety standards, as *approved,* shall be deemed as prima facie evidence of compliance with the intent of this code. Nothing herein shall derogate from the authority of the *fire code official* to determine compliance with codes or standards for those activities or installations within the *fire code official's* jurisdiction or responsibility.

[A] 102.9 Matters not provided for. Requirements that are essential for the public safety of an existing or proposed activity, building or structure, or for the safety of the occupants thereof, that are not specifically provided for by this code, shall be determined by the *fire code official.*

[A] 102.10 Conflicting provisions. Where there is a conflict between a general requirement and a specific requirement, the specific requirement shall be applicable. Where, in a specific case, different sections of this code specify different materials, methods of construction or other requirements, the most restrictive shall govern.

[A] 102.11 Other laws. The provisions of this code shall not be deemed to nullify any provisions of local, state or federal law.

[A] 102.12 Application of references. References to chapter or section numbers, or to provisions not specifically identified by number, shall be construed to refer to such chapter, section or provision of this code.

PART 2—ADMINISTRATIVE PROVISIONS

SECTION 103
DEPARTMENT OF FIRE PREVENTION

[A] 103.1 General. The department of fire prevention is established within the jurisdiction under the direction of the *fire code official.* The function of the department shall be the implementation, administration and enforcement of the provisions of this code.

[A] 103.2 Appointment. The *fire code official* shall be appointed by the chief appointing authority of the jurisdiction; and the *fire code official* shall not be removed from office except for cause and after full opportunity to be heard on specific and relevant charges by and before the appointing authority.

[A] 103.3 Deputies. In accordance with the prescribed procedures of this jurisdiction and with the concurrence of the appointing authority, the *fire code official* shall have the authority to appoint a deputy *fire code official,* other related technical officers, inspectors and other employees.

[A] 103.4 Liability. The *fire code official*, member of the board of appeals, officer or employee charged with the enforcement of this code, while acting for the jurisdiction, in good faith and without malice in the discharge of the duties required by this code or other pertinent law or ordinance, shall not thereby be rendered civilly or criminally liable personally, and is hereby relieved from all personal liability for any damage accruing to persons or property as a result of an act or by reason of an act or omission in the discharge of official duties.

[A] 103.4.1 Legal defense. Any suit or criminal complaint instituted against any officer or employee because of an act performed by that officer or employee in the lawful discharge of duties and under the provisions of this code shall be defended by the legal representatives of the jurisdiction until the final termination of the proceedings. The *fire code official* or any subordinate shall not be liable for costs in an action, suit or proceeding that is instituted in pursuance of the provisions of this code; and any officer of the department of fire prevention, acting in good faith and without malice, shall be free from liability for acts performed under any of its provisions or by reason of any act or omission in the performance of official duties in connection therewith.

SECTION 104
GENERAL AUTHORITY AND RESPONSIBILITIES

[A] 104.1 General. The *fire code official* is hereby authorized to enforce the provisions of this code and shall have the authority to render interpretations of this code, and to adopt policies, procedures, rules and regulations in order to clarify the application of its provisions. Such interpretations, policies, procedures, rules and regulations shall be in compliance with the intent and purpose of this code and shall not have the effect of waiving requirements specifically provided for in this code.

[A] 104.2 Applications and permits. The *fire code official* is authorized to receive applications, review *construction documents* and issue permits for construction regulated by this code, issue permits for operations regulated by this code, inspect the premises for which such permits have been issued and enforce compliance with the provisions of this code.

[A] 104.3 Right of entry. Where it is necessary to make an inspection to enforce the provisions of this code, or where the *fire code official* has reasonable cause to believe that there exists in a building or upon any premises any conditions or violations of this code that make the building or premises unsafe, dangerous or hazardous, the *fire code official* shall have the authority to enter the building or premises at all reasonable times to inspect or to perform the duties imposed upon the *fire code official* by this code. If such building or premises is occupied, the *fire code official* shall present credentials to the occupant and request entry. If such building or premises is unoccupied, the *fire code official* shall first make a reasonable effort to locate the *owner*, the owner's authorized agent or other person having charge or control of the building or premises and request entry. If entry is refused, the *fire code official* has recourse to every remedy provided by law to secure entry.

[A] 104.3.1 Warrant. Where the *fire code official* has first obtained a proper inspection warrant or other remedy provided by law to secure entry, an *owner*, the *owner's* authorized agent or occupant or person having charge, care or control of the building or premises shall not fail or neglect, after proper request is made as herein provided, to permit entry therein by the *fire code official* for the purpose of inspection and examination pursuant to this code.

[A] 104.4 Identification. The *fire code official* shall carry proper identification when inspecting structures or premises in the performance of duties under this code.

[A] 104.5 Notices and orders. The *fire code official* is authorized to issue such notices or orders as are required to affect compliance with this code in accordance with Sections 109.1 and 109.2.

[A] 104.6 Official records. The *fire code official* shall keep official records as required by Sections 104.6.1 through 104.6.4. Such official records shall be retained for not less than 5 years or for as long as the structure or activity to which such records relate remains in existence, unless otherwise provided by other regulations.

[A] 104.6.1 Approvals. A record of approvals shall be maintained by the *fire code official* and shall be available for public inspection during business hours in accordance with applicable laws.

[A] 104.6.2 Inspections. The *fire code official* shall keep a record of each inspection made, including notices and orders issued, showing the findings and disposition of each.

104.6.3 Fire records. The fire department shall keep a record of fires occurring within its jurisdiction and of facts concerning the same, including statistics as to the extent of such fires and the damage caused thereby, together with other information as required by the *fire code official*.

[A] 104.6.4 Administrative. Application for modification, alternative methods or materials and the final decision of the *fire code official* shall be in writing and shall be officially recorded in the permanent records of the *fire code official*.

[A] 104.7 Approved materials and equipment. Materials, equipment and devices *approved* by the *fire code official* shall be constructed and installed in accordance with such approval.

[A] 104.7.1 Material and equipment reuse. Materials, equipment and devices shall not be reused or reinstalled unless such elements have been reconditioned, tested and placed in good and proper working condition and *approved*.

[A] 104.7.2 Technical assistance. To determine the acceptability of technologies, processes, products, facilities, materials and uses attending the design, operation or use of a building or premises subject to inspection by the *fire code official*, the *fire code official* is authorized to require the *owner* or owner's authorized agent to provide, without charge to the jurisdiction, a technical opinion and report. The opinion and report shall be prepared by a qualified engineer, specialist, laboratory or fire safety specialty organization acceptable to the *fire code official* and shall analyze the fire safety properties of the design, operation or use of the building or premises and the facilities and appurtenances situated thereon, to recommend necessary changes. The *fire code official* is authorized to require design submittals to be prepared by, and bear the stamp of, a registered design professional.

[A] 104.8 Modifications. Where there are practical difficulties involved in carrying out the provisions of this code, the *fire code official* shall have the authority to grant modifications for individual cases, provided the *fire code official* shall first find that special individual reason makes the strict letter of this code impractical and the modification is in compliance with the intent and purpose of this code and that such modification does not lessen health, life and fire safety requirements. The details of action granting modifications shall be recorded and entered in the files of the department of fire prevention.

[A] 104.9 Alternative materials and methods. The provisions of this code are not intended to prevent the installation of any material or to prohibit any method of construction not specifically prescribed by this code, provided that any such alternative has been *approved*. The *fire code official* is authorized to approve an alternative material or method of construction where the *fire code official* finds that the proposed design is satisfactory and complies with the intent of the provisions of this code, and that the material, method or work offered is, for the purpose intended, at least the equivalent of that prescribed in this code in quality, strength, effectiveness, *fire resistance*, durability and safety. Where the alternative material, design or method of construction is not approved, the *fire code official* shall respond in writing, stating the reasons why the alternative was not approved.

[A] 104.9.1 Research reports. Supporting data, where necessary to assist in the approval of materials or assemblies not specifically provided for in this code, shall consist of valid research reports from *approved* sources.

[A] 104.9.2 Tests. Where there is insufficient evidence of compliance with the provisions of this code, or evidence that a material or method does not conform to the requirements of this code, or in order to substantiate claims for alternative materials or methods, the *fire code official* shall have the authority to require tests as evidence of compliance to be made at no expense to the jurisdiction. Test methods shall be as specified in this code or by other recognized test standards. In the absence of recognized and accepted test methods, the *fire code official* shall approve the testing procedures. Tests shall be performed by an *approved* agency. Reports of such tests shall be retained by the *fire code official* for the period required for retention of public records.

104.10 Fire investigations. The *fire code official*, the fire department or other responsible authority shall have the authority to investigate the cause, origin and circumstances of any fire, explosion or other hazardous condition. Information that could be related to trade secrets or processes shall not be made part of the public record, except as directed by a court of law.

104.10.1 Assistance from other agencies. Police and other enforcement agencies shall have authority to render necessary assistance in the investigation of fires when requested to do so.

104.11 Authority at fires and other emergencies. The fire chief or officer of the fire department in charge at the scene of a fire or other emergency involving the protection of life or property, or any part thereof, shall have the authority to direct such operation as necessary to extinguish or control any fire, perform any rescue operation, investigate the existence of suspected or reported fires, gas leaks or other hazardous conditions or situations, or take any other action necessary in the reasonable performance of duty. In the exercise of such power, the fire chief is authorized to prohibit any person, vehicle, vessel or thing from approaching the scene, and is authorized to remove, or cause to be removed or kept away from the scene, any vehicle, vessel or thing that could impede or interfere with the operations of the fire department and, in the judgment of the fire chief, any person not actually and usefully employed in the extinguishing of such fire or in the preservation of property in the vicinity thereof.

104.11.1 Barricades. The fire chief or officer of the fire department in charge at the scene of an emergency is authorized to place ropes, guards, barricades or other obstructions across any street, alley, place or private property in the vicinity of such operation so as to prevent accidents or interference with the lawful efforts of the fire department to manage and control the situation and to handle fire apparatus.

104.11.2 Obstructing operations. Persons shall not obstruct the operations of the fire department in connection with extinguishment or control of any fire, or actions relative to other emergencies, or disobey any lawful command of the fire chief or officer of the fire department in charge of the emergency, or any part thereof, or any lawful order of a police officer assisting the fire department.

104.11.3 Systems and devices. Persons shall not render a system or device inoperative during an emergency unless by direction of the fire chief or fire department official in charge of the incident.

SECTION 105
PERMITS

[A] 105.1 General. Permits shall be in accordance with Sections 105.1.1 through 105.7.18.

[A] 105.1.1 Permits required. A property owner or owner's authorized agent who intends to conduct an operation or business, or install or modify systems and equipment that are regulated by this code, or to cause any such work to be performed, shall first make application to the *fire code official* and obtain the required permit.

105.1.2 Types of permits. There shall be two types of permits as follows:

1. Operational permit. An operational permit allows the applicant to conduct an operation or a business for which a permit is required by Section 105.6 for either:
   1.1. A prescribed period.
   1.2. Until renewed or revoked.
2. Construction permit. A construction permit allows the applicant to install or modify systems and equipment for which a permit is required by Section 105.7.

105.1.3 Multiple permits for the same location. Where more than one permit is required for the same location, the *fire code official* is authorized to consolidate such permits into a single permit provided that each provision is listed in the permit.

[A] 105.1.4 Emergency repairs. Where equipment replacement and repairs must be performed in an emergency situation, the permit application shall be submitted within the next working business day to the *fire code official*.

[A] 105.1.5 Repairs. Application or notice to the *fire code official* is not required for ordinary repairs to structures, equipment or systems. Such repairs shall not include the cutting away of any wall, partition or portion thereof, the removal or change of any required *means of egress*, or rearrangement of parts of a structure affecting the egress requirements; nor shall any repairs include addition to, alteration of, replacement or relocation of any standpipe, fire protection water supply, *automatic sprinkler system*, fire alarm system or other work affecting fire protection or life safety.

[A] 105.1.6 Annual permit. Instead of an individual construction permit for each alteration to an already *approved* system or equipment installation, the *fire code official* is authorized to issue an annual permit upon application therefor to any person, firm or corporation regularly employing one or more qualified tradespersons in the building, structure or on the premises owned or operated by the applicant for the permit.

[A] 105.1.6.1 Annual permit records. The person to whom an annual permit is issued shall keep a detailed record of alterations made under such annual permit. The *fire code official* shall have access to such records at all times or such records shall be filed with the *fire code official* as designated.

[A] 105.2 Application. Application for a permit required by this code shall be made to the *fire code official* in such form and detail as prescribed by the *fire code official*. Applications for permits shall be accompanied by such plans as prescribed by the *fire code official*.

[A] 105.2.1 Refusal to issue permit. If the application for a permit describes a use that does not conform to the requirements of this code and other pertinent laws and ordinances, the *fire code official* shall not issue a permit, but shall return the application to the applicant with the refusal to issue such permit. Such refusal shall, where requested, be in writing and shall contain the reasons for refusal.

[A] 105.2.2 Inspection authorized. Before a new operational permit is *approved*, the *fire code official* is authorized to inspect the receptacles, vehicles, buildings, devices, premises, storage spaces or areas to be used to determine compliance with this code or any operational constraints required.

[A] 105.2.3 Time limitation of application. An application for a permit for any proposed work or operation shall be deemed to have been abandoned 180 days after the date of filing, unless such application has been diligently prosecuted or a permit shall have been issued; except that the *fire code official* is authorized to grant one or more extensions of time for additional periods not exceeding 90 days each. The extension shall be requested in writing and justifiable cause demonstrated.

[A] 105.2.4 Action on application. The *fire code official* shall examine or cause to be examined applications for permits and amendments thereto within a reasonable time after filing. If the application or the *construction documents* do not conform to the requirements of pertinent laws, the *fire code official* shall reject such application in writing, stating the reasons therefor. If the *fire code official* is satisfied that the proposed work or operation conforms to the requirements of this code and laws and ordinances applicable thereto, the *fire code official* shall issue a permit therefor as soon as practicable.

[A] 105.3 Conditions of a permit. A permit shall constitute permission to maintain, store or handle materials; or to conduct processes that produce conditions hazardous to life or property; or to install equipment utilized in connection with such activities; or to install or modify any *fire protection system* or equipment or any other construction, equipment installation or modification in accordance with the provisions of this code where a permit is required by Section 105.6 or 105.7. Such permission shall not be construed as authority to violate, cancel or set aside any of the provisions of this code or other applicable regulations or laws of the jurisdiction.

[A] 105.3.1 Expiration. An operational permit shall remain in effect until reissued, renewed or revoked, or for such a period of time as specified in the permit. Construction permits shall automatically become invalid unless the work authorized by such permit is commenced within 180 days after its issuance, or if the work authorized by such permit is suspended or abandoned for a period of 180 days after the time the work is commenced. Before such work recommences, a new permit shall be first obtained and the fee to recommence work, if any, shall be one-half the amount required for a new permit for such work, provided that changes have not been made and will not be made in the original construction documents for such work, and provided further that such suspension or abandonment has not exceeded one year. Permits are not transferable and any change in occupancy, operation, tenancy or ownership shall require that a new permit be issued.

[A] 105.3.2 Extensions. A permittee holding an unexpired permit shall have the right to apply for an extension of the time within which the permittee will commence work under that permit where work is unable to be commenced within the time required by this section for good and satisfactory reasons. The *fire code official* is authorized to grant, in writing, one or more extensions of the time period of a permit for periods of not more than 180 days each. Such extensions shall be requested by the permit holder in writing and justifiable cause demonstrated.

[A] 105.3.3 Occupancy prohibited before approval. The building or structure shall not be occupied prior to the *fire code official* issuing a permit and conducting associated inspections indicating the applicable provisions of this code have been met.

[A] 105.3.4 Conditional permits. Where permits are required and upon the request of a permit applicant, the *fire code official* is authorized to issue a conditional permit to occupy the premises or portion thereof before the entire work or operations on the premises is completed, provided that such portion or portions will be occupied safely prior to full completion or installation of equipment and operations without endangering life or public welfare. The *fire code official* shall notify the permit applicant in writing of any limitations or restrictions necessary to keep the permit area safe. The holder of a conditional permit shall proceed only to the point for which approval has been given, at the permit holder's own risk and without assurance that approval for the occupancy or the utilization of the entire premises, equipment or operations will be granted.

[A] 105.3.5 Posting the permit. Issued permits shall be kept on the premises designated therein at all times and shall be readily available for inspection by the *fire code official*.

[A] 105.3.6 Compliance with code. The issuance or granting of a permit shall not be construed to be a permit for, or an approval of, any violation of any of the provisions of this code or of any other ordinance of the jurisdiction. Permits presuming to give authority to violate or cancel the provisions of this code or other ordinances of the jurisdiction shall not be valid. The issuance of a permit based on *construction documents* and other data shall not prevent the *fire code official* from requiring the correction of errors in the *construction documents* and other data. Any addition to or alteration of *approved construction documents* shall be *approved* in advance by the *fire code official*, as evidenced by the issuance of a new or amended permit.

[A] 105.3.7 Information on the permit. The *fire code official* shall issue all permits required by this code on an *approved* form furnished for that purpose. The permit shall contain a general description of the operation or occupancy and its location and any other information required by the *fire code official*. Issued permits shall bear the signature of the *fire code official* or other *approved* legal authorization.

[A] 105.3.8 Validity of permit. The issuance or granting of a permit shall not be construed to be a permit for, or an approval of, any violation of any of the provisions of this code or of any other ordinances of the jurisdiction. Permits presuming to give authority to violate or cancel the provisions of this code or other ordinances of the jurisdiction shall not be valid. The issuance of a permit based on *construction documents*, operational documents and other data shall not prevent the *fire code official* from requiring correction of errors in the documents or other data.

[A] 105.4 Construction documents. *Construction documents* shall be in accordance with Sections 105.4.1 through 105.4.6.

[A] 105.4.1 Submittals. *Construction documents* and supporting data shall be submitted in two or more sets with each application for a permit and in such form and detail as required by the *fire code official*. The *construction documents* shall be prepared by a registered design professional where required by the statutes of the jurisdiction in which the project is to be constructed.

Exception: The *fire code official* is authorized to waive the submission of *construction documents* and supporting data not required to be prepared by a registered design professional if it is found that the nature of the work applied for is such that review of *construction documents* is not necessary to obtain compliance with this code.

[A] 105.4.1.1 Examination of documents. The *fire code official* shall examine or cause to be examined the accompanying *construction documents* and shall ascertain by such examinations whether the work indicated and described is in accordance with the requirements of this code.

[A] 105.4.2 Information on construction documents. *Construction documents* shall be drawn to scale upon suitable material. Electronic media documents are allowed to be submitted where *approved* by the *fire code official*. *Construction documents* shall be of sufficient clarity to indicate the location, nature and extent of the work proposed and show in detail that it will conform to the provisions of this code and relevant laws, ordinances, rules and regulations as determined by the *fire code official*.

[A] 105.4.2.1 Fire protection system shop drawings. Shop drawings for the fire protection system(s) shall be submitted to indicate compliance with this code and the *construction documents*, and shall be *approved* prior to the start of installation. Shop drawings shall contain all information as required by the referenced installation standards in Chapter 9.

[A] 105.4.3 Applicant responsibility. It shall be the responsibility of the applicant to ensure that the *construction documents* include all of the fire protection requirements and the shop drawings are complete and in compliance with the applicable codes and standards.

[A] 105.4.4 Approved documents. *Construction documents approved* by the *fire code official* are *approved* with the intent that such *construction documents* comply in all respects with this code. Review and approval by the *fire code official* shall not relieve the applicant of the responsibility of compliance with this code.

[A] 105.4.4.1 Phased approval. The *fire code official* is authorized to issue a permit for the construction of part of a structure, system or operation before the *construction documents* for the whole structure, system or operation have been submitted, provided that adequate information and detailed statements have been filed complying with pertinent requirements of this code. The holder of such permit for parts of a structure, system or operation shall proceed at the holder's own risk with the building operation and without assurance that a permit for the entire structure, system or operation will be granted.

[A] 105.4.5 Amended construction documents. Work shall be installed in accordance with the *approved construction documents*, and any changes made during construction that are not in compliance with the *approved construction documents* shall be resubmitted for approval as an amended set of *construction documents*.

[A] 105.4.6 Retention of construction documents. One set of *construction documents* shall be retained by the *fire code official* for a period of not less than 180 days from date of completion of the permitted work, or as required by state or local laws. One set of *approved construction documents* shall be returned to the applicant, and said set shall be kept on the site of the building or work at all times during which the work authorized thereby is in progress.

[A] 105.5 Revocation. The *fire code official* is authorized to revoke a permit issued under the provisions of this code where it is found by inspection or otherwise that there has been a false statement or misrepresentation as to the material facts in the application or *construction documents* on which the permit or approval was based including, but not limited to, any one of the following:

1. The permit is used for a location or establishment other than that for which it was issued.
2. The permit is used for a condition or activity other than that listed in the permit.
3. Conditions and limitations set forth in the permit have been violated.
4. There have been any false statements or misrepresentations as to the material fact in the application for permit or plans submitted or a condition of the permit.
5. The permit is used by a different person or firm than the name for which it was issued.
6. The permittee failed, refused or neglected to comply with orders or notices duly served in accordance with the provisions of this code within the time provided therein.
7. The permit was issued in error or in violation of an ordinance, regulation or this code.

105.6 Required operational permits. The *fire code official* is authorized to issue operational permits for the operations set forth in Sections 105.6.1 through 105.6.48.

105.6.1 Aerosol products. An operational permit is required to manufacture, store or handle an aggregate quantity of Level 2 or Level 3 aerosol products in excess of 500 pounds (227 kg) net weight.

105.6.2 Amusement buildings. An operational permit is required to operate a special amusement building.

105.6.3 Aviation facilities. An operational permit is required to use a Group H or Group S occupancy for aircraft servicing or repair and aircraft fuel-servicing vehicles. Additional permits required by other sections of this code include, but are not limited to, hot work, hazardous materials and flammable or combustible finishes.

105.6.4 Carbon dioxide systems used in beverage dispensing applications. An operational permit is required for carbon dioxide systems used in beverage dispensing applications having more than 100 pounds of carbon dioxide.

105.6.5 Carnivals and fairs. An operational permit is required to conduct a carnival or fair.

105.6.6 Cellulose nitrate film. An operational permit is required to store, handle or use cellulose nitrate film in a Group A occupancy.

105.6.7 Combustible dust-producing operations. An operational permit is required to operate a grain elevator, flour starch mill, feed mill, or a plant pulverizing aluminum, coal, cocoa, magnesium, spices or sugar, or other operations producing *combustible dusts* as defined in Chapter 2.

105.6.8 Combustible fibers. An operational permit is required for the storage and handling of *combustible fibers* in quantities greater than 100 cubic feet (2.8 m^3).

Exception: A permit is not required for agricultural storage.

105.6.9 Compressed gases. An operational permit is required for the storage, use or handling at *normal temperature and pressure* (NTP) of *compressed gases* in excess of the amounts listed in Table 105.6.9.

Exception: Vehicles equipped for and using *compressed gas* as a fuel for propelling the vehicle.

TABLE 105.6.9
PERMIT AMOUNTS FOR COMPRESSED GASES

TYPE OF GAS	AMOUNT (cubic feet at NTP)
Corrosive	200
Flammable (except cryogenic fluids and liquefied petroleum gases)	200
Highly toxic	Any Amount
Inert and simple asphyxiant[a]	6,000
Oxidizing (including oxygen)	504
Pyrophoric	Any Amount
Toxic	Any Amount

For SI: 1 cubic foot = 0.02832 m^3.

a. For carbon dioxide used in beverage dispensing applications, see Section 105.6.4.

105.6.10 Covered and open mall buildings. An operational permit is required for:

1. The placement of retail fixtures and displays, concession equipment, displays of highly combustible goods and similar items in the mall.
2. The display of liquid- or gas-fired equipment in the mall.
3. The use of open-flame or flame-producing equipment in the mall.

105.6.11 Cryogenic fluids. An operational permit is required to produce, store, transport on site, use, handle or dispense *cryogenic fluids* in excess of the amounts listed in Table 105.6.11.

Exception: Permits are not required for vehicles equipped for and using *cryogenic fluids* as a fuel for propelling the vehicle or for refrigerating the lading.

TABLE 105.6.11
PERMIT AMOUNTS FOR CRYOGENIC FLUIDS

TYPE OF CRYOGENIC FLUID	INSIDE BUILDING (gallons)	OUTSIDE BUILDING (gallons)
Flammable	More than 1	60
Inert	60	500
Oxidizing (includes oxygen)	10	50
Physical or health hazard not indicated above	Any Amount	Any Amount

For SI: 1 gallon = 3.785 L.

105.6.12 Cutting and welding. An operational permit is required to conduct cutting or welding operations within the jurisdiction.

105.6.13 Dry cleaning. An operational permit is required to engage in the business of dry cleaning or to change to a more hazardous cleaning solvent used in existing dry cleaning equipment.

105.6.14 Exhibits and trade shows. An operational permit is required to operate exhibits and trade shows.

105.6.15 Explosives. An operational permit is required for the manufacture, storage, handling, sale or use of any quantity of *explosives*, *explosive materials*, fireworks or pyrotechnic special effects within the scope of Chapter 56.

> **Exception:** Storage in Group R-3 occupancies of smokeless propellant, black powder and small arms primers for personal use, not for resale and in accordance with Section 5606.

105.6.16 Fire hydrants and valves. An operational permit is required to use or operate fire hydrants or valves intended for fire suppression purposes that are installed on water systems and accessible to a fire apparatus access road that is open to or generally used by the public.

> **Exception:** A permit is not required for authorized employees of the water company that supplies the system or the fire department to use or operate fire hydrants or valves.

105.6.17 Flammable and combustible liquids. An operational permit is required:

1. To use or operate a pipeline for the transportation within facilities of flammable or *combustible liquids*. This requirement shall not apply to the offsite transportation in pipelines regulated by the Department of Transportation (DOTn) nor does it apply to piping systems.
2. To store, handle or use Class I liquids in excess of 5 gallons (19 L) in a building or in excess of 10 gallons (37.9 L) outside of a building, except that a permit is not required for the following:
   - 2.1. The storage or use of Class I liquids in the fuel tank of a motor vehicle, aircraft, motorboat, mobile power plant or mobile heating plant, unless such storage, in the opinion of the *fire code official*, would cause an unsafe condition.
   - 2.2. The storage or use of paints, oils, varnishes or similar flammable mixtures where such liquids are stored for maintenance, painting or similar purposes for a period of not more than 30 days.
3. To store, handle or use Class II or Class IIIA liquids in excess of 25 gallons (95 L) in a building or in excess of 60 gallons (227 L) outside a building, except for fuel oil used in connection with oil-burning equipment.
4. To store, handle or use Class IIIB liquids in tanks or portable tanks for fueling motor vehicles at motor fuel-dispensing facilities or where connected to fuel-burning equipment.

   Exception: Fuel oil and used motor oil used for space heating or water heating.
5. To remove Class I or II liquids from an underground storage tank used for fueling motor vehicles by any means other than the *approved*, stationary on-site pumps normally used for dispensing purposes.
6. To operate tank vehicles, equipment, tanks, plants, terminals, wells, fuel-dispensing stations, refineries, distilleries and similar facilities where flammable and *combustible liquids* are produced, processed, transported, stored, dispensed or used.
7. To place temporarily out of service (for more than 90 days) an underground, protected above-ground or above-ground flammable or *combustible liquid* tank.
8. To change the type of contents stored in a flammable or *combustible liquid* tank to a material that poses a greater hazard than that for which the tank was designed and constructed.
9. To manufacture, process, blend or refine flammable or *combustible liquids*.
10. To engage in the dispensing of liquid fuels into the fuel tanks of motor vehicles at commercial, industrial, governmental or manufacturing establishments.
11. To utilize a site for the dispensing of liquid fuels from tank vehicles into the fuel tanks of motor vehicles, marine craft and other special equipment at commercial, industrial, governmental or manufacturing establishments.

105.6.18 Floor finishing. An operational permit is required for floor finishing or surfacing operations exceeding 350 square feet (33 m^2) using Class I or Class II liquids.

105.6.19 Fruit and crop ripening. An operational permit is required to operate a fruit- or crop-ripening facility or conduct a fruit-ripening process using ethylene gas.

105.6.20 Fumigation and insecticidal fogging. An operational permit is required to operate a business of fumigation or insecticidal fogging, and to maintain a room, vault or chamber in which a toxic or flammable fumigant is used.

105.6.21 Hazardous materials. An operational permit is required to store, transport on site, dispense, use or handle hazardous materials in excess of the amounts listed in Table 105.6.21.

105.6.22 HPM facilities. An operational permit is required to store, handle or use hazardous production materials.

**TABLE 105.6.21
PERMIT AMOUNTS FOR HAZARDOUS MATERIALS**

TYPE OF MATERIAL	AMOUNT
Combustible liquids	See Section 105.6.17
Corrosive materials	
Gases	See Section 105.6.9
Liquids	55 gallons
Solids	1000 pounds
Explosive materials	See Section 105.6.15
Flammable materials	
Gases	See Section 105.6.9
Liquids	See Section 105.6.17
Solids	100 pounds
Highly toxic materials	
Gases	See Section 105.6.9
Liquids	Any Amount
Solids	Any Amount
Oxidizing materials	
Gases	See Section 105.6.9
Liquids	
Class 4	Any Amount
Class 3	1 gallon[a]
Class 2	10 gallons
Class 1	55 gallons
Solids	
Class 4	Any Amount
Class 3	10 pounds[b]
Class 2	100 pounds
Class 1	500 pounds
Organic peroxides	
Liquids	
Class I	Any Amount
Class II	Any Amount
Class III	1 gallon
Class IV	2 gallons
Class V	No Permit Required
Solids	
Class I	Any Amount
Class II	Any Amount
Class III	10 pounds
Class IV	20 pounds
Class V	No Permit Required
Pyrophoric materials	
Gases	Any Amount
Liquids	Any Amount
Solids	Any Amount
Toxic materials	
Gases	See Section 105.6.9
Liquids	10 gallons
Solids	100 pounds
Unstable (reactive) materials	
Liquids	
Class 4	Any Amount
Class 3	Any Amount
Class 2	5 gallons
Class 1	10 gallons
Solids	
Class 4	Any Amount
Class 3	Any Amount
Class 2	50 pounds
Class 1	100 pounds

(continued)

**TABLE 105.6.21—continued
PERMIT AMOUNTS FOR HAZARDOUS MATERIALS**

TYPE OF MATERIAL	AMOUNT
Water-reactive materials	
Liquids	
Class 3	Any Amount
Class 2	5 gallons
Class 1	55 gallons
Solids	
Class 3	Any Amount
Class 2	50 pounds
Class 1	500 pounds

For SI: 1 gallon = 3.785 L, 1 pound = 0.454 kg.

a. 20 gallons where Table 5003.1.1(1) Note k applies and hazard identification signs in accordance with Section 5003.5 are provided for quantities of 20 gallons or less.

b. 200 pounds where Table 5003.1.1(1) Note k applies and hazard identification signs in accordance with Section 5003.5 are provided for quantities of 200 pounds or less.

105.6.23 High-piled storage. An operational permit is required to use a building or portion thereof as a *high-piled storage area* exceeding 500 square feet (46 m^2).

105.6.24 Hot work operations. An operational permit is required for hot work including, but not limited to:

1. Public exhibitions and demonstrations where hot work is conducted.
2. Use of portable hot work equipment inside a structure.

   Exception: Work that is conducted under a construction permit.
3. Fixed-site hot work equipment, such as welding booths.
4. Hot work conducted within a wildfire risk area.
5. Application of roof coverings with the use of an open-flame device.
6. Where *approved*, the *fire code official* shall issue a permit to carry out a hot work program. This program allows *approved* personnel to regulate their facility's hot work operations. The *approved* personnel shall be trained in the fire safety aspects denoted in this chapter and shall be responsible for issuing permits requiring compliance with the requirements found in Chapter 35. These permits shall be issued only to their employees or hot work operations under their supervision.

105.6.25 Industrial ovens. An operational permit is required for operation of industrial ovens regulated by Chapter 30.

105.6.26 Lumber yards and woodworking plants. An operational permit is required for the storage or processing of lumber exceeding 100,000 board feet (8,333 ft^3) (236 m^3).

105.6.27 Liquid- or gas-fueled vehicles or equipment in assembly buildings. An operational permit is required to display, operate or demonstrate liquid- or gas-fueled vehicles or equipment in assembly buildings.

105.6.28 LP-gas. An operational permit is required for:

1. Storage and use of LP-gas.

   Exception: A permit is not required for individual containers with a 500-gallon (1893 L) water capacity or less or multiple container systems having an aggregate quantity not exceeding 500 gallons (1893 L), serving occupancies in Group R-3.

2. Operation of cargo tankers that transport LP-gas.

105.6.29 Magnesium. An operational permit is required to melt, cast, heat treat or grind more than 10 pounds (4.54 kg) of magnesium.

105.6.30 Miscellaneous combustible storage. An operational permit is required to store in any building or upon any premises in excess of 2,500 cubic feet (71 m^3) gross volume of combustible empty packing cases, boxes, barrels or similar containers, rubber tires, rubber, cork or similar combustible material.

105.6.31 Motor fuel-dispensing facilities. An operational permit is required for the operation of automotive, marine and fleet motor fuel-dispensing facilities.

105.6.32 Open burning. An operational permit is required for the kindling or maintaining of an open fire or a fire on any public street, alley, road, or other public or private ground. Instructions and stipulations of the permit shall be adhered to.

Exception: *Recreational fires.*

105.6.33 Open flames and torches. An operational permit is required to remove paint with a torch; or to use a torch or open-flame device in a wildfire risk area.

105.6.34 Open flames and candles. An operational permit is required to use open flames or candles in connection with assembly areas, dining areas of restaurants or drinking establishments.

105.6.35 Organic coatings. An operational permit is required for any organic-coating manufacturing operation producing more than 1 gallon (4 L) of an organic coating in one day.

105.6.36 Places of assembly. An operational permit is required to operate a place of assembly.

105.6.37 Private fire hydrants. An operational permit is required for the removal from service, use or operation of private fire hydrants.

Exception: A permit is not required for private industry with trained maintenance personnel, private fire brigade or fire departments to maintain, test and use private hydrants.

105.6.38 Pyrotechnic special effects material. An operational permit is required for use and handling of pyrotechnic special effects material.

105.6.39 Pyroxylin plastics. An operational permit is required for storage or handling of more than 25 pounds (11 kg) of cellulose nitrate (pyroxylin) plastics, and for the assembly or manufacture of articles involving pyroxylin plastics.

105.6.40 Refrigeration equipment. An operational permit is required to operate a mechanical refrigeration unit or system regulated by Chapter 6.

105.6.41 Repair garages and motor fuel-dispensing facilities. An operational permit is required for operation of repair garages.

105.6.42 Rooftop heliports. An operational permit is required for the operation of a rooftop heliport.

105.6.43 Spraying or dipping. An operational permit is required to conduct a spraying or dipping operation utilizing flammable or *combustible liquids*, or the application of combustible powders regulated by Chapter 24.

105.6.44 Storage of scrap tires and tire byproducts. An operational permit is required to establish, conduct or maintain storage of scrap tires and tire byproducts that exceeds 2,500 cubic feet (71 m^3) of total volume of scrap tires, and for indoor storage of tires and tire byproducts.

105.6.45 Temporary membrane structures and tents. An operational permit is required to operate an air-supported temporary membrane structure, a temporary stage canopy or a tent having an area in excess of 400 square feet (37 m^2).

Exceptions:

1. Tents used exclusively for recreational camping purposes.
2. Tents open on all sides, which comply with all of the following:
   - 2.1. Individual tents having a maximum size of 700 square feet (65 m^2).
   - 2.2. The aggregate area of multiple tents placed side by side without a fire break clearance of not less than 12 feet (3658 mm) shall not exceed 700 square feet (65 m^2) total.
   - 2.3. A minimum clearance of 12 feet (3658 mm) to structures and other tents shall be provided.

105.6.46 Tire-rebuilding plants. An operational permit is required for the operation and maintenance of a tire-rebuilding plant.

105.6.47 Waste handling. An operational permit is required for the operation of wrecking yards, junk yards and waste material-handling facilities.

105.6.48 Wood products. An operational permit is required to store chips, hogged material, lumber or plywood in excess of 200 cubic feet (6 m^3).

[A] 105.7 Required construction permits. The *fire code official* is authorized to issue construction permits for work as set forth in Sections 105.7.1 through 105.7.18.

[A] 105.7.1 Automatic fire-extinguishing systems. A construction permit is required for installation of or modification to an automatic fire-extinguishing system. Maintenance performed in accordance with this code is not considered to be a modification and does not require a permit.

[A] 105.7.2 Battery systems. A permit is required to install stationary storage battery systems having a liquid capacity of more than 50 gallons (189 L).

[A] 105.7.3 Compressed gases. Where the compressed gases in use or storage exceed the amounts listed in Table 105.6.9, a construction permit is required to install, repair damage to, abandon, remove, place temporarily out of service, or close or substantially modify a *compressed gas* system.

Exceptions:

1. Routine maintenance.
2. For emergency repair work performed on an emergency basis, application for permit shall be made within two working days of commencement of work.

[A] 105.7.4 Cryogenic fluids. A construction permit is required for installation of or *alteration* to outdoor stationary *cryogenic fluid* storage systems where the system capacity exceeds the amounts listed in Table 105.6.11. Maintenance performed in accordance with this code is not considered to be an *alteration* and does not require a construction permit.

[A] 105.7.5 Emergency responder radio coverage system. A construction permit is required for installation of or modification to emergency responder radio coverage systems and related equipment. Maintenance performed in accordance with this code is not considered to be a modification and does not require a construction permit.

[A] 105.7.6 Fire alarm and detection systems and related equipment. A construction permit is required for installation of or modification to fire alarm and detection systems and related equipment. Maintenance performed in accordance with this code is not considered to be a modification and does not require a construction permit.

[A] 105.7.7 Fire pumps and related equipment. A construction permit is required for installation of or modification to fire pumps and related fuel tanks, jockey pumps, controllers and generators. Maintenance performed in accordance with this code is not considered to be a modification and does not require a construction permit.

[A] 105.7.8 Flammable and combustible liquids. A construction permit is required:

1. To install, repair or modify a pipeline for the transportation of flammable or *combustible liquids*.
2. To install, construct or alter tank vehicles, equipment, tanks, plants, terminals, wells, fuel-dispensing stations, refineries, distilleries and similar facilities where flammable and *combustible liquids* are produced, processed, transported, stored, dispensed or used.
3. To install, alter, remove, abandon or otherwise dispose of a flammable or *combustible liquid* tank.

[A] 105.7.9 Gates and barricades across fire apparatus access roads. A construction permit is required for the installation of or modification to a gate or barricade across a fire apparatus access road.

[A] 105.7.10 Hazardous materials. A construction permit is required to install, repair damage to, abandon, remove, place temporarily out of service, or close or substantially modify a storage facility or other area regulated by Chapter 50 where the hazardous materials in use or storage exceed the amounts listed in Table 105.6.21.

Exceptions:

1. Routine maintenance.
2. For repair work performed on an emergency basis, application for permit shall be made within two working days of commencement of work.

[A] 105.7.11 Industrial ovens. A construction permit is required for installation of industrial ovens covered by Chapter 30.

Exceptions:

1. Routine maintenance.
2. For repair work performed on an emergency basis, application for permit shall be made within two working days of commencement of work.

[A] 105.7.12 LP-gas. A construction permit is required for installation of or modification to an LP-gas system. Maintenance performed in accordance with this code is not considered to be a modification and does not require a permit.

[A] 105.7.13 Private fire hydrants. A construction permit is required for the installation or modification of private fire hydrants. Maintenance performed in accordance with this code is not considered to be a modification and does not require a permit.

[A] 105.7.14 Smoke control or smoke exhaust systems. Construction permits are required for installation of or alteration to smoke control or smoke exhaust systems. Maintenance performed in accordance with this code is not considered to be an alteration and does not require a permit.

[A] 105.7.15 Solar photovoltaic power systems. A construction permit is required to install or modify solar photovoltaic power systems. Maintenance performed in accordance with this code is not considered to be a modification and does not require a permit.

[A] 105.7.16 Spraying or dipping. A construction permit is required to install or modify a spray room, dip tank or booth. Maintenance performed in accordance with this code is not considered to be a modification and does not require a permit.

[A] 105.7.17 Standpipe systems. A construction permit is required for the installation, modification or removal from service of a standpipe system. Maintenance performed in accordance with this code is not considered to be a modification and does not require a permit.

[A] 105.7.18 Temporary membrane structures and tents. A construction permit is required to erect an air-supported temporary membrane structure, a temporary stage canopy or a tent having an area in excess of 400 square feet (37 m^2).

Exceptions:

1. Tents used exclusively for recreational camping purposes.
2. Funeral tents and curtains, or extensions attached thereto, when used for funeral services.
3. Tents and awnings open on all sides, which comply with all of the following:
   - 3.1. Individual tents shall have a maximum size of 700 square feet (65 m^2).
   - 3.2. The aggregate area of multiple tents placed side by side without a fire break clearance of not less than 12 feet (3658 mm) shall not exceed 700 square feet (65 m^2) total.
   - 3.3. A minimum clearance of 12 feet (3658 mm) to structures and other tents shall be maintained.

SECTION 106 INSPECTIONS

[A] 106.1 Inspection authority. The *fire code official* is authorized to enter and examine any building, structure, marine vessel, vehicle or premises in accordance with Section 104.3 for the purpose of enforcing this code.

[A] 106.2 Inspections. The *fire code official* is authorized to conduct such inspections as are deemed necessary to determine the extent of compliance with the provisions of this code and to approve reports of inspection by *approved* agencies or individuals. Reports of such inspections shall be prepared and submitted in writing for review and approval. Inspection reports shall be certified by a responsible officer of such *approved* agency or by the responsible individual. The *fire code official* is authorized to engage such expert opinion as deemed necessary to report upon unusual, detailed or complex technical issues subject to the approval of the governing body.

[A] 106.2.1 Inspection requests. It shall be the duty of the holder of the permit or their duly authorized agent to notify the *fire code official* when work is ready for inspection. It shall be the duty of the permit holder to provide access to and means for inspections of such work that are required by this code.

[A] 106.2.2 Approval required. Work shall not be done beyond the point indicated in each successive inspection without first obtaining the approval of the *fire code official.* The *fire code official,* upon notification, shall make the requested inspections and shall either indicate the portion of the construction that is satisfactory as completed, or notify the permit holder or his or her agent wherein the same fails to comply with this code. Any portions that do not comply shall be corrected, and such portion shall not be covered or concealed until authorized by the *fire code official.*

[A] 106.3 Concealed work. It shall be the duty of the permit applicant to cause the work to remain accessible and exposed for inspection purposes. Where any installation subject to inspection prior to use is covered or concealed without having first been inspected, the *fire code official* shall have the authority to require that such work be exposed for inspection. Neither the *fire code official* nor the jurisdiction shall be liable for expense entailed in the removal or replacement of any material required to allow inspection.

[A] 106.4 Approvals. Approval as the result of an inspection shall not be construed to be an approval of a violation of the provisions of this code or of other ordinances of the jurisdiction. Inspections presuming to give authority to violate or cancel provisions of this code or of other ordinances of the jurisdiction shall not be valid.

SECTION 107 MAINTENANCE

[A] 107.1 Maintenance of safeguards. Where any device, equipment, system, condition, arrangement, level of protection, or any other feature is required for compliance with the provisions of this code, or otherwise installed, such device, equipment, system, condition, arrangement, level of protection, or other feature shall thereafter be continuously maintained in accordance with this code and applicable referenced standards.

[A] 107.2 Testing and operation. Equipment requiring periodic testing or operation to ensure maintenance shall be tested or operated as specified in this code.

[A] 107.2.1 Reinspection and testing. Where any work or installation does not pass an initial test or inspection, the necessary corrections shall be made so as to achieve compliance with this code. The work or installation shall then be resubmitted to the *fire code official* for inspection and testing.

[A] 107.3 Recordkeeping. A record of periodic inspections, tests, servicing and other operations and maintenance shall be maintained on the premises or other *approved* location for not less than 3 years, or a different period of time where specified in this code or referenced standards. Records shall be made available for inspection by the *fire code official,* and a copy of the records shall be provided to the *fire code official* upon request.

The *fire code official* is authorized to prescribe the form and format of such recordkeeping. The *fire code official* is authorized to require that certain required records be filed with the *fire code official.*

[A] 107.4 Supervision. Maintenance and testing shall be under the supervision of a responsible person who shall ensure that such maintenance and testing are conducted at specified intervals in accordance with this code.

107.5 Rendering equipment inoperable. Portable or fixed fire-extinguishing systems or devices, and fire-warning systems, shall not be rendered inoperative or inaccessible, except as necessary during emergencies, maintenance, repairs, *alterations,* drills or prescribed testing.

[A] 107.6 Overcrowding. Overcrowding or admittance of any person beyond the *approved* capacity of a building or a portion thereof shall not be allowed. The *fire code official,* upon finding any overcrowding conditions or obstructions in *aisles,* passageways or other *means of egress,* or upon finding any condition that constitutes a life safety hazard, shall be authorized to cause the event to be stopped until such condition or obstruction is corrected.

SECTION 108 BOARD OF APPEALS

[A] 108.1 Board of appeals established. In order to hear and decide appeals of orders, decisions or determinations made by the *fire code official* relative to the application and interpretation of this code, there shall be and is hereby created a board of appeals. The board of appeals shall be appointed by the governing body and shall hold office at its pleasure. The *fire code official* shall be an ex officio member of said board but shall not have a vote on any matter before the board. The board shall adopt rules of procedure for conducting its business, and shall render all decisions and findings in writing to the appellant with a duplicate copy to the *fire code official.*

[A] 108.2 Limitations on authority. An application for appeal shall be based on a claim that the intent of this code or the rules legally adopted hereunder have been incorrectly interpreted, the provisions of this code do not fully apply, or an equivalent method of protection or safety is proposed. The board shall not have authority to waive requirements of this code.

[A] 108.3 Qualifications. The board of appeals shall consist of members who are qualified by experience and training to pass on matters pertaining to hazards of fire, explosions, hazardous conditions or *fire protection systems*, and are not employees of the jurisdiction.

SECTION 109 VIOLATIONS

[A] 109.1 Unlawful acts. It shall be unlawful for a person, firm or corporation to erect, construct, alter, repair, remove, demolish or utilize a building, occupancy, premises or system regulated by this code, or cause same to be done, in conflict with or in violation of any of the provisions of this code.

[A] 109.2 Owner/occupant responsibility. Correction and abatement of violations of this code shall be the responsibility of the *owner* or the owner's authorized agent. Where an occupant creates, or allows to be created, hazardous conditions in violation of this code, the occupant shall be held responsible for the abatement of such hazardous conditions.

[A] 109.3 Notice of violation. Where the *fire code official* finds a building, premises, vehicle, storage facility or outdoor area that is in violation of this code, the *fire code official* is authorized to prepare a written notice of violation describing the conditions deemed unsafe and, where compliance is not immediate, specifying a time for reinspection.

[A] 109.3.1 Service. A notice of violation issued pursuant to this code shall be served upon the *owner*, the owner's authorized agent, operator, occupant or other person responsible for the condition or violation, either by personal service, mail or by delivering the same to, and leaving it with, some person of responsibility upon the premises. For unattended or abandoned locations, a copy of such notice of violation shall be posted on the premises in a conspicuous place at or near the entrance to such premises and the notice of violation shall be mailed by certified mail with return receipt requested or a certificate of mailing, to the last known address of the *owner*, the owner's authorized agent, or occupant.

[A] 109.3.2 Compliance with orders and notices. A notice of violation issued or served as provided by this code shall be complied with by the *owner*, the owner's authorized agent, operator, occupant or other person responsible for the condition or violation to which the notice of violation pertains.

[A] 109.3.3 Prosecution of violations. If the notice of violation is not complied with promptly, the *fire code official* is authorized to request the legal counsel of the jurisdiction to institute the appropriate legal proceedings at law or in equity to restrain, correct or abate such violation or to require removal or termination of the unlawful occupancy of the structure in violation of the provisions of this code or of the order or direction made pursuant hereto.

[A] 109.3.4 Unauthorized tampering. Signs, tags or seals posted or affixed by the *fire code official* shall not be mutilated, destroyed or tampered with, or removed, without authorization from the *fire code official.*

[A] 109.4 Violation penalties. Persons who shall violate a provision of this code or shall fail to comply with any of the requirements thereof or who shall erect, install, alter, repair or do work in violation of the *approved construction documents* or directive of the *fire code official*, or of a permit or certificate used under provisions of this code, shall be guilty of a [SPECIFY OFFENSE], punishable by a fine of not more than [AMOUNT] dollars or by imprisonment not exceeding [NUMBER OF DAYS], or both such fine and imprisonment. Each day that a violation continues after due notice has been served shall be deemed a separate offense.

[A] 109.4.1 Abatement of violation. In addition to the imposition of the penalties herein described, the *fire code official* is authorized to institute appropriate action to prevent unlawful construction or to restrain, correct or abate a violation; or to prevent illegal occupancy of a structure or premises; or to stop an illegal act, conduct of business or occupancy of a structure on or about any premises.

SECTION 110 UNSAFE BUILDINGS

[A] 110.1 General. If during the inspection of a premises, a building or structure, or any building system, in whole or in part, constitutes a clear and inimical threat to human life, safety or health, the *fire code official* shall issue such notice or orders to remove or remedy the conditions as shall be deemed necessary in accordance with this section, and shall refer the building to the building department for any repairs, *alterations*, remodeling, removing or demolition required.

[A] 110.1.1 Unsafe conditions. Structures or existing equipment that are or hereafter become unsafe or deficient because of inadequate *means of egress* or which constitute a fire hazard, or are otherwise dangerous to human life or the public welfare, or which involve illegal or improper occupancy or inadequate maintenance, shall be deemed an unsafe condition. A vacant structure that is not secured against unauthorized entry as required by Section 311 shall be deemed unsafe.

[A] 110.1.2 Structural hazards. Where an apparent structural hazard is caused by the faulty installation, operation or malfunction of any of the items or devices governed by this code, the *fire code official* shall immediately notify the building code official in accordance with Section 110.1.

[A] 110.2 Evacuation. The *fire code official* or the fire department official in charge of an incident shall be authorized to order the immediate evacuation of any occupied building deemed unsafe where such building has hazardous conditions that present imminent danger to building occupants. Persons so notified shall immediately leave the structure or premises and shall not enter or re-enter until authorized to do so by the *fire code official* or the fire department official in charge of the incident.

[A] 110.3 Summary abatement. Where conditions exist that are deemed hazardous to life and property, the *fire code official* or fire department official in charge of the incident is authorized to abate summarily such hazardous conditions that are in violation of this code.

[A] 110.4 Abatement. The *owner*, the owner's authorized agent, operator or occupant of a building or premises deemed unsafe by the *fire code official* shall abate or cause to be abated or corrected such unsafe conditions either by repair, rehabilitation, demolition or other *approved* corrective action.

SECTION 111
STOP WORK ORDER

[A] 111.1 Order. Where the *fire code official* finds any work regulated by this code being performed in a manner contrary to the provisions of this code, or in a dangerous or unsafe manner, the *fire code official* is authorized to issue a stop work order.

[A] 111.2 Issuance. A stop work order shall be in writing and shall be given to the *owner* of the property, or to the *owner's* authorized agent, or to the person doing the work. Upon issuance of a stop work order, the cited work shall immediately cease. The stop work order shall state the reason for the order, and the conditions under which the cited work is authorized to resume.

[A] 111.3 Emergencies. Where an emergency exists, the *fire code official* shall not be required to give a written notice prior to stopping the work.

[A] 111.4 Failure to comply. Any person who shall continue any work after having been served with a stop work order, except such work as that person is directed to perform to remove a violation or unsafe condition, shall be liable to a fine of not less than [AMOUNT] dollars or more than [AMOUNT] dollars.

SECTION 112
SERVICE UTILITIES

[A] 112.1 Authority to disconnect service utilities. The *fire code official* shall have the authority to authorize disconnection of utility service to the building, structure or system in order to safely execute emergency operations or to eliminate an immediate hazard. The *fire code official* shall notify the serving utility and, where possible, the *owner* or the owner's authorized agent and the occupant of the building, structure or service system of the decision to disconnect prior to taking such action. If not notified prior to disconnection, then the *owner*, the owner's authorized agent or occupant of the building, structure or service system shall be notified in writing as soon as practical thereafter.

SECTION 113
FEES

[A] 113.1 Fees. A permit shall not be issued until the fees have been paid, nor shall an amendment to a permit be released until the additional fee, if any, has been paid.

[A] 113.2 Schedule of permit fees. A fee for each permit shall be paid as required, in accordance with the schedule as established by the applicable governing authority.

[A] 113.3 Work commencing before permit issuance. A person who commences any work, activity or operation regulated by this code before obtaining the necessary permits shall be subject to an additional fee established by the applicable governing authority, which shall be in addition to the required permit fees.

[A] 113.4 Related fees. The payment of the fee for the construction, *alteration*, removal or demolition of work done in connection to or concurrently with the work or activity authorized by a permit shall not relieve the applicant or holder of the permit from the payment of other fees that are prescribed by law.

[A] 113.5 Refunds. The applicable governing authority is authorized to establish a refund policy.

CHAPTER 2

DEFINITIONS

SECTION 201 GENERAL

201.1 Scope. Unless otherwise expressly stated, the following words and terms shall, for the purposes of this code, have the meanings shown in this chapter.

201.2 Interchangeability. Words used in the present tense include the future; words stated in the masculine gender include the feminine and neuter; the singular number includes the plural and the plural, the singular.

201.3 Terms defined in other codes. Where terms are not defined in this code and are defined in the *International Building Code*, *International Fuel Gas Code*, *International Mechanical Code* or *International Plumbing Code*, such terms shall have the meanings ascribed to them as in those codes.

201.4 Terms not defined. Where terms are not defined through the methods authorized by this section, such terms shall have ordinarily accepted meanings such as the context implies. *Merriam Webster's Collegiate Dictionary, 11th Edition*, shall be considered as providing ordinarily accepted meanings.

SECTION 202 GENERAL DEFINITIONS

[BG] 24-HOUR BASIS. The actual time that a person is an occupant within a facility for the purpose of receiving care. It shall not include a facility that is open for 24 hours and is capable of providing care to someone visiting the facility during any segment of the 24 hours.

[BE] ACCESSIBLE MEANS OF EGRESS. A continuous and unobstructed way of egress travel from any *accessible* point in a building or facility to a *public way*.

[BE] ACCESSIBLE ROUTE. A continuous, unobstructed path that complies with Chapter 11 of the *International Building Code*.

AEROSOL. A product that is dispensed from an aerosol container by a propellant.

Aerosol products shall be classified by means of the calculation of their chemical heats of combustion and shall be designated Level 1, Level 2 or Level 3.

Level 1 aerosol products. Those with a total chemical heat of combustion that is less than or equal to 8,600 British thermal units per pound (Btu/lb) (20 kJ/g).

Level 2 aerosol products. Those with a total chemical heat of combustion that is greater than 8,600 Btu/lb (20 kJ/g), but less than or equal to 13,000 Btu/lb (30 kJ/g).

Level 3 aerosol products. Those with a total chemical heat of combustion that is greater than 13,000 Btu/lb (30 kJ/g).

AEROSOL CONTAINER. A metal can, or a glass or plastic bottle designed to dispense an aerosol.

AEROSOL WAREHOUSE. A building used for warehousing aerosol products.

AGENCY. Any emergency responder department within the jurisdiction that utilizes radio frequencies for communication. This could include, but not be limited to, various public safety agencies such as fire departments, emergency medical services and law enforcement.

AGENT. A person who shall have charge, care or control of any structure as *owner*, or agent of the *owner*, or as executor, executrix, administrator, administratrix, trustee or guardian of the estate of the *owner*. Any such person representing the actual *owner* shall be bound to comply with the provisions of this code to the same extent as if that person was the *owner*.

[BG] AGRICULTURAL BUILDING. A structure designed and constructed to house farm implements, hay, grain, poultry, livestock or other horticultural products. This structure shall not be a place of human habitation or a place of employment where agricultural products are processed, treated or packaged, nor shall it be a place used by the public.

AGRO-INDUSTRIAL. A facility, or portion thereof, housing operations involving the transforming of raw agricultural products into intermediate or consumable products.

[BG] AIR-INFLATED STRUCTURE. A structure that uses air-pressurized membrane beams, arches or other elements to enclose space. Occupants of such a structure do not occupy the pressurized areas used to support the structure.

[BG] AIR-SUPPORTED STRUCTURE. A structure wherein the shape of the structure is attained by air pressure, and occupants of the structure are within the elevated pressure area. Air supported structures are of two basic types:

Double skin. Similar to a single skin, but with an attached liner that is separated from the outer skin and provides an airspace which serves for insulation, acoustic, aesthetic or similar purposes.

Single skin. Where there is only the single outer skin and the air pressure is directly against that skin.

AIRCRAFT MOTOR-VEHICLE FUEL-DISPENSING FACILITY. That portion of property where flammable or *combustible liquids* or gases used as motor fuels are stored and dispensed from fixed automotive-type equipment into the fuel tanks of aircraft.

AIRCRAFT OPERATION AREA (AOA). Any area used or intended for use for the parking, taxiing, takeoff, landing or other ground-based aircraft activity.

AIRPORT. An area of land or structural surface that is used, or intended for use, for the landing and taking off of aircraft with an overall length greater than 39 feet (11 887 mm) and an overall exterior fuselage width greater than 6.6 feet (2012 mm), and any appurtenant areas that are used or intended for use for airport buildings and other airport facilities.

[BE] AISLE. An unenclosed *exit access* component that defines and provides a path of egress travel.

[BE] AISLE ACCESSWAY. That portion of an *exit access* that leads to an *aisle*.

ALARM, NUISANCE. See "Nuisance alarm."

ALARM DEVICE, MULTIPLE STATION. See "Multiple-station alarm device."

ALARM NOTIFICATION APPLIANCE. A fire alarm system component such as a bell, horn, speaker, light or text display that provides audible, tactile or visible outputs, or any combination thereof. See also "Audible alarm notification appliance" or "Visible alarm notification appliance."

ALARM SIGNAL. A signal indicating an emergency requiring immediate action, such as a signal indicative of fire.

ALARM VERIFICATION FEATURE. A feature of automatic fire detection and alarm systems to reduce unwanted alarms wherein smoke detectors report alarm conditions for a minimum period of time, or confirm alarm conditions within a given time period, after being automatically reset, in order to be accepted as a valid alarm-initiation signal.

ALCOHOL-BASED HAND RUB. An alcohol-containing preparation designed for application to the hands for reducing the number of viable microorganisms on the hands and containing ethanol or isopropanol in an amount not exceeding 95-percent by volume.

ALCOHOL-BLENDED FUELS. Flammable liquids consisting of 10-percent or greater, by volume, ethanol or other alcohols blended with gasoline.

[A] ALTERATION. Any construction or renovation to an existing structure other than a repair or addition.

[BE] ALTERNATING TREAD DEVICE. A device that has a series of steps between 50 and 70 degrees (0.87 and 1.22 rad) from horizontal, usually attached to a center support rail in an alternating manner so that the user does not have both feet on the same level at the same time.

[BG] AMBULATORY CARE FACILITY. Buildings or portions thereof used to provide medical, surgical, psychiatric, nursing or similar care on a less-than-24-hour basis to persons who are rendered incapable of self-preservation by the services provided.

AMMONIUM NITRATE. A chemical compound represented by the formula NH_4NO_3.

ANNUNCIATOR. A unit containing one or more indicator lamps, alphanumeric displays or other equivalent means in which each indication provides status information about a circuit, condition or location.

[A] APPROVED. Acceptable to the *fire code official*.

[BG] AREA, BUILDING. The area included within surrounding *exterior walls* (or *exterior walls* and *fire walls*) exclusive of vent shafts and *courts*. Areas of the building not provided with surrounding walls shall be included in the building area if such areas are included within the horizontal projection of the roof or floor above.

[BE] AREA OF REFUGE. An area where persons unable to use *stairways* can remain temporarily to await instructions or assistance during emergency evacuation.

ARRAY. The configuration of storage. Characteristics considered in defining an array include the type of packaging, flue spaces, height of storage and compactness of storage.

ARRAY, CLOSED. A storage configuration having a 6-inch (152 mm) or smaller width vertical flue space that restricts air movement through the stored commodity.

[BG] ATRIUM. An opening connecting two or more stories other than enclosed *stairways*, elevators, hoistways, escalators, plumbing, electrical, air-conditioning or other equipment, which is closed at the top and not defined as a mall. Stories, as used in this definition, do not include balconies within assembly groups or mezzanines that comply with Section 505 of the *International Building Code*.

[BG] ATTIC. The space between the ceiling beams of the top story and the roof rafters.

AUDIBLE ALARM NOTIFICATION APPLIANCE. A notification appliance that alerts by the sense of hearing.

AUTOMATED RACK STORAGE. Automated rack storage is a stocking method whereby the movement of pallets, products, apparatus or systems are automatically controlled by mechanical or electronic devices.

AUTOMATIC. As applied to fire protection devices, a device or system providing an emergency function without the necessity for human intervention and activated as a result of a predetermined temperature rise, rate of temperature rise or combustion products.

AUTOMATIC FIRE-EXTINGUISHING SYSTEM. An *approved* system of devices and equipment which automatically detects a fire and discharges an *approved* fire-extinguishing agent onto or in the area of a fire.

AUTOMATIC SMOKE DETECTION SYSTEM. A fire alarm system that has initiation devices that utilize smoke detectors for protection of an area such as a room or space with detectors to provide early warning of fire.

AUTOMATIC SPRINKLER SYSTEM. An *automatic sprinkler system*, for fire protection purposes, is an integrated system of underground and overhead piping designed in accordance with fire protection engineering standards. The system includes a suitable water supply. The portion of the system above the ground is a network of specially sized or hydraulically designed piping installed in a structure or area, generally overhead, and to which automatic sprinklers are connected in a systematic pattern. The system is usually activated by heat from a fire and discharges water over the fire area.

AUTOMATIC WATER MIST SYSTEM. A system consisting of a water supply, a pressure source and a distribution piping system with attached nozzles which, at or above a minimum operating pressure, defined by its listing, discharges water in fine droplets meeting the requirements of NFPA 750 for the purpose of the control, suppression or extinguishment of a fire. Such systems include wet-pipe, dry-pipe and pre-action types. The systems are designed as engineered, preengineered, local-application or total flooding systems.

AUTOMOTIVE MOTOR FUEL-DISPENSING FACILITY. That portion of property where flammable or *combustible liquids* or gases used as motor fuels are stored and dispensed from fixed equipment into the fuel tanks of motor vehicles.

AVERAGE AMBIENT SOUND LEVEL. The root mean square, A-weighted sound pressure level measured over a 24-hour period, or the time any person is present, whichever time period is less.

[BG] AWNING. An architectural projection that provides weather protection, identity or decoration and is partially or wholly supported by the building to which it is attached. An awning is comprised of a lightweight frame structure over which a covering is attached.

[BE] BALANCED DOOR. A door equipped with double-pivoted hardware so designed as to cause a semicounter balanced swing action when opening.

BALED COTTON. See "Cotton."

BALED COTTON, DENSELY PACKED. See "Cotton."

BARRICADE. A structure that consists of a combination of walls, floor and roof, which is designed to withstand the rapid release of energy in an explosion and which is fully confined, partially vented or fully vented; or other effective method of shielding from *explosive materials* by a natural or artificial barrier.

Artificial barricade. An artificial mound or revetment with a minimum thickness of 3 feet (914 mm).

Natural barricade. Natural features of the ground, such as hills, or timber of sufficient density that the surrounding exposures that require protection cannot be seen from the magazine or building containing *explosives* when the trees are bare of leaves.

BARRICADED. The effective screening of a building containing *explosive materials* from the magazine or other building, railway or highway by a natural or an artificial barrier. A straight line from the top of any sidewall of the building containing *explosive materials* to the eave line of any magazine or other building or to a point 12 feet (3658 mm) above the center of a railway or highway shall pass through such barrier.

[BG] BASEMENT. A story that is not a story above grade plane.

BATTERY SYSTEM, STATIONARY LEAD ACID. A system which consists of three interconnected subsystems:

1. A lead-acid battery.
2. A battery charger.
3. A collection of rectifiers, inverters, converters and associated electrical equipment as required for a particular application.

BATTERY TYPES.

Lithium-ion battery. A storage battery that consists of lithium ions embedded in a carbon graphite or nickel metal-oxide substrate. The electrolyte is a carbonate mixture or a gelled polymer. The lithium ions are the charge carriers of the battery.

Lithium metal polymer battery. A storage battery that is comprised of nonaqueous liquid or polymerized electrolytes, which provide ionic conductivity between lithiated positive active material electrically separated from metallic lithium or lithiated negative active material.

Nickel cadmium (Ni-Cd) battery. An alkaline storage battery in which the positive active material is nickel oxide, the negative contains cadmium and the electrolyte is potassium hydroxide.

Nonrecombinant battery. A storage battery in which, under conditions of normal use, hydrogen and oxygen gasses created by electrolysis are vented into the air outside of the battery.

Recombinant battery. A storage battery in which, under conditions of normal use, hydrogen and oxygen gases created by electrolysis are converted back into water inside the battery instead of venting into the air outside of the battery.

Stationary storage battery. A group of electrochemical cells interconnected to supply a nominal voltage of DC power to a suitably connected electrical load, designed for service in a permanent location. The number of cells connected in a series determines the nominal voltage rating of the battery. The size of the cells determines the discharge capacity of the entire battery. After discharge, it may be restored to a fully charged condition by an electric current flowing in a direction opposite to the flow of current when the battery is discharged.

Valve-regulated lead-acid (VRLA) battery. A lead-acid battery consisting of sealed cells furnished with a valve that opens to vent the battery whenever the internal pressure of the battery exceeds the ambient pressure by a set amount. In VRLA batteries, the liquid electrolyte in the cells is immobilized in an absorptive glass mat (AGM cells or batteries) or by the addition of a gelling agent (gel cells or gelled batteries).

Vented (flooded) lead-acid battery. A lead-acid battery consisting of cells that have electrodes immersed in liquid electrolyte. Flooded lead-acid batteries have a provision for the user to add water to the cell and are equipped with a flame-arresting vent which permits the escape of hydrogen and oxygen gas from the cell in a diffused manner such that a spark, or other ignition source, outside the cell will not ignite the gases inside the cell.

BIN BOX. A five-sided container with the open side facing an aisle. Bin boxes are self-supporting or supported by a structure designed so that little or no horizontal or vertical space exists around the boxes.

BIOMASS. Plant- or animal-based material of biological origin excluding material embedded in geologic formations or transformed into fossils.

BLAST AREA. The area including the blast site and the immediate adjacent area within the influence of flying rock, missiles and concussion.

BLAST SITE. The area in which *explosive materials* are being or have been loaded and which includes all holes loaded or to be loaded for the same blast and a distance of 50 feet (15 240 mm) in all directions.

BLASTER. A person qualified in accordance with Section 3301.4 to be in charge of and responsible for the loading and firing of a blast.

BLASTING AGENT. A material or mixture consisting of fuel and oxidizer, intended for blasting provided that the finished product, as mixed for use or shipment, cannot be detonated by means of a No. 8 test detonator when unconfined. Blasting agents are labeled and placarded as Class 1.5 material by US DOTn.

[BE] BLEACHERS. Tiered seating supported on a dedicated structural system and two or more rows high and is not a building element (see "*Grandstand*").

[BG] BOARDING HOUSE. A building arranged or used for lodging for compensation, with or without meals, and not occupied as a single-family unit.

BOILING POINT. The temperature at which the vapor pressure of a liquid equals the atmospheric pressure of 14.7 pounds per square inch absolute (psia) (101 kPa) or 760 mm of mercury. Where an accurate boiling point is unavailable for the material in question, or for mixtures which do not have a constant boiling point, for the purposes of this classification, the 20-percent evaporated point of a distillation performed in accordance with ASTM D 86 shall be used as the boiling point of the liquid.

BONFIRE. An outdoor fire utilized for ceremonial purposes.

[BE] BREAKOUT. For revolving doors, a process whereby wings or door panels can be pushed open manually for *means of egress* travel.

BRITISH THERMAL UNIT (BTU). The heat necessary to raise the temperature of 1 pound (0.454 kg) of water by 1°F (0.5565°C).

[A] BUILDING. Any structure used or intended for supporting or sheltering any use or occupancy.

BUILDING AREA. See "Area, building."

BUILDING HEIGHT. See "Height, building."

[A] BUILDING OFFICIAL. The officer or other designated authority charged with the administration and enforcement of the *International Building Code*, or a duly authorized representative.

BULK OXYGEN SYSTEM. An assembly of equipment, such as oxygen storage containers, pressure regulators, safety devices, vaporizers, manifolds and interconnecting piping, that has a storage capacity of more than 20,000 cubic feet (566 m^3) of oxygen at *normal temperature and pressure (NTP)* including unconnected reserves on hand at the site. The bulk oxygen system terminates at the point where oxygen at service pressure first enters the supply line. The oxygen containers can be stationary or movable, and the oxygen can be stored as a gas or liquid.

BULK PLANT OR TERMINAL. That portion of a property where flammable or *combustible liquids* are received by tank vessel, pipelines, tank car or tank vehicle and are stored or blended in bulk for the purpose of distributing such liquids by tank vessel, pipeline, tank car, tank vehicle, portable tank or container.

BULK TRANSFER. The loading or unloading of flammable or *combustible liquids* from or between tank vehicles, tank cars or storage tanks.

BULLET RESISTANT. Constructed so as to resist penetration of a bullet of 150-grain M2 ball ammunition having a nominal muzzle velocity of 2,700 feet per second (fps) (824 mps) when fired from a 30-caliber rifle at a distance of 100 feet (30 480 mm), measured perpendicular to the target.

CANOPY. A structure or architectural projection of rigid construction over which a covering is attached that provides weather protection, identity or decoration, and may be structurally independent or supported by attachment to a building on one end and by not less than one stanchion on the outer end.

CARBON DIOXIDE EXTINGUISHING SYSTEM. A system supplying carbon dioxide (CO_2) from a pressurized vessel through fixed pipes and nozzles. The system includes a manual- or automatic-actuating mechanism.

[BG] CARE SUITE. In Group I-2 occupancies, a group of treatment rooms, care recipient sleeping rooms and the support rooms or spaces and circulation space within the suite where staff are in attendance for supervision of all care recipients within the suite, and the suite is in compliance with the requirements of Section 407.4.4 of the *International Building Code*.

CARTON. A cardboard or fiberboard box enclosing a product.

CEILING LIMIT. The maximum concentration of an airborne contaminant to which one may be exposed. The ceiling limits utilized are those published in DOL 29 CFR Part 1910.1000. The ceiling Recommended Exposure Limit (REL-C) concentrations published by the U.S. National Institute for Occupational Safety and Health (NIOSH), Threshold Limit Value-Ceiling (TLV-C) concentrations published by the American Conference of Governmental Industrial Hygienists (ACGIH), Ceiling Workplace Environmental Exposure Level (WEEL-Ceiling) Guides published by the American Industrial Hygiene Association (AIHA), and other *approved*, consistent measures are allowed as surrogates for hazardous substances not listed in DOL 29 CFR Part 1910.1000.

[A] CHANGE OF OCCUPANCY. A change in the use of a building or a portion of a building. A change of occupancy shall include any change of occupancy classification, any change from one group to another group within an occupancy classification or any change in use within a group for a specific occupancy classification.

CHEMICAL. An element, chemical compound or mixture of elements or compounds or both.

CHEMICAL NAME. The scientific designation of a chemical in accordance with the nomenclature system developed by the International Union of Pure and Applied Chemistry, the Chemical Abstracts Service rules of nomenclature, or a name which will clearly identify a chemical for the purpose of conducting an evaluation.

[M] CHIMNEY. A primarily vertical structure containing one or more flues for the purpose of carrying gaseous products of combustion and air from a fuel-burning appliance to the outdoor atmosphere.

Factory-built chimney. A listed and labeled chimney composed of factory-made components, assembled in the field in accordance with manufacturer's instructions and the conditions of the listing.

Masonry chimney. A field-constructed chimney composed of solid masonry units, bricks, stones, or concrete.

Metal chimney. A field-constructed chimney of metal.

CLEAN AGENT. Electrically nonconducting, volatile or gaseous fire extinguishant that does not leave a residue upon evaporation.

[BG] CLINIC, OUTPATIENT. Buildings or portions thereof used to provide medical care on a less-than-24-hour basis to persons who are not rendered incapable of self-preservation by the services provided.

CLOSED CONTAINER. A container sealed by means of a lid or other device such that liquid, vapor or dusts will not escape from it under ordinary conditions of use or handling.

CLOSED SYSTEM. The use of a solid or liquid hazardous material involving a closed vessel or system that remains closed during normal operations where vapors emitted by the product are not liberated outside of the vessel or system and the product is not exposed to the atmosphere during normal operations; and all uses of *compressed gases*. Examples of closed systems for solids and liquids include product conveyed through a piping system into a closed vessel, system or piece of equipment.

COLD DECK. A pile of unfinished cut logs.

COMBUSTIBLE DUST. Finely divided solid material which is 420 microns or less in diameter and which, when dispersed in air in the proper proportions, could be ignited by a flame, spark or other source of ignition. Combustible dust will pass through a U.S. No. 40 standard sieve.

COMBUSTIBLE FIBERS. Readily ignitable and free-burning materials in a fibrous or shredded form, such as cocoa fiber, cloth, cotton, excelsior, hay, hemp, henequen, istle, jute, kapok, oakum, rags, sisal, Spanish moss, straw, tow, wastepaper, certain synthetic fibers or other like materials. This definition does not include densely packed baled cotton.

COMBUSTIBLE GAS DETECTOR. An instrument that samples the local atmosphere and indicates the presence of ignitable vapors or gases within the flammable or explosive range expressed as a volume percent in air.

COMBUSTIBLE LIQUID. A liquid having a closed cup flash point at or above 100°F (38°C). Combustible liquids shall be subdivided as follows:

Class II. Liquids having a closed cup flash point at or above 100°F (38°C) and below 140°F (60°C).

Class IIIA. Liquids having a closed cup flash point at or above 140°F (60°C) and below 200°F (93°C).

Class IIIB. Liquids having closed cup *flash points* at or above 200°F (93°C).

The category of combustible liquids does not include *compressed gases* or *cryogenic fluids*.

[M] COMMERCIAL COOKING APPLIANCES. Appliances used in a commercial food service establishment for heating or cooking food and which produce grease vapors, steam, fumes, smoke or odors that are required to be removed through a local exhaust ventilation system. Such appliances include deep fat fryers, upright broilers, griddles, broilers, steam-jacketed kettles, hot-top ranges, under-fired broilers (charbroilers), ovens, barbecues, rotisseries, and similar appliances. For the purpose of this definition, a food service establishment shall include any building or a portion thereof used for the preparation and serving of food.

COMMERCIAL MOTOR VEHICLE. A motor vehicle used to transport passengers or property where the motor vehicle:

1. Has a gross vehicle weight rating of 10,000 pounds (454 kg) or more; or
2. Is designed to transport 16 or more passengers, including the driver.

COMMODITY. A combination of products, packing materials and containers.

[BE] COMMON PATH OF EGRESS TRAVEL. That portion of the *exit access* travel distance measured from the most remote point within a story to that point where the occupants have separate and distinct access to two *exits* or *exit access* doorways.

[BE] COMMON USE. Interior or exterior circulation paths, rooms, spaces or elements that are not for public use and are made available for the shared use of two or more people.

COMPRESSED GAS. A material, or mixture of materials that:

1. Is a gas at 68°F (20°C) or less at 14.7 psia (101 kPa) of pressure; and
2. Has a *boiling point* of 68°F (20°C) or less at 14.7 psia (101 kPa) which is either liquefied, nonliquefied or in solution, except those gases which have no other health- or physical-hazard properties are not considered to be compressed until the pressure in the packaging exceeds 41 psia (282 kPa) at 68°F (20°C).

The states of a compressed gas are categorized as follows:

1. Nonliquefied compressed gases are gases, other than those in solution, which are in a packaging under the charged pressure and are entirely gaseous at a temperature of 68°F (20°C).

2. Liquefied compressed gases are gases that, in a packaging under the charged pressure, are partially liquid at a temperature of 68°F (20°C).
3. Compressed gases in solution are nonliquefied gases that are dissolved in a solvent.
4. Compressed gas mixtures consist of a mixture of two or more compressed gases contained in a packaging, the hazard properties of which are represented by the properties of the mixture as a whole.

COMPRESSED GAS CONTAINER. A pressure vessel designed to hold *compressed gases* at pressures greater than one atmosphere at 68°F (20°C) and includes cylinders, containers and tanks.

COMPRESSED GAS SYSTEM. An assembly of equipment designed to contain, distribute or transport *compressed gases*. It can consist of a *compressed gas* container or containers, reactors and appurtenances, including pumps, compressors and connecting piping and tubing.

[BG] CONGREGATE LIVING FACILITIES. A building or part thereof that contains sleeping units where residents share bathroom and/or kitchen facilities.

CONSTANTLY ATTENDED LOCATION. A designated location at a facility staffed by trained personnel on a continuous basis where alarm or supervisory signals are monitored and facilities are provided for notification of the fire department or other emergency services.

[A] CONSTRUCTION DOCUMENTS. The written, graphic and pictorial documents prepared or assembled for describing the design, location and physical characteristics of the elements of the project necessary for obtaining a permit.

CONTAINER. A vessel of 60 gallons (227 L) or less in capacity used for transporting or storing hazardous materials. Pipes, piping systems, engines and engine fuel tanks are not considered to be containers.

CONTAINMENT SYSTEM. A gas-tight recovery system comprised of equipment or devices which can be placed over a leak in a *compressed gas* container, thereby stopping or controlling the escape of gas from the leaking container.

CONTAINMENT VESSEL. A gas-tight recovery vessel designed so that a leaking *compressed gas* container can be placed within its confines thereby encapsulating the leaking container.

CONTINUOUS GAS DETECTION SYSTEM. A gas detection system where the analytical instrument is maintained in continuous operation and sampling is performed without interruption. Analysis is allowed to be performed on a cyclical basis at intervals not to exceed 30 minutes.

CONTROL AREA. Spaces within a building where quantities of hazardous materials not exceeding the *maximum allowable quantities per control area* are stored, dispensed, used or handled. See also the definition of "Outdoor control area."

[BE] CORRIDOR. An enclosed *exit access* component that defines and provides a path of egress travel.

CORRIDOR, OPEN-ENDED. See "Open-ended corridor."

CORROSIVE. A chemical that causes visible destruction of, or irreversible alterations in, living tissue by chemical action at the point of contact. A chemical shall be considered corrosive if, when tested on the intact skin of albino rabbits by the method described in DOTn 49 CFR 173.137, such chemical destroys or changes irreversibly the structure of the tissue at the point of contact following an exposure period of 4 hours. This term does not refer to action on inanimate surfaces.

COTTON.

Baled cotton. A natural seed fiber wrapped in and secured with industry-accepted materials, usually consisting of burlap, woven polypropylene, polyethylene or cotton or sheet polyethylene, and secured with steel, synthetic or wire bands, or wire; also includes linters (lint removed from the cottonseed) and motes (residual materials from the ginning process).

Baled cotton, densely packed. Cotton, made into banded bales, with a packing density of not less than 22 pounds per cubic foot (360 kg/m^3), and dimensions complying with the following: a length of 55 inches (1397 mm), a width of 21 inches (533.4 mm) and a height of 27.6 to 35.4 inches (701 to 899 mm).

Seed cotton. Perishable raw agricultural commodity consisting of cotton fiber (lint) attached to the seed of the cotton plant, which requires ginning to become a commercial product.

[BG] COURT. An open, uncovered space, unobstructed to the sky, bounded on three or more sides by exterior building walls or other enclosing devices.

[BG] COVERED MALL BUILDING. A single building enclosing a number of tenants and occupants such as retail stores, drinking and dining establishments, entertainment and amusement facilities, passenger transportation terminals, offices, and other similar uses wherein two or more tenants have a main entrance into one or more malls. Anchor buildings shall not be considered as a part of the covered mall building. The term "covered mall building" shall include open mall buildings as defined below.

Mall. A roofed or covered common pedestrian area within a covered mall building that serves as access for two or more tenants and not to exceed three levels that are open to each other. The term "mall" shall include open malls as defined below.

Open mall. An unroofed common pedestrian way serving a number of tenants not exceeding three levels. Circulation at levels above grade shall be permitted to include open exterior balconies leading to *exits* discharging at grade.

Open mall building. Several structures housing a number of tenants such as retail stores, drinking and dining establishments, entertainment and amusement facilities, offices, and other similar uses wherein two or more tenants have a main entrance into one or more open malls. Anchor buildings are not considered as a part of the open mall building.

CRITICAL CIRCUIT. A circuit that requires continuous operation to ensure safety of the structure and occupants.

CRYOGENIC CONTAINER. A cryogenic vessel of any size used for the transportation, handling or storage of *cryogenic fluids*.

CRYOGENIC FLUID. A fluid having a *boiling point* lower than -130°F (-89.9°C) at 14.7 pounds per square inch atmosphere (psia) (an absolute pressure of 101.3 kPa).

CRYOGENIC VESSEL. A pressure vessel, low-pressure tank or atmospheric tank designed to contain a *cryogenic fluid* on which venting, insulation, refrigeration or a combination of these is used in order to maintain the operating pressure within the design pressure and the contents in a liquid phase.

[BG] CUSTODIAL CARE. Assistance with day-to-day living tasks; such as assistance with cooking, taking medication, bathing, using toilet facilities and other tasks of daily living. Custodial care includes persons receiving care who have the ability to respond to emergency situations and evacuate at a slower rate and/or who have mental and psychiatric complications.

CYLINDER. A pressure vessel designed for pressures higher than 40 psia (275.6 kPa) and having a circular cross section. It does not include a portable tank, multiunit tank car tank, cargo tank or tank car.

DAMPER. See "Fire damper" and "Smoke damper."

DAY BOX. A portable magazine designed to hold *explosive* materials and constructed in accordance with the requirements for a Type 3 magazine as defined and classified in Chapter 56.

DECORATIVE MATERIALS. All materials applied over the building interior finish for decorative, acoustical or other effect including, but not limited to, curtains, draperies, fabrics, streamers and all other materials utilized for decorative effect including, but not limited to, bulletin boards, artwork, posters, photographs, paintings, batting, cloth, cotton, hay, stalks, straw, vines, leaves, trees, moss and similar items, foam plastics and materials containing foam plastics. Decorative materials do not include wall coverings, ceiling coverings, floor coverings, ordinary window shades, interior finish and materials 0.025 inch (0.64 mm) or less in thickness applied directly to and adhering tightly to a substrate.

DEFLAGRATION. An exothermic reaction, such as the extremely rapid oxidation of a flammable dust or vapor in air, in which the reaction progresses through the unburned material at a rate less than the velocity of sound. A deflagration can have an explosive effect.

DELUGE SYSTEM. A sprinkler system employing open sprinklers attached to a piping system connected to a water supply through a valve that is opened by the operation of a detection system installed in the same area as the sprinklers. When this valve opens, water flows into the piping system and discharges from all sprinklers attached thereto.

DESIGN PRESSURE. The maximum gauge pressure that a pressure vessel, device, component or system is designed to withstand safely under the temperature and conditions of use expected.

DETACHED BUILDING. A separate single-story building, without a *basement* or crawl space, used for the storage or use of hazardous materials and located an *approved* distance from all structures.

DETEARING. A process for rapidly removing excess wet coating material from a dipped or coated object or material by passing it through an electrostatic field.

DETECTOR, HEAT. A fire detector that senses heat, either abnormally high temperature or rate of rise, or both.

DETONATING CORD. A flexible cord containing a center core of high *explosive* used to initiate other *explosives*.

DETONATION. An exothermic reaction characterized by the presence of a shock wave in the material which establishes and maintains the reaction. The reaction zone progresses through the material at a rate greater than the velocity of sound. The principal heating mechanism is one of shock compression. *Detonations* have an *explosive* effect.

DETONATOR. A device containing any initiating or primary *explosive* that is used for initiating *detonation*. A detonator shall not contain more than 154.32 grains (10 grams) of total *explosives* by weight, excluding ignition or delay charges. The term includes, but is not limited to, electric blasting caps of instantaneous and delay types, blasting caps for use with safety fuses, detonating cord delay connectors, and noninstantaneous and delay blasting caps which use detonating cord, shock tube or any other replacement for electric leg wires. All types of detonators in strengths through No. 8 cap should be rated at $1^1/_2$ pounds (0.68 kg) of explosives per 1,000 caps. For strengths higher than No. 8 cap, consult the manufacturer.

[BG] DETOXIFICATION FACILITIES. Facilities that provide treatment for substance abuse serving care recipients who are incapable of self-preservation or who are harmful to themselves or others.

DIP TANK. A tank, vat or container of flammable or combustible liquid in which articles or materials are immersed for the purpose of coating, finishing, treating and similar processes.

DISCHARGE SITE. The immediate area surrounding the fireworks mortars used for an outdoor fireworks display.

DISPENSING. The pouring or transferring of any material from a container, tank or similar vessel, whereby vapors, dusts, fumes, mists or gases are liberated to the atmosphere.

DISPENSING DEVICE, OVERHEAD TYPE. A dispensing device that consists of one or more individual units intended for installation in conjunction with each other, mounted above a dispensing area typically within the motor fuel-dispensing facility canopy structure, and characterized by the use of an overhead hose reel.

DISPLAY SITE. The immediate area where a fireworks display is conducted. The display area includes the discharge site, the fallout area and the required separation distance from the mortars to spectator viewing areas. The display area does not include spectator viewing areas or vehicle parking areas.

DOOR, BALANCED. See "Balanced door."

DOOR, DUTCH. See "Dutch door."

DOOR, LOW ENERGY POWER-OPERATED. See "Low energy power-operated door."

DOOR, POWER-ASSISTED. See "Power-assisted door."

DOOR, POWER-OPERATED. See "Power-operated door."

DOORWAY, EXIT ACCESS. See "Exit access doorway."

[BG] DORMITORY. A space in a building where group sleeping accommodations are provided in one room, or in a series of closely associated rooms, for persons not members of the same family group, under joint occupancy and single management, as in college dormitories or fraternity houses.

DRAFT CURTAIN. A structure arranged to limit the spread of smoke and heat along the underside of the ceiling or roof.

[BF] DRAFTSTOP. A material, device or construction installed to restrict the movement of air within open spaces of concealed areas of building components such as crawl spaces, floor/ceiling assemblies, roof/ceiling assemblies and attics.

DRY-CHEMICAL EXTINGUISHING AGENT. A powder composed of small particles, usually of sodium bicarbonate, potassium bicarbonate, urea-potassium-based bicarbonate, potassium chloride or monoammonium phosphate, with added particulate material supplemented by special treatment to provide resistance to packing, resistance to moisture absorption (caking) and the proper flow capabilities.

DRY CLEANING. The process of removing dirt, grease, paints and other stains from such items as wearing apparel, textiles, fabrics and rugs by use of nonaqueous liquids (solvents).

DRY CLEANING PLANT. A facility in which dry cleaning and associated operations are conducted, including the office, receiving area and storage rooms.

DRY CLEANING ROOM. An occupiable space within a building used for performing dry cleaning operations, the installation of solvent-handling equipment or the storage of dry cleaning solvents.

DRY CLEANING SYSTEM. Machinery or equipment in which textiles are immersed or agitated in solvent or in which dry cleaning solvent is extracted from textiles.

DUTCH DOOR. A door divided horizontally so that the top can be operated independently from the bottom.

[BG] DWELLING. A building that contains one or two *dwelling units* used, intended or designed to be used, rented, leased, let or hired out to be occupied for living purposes.

[BG] DWELLING UNIT. A single unit providing complete, independent living facilities for one or more persons, including permanent provisions for living, sleeping, eating, cooking and sanitation.

EARLY SUPPRESSION FAST-RESPONSE (ESFR) SPRINKLER. A sprinkler *listed* for early suppression fast-response performance.

[BE] EGRESS COURT. A court or *yard* which provides access to a *public way* for one or more *exits*.

ELECTROSTATIC FLUIDIZED BED. A container holding powder coating material that is aerated from below so as to form an air-supported expanded cloud of such material that is electrically charged with a charge opposite to that of the object to be coated. Such object is transported through the container immediately above the charged and aerated materials in order to be coated.

ELEVATOR GROUP. A grouping of elevators in a building located adjacent or directly across from one another that respond to a common hall call button(s).

EMERGENCY ALARM SYSTEM. A system to provide indication and warning of emergency situations involving hazardous materials.

EMERGENCY CONTROL STATION. An *approved* location on the premises where signals from emergency equipment are received and which is staffed by trained personnel.

[BE] EMERGENCY ESCAPE AND RESCUE OPENING. An operable window, door or other similar device that provides for a means of escape and access for rescue in the event of an emergency.

EMERGENCY EVACUATION DRILL. An exercise performed to train staff and occupants and to evaluate their efficiency and effectiveness in carrying out emergency evacuation procedures.

EMERGENCY POWER SYSTEM. A source of automatic electric power of a required capacity and duration to operate required life safety, fire alarm, detection and ventilation systems in the event of a failure of the primary power. Emergency power systems are required for electrical loads where interruption of the primary power could result in loss of human life or serious injuries.

EMERGENCY SHUTOFF VALVE. A valve designed to shut off the flow of gases or liquids.

EMERGENCY SHUTOFF VALVE, AUTOMATIC. A fail-safe automatic-closing valve designed to shut off the flow of gases or liquids initiated by a control system that is activated by automatic means.

EMERGENCY SHUTOFF VALVE, MANUAL. A manually operated valve designed to shut off the flow of gases or liquids.

EMERGENCY VOICE/ALARM COMMUNICATIONS. Dedicated manual or automatic facilities for originating and distributing voice instructions, as well as alert and evacuation signals pertaining to a fire emergency, to the occupants of a building.

[BG] EMPLOYEE WORK AREA. All or any portion of a space used only by employees and only for work. *Corridors*, toilet rooms, kitchenettes and break rooms are not employee work areas.

[BG] EQUIPMENT PLATFORM. An unoccupied, elevated platform used exclusively for mechanical systems or industrial process equipment, including the associated elevated walkways, *stairways*, *alternating tread devices* and ladders necessary to access the platform (see Section 505.3 of the *International Building Code*).

EXCESS FLOW CONTROL. A fail-safe system or other *approved* means designed to shut off flow caused by a rupture in pressurized piping systems.

EXCESS FLOW VALVE. A valve inserted into a *compressed gas* cylinder, portable tank or stationary tank that is designed to positively shut off the flow of gas in the event that its predetermined flow is exceeded.

EXHAUSTED ENCLOSURE. An appliance or piece of equipment which consists of a top, a back and two sides providing a means of local exhaust for capturing gases, fumes, vapors and mists. Such enclosures include laboratory hoods, exhaust fume hoods and similar appliances and equipment used to retain and exhaust locally the gases, fumes, vapors and mists that could be released. Rooms or areas provided with general ventilation, in themselves, are not exhausted enclosures.

EXISTING. Buildings, facilities or conditions that are already in existence, constructed or officially authorized prior to the adoption of this code.

[BE] EXIT. That portion of a *means of egress* system between the *exit access* and the *exit discharge* or *public way*. Exit components include exterior exit doors at the *level of exit discharge*, *interior exit stairways* and *ramps*, *exit passageways*, *exterior exit stairways* and *ramps* and *horizontal exits*.

[BE] EXIT ACCESS. That portion of a *means of egress* system that leads from any occupied portion of a building or structure to an *exit*.

[BE] EXIT ACCESS DOORWAY. A door or access point along the path of egress travel from an occupied room, area or space where the path of egress enters an intervening room, *corridor*, *exit access stairway* or *ramp*.

[BE] EXIT ACCESS RAMP. A *ramp* within the *exit access* portion of the *means of egress* system.

[BE] EXIT ACCESS STAIRWAY. A *stairway* within the *exit access* portion of the *means of egress* system.

[BE] EXIT DISCHARGE. That portion of a *means of egress* system between the termination of an *exit* and a *public way*.

[BE] EXIT DISCHARGE, LEVEL OF. The *story* at the point at which an *exit* terminates and an *exit discharge* begins.

[BE] EXIT PASSAGEWAY. An *exit* component that is separated from other interior spaces of a building or structure by fire-resistance-rated construction and opening protectives, and provides for a protected path of egress travel in a horizontal direction to the *exit discharge*.

EXPANDED PLASTIC. A foam or cellular plastic material having a reduced density based on the presence of numerous small cavities or cells dispersed throughout the material.

EXPLOSION. An effect produced by the sudden violent expansion of gases, which may be accompanied by a shock wave or disruption, or both, of enclosing materials or structures. An explosion could result from any of the following:

1. Chemical changes such as rapid oxidation, *deflagration* or *detonation*, decomposition of molecules and runaway polymerization (usually *detonations*).
2. Physical changes such as pressure tank ruptures.
3. Atomic changes (nuclear fission or fusion).

EXPLOSIVE. A chemical compound, mixture or device, the primary or common purpose of which is to function by explosion. The term includes, but is not limited to, dynamite, black powder, pellet powder, initiating explosives, detonators, safety fuses, squibs, detonating cord, igniter cord, igniters and display fireworks, 1.3G.

The term "Explosive" includes any material determined to be within the scope of USC Title 18: Chapter 40 and also includes any material classified as an explosive other than consumer fireworks, 1.4G by the hazardous materials regulations of DOTn 49 CFR Parts 100-185.

High explosive. *Explosive material*, such as dynamite, which can be caused to detonate by means of a No. 8 test blasting cap where unconfined.

Low explosive. *Explosive material* that will burn or deflagrate when ignited. It is characterized by a rate of reaction that is less than the speed of sound. Examples of low *explosives* include, but are not limited to, black powder, safety fuse, igniters, igniter cord, fuse lighters, fireworks, 1.3G and propellants, 1.3C.

Mass-detonating explosives. Division 1.1, 1.2 and 1.5 *explosives* alone or in combination, or loaded into various types of ammunition or containers, most of which can be expected to explode virtually instantaneously when a small portion is subjected to fire, severe concussion, impact, the impulse of an initiating agent or the effect of a considerable discharge of energy from without. Materials that react in this manner represent a mass explosion hazard. Such an *explosive* will normally cause severe structural damage to adjacent objects. Explosive propagation could occur immediately to other items of ammunition and *explosives* stored sufficiently close to and not adequately protected from the initially exploding pile with a time interval short enough so that two or more quantities must be considered as one for quantity-distance purposes.

UN/DOTn Class 1 explosives. The former classification system used by DOTn included the terms "high" and "low" *explosives* as defined herein. The following terms further define *explosives* under the current system applied by DOTn for all *explosive materials* defined as hazard Class 1 materials. Compatibility group letters are used in concert with the division to specify further limitations on each division noted (for example, the letter G identifies the material as a pyrotechnic substance or article containing a pyrotechnic substance and similar materials).

Division 1.1. *Explosives* that have a mass explosion hazard. A mass explosion is one which affects almost the entire load instantaneously.

Division 1.2. *Explosives* that have a projection hazard but not a mass explosion hazard.

Division 1.3. *Explosives* that have a fire hazard and either a minor blast hazard or a minor projection hazard or both, but not a mass explosion hazard.

Division 1.4. *Explosives* that pose a minor explosion hazard. The explosive effects are largely confined to the package and no projection of fragments of appreciable size or range is to be expected. An external fire must not cause virtually instantaneous explosion of almost the entire contents of the package.

Division 1.5. Very insensitive *explosives*. This division is comprised of substances that have a mass explosion hazard but which are so insensitive that there is very little probability of initiation or of transition from burning to *detonation* under normal conditions of transport.

Division 1.6. Extremely insensitive articles which do not have a mass explosion hazard. This division is comprised of articles that contain only extremely insensitive detonating substances and which demonstrate a negligible probability of accidental initiation or propagation.

EXPLOSIVE MATERIAL. The term "explosive" material means *explosives*, blasting agents and detonators.

[BE] EXTERIOR EXIT RAMP. An *exit* component that serves to meet one or more *means of egress* design requirements, such as required number of *exits* or *exit access* travel distance, and is open to yards, courts or *public ways*.

[BE] EXTERIOR EXIT STAIRWAY. An *exit* component that serves to meet one or more *means of egress* design requirements, such as required number of *exits* or *exit access* travel distance, and is open to yards, courts or *public ways*.

[BF] EXTERIOR WALL. A wall, bearing or nonbearing, that is used as an enclosing wall for a building, other than a *fire wall*, and that has a slope of 60 degrees (1.05 rad) or greater with the horizontal plane.

EXTRA-HIGH-RACK COMBUSTIBLE STORAGE. Storage on racks of Class I, II, III or IV commodities that exceed 40 feet (12 192 mm) in height and storage on racks of high-hazard commodities that exceed 30 feet (9144 mm) in height.

FABRICATION AREA. An area within a semiconductor fabrication facility and related research and development areas in which there are processes using hazardous production materials. Such areas are allowed to include ancillary rooms or areas such as dressing rooms and offices that are directly related to the fabrication area processes.

[A] FACILITY. A building or use in a fixed location including exterior storage areas for flammable and combustible substances and hazardous materials, piers, wharves, tank farms and similar uses. This term includes recreational vehicles, mobile home and manufactured housing parks, sales and storage lots.

FAIL-SAFE. A design condition incorporating a feature for automatically counteracting the effect of an anticipated possible source of failure; also, a design condition eliminating or mitigating a hazardous condition by compensating automatically for a failure or malfunction.

FALLOUT AREA. The area over which aerial shells are fired. The shells burst over the area, and unsafe debris and malfunctioning aerial shells fall into this area. The fallout area is the location where a typical aerial shell dud falls to the ground depending on the wind and the angle of mortar placement.

FALSE ALARM. The willful and knowing initiation or transmission of a signal, message or other notification of an event of fire when no such danger exists.

FINES. Small pieces or splinters of wood byproducts that will pass through a 0.25-inch (6.4 mm) screen.

FIRE ALARM. The giving, signaling or transmission to any public fire station, or company or to any officer or employee thereof, whether by telephone, spoken word or otherwise, of information to the effect that there is a fire at or near the place indicated by the person giving, signaling or transmitting such information.

FIRE ALARM BOX, MANUAL. See "Manual fire alarm box."

FIRE ALARM CONTROL UNIT. A system component that receives inputs from automatic and manual fire alarm devices and may be capable of supplying power to detection devices and transponder(s) or off-premises transmitter(s). The control unit may be capable of providing a transfer of power to the notification appliances and transfer of condition to relays or devices.

FIRE ALARM SIGNAL. A signal initiated by a fire alarm-initiating device such as a manual fire alarm box, automatic fire detector, waterflow switch or other device whose activation is indicative of the presence of a fire or fire signature.

FIRE ALARM SYSTEM. A system or portion of a combination system consisting of components and circuits arranged to monitor and annunciate the status of fire alarm or supervisory signal-initiating devices and to initiate the appropriate response to those signals.

FIRE APPARATUS ACCESS ROAD. A road that provides fire apparatus access from a fire station to a facility, building or portion thereof. This is a general term inclusive of all other terms such as *fire lane*, public street, private street, parking lot lane and access roadway.

[BF] FIRE AREA. The aggregate floor area enclosed and bounded by *fire walls*, *fire barriers*, *exterior walls* or *horizontal assemblies* of a building. Areas of the building not provided with surrounding walls shall be included in the fire area if such areas are included within the horizontal projection of the roof or floor next above.

[BF] FIRE BARRIER. A fire-resistance-rated wall assembly of materials designed to restrict the spread of fire in which continuity is maintained.

FIRE CHIEF. The chief officer of the fire department serving the jurisdiction, or a duly authorized representative.

FIRE CODE OFFICIAL. The fire chief or other designated authority charged with the administration and enforcement of the code, or a duly authorized representative.

FIRE COMMAND CENTER. The principal attended or unattended location where the status of detection, alarm communications and control systems is displayed, and from which the system(s) can be manually controlled.

[BF] FIRE DAMPER. A *listed* device installed in ducts and air transfer openings designed to close automatically upon detection of heat and resist the passage of flame. Fire dampers are classified for use in either static systems that will automatically shut down in the event of a fire, or in dynamic systems that continue to operate during a fire. A dynamic fire damper is tested and rated for closure under elevated temperature airflow.

FIRE DEPARTMENT MASTER KEY. A limited issue key of special or controlled design to be carried by fire department officials in command which will open key boxes on specified properties.

FIRE DETECTOR, AUTOMATIC. A device designed to detect the presence of a fire signature and to initiate action.

[BF] FIRE DOOR. The door component of a fire door assembly.

[BF] FIRE DOOR ASSEMBLY. Any combination of a fire door, frame, hardware and other accessories that together provide a specific degree of fire protection to the opening.

[BF] FIRE EXIT HARDWARE. Panic hardware that is *listed* for use on *fire door assemblies*.

FIRE LANE. A road or other passageway developed to allow the passage of fire apparatus. A fire lane is not necessarily intended for vehicular traffic other than fire apparatus.

[BF] FIRE PARTITION. A vertical assembly of materials designed to restrict the spread of fire in which openings are protected.

FIRE POINT. The lowest temperature at which a liquid will ignite and achieve sustained burning when exposed to a test flame in accordance with ASTM D 92.

[BF] FIRE PROTECTION RATING. The period of time that an opening protective assembly will maintain the ability to confine a fire as determined by tests prescribed in Section 716 of the *International Building Code*. Ratings are stated in hours or minutes.

FIRE PROTECTION SYSTEM. *Approved* devices, equipment and systems or combinations of systems used to detect a fire, activate an alarm, extinguish or control a fire, control or manage smoke and products of a fire or any combination thereof.

[BF] FIRE RESISTANCE. That property of materials or their assemblies that prevents or retards the passage of excessive heat, hot gases or flames under conditions of use.

[BF] FIRE-RESISTANCE RATING. The period of time a building element, component or assembly maintains the ability to confine a fire, continues to perform a given structural function, or both, as determined by the tests, or the methods based on tests, prescribed in Section 703 of the *International Building Code*.

[BF] FIRE-RESISTANT JOINT SYSTEM. An assemblage of specific materials or products that are designed, tested and fire-resistance rated in accordance with either ASTM E 1966 or UL 2079 to resist for a prescribed period of time the passage of fire through joints made in or between fire-resistance-rated assemblies.

FIRE SAFETY FUNCTIONS. Building and fire control functions that are intended to increase the level of life safety for occupants or to control the spread of the harmful effects of fire.

[BF] FIRE SEPARATION DISTANCE. The distance measured from the building face to one of the following:

1. The closest interior *lot line*.
2. To the centerline of a street, an alley or *public way*.
3. To an imaginary line between two buildings on the lot.

The distance shall be measured at right angles from the face of the wall.

[BF] FIRE WALL. A fire-resistance-rated wall having protected openings, which restricts the spread of fire and extends continuously from the foundation to or through the roof, with sufficient structural stability under fire conditions to allow collapse of construction on either side without collapse of the wall.

FIRE WATCH. A temporary measure intended to ensure continuous and systematic surveillance of a building or portion thereof by one or more qualified individuals for the purposes of identifying and controlling fire hazards, detecting early signs of unwanted fire, raising an alarm of fire and notifying the fire department.

[BF] FIREBLOCKING. Building materials, or materials *approved* for use as fireblocking, installed to resist the free passage of flame to other areas of the building through concealed spaces.

FIREWORKS. Any composition or device for the purpose of producing a visible or an audible effect for entertainment purposes by combustion, *deflagration* or *detonation* that meets the definition of 1.4G fireworks or 1.3G fireworks.

Fireworks, 1.4G. Small fireworks devices containing restricted amounts of pyrotechnic composition designed primarily to produce visible or audible effects by combustion. Such 1.4G fireworks which comply with the construction, chemical composition and labeling regulations of the DOTn for Fireworks, UN 0336, and the U.S. Consumer Product Safety Commission as set forth in CPSC 16 CFR Parts 1500 and 1507, are not *explosive materials* for the purpose of this code.

Fireworks, 1.3G. Large fireworks devices, which are *explosive materials*, intended for use in fireworks displays and designed to produce audible or visible effects by combustion, *deflagration* or *detonation*. Such 1.3G fireworks include, but are not limited to, firecrackers containing more than 130 milligrams (2 grains) of explosive composition, aerial shells containing more than 40 grams of pyrotechnic composition and other display pieces which exceed the limits for classification as 1.4G fireworks. Such 1.3G fireworks are also described as Fireworks, UN 0335 by the DOTn.

FIREWORKS DISPLAY. A presentation of fireworks for a public or private gathering.

[BG] FIXED BASE OPERATOR (FBO). A commercial business granted the right by the airport sponsor to operate on an airport and provide aeronautical services such as fueling, hangaring, tie-down and parking, aircraft rental, aircraft maintenance and flight instruction.

[BE] FIXED SEATING. Furniture or fixtures designed and installed for the use of sitting and secured in place including bench-type seats and seats with or without back or arm rests.

[BF] FLAME SPREAD. The propagation of flame over a surface.

[BF] FLAME SPREAD INDEX. A comparative measure, expressed as a dimensionless number, derived from visual measurements of the spread of flame versus time for a material tested in accordance with ASTM E 84 or UL 723.

FLAMMABLE CRYOGENIC FLUID. A *cryogenic fluid* that is flammable in its vapor state.

FLAMMABLE FINISHES. Coatings to articles or materials in which the material being applied is a flammable liquid, combustible liquid, combustible powder, fiberglass resin or flammable or combustible gel coating.

FLAMMABLE GAS. A material which is a gas at 68°F (20°C) or less at 14.7 pounds per square inch atmosphere (psia) (101 kPa) of pressure [a material that has a *boiling point* of 68°F (20°C) or less at 14.7 psia (101 kPa)] which:

1. Is ignitable at 14.7 psia (101 kPa) when in a mixture of 13 percent or less by volume with air; or
2. Has a flammable range at 14.7 psia (101 kPa) with air of not less than 12 percent, regardless of the lower limit.

The limits specified shall be determined at 14.7 psi (101 kPa) of pressure and a temperature of 68°F (20°C) in accordance with ASTM E 681.

FLAMMABLE LIQUEFIED GAS. A liquefied *compressed gas* which, under a charged pressure, is partially liquid at a temperature of 68°F (20°C) and which is flammable.

FLAMMABLE LIQUID. A liquid having a closed cup flash point below 100°F (38°C). Flammable liquids are further categorized into a group known as Class I liquids. The Class I category is subdivided as follows:

Class IA. Liquids having a flash point below 73°F (23°C) and having a *boiling point* below 100°F (38°C).

Class IB. Liquids having a *flash point* below 73°F (23°C) and having a *boiling point* at or above 100°F (38°C).

Class IC. Liquids having a *flash point* at or above 73°F (23°C) and below 100°F (38°C).

The category of flammable liquids does not include *compressed gases* or *cryogenic fluids*.

FLAMMABLE MATERIAL. A material capable of being readily ignited from common sources of heat or at a temperature of 600°F (316°C) or less.

FLAMMABLE SOLID. A solid, other than a blasting agent or *explosive*, that is capable of causing fire through friction, absorption of moisture, spontaneous chemical change or retained heat from manufacturing or processing, or which has an ignition temperature below 212°F (100°C) or which burns so vigorously and persistently when ignited as to create a serious hazard. A chemical shall be considered a flammable solid as determined in accordance with the test method of CPSC 16 CFR Part 1500.44, if it ignites and burns with a self-sustained flame at a rate greater than 0.0866 inch (2.2 mm) per second along its major axis.

FLAMMABLE VAPOR AREA. An area in which the concentration of flammable constituents (vapor, gas, fume, mist or dust) in air exceeds 25 percent of their lower flammable limit (LFL) because of the flammable finish processes operation. It shall include:

1. The interior of spray booths.
2. The interior of ducts exhausting from spraying processes.
3. Any area in the direct path of spray or any area containing dangerous quantities of air-suspended powder, combustible residue, dust, deposits, vapor or mists as a result of spraying operations.
4. The area in the vicinity of dip tanks, drain boards or associated drying, conveying or other equipment during operation or shutdown periods.

The *fire code official* is authorized to determine the extent of the flammable vapor area, taking into consideration the material characteristics of the flammable materials, the degree of sustained ventilation and the nature of the operations.

FLAMMABLE VAPORS OR FUMES. The concentration of flammable constituents in air that exceeds 25 percent of their lower flammable limit (LFL).

FLASH POINT. The minimum temperature in degrees Fahrenheit at which a liquid will give off sufficient vapors to form an ignitable mixture with air near the surface or in the container, but will not sustain combustion. The flash point of a liquid shall be determined by appropriate test procedure and apparatus as specified in ASTM D 56, ASTM D 93 or ASTM D 3278.

FLEET VEHICLE MOTOR FUEL-DISPENSING FACILITY. That portion of a commercial, industrial, governmental or manufacturing property where liquids used as fuels are stored and dispensed into the fuel tanks of motor vehicles that are used in connection with such businesses, by persons within the employ of such businesses.

[BE] FLIGHT. A continuous run of rectangular treads, *winders* or combination thereof from one landing to another.

FLOAT. A floating structure normally used as a point of transfer for passengers and goods, or both, for mooring purposes.

[BE] FLOOR AREA, GROSS. The floor area within the inside perimeter of the *exterior walls* of the building under consideration, exclusive of vent shafts and courts, without deduction for *corridors*, *stairways*, *ramps*, closets, the thickness of interior walls, columns or other features. The floor area of a building, or portion thereof, not provided with surrounding *exterior walls* shall be the usable area under the horizontal projection of the roof or floor above. The *gross floor area* shall not include shafts with no openings or interior courts.

[BE] FLOOR AREA, NET. The actual occupied area not including unoccupied accessory areas such as corridors, *stairways*, *ramps*, toilet rooms, mechanical rooms and closets.

FLUE SPACES.

Longitudinal flue space. The flue space between rows of storage perpendicular to the direction of loading.

Transverse flue space. The space between rows of storage parallel to the direction of loading.

FLUIDIZED BED. A container holding powder coating material that is aerated from below so as to form an air-supported expanded cloud of such material through which the preheated object to be coated is immersed and transported.

FOAM-EXTINGUISHING SYSTEM. A special system discharging a foam made from concentrates, either mechanically or chemically, over the area to be protected.

[BE] FOLDING AND TELESCOPIC SEATING. Tiered seating having an overall shape and size that is capable of being reduced for purposes of moving or storing and is not a building element.

[BG] FOSTER CARE FACILITIES. Facilities that provide care to more than five children, $2^1/_2$ years of age or less.

FUEL LIMIT SWITCH. A mechanism, located on a tank vehicle, that limits the quantity of product dispensed at one time.

FUMIGANT. A substance which by itself or in combination with any other substance emits or liberates a gas, fume or vapor utilized for the destruction or control of insects, fungi, vermin, germs, rats or other pests, and shall be distinguished from insecticides and disinfectants which are essentially effective in the solid or liquid phases. Examples are methyl bromide, ethylene dibromide, hydrogen cyanide, carbon disulfide and sulfuryl fluoride.

FUMIGATION. The utilization within an enclosed space of a fumigant in concentrations that are hazardous or acutely toxic to humans.

FURNACE CLASS A. An oven or furnace that has heat utilization equipment operating at approximately atmospheric pressure wherein there is a potential explosion or fire hazard that could be occasioned by the presence of flammable volatiles or combustible materials processed or heated in the furnace.

Note: Such flammable volatiles or combustible materials can, for instance, originate from the following:

1. Paints, powders, inks, and adhesives from finishing processes, such as dipped, coated, sprayed and impregnated materials.
2. The substrate material.
3. Wood, paper and plastic pallets, spacers or packaging materials.
4. Polymerization or other molecular rearrangements.

Potentially flammable materials, such as quench oil, water-borne finishes, cooling oil or cooking oils, that present a hazard are ventilated according to Class A standards.

FURNACE CLASS B. An oven or furnace that has heat utilization equipment operating at approximately atmospheric pressure wherein there are no flammable volatiles or combustible materials being heated.

FURNACE CLASS C. An oven or furnace that has a potential hazard due to a flammable or other special atmosphere being used for treatment of material in process. This type of furnace can use any type of heating system and includes a special atmosphere supply system. Also included in the Class C classification are integral quench furnaces and molten salt bath furnaces.

FURNACE CLASS D. An oven or furnace that operates at temperatures from above ambient to over 5,000°F (2760°C) and at pressures normally below atmospheric using any type of heating system. These furnaces can include the use of special processing atmospheres.

GAS CABINET. A fully enclosed, ventilated, noncombustible enclosure used to provide an isolated environment for *compressed gas* cylinders in storage or use. Doors and access ports for exchanging cylinders and accessing pressure-regulating controls are allowed to be included.

GAS DETECTION SYSTEM, CONTINUOUS. See "Continuous gas detection system."

GAS ROOM. A separately ventilated, fully enclosed room in which only *compressed gases* and associated equipment and supplies are stored or used.

GAS ROOM, HYDROGEN FUEL. See "Hydrogen fuel gas room."

GASEOUS HYDROGEN SYSTEM. An assembly of piping, devices and apparatus designed to generate, store, contain, distribute or transport a nontoxic, gaseous hydrogen-containing mixture having not less than 95-percent hydrogen gas by volume and not more than 1-percent oxygen by volume. Gaseous hydrogen systems consist of items such as *compressed gas* containers, reactors and appurtenances, including pressure regulators, pressure relief devices, manifolds, pumps, compressors and interconnecting piping and tubing and controls.

[BG] GRADE FLOOR OPENING. A window or other opening located such that the sill height of the opening is not more than 44 inches (1118 mm) above or below the finished ground level adjacent to the opening.

[BG] GRADE PLANE. A reference plane representing the average of finished ground level adjoining the building at exterior walls. Where the finished ground level slopes away from the exterior walls, the reference plane shall be established by the lowest points within the area between the building and the *lot line* or, where the *lot line* is more than 6 feet (1829 mm) from the building, between the building and a point 6 feet (1829 mm) from the building.

[BE] GRANDSTAND. Tiered seating supported on a dedicated structural system and two or more rows high and is not a building element (see "*Bleachers*").

[BG] GROUP HOME. A facility for social rehabilitation, substance abuse or mental health problems that contains a group housing arrangement that provides custodial care but does not provide medical care.

[BE] GUARD. A building component or a system of building components located at or near the open sides of elevated walking surfaces that minimizes the possibility of a fall from the walking surface to a lower level.

[BG] GUEST ROOM. A room used or intended to be used by one or more guests for living or sleeping purposes.

[BS] GYPSUM BOARD. Gypsum wallboard, gypsum sheathing, gypsum base for gypsum veneer plaster, exterior gypsum soffit board, predecorated gypsum board or water-resistant gypsum backing board complying with the standards listed in Tables 2506.2 and 2507.2 and Chapter 35 of the *International Building Code.*

[BG] HABITABLE SPACE. A space in a building for living, sleeping, eating or cooking. Bathrooms, toilet rooms, closets, halls, storage or utility spaces and similar areas are not considered habitable spaces.

HALOGENATED EXTINGUISHING SYSTEM. A fire-extinguishing system using one or more atoms of an element from the halogen chemical series: fluorine, chlorine, bromine and iodine.

HANDLING. The deliberate transport by any means to a point of storage or use.

[BE] HANDRAIL. A horizontal or sloping rail intended for grasping by the hand for guidance or support.

HAZARDOUS MATERIALS. Those chemicals or substances which are *physical hazards* or *health hazards* as defined and classified in this chapter, whether the materials are in usable or waste condition.

HAZARDOUS PRODUCTION MATERIAL (HPM). A solid, liquid or gas associated with semiconductor manufacturing that has a degree-of-hazard rating in health, flammability or instability of Class 3 or 4 as ranked by NFPA 704 and which is used directly in research, laboratory or production processes which have, as their end product, materials that are not hazardous.

HEALTH HAZARD. A classification of a chemical for which there is statistically significant evidence that acute or chronic health effects are capable of occurring in exposed persons. The term "health hazard" includes chemicals that are toxic, highly toxic and *corrosive.*

HEAT DETECTOR. See "Detector, heat."

[BG] HEIGHT, BUILDING. The vertical distance from grade plane to the average height of the highest roof surface.

HELIPORT. An area of land or water or a structural surface that is used, or intended for use, for the landing and taking off of helicopters, and any appurtenant areas which are used, or intended for use, for heliport buildings and other heliport facilities.

HELISTOP. The same as "Heliport," except that fueling, defueling, maintenance, repairs or storage of helicopters is not permitted.

HI-BOY. A cart used to transport hot roofing materials on a roof.

HIGH-PILED COMBUSTIBLE STORAGE. Storage of combustible materials in closely packed piles or combustible materials on pallets, in racks or on shelves where the top of storage is greater than 12 feet (3658 mm) in height. Where required by the *fire code official, high-piled combustible storage* also includes certain high-hazard commodities, such as rubber tires, Group A plastics, flammable liquids, idle pallets and similar commodities, where the top of storage is greater than 6 feet (1829 mm) in height.

HIGH-PILED STORAGE AREA. An area within a building which is designated, intended, proposed or actually used for *high-piled combustible storage.*

[BG] HIGH-RISE BUILDING. A building with an occupied floor located more than 75 feet (22 860 mm) above the lowest level of fire department vehicle access.

HIGH-VOLTAGE TRANSMISSION LINE. An electrical power transmission line operating at or above 66 kilovolts.

HIGHLY TOXIC. A material which produces a lethal dose or lethal concentration which falls within any of the following categories:

1. A chemical that has a median lethal dose (LD_{50}) of 50 milligrams or less per kilogram of body weight when administered orally to albino rats weighing between 200 and 300 grams each.
2. A chemical that has a median lethal dose (LD_{50}) of 200 milligrams or less per kilogram of body weight when administered by continuous contact for 24 hours (or less if death occurs within 24 hours) with the bare skin of albino rabbits weighing between 2 and 3 kilograms each.
3. A chemical that has a median lethal concentration (LC_{50}) in air of 200 parts per million by volume or less of gas or vapor, or 2 milligrams per liter or less of mist, fume or dust, when administered by continuous inhalation for one hour (or less if death occurs within 1 hour) to albino rats weighing between 200 and 300 grams each.

Mixtures of these materials with ordinary materials, such as water, might not warrant classification as highly toxic. While this system is basically simple in application, any hazard evaluation that is required for the precise categorization of this type of material shall be performed by experienced, technically competent persons.

HIGHLY VOLATILE LIQUID. A liquefied *compressed gas* with a *boiling point* of less than 68°F (20°C).

HIGHWAY. A public street, public alley or public road.

[A] HISTORIC BUILDINGS. Buildings that are listed in or eligible for listing in the National Register of Historic Places, or designated as historic under an appropriate state or local law.

HOGGED MATERIALS. Wood waste materials produced from the lumber production process.

[M] HOOD. An air-intake device used to capture by entrapment, impingement, adhesion or similar means, grease and similar contaminants before they enter a duct system.

Type I. A kitchen hood for collecting and removing grease vapors and smoke.

Type II. A general kitchen hood for collecting and removing steam vapor, heat, odors and products of combustion.

[BF] HORIZONTAL ASSEMBLY. A fire-resistance-rated floor or roof assembly of materials designed to restrict the spread of fire in which continuity is maintained.

[BE] HORIZONTAL EXIT. An *exit* component consisting of fire-resistance-rated construction and opening protectives intended to compartmentalize portions of a building thereby creating refuge areas that afford safety from fire and smoke from the area of fire origin.

[BG] HOSPITALS AND PSYCHIATRIC HOSPITALS. Facilities that provide care or treatment for the medical, psychiatric, obstetrical, or surgical treatment of inpatient care recipients that are incapable of self-preservation.

HOT WORK. Operations including cutting, welding, Thermit welding, brazing, soldering, grinding, thermal spraying, thawing pipe, installation of torch-applied roof systems or any other similar activity.

HOT WORK AREA. The area exposed to sparks, hot slag, radiant heat, or convective heat as a result of the hot work.

HOT WORK EQUIPMENT. Electric or gas welding or cutting equipment used for hot work.

HOT WORK PERMITS. Permits issued by the responsible person at the facility under the hot work permit program permitting welding or other hot work to be done in locations referred to in Section 3503.3 and prepermitted by the *fire code official*.

HOT WORK PROGRAM. A permitted program, carried out by *approved* facilities-designated personnel, allowing them to oversee and issue permits for hot work conducted by their personnel or at their facility. The intent is to have trained, on-site, responsible personnel ensure that required hot work safety measures are taken to prevent fires and fire spread.

HPM FACILITY. See "Semiconductor fabrication facility."

HPM ROOM. A room used in conjunction with or serving a Group H-5 occupancy, where HPM is stored or used and which is classified as a Group H-2, H-3 or H-4 occupancy.

HYDROGEN FUEL GAS ROOM. A room or space that is intended exclusively to house a *gaseous hydrogen system*.

IMMEDIATELY DANGEROUS TO LIFE AND HEALTH (IDLH). The concentration of air-borne contaminants that poses a threat of death, immediate or delayed permanent adverse health effects, or effects that could prevent escape from such an environment. This contaminant concentration level is established by the National Institute of Occupational Safety and Health (NIOSH) based on both toxicity and flammability. It generally is expressed in parts per million by volume (ppm v/v) or milligrams per cubic meter (mg/m^3). Where adequate data do not exist for precise establishment of IDLH concentrations, an independent certified industrial hygienist, industrial toxicologist, appropriate regulatory agency or other source *approved* by the *fire code official* shall make such determination.

IMPAIRMENT COORDINATOR. The person responsible for the maintenance of a particular *fire protection system*.

[BG] INCAPABLE OF SELF-PRESERVATION. Persons who, because of age, physical limitations, mental limitations, chemical dependency or medical treatment, cannot respond as an individual to an emergency situation.

INCOMPATIBLE MATERIALS. Materials that, when mixed, have the potential to react in a manner which generates heat, fumes, gases or byproducts which are hazardous to life or property.

INERT GAS. A gas that is capable of reacting with other materials only under abnormal conditions such as high temperatures, pressures and similar extrinsic physical forces. Within the context of the code, inert gases do not exhibit either physical or *health hazard* properties as defined (other than acting as a simple asphyxiant) or hazard properties other than those of a *compressed gas*. Some of the more common inert gases include argon, helium, krypton, neon, nitrogen and xenon.

INHABITED BUILDING. A building regularly occupied in whole or in part as a habitation for people, or any place of religious worship, schoolhouse, railroad station, store or other structure where people are accustomed to assemble, except any building or structure occupied in connection with the manufacture, transportation, storage or use of *explosive materials*.

INITIATING DEVICE. A system component that originates transmission of a change-of-state condition, such as in a smoke detector, manual fire alarm box, or supervisory switch.

INSECTICIDAL FOGGING. The utilization of insecticidal liquids passed through fog-generating units where, by means of pressure and turbulence, with or without the application of heat, such liquids are transformed and discharged in the form of fog or mist blown into an area to be treated.

[BE] INTERIOR EXIT RAMP. An exit component that serves to meet one or more means of egress design requirements, such as required number of exits or exit access travel distance, and provides for a protected path of egress travel to the exit discharge or public way.

[BE] INTERIOR EXIT STAIRWAY. An exit component that serves to meet one or more means of egress design requirements, such as required number of exits or exit access travel distance, and provides for a protected path of egress travel to the exit discharge or public way.

[BG] INTERIOR FINISH. Interior finish includes interior wall and ceiling finish and interior floor finish.

[BG] INTERIOR FLOOR-WALL BASE. Interior floor finish trim used to provide a functional or decorative border at the intersection of walls and floors.

[BG] INTERIOR WALL AND CEILING FINISH. The exposed interior surfaces of buildings, including but not limited to: fixed or movable walls and partitions; toilet room privacy partitions; columns; ceilings; and interior wainscoting, paneling or other finish applied structurally or for decoration, acoustical correction, surface insulation, structural *fire resistance* or similar purposes, but not including trim.

IRRITANT. A chemical which is not *corrosive*, but which causes a reversible inflammatory effect on living tissue by chemical action at the site of contact. A chemical is a skin irritant if, when tested on the intact skin of albino rabbits by the methods of CPSC 16 CFR Part 1500.41 for an exposure of four or more hours or by other appropriate techniques, it results in an empirical score of 5 or more. A chemical is classified as an eye irritant if so determined under the procedure listed in CPSC 16 CFR Part 1500.42 or other *approved* techniques.

[A] JURISDICTION. The governmental unit that has adopted this code under due legislative authority.

KEY BOX. A secure device with a lock operable only by a fire department master key, and containing building entry keys and other keys that may be required for access in an emergency.

[A] LABELED. Equipment, materials or products to which have been affixed a label, seal, symbol or other identifying mark of a nationally recognized testing laboratory, *approved* agency or other organization concerned with product evaluation that maintains periodic inspection of the production of the labeled items and whose labeling indicates either that the equipment, material or product meets identified standards or has been tested and found suitable for a specified purpose.

LEVEL OF EXIT DISCHARGE. See "Exit discharge, level of."

LIMITED SPRAYING SPACE. An area in which operations for touch-up or spot painting of a surface area of 9 square feet (0.84 m^2) or less are conducted.

LIQUEFIED NATURAL GAS (LNG). A fluid in the liquid state composed predominantly of methane and which may contain minor quantities of ethane, propane, nitrogen or other components normally found in natural gas.

LIQUEFIED PETROLEUM GAS (LP-gas). A material which is composed predominantly of the following hydrocarbons or mixtures of them: propane, propylene, butane (normal butane or isobutane) and butylenes.

LIQUID. A material having a melting point that is equal to or less than 68°F (20°C) and a *boiling point* which is greater than 68°F (20°C) at 14.7 pounds per square inch absolute (psia) (101 kPa). Where not otherwise identified, the term "liquid" includes both flammable and *combustible liquids*.

LIQUID OXYGEN AMBULATORY CONTAINER. A container used for liquid oxygen not exceeding 0.396 gallons (1.5 liters) specifically designed for use as a medical device as defined by 21 USC Chapter 9 that is intended for portable therapeutic use and to be filled from its companion base unit, a liquid oxygen home care container.

LIQUID OXYGEN HOME CARE CONTAINER. A container used for liquid oxygen not exceeding 15.8 gallons (60 liters) specifically designed for use as a medical device as defined by 21 USC Chapter 9 that is intended to deliver gaseous oxygen for therapeutic use in a home environment.

LIQUID STORAGE ROOM. A room classified as a Group H-3 occupancy used for the storage of flammable or *combustible liquids* in a closed condition.

LIQUID STORAGE WAREHOUSE. A building classified as a Group H-2 or H-3 occupancy used for the storage of flammable or *combustible liquids* in a closed condition.

[A] LISTED. Equipment, materials, products or services included in a list published by an organization acceptable to the *fire code official* and concerned with evaluation of products or services that maintains periodic inspection of production of listed equipment or materials or periodic evaluation of services and whose listing states either that the equipment, material, product or service meets identified standards or has been tested and found suitable for a specified purpose.

LOCKDOWN. An emergency situation, in other than a Group I-3 occupancy, requiring that the occupants be sheltered and secured in place within a building when normal evacuation would put occupants at risk.

[BG] LODGING HOUSE. A one-family dwelling where one or more occupants are primarily permanent in nature and rent is paid for guest rooms.

LONGITUDINAL FLUE SPACE. See "Flue space—longitudinal."

[A] LOT. A portion or parcel of land considered as a unit.

[A] LOT LINE. A line dividing one lot from another, or from a street or any public place.

[BE] LOW ENERGY POWER-OPERATED DOOR. Swinging door which opens automatically upon an action by a pedestrian such as pressing a push plate or waving a hand in front of a sensor. The door closes automatically, and operates with decreased forces and decreased speeds. See also "Power-assisted door" and "Power-operated door."

LOW-PRESSURE TANK. A storage tank designed to withstand an internal pressure greater than 0.5 pound per square inch gauge (psig) (3.4 kPa) but not greater than 15 psig (103.4 kPa).

LOWER EXPLOSIVE LIMIT (LEL). See "Lower flammable limit."

LOWER FLAMMABLE LIMIT (LFL). The minimum concentration of vapor in air at which propagation of flame will occur in the presence of an ignition source. The LFL is sometimes referred to as LEL or lower explosive limit.

LP-GAS CONTAINER. Any vessel, including cylinders, tanks, portable tanks and cargo tanks, used for transporting or storing LP-gases.

MAGAZINE. A building, structure or container, other than an operating building, *approved* for storage of *explosive materials*.

Indoor. A portable structure, such as a box, bin or other container, constructed as required for Type 2, 4 or 5 magazines in accordance with NFPA 495, NFPA 1124 or DOTy 27 CFR Part 55 so as to be fire resistant and theft resistant.

Type 1. A permanent structure, such as a building or igloo, that is bullet resistant, fire resistant, theft resistant, weather resistant and ventilated in accordance with the requirements of NFPA 495, NFPA 1124 or DOTy 27 CFR Part 55.

Type 2. A portable or mobile structure, such as a box, skid-magazine, trailer or semitrailer, constructed in accordance with the requirements of NFPA 495, NFPA 1124 or DOTy 27 CFR Part 55 that is fire resistant, theft resistant, weather resistant and ventilated. If used outdoors, a Type 2 magazine is also bullet resistant.

Type 3. A fire resistant, theft resistant and weather resistant "day box" or portable structure constructed in accordance with NFPA 495, NFPA 1124 or DOTy 27 CFR Part 55 used for the temporary storage of *explosive materials*.

Type 4. A permanent, portable or mobile structure such as a building, igloo, box, semitrailer or other mobile container that is fire resistant, theft resistant and weather resistant and constructed in accordance with NFPA 495, NFPA 1124 or DOTy 27 CFR Part 55.

Type 5. A permanent, portable or mobile structure such as a building, igloo, box, bin, tank, semitrailer, bulk trailer, tank trailer, bulk truck, tank truck or other mobile container that is theft resistant, which is constructed in accordance with NFPA 495, NFPA 1124 or DOTy 27 CFR Part 55.

MAGNESIUM. The pure metal and alloys, of which the major part is magnesium.

MALL. See "Covered mall building."

MANUAL FIRE ALARM BOX. A manually operated device used to initiate an alarm signal.

MANUAL STOCKING METHODS. Stocking methods utilizing ladders or other nonmechanical equipment to move stock.

MARINA. Any portion of the ocean or inland water, either naturally or artificially protected, for the mooring, servicing or safety of vessels and shall include artificially protected works, the public or private lands ashore, and structures or facilities provided within the enclosed body of water and ashore for the mooring or servicing of vessels or the servicing of their crews or passengers.

MARINE MOTOR FUEL-DISPENSING FACILITY. That portion of property where flammable or *combustible liquids* or gases used as fuel for watercraft are stored and dispensed from fixed equipment on shore, piers, wharves, floats or barges into the fuel tanks of watercraft and shall include all other facilities used in connection therewith.

MATERIAL SAFETY DATA SHEET (MSDS). Information concerning a hazardous material which is prepared in accordance with the provisions of DOL 29 CFR Part 1910.1200 or in accordance with the provisions of a federally *approved* state OSHA plan.

MAXIMUM ALLOWABLE QUANTITY PER CONTROL AREA. The maximum amount of a hazardous material allowed to be stored or used within a *control area* inside a building or an outdoor *control area*. The maximum allowable quantity per control area is based on the material state (solid, liquid or gas) and the material storage or use conditions.

[BE] MEANS OF EGRESS. A continuous and unobstructed path of vertical and horizontal egress travel from any occupied portion of a building or structure to a *public way*. A means of egress consists of three separate and distinct parts: the *exit access*, the *exit* and the *exit discharge*.

MECHANICAL STOCKING METHODS. Stocking methods utilizing motorized vehicles or hydraulic jacks to move stock.

[BG] MEDICAL CARE. Care involving medical or surgical procedures, nursing or for psychiatric purposes.

MEMBRANE STRUCTURE. An air-inflated, air-supported, cable or frame-covered structure as defined by the *International Building Code* and not otherwise defined as a tent. See Chapter 31 of the *International Building Code*.

[BE] MERCHANDISE PAD. A merchandise pad is an area for display of merchandise surrounded by *aisles*, permanent fixtures or walls. Merchandise pads contain elements such as nonfixed and moveable fixtures, cases, racks, counters and partitions as indicated in Section 105.2 of the *International Building Code* from which customers browse or shop.

METAL HYDRIDE. A generic name for compounds composed of metallic element(s) and hydrogen.

METAL HYDRIDE STORAGE SYSTEM. A *closed system* consisting of a group of components assembled as a package to contain metal-hydrogen compounds for which there exists an equilibrium condition where the hydrogen-absorbing metal alloy(s), hydrogen gas and the metal-hydrogen compound(s) coexist and where only hydrogen gas is released from the system in normal use.

[BG] MEZZANINE. An intermediate level or levels between the floor and ceiling of any story and in accordance with Section 505 of the *International Building Code*.

MOBILE FUELING. The operation of dispensing liquid fuels from tank vehicles into the fuel tanks of motor vehicles. Mobile fueling may also be known by the terms "Mobile fleet fueling," "Wet fueling" and "Wet hosing."

MORTAR. A tube from which fireworks shells are fired into the air.

MULTIPLE-STATION ALARM DEVICE. Two or more single-station alarm devices that can be interconnected such that actuation of one causes all integral or separate audible alarms to operate. A multiple-station alarm device can consist of one single-station alarm device having connections to other detectors or to a manual fire alarm box.

MULTIPLE-STATION SMOKE ALARM. Two or more single-station alarm devices that are capable of interconnection such that actuation of one causes the appropriate alarm signal to operate in all interconnected alarms.

NESTING. A method of securing flat-bottomed *compressed gas* cylinders upright in a tight mass using a contiguous three-point contact system whereby all cylinders within a group have not less than three points of contact with other cylinders, walls or bracing.

NET EXPLOSIVE WEIGHT (net weight). The weight of *explosive material* expressed in pounds. The net explosive weight is the aggregate amount of *explosive material* contained within buildings, magazines, structures or portions thereof, used to establish quantity-distance relationships.

NORMAL TEMPERATURE AND PRESSURE (NTP). A temperature of 70°F (21°C) and a pressure of 1 atmosphere [14.7 psia (101 kPa)].

[BE] NOSING. The leading edge of treads of *stairs* and of landings at the top of *stairway flights*.

NOTIFICATION ZONE. See "Zone, notification."

NUISANCE ALARM. An alarm caused by mechanical failure, malfunction, improper installation or lack of proper maintenance, or an alarm activated by a cause that cannot be determined.

[BG] NURSING HOMES. Facilities that provide care, including both intermediate care facilities and skilled nursing facilities, where any of the persons are incapable of self-preservation.

OCCUPANCY CLASSIFICATION. For the purposes of this code, certain occupancies are defined as follows:

[BG] Assembly Group A. Assembly Group A occupancy includes, among others, the use of a building or structure, or a portion thereof, for the gathering of persons for purposes such as civic, social or religious functions; recreation, food or drink consumption; or awaiting transportation.

[BG] Small buildings and tenant spaces. A building or tenant space used for assembly purposes with an *occupant load* of less than 50 persons shall be classified as a Group B occupancy.

[BG] Small assembly spaces. The following rooms and spaces shall not be classified as assembly occupancies:

1. A room or space used for assembly purposes with an *occupant load* of less than 50 persons and accessory to another occupancy shall be classified as a Group B occupancy or as part of that occupancy.
2. A room or space used for assembly purposes that is less than 750 square feet (70 m^2) in area and accessory to another occupancy shall be classified as a Group B occupancy or as part of that occupancy.

[BG] Associated with Group E occupancies. A room or space used for assembly purposes that is associated with a Group E occupancy is not considered a separate occupancy.

[BG] Accessory with places of religious worship. Accessory religious educational rooms and religious auditoriums with *occupant loads* of less than 100 per room or space are not considered separate occupancies

[BG] Assembly Group A-1. Group A occupancy includes assembly uses, usually with fixed seating, intended for the production and viewing of performing arts or motion pictures including, but not limited to:

Motion picture theaters
Symphony and concert halls
Television and radio studios admitting an audience
Theaters

[BG] Assembly Group A-2. Group A-2 occupancy includes assembly uses intended for food and/or drink consumption including, but not limited to:

Banquet halls
Casinos (gaming areas)
Night clubs
Restaurants, cafeterias and similar dining facilities (including associated commercial kitchens)
Taverns and bars

[BG] Assembly Group A-3. Group A-3 occupancy includes assembly uses intended for worship, recreation or amusement and other assembly uses not classified elsewhere in Group A, including, but not limited to:

Amusement arcades
Art galleries
Bowling alleys
Community halls
Courtrooms
Dance halls (not including food or drink consumption)
Exhibition halls
Funeral parlors
Gymnasiums (without spectator seating)
Indoor swimming pools (without spectator seating)
Indoor tennis courts (without spectator seating)
Lecture halls
Libraries
Museums
Places of religious worship
Pool and billiard parlors
Waiting areas in transportation terminals

[BG] Assembly Group A-4. Group A-4 occupancy includes assembly uses intended for viewing of indoor sporting events and activities with spectator seating including, but not limited to:

Arenas
Skating rinks
Swimming pools
Tennis courts

[BG] Assembly Group A-5. Group A-5 occupancy includes assembly uses intended for participation in or viewing outdoor activities including, but not limited to:

Amusement park structures
Bleachers
Grandstands
Stadiums

[BG] Business Group B. Business Group B occupancy includes, among others, the use of a building or structure, or a portion thereof, for office, professional or service-type transactions, including storage of records and accounts. Business occupancies shall include, but not be limited to, the following:

- Airport traffic control towers
- Ambulatory care facilities
- Animal hospitals, kennels and pounds
- Banks
- Barber and beauty shops
- Car wash
- Civic administration
- Clinic-outpatient
- Dry cleaning and laundries: pick-up and delivery stations and self-service
- Educational occupancies for students above the 12th grade
- Electronic data processing
- Food processing establishments and commercial kitchens not associated with restaurants, cafeterias and similar dining facilities not more than 2,500 square feet (232 m^2) in area.
- Laboratories: testing and research
- Motor vehicle showrooms
- Post offices
- Print shops
- Professional services (architects, attorneys, dentists, physicians, engineers, etc.)
- Radio and television stations
- Telephone exchanges
- Training and skill development not in a school or academic program (This shall include, but not be limited to, tutoring centers, martial arts studios, gymnastics and similar uses regardless of the ages served, and where not classified as a Group A occupancy).

[BG] Educational Group E. Educational Group E occupancy includes, among others, the use of a building or structure, or a portion thereof, by six or more persons at any one time for educational purposes through the 12th grade.

[BG] Accessory to places of religious worship. Religious educational rooms and religious auditoriums, which are accessory to places of religious worship in accordance with Section 508.3.1 of the *International Building Code* and have *occupant loads* of less than 100 per room or space shall be classified as Group A-3 occupancies.

[BG] Group E, day care facilities. This group includes buildings and structures or portions thereof occupied by more than five children older than $2^1/_2$ years of age who receive educational, supervision or *personal care services* for less than 24 hours per day.

[BG] Within places of worship. Rooms and spaces within places of worship providing such care during religious functions shall be classified as part of the primary occupancy.

[BG] Five or fewer children. A facility having five or fewer children receiving such care shall be classified as part of the primary occupancy.

[BG] Five or fewer children in a dwelling unit. A facility such as the above within a dwelling unit and having five or fewer children receiving such care shall be classified as a Group R-3 occupancy or shall comply with the *International Residential Code.*

[BG] Factory Industrial Group F. Factory Industrial Group F occupancy includes, among others, the use of a building or structure, or a portion thereof, for assembling, disassembling, fabricating, finishing, manufacturing, packaging, repair or processing operations that are not classified as a Group H high-hazard or Group S storage occupancy.

[BG] Factory Industrial F-1 Moderate-hazard occupancy. Factory industrial uses that are not classified as Factory Industrial F-2 Low Hazard shall be classified as F-1 Moderate Hazard and shall include, but not be limited to, the following:

- Aircraft (manufacturing, not to include repair)
- Appliances
- Athletic equipment
- Automobiles and other motor vehicles
- Bakeries
- Beverages; over 16-percent alcohol content
- Bicycles
- Boats
- Brooms or brushes
- Business machines
- Cameras and photo equipment
- Canvas or similar fabric
- Carpets and rugs (includes cleaning)
- Clothing
- Construction and agricultural machinery
- Disinfectants
- Dry cleaning and dyeing
- Electric generation plants
- Electronics
- Engines (including rebuilding)
- Food processing and commercial kitchens not associated with restaurants, cafeterias and similar dining facilities more than 2,500 square feet (232 m^2) in area.
- Furniture
- Hemp products
- Jute products
- Laundries
- Leather products
- Machinery
- Metals
- Millwork (sash and door)
- Motion pictures and television filming (without spectators)
- Musical instruments
- Optical goods
- Paper mills or products
- Photographic film
- Plastic products
- Printing or publishing
- Refuse incineration
- Shoes

Soaps and detergents
Textiles
Tobacco
Trailers
Upholstering
Wood; distillation
Woodworking (cabinet)

[BG] Factory Industrial F-2 Low-hazard Occupancy. Factory industrial uses involving the fabrication or manufacturing of noncombustible materials that, during finishing, packaging or processing do not involve a significant fire hazard, shall be classified as Group F-2 occupancies and shall include, but not be limited to, the following:

Beverages; up to and including 16-percent alcohol content
Brick and masonry
Ceramic products
Foundries
Glass products
Gypsum
Ice
Metal products (fabrication and assembly)

High-hazard Group H. High-hazard Group H occupancy includes, among others, the use of a building or structure, or a portion thereof, that involves the manufacturing, processing, generation or storage of materials that constitute a physical or *health hazard* in quantities in excess of those allowed in *control areas* complying with Section 5003.8.3, based on the maximum allowable quantity limits for *control areas* set forth in Tables 5003.1.1(1) and 5003.1.1(2). Hazardous occupancies are classified in Groups H-1, H-2, H-3, H-4 and H-5 and shall be in accordance with this code and the requirements of Section 415 of the *International Building Code*. Hazardous materials stored or used on top of roofs or canopies shall be classified as outdoor storage or use and shall comply with this code.

Uses other than Group H. The storage, use or handling of hazardous materials as described in one or more of the following items shall not cause the occupancy to be classified as Group H, but it shall be classified as the occupancy that it most nearly resembles:

1. Buildings and structures occupied for the application of flammable finishes, provided that such buildings or areas conform to the requirements of Chapter 24 of this code and Section 416 of the *International Building Code*.
2. Wholesale and retail sales and storage of flammable and *combustible liquids* in mercantile occupancies conforming to Chapter 57.
3. Closed piping system containing flammable or *combustible liquids* or gases utilized for the operation of machinery or equipment.
4. Cleaning establishments that utilize *combustible liquid* solvents having a *flash point* of 140°F (60°C) or higher in *closed systems* employing equipment *listed* by an *approved* testing agency, provided that this occupancy is separated from all other areas of the building by 1-hour *fire barriers* in accordance with Section 707 of the *International Building Code* or 1-hour *horizontal assemblies* in accordance with Section 711 of the *International Building Code*, or both.
5. Cleaning establishments that utilize a liquid solvent having a *flash point* at or above 200°F (93°C).
6. Liquor stores and distributors without bulk storage.
7. Refrigeration systems.
8. The storage or utilization of materials for agricultural purposes on the premises.
9. Stationary batteries utilized for facility emergency power, uninterruptible power supply or telecommunication facilities, provided that the batteries are equipped with safety venting caps and ventilation is provided in accordance with the *International Mechanical Code*.
10. *Corrosive* personal or household products in their original packaging used in retail display.
11. Commonly used *corrosive* building materials.
12. Buildings and structures occupied for aerosol storage shall be classified as Group S-1, provided that such buildings conform to the requirements of Chapter 51.
13. Display and storage of nonflammable solid and nonflammable or noncombustible liquid hazardous materials in quantities not exceeding the *maximum allowable quantity per control area* in Group M or S occupancies complying with Section 5003.8.3.5.
14. The storage of black powder, smokeless propellant and small arms primers in Groups M and R-3 and special industrial explosive devices in Groups B, F, M and S, provided such storage conforms to the quantity limits and requirements of this code.

High-hazard Group H-1. Buildings and structures containing materials that pose a *detonation* hazard shall be classified as Group H-1. Such materials shall include, but not be limited to, the following:

Detonable pyrophoric materials

Explosives:

Division 1.1
Division 1.2
Division 1.3
Division 1.4
Division 1.5
Division 1.6

Organic peroxides, unclassified detonable
Oxidizers, Class 4
Unstable (reactive) materials, Class 3 detonable, and Class 4

Occupancies containing explosives not classified as H-1. The following occupancies containing explosive materials shall be classified as follows:

1. Division 1.3 explosive materials that are used and maintained in a form where either confinement or configuration will not elevate the hazard from a mass fire hazard to mass explosion hazard shall be allowed in Group H-2 occupancies.
2. Articles, including articles packaged for shipment, that are not regulated as a Division 1.4 explosive under Bureau of Alcohol, Tobacco, Firearms and Explosives regulations, or unpackaged articles used in process operations that do not propagate a *detonation* or deflagration between articles shall be allowed in H-3 occupancies.

High-hazard Group H-2. Buildings and structures containing materials that pose a *deflagration* hazard or a hazard from accelerated burning shall be classified as Group H-2. Such materials shall include, but not be limited to, the following:

Class I, II or IIIA flammable or *combustible liquids* that are used or stored in normally open containers or systems, or in closed containers or systems pressurized at more than 15 pounds per square inch gauge (103.4 kPa)
Combustible dusts where manufactured, generated or used in such a manner that the concentration and conditions create a fire or explosion hazard based on information prepared in accordance with Section 414.1.3 of the *International Building Code*
Cryogenic fluids, flammable
Flammable gases
Organic peroxides, Class I
Oxidizers, Class 3, that are used or stored in normally open containers or systems, or in closed containers or systems pressurized at more than 15 pounds per square inch gauge (103.4 kPa)
Pyrophoric liquids, solids and gases, nondetonable
Unstable (reactive) materials, Class 3, nondetonable
Water-reactive materials, Class 3

High-hazard Group H-3. Buildings and structures containing materials that readily support combustion or that pose a *physical hazard* shall be classified as Group H-3. Such materials shall include, but not be limited to, the following:

Combustible fibers, other than densely packed baled cotton, where manufactured, generated or used in such a manner that the concentration and conditions create a fire or explosion hazard based on information prepared in accordance with Section 414.1.3 of the *International Building Code*
Consumer fireworks, 1.4G (Class C, Common)
Cryogenic fluids, oxidizing
Flammable solids
Organic peroxides, Class II and III
Oxidizers, Class 2
Oxidizers, Class 3, that are used or stored in normally closed containers or systems pressurized at 15 pounds per square inch gauge (103 kPa) or less
Oxidizing gases
Unstable (reactive) materials, Class 2
Water-reactive materials, Class 2

High-hazard Group H-4. Buildings and structures containing materials that are *health hazards* shall be classified as Group H-4. Such materials shall include, but not be limited to, the following:

Corrosives
Highly toxic materials
Toxic materials

High-hazard Group H-5. Semiconductor fabrication facilities and comparable research and development areas in which hazardous production materials (HPM) are used and the aggregate quantity of materials is in excess of those listed in Tables 5003.1.1(1) and 5003.1.1(2) shall be classified as Group H-5. Such facilities and areas shall be designed and constructed in accordance with Section 415.11 of the *International Building Code*.

[BG] Institutional Group I. Institutional Group I occupancy includes, among others, the use of a building or structure, or a portion thereof, in which care or supervision is provided to persons who are or are not capable of self-preservation without physical assistance or in which persons are detained for penal or correctional purposes or in which the liberty of the occupants is restricted. Institutional occupancies shall be classified as Group I-1, I-2, I-3 or I-4.

[BG] Institutional Group I-1. Institutional Group I-1 occupancy shall include buildings, structures or portions thereof for more than 16 persons, excluding staff, who reside on a 24-hour basis in a supervised environment and receive custodial care. Buildings of Group I-1 shall be classified as one of the occupancy conditions indicated below. This group shall include, but not be limited to, the following:

Alcohol and drug centers
Assisted living facilities
Congregate care facilities
Group homes
Halfway houses
Residential board and care facilities
Residential board and custodial care facilities
Social rehabilitation facilities

[BG] Condition 1. This occupancy condition shall include buildings in which all persons receiving custodial care who, without any assistance, are capable of responding to an emergency situation to complete building evacuation.

[BG] Condition 2. This occupancy condition shall include buildings in which there are any persons receiving custodial care who require limited verbal or physical assistance while responding to an emergency situation to complete building evacuation.

[BG] Six to 16 persons receiving custodial care. A facility housing not fewer than six and not more than

16 persons receiving custodial care shall be classified as Group R-4.

[BG] Five or fewer persons receiving custodial care. A facility with five or fewer persons receiving custodial care shall be classified as Group R-3 or shall comply with the *International Residential Code* provided an *automatic sprinkler system* is installed in accordance with Section 903.3.1.3 or with Section P2904 of the *International Residential Code*.

[BG] Institutional Group I-2. Institutional Group I-2 occupancy shall include buildings and structures used for medical care on a 24-hour basis for more than five persons who are not capable of self-preservation. This group shall include, but not be limited to, the following:

Foster care facilities
Detoxification facilities
Hospitals
Nursing homes
Psychiatric hospitals

[BG] Occupancy Conditions. Buildings of Group I-2 shall be classified as one of the following occupancy conditions:

[BG] Condition 1. This occupancy condition shall include facilities that provide nursing and medical care but do not provide emergency care, surgery, obstetrics, or in-patient stabilization units for psychiatric or detoxification, including, but not limited to, nursing homes and foster care facilities.

[BG] Condition 2. This occupancy condition shall include facilities that provide nursing and medical care and could provide emergency care, surgery, obstetrics, or inpatient stabilization units for psychiatric or detoxification, including, but not limited to, hospitals.

[BG] Five or fewer persons receiving medical care. A facility with five or fewer persons receiving medical care shall be classified as Group R-3 or shall comply with the *International Residential Code* provided an *automatic sprinkler system* is installed in accordance with Section 903.3.1.3 or with Section P2904 of the *International Residential Code*.

[BG] Institutional Group I-3. Institutional Group I-3 occupancy shall include buildings and structures which are inhabited by more than five persons who are under restraint or security. A Group I-3 facility is occupied by persons who are generally incapable of self-preservation due to security measures not under the occupants' control. This group shall include, but not be limited to, the following:

Correctional centers
Detention centers
Jails
Prerelease centers
Prisons
Reformatories

Buildings of Group I-3 shall be classified as one of the following occupancy conditions:

[BG] Condition 1. This occupancy condition shall include buildings in which free movement is allowed from sleeping areas and other spaces where access or occupancy is permitted to the exterior via *means of egress* without restraint. A Condition 1 facility is permitted to be constructed as Group R.

[BG] Condition 2. This occupancy condition shall include buildings in which free movement is allowed from sleeping areas and any other occupied smoke compartment to one or more other smoke compartments. Egress to the exterior is impeded by locked *exits*.

[BG] Condition 3. This occupancy condition shall include buildings in which free movement is allowed within individual smoke compartments, such as within a residential unit comprised of individual *sleeping units* and group activity spaces, where egress is impeded by remote-controlled release of *means of egress* from such smoke compartment to another smoke compartment.

[BG] Condition 4. This occupancy condition shall include buildings in which free movement is restricted from an occupied space. Remote-controlled release is provided to permit movement from *sleeping units*, activity spaces and other occupied areas within the smoke compartment to other smoke compartments.

[BG] Condition 5. This occupancy condition shall include buildings in which free movement is restricted from an occupied space. Staff-controlled manual release is provided to permit movement from *sleeping units*, activity spaces and other occupied areas within the smoke compartment to other smoke compartments.

[BG] Institutional Group I-4, day care facilities. Institutional Group I-4 shall include buildings and structures occupied by more than five persons of any age who receive custodial care for less than 24 hours by persons other than parents or guardians, relatives by blood, marriage, or adoption, and in a place other than the home of the person cared for. This group shall include, but not be limited to, the following:

Adult day care
Child day care

[BG] Classification as Group E. A child day care facility that provides care for more than five but not more than 100 children $2^1/_2$ years or less of age, where the rooms in which the children are cared for are located on a *level of exit discharge* serving such rooms and each of these child care rooms has an *exit* door directly to the exterior, shall be classified as Group E.

[BG] Within a place of religious worship. Rooms and spaces within places of religious worship providing such care during religious functions shall be classified as part of the primary occupancy.

[BG] Five or fewer occupants receiving care. A facility having five or fewer persons receiving custodial care shall be classified as part of the primary occupancy.

[BG] Five or fewer occupants receiving care in a dwelling unit. A facility such as the above within a *dwelling unit* and having five or fewer persons receiving custodial care shall be classified as a Group R-3 occupancy or shall comply with the *International Residential Code.*

[BG] Mercantile Group M. Mercantile Group M occupancy includes, among others, the use of a building or structure or a portion thereof, for the display and sale of merchandise, and involves stocks of goods, wares or merchandise incidental to such purposes and accessible to the public. Mercantile occupancies shall include, but not be limited to, the following:

Department stores
Drug stores
Markets
Motor fuel-dispensing facilities
Retail or wholesale stores
Sales rooms

[BG] Residential Group R. Residential Group R includes, among others, the use of a building or structure, or a portion thereof, for sleeping purposes when not classified as an Institutional Group I or when not regulated by the *International Residential Code* in accordance with Section 101.2 of the *International Building Code.*

[BG] Residential Group R-1. Residential Group R-1 occupancies containing *sleeping units* where the occupants are primarily transient in nature, including:

Boarding houses (transient) with more than 10 occupants
Congregate living facilities (transient) with more than 10 occupants
Hotels (transient)
Motels (transient)

[BG] Residential Group R-2. Residential Group R-2 occupancies containing *sleeping units* or more than two *dwelling units* where the occupants are primarily permanent in nature, including:

Apartment houses
Boarding houses (nontransient) with more than 16 occupants
Congregate living facilities (nontransient) with more than 16 occupants
Convents
Dormitories
Fraternities and sororities
Hotels (nontransient)
Live/work units
Monasteries
Motels (nontransient)
Vacation timeshare properties

[BG] Residential Group R-3. Residential Group R-3 occupancies where the occupants are primarily permanent in nature and not classified as Group R-1, R-2, R-4 or I, including:

Boarding houses (nontransient) with 16 or fewer occupants
Boarding houses (transient) with 10 or fewer occupants
Buildings that do not contain more than two *dwelling units*
Care facilities that provide accommodations for five or fewer persons receiving care
Congregate living facilities (nontransient) with 16 or fewer occupants
Congregate living facilities (transient) with 10 or fewer occupants
Lodging houses with five or fewer guest rooms

[BG] Care facilities within a dwelling. Care facilities for five or fewer persons receiving care that are within a single-family dwelling are permitted to comply with the *International Residential Code* provided an *automatic sprinkler system* is installed in accordance with Section 903.3.1.3 or Section P2904 of the *International Residential Code.*

[BG] Lodging houses. Owner-occupied *lodging houses* with five or fewer guest rooms shall be permitted to be constructed in accordance with the *International Residential Code.*

[BG] Residential Group R-4. Residential Group R-4 shall include buildings, structures or portions thereof for more than five but not more than 16 persons, excluding staff, who reside on a 24-hour basis in a supervised residential environment and receive custodial care. Buildings of Group R-4 shall be classified as one of the occupancy conditions indicated below. This group shall include, but not be limited to, the following:

Alcohol and drug centers
Assisted living facilities
Congregate care facilities
Group homes
Halfway houses
Residential board and care facilities
Social rehabilitation facilities

Group R-4 occupancies shall meet the requirements for construction as defined for Group R-3, except as otherwise provided for in the *International Building Code.*

[BG] Condition 1. This occupancy condition shall include buildings in which all persons receiving custodial care, without any assistance, are capable of responding to an emergency situation to complete building evacuation.

[BG] Condition 2. This occupancy condition shall include buildings in which there are any persons receiving custodial care who require limited verbal or physical assistance while responding to an emergency situation to complete building evacuation.

[BG] Storage Group S. Storage Group S occupancy includes, among others, the use of a building or structure, or a portion thereof, for storage that is not classified as a hazardous occupancy.

[BG] Accessory storage spaces. A room or space used for storage purposes that is less than 100 square feet (9.3 m^2) in area and accessory to another occupancy shall be classified as part of that occupancy. The aggregate area of such rooms or spaces shall not exceed the allowable area limits of Section 508.2 of the *International Building Code*.

[BG] Moderate-hazard storage, Group S-1. Storage Group S-1 occupancies are buildings occupied for storage uses that are not classified as Group S-2, including, but not limited to, storage of the following:

Aerosols, Levels 2 and 3
Aircraft hangar (storage and repair)
Bags: cloth, burlap and paper
Bamboos and rattan
Baskets
Belting: canvas and leather
Books and paper in rolls or packs
Boots and shoes
Buttons, including cloth covered, pearl or bone
Cardboard and cardboard boxes
Clothing, woolen wearing apparel
Cordage
Dry boat storage (indoor)
Furniture
Furs
Glues, mucilage, pastes and size
Grains
Horns and combs, other than celluloid
Leather
Linoleum
Lumber
Motor vehicle repair garages complying with the maximum allowable quantities of hazardous materials listed in Table 5003.1.1(1) (see Section 406.8 of the *International Building Code*)
Photo engravings
Resilient flooring
Silks
Soaps
Sugar
Tires, bulk storage of
Tobacco, cigars, cigarettes and snuff
Upholstery and mattresses
Wax candles

[BG] Low-hazard storage, Group S-2. Storage Group S-2 occupancies include, among others, buildings used for the storage of noncombustible materials such as products on wood pallets or in paper cartons with or without single thickness divisions; or in paper wrappings. Such products are permitted to have a negligible amount of plastic trim, such as knobs, handles or film wrapping. Storage uses shall include, but not be limited to, storage of the following:

Asbestos
Beverages up to and including 16-percent alcohol in metal, glass or ceramic containers
Cement in bags
Chalk and crayons
Dairy products in nonwaxed coated paper containers
Dry cell batteries
Electrical coils
Electrical motors
Empty cans
Food products
Foods in noncombustible containers
Fresh fruits and vegetables in nonplastic trays or containers
Frozen foods
Glass
Glass bottles, empty or filled with noncombustible liquids
Gypsum board
Inert pigments
Ivory
Meats
Metal cabinets
Metal desks with plastic tops and trim
Metal parts
Metals
Mirrors
Oil-filled and other types of distribution transformers
Parking garages, open or enclosed
Porcelain and pottery
Stoves
Talc and soapstones
Washers and dryers

[BG] Miscellaneous Group U. Buildings and structures of an accessory character and miscellaneous structures not classified in any specific occupancy shall be constructed, equipped and maintained to conform to the requirements of this code commensurate with the fire and life hazard incidental to their occupancy. Group U shall include, but not be limited to, the following:

Agricultural buildings
Aircraft hangar, accessory to a one- or two-family residence (see Section 412.5 of the *International Building Code*)
Barns
Carports
Fences more than 6 feet (1829 mm) high
Grain silos, accessory to a residential occupancy
Greenhouses
Livestock shelters
Private garages
Retaining walls
Sheds
Stables
Tanks
Towers

[BG] OCCUPANT LOAD. The number of persons for which the *means of egress* of a building or portion thereof is designed.

OPEN BURNING. The burning of materials wherein products of combustion are emitted directly into the ambient air without passing through a stack or chimney from an enclosed chamber. Open burning does not include road flares, smudgepots and similar devices associated with safety or occupational uses typically considered open flames, *recreational fires* or use of portable outdoor fireplaces. For the purpose of this definition, a chamber shall be regarded as enclosed when, during the time combustion occurs, only apertures, ducts, stacks, flues or chimneys necessary to provide combustion air and permit the escape of exhaust gas are open.

[BE] OPEN-ENDED CORRIDOR. An interior *corridor* that is open on each end and connects to an exterior *stairway* or *ramp* at each end with no intervening doors or separation from the *corridor*.

OPEN MALL. See "Covered mall building."

OPEN MALL BUILDING. See "Covered mall building."

[BG] OPEN PARKING GARAGE. A structure or portion of a structure with the openings as described in Section 406.5.2 of the *International Building Code* on two or more sides that is used for the parking or storage of private motor vehicles as described in Section 406.5 of the *International Building Code.*

OPEN SYSTEM. The use of a solid or liquid hazardous material involving a vessel or system that is continuously open to the atmosphere during normal operations and where vapors are liberated, or the product is exposed to the atmosphere during normal operations. Examples of open systems for solids and liquids include dispensing from or into open beakers or containers, dip tank and plating tank operations.

OPERATING BUILDING. A building occupied in conjunction with the manufacture, transportation or use of *explosive materials*. Operating buildings are separated from one another with the use of intraplant or intraline distances.

OPERATING LINE. A group of buildings, facilities or workstations so arranged as to permit performance of the steps in the manufacture of an *explosive* or in the loading, assembly, modification and maintenance of ammunition or devices containing *explosive materials*.

OPERATING PRESSURE. The pressure at which a system operates.

ORGANIC COATING. A liquid mixture of binders such as alkyd, nitrocellulose, acrylic or oil, and flammable and combustible solvents such as hydrocarbon, ester, ketone or alcohol, which, when spread in a thin film, convert to a durable protective and decorative finish.

ORGANIC PEROXIDE. An organic compound that contains the bivalent -O-O- structure and which may be considered to be a structural derivative of hydrogen peroxide where one or both of the hydrogen atoms have been replaced by an organic radical. Organic peroxides can present an explosion hazard (*detonation* or *deflagration*) or they can be shock sensitive. They can also decompose into various unstable compounds over an extended period of time.

Class I. Describes those formulations that are capable of *deflagration* but not *detonation*.

Class II. Describes those formulations that burn very rapidly and that pose a moderate reactivity hazard.

Class III. Describes those formulations that burn rapidly and that pose a moderate reactivity hazard.

Class IV. Describes those formulations that burn in the same manner as ordinary combustibles and that pose a minimal reactivity hazard.

Class V. Describes those formulations that burn with less intensity than ordinary combustibles or do not sustain combustion and that pose no reactivity hazard.

Unclassified detonable. Organic peroxides that are capable of *detonation*. These peroxides pose an extremely high-explosion hazard through rapid explosive decomposition.

OUTDOOR CONTROL AREA. An outdoor area that contains hazardous materials in amounts not exceeding the maximum allowable quantities of Table 5003.1.1(3) or Table 5003.1.1(4).

OUTPATIENT CLINIC. See "Clinic, outpatient."

OVERCROWDING. A condition that exists when either there are more people in a building, structure or portion thereof than have been authorized or posted by the *fire code official*, or when the *fire code official* determines that a threat exists to the safety of the occupants due to persons sitting and/or standing in locations that may obstruct or impede the use of *aisles*, passages, *corridors*, *stairways*, *exits* or other components of the *means of egress*.

[A] OWNER. Any person, agent, operator, entity, firm or corporation having any legal or equitable interest in the property; or recorded in the official records of the state, county or municipality as holding an interest or title to the property; or otherwise having possession or control of the property, including the guardian of the estate of any such person, and the executor or administrator of the estate of such person if ordered to take possession of real property by a court.

OXIDIZER. A material that readily yields oxygen or other oxidizing gas, or that readily reacts to promote or initiate combustion of combustible materials and, if heated or contaminated, can result in vigorous self-sustained decomposition.

Class 4. An oxidizer that can undergo an explosive reaction due to contamination or exposure to thermal or physical shock and that causes a severe increase in the burning rate of combustible materials with which it comes into contact. Additionally, the oxidizer causes a severe increase in the burning rate and can cause spontaneous ignition of combustibles.

Class 3. An oxidizer that causes a severe increase in the burning rate of combustible materials with which it comes in contact.

Class 2. An oxidizer that will cause a moderate increase in the burning rate of combustible materials with which it comes in contact.

Class 1. An oxidizer that does not moderately increase the burning rate of combustible materials.

OXIDIZING CRYOGENIC FLUID. An oxidizing gas in the cryogenic state.

OXIDIZING GAS. A gas that can support and accelerate combustion of other materials more than air does.

OZONE-GAS GENERATOR. Equipment which causes the production of ozone.

[BE] PANIC HARDWARE. A door-latching assembly incorporating a device that releases the latch upon the application of a force in the direction of egress travel. See also "Fire exit hardware."

PASS-THROUGH. An enclosure installed in a wall with a door on each side that allows chemicals, HPM, equipment, and parts to be transferred from one side of the wall to the other.

[BG] PENTHOUSE. An enclosed, unoccupied rooftop structure used for sheltering mechanical and electrical equipment, tanks, elevators and related machinery, and vertical shaft openings.

PERMISSIBLE EXPOSURE LIMIT (PEL). The maximum permitted 8-hour time-weighted-average concentration of an air-borne contaminant. The exposure limits to be utilized are those published in DOL 29 CFR Part 1910.1000. The Recommended Exposure Limit (REL) concentrations published by the U.S. National Institute for Occupational Safety and Health (NIOSH), Threshold Limit Value-Time Weighted Average (TLV-TWA) concentrations published by the American Conference of Governmental Industrial Hygienists (ACGIH), Workplace Environmental Exposure Level (WEEL) Guides published by the American Industrial Hygiene Association (AIHA), and other *approved*, consistent measures are allowed as surrogates for hazardous substances not *listed* in DOL 29 CFR Part 1910.1000.

[A] PERMIT. An official document or certificate issued by the *fire code official* that authorizes performance of a specified activity.

[A] PERSON. An individual, heirs, executors, administrators or assigns, and also includes a firm, partnership or corporation, its or their successors or assigns, or the agent of any of the aforesaid.

[BG] PERSONAL CARE SERVICE. The care of persons who do not require medical care. Personal care involves responsibility for the safety of the persons while inside the building.

PESTICIDE. A substance or mixture of substances, including fungicides, intended for preventing, destroying, repelling or mitigating pests and substances or a mixture of substances intended for use as a plant regulator, defoliant or desiccant. Products defined as drugs in the Federal Food, Drug and Cosmetic Act are not pesticides.

[BE] PHOTOLUMINESCENT. Having the property of emitting light that continues for a length of time after excitation by visible or invisible light has been removed.

PHYSICAL HAZARD. A chemical for which there is evidence that it is a *combustible liquid*, *cryogenic fluid*, *explosive*, flammable (solid, liquid or gas), organic peroxide (solid or liquid), oxidizer (solid or liquid), oxidizing gas, pyrophoric (solid, liquid or gas), unstable (reactive) material (solid, liquid or gas) or water-reactive material (solid or liquid).

PHYSIOLOGICAL WARNING THRESHOLD. A concentration of air-borne contaminants, normally expressed in parts per million (ppm) or milligrams per cubic meter (mg/m^3), that represents the concentration at which persons can sense the presence of the contaminant due to odor, irritation or other quick-acting physiological responses. When used in conjunction with the permissible exposure limit (PEL), the physiological warning threshold levels are those consistent with the classification system used to establish the PEL. See the definition of "Permissible exposure limit (PEL)."

PIER. A structure built over the water, supported by pillars or piles, and used as a landing place, pleasure pavilion or similar purpose.

PLACE OF RELIGIOUS WORSHIP. See "Religious worship, place of."

[M] PLENUM. An enclosed portion of the building structure, other than an occupiable space being conditioned, that is designed to allow air movement and thereby serve as part of an air distribution system.

PLOSOPHORIC MATERIAL. Two or more unmixed, commercially manufactured, prepackaged chemical substances including oxidizers, flammable liquids or solids, or similar substances that are not independently classified as *explosives* but which, when mixed or combined, form an *explosive* that is intended for blasting.

PLYWOOD AND VENEER MILLS. Facilities where raw wood products are processed into finished wood products, including waferboard, oriented strandboard, fiberboard, composite wood panels and plywood.

PORTABLE OUTDOOR FIREPLACE. A portable, outdoor, solid-fuel-burning fireplace that may be constructed of steel, concrete, clay or other noncombustible material. A portable outdoor fireplace may be open in design, or may be equipped with a small hearth opening and a short chimney or chimney opening in the top.

POWERED INDUSTRIAL TRUCK. A forklift, tractor, platform lift truck or motorized hand truck powered by an electrical motor or internal combustion engine. Powered industrial trucks do not include farm vehicles or automotive vehicles for highway use.

[BE] POWER-ASSISTED DOOR. Swinging door that opens by reduced pushing or pulling force on the door-operating hardware. The door closes automatically after the pushing or pulling force is released, and functions with decreased forces. See also "Low energy power-operated door" and "Power-operated door."

[BE] POWER-OPERATED DOOR. Swinging, sliding, or folding door that opens automatically when approached by a pedestrian or opens automatically upon an action by a pedestrian. The door closes automatically and includes provisions such as presence sensors to prevent entrapment. See also "Low energy power-operated door" and "Power-assisted door."

PRESSURE VESSEL. A closed vessel designed to operate at pressures above 15 psig (103 kPa).

PRIMARY CONTAINMENT. The first level of containment, consisting of the inside portion of that container which comes into immediate contact on its inner surface with the material being contained.

[BG] PRIVATE GARAGE. A building or portion of a building in which motor vehicles used by the tenants of the building or buildings on the premises are stored or kept, without provisions for repairing or servicing such vehicles for profit.

PROCESS TRANSFER. The transfer of flammable or *combustible liquids* between tank vehicles or tank cars and process operations. Process operations may include containers, tanks, piping and equipment.

PROPELLANT. The liquefied or *compressed gas* in an aerosol container that expels the contents from an aerosol container when the valve is actuated. A propellant is considered flammable if it forms a flammable mixture with air, or if a flame is self-propagating in a mixture with air.

PROXIMATE AUDIENCE. An audience closer to pyrotechnic devices than allowed by NFPA 1123.

[B] PSYCHIATRIC HOSPITALS. See "Hospitals."

PUBLIC TRAFFIC ROUTE (PTR). Any public street, road, highway, navigable stream or passenger railroad that is used for through traffic by the general public.

[BE] PUBLIC-USE AREAS. Interior or exterior rooms or spaces that are made available to the general public.

[A] PUBLIC WAY. A street, alley or other parcel of land open to the outside air leading to a street, that has been deeded, dedicated or otherwise permanently appropriated to the public for public use and which has a clear width and height of not less than 10 feet (3048 mm).

PYROPHORIC. A chemical with an autoignition temperature in air, at or below a temperature of 130°F (54°C).

PYROTECHNIC ARTICLE. A pyrotechnic device for use in the entertainment industry, which is not classified as fireworks.

PYROTECHNIC COMPOSITION. A chemical mixture that produces visible light displays or sounds through a self-propagating, heat-releasing chemical reaction which is initiated by ignition.

PYROTECHNIC SPECIAL EFFECT. A visible or audible effect for entertainment created through the use of pyrotechnic materials and devices.

PYROTECHNIC SPECIAL-EFFECT MATERIAL. A chemical mixture used in the entertainment industry to produce visible or audible effects by combustion, *deflagration* or *detonation*. Such a chemical mixture predominantly consists of solids capable of producing a controlled, self-sustaining and self-contained exothermic chemical reaction that results in heat, gas sound, light or a combination of these effects. The chemical reaction functions without external oxygen.

PYROTECHNICS. Controlled exothermic chemical reactions timed to create the effects of heat, hot gas, sound, dispersion of aerosols, emission of visible light or a combination of such effects to achieve the maximum effect from the least volume of pyrotechnic composition.

QUANTITY-DISTANCE (Q-D). The quantity of *explosive material* and separation distance relationships providing protection. These relationships are based on levels of risk considered acceptable for the stipulated exposures and are tabulated in the appropriate Q-D tables. The separation distances specified afford less than absolute safety:

Inhabited building distance (IBD). The minimum separation distance between an operating building or magazine containing *explosive materials* and an inhabited building or site boundary.

Intermagazine distance (IMD). The minimum separation distance between magazines.

Intraline distance (ILD) or Intraplant distance (IPD). The distance to be maintained between any two operating buildings on an *explosives* manufacturing site when at least one contains or is designed to contain *explosives*, or the distance between a magazine and an operating building.

Minimum separation distance (D_o). The minimum separation distance between adjacent buildings occupied in conjunction with the manufacture, transportation, storage or use of *explosive materials* where one of the buildings contains *explosive materials* and the other building does not.

RAILWAY. A steam, electric or other railroad or railway that carriers passengers for hire.

[BE] RAMP. A walking surface that has a running slope steeper than one unit vertical in 20 units horizontal (5-percent slope).

RAMP, EXIT ACCESS. See "Exit access ramp."

RAMP, EXTERIOR EXIT. See "Exterior exit ramp."

RAMP, INTERIOR EXIT. See "Interior exit ramp."

RAW PRODUCT. A mixture of natural materials such as tree, brush trimmings, or waste logs and stumps.

READY BOX. A weather-resistant container with a self-closing or automatic-closing cover that protects fireworks shells from burning debris. Tarpaulins shall not be considered as ready boxes.

[A] RECORD DRAWINGS. Drawings ("as builts") that document the location of all devices, appliances, wiring, sequences, wiring methods and connections of the components of a fire alarm system as installed.

RECREATIONAL FIRE. An outdoor fire burning materials other than rubbish where the fuel being burned is not contained in an incinerator, outdoor fireplace, portable outdoor fireplace, barbeque grill or barbeque pit and has a total fuel area of 3 feet (914 mm) or less in diameter and 2 feet (610 mm) or less in height for pleasure, religious, ceremonial, cooking, warmth or similar purposes.

REDUCED FLOW VALVE. A valve equipped with a restricted flow orifice and inserted into a *compressed gas* cylinder, portable tank or stationary tank that is designed to reduce the maximum flow from the valve under full-flow conditions. The maximum flow rate from the valve is determined with the valve allowed to flow to atmosphere with no other piping or fittings attached.

REFINERY. A plant in which flammable or *combustible liquids* are produced on a commercial scale from crude petroleum, natural gasoline or other hydrocarbon sources.

REFRIGERANT. The fluid used for heat transfer in a refrigeration system; the refrigerant absorbs heat and transfers it at a higher temperature and a higher pressure, usually with a change of state.

[M] REFRIGERATING (REFRIGERATION) SYSTEM. A combination of interconnected refrigerant-containing parts constituting one closed refrigerant circuit in which a refrigerant is circulated for the purpose of extracting heat.

[A] REGISTERED DESIGN PROFESSIONAL. An architect or engineer, registered or licensed to practice professional architecture or engineering, as defined by the statutory requirements of the professional registration laws of the state in which the project is to be constructed.

[BG] RELIGIOUS WORSHIP, PLACE OF. A building or portion thereof intended for the performance of religious services.

REMOTE EMERGENCY SHUTOFF DEVICE. The combination of an operator-carried signaling device and a mechanism on the tank vehicle. Activation of the remote emergency shutoff device sends a signal to the tanker-mounted mechanism and causes fuel flow to cease.

REMOTE SOLVENT RESERVOIR. A liquid solvent container enclosed against evaporative losses to the atmosphere during periods when the container is not being utilized, except for a solvent return opening not larger than 16 square inches (10 322 mm^2). Such return allows pump-cycled used solvent to drain back into the reservoir from a separate solvent sink or work area.

REMOTELY LOCATED, MANUALLY ACTIVATED SHUTDOWN CONTROL. A control system that is designed to initiate shutdown of the flow of gases or liquids that is manually activated from a point located some distance from the delivery system.

REPAIR GARAGE. A building, structure or portion thereof used for servicing or repairing motor vehicles.

RESIN APPLICATION AREA. An area where reinforced plastics are used to manufacture products by hand lay-up or spray-fabrication methods.

RESPONSIBLE PERSON. A person trained in the safety and fire safety considerations concerned with hot work. Responsible for reviewing the sites prior to issuing permits as part of the hot work permit program and following up as the job progresses.

RETAIL DISPLAY AREA. The area of a Group M occupancy open for the purpose of viewing or purchasing merchandise offered for sale. Individuals in such establishments are free to circulate among the items offered for sale which are typically displayed on shelves, racks or the floor.

ROLL COATING. The process of coating, spreading and impregnating fabrics, paper or other materials as they are passed directly through a tank or trough containing flammable or *combustible liquids*, or over the surface of a roller revolving partially submerged in a flammable or *combustible liquid*.

RUBBISH (TRASH). Combustible and noncombustible waste materials, including residue from the burning of coal, wood, coke or other combustible material, paper, rags, cartons, tin cans, metals, mineral matter, glass crockery, dust and discarded refrigerators, and heating, cooking or incinerator-type appliances.

SAFETY CAN. An *approved* container of not more than 5-gallon (19 L) capacity having a spring-closing lid and spout cover so designed that it will relieve internal pressure when subjected to fire exposure.

[BE] SCISSOR STAIRWAY. Two interlocking *stairways* providing two separate paths of egress located within one *exit* enclosure.

SECONDARY CONTAINMENT. That level of containment that is external to and separate from primary containment.

SEED COTTON. See "Cotton."

SEGREGATED. Storage in the same room or inside area, but physically separated by distance from *incompatible materials*.

[BF] SELF-CLOSING. As applied to a fire door or other opening, means equipped with an *approved* device that will ensure closing after having been opened.

[BE] SELF-LUMINOUS. Illuminated by a self-contained power source, other than batteries, and operated independently of external power sources.

SELF-PRESERVATION, INCAPABLE OF. See "Incapable of self-preservation."

SELF-SERVICE MOTOR FUEL-DISPENSING FACILITY. That portion of motor fuel-dispensing facility where liquid motor fuels are dispensed from fixed *approved* dispensing equipment into the fuel tanks of motor vehicles by persons other than a motor fuel-dispensing facility attendant.

SEMICONDUCTOR FABRICATION FACILITY. A building or a portion of a building in which electrical circuits or devices are created on solid crystalline substances having electrical conductivity greater than insulators but less than conductors. These circuits or devices are commonly known as semiconductors.

SERVICE CORRIDOR. A fully enclosed passage used for transporting HPM and purposes other than required *means of egress*.

SHELF STORAGE. Storage on shelves less than 30 inches (762 mm) deep with the distance between shelves not exceeding 3 feet (914 mm) vertically. For other shelving arrangements, see the requirements for rack storage.

SINGLE-STATION SMOKE ALARM. An assembly incorporating the detector, the control equipment and the alarm-sounding device in one unit, operated from a power supply either in the unit or obtained at the point of installation.

[BG] SITE. A parcel of land bounded by a *lot line* or a designated portion of a public right-of-way.

[BG] SITE-FABRICATED STRETCH SYSTEM. A system, fabricated on site and intended for acoustical, tackable or aesthetic purposes, that is composed of three elements:

1. A frame constructed of plastic, wood, metal or other material used to hold fabric in place;
2. A core material (infill, with the correct properties for the application); and
3. An outside layer, comprised of a textile, fabric or vinyl, that is stretched taut and held in place by tension or mechanical fasteners via the frame.

SKY LANTERN. An unmanned device with a fuel source that incorporates an open flame in order to make the device airborne.

[BG] SLEEPING UNIT. A room or space in which people sleep, which can also include permanent provisions for living, eating, and either sanitation or kitchen facilities but not both. Such rooms and spaces that are also part of a *dwelling unit* are not sleeping units.

SMALL ARMS AMMUNITION. A shotgun, rifle or pistol cartridge and any cartridge for propellant-actuated devices. This definition does not include military ammunition containing bursting charges or incendiary, trace, spotting or pyrotechnic projectiles.

SMALL ARMS PRIMERS. Small percussion-sensitive *explosive* charges, encased in a cap, used to ignite propellant powder.

SMOKE ALARM. A single- or multiple-station alarm responsive to smoke. See also "Single-station smoke alarm" and "Multiple-station smoke alarm."

[BF] SMOKE BARRIER. A continuous membrane, either vertical or horizontal, such as a wall, floor, or ceiling assembly, that is designed and constructed to restrict the movement of smoke.

[BG] SMOKE COMPARTMENT. A space within a building enclosed by *smoke barriers* on all sides, including the top and bottom.

[BF] SMOKE DAMPER. A *listed* device installed in ducts and air transfer openings designed to resist the passage of smoke. The device is installed to operate automatically, controlled by a smoke detection system, and where required, is capable of being positioned from a *fire command center*.

SMOKE DETECTOR. A *listed* device that senses visible or invisible particles of combustion.

[BG] SMOKE-DEVELOPED INDEX. A comparative measure, expressed as a dimensionless number, derived from measurements of smoke obscuration versus time for a material tested in accordance with ASTM E 84.

[BE] SMOKE-PROTECTED ASSEMBLY SEATING. Seating served by means of egress that is not subject to smoke accumulation within or under a structure.

SMOKELESS PROPELLANTS. Solid propellants, commonly referred to as smokeless powders, used in small arms ammunition, cannons, rockets, propellant-actuated devices and similar articles.

[BF] SMOKEPROOF ENCLOSURE. An *interior exit stairway* designed and constructed so that the movement of the products of combustion produced by a fire occurring in any part of the building into the enclosure is limited.

SOLID. A material that has a melting point and decomposes or sublimes at a temperature greater than 68°F (20°C).

SOLID BIOFUEL. Densified biomass made in the form of cubiform, polyhedral, polyhydric or cylindrical units, produced by compressing milled biomass.

SOLID BIOMASS FEEDSTOCK. The basic materials of which solid biofuel is composed, manufactured or made.

SOLID SHELVING. Shelving that is solid, slatted or of other construction located in racks and which obstructs sprinkler discharge down into the racks.

SOLVENT DISTILLATION UNIT. An appliance that receives contaminated flammable or *combustible liquids* and which distills the contents to remove contaminants and recover the solvents.

SOLVENT OR LIQUID CLASSIFICATIONS. A method for classifying solvents or liquids according to the following classes:

Class I solvents. Liquids having a *flash point* below 100°F (38°C).

Class II solvents. Liquids having a *flash point* at or above 100°F (38°C) and below 140°F (60°C).

Class IIIA solvents. Liquids having a *flash point* at or above 140°F (60°C) and below 200°F (93°C).

Class IIIB solvents. Liquids having a *flash point* at or above 200°F (93°C).

Class IV solvents. Liquids classified as nonflammable.

SPECIAL AMUSEMENT BUILDING. A building that is temporary, permanent or mobile that contains a device or system that conveys passengers or provides a walkway along, around or over a course in any direction as a form of amusement arranged so that the egress path is not readily apparent due to visual or audio distractions or an intentionally confounded egress path, or is not readily available because of the mode of conveyance through the building or structure.

SPECIAL INDUSTRIAL EXPLOSIVE DEVICE. An explosive power pack containing an *explosive* charge in the form of a cartridge or construction device. The term includes but is not limited to explosive rivets, explosive bolts, *explosive* charges for driving pins or studs, cartridges for *explosive*-actuated power tools and charges of *explosives* used in automotive air bag inflators, jet tapping of open hearth furnaces and jet perforation of oil well casings.

SPRAY BOOTH. A mechanically ventilated appliance of varying dimensions and construction provided to enclose or accommodate a spraying operation and to confine and limit the escape of spray vapor and residue and to exhaust it safely.

SPRAY ROOM. A room designed to accommodate spraying operations, constructed in accordance with the *International*

Building Code and separated from the remainder of the building by a minimum 1-hour *fire barrier*.

SPRAYING SPACE. An area in which dangerous quantities of flammable vapors or combustible residues, dusts or deposits are present due to the operation of spraying processes. The *fire code official* is authorized to define the limits of the spraying space in any specific case.

[BE] STAIR. A change in elevation, consisting of one or more risers.

[BE] STAIRWAY. One or more *flights* of *stairs*, either exterior or interior, with the necessary landings and platforms connecting them, to form a continuous and uninterrupted passage from one level to another.

STAIRWAY, EXIT ACCESS. See "Exit access stairway."

STAIRWAY, EXTERIOR EXIT. See "Exterior exit stairway."

STAIRWAY, INTERIOR EXIT. See "Interior Exit Stairway."

STAIRWAY, SCISSOR. See "Scissor stairway."

[BE] STAIRWAY, SPIRAL. A *stairway* having a closed circular form in its plan view with uniform section-shaped treads attached to and radiating from a minimum-diameter supporting column.

STANDBY POWER SYSTEM. A source of automatic electric power of a required capacity and duration to operate required building, hazardous materials or ventilation systems in the event of a failure of the primary power. Standby power systems are required for electrical loads where interruption of the primary power could create hazards or hamper rescue or fire-fighting operations.

STANDPIPE SYSTEM, CLASSES OF. Standpipe system classes are as follows:

Class I system. A system providing $2^1/_2$-inch (64 mm) hose connections to supply water for use by fire departments and those trained in handling heavy fire streams.

Class II system. A system providing $1^1/_2$-inch (38 mm) hose stations to supply water for use primarily by the building occupants or by the fire department during initial response.

Class III system. A system providing $1^1/_2$-inch (38 mm) hose stations to supply water for use by building occupants and $2^1/_2$-inch (64 mm) hose connections to supply a larger volume of water for use by fire departments and those trained in handling heavy fire streams.

STANDPIPE, TYPES OF. Standpipe types are as follows:

Automatic dry. A dry standpipe system, normally filled with pressurized air, that is arranged through the use of a device, such as a dry pipe valve, to admit water into the system piping automatically upon the opening of a hose valve. The water supply for an automatic dry standpipe system shall be capable of supplying the system demand.

Automatic wet. A wet standpipe system that has a water supply that is capable of supplying the system demand automatically.

Manual dry. A dry standpipe system that does not have a permanent water supply attached to the system. Manual dry standpipe systems require water from a fire department pumper to be pumped into the system through the fire department connection in order to supply the system demand.

Manual wet. A wet standpipe system connected to a water supply for the purpose of maintaining water within the system but which does not have a water supply capable of delivering the system demand attached to the system. Manual wet standpipe systems require water from a fire department pumper (or the like) to be pumped into the system in order to supply the system demand.

Semiautomatic dry. A dry standpipe system that is arranged through the use of a device, such as a deluge valve, to admit water into the system piping upon activation of a remote control device located at a hose connection. A remote control activation device shall be provided at each hose connection. The water supply for a semiautomatic dry standpipe system shall be capable of supplying the system demand.

STATIC PILES. Piles in which processed wood product or solid biomass feedstock is mounded and is not being turned or moved.

STEEL. Hot- or cold-rolled as defined by the *International Building Code*.

STORAGE, HAZARDOUS MATERIALS. The keeping, retention or leaving of hazardous materials in closed containers, tanks, cylinders, or similar vessels; or vessels supplying operations through closed connections to the vessel.

[BG] STORY. That portion of a building included between the upper surface of a floor and the upper surface of the floor or roof next above (see "Basement," "Building height," "Grade plane" and "Mezzanine"). A story is measured as the vertical distance from top to top of two successive tiers of beams or finished floor surfaces and, for the topmost story, from the top of the floor finish to the top of the ceiling joists or, where there is not a ceiling, to the top of the roof rafters.

[BG] STORY ABOVE GRADE PLANE. Any story having its finished floor surface entirely above grade plane, or in which the finished surface of the floor next above is:

1. More than 6 feet (1829 mm) above grade plane; or
2. More than 12 feet (3658 mm) above the finished ground level at any point.

SUPERVISING STATION. A facility that receives signals and at which personnel are in attendance at all times to respond to these signals.

SUPERVISORY SERVICE. The service required to monitor performance of guard tours and the operative condition of fixed suppression systems or other systems for the protection of life and property.

SUPERVISORY SIGNAL. A signal indicating the need of action in connection with the supervision of guard tours, the fire suppression systems or equipment, or the maintenance features of related systems.

SUPERVISORY SIGNAL-INITIATING DEVICE. An initiating device such as a valve supervisory switch, water level indicator, or low-air pressure switch on a dry-pipe sprinkler system whose change of state signals an off-normal condition and its restoration to normal of a fire protection or life safety system; or a need for action in connection with guard tours, fire suppression systems or equipment, or maintenance features of related systems.

SYSTEM. An assembly of equipment consisting of a tank, container or containers, appurtenances, pumps, compressors and connecting piping.

TANK. A vessel containing more than 60 gallons (227 L).

TANK, ATMOSPHERIC. A storage tank designed to operate at pressures from atmospheric through 1.0 pound per square inch gauge (760 mm Hg through 812 mm Hg) measured at the top of the tank.

TANK, PORTABLE. A packaging of more than 60-gallon (227 L) capacity and designed primarily to be loaded into or on or temporarily attached to a transport vehicle or ship and equipped with skids, mountings or accessories to facilitate handling of the tank by mechanical means. It does not include any cylinder having less than a 1,000-pound (454 kg) water capacity, cargo tank, tank car tank or trailers carrying cylinders of more than 1,000-pound (454 kg) water capacity.

TANK, PRIMARY. A *listed* atmospheric tank used to store liquid. See "Primary containment."

TANK, PROTECTED ABOVE GROUND. A tank *listed* in accordance with UL 2085 consisting of a primary tank provided with protection from physical damage and fire-resistive protection from a high-intensity liquid pool fire exposure. The tank may provide protection elements as a unit or may be an assembly of components, or a combination thereof.

TANK, STATIONARY. Packaging designed primarily for stationary installations not intended for loading, unloading or attachment to a transport vehicle as part of its normal operation in the process of use. It does not include cylinders having less than a 1,000-pound (454 kg) water capacity.

TANK VEHICLE. A vehicle other than a railroad tank car or boat, with a cargo tank mounted thereon or built as an integral part thereof, used for the transportation of flammable or *combustible liquids*, LP-gas or hazardous chemicals. Tank vehicles include self-propelled vehicles and full trailers and semitrailers, with or without motive power, and carrying part or all of the load.

TEMPORARY STAGE CANOPY. A temporary ground-supported membrane-covered frame structure used to cover stage areas and support equipment in the production of outdoor entertainment events.

[BG] TENT. A structure, enclosure or shelter, with or without sidewalls or drops, constructed of fabric or pliable material supported by any manner except by air or the contents that it protects.

THEFT RESISTANT. Construction designed to deter illegal entry into facilities for the storage of *explosive materials*.

TIMBER AND LUMBER PRODUCTION FACILITIES. Facilities where raw wood products are processed into finished wood products.

TIRES, BULK STORAGE OF. Storage of tires where the area available for storage exceeds 20,000 cubic feet (566 m^3).

TOOL. A device, storage container, workstation or process machine used in a fabrication area.

TORCH-APPLIED ROOF SYSTEM. Bituminous roofing systems using membranes that are adhered by heating with a torch and melting asphalt back coating instead of mopping hot asphalt for adhesion.

[A] TOWNHOUSE. A single-family *dwelling unit* constructed in a group of three or more attached units in which each unit extends from the foundation to roof and with open space on not less than two sides.

TOXIC. A chemical falling within any of the following categories:

1. A chemical that has a median lethal dose (LD_{50}) of more than 50 milligrams per kilogram, but not more than 500 milligrams per kilogram of body weight when administered orally to albino rats weighing between 200 and 300 grams each.
2. A chemical that has a median lethal dose (LD_{50}) of more than 200 milligrams per kilogram but not more than 1,000 milligrams per kilogram of body weight when administered by continuous contact for 24 hours (or less if death occurs within 24 hours) with the bare skin of albino rabbits weighing between 2 and 3 kilograms each.
3. A chemical that has a median lethal concentration (LC_{50}) in air of more than 200 parts per million but not more than 2,000 parts per million by volume of gas or vapor, or more than 2 milligrams per liter but not more than 20 milligrams per liter of mist, fume or dust, when administered by continuous inhalation for 1 hour (or less if death occurs within 1 hour) to albino rats weighing between 200 and 300 grams each.

TRAFFIC CALMING DEVICES. Traffic calming devices are design elements of fire apparatus access roads such as street alignment, installation of barriers, and other physical measures intended to reduce traffic and cut-through volumes, and slow vehicle speeds.

[BG] TRANSIENT. Occupancy of a dwelling unit or sleeping unit for not more than 30 days.

[BG] TRANSIENT AIRCRAFT. Aircraft based at another location and that is at the transient location for not more than 90 days.

TRANSVERSE FLUE SPACE. See "Flue space—Transverse."

TRASH. See "Rubbish."

TROUBLE SIGNAL. A signal initiated by the fire alarm system or device indicative of a fault in a monitored circuit or component.

TUBE TRAILER. A semitrailer on which a number of tubular gas cylinders have been mounted. A manifold is typically provided that connects the cylinder valves enabling gas to be discharged from one or more tubes or cylinders through a piping and control system.

TWENTY-FOUR HOUR BASIS. See "24-hour basis" before the "A" entries.

UNAUTHORIZED DISCHARGE. A release or emission of materials in a manner which does not conform to the provisions of this code or applicable public health and safety regulations.

UNSTABLE (REACTIVE) MATERIAL. A material, other than an *explosive*, which in the pure state or as commercially produced, will vigorously polymerize, decompose, condense or become self-reactive and undergo other violent chemical changes, including explosion, when exposed to heat, friction or shock, or in the absence of an inhibitor, or in the presence of contaminants, or in contact with *incompatible materials*. Unstable (reactive) materials are subdivided as follows:

Class 4. Materials that in themselves are readily capable of *detonation* or explosive decomposition or explosive reaction at *normal temperatures and pressures*. This class includes materials that are sensitive to mechanical or localized thermal shock at *normal temperatures and pressures*.

Class 3. Materials that in themselves are capable of *detonation* or of explosive decomposition or explosive reaction but which require a strong initiating source or which must be heated under confinement before initiation. This class includes materials that are sensitive to thermal or mechanical shock at elevated temperatures and pressures.

Class 2. Materials that in themselves are normally unstable and readily undergo violent chemical change but do not detonate. This class includes materials that can undergo chemical change with rapid release of energy at *normal temperatures and pressures*, and that can undergo violent chemical change at elevated temperatures and pressures.

Class 1. Materials that in themselves are normally stable but which can become unstable at elevated temperatures and pressure.

UNWANTED FIRE. A fire not used for cooking, heating or recreational purposes or one not incidental to the normal operations of the property.

USE (MATERIAL). Placing a material into action, including solids, liquids and gases.

VAPOR PRESSURE. The pressure exerted by a volatile fluid as determined in accordance with ASTM D 323.

[M] VENTILATION. The natural or mechanical process of supplying conditioned or unconditioned air to, or removing such air from, any space.

VESSEL. A motorized watercraft, other than a seaplane on the water, used or capable of being used as a means of transportation. Nontransportation vessels, such as houseboats and boathouses, are included in this definition.

VISIBLE ALARM NOTIFICATION APPLIANCE. A notification appliance that alerts by the sense of sight.

WATER MIST SYSTEM, AUTOMATIC. See "Automatic water mist system."

WATER-REACTIVE MATERIAL. A material that explodes; violently reacts; produces flammable, toxic or other hazardous gases; or evolves enough heat to cause autoignition or ignition of combustibles upon exposure to water or moisture. Water-reactive materials are subdivided as follows:

Class 3. Materials that react explosively with water without requiring heat or confinement.

Class 2. Materials that react violently with water or have the ability to boil water. Materials that produce flammable, toxic or other hazardous gases, or evolve enough heat to cause autoignition or ignition of combustibles upon exposure to water or moisture.

Class 1. Materials that react with water with some release of energy, but not violently.

WET-CHEMICAL EXTINGUISHING AGENT. A solution of water and potassium-carbonate-based chemical, potassium-acetate-based chemical or a combination thereof, forming an extinguishing agent.

WET FUELING. See "Mobile fueling."

WET HOSING. See "Mobile fueling."

WHARF. A structure or bulkhead constructed of wood, stone, concrete or similar material built at the shore of a harbor, lake or river for vessels to lie alongside of, and to anchor piers or floats.

WILDFIRE RISK AREA. Land that is covered with grass, grain, brush or forest, whether privately or publicly owned, which is so situated or is of such inaccessible location that a fire originating upon it would present an abnormally difficult job of suppression or would result in great or unusual damage through fire or such areas designated by the *fire code official*.

[BE] WINDER. A tread with nonparallel edges.

WIRELESS PROTECTION SYSTEM. A system or a part of a system that can transmit and receive signals without the aid of wire.

WORKSTATION. A defined space or an independent principal piece of equipment using HPM within a fabrication area where a specific function, laboratory procedure or research activity occurs. *Approved* or *listed* hazardous materials storage cabinets, flammable liquid storage cabinets or gas cabinets serving a workstation are included as part of the workstation. A workstation is allowed to contain ventilation equipment, fire protection devices, detection devices, electrical devices and other processing and scientific equipment.

[BG] YARD. An open space, other than a *court*, unobstructed from the ground to the sky, except where specifically provided by the *International Building Code*, on the lot on which a building is situated.

ZONE. A defined area within the protected premises. A zone can define an area from which a signal can be received, an area to which a signal can be sent or an area in which a form of control can be executed.

ZONE, NOTIFICATION. An area within a building or facility covered by notification appliances which are activated simultaneously.

Part II—General Safety Provisions

CHAPTER 3

GENERAL REQUIREMENTS

SECTION 301 GENERAL

301.1 Scope. The provisions of this chapter shall govern the occupancy and maintenance of all structures and premises for precautions against fire and the spread of fire and general requirements of fire safety.

301.2 Permits. Permits shall be required as set forth in Section 105.6 for the activities or uses regulated by Sections 306, 307, 308 and 315.

SECTION 302 DEFINITIONS

302.1 Definitions. The following terms are defined in Chapter 2:

BONFIRE.

HI-BOY.

HIGH-VOLTAGE TRANSMISSION LINE.

OPEN BURNING.

PORTABLE OUTDOOR FIREPLACE.

POWERED INDUSTRIAL TRUCK.

RECREATIONAL FIRE.

SKY LANTERN.

SECTION 303 ASPHALT KETTLES

303.1 Transporting. Asphalt (tar) kettles shall not be transported over any highway, road or street when the heat source for the kettle is operating.

> **Exception:** Asphalt (tar) kettles in the process of patching road surfaces.

303.2 Location. Asphalt (tar) kettles shall not be located within 20 feet (6096 mm) of any combustible material, combustible building surface or any building opening and within a controlled area identified by the use of traffic cones, barriers or other *approved* means. Asphalt (tar) kettles and pots shall not be utilized inside or on the roof of a building or structure. Roofing kettles and operating asphalt (tar) kettles shall not block *means of egress*, gates, roadways or entrances.

303.3 Location of fuel containers. Fuel containers shall be located not less than 10 feet (3048 mm) from the burner.

> **Exception:** Containers properly insulated from heat or flame are allowed to be within 2 feet (610 mm) of the burner.

303.4 Attendant. An operating kettle shall be attended by not less than one employee knowledgeable of the operations and hazards. The employee shall be within 100 feet (30 480 mm) of the kettle and have the kettle within sight. Ladders or similar obstacles shall not form a part of the route between the attendant and the kettle.

303.5 Fire extinguishers. There shall be a portable fire extinguisher complying with Section 906 and with a minimum 40-B:C rating within 25 feet (7620 mm) of each asphalt (tar) kettle during the period such kettle is being utilized. Additionally, there shall be one portable fire extinguisher with a minimum 3-A:40-B:C rating on the roof being covered.

303.6 Lids. Asphalt (tar) kettles shall be equipped with tight-fitting lids.

303.7 Hi-boys. Hi-boys shall be constructed of noncombustible materials. Hi-boys shall be limited to a capacity of 55 gallons (208 L). Fuel sources or heating elements shall not be allowed as part of a hi-boy.

303.8 Roofing kettles. Roofing kettles shall be constructed of noncombustible materials.

303.9 Fuel containers under air pressure. Fuel containers that operate under air pressure shall not exceed 20 gallons (76 L) in capacity and shall be *approved*.

SECTION 304 COMBUSTIBLE WASTE MATERIAL

304.1 Waste accumulation prohibited. Combustible waste material creating a fire hazard shall not be allowed to accumulate in buildings or structures or upon premises.

304.1.1 Waste material. Accumulations of wastepaper, wood, hay, straw, weeds, litter or combustible or flammable waste or rubbish of any type shall not be permitted to remain on a roof or in any *court*, yard, vacant lot, alley, parking lot, open space, or beneath a grandstand, *bleacher*, pier, wharf, manufactured home, recreational vehicle or other similar structure.

304.1.2 Vegetation. Weeds, grass, vines or other growth that is capable of being ignited and endangering property, shall be cut down and removed by the *owner* or occupant of the premises. Vegetation clearance requirements in urban-wildland interface areas shall be in accordance with the *International Wildland-Urban Interface Code*.

304.1.3 Space underneath seats. Spaces underneath grandstand and bleacher seats shall be kept free from combustible and flammable materials. Except where enclosed in not less than 1-hour fire-resistance-rated construction in accordance with the *International Building Code*, spaces underneath grandstand and bleacher seats shall not be occupied or utilized for purposes other than *means of egress*.

304.2 Storage. Storage of combustible rubbish shall not produce conditions that will create a nuisance or a hazard to the public health, safety or welfare.

304.3 Containers. Combustible rubbish, and waste material kept within or near a structure shall be stored in accordance with Sections 304.3.1 through 304.3.4.

304.3.1 Spontaneous ignition. Materials susceptible to spontaneous ignition, such as oily rags, shall be stored in a *listed* disposal container. Contents of such containers shall be removed and disposed of daily.

304.3.2 Capacity exceeding 5.33 cubic feet. Containers with a capacity exceeding 5.33 cubic feet (40 gallons) (0.15 m^3) shall be provided with lids. Containers and lids shall be constructed of noncombustible materials or of combustible materials with a peak rate of heat release not exceeding 300 kW/m^2 where tested in accordance with ASTM E 1354 at an incident heat flux of 50 kW/m^2 in the horizontal orientation.

Exception: Wastebaskets complying with Section 808.

304.3.3 Capacity exceeding 1.5 cubic yards. Dumpsters and containers with an individual capacity of 1.5 cubic yards [40.5 cubic feet (1.15 m^3)] or more shall not be stored in buildings or placed within 5 feet (1524 mm) of combustible walls, openings or combustible roof eave lines.

Exceptions:

1. Dumpsters or containers in areas protected by an *approved automatic sprinkler system* installed throughout in accordance with Section 903.3.1.1, 903.3.1.2 or 903.3.1.3.
2. Storage in a structure shall not be prohibited where the structure is of Type I or IIA construction, located not less than 10 feet (3048 mm) from other buildings and used exclusively for dumpster or container storage.

304.3.4 Capacity of 1 cubic yard or more. Dumpsters with an individual capacity of 1.0 cubic yard [200 gallons (0.76 m^3)] or more shall not be stored in buildings or placed within 5 feet (1524 mm) of combustible walls, openings or combustible roof eave lines unless the dumpsters are constructed of noncombustible materials or of combustible materials with a peak rate of heat release not exceeding 300 kW/m^2 where tested in accordance with ASTM E 1354 at an incident heat flux of 50 kW/m^2 in the horizontal orientation.

Exceptions:

1. Dumpsters in areas protected by an *approved automatic sprinkler system* installed throughout in accordance with Section 903.3.1.1, 903.3.1.2 or 903.3.1.3.
2. Storage in a structure shall not be prohibited where the structure is of Type I or IIA construction, located not less than 10 feet (3048 mm) from other buildings and used exclusively for dumpster or container storage.

SECTION 305
IGNITION SOURCES

305.1 Clearance from ignition sources. Clearance between ignition sources, such as luminaires, heaters, flame-producing devices and combustible materials, shall be maintained in an *approved* manner.

305.2 Hot ashes and spontaneous ignition sources. Hot ashes, cinders, smoldering coals or greasy or oily materials subject to spontaneous ignition shall not be deposited in a combustible receptacle, within 10 feet (3048 mm) of other combustible material including combustible walls and partitions or within 2 feet (610 mm) of openings to buildings.

Exception: The minimum required separation distance to other combustible materials shall be 2 feet (610 mm) where the material is deposited in a covered, noncombustible receptacle placed on a noncombustible floor, ground surface or stand.

305.3 Open-flame warning devices. Open-flame warning devices shall not be used along an excavation, road, or any place where the dislodgment of such device might permit the device to roll, fall or slide on to any area or land containing combustible material.

305.4 Deliberate or negligent burning. It shall be unlawful to deliberately or through negligence set fire to or cause the burning of combustible material in such a manner as to endanger the safety of persons or property.

305.5 Unwanted fire ignitions. Acts or processes that have caused repeated ignition of unwanted fires shall be modified to prevent future ignition.

SECTION 306
MOTION PICTURE PROJECTION ROOMS AND FILM

306.1 Motion picture projection rooms. Electric arc, xenon or other light source projection equipment that develops hazardous gases, dust or radiation and the projection of ribbon-type cellulose nitrate film, regardless of the light source used in projection, shall be operated within a motion picture projection room complying with Section 409 of the *International Building Code*.

306.2 Cellulose nitrate film storage. Storage of cellulose nitrate film shall be in accordance with NFPA 40.

SECTION 307
OPEN BURNING, RECREATIONAL FIRES AND PORTABLE OUTDOOR FIREPLACES

307.1 General. A person shall not kindle or maintain or authorize to be kindled or maintained any *open burning* unless conducted and *approved* in accordance with Sections 307.1.1 through 307.5.

307.1.1 Prohibited open burning. Open burning shall be prohibited when atmospheric conditions or local circumstances make such fires hazardous.

Exception: Prescribed burning for the purpose of reducing the impact of wildland fire when authorized by the *fire code official*.

307.2 Permit required. A permit shall be obtained from the *fire code official* in accordance with Section 105.6 prior to kindling a fire for recognized silvicultural or range or wildlife management practices, prevention or control of disease or pests, or a bonfire. Application for such approval shall only be presented by and permits issued to the *owner* of the land upon which the fire is to be kindled.

307.2.1 Authorization. Where required by state or local law or regulations, *open burning* shall only be permitted with prior approval from the state or local air and water quality management authority, provided that all conditions specified in the authorization are followed.

307.3 Extinguishment authority. Where open burning creates or adds to a hazardous situation, or a required permit for open burning has not been obtained, the *fire code official* is authorized to order the extinguishment of the open burning operation.

307.4 Location. The location for *open burning* shall be not less than 50 feet (15 240 mm) from any structure, and provisions shall be made to prevent the fire from spreading to within 50 feet (15 240 mm) of any structure.

Exceptions:

1. Fires in *approved* containers that are not less than 15 feet (4572 mm) from a structure.
2. The minimum required distance from a structure shall be 25 feet (7620 mm) where the pile size is 3 feet (914 mm) or less in diameter and 2 feet (610 mm) or less in height.

307.4.1 Bonfires. A bonfire shall not be conducted within 50 feet (15 240 mm) of a structure or combustible material unless the fire is contained in a barbecue pit. Conditions that could cause a fire to spread within 50 feet (15 240 mm) of a structure shall be eliminated prior to ignition.

307.4.2 Recreational fires. *Recreational fires* shall not be conducted within 25 feet (7620 mm) of a structure or combustible material. Conditions that could cause a fire to spread within 25 feet (7620 mm) of a structure shall be eliminated prior to ignition.

307.4.3 Portable outdoor fireplaces. Portable outdoor fireplaces shall be used in accordance with the manufacturer's instructions and shall not be operated within 15 feet (3048 mm) of a structure or combustible material.

Exception: Portable outdoor fireplaces used at one- and two-family *dwellings*.

307.5 Attendance. *Open burning*, bonfires, *recreational fires* and use of portable outdoor fireplaces shall be constantly attended until the fire is extinguished. A minimum of one portable fire extinguisher complying with Section 906 with a minimum 4-A rating or other *approved* on-site fire-extinguishing equipment, such as dirt, sand, water barrel, garden hose or water truck, shall be available for immediate utilization.

SECTION 308
OPEN FLAMES

308.1 General. Open flame, fire and burning on all premises shall be in accordance with Sections 308.1.1 through 308.4.1 and with other applicable sections of this code.

308.1.1 Where prohibited. A person shall not take or utilize an open flame or light in a structure, vessel, boat or other place where highly flammable, combustible or explosive material is utilized or stored. Lighting appliances shall be well-secured in a glass globe and wire mesh cage or a similar *approved* device.

308.1.2 Throwing or placing sources of ignition. A person shall not throw or place, or cause to be thrown or placed, a lighted match, cigar, cigarette, matches, or other flaming or glowing substance or object on any surface or article where it can cause an unwanted fire.

308.1.3 Torches for removing paint. A person utilizing a torch or other flame-producing device for removing paint from a structure shall provide not less than one portable fire extinguisher complying with Section 906 and with a minimum 4-A rating, two portable fire extinguishers, each with a minimum 2-A rating, or a water hose connected to the water supply on the premises where such burning is done. The person doing the burning shall remain on the premises 1 hour after the torch or flame-producing device is utilized.

308.1.4 Open-flame cooking devices. Charcoal burners and other open-flame cooking devices shall not be operated on combustible balconies or within 10 feet (3048 mm) of combustible construction.

Exceptions:

1. One- and two-family *dwellings*.
2. Where buildings, balconies and decks are protected by an *automatic sprinkler system*.
3. LP-gas cooking devices having LP-gas container with a water capacity not greater than $2^1/_2$ pounds [nominal 1 pound (0.454 kg) LP-gas capacity].

308.1.5 Location near combustibles. Open flames such as from candles, lanterns, kerosene heaters and gas-fired heaters shall not be located on or near decorative material or similar combustible materials.

308.1.6 Open-flame devices. Torches and other devices, machines or processes liable to start or cause fire shall not be operated or used in or upon wildfire risk areas, except by a permit in accordance with Section 105.6 secured from the *fire code official*.

Exception: Use within inhabited premises or designated campsites that are not less than 30 feet (9144 mm) from grass-, grain-, brush- or forest-covered areas.

308.1.6.1 Signals and markers. Flame-employing devices, such as lanterns or kerosene road flares, shall not be operated or used as a signal or marker in or upon wildfire risk areas.

Exception: The proper use of fusees at the scenes of emergencies or as required by standard railroad operating procedures.

308.1.6.2 Portable fueled open-flame devices. Portable open-flame devices fueled by flammable or combustible gases or liquids shall be enclosed or installed in such a manner as to prevent the flame from contacting combustible material.

Exceptions:

1. LP-gas-fueled devices used for sweating pipe joints or removing paint in accordance with Chapter 61.
2. Cutting and welding operations in accordance with Chapter 35.
3. Torches or flame-producing devices in accordance with Section 308.4.
4. Candles and open-flame decorative devices in accordance with Section 308.3.

308.1.6.3 Sky lanterns. A person shall not release or cause to be released an untethered sky lantern.

308.1.7 Religious ceremonies. When, in the opinion of the *fire code official*, adequate safeguards have been taken, participants in religious ceremonies are allowed to carry hand-held candles. Hand-held candles shall not be passed from one person to another while lighted.

308.1.7.1 Aisles and exits. Candles shall be prohibited in areas where occupants stand, or in an *aisle* or *exit*.

308.1.8 Flaming food and beverage preparation. The preparation of flaming foods or beverages in places of assembly and drinking or dining establishments shall be in accordance with Sections 308.1.8.1 through 308.1.8.5.

308.1.8.1 Dispensing. Flammable or *combustible liquids* used in the preparation of flaming foods or beverages shall be dispensed from one of the following:

1. A 1-ounce (29.6 ml) container.
2. A container not exceeding 1-quart (946.5 ml) capacity with a controlled pouring device that will limit the flow to a 1-ounce (29.6 ml) serving.

308.1.8.2 Containers not in use. Containers shall be secured to prevent spillage when not in use.

308.1.8.3 Serving of flaming food. The serving of flaming foods or beverages shall be done in a safe manner and shall not create high flames. The pouring, ladling or spooning of liquids is restricted to a maximum height of 8 inches (203 mm) above the receiving receptacle.

308.1.8.4 Location. Flaming foods or beverages shall be prepared only in the immediate vicinity of the table being serviced. They shall not be transported or carried while burning.

308.1.8.5 Fire protection. The person preparing the flaming foods or beverages shall have a wet cloth towel immediately available for use in smothering the flames in the event of an emergency.

308.2 Permits required. Permits shall be obtained from the *fire code official* in accordance with Section 105.6 prior to engaging in the following activities involving open flame, fire and burning:

1. Use of a torch or flame-producing device to remove paint from a structure.
2. Use of open flame, fire or burning in connection with Group A or E occupancies.
3. Use or operation of torches and other devices, machines or processes liable to start or cause fire in or upon wildfire risk areas.

308.3 Group A occupancies. Open-flame devices shall not be used in a Group A occupancy.

Exceptions:

1. Open-flame devices are allowed to be used in the following situations, provided *approved* precautions are taken to prevent ignition of a combustible material or injury to occupants:
   1.1. Where necessary for ceremonial or religious purposes in accordance with Section 308.1.7.
   1.2. On stages and platforms as a necessary part of a performance in accordance with Section 308.3.2.
   1.3. Where candles on tables are securely supported on substantial noncombustible bases and the candle flames are protected.
2. Heat-producing equipment complying with Chapter 6 and the *International Mechanical Code*.
3. Gas lights are allowed to be used provided adequate precautions satisfactory to the *fire code official* are taken to prevent ignition of combustible materials.

308.3.1 Open-flame decorative devices. Open-flame decorative devices shall comply with all of the following restrictions:

1. Class I and Class II liquids and LP-gas shall not be used.
2. Liquid- or solid-fueled lighting devices containing more than 8 ounces (237 ml) of fuel must self-extinguish and not leak fuel at a rate of more than 0.25 teaspoon per minute (1.26 ml per minute) if tipped over.
3. The device or holder shall be constructed to prevent the spillage of liquid fuel or wax at the rate of more than 0.25 teaspoon per minute (1.26 ml per minute) when the device or holder is not in an upright position.

4. The device or holder shall be designed so that it will return to the upright position after being tilted to an angle of 45 degrees (0.79 rad) from vertical.

   Exception: Devices that self-extinguish if tipped over and do not spill fuel or wax at the rate of more than 0.25 teaspoon per minute (1.26 ml per minute) if tipped over.

5. The flame shall be enclosed except where openings on the side are not more than 0.375-inch (9.5 mm) diameter or where openings are on the top and the distance to the top is such that a piece of tissue paper placed on the top will not ignite in 10 seconds.

6. Chimneys shall be made of noncombustible materials and securely attached to the open-flame device.

   Exception: A chimney is not required to be attached to any open-flame device that will self-extinguish if the device is tipped over.

7. Fuel canisters shall be safely sealed for storage.

8. Storage and handling of *combustible liquids* shall be in accordance with Chapter 57.

9. Shades, where used, shall be made of noncombustible materials and securely attached to the open-flame device holder or chimney.

10. Candelabras with flame-lighted candles shall be securely fastened in place to prevent overturning, and shall be located away from occupants using the area and away from possible contact with drapes, curtains or other combustibles.

308.3.2 Theatrical performances. Where *approved*, open-flame devices used in conjunction with theatrical performances are allowed to be used when adequate safety precautions have been taken in accordance with NFPA 160.

308.4 Group R occupancies. Open flame, fire and burning in Group R occupancies shall comply with the requirements of Sections 308.1 through 308.1.6.3 and Section 308.4.1.

308.4.1 Group R-2 dormitories. Candles, incense and similar open-flame-producing items shall not be allowed in sleeping units in Group R-2 dormitory occupancies.

SECTION 309
POWERED INDUSTRIAL TRUCKS AND EQUIPMENT

309.1 General. Powered industrial trucks and similar equipment including, but not limited to, floor scrubbers and floor buffers, shall be operated and maintained in accordance with Section 309.2 through 309.6.

309.2 Battery chargers. Battery chargers shall be of an *approved* type. Combustible storage shall be kept not less than 3 feet (915 mm) from battery chargers. Battery charging shall not be conducted in areas accessible to the public.

309.3 Ventilation. Ventilation shall be provided in an *approved* manner in battery-charging areas to prevent a dangerous accumulation of flammable gases.

309.4 Fire extinguishers. Battery-charging areas shall be provided with a fire extinguisher complying with Section 906 having a minimum 4-A:20-B:C rating within 20 feet (6096 mm) of the battery charger.

309.5 Refueling. Powered industrial trucks using liquid fuel, LP-gas or hydrogen shall be refueled outside of buildings or in areas specifically *approved* for that purpose. Fixed fuel-dispensing equipment and associated fueling operations shall be in accordance with Chapter 23. Other fuel-dispensing equipment and operations, including cylinder exchange for LP-gas-fueled vehicles, shall be in accordance with Chapter 57 for flammable and *combustible liquids* or Chapter 61 for LP-gas.

309.6 Repairs. Repairs to fuel systems, electrical systems and repairs utilizing open flame or welding shall be done in *approved* locations outside of buildings or in areas specifically *approved* for that purpose.

SECTION 310
SMOKING

310.1 General. The smoking or carrying of a lighted pipe, cigar, cigarette or any other type of smoking paraphernalia or material is prohibited in the areas indicated in Sections 310.2 through 310.8.

310.2 Prohibited areas. Smoking shall be prohibited where conditions are such as to make smoking a hazard, and in spaces where flammable or combustible materials are stored or handled.

310.3 "No Smoking" signs. The *fire code official* is authorized to order the posting of "No Smoking" signs in a conspicuous location in each structure or location in which smoking is prohibited. The content, lettering, size, color and location of required "No Smoking" signs shall be *approved.*

Exception: In Group I-2 occupancies where smoking is prohibited, "No Smoking" signs are not required in interior locations of the facility where signs are displayed at all major entrances into the facility.

310.4 Removal of signs prohibited. A posted "No Smoking" sign shall not be obscured, removed, defaced, mutilated or destroyed.

310.5 Compliance with "No Smoking" signs. Smoking shall not be permitted nor shall a person smoke, throw or deposit any lighted or smoldering substance in any place where "No Smoking" signs are posted.

310.6 Ash trays. Where smoking is permitted, suitable noncombustible ash trays or match receivers shall be provided on each table and at other appropriate locations.

310.7 Burning objects. Lighted matches, cigarettes, cigars or other burning object shall not be discarded in such a manner that could cause ignition of other combustible material.

310.8 Hazardous environmental conditions. Where the *fire code official* determines that hazardous environmental conditions necessitate controlled use of smoking materials, the ignition or use of such materials in mountainous, brush-covered or forest-covered areas or other designated areas is prohibited except in *approved* designated smoking areas.

SECTION 311
VACANT PREMISES

311.1 General. Temporarily unoccupied buildings, structures, premises or portions thereof, including tenant spaces, shall be safeguarded and maintained in accordance with Sections 311.1.1 through 311.6.

311.1.1 Abandoned premises. Buildings, structures and premises for which an *owner* cannot be identified or located by dispatch of a certificate of mailing to the last known or registered address, which persistently or repeatedly become unprotected or unsecured, which have been occupied by unauthorized persons or for illegal purposes, or which present a danger of structural collapse or fire spread to adjacent properties shall be considered abandoned, declared unsafe and abated by demolition or rehabilitation in accordance with the *International Property Maintenance Code* and the *International Building Code*.

311.1.2 Tenant spaces. Storage and lease plans required by this code shall be revised and updated to reflect temporary or partial vacancies.

311.2 Safeguarding vacant premises. Temporarily unoccupied buildings, structures, premises or portions thereof shall be secured and protected in accordance with Sections 311.2.1 through 311.2.3.

311.2.1 Security. Exterior and interior openings accessible to other tenants or unauthorized persons shall be boarded, locked, blocked or otherwise protected to prevent entry by unauthorized individuals. The *fire code official* is authorized to placard, post signs, erect barrier tape or take similar measures as necessary to secure public safety.

311.2.2 Fire protection. Fire alarm, sprinkler and standpipe systems shall be maintained in an operable condition at all times.

Exceptions:

1. Where the premises have been cleared of all combustible materials and debris and, in the opinion of the *fire code official*, the type of construction, *fire separation distance* and security of the premises do not create a fire hazard.
2. Where *approved* by the fire chief, buildings that will not be heated and where *fire protection systems* will be exposed to freezing temperatures, fire alarm and sprinkler systems are permitted to be placed out of service and standpipes are permitted to be maintained as dry systems (without an automatic water supply), provided the building has no contents or storage, and windows, doors and other openings are secured to prohibit entry by unauthorized persons.

311.2.3 Fire separation. Fire-resistance-rated partitions, *fire barriers* and *fire walls* separating vacant tenant spaces from the remainder of the building shall be maintained. Openings, joints and penetrations in fire-resistance-rated assemblies shall be protected in accordance with Chapter 7.

311.3 Removal of combustibles. Persons owning, or in charge or control of, a vacant building or portion thereof, shall remove therefrom all accumulations of combustible materials, flammable or combustible waste or rubbish and shall securely lock or otherwise secure doors, windows and other openings to prevent entry by unauthorized persons. The premises shall be maintained clear of waste or hazardous materials.

Exceptions:

1. Buildings or portions of buildings undergoing additions, *alterations*, repairs or change of occupancy in accordance with the *International Building Code*, where waste is controlled and removed as required by Section 304.
2. Seasonally occupied buildings.

311.4 Removal of hazardous materials. Persons owning or having charge or control of a vacant building containing hazardous materials regulated by Chapter 50 shall comply with the facility closure requirements of Section 5001.6.

311.5 Placards. Any vacant or abandoned buildings or structures determined to be unsafe pursuant to Section 110 of this code relating to structural or interior hazards shall be marked as required by Sections 311.5.1 through 311.5.5.

311.5.1 Placard location. Placards shall be applied on the front of the structure and be visible from the street. Additional placards shall be applied to the side of each entrance to the structure and on penthouses.

311.5.2 Placard size and color. Placards shall be 24 inches by 24 inches (610 mm by 610 mm) minimum in size with a red background, white reflective stripes and a white reflective border. The stripes and border shall have a 2-inch (51 mm) minimum stroke.

311.5.3 Placard date. Placards shall bear the date of their application to the building and the date of the most recent inspection.

311.5.4 Placard symbols. The design of the placards shall use the following symbols:

1. ☐ This symbol shall mean that the structure had normal structural conditions at the time of marking.
2. ◹ This symbol shall mean that structural or interior hazards exist and interior fire-fighting or rescue operations should be conducted with extreme caution.
3. ☒ This symbol shall mean that structural or interior hazards exist to a degree that consideration should be given to limit fire fighting to exterior operations only, with entry only occurring for known life hazards.
4. Vacant marker hazard identification symbols: The following symbols shall be used to designate known

hazards on the vacant building marker. They shall be placed directly above the symbol.

4.1. R/O—Roof open

4.2. S/M—Stairs, steps and landing missing

4.3. F/E—Avoid fire escapes

4.4. H/F—Holes in floor

311.5.5 Informational use. The use of these symbols shall be informational only and shall not in any way limit the discretion of the on-scene incident commander.

** **311.6 Unoccupied tenant spaces in mall buildings.** Unoccupied tenant spaces in covered and open mall buildings shall be:

1. Kept free from the storage of any materials.
2. Separated from the remainder of the building by partitions of not less than 0.5-inch-thick (12.7 mm) gypsum board or an *approved* equivalent to the underside of the ceiling of the adjoining tenant spaces.
3. Without doors or other access openings other than one door that shall be kept key locked in the closed position except during that time when opened for inspection.
4. Kept free from combustible waste and be broomswept clean.

SECTION 312 VEHICLE IMPACT PROTECTION

312.1 General. Vehicle impact protection required by this code shall be provided by posts that comply with Section 312.2 or by other *approved* physical barriers that comply with Section 312.3.

312.2 Posts. Guard posts shall comply with all of the following requirements:

1. Constructed of steel not less than 4 inches (102 mm) in diameter and concrete filled.
2. Spaced not more than 4 feet (1219 mm) betwe

en posts on center.

3. Set not less than 3 feet (914 mm) deep in a concrete footing of not less than a 15-inch (381 mm) diameter.
4. Set with the top of the posts not less than 3 feet (914 mm) above ground.
5. Located not less than 3 feet (914 mm) from the protected object.

312.3 Other barriers. Barriers, other than posts specified in Section 312.2, that are designed to resist, deflect or visually deter vehicular impact commensurate with an anticipated impact scenario shall be permitted where *approved.*

SECTION 313 FUELED EQUIPMENT

313.1 General. Fueled equipment including, but not limited to, motorcycles, mopeds, lawn-care equipment, portable generators and portable cooking equipment, shall not be stored, operated or repaired within a building.

Exceptions:

1. Buildings or rooms constructed for such use in accordance with the *International Building Code*.
2. Where allowed by Section 314.
3. Storage of equipment utilized for maintenance purposes is allowed in *approved* locations where the aggregate fuel capacity of the stored equipment does not exceed 10 gallons (38 L) and the building is equipped throughout with an *automatic sprinkler system* installed in accordance with Section 903.3.1.1.

313.1.1 Removal. The *fire code official* is authorized to require removal of fueled equipment from locations where the presence of such equipment is determined by the *fire code official* to be hazardous.

313.2 Group R occupancies. Vehicles powered by flammable liquids, Class II *combustible liquids* or compressed flammable gases shall not be stored within the living space of Group R buildings.

SECTION 314 INDOOR DISPLAYS

314.1 General. Indoor displays constructed within any occupancy shall comply with Sections 314.2 through 314.4.

314.2 Fixtures and displays. Fixtures and displays of goods for sale to the public shall be arranged so as to maintain free, immediate and unobstructed access to exits as required by Chapter 10.

314.3 Highly combustible goods. The display of highly combustible goods, including but not limited to fireworks, flammable or *combustible liquids*, liquefied flammable gases, oxidizing materials, pyroxylin plastics and agricultural goods, in main *exit access aisles*, *corridors*, covered and open malls, or within 5 feet (1524 mm) of entrances to *exits* and exterior exit doors is prohibited where a fire involving such goods would rapidly prevent or obstruct egress.

314.4 Vehicles. Liquid- or gas-fueled vehicles, boats or other motorcraft shall not be located indoors except as follows:

1. Batteries are disconnected.
2. Fuel in fuel tanks does not exceed one-quarter tank or 5 gallons (19 L) (whichever is least).
3. Fuel tanks and fill openings are closed and sealed to prevent tampering.
4. Vehicles, boats or other motorcraft equipment are not fueled or defueled within the building.

SECTION 315 GENERAL STORAGE

315.1 General. Storage shall be in accordance with Sections 315.2 through 315.5.

315.2 Permit required. A permit for miscellaneous combustible storage shall be required as set forth in Section 105.6.

315.3 Storage in buildings. Storage of materials in buildings shall be orderly and stacks shall be stable. Storage of combustible materials shall be separated from heaters or heating devices by distance or shielding so that ignition cannot occur.

315.3.1 Ceiling clearance. Storage shall be maintained 2 feet (610 mm) or more below the ceiling in nonsprinklered areas of buildings or not less than 18 inches (457 mm) below sprinkler head deflectors in sprinklered areas of buildings.

315.3.2 Means of egress. Combustible materials shall not be stored in exits or enclosures for stairways and ramps.

315.3.3 Equipment rooms. Combustible material shall not be stored in boiler rooms, mechanical rooms, electrical equipment rooms or in *fire command centers* as specified in Section 508.1.5.

315.3.4 Attic, under-floor and concealed spaces. Attic, under-floor and concealed spaces used for storage of combustible materials shall be protected on the storage side as required for 1-hour fire-resistance-rated construction. Openings shall be protected by assemblies that are self-closing and are of noncombustible construction or solid wood core not less than $1^3/_4$ inches (44.5 mm) in thickness. Storage shall not be placed on exposed joists.

Exceptions:

1. Areas protected by *approved automatic sprinkler systems*.
2. Group R-3 and Group U occupancies.

315.4 Outside storage. Outside storage of combustible materials shall not be located within 10 feet (3048 mm) of a lot line.

Exceptions:

1. The separation distance is allowed to be reduced to 3 feet (914 mm) for storage not exceeding 6 feet (1829 mm) in height.
2. The separation distance is allowed to be reduced where the *fire code official* determines that no hazard to the adjoining property exists.

315.4.1 Storage beneath overhead projections from buildings. Where buildings are protected by automatic sprinklers, the outdoor storage, display and handling of combustible materials under eaves, canopies or other projections or overhangs are prohibited except where automatic sprinklers are installed under such eaves, canopies or other projections or overhangs.

315.4.2 Height. Storage in the open shall not exceed 20 feet (6096 mm) in height.

315.5 Storage underneath high-voltage transmission lines. Storage located underneath high-voltage transmission lines shall be in accordance with Section 316.6.2.

315.6 Storage in plenums. Storage shall not be permitted in plenums. Abandoned material in plenums shall be deemed to be storage and shall be removed. Where located in plenums, the accessible portion of abandoned cables that are not identified for future use with a tag shall be deemed storage and shall be removed.

SECTION 316
HAZARDS TO FIRE FIGHTERS

316.1 Trapdoors to be closed. Trapdoors and scuttle covers, other than those that are within a *dwelling unit* or automatically operated, shall be kept closed at all times except when in use.

316.2 Shaftway markings. Vertical shafts shall be identified as required by this section.

316.2.1 Exterior access to shaftways. Outside openings accessible to the fire department and that open directly on a hoistway or shaftway communicating between two or more floors in a building shall be plainly marked with the word SHAFTWAY in red letters not less than 6 inches (152 mm) high on a white background. Such warning signs shall be placed so as to be readily discernible from the outside of the building.

316.2.2 Interior access to shaftways. Door or window openings to a hoistway or shaftway from the interior of the building shall be plainly marked with the word SHAFTWAY in red letters not less than 6 inches (152 mm) high on a white background. Such warning signs shall be placed so as to be readily discernible.

Exception: Marking shall not be required on shaftway openings that are readily discernible as openings onto a shaftway by the construction or arrangement.

316.3 Pitfalls. The intentional design or *alteration* of buildings to disable, injure, maim or kill intruders is prohibited. A person shall not install and use firearms, sharp or pointed objects, razor wire, *explosives*, flammable or *combustible liquid* containers, or dispensers containing highly toxic, toxic, irritant or other hazardous materials in a manner that could passively or actively disable, injure, maim or kill a fire fighter who forcibly enters a building for the purpose of controlling or extinguishing a fire, rescuing trapped occupants or rendering other emergency assistance.

316.4 Obstructions on roofs. Wires, cables, ropes, antennas, or other suspended obstructions installed on the roof of a building having a roof slope of less than 30 degrees (0.52 rad) shall not create an obstruction that is less than 7 feet (2133 mm) high above the surface of the roof.

Exceptions:

1. Such obstruction shall be permitted where the wire, cable, rope, antenna or suspended obstruction is encased in a white, 2-inch (51 mm) minimum diameter plastic pipe or an approved equivalent.
2. Such obstruction shall be permitted where there is a solid obstruction below such that accidentally walking into the wire, cable, rope, antenna or suspended obstruction is not possible.

316.5 Security device. Any security device or system that emits any medium that could obscure a *means of egress* in any building, structure or premise shall be prohibited.

316.6 Structures and outdoor storage underneath high-voltage transmission lines. Structures and outdoor storage underneath high-voltage transmission lines shall comply with Sections 316.6.1 and 316.6.2, respectively.

316.6.1 Structures. Structures shall not be constructed within the utility easement beneath high-voltage transmission lines.

Exception: Restrooms and unoccupied telecommunication structures of noncombustible construction less than 15 feet (4572 mm) in height.

316.6.2 Outdoor storage. Outdoor storage within the utility easement underneath high-voltage transmission lines shall be limited to noncombustible material. Storage of hazardous materials including, but not limited to, flammable and *combustible liquids* is prohibited.

Exception: Combustible storage, including vehicles and fuel storage for backup power equipment serving public utility equipment, is allowed, provided that a plan indicating the storage configuration is submitted and *approved*.

SECTION 317 ROOFTOP GARDENS AND LANDSCAPED ROOFS

317.1 General. Rooftop gardens and landscaped roofs shall be installed and maintained in accordance with Sections 317.2 through 317.5 and Sections 1505 and 1507.16 of the *International Building Code.*

317.2 Rooftop garden or landscaped roof size. Rooftop garden or landscaped roof areas shall not exceed 15,625 square feet (1450 m^2) in size for any single area with a maximum dimension of 125 feet (39 m) in length or width. A minimum 6-foot-wide (1.8 m) clearance consisting of a Class A-rated roof system complying with ASTM E 108 or UL 790 shall be provided between adjacent rooftop gardens or landscaped roof areas.

317.3 Rooftop structure and equipment clearance. For all vegetated roofing systems abutting combustible vertical surfaces, a Class A-rated roof system complying with ASTM E 108 or UL 790 shall be achieved for a minimum 6-foot-wide (1829 mm) continuous border placed around rooftop structures and all rooftop equipment including, but not limited to, mechanical and machine rooms, penthouses, skylights, roof vents, solar panels, antenna supports and building service equipment.

317.4 Vegetation. Vegetation shall be maintained in accordance with Sections 317.4.1 and 317.4.2.

317.4.1 Irrigation. Supplemental irrigation shall be provided to maintain levels of hydration necessary to keep green roof plants alive and to keep dry foliage to a minimum.

317.4.2 Dead foliage. Excess biomass, such as overgrown vegetation, leaves and other dead and decaying material, shall be removed at regular intervals not less than two times per year.

317.4.3 Maintenance plan. The *fire code official* is authorized to require a maintenance plan for vegetation placed on roofs due to the size of a roof garden, materials used or where a fire hazard exists to the building or exposures due to the lack of maintenance.

317.5 Maintenance equipment. Fueled equipment stored on roofs and used for the care and maintenance of vegetation on roofs shall be stored in accordance with Section 313.

SECTION 318 LAUNDRY CARTS

318.1 Laundry carts with a capacity of 1 cubic yard or more. Laundry carts with an individual capacity of 1 cubic yard [200 gallons (0.76 m^3)] or more, used in laundries within Group B, E, F-1, I, M and R-1 occupancies, shall be constructed of noncombustible materials or materials having a peak rate of heat release not exceeding 300 kW/m^2 at a flux of 50 kW/m^2 where tested in a horizontal orientation in accordance with ASTM E 1354.

Exceptions:

1. Laundry carts in areas protected by an *approved automatic sprinkler system* installed throughout in accordance with Section 903.3.1.1.
2. Laundry carts in coin-operated laundries.

CHAPTER 4

EMERGENCY PLANNING AND PREPAREDNESS

SECTION 401 GENERAL

401.1 Scope. Reporting of emergencies, coordination with emergency response forces, emergency plans and procedures for managing or responding to emergencies shall comply with the provisions of this section.

Exception: Firms that have *approved* on-premises firefighting organizations and that are in compliance with *approved* procedures for fire reporting.

401.2 Approval. Where required by this code, fire safety plans, emergency procedures and employee training programs shall be *approved* by the *fire code official.*

401.3 Emergency responder notification. Notification of emergency responders shall be in accordance with Sections 401.3.1 through 401.3.3.

401.3.1 Fire events. In the event an unwanted fire occurs on a property, the *owner* or occupant shall immediately report such condition to the fire department.

401.3.2 Alarm activations. Upon activation of a fire alarm signal, employees or staff shall immediately notify the fire department.

401.3.3 Delayed notification. A person shall not, by verbal or written directive, require any delay in the reporting of a fire to the fire department.

401.4 Required plan implementation. In the event an unwanted fire is detected in a building or a fire alarm activates, the emergency plan shall be implemented.

401.5 Making false report. A person shall not give, signal or transmit a false alarm.

401.6 Emergency evacuation drills. The sounding of a fire alarm signal and the carrying out of an emergency evacuation drill in accordance with the provisions of Section 405 shall be allowed.

401.7 Unplanned evacuation. Evacuations made necessary by the unplanned activation of a fire alarm system or by any other emergency shall not be substituted for a required evacuation drill.

401.8 Interference with fire department operations. It shall be unlawful to interfere with, attempt to interfere with, conspire to interfere with, obstruct or restrict the mobility of or block the path of travel of a fire department emergency vehicle in any way, or to interfere with, attempt to interfere with, conspire to interfere with, obstruct or hamper any fire department operation.

SECTION 402 DEFINITIONS

402.1 Definitions. The following terms are defined in Chapter 2:

EMERGENCY EVACUATION DRILL.

LOCKDOWN.

SECTION 403 EMERGENCY PREPAREDNESS REQUIREMENTS **

403.1 General. In addition to the requirements of Section 401, occupancies, uses and outdoor locations shall comply with the emergency preparedness requirements set forth in Sections 403.2 through 403.12.3.3. Where a fire safety and evacuation plan is required by Sections 403.2 through 403.11.4, evacuation drills shall be in accordance with Section 405 and employee training shall be in accordance with Section 406.

403.2 Group A occupancies. An *approved* fire safety and evacuation plan in accordance with Section 404 shall be prepared and maintained for Group A *occupancies*, other than those occupancies used exclusively for purposes of religious worship with an occupant load less than 2,000, and for buildings containing both a Group A occupancy and an atrium. Group A occupancies shall comply with Sections 403.2.1 through 403.2.4.

403.2.1 Seating plan. In addition to the requirements of Section 404.2, the fire safety and evacuation plans for assembly occupancies shall include a detailed seating plan, *occupant load* and *occupant load* limit. Deviations from the *approved* plans shall be allowed provided the *occupant load* limit for the occupancy is not exceeded and the *aisles* and exit accessways remain unobstructed.

403.2.2 Announcements. In theaters, motion picture theaters, auditoriums and similar assembly occupancies in Group A used for noncontinuous programs, an audible announcement shall be made not more than 10 minutes prior to the start of each program to notify the occupants of the location of the exits to be used in the event of a fire or other emergency.

Exception: In motion picture theaters, the announcement is allowed to be projected upon the screen in a manner *approved* by the *fire code official.*

403.2.3 Fire watch personnel. Fire watch personnel shall be provided where required by Section 403.12.1.

403.2.4 Crowd managers. Crowd managers shall be provided where required by Section 403.12.3.

403.3 Ambulatory care facilities. Ambulatory care facilities shall comply with the requirements of Sections 403.3.1 through 403.3.3 as well as 401 and 404 through 406.

403.3.1 Fire evacuation plan. The fire safety and evacuation plan required by Section 404 shall include a description of special staff actions. This shall include procedures for stabilizing patients in a defend-in-place response, staged evacuation, or full evacuation in conjunction with the entire building if part of a multitenant facility.

403.3.2 Fire safety plan. A copy of the plan shall be maintained at the facility at all times. The plan shall include all of the following in addition to the requirements of Section 404:

1. Locations of patients who are rendered incapable of self-preservation.
2. Maximum number of patients rendered incapable of self-preservation.
3. Area and extent of each ambulatory care facility.
4. Location of adjacent smoke compartments or refuge areas, where required.
5. Path of travel to adjacent smoke compartments.
6. Location of any special locking, delayed egress or access control arrangements.

403.3.3 Staff training. Employees shall be periodically instructed and kept informed of their duties and responsibilities under the plan. Records of instruction shall be maintained. Such instruction shall be reviewed by the staff not less than every two months. A copy of the plan shall be readily available at all times within the facility.

403.3.4 Emergency evacuation drills. Emergency evacuation drills shall comply with Section 405. Emergency evacuation drills shall be conducted not less than four times per year.

Exceptions: The movement of patients to safe areas or to the exterior of the building is not required.

403.4 Group B occupancies. An *approved* fire safety and evacuation plan in accordance with Section 404 shall be prepared and maintained for buildings containing a Group B occupancy where the Group B occupancy has an *occupant load* of 500 or more persons or more than 100 persons above or below the lowest *level of exit discharge* and for buildings having an ambulatory care facility.

403.5 Group E occupancies. An *approved* fire safety and evacuation plan in accordance with Section 404 shall be prepared and maintained for Group E occupancies and for buildings containing both a Group E occupancy and an atrium. Group E occupancies shall comply with Sections 403.5.1 through 403.5.3.

403.5.1 First emergency evacuation drill. The first emergency evacuation drill of each school year shall be conducted within 10 days of the beginning of classes.

403.5.2 Time of day. Emergency evacuation drills shall be conducted at different hours of the day or evening, during the changing of classes, when the school is at assembly, during the recess or gymnastic periods, or during other times to avoid distinction between drills and actual fires.

403.5.3 Assembly points. Outdoor assembly areas shall be designated and shall be located a safe distance from the building being evacuated so as to avoid interference with fire department operations. The assembly areas shall be arranged to keep each class separate to provide accountability of all individuals.

403.6 Group F occupancies. An *approved* fire safety and evacuation plan in accordance with Section 404 shall be prepared and maintained for buildings containing a Group F occupancy where the Group F occupancy has an *occupant load* of 500 or more persons or more than 100 persons above or below the lowest *level of exit discharge*.

403.7 Group H occupancies. An *approved* fire safety and evacuation plan in accordance with Section 404 shall be prepared and maintained for Group H occupancies.

403.7.1 Group H-5 occupancies. Group H-5 occupancies shall comply with Sections 403.7.1.1 through 403.7.1.4.

403.7.1.1 Plans and diagrams. In addition to the requirements of Section 404 and Section 407.6, plans and diagrams shall be maintained in *approved* locations indicating the approximate plan for each area, the amount and type of HPM stored, handled and used, locations of shutoff valves for HPM supply piping, emergency telephone locations and locations of exits.

403.7.1.2 Plan updating. The plans and diagrams required by Sections 404, 403.7.1.1 and 407.6 shall be maintained up to date and the *fire code official* and fire department shall be informed of major changes.

403.7.1.3 Emergency response team. Responsible persons shall be designated as an on-site emergency response team and trained to be liaison personnel for the fire department. These persons shall aid the fire department in preplanning emergency responses, identifying locations where HPM is stored, handled and used, and be familiar with the chemical nature of such material. An adequate number of personnel for each work shift shall be designated.

403.7.1.4 Emergency drills. Emergency drills of the on-site emergency response team shall be conducted on a regular basis but not less than once every three months. Records of drills conducted shall be maintained.

403.8 Group I occupancies. An *approved* fire safety and evacuation plan in accordance with Section 404 shall be prepared and maintained for Group I occupancies. Group I occupancies shall comply with Sections 403.8.1 through 403.8.3.4.

403.8.1 Group I-1 occupancies. Group I-1 occupancies shall comply with Sections 403.8.1.1 through 403.8.1.7.

403.8.1.1 Fire safety and evacuation plan. The fire safety and evacuation plan required by Section 404 shall include special employee actions, including fire protection procedures necessary for residents, and shall be amended or revised upon admission of any resident with unusual needs.

403.8.1.1.1 Fire evacuation plan. The fire evacuation plan required by Section 404 shall include a description of special staff actions. In addition to the requirements of Section 404, plans in Group I-1 Condition 2 occupancies shall include procedures for evacuation through a refuge area in an adjacent smoke compartment and then to an exterior assembly point.

403.8.1.1.2 Fire safety plans. A copy of the fire safety plan shall be maintained at the facility at all times. Plans shall include the following in addition to the requirements of Section 404:

1. Location and number of resident sleeping rooms.
2. Location of special locking or egress control arrangements.

403.8.1.2 Employee training. Employees shall be periodically instructed and kept informed of their duties and responsibilities under the plan. Such instruction shall be reviewed by employees at intervals not exceeding two months. A copy of the plan shall be readily available at all times within the facility.

403.8.1.3 Resident training. Residents capable of assisting in their own evacuation shall be trained in the proper actions to take in the event of a fire. In Group I-1 Condition 2 occupancies, training shall include evacuation through an adjacent smoke compartment and then to an exterior assembly point. The training shall include actions to take if the primary escape route is blocked. Where the resident is given rehabilitation or habilitation training, methods of fire prevention and actions to take in the event of a fire shall be a part of the rehabilitation training program. Residents shall be trained to assist each other in case of fire to the extent their physical and mental abilities permit them to do so without additional personal risk.

403.8.1.4 Drill frequency. In addition to the evacuation drills required in Section 405.2, employees shall participate in drills an additional two times a year on each shift. Twelve drills with all occupants shall be conducted in the first year of operation. Drills are not required to comply with the time requirements of Section 405.4.

403.8.1.5 Drill times. Drill times are not required to comply with Section 405.4.

403.8.1.6 Resident participation in drills. Emergency evacuation drills shall involve the actual evacuation of residents to a selected assembly point and shall provide residents with experience in exiting through all required exits. All required exits shall be used during emergency evacuation drills.

403.8.1.7 Emergency evacuation drill deferral. In severe climates, the *fire code official* shall have the authority to modify the emergency evacuation drill frequency specified in Section 405.2.

403.8.2 Group I-2 occupancies. Group I-2 occupancies shall comply with Sections 403.8.2.1 through 403.8.2.3 as well as 401 and 404 through 406.

403.8.2.1 Fire evacuation plans. The fire safety and evacuation plans required by Section 404 shall include a description of special staff *actions*. Plans shall include all of the following in addition to the requirements of Section 404.

1. Procedures for evacuation for patients with needs for containment or restraint and post-evacuation containment, where present.
2. A written plan for maintenance of the means of egress.
3. Procedure for a defend-in-place strategy.
4. Procedures for a full-floor or building evacuation, where necessary.

403.8.2.2 Fire safety plans. A copy of the plan shall be maintained at the facility at all times. Plans shall include all of the following in addition to the requirements of Section 404:

1. Location and number of patient sleeping rooms and operating rooms.
2. Location of adjacent smoke compartments or refuge areas.
3. Path of travel to adjacent smoke compartments.
4. Location of special locking, delayed egress or access control arrangements.
5. Location of elevators utilized for patient movement in accordance with the fire safety plan, where provided.

403.8.2.3 Emergency evacuation drills. Emergency evacuation drills shall comply with Section 405.

Exceptions:

1. The movement of patients to safe areas or to the exterior of the building is not required.
2. Where emergency evacuation drills are conducted after visiting hours or where patients or residents are expected to be asleep, a coded announcement shall be an acceptable alternative to audible alarms.

403.8.3 Group I-3 occupancies. Group I-3 occupancies shall comply with Sections 403.8.3.1 through 403.8.3.4.

403.8.3.1 Employee training. Employees shall be instructed in the proper use of portable fire extinguishers and other manual fire suppression equipment. Training of new employees shall be provided promptly upon entrance to duty. Refresher training shall be provided not less than annually.

403.8.3.2 Employee staffing. Group I-3 occupancies shall be provided with 24-hour staffing. An employee shall be within three floors or 300 feet (91 440 mm) horizontal distance of the access door of each resident housing area. In Group I-3 Conditions 3, 4 and 5, as defined in Chapter 2, the arrangement shall be such that the employee involved can start release of locks necessary for emergency evacuation or rescue and initiate other necessary emergency actions within 2 minutes of an alarm.

Exception: An employee shall not be required to be within three floors or 300 feet (91 440 mm) horizontal distance of the access door of each resident housing area in areas in which all locks are unlocked remotely and automatically in accordance with Section 408.4 of the *International Building Code*.

403.8.3.3 Notification. Provisions shall be made for residents in Group I-3 Conditions 3, 4 and 5, as defined in Chapter 2, to readily notify an employee of an emergency.

403.8.3.4 Keys. Keys necessary for unlocking doors installed in a *means of egress* shall be individually identifiable by both touch and sight.

403.9 Group M occupancies. An *approved* fire safety and evacuation plan in accordance with Section 404 shall be prepared and maintained for buildings containing a Group M occupancy where the Group M occupancy has an *occupant load* of 500 or more persons or more than 100 persons above or below the lowest *level of exit discharge* and for buildings containing both a Group M occupancy and an atrium.

403.10 Group R occupancies. Group R occupancies shall comply with Sections 403.10.1 through 403.10.3.6.

403.10.1 Group R-1 occupancies. An approved fire safety and evacuation plan in accordance with Section 404 shall be prepared and maintained for Group R-1 occupancies. Group R-1 occupancies shall comply with Sections 403.10.1.1 through 403.10.1.3.

403.10.1.1 Evacuation diagrams. A diagram depicting two evacuation routes shall be posted on or immediately adjacent to every required egress door from each hotel or motel sleeping unit.

403.10.1.2 Emergency duties. Upon discovery of a fire or suspected fire, hotel and motel employees shall perform the following duties:

1. Activate the fire alarm system, where provided.
2. Notify the public fire department.
3. Take other action as previously instructed.

403.10.1.3 Fire safety and evacuation instructions. Information shall be provided in the fire safety and evacuation plan required by Section 404 to allow guests to decide whether to evacuate to the outside, evacuate to an *area of refuge,* remain in place, or any combination of the three.

403.10.2 Group R-2 occupancies. Group R-2 occupancies shall comply with Sections 403.10.2.1 through 403.10.2.3.

403.10.2.1 College and university buildings. An *approved* fire safety and evacuation plan in accordance with Section 404 shall be prepared and maintained for Group R-2 college and university buildings. Group R-2 college and university buildings shall comply with Sections 403.10.2.1.1 and 403.10.2.1.2.

403.10.2.1.1 First emergency evacuation drill. The first emergency evacuation drill of each school year shall be conducted within 10 days of the beginning of classes.

403.10.2.1.2 Time of day. Emergency evacuation drills shall be conducted at different hours of the day or evening, during the changing of classes, when school is at assembly, during recess or gymnastic periods or during other times to avoid distinction between drills and actual fires. One required drill shall be held during hours after sunset or before sunrise.

403.10.2.2 Emergency guide. Fire emergency guides shall be provided for Group R-2 occupancies. Guide contents, maintenance and distribution shall comply with Sections 403.10.2.2.1 through 403.10.2.2.3.

403.10.2.2.1 Guide contents. A fire emergency guide shall describe the location, function and use of fire protection equipment and appliances accessible to residents, including fire alarm systems, smoke alarms and portable fire extinguishers. Guides shall include an emergency evacuation plan for each *dwelling unit.*

403.10.2.2.2 Emergency guide maintenance. Emergency guides shall be reviewed and approved by the *fire code official.*

403.10.2.2.3 Emergency guide distribution. A copy of the emergency guide shall be given to each tenant prior to initial occupancy.

403.10.2.3 Evacuation diagrams for dormitories. A diagram depicting two evacuation routes shall be posted on or immediately adjacent to every required egress door from each dormitory *sleeping unit.* Evacuation diagrams shall be reviewed and updated as needed to maintain accuracy.

403.10.3 Group R-4 occupancies. An *approved* fire safety and evacuation plan in accordance with Section 404 shall be prepared and maintained for Group R-4 occupancies. Group R-4 occupancies shall comply with Sections 403.10.3.1 through 403.10.3.6.

403.10.3.1 Fire safety and evacuation plan. The fire safety and evacuation plan required by Section 404 shall include special employee actions, including fire protection procedures necessary for residents, and shall be amended or revised upon admission of a resident with unusual needs.

403.10.3.1.1 Fire safety plans. A copy of the plan shall be maintained at the facility at all times. Plans shall include the following in addition to the requirements of Section 404:

1. Location and number of resident sleeping rooms.
2. Location of special locking or egress control arrangements.

403.10.3.2 Employee training. Employees shall be periodically instructed and kept informed of their duties and responsibilities under the plan. Records of instruction shall be maintained. Such instruction shall be reviewed by employees at intervals not exceeding two months. A copy of the plan shall be readily available at all times within the facility.

403.10.3.3 Resident training. Residents capable of assisting in their own evacuation shall be trained in the proper actions to take in the event of a fire. The training shall include actions to take if the primary escape route is blocked. Where the resident is given rehabilitation or habilitation training, methods of fire prevention and actions to take in the event of a fire shall be a part of the rehabilitation training program. Residents shall be trained to assist each other in case of fire to the extent their physical and mental abilities permit them to do so without additional personal risk.

403.10.3.4 Drill frequency. In addition to the evacuation drills required in Section 405.2, employees shall participate in drills an additional two times a year on each shift. Twelve drills with all occupants shall be conducted in the first year of operation.

403.10.3.5 Drill times. Drill times are not required to comply with Section 405.4.

403.10.3.6 Resident participation in drills. Emergency evacuation drills shall involve the actual evacuation of residents to a selected assembly point and shall provide residents with experience in exiting through all required exits. All required exits shall be used during emergency evacuation drills.

Exception: Actual exiting from emergency escape and rescue windows shall not be required. Opening the emergency escape and rescue window and signaling for help shall be an acceptable alternative.

403.11 Special uses. Special uses shall be in accordance with Sections 403.11.1 through 403.11.4.

403.11.1 Covered and open mall buildings. Covered and open mall buildings shall comply with the requirements of Sections 403.11.1.1 through 403.11.1.6.

403.11.1.1 Malls and mall buildings exceeding 50,000 square feet. An *approved* fire safety and evacuation plan in accordance with Section 404 shall be prepared and maintained for covered malls exceeding 50,000 square feet (4645 m^2) in aggregate floor area and for open mall buildings exceeding 50,000 square feet (4645 m^2) in aggregate area within the perimeter line.

403.11.1.2 Lease plan. In addition to the requirements of Section 404.2.2, a lease plan that includes the following information shall be prepared for each covered and open mall building:

1. Each occupancy, including identification of tenant.
2. *Exits* from each tenant space.
3. Fire protection features, including the following:

   3.1. Fire department connections.

   3.2. *Fire command center.*

   3.3. Smoke management system controls.

   3.4. Elevators, elevator machine rooms and controls.

   3.5. Hose valve outlets.

   3.6. Sprinkler and standpipe control valves.

   3.7. Automatic fire-extinguishing system areas.

   3.8. Automatic fire detector zones.

   3.9. *Fire barriers.*

403.11.1.3 Lease plan approval. The lease plan shall be submitted to the *fire code official* for approval, and shall be maintained on site for immediate reference by responding fire service personnel.

403.11.1.4 Lease plan revisions. The lease plans shall be revised annually or as often as necessary to keep them current. Modifications or changes in tenants or occupancies shall not be made without prior approval of the *fire code official* and building official.

403.11.1.5 Tenant identification. Tenant identification shall be provided for secondary *exits* from occupied tenant spaces that lead to an *exit corridor* or directly to the exterior of the building. Tenant identification shall be posted on the exterior side of the *exit* or exit access door and shall identify the business name and address using plainly legible letters and numbers that contrast with their background.

Exception: Tenant identification is not required for anchor stores.

403.11.1.6 Unoccupied tenant spaces. The fire safety and evacuation plan shall provide for compliance with the requirements for unoccupied tenant spaces in Section 311.

403.11.2 High-rise buildings. An *approved* fire safety and evacuation plan in accordance with Section 404 shall be prepared and maintained for high-rise buildings.

403.11.3 Underground buildings. An *approved* fire safety and evacuation plan in accordance with Section 404 shall be prepared and maintained for underground buildings.

403.11.4 Buildings using occupant evacuation elevators. In buildings using occupant evacuation elevators in accordance with Section 3008 of the *International Building Code*, the fire safety and evacuation plan and the training required by Sections 404 and 406, respectively, shall incorporate specific procedures for the occupants using such elevators.

403.12 Special requirements for public safety. Special requirements for public safety shall be in accordance with Sections 403.12.1 through 403.12.3.3.

403.12.1 Fire watch personnel. Where, in the opinion of the *fire code official*, it is essential for public safety in a place of assembly or any other place where people congregate, because of the number of persons, or the nature of the performance, exhibition, display, contest or activity, the *owner*, agent or lessee shall provide one or more fire watch personnel, as required and *approved*. Fire watch personnel shall comply with Sections 403.12.1.1 and 403.12.1.2.

403.12.1.1 Duty times. Fire watch personnel shall remain on duty while places requiring a fire watch are open to the public, or when an activity requiring a fire watch is being conducted.

403.12.1.2 Duties. On-duty fire watch personnel shall have the following responsibilities:

1. Keep diligent watch for fires, obstructions to *means of egress* and other hazards.
2. Take prompt measures for remediation of hazards and extinguishment of fires that occur.
3. Take prompt measures to assist in the evacuation of the public from the structures.

403.12.2 Public safety plan for gatherings. Where the *fire code official* determines that an indoor or outdoor gathering of persons has an adverse impact on public safety through diminished access to buildings, structures, fire hydrants and fire apparatus access roads or where such gatherings adversely affect public safety services of any kind, the *fire code official* shall have the authority to order the development of or prescribe a public safety plan that provides an *approved* level of public safety and addresses the following items:

1. Emergency vehicle ingress and egress.
2. Fire protection.
3. Emergency egress or escape routes.
4. Emergency medical services.
5. Public assembly areas.
6. The directing of both attendees and vehicles, including the parking of vehicles.
7. Vendor and food concession distribution.
8. The need for the presence of law enforcement.
9. The need for fire and emergency medical services personnel.

403.12.3 Crowd managers for gatherings exceeding 1,000 people. Where facilities or events involve a gathering of more than 1,000 people, crowd managers shall be provided in accordance with Sections 403.12.3.1 through 403.12.3.3.

403.12.3.1 Number of crowd managers. The minimum number of crowd managers shall be established at a ratio of one crowd manager for every 250 persons.

Exception: Where approved by the *fire code official*, the number of crowd managers shall be permitted to be reduced where the facility is equipped throughout with an *approved automatic sprinkler system* or based upon the nature of the event.

403.12.3.2 Training. Training for crowd managers shall be *approved*.

403.12.3.3 Duties. The duties of crowd managers shall include, but not be limited to:

1. Conduct an inspection of the area of responsibility and identify and address any egress barriers.
2. Conduct an inspection of the area of responsibility to identify and mitigate any fire hazards.
3. Verify compliance with all permit conditions, including those governing pyrotechnics and other special effects.
4. Direct and assist the event attendees in evacuation during an emergency.
5. Assist emergency response personnel where requested.
6. Other duties required by the *fire code official*.
7. Other duties as specified in the fire safety plan.

SECTION 404
FIRE SAFETY, EVACUATION AND LOCKDOWN PLANS

404.1 General. Where required by Section 403, fire safety, evacuation and lockdown plans shall comply with Sections 404.2 through 404.4.1.

404.2 Contents. Fire safety and evacuation plan contents shall be in accordance with Sections 404.2.1 and 404.2.2.

404.2.1 Fire evacuation plans. Fire evacuation plans shall include the following:

1. Emergency egress or escape routes and whether evacuation of the building is to be complete by selected floors or areas only or with a defend-in-place response.
2. Procedures for employees who must remain to operate critical equipment before evacuating.
3. Procedures for the use of elevators to evacuate the building where occupant evacuation elevators complying with Section 3008 of the *International Building Code* are provided.
4. Procedures for assisted rescue for persons unable to use the general *means of egress* unassisted.
5. Procedures for accounting for employees and occupants after evacuation has been completed.
6. Identification and assignment of personnel responsible for rescue or emergency medical aid.
7. The preferred and any alternative means of notifying occupants of a fire or emergency.
8. The preferred and any alternative means of reporting fires and other emergencies to the fire department or designated emergency response organization.
9. Identification and assignment of personnel who can be contacted for further information or explanation of duties under the plan.

10. A description of the emergency voice/alarm communication system alert tone and preprogrammed voice messages, where provided.

404.2.2 Fire safety plans. Fire safety plans shall include the following:

1. The procedure for reporting a fire or other emergency.
2. The life safety strategy including the following:
   - 2.1. Procedures for notifying occupants, including areas with a private mode alarm system.
   - 2.2. Procedures for occupants under a defend-in-place response.
   - 2.3. Procedures for evacuating occupants, including those who need evacuation assistance.
3. Site plans indicating the following:
   - 3.1. The occupancy assembly point.
   - 3.2. The locations of fire hydrants.
   - 3.3. The normal routes of fire department vehicle access.
4. Floor plans identifying the locations of the following:
   - 4.1. Exits.
   - 4.2. Primary evacuation routes.
   - 4.3. Secondary evacuation routes.
   - 4.4. Accessible egress routes.
     - 4.4.1. Areas of refuge.
     - 4.4.2. Exterior areas for assisted rescue.
   - 4.5. Refuge areas associated with *smoke barriers* and *horizontal exits*.
   - 4.6. Manual fire alarm boxes.
   - 4.7. Portable fire extinguishers.
   - 4.8. Occupant-use hose stations.
   - 4.9. Fire alarm annunciators and controls.
5. A list of major fire hazards associated with the normal use and occupancy of the premises, including maintenance and housekeeping procedures.
6. Identification and assignment of personnel responsible for maintenance of systems and equipment installed to prevent or control fires.
7. Identification and assignment of personnel responsible for maintenance, housekeeping and controlling fuel hazard sources.

404.2.3 Lockdown plans. Where facilities develop a lockdown plan, it shall be in accordance with Sections 404.2.3.1 through 404.2.3.3.

404.2.3.1 Lockdown plan contents. Lockdown plans shall be *approved* by the *fire code official* and shall include the following:

1. Initiation. The plan shall include instructions for reporting an emergency that requires a lockdown.
2. Accountability. The plan shall include accountability procedures for staff to report the presence or absence of occupants.
3. Recall. The plan shall include a prearranged signal for returning to normal activity.
4. Communication and coordination. The plan shall include an *approved* means of two-way communication between a central location and each secured area.

404.2.3.2 Training frequency. The training frequency shall be included in the lockdown plan. The lockdown drills shall not substitute for any of the fire and evacuation drills required in Section 405.2.

404.2.3.3 Lockdown notification. The method of notifying building occupants of a lockdown shall be included in the plan. The method of notification shall be separate and distinct from the fire alarm signal.

404.3 Maintenance. Fire safety and evacuation plans shall be reviewed or updated annually or as necessitated by changes in staff assignments, occupancy or the physical arrangement of the building.

404.4 Availability. Fire safety and evacuation plans shall be available in the workplace for reference and review by employees, and copies shall be furnished to the *fire code official* for review upon request.

404.4.1 Distribution. The fire safety and evacuation plans shall be distributed to the tenants and building service employees by the *owner* or *owner's* agent. Tenants shall distribute to their employees applicable parts of the fire safety plan affecting the employees' actions in the event of a fire or other emergency.

SECTION 405
EMERGENCY EVACUATION DRILLS

405.1 General. Emergency evacuation drills complying with Sections 405.2 through 405.9 shall be conducted not less than annually where fire safety and evacuation plans are required by Section 403 or where required by the *fire code official.* Drills shall be designed in cooperation with the local authorities.

405.2 Frequency. Required emergency evacuation drills shall be held at the intervals specified in Table 405.2 or more frequently where necessary to familiarize all occupants with the drill procedure.

TABLE 405.2
FIRE AND EVACUATION DRILL
FREQUENCY AND PARTICIPATION

GROUP OR OCCUPANCY	FREQUENCY	PARTICIPATION
Group A	Quarterly	Employees
Group B[b]	Annually	All occupants
Group B[b, c] (Ambulatory care facilities)	Annually	Employees
Group B[b] (Clinic, outpatient)	Annually	Employees
Group E	Monthly[a]	All occupants
Group F	Annually	Employees
Group I-1	Semiannually on each shift	All occupants
Group I-2	Quarterly on each shift[a]	Employees
Group I-3	Quarterly on each shift[a]	Employees
Group I-4	Monthly on each shift[a]	All occupants
Group R-1	Quarterly on each shift	Employees
Group R-2[d]	Four annually	All occupants
Group R-4	Semiannually on each shift[a]	All occupants

a. In severe climates, the *fire code official* shall have the authority to modify the emergency evacuation drill frequency.

b. Emergency evacuation drills are required in Group B buildings having an *occupant load* of 500 or more persons or more than 100 persons above or below the lowest *level of exit discharge*.

c. Emergency evacuation drills are required in ambulatory care facilities in accordance with Section 403.3.

d. Emergency evacuation drills in Group R-2 college and university buildings shall be in accordance with Section 403.10.2.1. Other Group R-2 occupancies shall be in accordance with Section 403.10.2.2.

405.3 Leadership. Responsibility for the planning and conduct of drills shall be assigned to competent persons designated to exercise leadership.

405.4 Time. Drills shall be held at unexpected times and under varying conditions to simulate the unusual conditions that occur in case of fire.

405.5 Record keeping. Records shall be maintained of required emergency evacuation drills and include the following information:

1. Identity of the person conducting the drill.
2. Date and time of the drill.
3. Notification method used.
4. Employees on duty and participating.
5. Number of occupants evacuated.
6. Special conditions simulated.
7. Problems encountered.
8. Weather conditions when occupants were evacuated.
9. Time required to accomplish complete evacuation.

405.6 Notification. Where required by the *fire code official*, prior notification of emergency evacuation drills shall be given to the *fire code official*.

405.7 Initiation. Where a fire alarm system is provided, emergency evacuation drills shall be initiated by activating the fire alarm system.

405.8 Accountability. As building occupants arrive at the assembly point, efforts shall be made to determine if all occupants have been successfully evacuated or have been accounted for.

405.9 Recall and reentry. An electrically or mechanically operated signal used to recall occupants after an evacuation shall be separate and distinct from the signal used to initiate the evacuation. The recall signal initiation means shall be manually operated and under the control of the person in charge of the premises or the official in charge of the incident. Persons shall not reenter the premises until authorized to do so by the official in charge.

SECTION 406
EMPLOYEE TRAINING AND RESPONSE PROCEDURES

406.1 General. Where fire safety and evacuation plans are required by Section 403, employees shall be trained in fire emergency procedures based on plans prepared in accordance with Section 404.

406.2 Frequency. Employees shall receive training in the contents of fire safety and evacuation plans and their duties as part of new employee orientation and not less than annually thereafter. Records of training shall be maintained.

406.3 Employee training program. Employees shall be trained in fire prevention, evacuation and fire safety in accordance with Sections 406.3.1 through 406.3.4.

406.3.1 Fire prevention training. Employees shall be apprised of the fire hazards of the materials and processes to which they are exposed. Each employee shall be instructed in the proper procedures for preventing fires in the conduct of their assigned duties.

406.3.2 Evacuation training. Employees shall be familiarized with the fire alarm and evacuation signals, their assigned duties in the event of an alarm or emergency, evacuation routes, areas of refuge, exterior assembly areas and procedures for evacuation.

406.3.3 Fire safety training. Employees assigned fire-fighting duties shall be trained to know the locations and proper use of portable fire extinguishers or other manual fire-fighting equipment and the protective clothing or equipment required for its safe and proper use.

406.4 Emergency lockdown training. Where a facility has a lockdown plan, employees shall be trained on their assigned duties and procedures in the event of an emergency lockdown.

SECTION 407
HAZARD COMMUNICATION

407.1 General. The provisions of Sections 407.2 through 407.7 shall be applicable where hazardous materials subject to permits under Section 5001.5 are located on the premises or where required by the *fire code official.*

407.2 Material Safety Data Sheets. Material Safety Data Sheets (MSDS) for all hazardous materials shall be either readily available on the premises as a paper copy, or where *approved*, shall be permitted to be readily retrievable by electronic access.

407.3 Identification. Individual containers of hazardous materials, cartons or packages shall be marked or labeled in accordance with applicable federal regulations. Buildings, rooms and spaces containing hazardous materials shall be identified by hazard warning signs in accordance with Section 5003.5.

407.4 Training. Persons responsible for the operation of areas in which hazardous materials are stored, dispensed, handled or used shall be familiar with the chemical nature of the materials and the appropriate mitigating actions necessary in the event of a fire, leak or spill. Responsible persons shall be designated and trained to be liaison personnel for the fire department. These persons shall aid the fire department in preplanning emergency responses and identification of where hazardous materials are located, and shall have access to Material Safety Data Sheets and be knowledgeable in the site emergency response procedures.

407.5 Hazardous Materials Inventory Statement. Where required by the *fire code official*, each application for a permit shall include a Hazardous Materials Inventory Statement (HMIS) in accordance with Section 5001.5.2.

407.6 Hazardous Materials Management Plan. Where required by the *fire code official*, each application for a permit shall include a Hazardous Materials Management Plan (HMMP) in accordance with Section 5001.5.1. The *fire code official* is authorized to accept a similar plan required by other regulations.

407.7 Facility closure plans. The permit holder or applicant shall submit to the *fire code official* a facility closure plan in accordance with Section 5001.6.3 to terminate storage, dispensing, handling or use of hazardous materials.

*

Part III—Building and Equipment Design Features

CHAPTER 5

FIRE SERVICE FEATURES

SECTION 501 GENERAL

501.1 Scope. Fire service features for buildings, structures and premises shall comply with this chapter.

501.2 Permits. A permit shall be required as set forth in Sections 105.6 and 105.7.

501.3 Construction documents. *Construction documents* for proposed fire apparatus access, location of *fire lanes*, security gates across fire apparatus access roads and *construction documents* and hydraulic calculations for fire hydrant systems shall be submitted to the fire department for review and approval prior to construction.

501.4 Timing of installation. Where fire apparatus access roads or a water supply for fire protection are required to be installed, such protection shall be installed and made serviceable prior to and during the time of construction except when *approved* alternative methods of protection are provided. Temporary street signs shall be installed at each street intersection where construction of new roadways allows passage by vehicles in accordance with Section 505.2.

SECTION 502 DEFINITIONS

502.1 Definitions. The following terms are defined in Chapter 2:

AGENCY.

FIRE APPARATUS ACCESS ROAD.

FIRE COMMAND CENTER.

FIRE DEPARTMENT MASTER KEY.

FIRE LANE.

KEY BOX.

TRAFFIC CALMING DEVICES.

SECTION 503 FIRE APPARATUS ACCESS ROADS

503.1 Where required. Fire apparatus access roads shall be provided and maintained in accordance with Sections 503.1.1 through 503.1.3.

503.1.1 Buildings and facilities. *Approved* fire apparatus access roads shall be provided for every facility, building or portion of a building hereafter constructed or moved into or within the jurisdiction. The fire apparatus access road shall comply with the requirements of this section and shall extend to within 150 feet (45 720 mm) of all portions of the facility and all portions of the *exterior walls* of the first story of the building as measured by an *approved* route around the exterior of the building or facility.

Exceptions:

1. The *fire code official* is authorized to increase the dimension of 150 feet (45 720 mm) where any of the following conditions occur:
   - 1.1. The building is equipped throughout with an *approved automatic sprinkler system* installed in accordance with Section 903.3.1.1, 903.3.1.2 or 903.3.1.3.
   - 1.2. Fire apparatus access roads cannot be installed because of location on property, topography, waterways, nonnegotiable grades or other similar conditions, and an *approved* alternative means of fire protection is provided.
   - 1.3. There are not more than two Group R-3 or Group U occupancies.
2. Where approved by the *fire code official*, fire apparatus access roads shall be permitted to be exempted or modified for solar photovoltaic power generation facilities.

503.1.2 Additional access. The *fire code official* is authorized to require more than one fire apparatus access road based on the potential for impairment of a single road by vehicle congestion, condition of terrain, climatic conditions or other factors that could limit access.

503.1.3 High-piled storage. Fire department vehicle access to buildings used for *high-piled combustible storage* shall comply with the applicable provisions of Chapter 32.

503.2 Specifications. Fire apparatus access roads shall be installed and arranged in accordance with Sections 503.2.1 through 503.2.8.

503.2.1 Dimensions. Fire apparatus access roads shall have an unobstructed width of not less than 20 feet (6096 mm), exclusive of shoulders, except for *approved* security gates in accordance with Section 503.6, and an unobstructed vertical clearance of not less than 13 feet 6 inches (4115 mm).

503.2.2 Authority. The *fire code official* shall have the authority to require or permit modifications to the required access widths where they are inadequate for fire or rescue operations or where necessary to meet the public safety objectives of the jurisdiction.

503.2.3 Surface. Fire apparatus access roads shall be designed and maintained to support the imposed loads of fire apparatus and shall be surfaced so as to provide all-weather driving capabilities.

503.2.4 Turning radius. The required turning radius of a fire apparatus access road shall be determined by the *fire code official.*

503.2.5 Dead ends. Dead-end fire apparatus access roads in excess of 150 feet (45 720 mm) in length shall be provided with an *approved* area for turning around fire apparatus.

503.2.6 Bridges and elevated surfaces. Where a bridge or an elevated surface is part of a fire apparatus access road, the bridge shall be constructed and maintained in accordance with AASHTO HB-17. Bridges and elevated surfaces shall be designed for a live load sufficient to carry the imposed loads of fire apparatus. Vehicle load limits shall be posted at both entrances to bridges where required by the *fire code official.* Where elevated surfaces designed for emergency vehicle use are adjacent to surfaces that are not designed for such use, *approved* barriers, *approved* signs or both shall be installed and maintained where required by the *fire code official.*

503.2.7 Grade. The grade of the fire apparatus access road shall be within the limits established by the *fire code official* based on the fire department's apparatus.

503.2.8 Angles of approach and departure. The angles of approach and departure for fire apparatus access roads shall be within the limits established by the *fire code official* based on the fire department's apparatus.

503.3 Marking. Where required by the *fire code official, approved* signs or other *approved* notices or markings that include the words NO PARKING—FIRE LANE shall be provided for fire apparatus access roads to identify such roads or prohibit the obstruction thereof. The means by which *fire lanes* are designated shall be maintained in a clean and legible condition at all times and be replaced or repaired when necessary to provide adequate visibility.

503.4 Obstruction of fire apparatus access roads. Fire apparatus access roads shall not be obstructed in any manner, including the parking of vehicles. The minimum widths and clearances established in Sections 503.2.1 and 503.2.2 shall be maintained at all times.

503.4.1. Traffic calming devices. Traffic calming devices shall be prohibited unless *approved* by the *fire code official.*

503.5 Required gates or barricades. The *fire code official* is authorized to require the installation and maintenance of gates or other *approved* barricades across fire apparatus access roads, trails or other accessways, not including public streets, alleys or highways. Electric gate operators, where provided, shall be *listed* in accordance with UL 325. Gates intended for automatic operation shall be designed, constructed and installed to comply with the requirements of ASTM F 2200.

503.5.1 Secured gates and barricades. Where required, gates and barricades shall be secured in an *approved* manner. Roads, trails and other accessways that have been closed and obstructed in the manner prescribed by Section 503.5 shall not be trespassed on or used unless authorized by the *owner* and the *fire code official.*

Exception: The restriction on use shall not apply to public officers acting within the scope of duty.

503.6 Security gates. The installation of security gates across a fire apparatus access road shall be *approved* by the fire chief. Where security gates are installed, they shall have an *approved* means of emergency operation. The security gates and the emergency operation shall be maintained operational at all times. Electric gate operators, where provided, shall be *listed* in accordance with UL 325. Gates intended for automatic operation shall be designed, constructed and installed to comply with the requirements of ASTM F 2200.

SECTION 504
ACCESS TO BUILDING OPENINGS AND ROOFS

504.1 Required access. Exterior doors and openings required by this code or the *International Building Code* shall be maintained readily accessible for emergency access by the fire department. An *approved* access walkway leading from fire apparatus access roads to exterior openings shall be provided when required by the *fire code official.*

504.2 Maintenance of exterior doors and openings. Exterior doors and their function shall not be eliminated without prior approval. Exterior doors that have been rendered nonfunctional and that retain a functional door exterior appearance shall have a sign affixed to the exterior side of the door with the words THIS DOOR BLOCKED. The sign shall consist of letters having a principal stroke of not less than $^3/_4$ inch (19.1 mm) wide and not less than 6 inches (152 mm) high on a contrasting background. Required fire department access doors shall not be obstructed or eliminated. Exit and *exit access* doors shall comply with Chapter 10. Access doors for *high-piled combustible storage* shall comply with Section 3206.6.1.

504.3 Stairway access to roof. New buildings four or more stories above grade plane, except those with a roof slope greater than four units vertical in 12 units horizontal (33.3-percent slope), shall be provided with a *stairway* to the roof. *Stairway* access to the roof shall be in accordance with Section 1011.12. Such *stairway* shall be marked at street and floor levels with a sign indicating that the *stairway* continues to the roof. Where roofs are used for roof gardens or for other purposes, *stairways* shall be provided as required for such occupancy classification.

SECTION 505
PREMISES IDENTIFICATION

505.1 Address identification. New and existing buildings shall be provided with *approved* address identification. The address identification shall be legible and placed in a position that is visible from the street or road fronting the property. Address identification characters shall contrast with their background. Address numbers shall be Arabic numbers or alphabetical letters. Numbers shall not be spelled out. Each character shall be not less than 4 inches (102 mm) high with a minimum stroke width of $^1/_2$ inch (12.7 mm). Where required by the *fire code official*, address identification shall be provided in additional *approved* locations to facilitate emergency response. Where access is by means of a private road and the building cannot be viewed from the *public way*, a monument, pole or other sign or means shall be used to identify the structure. Address identification shall be maintained.

505.2 Street or road signs. Streets and roads shall be identified with *approved* signs. Temporary signs shall be installed at each street intersection when construction of new roadways allows passage by vehicles. Signs shall be of an *approved* size, weather resistant and be maintained until replaced by permanent signs.

SECTION 506
KEY BOXES

506.1 Where required. Where access to or within a structure or an area is restricted because of secured openings or where immediate access is necessary for life-saving or fire-fighting purposes, the *fire code official* is authorized to require a key box to be installed in an *approved* location. The key box shall be of an *approved* type listed in accordance with UL 1037, and shall contain keys to gain necessary access as required by the *fire code official*.

506.1.1 Locks. An *approved* lock shall be installed on gates or similar barriers where required by the *fire code official*.

506.1.2 Key boxes for nonstandardized fire service elevator keys. Key boxes provided for nonstandardized fire service elevator keys shall comply with Section 506.1 and all of the following:

1. The key box shall be compatible with an existing rapid entry key box system in use in the jurisdiction and *approved* by the *fire code official*.
2. The front cover shall be permanently labeled with the words "Fire Department Use Only—Elevator Keys."
3. The key box shall be mounted at each elevator bank at the lobby nearest to the lowest level of fire department access.
4. The key box shall be mounted 5 feet 6 inches (1676 mm) above the finished floor to the right side of the elevator bank.
5. Contents of the key box are limited to fire service elevator keys. Additional elevator access tools, keys and information pertinent to emergency planning or elevator access shall be permitted where authorized by the *fire code official*.
6. In buildings with two or more elevator banks, a single key box shall be permitted to be used where such elevator banks are separated by not more than 30 feet (9144 mm). Additional key boxes shall be provided for each individual elevator or elevator bank separated by more than 30 feet (9144 mm).

Exception: A single key box shall be permitted to be located adjacent to a *fire command center* or the nonstandard fire service elevator key shall be permitted to be secured in a key box used for other purposes and located in accordance with Section 506.1.

506.2 Key box maintenance. The operator of the building shall immediately notify the *fire code official* and provide the new key where a lock is changed or rekeyed. The key to such lock shall be secured in the key box.

SECTION 507
FIRE PROTECTION WATER SUPPLIES

507.1 Required water supply. An *approved* water supply capable of supplying the required fire flow for fire protection shall be provided to premises upon which facilities, buildings or portions of buildings are hereafter constructed or moved into or within the jurisdiction.

507.2 Type of water supply. A water supply shall consist of reservoirs, pressure tanks, elevated tanks, water mains or other fixed systems capable of providing the required fire flow.

507.2.1 Private fire service mains. Private fire service mains and appurtenances shall be installed in accordance with NFPA 24.

507.2.2 Water tanks. Water tanks for private fire protection shall be installed in accordance with NFPA 22.

507.3 Fire flow. Fire flow requirements for buildings or portions of buildings and facilities shall be determined by an *approved* method.

507.4 Water supply test. The *fire code official* shall be notified prior to the water supply test. Water supply tests shall be witnessed by the *fire code official* or *approved* documentation of the test shall be provided to the *fire code official* prior to final approval of the water supply system.

507.5 Fire hydrant systems. Fire hydrant systems shall comply with Sections 507.5.1 through 507.5.6.

507.5.1 Where required. Where a portion of the facility or building hereafter constructed or moved into or within the jurisdiction is more than 400 feet (122 m) from a hydrant on a fire apparatus access road, as measured by an *approved* route around the exterior of the facility or building, on-site fire hydrants and mains shall be provided where required by the *fire code official*.

Exceptions:

1. For Group R-3 and Group U occupancies, the distance requirement shall be 600 feet (183 m).
2. For buildings equipped throughout with an *approved automatic sprinkler system* installed in accordance with Section 903.3.1.1 or 903.3.1.2, the distance requirement shall be 600 feet (183 m).

507.5.1.1 Hydrant for standpipe systems. Buildings equipped with a standpipe system installed in accordance with Section 905 shall have a fire hydrant within 100 feet (30 480 mm) of the fire department connections.

Exception: The distance shall be permitted to exceed 100 feet (30 480 mm) where *approved* by the *fire code official.*

507.5.2 Inspection, testing and maintenance. Fire hydrant systems shall be subject to periodic tests as required by the *fire code official.* Fire hydrant systems shall be maintained in an operative condition at all times and shall be repaired where defective. Additions, repairs, *alterations* and servicing shall comply with *approved* standards. Records of tests and required maintenance shall be maintained.

507.5.3 Private fire service mains and water tanks. Private fire service mains and water tanks shall be periodically inspected, tested and maintained in accordance with NFPA 25 at the following intervals:

1. Private fire hydrants of all types: Inspection annually and after each operation; flow test and maintenance annually.
2. Fire service main piping: Inspection of exposed, annually; flow test every 5 years.
3. Fire service main piping strainers: Inspection and maintenance after each use.

Records of inspections, testing and maintenance shall be maintained.

507.5.4 Obstruction. Unobstructed access to fire hydrants shall be maintained at all times. The fire department shall not be deterred or hindered from gaining immediate access to fire protection equipment or fire hydrants.

507.5.5 Clear space around hydrants. A 3-foot (914 mm) clear space shall be maintained around the circumference of fire hydrants, except as otherwise required or *approved.*

507.5.6 Physical protection. Where fire hydrants are subject to impact by a motor vehicle, guard posts or other *approved* means shall comply with Section 312.

SECTION 508
FIRE COMMAND CENTER

508.1 General. Where required by other sections of this code and in all buildings classified as high-rise buildings by the *International Building Code*, a *fire command center* for fire department operations shall be provided and shall comply with Sections 508.1.1 through 508.1.6.

508.1.1 Location and access. The location and accessibility of the *fire command center* shall be *approved* by the fire chief.

508.1.2 Separation. The *fire command center* shall be separated from the remainder of the building by not less than a 1-hour *fire barrier* constructed in accordance with Section 707 of the *International Building Code* or *horizontal assembly* constructed in accordance with Section 711 of the *International Building Code*, or both.

508.1.3 Size. The *fire command center* shall be not less than 200 square feet (19 m^2) in area with a minimum dimension of 10 feet (3048 mm).

508.1.4 Layout approval. A layout of the *fire command center* and all features required by this section to be contained therein shall be submitted for approval prior to installation.

508.1.5 Storage. Storage unrelated to operation of the *fire command center* shall be prohibited.

508.1.6 Required features. The *fire command center* shall comply with NFPA 72 and shall contain the following features:

1. The emergency voice/alarm communication system control unit.
2. The fire department communications system.
3. Fire detection and alarm system annunciator.
4. Annunciator unit visually indicating the location of the elevators and whether they are operational.
5. Status indicators and controls for air distribution systems.
6. The fire fighter's control panel required by Section 909.16 for smoke control systems installed in the building.
7. Controls for unlocking *stairway* doors simultaneously.
8. Sprinkler valve and water-flow detector display panels.
9. Emergency and standby power status indicators.
10. A telephone for fire department use with controlled access to the public telephone system.
11. Fire pump status indicators.
12. Schematic building plans indicating the typical floor plan and detailing the building core, *means of egress*, *fire protection systems*, fire-fighter air-replenishment systems, fire-fighting equipment and fire department access, and the location of *fire walls*, *fire barriers*, *fire partitions*, *smoke barriers* and smoke partitions.
13. An *approved* Building Information Card that includes, but is not limited to, all of the following information:
    13.1. General building information that includes: property name, address, the number of floors in the building above and below grade, use and occupancy classification (for mixed uses, identify the different types of occupancies on each floor) and the estimated building population during the day, night and weekend;

13.2. Building emergency contact information that includes: a list of the building's emergency contacts including but not limited to building manager, building engineer and their respective work phone number, cell phone number and e-mail address;

13.3. Building construction information that includes: the type of building construction including but not limited to floors, walls, columns and roof assembly;

13.4. *Exit access stairway* and *exit stairway* information that includes: number of *exit access stairways* and *exit stairways* in building; each *exit access stairway* and *exit stairway* designation and floors served; location where each *exit access stairway* and *exit stairway* discharges, *interior exit stairways* that are pressurized; *exit stairways* provided with emergency lighting; each *exit stairway* that allows reentry; *exit stairways* providing roof access; elevator information that includes: number of elevator banks, elevator bank designation, elevator car numbers and respective floors that they serve; location of elevator machine rooms, control rooms and control spaces; location of sky lobby; and location of freight elevator banks;

13.5. Building services and system information that includes: location of mechanical rooms, location of building management system, location and capacity of all fuel oil tanks, location of emergency generator and location of natural gas service;

13.6. *Fire protection system* information that includes: location of standpipes, location of fire pump room, location of fire department connections, floors protected by automatic sprinklers and location of different types of *automatic sprinkler systems* installed including but not limited to dry, wet and pre-action;

13.7. Hazardous material information that includes: location and quantity of hazardous material.

14. Work table.
15. Generator supervision devices, manual start and transfer features.
16. Public address system, where specifically required by other sections of this code.
17. Elevator fire recall switch in accordance with ASME A17.1.
18. Elevator emergency or standby power selector switch(es), where emergency or standby power is provided.

SECTION 509 FIRE PROTECTION AND UTILITY EQUIPMENT IDENTIFICATION AND ACCESS

509.1 Identification. Fire protection equipment shall be identified in an *approved* manner. Rooms containing controls for air-conditioning systems, sprinkler risers and valves, or other fire detection, suppression or control elements shall be identified for the use of the fire department. *Approved* signs required to identify fire protection equipment and equipment location shall be constructed of durable materials, permanently installed and readily visible.

509.1.1 Utility identification. Where required by the *fire code official*, gas shutoff valves, electric meters, service switches and other utility equipment shall be clearly and legibly marked to identify the unit or space that it serves. Identification shall be made in an *approved* manner, readily visible and shall be maintained.

509.2 Equipment access. *Approved* access shall be provided and maintained for all fire protection equipment to permit immediate safe operation and maintenance of such equipment. Storage, trash and other materials or objects shall not be placed or kept in such a manner that would prevent such equipment from being readily accessible.

SECTION 510 EMERGENCY RESPONDER RADIO COVERAGE

510.1 Emergency responder radio coverage in new buildings. All new buildings shall have *approved* radio coverage for emergency responders within the building based upon the existing coverage levels of the public safety communication systems of the jurisdiction at the exterior of the building. This section shall not require improvement of the existing public safety communication systems.

Exceptions:

1. Where *approved* by the building official and the *fire code official*, a wired communication system in accordance with Section 907.2.13.2 shall be permitted to be installed or maintained instead of an *approved* radio coverage system.
2. Where it is determined by the *fire code official* that the radio coverage system is not needed.
3. In facilities where emergency responder radio coverage is required and such systems, components or equipment required could have a negative impact on the normal operations of that facility, the *fire code official* shall have the authority to accept an automatically activated emergency responder radio coverage system.

510.2 Emergency responder radio coverage in existing buildings. Existing buildings shall be provided with *approved* radio coverage for emergency responders as required in Chapter 11.

510.3 Permit required. A construction permit for the installation of or modification to emergency responder radio coverage systems and related equipment is required as specified in Section 105.7.5. Maintenance performed in accordance with this code is not considered a modification and does not require a permit.

510.4 Technical requirements. Systems, components and equipment required to provide the emergency responder radio coverage system shall comply with Sections 510.4.1 through 510.4.2.5.

510.4.1 Radio signal strength. The building shall be considered to have acceptable emergency responder radio coverage when signal strength measurements in 95 percent of all areas on each floor of the building meet the signal strength requirements in Sections 510.4.1.1 and 510.4.1.2.

510.4.1.1 Minimum signal strength into the building. A minimum signal strength of -95 dBm shall be receivable within the building.

510.4.1.2 Minimum signal strength out of the building. A minimum signal strength of -95 dBm shall be received by the agency's radio system when transmitted from within the building.

510.4.2 System design. The emergency responder radio coverage system shall be designed in accordance with Sections 510.4.2.1 through 510.4.2.5.

510.4.2.1 Amplification systems allowed. Buildings and structures that cannot support the required level of radio coverage shall be equipped with a radiating cable system, a distributed antenna system with Federal Communications Commission (FCC)-certified signal boosters, or other system approved by the *fire code official* in order to achieve the required adequate radio coverage.

510.4.2.2 Technical criteria. The *fire code official* shall maintain a document providing the specific technical information and requirements for the emergency responder radio coverage system. This document shall contain, but not be limited to, the various frequencies required, the location of radio sites, effective radiated power of radio sites, and other supporting technical information.

510.4.2.3 Standby power. Emergency responder radio coverage systems shall be provided with standby power in accordance with Section 604. The standby power supply shall be capable of operating the emergency responder radio coverage system for a duration of not less than 24 hours.

510.4.2.4 Signal booster requirements. If used, signal boosters shall meet the following requirements:

1. All signal booster components shall be contained in a National Electrical Manufacturer's Association (NEMA) 4-type waterproof cabinet.
2. Battery systems used for the emergency power source shall be contained in a NEMA 4-type waterproof cabinet.
3. The signal booster system and battery system shall be electrically supervised and monitored by a supervisory service, or when *approved* by the *fire code official*, shall sound an audible signal at a constantly attended location
4. Equipment shall have FCC certification prior to installation.

510.4.2.5 Additional frequencies and change of frequencies. The emergency responder radio coverage system shall be capable of modification or expansion in the event frequency changes are required by the FCC or additional frequencies are made available by the FCC.

510.5 Installation requirements. The installation of the public safety radio coverage system shall be in accordance with Sections 510.5.1 through 510.5.4.

510.5.1 Approval prior to installation. Amplification systems capable of operating on frequencies licensed to any public safety agency by the FCC shall not be installed without prior coordination and approval of the *fire code official.*

510.5.2 Minimum qualifications of personnel. The minimum qualifications of the system designer and lead installation personnel shall include both of the following:

1. A valid FCC-issued general radio operators license.
2. Certification of in-building system training issued by a nationally recognized organization, school or a certificate issued by the manufacturer of the equipment being installed.

These qualifications shall not be required where demonstration of adequate skills and experience satisfactory to the *fire code official* is provided.

510.5.3 Acceptance test procedure. Where an emergency responder radio coverage system is required, and upon completion of installation, the building *owner* shall have the radio system tested to verify that two-way coverage on each floor of the building is not less than 90 percent. The test procedure shall be conducted as follows:

1. Each floor of the building shall be divided into a grid of 20 approximately equal test areas.
2. The test shall be conducted using a calibrated portable radio of the latest brand and model used by the agency talking through the agency's radio communications system.
3. Failure of not more than two nonadjacent test areas shall not result in failure of the test.
4. In the event that three of the test areas fail the test, in order to be more statistically accurate, the floor shall be permitted to be divided into 40 equal test areas. Failure of not more than four nonadjacent test areas shall not result in failure of the test. If the system fails the 40-area test, the system shall be altered to meet the 90-percent coverage requirement.
5. A test location approximately in the center of each test area shall be selected for the test, with the radio enabled to verify two-way communications to and from the outside of the building through the public agency's radio communications system. Once the test location has been selected, that location shall represent the entire test area. Failure in the selected test location shall be considered failure of that test area. Additional test locations shall not be permitted.

6. The gain values of all amplifiers shall be measured and the test measurement results shall be kept on file with the building *owner* so that the measurements can be verified during annual tests. In the event that the measurement results become lost, the building *owner* shall be required to rerun the acceptance test to reestablish the gain values.
7. As part of the installation a spectrum analyzer or other suitable test equipment shall be utilized to ensure spurious oscillations are not being generated by the subject signal booster. This test shall be conducted at the time of installation and subsequent annual inspections.

510.5.4 FCC compliance. The emergency responder radio coverage system installation and components shall also comply with all applicable federal regulations including, but not limited to, FCC 47 CFR Part 90.219.

510.6 Maintenance. The emergency responder radio coverage system shall be maintained operational at all times in accordance with Sections 510.6.1 through 510.6.3.

510.6.1 Testing and proof of compliance. The emergency responder radio coverage system shall be inspected and tested annually or where structural changes occur including additions or remodels that could materially change the original field performance tests. Testing shall consist of the following:

1. In-building coverage test as described in Section 510.5.3.
2. Signal boosters shall be tested to verify that the gain is the same as it was upon initial installation and acceptance.
3. Backup batteries and power supplies shall be tested under load of a period of 1 hour to verify that they will properly operate during an actual power outage. If within the 1-hour test period the battery exhibits symptoms of failure, the test shall be extended for additional 1-hour periods until the integrity of the battery can be determined.
4. Other active components shall be checked to verify operation within the manufacturer's specifications.
5. At the conclusion of the testing, a report, which shall verify compliance with Section 510.5.3, shall be submitted to the *fire code official*.

510.6.2 Additional frequencies. The building *owner* shall modify or expand the emergency responder radio coverage system at his or her expense in the event frequency changes are required by the FCC or additional frequencies are made available by the FCC. Prior approval of a public safety radio coverage system on previous frequencies does not exempt this section.

510.6.3 Field testing. Agency personnel shall have the right to enter onto the property at any reasonable time to conduct field testing to verify the required level of radio coverage.

CHAPTER 6

BUILDING SERVICES AND SYSTEMS

SECTION 601 GENERAL

601.1 Scope. The provisions of this chapter shall apply to the installation, operation and maintenance of fuel-fired appliances and heating systems, emergency and standby power systems, electrical systems and equipment, mechanical refrigeration systems, elevator recall, stationary storage battery systems and commercial kitchen equipment.

601.2 Permits. Permits shall be obtained for refrigeration systems, battery systems and solar photovoltaic power systems as set forth in Sections 105.6 and 105.7.

SECTION 602 DEFINITIONS

602.1 Definitions. The following terms are defined in Chapter 2:

BATTERY SYSTEM, STATIONARY LEAD-ACID.

BATTERY TYPES.

COMMERCIAL COOKING APPLIANCES.

CRITICAL CIRCUIT.

EMERGENCY POWER SYSTEM.

HOOD.

Type I.

Type II.

REFRIGERANT.

REFRIGERATION SYSTEM.

STANDBY POWER SYSTEM.

SECTION 603 FUEL-FIRED APPLIANCES

603.1 Installation. The installation of nonportable fuel gas appliances and systems shall comply with the *International Fuel Gas Code*. The installation of all other fuel-fired appliances, other than internal combustion engines, oil lamps and portable devices such as blow torches, melting pots and weed burners, shall comply with this section and the *International Mechanical Code*.

603.1.1 Manufacturer's instructions. The installation shall be made in accordance with the manufacturer's instructions and applicable federal, state and local rules and regulations. Where it becomes necessary to change, modify or alter a manufacturer's instructions in any way, written approval shall first be obtained from the manufacturer.

603.1.2 Approval. The design, construction and installation of fuel-fired appliances shall be in accordance with the *International Fuel Gas Code* and the *International Mechanical Code*.

603.1.3 Electrical wiring and equipment. Electrical wiring and equipment used in connection with oil-burning equipment shall be installed and maintained in accordance with Section 605 and NFPA 70.

603.1.4 Fuel oil. The grade of fuel oil used in a burner shall be that for which the burner is *approved* and as stipulated by the burner manufacturer. Oil containing gasoline shall not be used. Waste crankcase oil shall be an acceptable fuel in Group F, M and S occupancies where utilized in equipment *listed* for use with waste oil and where such equipment is installed in accordance with the manufacturer's instructions and the terms of its listing.

603.1.5 Access. The installation shall be readily accessible for cleaning hot surfaces; removing burners; replacing motors, controls, air filters, chimney connectors, draft regulators and other working parts; and for adjusting, cleaning and lubricating parts.

603.1.6 Testing, diagrams and instructions. After installation of the oil-burning equipment, operation and combustion performance tests shall be conducted to determine that the burner is in proper operating condition and that all accessory equipment, controls, and safety devices function properly.

603.1.6.1 Diagrams. Contractors installing industrial oil-burning systems shall furnish not less than two copies of diagrams showing the main oil lines and controlling valves, one copy of which shall be posted at the oil-burning equipment and another at an *approved* location that will be accessible in case of emergency.

603.1.6.2 Instructions. After completing the installation, the installer shall instruct the *owner* or operator in the proper operation of the equipment. The installer shall furnish the *owner* or operator with the name and telephone number of persons to contact for technical information or assistance and routine or emergency services.

603.1.7 Clearances. Working clearances between oil-fired appliances and electrical panelboards and equipment shall be in accordance with NFPA 70. Clearances between oil-fired equipment and oil supply tanks shall be in accordance with NFPA 31.

603.2 Chimneys. Masonry chimneys shall be constructed in accordance with the *International Building Code*. Factory-built chimneys shall be installed in accordance with the *International Mechanical Code*. Metal chimneys shall be constructed and installed in accordance with NFPA 211.

603.3 Fuel oil storage systems. Fuel oil storage systems shall be installed in accordance with this code. Fuel-oil piping systems shall be installed in accordance with the *International Mechanical Code*.

603.3.1 Fuel oil storage in outside, above-ground tanks. Where connected to a fuel-oil piping system, the maximum amount of fuel oil storage allowed outside above ground without additional protection shall be 660 gallons (2498 L). The storage of fuel oil above ground in quantities exceeding 660 gallons (2498 L) shall comply with NFPA 31.

603.3.2 Fuel oil storage inside buildings. Fuel oil storage inside buildings shall comply with Sections 603.3.2.1 through 603.3.2.5 or Chapter 57.

603.3.2.1 Quantity limits. One or more fuel oil storage tanks containing Class II or III *combustible liquid* shall be permitted in a building. The aggregate capacity of all such tanks shall not exceed 660 gallons (2498 L).

Exception: The aggregate capacity limit shall be permitted to be increased to 3,000 gallons (11 356 L) of Class II or III liquid for storage in protected above-ground tanks complying with Section 5704.2.9.7, where all of the following conditions are met:

1. The entire 3,000-gallon (11 356 L) quantity shall be stored in protected above-ground tanks.
2. The 3,000-gallon (11 356 L) capacity shall be permitted to be stored in a single tank or multiple smaller tanks.
3. The tanks shall be located in a room protected by an *automatic sprinkler system* complying with Section 903.3.1.1.

603.3.2.2 Restricted use and connection. Tanks installed in accordance with Section 603.3.2 shall be used only to supply fuel oil to fuel-burning or generator equipment installed in accordance with Section 603.3.2.4. Connections between tanks and equipment supplied by such tanks shall be made using closed piping systems.

603.3.2.3 Applicability of maximum allowable quantity and control area requirements. The quantity of *combustible liquid* stored in tanks complying with Section 603.3.2 shall not be counted towards the maximum allowable quantity set forth in Table 5003.1.1(1), and such tanks shall not be required to be located in a *control area*.

603.3.2.4 Installation. Tanks and piping systems shall be installed and separated from other uses in accordance with Section 915 and Chapter 13, both of the *International Mechanical Code*, as applicable.

Exception: Protected above-ground tanks complying with Section 5704.2.9.7 shall not be required to be separated from surrounding areas.

603.3.2.5 Tanks in basements. Tanks in *basements* shall be located not more than two stories below grade plane.

603.3.3 Underground storage of fuel oil. The storage of fuel oil in underground storage tanks shall comply with NFPA 31.

603.4 Portable unvented heaters. Portable unvented fuel-fired heating equipment shall be prohibited in occupancies in Groups A, E, I, R-1, R-2, R-3 and R-4.

Exceptions:

1. *Listed* and *approved* unvented fuel-fired heaters, including portable outdoor gas-fired heating appliances, in one- and two-family *dwellings*.
2. Portable outdoor gas-fired heating appliances shall be allowed in accordance with Section 603.4.2.

603.4.1 Prohibited locations. Unvented fuel-fired heating equipment shall not be located in, or obtain combustion air from, any of the following rooms or spaces: sleeping rooms, bathrooms, toilet rooms or storage closets.

603.4.2 Portable outdoor gas-fired heating appliances. Portable gas-fired heating appliances located outdoors shall be in accordance with Sections 603.4.2.1 through 603.4.2.3.4.

603.4.2.1 Location. Portable outdoor gas-fired heating appliances shall be located in accordance with Sections 603.4.2.1.1 through 603.4.2.1.4.

603.4.2.1.1 Prohibited locations. The storage or use of portable outdoor gas-fired heating appliances is prohibited in any of the following locations:

1. Inside of any occupancy where connected to the fuel gas container.
2. Inside of tents, canopies and membrane structures.
3. On exterior balconies.

Exception: As allowed in Section 6.20 of NFPA 58.

603.4.2.1.2 Clearance to buildings. Portable outdoor gas-fired heating appliances shall be located not less than 5 feet (1524 mm) from buildings.

603.4.2.1.3 Clearance to combustible materials. Portable outdoor gas-fired heating appliances shall not be located beneath, or closer than 5 feet (1524 mm) to combustible decorations and combustible overhangs, awnings, sunshades or similar combustible attachments to buildings.

603.4.2.1.4 Proximity to exits. Portable outdoor gas-fired heating appliances shall not be located within 5 feet (1524 mm) of *exits* or *exit discharges*.

603.4.2.2 Installation and operation. Portable outdoor gas-fired heating appliances shall be installed and operated in accordance with Sections 603.4.2.2.1 through 603.4.2.2.4.

603.4.2.2.1 Listing and approval. Only *listed* and *approved* portable outdoor gas-fired heating appliances utilizing a fuel gas container that is integral to the appliance shall be used.

603.4.2.2.2 Installation and maintenance. Portable outdoor gas-fired heating appliances shall be installed and maintained in accordance with the manufacturer's instructions.

603.4.2.2.3 Tip-over switch. Portable outdoor gas-fired heating appliances shall be equipped with a tilt or tip-over switch that automatically shuts off the flow of gas if the appliance is tilted more than 15 degrees (0.26 rad) from the vertical.

603.4.2.2.4 Guard against contact. The heating element or combustion chamber of portable outdoor gas-fired heating appliances shall be permanently guarded so as to prevent accidental contact by persons or material.

603.4.2.3 Gas containers. Fuel gas containers for portable outdoor gas-fired heating appliances shall comply with Sections 603.4.2.3.1 through 603.4.2.3.4.

603.4.2.3.1 Approved containers. Only *approved* DOTn or ASME gas containers shall be used.

603.4.2.3.2 Container replacement. Replacement of fuel gas containers in portable outdoor gas-fired heating appliances shall not be conducted while the public is present.

603.4.2.3.3 Container capacity. The maximum individual capacity of gas containers used in connection with portable outdoor gas-fired heating appliances shall not exceed 20 pounds (9 kg).

603.4.2.3.4 Indoor storage prohibited. Gas containers shall not be stored inside of buildings except in accordance with Section 6109.9.

603.5 Heating appliances. Heating appliances shall be *listed* and shall comply with Sections 603.5.1 and 603.5.2.

603.5.1 Guard against contact. The heating element or combustion chamber shall be permanently guarded so as to prevent accidental contact by persons or material.

603.5.2 Heating appliance installation and maintenance. Heating appliances shall be installed and maintained in accordance with the manufacturer's instructions, the *International Building Code*, the *International Mechanical Code*, the *International Fuel Gas Code* and NFPA 70.

603.6 Chimneys and appliances. Chimneys, incinerators, smokestacks or similar devices for conveying smoke or hot gases to the outer air and the stoves, furnaces, fireboxes or boilers to which such devices are connected, shall be maintained so as not to create a fire hazard.

603.6.1 Masonry chimneys. Masonry chimneys that, upon inspection, are found to be without a flue liner and that have open mortar joints which will permit smoke or gases to be discharged into the building, or which are cracked as to be dangerous, shall be repaired or relined with a *listed* chimney liner system installed in accordance with the manufacturer's instructions or a flue lining system installed in accordance with the requirements of the *International Building Code* and appropriate for the intended class of chimney service.

603.6.2 Metal chimneys. Metal chimneys which are corroded or improperly supported shall be repaired or replaced.

603.6.3 Decorative shrouds. Decorative shrouds installed at the termination of factory-built chimneys shall be removed except where such shrouds are *listed* and *labeled* for use with the specific factory-built chimney system and are installed in accordance with the chimney manufacturer's instructions.

603.6.4 Factory-built chimneys. Existing factory-built chimneys that are damaged, corroded or improperly supported shall be repaired or replaced.

603.6.5 Connectors. Existing chimney and vent connectors that are damaged, corroded or improperly supported shall be repaired or replaced.

603.7 Discontinuing operation of unsafe heating appliances. The *fire code official* is authorized to order that measures be taken to prevent the operation of any existing stove, oven, furnace, incinerator, boiler or any other heat-producing device or appliance found to be defective or in violation of code requirements for existing appliances after giving notice to this effect to any person, *owner*, firm or agent or operator in charge of the same. The *fire code official* is authorized to take measures to prevent the operation of any device or appliance without notice when inspection shows the existence of an immediate fire hazard or when imperiling human life. The defective device shall remain withdrawn from service until all necessary repairs or *alterations* have been made.

603.7.1 Unauthorized operation. It shall be a violation of this code for any person, user, firm or agent to continue the utilization of any device or appliance (the operation of which has been discontinued or ordered discontinued in accordance with Section 603.7) unless written authority to resume operation is given by the *fire code official*. Removing or breaking the means by which operation of the device is prevented shall be a violation of this code.

603.8 Incinerators. Commercial, industrial and residential-type incinerators and chimneys shall be constructed in accordance with the *International Building Code*, the *International Fuel Gas Code* and the *International Mechanical Code*.

603.8.1 Residential incinerators. Residential incinerators shall be of an *approved* type.

603.8.2 Spark arrestor. Incinerators shall be equipped with an effective means for arresting sparks.

603.8.3 Restrictions. Where the *fire code official* determines that burning in incinerators located within 500 feet (152 m) of mountainous, brush or grass-covered areas will create an undue fire hazard because of atmospheric conditions, such burning shall be prohibited.

603.8.4 Time of burning. Burning shall take place only during *approved* hours.

603.8.5 Discontinuance. The *fire code official* is authorized to require incinerator use to be discontinued immediately if the *fire code official* determines that smoke emissions are offensive to occupants of surrounding property or if the use of incinerators is determined by the *fire code official* to constitute a hazardous condition.

603.8.6 Flue-fed incinerators in Group I-2. In Group I-2 occupancies, the continued use of existing flue-fed incinerators is prohibited.

603.8.7 Incinerator inspections in Group I-2. Incinerators in Group I-2 occupancies shall be inspected not less than annually in accordance with the manufacturer's instructions. Inspection records shall be maintained on the premises and made available to the *fire code official* upon request

603.9 Gas meters. Above-ground gas meters, regulators and piping subject to damage shall be protected by a barrier complying with Section 312 or otherwise protected in an *approved* manner.

SECTION 604 EMERGENCY AND STANDBY POWER SYSTEMS

604.1 General. Emergency power systems and standby power systems required by this code or the *International Building Code* shall comply with Sections 604.1.1 through 604.1.8.

604.1.1 Stationary generators. Stationary emergency and standby power generators required by this code shall be *listed* in accordance with UL 2200.

604.1.2 Installation. Emergency power systems and standby power systems shall be installed in accordance with the *International Building Code,* NFPA 70, NFPA 110 and NFPA 111.

604.1.3 Load transfer. Emergency power systems shall automatically provide secondary power within 10 seconds after primary power is lost, unless specified otherwise in this code. Standby power systems shall automatically provide secondary power within 60 seconds after primary power is lost unless specified otherwise in this code.

604.1.4 Load duration. Emergency power systems and standby power systems shall be designed to provide the required power for a minimum duration of 2 hours without being refueled or recharged, unless specified otherwise in this code.

604.1.5 Uninterruptable power source. An uninterrupted source of power shall be provided for equipment where required by the manufacturer's instructions, the listing, this code or applicable referenced standards.

604.1.6 Interchangeability. Emergency power systems shall be an acceptable alternative for installations that require standby power systems.

604.1.7 Group I-2 occupancies. In Group I-2 occupancies, where an essential electrical system is located in flood hazard areas established in Section 1612.3 of the *International Building Code* and where new or replacement essential electrical system generators are installed, the system shall be located and installed in accordance with ASCE 24.

604.1.8 Maintenance. Existing installations shall be maintained in accordance with the original approval and Section 604.4.

604.2 Where required. Emergency and standby power systems shall be provided where required by Sections 604.2.1 through 604.2.16.

604.2.1 Elevators and platform lifts. Standby power shall be provided for elevators and platform lifts as required in Sections 607.2, 1009.4, and 1009.5.

604.2.2 Emergency alarm systems. Emergency power shall be provided for emergency alarm systems as required by Section 414 of the *International Building Code.*

604.2.3 Emergency responder radio coverage systems. Standby power shall be provided for emergency responder radio coverage systems as required in Section 510.4.2.3. The standby power supply shall be capable of operating the emergency responder radio coverage system for a duration of not less than 24 hours.

604.2.4 Emergency voice/alarm communication systems. Emergency power shall be provided for emergency voice/alarm communication systems as required in Section 907.5.2.2.5. The system shall be capable of powering the required load for a duration of not less than 24 hours, as required in NFPA 72.

604.2.5 Exit signs. Emergency power shall be provided for *exit* signs as required in Section 1013.6.3. The system shall be capable of powering the required load for a duration of not less than 90 minutes.

604.2.6 Group I-2 occupancies. Essential electrical systems for Group I-2 occupancies shall be in accordance with Section 407.10 of the *International Building Code.*

604.2.7 Group I-3 occupancies. Power-operated sliding doors or power-operated locks for swinging doors in Group I-3 occupancies shall be operable by a manual release mechanism at the door. Emergency power shall be provided for the doors and locks in accordance with Section 604.

Exceptions:

1. Emergency power is not required in facilities where provisions for remote locking and unlocking of occupied rooms in Occupancy Condition 4 are not required as set forth in the *International Building Code.*
2. Emergency power is not required where remote mechanical operating releases are provided.

604.2.8 Hazardous materials. Emergency and standby power shall be provided in occupancies with hazardous materials as required in the following sections:

1. Sections 5004.7 and 5005.1.5 for hazardous materials.
2. Sections 6004.2.2.8 and 6004.3.4.2 for highly toxic and toxic gases.
3. Section 6204.1.11 for organic peroxides.

604.2.9 High-rise buildings. Standby power and emergency power shall be provided for high-rise buildings as required in Section 403 of the *International Building Code*, and shall be in accordance with Section 604.

604.2.10 Horizontal sliding doors. Standby power shall be provided for horizontal sliding doors as required in Section 1010.1.4.3. The standby power supply shall have a capacity to operate not fewer than 50 closing cycles of the door.

604.2.11 Hydrogen fuel gas rooms. Standby power shall be provided for hydrogen fuel gas rooms as required by Section 5808.7.

604.2.12 Means of egress illumination. Emergency power shall be provided for *means of egress* illumination in accordance with Sections 1008.3 and 1104.5.1.

604.2.13 Membrane structures. Standby power shall be provided for auxiliary inflation systems in permanent membrane structures in accordance with Section 2702 of the *International Building Code*. Auxiliary inflation systems shall be provided in temporary air-supported and air-inflated membrane structures in accordance with Section 3103.10.4.

604.2.14 Semiconductor fabrication facilities. Emergency power shall be provided for semiconductor fabrication facilities as required in Section 2703.15.

604.2.15 Smoke control systems. Standby power shall be provided for smoke control systems as required in Section 909.11.

604.2.16 Underground buildings. Emergency and standby power shall be provided in underground buildings as required in Section 405 of the *International Building Code* and shall be in accordance with Section 604.

*

604.3 Critical circuits. Cables used for survivability of required critical circuits shall be listed in accordance with UL 2196. Electrical circuit protective systems shall be installed in accordance with their listing requirements.

604.4 Maintenance. Emergency and standby power systems shall be maintained in accordance with NFPA 110 and NFPA 111 such that the system is capable of supplying service within the time specified for the type and duration required.

604.4.1 Schedule. Inspection, testing and maintenance of emergency and standby power systems shall be in accordance with an approved schedule established upon completion and approval of the system installation.

604.4.2 Records. Records of the inspection, testing and maintenance of emergency and standby power systems shall include the date of service, name of the servicing technician, a summary of conditions noted and a detailed description of any conditions requiring correction and what corrective action was taken. Such records shall be maintained.

604.4.3 Switch maintenance. Emergency and standby power system transfer switches shall be included in the inspection, testing and maintenance schedule required by Section 604.4.1. Transfer switches shall be maintained free from accumulated dust and dirt. Inspection shall include examination of the transfer switch contacts for evidence of deterioration. When evidence of contact deterioration is detected, the contacts shall be replaced in accordance with the transfer switch manufacturer's instructions.

604.5 Operational inspection and testing. Emergency power systems, including all appurtenant components, shall be inspected and tested under load in accordance with NFPA 110 and NFPA 111.

Exception: Where the emergency power system is used for standby power or peak load shaving, such use shall be recorded and shall be allowed to be substituted for scheduled testing of the generator set, provided that appropriate records are maintained.

604.5.1 Transfer switch test. The test of the transfer switch shall consist of electrically operating the transfer switch from the normal position to the alternate position and then return to the normal position.

604.6 Emergency lighting equipment. Emergency lighting shall be inspected and tested in accordance with Sections 604.6.1 through 604.6.2.1.

604.6.1 Activation test. An activation test of the emergency lighting equipment shall be completed monthly. The activation test shall ensure the emergency lighting activates automatically upon normal electrical disconnect and stays sufficiently illuminated for not less than 30 seconds.

604.6.1.1 Activation test record. Records of tests shall be maintained. The record shall include the location of the emergency lighting tested, whether the unit passed or failed, the date of the test and the person completing the test.

604.6.2 Power test. For battery-powered emergency lighting, a power test of the emergency lighting equipment shall be completed annually. The power test shall operate the emergency lighting for not less than 90 minutes and shall remain sufficiently illuminated for the duration of the test.

604.6.2.1 Power test record. Records of tests shall be maintained. The record shall include the location of the emergency lighting tested, whether the unit passed or failed, the date of the test and the person completing the test.

604.7 Supervision of maintenance and testing. Routine maintenance, inspection and operational testing shall be overseen by a properly instructed individual.

SECTION 605
ELECTRICAL EQUIPMENT, WIRING AND HAZARDS

605.1 Abatement of electrical hazards. Identified electrical hazards shall be abated. Identified hazardous electrical conditions in permanent wiring shall be brought to the attention of the responsible code official. Electrical wiring, devices, appliances and other equipment that is modified or damaged and constitutes an electrical shock or fire hazard shall not be used.

605.2 Illumination. Illumination shall be provided for service equipment areas, motor control centers and electrical panelboards.

605.3 Working space and clearance. A working space of not less than 30 inches (762 mm) in width, 36 inches (914 mm) in depth and 78 inches (1981 mm) in height shall be provided in front of electrical service equipment. Where the electrical service equipment is wider than 30 inches (762 mm), the working space shall be not less than the width of the equipment. Storage of materials shall not be located within the designated working space.

Exceptions:

1. Where other dimensions are required or allowed by NFPA 70.
2. Access openings into attics or under-floor areas which provide a minimum clear opening of 22 inches (559 mm) by 30 inches (762 mm).

605.3.1 Labeling. Doors into electrical control panel rooms shall be marked with a plainly visible and legible sign stating ELECTRICAL ROOM or similar approved wording. The disconnecting means for each service, feeder or branch circuit originating on a switchboard or panelboard shall be legibly and durably marked to indicate its purpose unless such purpose is clearly evident.

605.4 Multiplug adapters. Multiplug adapters, such as cube adapters, unfused plug strips or any other device not complying with NFPA 70 shall be prohibited.

605.4.1 Power tap design. Relocatable power taps shall be of the polarized or grounded type, equipped with overcurrent protection, and shall be *listed* in accordance with UL 1363.

605.4.2 Power supply. Relocatable power taps shall be directly connected to a permanently installed receptacle.

605.4.3 Installation. Relocatable power tap cords shall not extend through walls, ceilings, floors, under doors or floor coverings, or be subject to environmental or physical damage.

605.5 Extension cords. Extension cords and flexible cords shall not be a substitute for permanent wiring. Extension cords and flexible cords shall not be affixed to structures, extended through walls, ceilings or floors, or under doors or floor coverings, nor shall such cords be subject to environmental damage or physical impact. Extension cords shall be used only with portable appliances.

605.5.1 Power supply. Extension cords shall be plugged directly into an *approved* receptacle, power tap or multiplug adapter and, except for *approved* multiplug extension cords, shall serve only one portable appliance.

605.5.2 Ampacity. The ampacity of the extension cords shall be not less than the rated capacity of the portable appliance supplied by the cord.

605.5.3 Maintenance. Extension cords shall be maintained in good condition without splices, deterioration or damage.

605.5.4 Grounding. Extension cords shall be grounded where serving grounded portable appliances.

605.6 Unapproved conditions. Open junction boxes and open-wiring splices shall be prohibited. *Approved* covers shall be provided for all switch and electrical outlet boxes.

605.7 Appliances. Electrical appliances and fixtures shall be tested and *listed* in published reports of inspected electrical equipment by an *approved* agency and installed and maintained in accordance with all instructions included as part of such listing.

605.8 Electrical motors. Electrical motors shall be maintained free from excessive accumulations of oil, dirt, waste and debris.

605.9 Temporary wiring. Temporary wiring for electrical power and lighting installations is allowed for a period not to exceed 90 days. Temporary wiring methods shall meet the applicable provisions of NFPA 70.

Exception: Temporary wiring for electrical power and lighting installations is allowed during periods of construction, remodeling, repair or demolition of buildings, structures, equipment or similar activities.

605.9.1 Attachment to structures. Temporary wiring attached to a structure shall be attached in an *approved* manner.

605.10 Portable, electric space heaters. Where not prohibited by other sections of this code, portable, electric space heaters shall be permitted to be used in all occupancies other than Group I-2 and in accordance with Sections 605.10.1 through 605.10.4.

Exception: The use of portable, electric space heaters in which the heating element cannot exceed a temperature of 212°F (100°C) shall be permitted in nonsleeping staff and employee areas in Group I-2 occupancies.

605.10.1 Listed and labeled. Only *listed* and *labeled* portable, electric space heaters shall be used.

605.10.2 Power supply. Portable, electric space heaters shall be plugged directly into an *approved* receptacle.

605.10.3 Extension cords. Portable, electric space heaters shall not be plugged into extension cords.

605.10.4 Prohibited areas. Portable, electric space heaters shall not be operated within 3 feet (914 mm) of any combustible materials. Portable, electric space heaters shall be operated only in locations for which they are *listed*.

605.11 Solar photovoltaic power systems. Solar photovoltaic power systems shall be installed in accordance with Sections 605.11.1 through 605.11.2, the *International Building Code* or *International Residential Code*, and NFPA 70.

605.11.1 Access and pathways. Roof access, pathways, and spacing requirements shall be provided in accordance with Sections 605.11.1.1 through 605.11.1.3.3.

Exceptions:

1. Detached, nonhabitable Group U structures including, but not limited to, parking shade structures, carports, solar trellises and similar structures.
2. Roof access, pathways and spacing requirements need not be provided where the fire chief has determined that rooftop operations will not be employed.

605.11.1.1 Roof access points. Roof access points shall be located in areas that do not require the placement of ground ladders over openings such as windows or doors, and located at strong points of building construction in locations where the access point does not conflict with overhead obstructions such as tree limbs, wires or signs.

605.11.1.2 Solar photovoltaic systems for Group R-3 buildings. Solar photovoltaic systems for Group R-3 buildings shall comply with Sections 605.11.1.2.1 through 605.11.1.2.5.

Exception: These requirements shall not apply to structures designed and constructed in accordance with the *International Residential Code*.

605.11.1.2.1 Size of solar photovoltaic array. Each photovoltaic array shall be limited to 150 feet (45 720 mm) by 150 feet (45 720 mm). Multiple arrays shall be separated by a 3-foot-wide (914 mm) clear access pathway.

605.11.1.2.2 Hip roof layouts. Panels and modules installed on Group R-3 buildings with hip roof layouts shall be located in a manner that provides a 3-foot-wide (914 mm) clear access pathway from the eave to the ridge on each roof slope where panels and modules are located. The access pathway shall be at a location on the building capable of supporting the fire fighters accessing the roof.

Exception: These requirements shall not apply to roofs with slopes of two units vertical in 12 units horizontal (2:12) or less.

605.11.1.2.3 Single-ridge roofs. Panels and modules installed on Group R-3 buildings with a single ridge shall be located in a manner that provides two, 3-foot-wide (914 mm) access pathways from the eave to the ridge on each roof slope where panels and modules are located.

Exception: This requirement shall not apply to roofs with slopes of two units vertical in 12 units horizontal (2:12) or less.

605.11.1.2.4 Roofs with hips and valleys. Panels and modules installed on Group R-3 buildings with roof hips and valleys shall not be located closer than 18 inches (457 mm) to a hip or a valley where panels/modules are to be placed on both sides of a hip or valley. Where panels are to be located on only one side of a hip or valley that is of equal length, the panels shall be permitted to be placed directly adjacent to the hip or valley.

Exception: These requirements shall not apply to roofs with slopes of two units vertical in 12 units horizontal (2:12) or less.

605.11.1.2.5 Allowance for smoke ventilation operations. Panels and modules installed on Group R-3 buildings shall be located not less than 3 feet (914 mm) from the ridge in order to allow for fire department smoke ventilation operations.

Exception: Panels and modules shall be permitted to be located up to the roof ridge where an alternative ventilation method *approved* by the fire chief has been provided or where the fire chief has determined vertical ventilation techniques will not be employed.

605.11.1.3 Other than Group R-3 buildings. Access to systems for buildings, other than those containing Group R-3 occupancies, shall be provided in accordance with Sections 605.11.1.3.1 through 605.11.1.3.3.

Exception: Where it is determined by the fire code official that the roof configuration is similar to that of a Group R-3 occupancy, the residential access and ventilation requirements in Sections 605.11.1.2.1 through 605.11.1.2.5 shall be permitted to be used.

605.11.1.3.1 Access. There shall be a minimum 6-foot-wide (1829 mm) clear perimeter around the edges of the roof.

Exception: Where either axis of the building is 250 feet (76 200 mm) or less, the clear perimeter around the edges of the roof shall be permitted to be reduced to a minimum 4 foot wide (1290 mm).

605.11.1.3.2 Pathways. The solar installation shall be designed to provide designated pathways. The pathways shall meet the following requirements:

1. The pathway shall be over areas capable of supporting fire fighters accessing the roof.
2. The centerline axis pathways shall be provided in both axes of the roof. Centerline axis pathways shall run where the roof structure is capable of supporting fire fighters accessing the roof.
3. Pathways shall be a straight line not less than 4 feet (1290 mm) clear to roof standpipes or ventilation hatches.
4. Pathways shall provide not less than 4 feet (1290 mm) clear around roof access hatch with not less than one singular pathway not less than 4 feet (1290 mm) clear to a parapet or roof edge.

605.11.1.3.3 Smoke ventilation. The solar installation shall be designed to meet the following requirements:

1. Arrays shall be not greater than 150 feet (45 720 mm) by 150 feet (45 720 mm) in distance in either axis in order to create opportunities for fire department smoke ventilation operations.
2. Smoke ventilation options between array sections shall be one of the following:
   - 2.1. A pathway 8 feet (2438 mm) or greater in width.
   - 2.2. A 4-foot (1290 mm) or greater in width pathway and bordering roof skylights or gravity-operated dropout smoke and heat vents on not less than one side.
   - 2.3. A 4-foot (1290 mm) or greater in width pathway and bordering all sides of nongravity-operated dropout smoke and heat vents.
   - 2.4. A 4-foot (1290 mm) or greater in width pathway and bordering 4-foot by 8-foot (1290 mm by 2438 mm) "venting cutouts" every 20 feet (6096 mm) on alternating sides of the pathway.

605.11.2 Ground-mounted photovoltaic arrays. Ground-mounted photovoltaic arrays shall comply with Section 605.11 and this section. Setback requirements shall not apply to ground-mounted, free-standing photovoltaic arrays. A clear, brush-free area of 10 feet (3048 mm) shall be required for ground-mounted photovoltaic arrays.

605.12 Abandoned wiring in plenums. Accessible portions of abandoned cables in air-handling plenums shall be removed. Cables that are unused and have not been tagged for future use shall be considered abandoned.

SECTION 606 MECHANICAL REFRIGERATION

[M] 606.1 Scope. Refrigeration systems shall be installed in accordance with the *International Mechanical Code.*

[M] 606.2 Refrigerants. The use and purity of new, recovered and reclaimed refrigerants shall be in accordance with the *International Mechanical Code.*

[M] 606.3 Refrigerant classification. Refrigerants shall be classified in accordance with the *International Mechanical Code.*

[M] 606.4 Change in refrigerant type. A change in the type of refrigerant in a refrigeration system shall be in accordance with the *International Mechanical Code.*

606.5 Access. Refrigeration systems having a refrigerant circuit containing more than 220 pounds (100 kg) of Group A1 or 30 pounds (14 kg) of any other group refrigerant shall be accessible to the fire department at all times as required by the *fire code official.*

606.6 Testing of equipment. Refrigeration equipment and systems having a refrigerant circuit containing more than 220 pounds (100 kg) of Group A1 or 30 pounds (14 kg) of any other group refrigerant shall be subject to periodic testing in accordance with Section 606.6.1. Records of tests shall be maintained. Tests of emergency devices or systems required by this chapter shall be conducted by persons trained and qualified in refrigeration systems.

606.6.1 Periodic testing. The following emergency devices or systems shall be periodically tested in accordance with the manufacturer's instructions and as required by the *fire code official.*

1. Treatment and flaring systems.
2. Valves and appurtenances necessary to the operation of emergency refrigeration control boxes.
3. Fans and associated equipment intended to operate emergency ventilation systems.
4. Detection and alarm systems.

606.7 Emergency signs. Refrigeration units or systems having a refrigerant circuit containing more than 220 pounds (100 kg) of Group A1 or 30 pounds (14 kg) of any other group refrigerant shall be provided with *approved* emergency signs, charts and labels in accordance with NFPA 704. Hazard signs shall be in accordance with the *International Mechanical Code* for the classification of refrigerants listed therein.

606.8 Refrigerant detector. Machinery rooms shall contain a refrigerant detector with an audible and visual alarm. The detector, or a sampling tube that draws air to the detector, shall be located in an area where refrigerant from a leak will concentrate. The alarm shall be actuated at a value not greater than the corresponding TLV-TWA values shown in the *International Mechanical Code* for the refrigerant classification. Detectors and alarms shall be placed in *approved* locations. The detector shall transmit a signal to an *approved* location.

606.9 Remote controls. Where flammable refrigerants are used and compliance with Section 1106 of the *International Mechanical Code* is required, remote control of the mechanical equipment and appliances located in the machinery room as required by Sections 606.9.1 and 606.9.2 shall be provided at an approved location immediately outside the machinery room and adjacent to its principal entrance.

606.9.1 Refrigeration system emergency shutoff. A clearly identified switch of the break-glass type or with an *approved* tamper-resistant cover shall provide off-only control of refrigerant compressors, refrigerant pumps and normally closed automatic refrigerant valves located in the machinery room. Additionally, this equipment shall be automatically shut off when the refrigerant vapor concentration in the machinery room exceeds the vapor detector's upper detection limit or 25 percent of the LEL, whichever is lower.

606.9.2 Ventilation system. A clearly identified switch of the break-glass type or with an approved tamper-resistant cover shall provide on-only control of the machinery room ventilation fans.

606.10 Emergency pressure control system. Permanently installed refrigeration systems containing more than 6.6 pounds (3 kg) of flammable, toxic or highly toxic refrigerant or ammonia shall be provided with an emergency pressure control system in accordance with Sections 606.10.1 and 606.10.2.

606.10.1 Automatic crossover valves. Each high- and intermediate-pressure zone in a refrigeration system shall be provided with a single automatic valve providing a crossover connection to a lower pressure zone. Automatic crossover valves shall comply with Sections 606.10.1.1 through 606.10.1.3.

606.10.1.1 Overpressure limit set point. Automatic crossover valves shall be arranged to automatically relieve excess system pressure to a lower pressure zone if the pressure in a high- or intermediate-pressure zone rises to within 90 percent of the set point for emergency pressure relief devices.

606.10.1.2 Manual operation. Where required by the *fire code official*, automatic crossover valves shall be capable of manual operation.

606.10.1.3 System design pressure. Refrigeration system zones that are connected to a higher pressure zone by an automatic crossover valve shall be designed to safely contain the maximum pressure that can be achieved by interconnection of the two zones.

606.10.2 Automatic emergency stop. An automatic emergency stop feature shall be provided in accordance with Sections 606.10.2.1 and 606.10.2.2.

606.10.2.1 Operation of an automatic crossover valve. Operation of an automatic crossover valve shall cause all compressors on the affected system to immediately stop. Dedicated pressure-sensing devices located immediately adjacent to crossover valves shall be permitted as a means for determining operation of a valve. To ensure that the automatic crossover valve system provides a redundant means of stopping compressors in an overpressure condition, high-pressure cutout sensors associated with compressors shall not be used as a basis for determining operation of a crossover valve.

606.10.2.2 Overpressure in low-pressure zone. The lowest pressure zone in a refrigeration system shall be provided with a dedicated means of determining a rise in system pressure to within 90 percent of the set point for emergency pressure relief devices. Activation of the overpressure sensing device shall cause all compressors on the affected system to immediately stop.

606.11 Storage, use and handling. Flammable and combustible materials shall not be stored in machinery rooms for refrigeration systems having a refrigerant circuit containing more than 220 pounds (100 kg) of Group A1 or 30 pounds (14 kg) of any other group refrigerant. Storage, use or handling of extra refrigerant or refrigerant oils shall be as required by Chapters 50, 53, 55 and 57.

Exception: This provision shall not apply to spare parts, tools and incidental materials necessary for the safe and proper operation and maintenance of the system.

606.12 Discharge and termination of pressure relief and purge systems. Pressure relief devices, fusible plugs and purge systems discharging to the atmosphere from refrigeration systems containing flammable, toxic or highly toxic refrigerants or ammonia shall comply with Sections 606.12.3 through 606.12.5.

606.12.1 Standards. Refrigeration systems and the buildings in which such systems are installed shall be in accordance with ASHRAE 15.

606.12.1.1 Ammonia refrigeration. Refrigeration systems using ammonia refrigerant and the buildings in which such systems are installed shall comply with IIAR-2 for system design and installation and IIAR-7 for operating procedures.

606.12.2 Fusible plugs and rupture members. Discharge piping and devices connected to the discharge side of a fusible plug or rupture member shall have provisions to prevent plugging the pipe in the event the fusible plug or rupture member functions.

606.12.3 Flammable refrigerants. Systems containing more than 6.6 pounds (3 kg) of flammable refrigerants having a density equal to or greater than the density of air shall discharge vapor to the atmosphere only through an *approved* treatment system in accordance with Section 606.12.6 or a flaring system in accordance with Section 606.12.7. Systems containing more than 6.6 pounds (3 kg) of flammable refrigerants having a density less than the density of air shall be permitted to discharge vapor to the atmosphere provided that the point of discharge is located outside of the structure at not less than 15 feet (4572 mm) above the adjoining grade level and not less than 20 feet (6096 mm) from any window, ventilation opening or *exit*.

606.12.4 Toxic and highly toxic refrigerants. Systems containing more than 6.6 pounds (3 kg) of toxic or highly toxic refrigerants shall discharge vapor to the atmosphere only through an *approved* treatment system in accordance with Section 606.12.6 or a flaring system in accordance with Section 606.12.7.

606.12.5 Ammonia refrigerant. Systems containing more than 6.6 pounds (3 kg) of ammonia refrigerant shall discharge vapor to the atmosphere in accordance with one of the following methods:

1. Directly to atmosphere where the *fire code official* determines, on review of an engineering analysis prepared in accordance with Section 104.7.2, that a fire, health or environmental hazard would not result from atmospheric discharge of ammonia.
2. Through an *approved* treatment system in accordance with Section 606.12.6.

3. Through a flaring system in accordance with Section 606.12.7.
4. Through an *approved* ammonia diffusion system in accordance with Section 606.12.8.
5. By other *approved* means.

Exception: Ammonia/water absorption systems containing less than 22 pounds (10 kg) of ammonia and for which the ammonia circuit is located entirely outdoors.

606.12.6 Treatment systems. Treatment systems shall be designed to reduce the allowable discharge concentration of the refrigerant gas to not more than 50 percent of the IDLH at the point of exhaust. Treatment systems shall be in accordance with Chapter 60.

606.12.7 Flaring systems. Flaring systems for incineration of flammable refrigerants shall be designed to incinerate the entire discharge. The products of refrigerant incineration shall not pose health or environmental hazards. Incineration shall be automatic upon initiation of discharge, shall be designed to prevent blowback and shall not expose structures or materials to threat of fire. Standby fuel, such as LP-gas, and standby power shall have the capacity to operate for one and one-half the required time for complete incineration of refrigerant in the system. Standby electrical power, where required to complete the incineration process, shall be in accordance with Section 604.

606.12.8 Ammonia diffusion systems. Ammonia diffusion systems shall include a tank containing 1 gallon of water for each pound of ammonia (8.3 L of water for each 1 kg of ammonia) that will be released in 1 hour from the largest relief device connected to the discharge pipe. The water shall be prevented from freezing. The discharge pipe from the pressure relief device shall distribute ammonia in the bottom of the tank, but not lower than 33 feet (10 058 mm) below the maximum liquid level. The tank shall contain the volume of water and ammonia without overflowing.

606.13 Discharge location for refrigeration machinery room ventilation. Exhaust from mechanical ventilation systems serving refrigeration machinery rooms containing flammable, toxic or highly toxic refrigerants, other than ammonia, capable of exceeding 25 percent of the LFL or 50 percent of the IDLH shall be equipped with *approved* treatment systems to reduce the discharge concentrations to those values or lower.

606.14 Notification of refrigerant discharges. The *fire code official* shall be notified immediately when a discharge becomes reportable under state, federal or local regulations in accordance with Section 5003.3.1.

606.15 Records. A record of refrigerant quantities brought into and removed from the premises shall be maintained.

606.16 Electrical equipment. Where refrigerants of Groups A2, A3, B2 and B3, as defined in the *International Mechanical Code*, are used, refrigeration machinery rooms shall conform to the Class I, Division 2 hazardous location classification requirements of NFPA 70.

Exception: Ammonia machinery rooms that are provided with ventilation in accordance with Section 1106.3 of the *International Mechanical Code*.

SECTION 607
ELEVATOR OPERATION, MAINTENANCE AND FIRE SERVICE KEYS

607.1 Emergency operation. Existing elevators with a travel distance of 25 feet (7620 mm) or more shall comply with the requirements in Chapter 11. New elevators shall be provided with Phase I emergency recall operation and Phase II emergency in-car operation in accordance with ASME A17.1.

607.2 Standby power. In buildings and structures where standby power is required or furnished to operate an elevator, standby power shall be provided in accordance with Section 604. Operation of the system shall be in accordance with Sections 607.2.1 through 607.2.4. **

607.2.1 Manual transfer. Standby power shall be manually transferable to all elevators in each bank.

607.2.2 One elevator. Where only one elevator is installed, the elevator shall automatically transfer to standby power within 60 seconds after failure of normal power.

607.2.3 Two or more elevators. Where two or more elevators are controlled by a common operating system, all elevators shall automatically transfer to standby power within 60 seconds after failure of normal power where the standby power source is of sufficient capacity to operate all elevators at the same time. Where the standby power source is not of sufficient capacity to operate all elevators at the same time, all elevators shall transfer to standby power in sequence, return to the designated landing and disconnect from the standby power source. After all elevators have been returned to the designated level, not less than one elevator shall remain operable from the standby power source.

607.2.4 Machine room ventilation. Where standby power is connected to elevators, the machine room ventilation or air conditioning shall be connected to the standby power source.

[BE] 607.3 Emergency signs. An *approved* pictorial sign of a standardized design shall be posted adjacent to each elevator call station on all floors instructing occupants to use the exit stairways and not to use the elevators in case of fire. The sign shall read: IN FIRE EMERGENCY, DO NOT USE ELEVATOR. USE EXIT STAIRS.

Exceptions:

1. The emergency sign shall not be required for elevators that are part of an accessible *means of egress* complying with Section 1009.4.

2. The emergency sign shall not be required for elevators that are used for occupant self-evacuation in accordance with Section 3008 of the *International Building Code*.

607.4 Fire service access elevator lobbies. Where fire service access elevators are required by Section 3007 of the *International Building Code*, fire service access elevator lobbies shall be maintained free of storage and furniture.

607.5 Occupant evacuation elevator lobbies. Where occupant evacuation elevators are provided in accordance with Section 3008 of the *International Building Code*, occupant evacuation elevator lobbies shall be maintained free of storage and furniture.

607.6 Water protection of hoistway enclosures. Methods to prevent water from infiltrating into a hoistway enclosure required by Section 3007.4 and Section 3008.4 of the *International Building Code* shall be maintained.

607.7 Elevator key location. Keys for the elevator car doors and fire-fighter service keys shall be kept in an *approved* location for immediate use by the fire department.

607.8 Standardized fire service elevator keys. Buildings with elevators equipped with Phase I emergency recall, Phase II emergency in-car operation, or a fire service access elevator shall be equipped to operate with a standardized fire service elevator key approved by the *fire code official*.

Exception: The owner shall be permitted to place the building's nonstandardized fire service elevator keys in a key box installed in accordance with Section 506.1.2.

607.8.1 Requirements for standardized fire service elevator keys. Standardized fire service elevator keys shall comply with all of the following:

1. All fire service elevator keys within the jurisdiction shall be uniform and specific for the jurisdiction. Keys shall be cut to a uniform key code.
2. Fire service elevator keys shall be of a patent-protected design to prevent unauthorized duplication.
3. Fire service elevator keys shall be factory restricted by the manufacturer to prevent the unauthorized distribution of key blanks. Uncut key blanks shall not be permitted to leave the factory.
4. Fire service elevator keys subject to these rules shall be engraved with the words "DO NOT DUPLICATE."

607.8.2 Access to standardized fire service keys. Access to standardized fire service elevator keys shall be restricted to the following:

1. Elevator owners or their authorized agents.
2. Elevator contractors.
3. Elevator inspectors of the jurisdiction.
4. *Fire code officials* of the jurisdiction.
5. The fire department and other emergency response agencies designated by the *fire code official*.

607.8.3 Duplication or distribution of keys. A person shall not duplicate a standardized fire service elevator key or issue, give, or sell a duplicated key unless in accordance with this code.

607.8.4 Responsibility to provide keys. The building owner shall provide up to three standardized fire service elevator keys where required by the *fire code official*, upon installation of a standardized fire service key switch or switches in the building.

SECTION 608
STATIONARY STORAGE BATTERY SYSTEMS

608.1 Scope. Stationary storage battery systems having an electrolyte capacity of more than 50 gallons (189 L) for flooded lead-acid, nickel cadmium (Ni-Cd) and valve-regulated lead-acid (VRLA), or more than 1,000 pounds (454 kg) for lithium-ion and lithium metal polymer, used for facility standby power, emergency power or uninterruptible power supplies shall comply with this section and Table 608.1.

608.2 Safety caps. Safety caps for stationary storage battery systems shall comply with Sections 608.2.1 and 608.2.2.

608.2.1 Nonrecombinant batteries. Vented lead-acid, nickel-cadmium or other types of nonrecombinant batteries shall be provided with safety venting caps.

608.2.2 Recombinant batteries. VRLA batteries shall be equipped with self-resealing flame-arresting safety vents.

608.3 Thermal runaway. VRLA and lithium metal polymer battery systems shall be provided with a *listed* device or other *approved* method to preclude, detect and control thermal runaway.

608.4 Room design and construction. Enclosure of stationary battery systems shall comply with the *International Building Code.* Battery systems shall be allowed to be in the same room with the equipment they support.

608.4.1 Separate rooms. Where stationary batteries are installed in a separate equipment room accessible only to authorized personnel, they shall be permitted to be installed on an open rack for ease of maintenance.

608.4.2 Occupied work centers. Where a system of VRLA, lithium-ion, or other type of sealed, nonventing batteries is situated in an occupied work center, it shall be allowed to be housed in a noncombustible cabinet or other enclosure to prevent access by unauthorized personnel.

608.4.3 Cabinets. Where stationary batteries are contained in cabinets in occupied work centers, the cabinet enclosures shall be located within 10 feet (3048 mm) of the equipment that they support.

608.5 Spill control and neutralization. An *approved* method and materials for the control and neutralization of a spill of electrolyte shall be provided in areas containing lead-acid, nickel-cadmium or other types of batteries with free-flowing liquid electrolyte. For purposes of this paragraph, a "spill" is defined as any unintentional release of electrolyte.

Exception: VRLA, lithium-ion, lithium metal polymer or other types of sealed batteries with immobilized electrolyte shall not require spill control.

TABLE 608.1
BATTERY REQUIREMENTS

REQUIREMENT	NONRECOMBINANT BATTERIES		RECOMBINANT BATTERIES		OTHER BATTERIES
	Vented (Flooded) Lead Acid Batteries	Vented (Flooded) Nickel-Cadmium (Ni-Cd) Batteries	Valve Regulated Lead-Acid (VRLA) Cells	Lithium-Ion Cells	Lithium Metal Cells
Safety caps	Venting caps (608.2.1)	Venting caps (608.2.1)	Self-resealing flame-arresting caps (608.2.2)	No caps	No caps
Thermal runaway management	Not required	Not required	Required (608.3)	Not required	Required (608.3)
Spill control	Required (608.5)	Required (608.5)	Not required	Not required	Not required
Neutralization	Required (608.5.1)	Required (608.5.1)	Required (608.5.2)	Not required	Not required
Ventilation	Required (608.6.1; 608.6.2)	Required (608.6.1; 608.6.2)	Required (608.6.1; 608.6.2)	Not required	Not required
Signage	Required (608.7)	Required (608.7)	Required (608.7)	Required (608.7)	Required (608.7)
Seismic protection	Required (608.8)	Required (608.8)	Required (608.8)	Required (608.8)	Required (608.8)
Smoke detection	Required (608.9)	Required (608.9)	Required (608.9)	Required (608.9)	Required (608.9)

608.5.1 Nonrecombinant battery neutralization. For battery systems containing lead acid, nickel cadmium or other types of batteries with free-flowing electrolyte, the method and materials shall be capable of neutralizing a spill of the total capacity from the largest cell or block to a pH between 5.0 and 9.0.

608.5.2 Recombinant battery neutralization. For VRLA or other types of batteries with immobilized electrolyte, the method and material shall be capable of neutralizing a spill of 3.0 percent of the capacity of the largest cell or block in the room to a pH between 5.0 and 9.0.

Exception: Lithium-ion and lithium metal polymer batteries shall not require neutralization.

608.6 Ventilation. Ventilation of stationary storage battery systems shall comply with Sections 608.6.1 and 608.6.2.

608.6.1 Room ventilation. Ventilation shall be provided in accordance with the *International Mechanical Code* and the following:

1. For flooded lead-acid, flooded Ni-Cd and VRLA batteries, the ventilation system shall be designed to limit the maximum concentration of hydrogen to 1.0 percent of the total volume of the room; or
2. Continuous ventilation shall be provided at a rate of not less than 1 cubic foot per minute per square foot (1 $ft^3/min/ft^2$) [0.0051 $m^3/s \cdot m^2$] of floor area of the room.

Exception: Lithium-ion and lithium metal polymer batteries shall not require additional ventilation beyond that which would normally be required for human occupancy of the space in accordance with the *International Mechanical Code*.

608.6.2 Cabinet ventilation. Where VRLA batteries are installed inside a cabinet, the cabinet shall be *approved* for use in occupied spaces and shall be mechanically or naturally vented by one of the following methods:

1. The cabinet ventilation shall limit the maximum concentration of hydrogen to 1 percent of the total volume of the cabinet during the worst-case event of simultaneous "boost" charging of all the batteries in the cabinet.
2. Where calculations are not available to substantiate the ventilation rate, continuous ventilation shall be provided at a rate of not less than 1 cubic foot per minute per square foot [1 $ft^3/min/ft^2$ or 0.0051 $m^3/(s \cdot m^2)$] of floor area covered by the cabinet. The room in which the cabinet is installed shall be ventilated as required in Section 608.6.1.

608.6.3 Supervision. Mechanical ventilation systems where required by Sections 608.6.1 and 608.6.2 shall be supervised by an *approved* central, proprietary or remote station service or shall initiate an audible and visual signal at a constantly attended on-site location.

608.7 Signage. Signs shall comply with Sections 608.7.1 and 608.7.2.

608.7.1 Equipment room and building signage. Doors into electrical equipment rooms or buildings containing stationary battery systems shall be provided with *approved* signs. The signs shall state that:

1. The room contains energized battery systems.
2. The room contains energized electrical circuits.
3. The battery electrolyte solutions, where present, are *corrosive* liquids.

608.7.2 Cabinet signage. Cabinets shall have exterior labels that identify the manufacturer and model number of the system and electrical rating (voltage and current) of the contained battery system. There shall be signs within the cabinet that indicate the relevant electrical, chemical and fire hazards.

608.8 Seismic protection. The battery systems shall be seismically braced in accordance with the *International Building Code.*

608.9 Smoke detection. An *approved* automatic smoke detection system shall be installed in accordance with Section 907.2 in rooms containing stationary battery systems.

SECTION 609 COMMERCIAL KITCHEN HOODS

[M] 609.1 General. Commercial kitchen exhaust hoods shall comply with the requirements of the *International Mechanical Code.*

[M] 609.2 Where required. A Type I hood shall be installed at or above all commercial cooking appliances and domestic cooking appliances used for commercial purposes that produce grease vapors.

Exception: A Type I hood shall not be required for an electric cooking appliance where an approved testing agency provides documentation that the appliance effluent contains 5 mg/m^3 or less of grease when tested at an exhaust flow rate of 500 cfm (0.236 m^3/s) in accordance with UL 710B.

609.3 Operations and maintenance. Commercial cooking systems shall be operated and maintained in accordance with Sections 609.3.1 through 609.3.4.

609.3.1 Ventilation system. The ventilation system in connection with hoods shall be operated at the required rate of air movement, and classified grease filters shall be in place when equipment under a kitchen grease hood is used.

609.3.2 Grease extractors. Where grease extractors are installed, they shall be operated when the commercial-type cooking equipment is used.

609.3.3 Cleaning. Hoods, grease-removal devices, fans, ducts and other appurtenances shall be cleaned at intervals as required by Sections 609.3.3.1 through 609.3.3.3.

609.3.3.1 Inspection. Hoods, grease-removal devices, fans, ducts and other appurtenances shall be inspected at intervals specified in Table 609.3.3.1 or as *approved* by the *fire code official.* Inspections shall be completed by qualified individuals.

TABLE 609.3.3.1
COMMERCIAL COOKING SYSTEM INSPECTION FREQUENCY

TYPE OF COOKING OPERATIONS	FREQUENCY OF INSPECTION
High-volume cooking operations such as 24-hour cooking, charbroiling or wok cooking	3 months
Low-volume cooking operations such as places of religious worship, seasonal businesses and senior centers	12 months
Cooking operations utilizing solid fuel-burning cooking appliances	1 month
All other cooking operations	6 months

609.3.3.2 Grease accumulation. If during the inspection it is found that hoods, grease-removal devices, fans, ducts or other appurtenances have an accumulation of grease, such components shall be cleaned in accordance with ANSI/IKECA C 10.

609.3.3.3 Records. Records for inspections shall state the individual and company performing the inspection, a description of the inspection and when the inspection took place. Records for cleanings shall state the individual and company performing the cleaning and when the cleaning took place. Such records shall be completed after each inspection or cleaning and maintained.

609.3.3.3.1 Tags. When a commercial kitchen hood or duct system is inspected, a tag containing the service provider name, address, telephone number and date of service shall be provided in a conspicuous location. Prior tags shall be covered or removed.

609.3.4 Extinguishing system service. Automatic fire-extinguishing systems protecting commercial cooking systems shall be serviced as required in Section 904.12.6.

609.4 Appliance connection to building piping. Gas-fired commercial cooking appliances installed on casters and appliances that are moved for cleaning and sanitation purposes shall be connected to the piping system with an appliance connector listed as complying with ANSI Z21.69. The commercial cooking appliance connector installation shall be configured in accordance with the manufacturer's installation instructions. Movement of appliances with casters shall be limited by a restraining device installed in accordance with the connector and appliance manufacturer's instructions.

SECTION 610 COMMERCIAL KITCHEN COOKING OIL STORAGE

610.1 General. Storage of cooking oil (grease) in commercial cooking operations utilizing above-ground tanks with a capacity greater than 60 gal (227 L) installed within a building shall comply with Sections 610.2 through 610.7 and NFPA 30. For purposes of this section, cooking oil shall be classified as a Class IIIB liquid unless otherwise determined by testing.

610.2 Metallic storage tanks. Metallic cooking oil storage tanks shall be listed in accordance with UL 142 or UL 80, and shall be installed in accordance with the tank manufacturer's instructions.

610.3 Nonmetallic storage tanks. Nonmetallic cooking oil storage tanks shall be installed in accordance with the tank manufacturer's instructions and shall also comply with all of the following:

1. Tanks shall be listed for use with cooking oil, including maximum temperature to which the tank will be exposed during use.
2. Tank capacity shall not exceed 200 gallons (757 L) per tank.

610.4 Cooking oil storage system components. Cooking oil storage system components shall include but are not limited to piping, connections, fittings, valves, tubing, hose, pumps, vents and other related components used for the transfer of cooking oil, and are permitted to be of either metallic or nonmetallic construction.

610.4.1 Design standards. The design, fabrication and assembly of system components shall be suitable for the working pressures, temperatures and structural stresses to be encountered by the components.

610.4.2 Components in contact with heated oil. System components that come in contact with heated cooking oil shall be rated for the maximum operating temperatures expected in the system.

610.5 Tank venting. Normal and emergency venting shall be provided for cooking oil storage tanks.

610.5.1 Normal vents. Normal vents shall be located above the maximum normal liquid line, and shall have a minimum effective area not smaller than the largest filling or withdrawal connection. Normal vents shall be permitted to vent inside the building.

610.5.2 Emergency vents. Emergency relief vents shall be located above the maximum normal liquid line, and shall be in the form of a device or devices that will relieve excessive internal pressure caused by an exposure fire. For nonmetallic tanks, the emergency relief vent shall be allowed to be in the form of construction. Emergency vents shall be permitted to vent inside the building.

610.6 Heating of cooking oil. Electrical equipment used for heating cooking oil in cooking oil storage systems shall be listed to UL 499 and shall comply with NFPA 70. Use of electrical immersion heaters shall be prohibited in nonmetallic tanks.

610.7 Electrical equipment. Electrical equipment used for the operation of cooking oil storage systems shall comply with NFPA 70.

SECTION 611
HYPERBARIC FACILITIES

611.1 General. Hyperbaric facilities shall be inspected, tested and maintained in accordance with NFPA 99.

611.2 Records. Records shall be maintained of all testing and repair conducted on the hyperbaric chamber and associated devices and equipment. Records shall be available to the *fire code official*.

CHAPTER 7

FIRE AND SMOKE PROTECTION FEATURES

SECTION 701 GENERAL

701.1 Scope. The provisions of this chapter shall govern maintenance of the materials, systems and assemblies used for structural fire resistance and fire-resistance-rated construction separation of adjacent spaces to safeguard against the spread of fire and smoke within a building and the spread of fire to or from buildings. New buildings shall comply with the *International Building Code*.

701.2 Unsafe conditions. Where any components in this chapter are not maintained and do not function as intended or do not have the *fire resistance* required by the code under which the building was constructed, remodeled or altered, such component(s) or portion thereof shall be deemed an unsafe condition, in accordance with Section 110.1.1. Components or portions thereof determined to be unsafe shall be repaired or replaced to conform to that code under which the building was constructed, remodeled, altered or this chapter, as deemed appropriate by the *fire code official*.

Where the extent of the conditions of components is such that any building, structure or portion thereof presents an imminent danger to the occupants of the building, structure or portion thereof, the *fire code official* shall act in accordance with Section 110.2.

SECTION 702 DEFINITIONS

702.1 Definitions. The following terms are defined in Chapter 2:

DRAFTSTOP.

FIRE-RESISTANT JOINT SYSTEM.

FIREBLOCKING.

SECTION 703 FIRE-RESISTANCE-RATED CONSTRUCTION

703.1 Maintenance. The required *fire-resistance rating* of fire-resistance-rated construction, including, but not limited to, walls, firestops, shaft enclosures, partitions, *smoke barriers*, floors, fire-resistive coatings and sprayed fire-resistant materials applied to structural members and fire-resistant joint systems, shall be maintained. Such elements shall be visually inspected by the *owner* annually and properly repaired, restored or replaced where damaged, altered, breached or penetrated. Records of inspections and repairs shall be maintained. Where concealed, such elements shall not be required to be visually inspected by the *owner* unless the concealed space is accessible by the removal or movement of a panel, access door, ceiling tile or similar movable entry to the space. Openings made therein for the passage of pipes, electrical conduit, wires, ducts, air transfer openings and holes made for any reason shall be protected with *approved* methods capable of resisting the passage of smoke and fire. Openings through fire-resistance-rated assemblies shall be protected by self- or automatic-closing doors of *approved* construction meeting the fire protection requirements for the assembly.

703.1.1 Fireblocking and draftstopping. Required *fireblocking* and draftstopping in combustible concealed spaces shall be maintained to provide continuity and integrity of the construction.

703.1.2 Smoke barriers and smoke partitions. Required *smoke barriers* and smoke partitions shall be maintained to prevent the passage of smoke. Openings protected with *approved* smoke barrier doors or smoke dampers shall be maintained in accordance with NFPA 105.

703.1.3 Fire walls, fire barriers and fire partitions. Required *fire walls*, *fire barriers* and *fire partitions* shall be maintained to prevent the passage of fire. Openings protected with *approved* doors or fire dampers shall be maintained in accordance with NFPA 80.

703.2 Opening protectives. Opening protectives shall be maintained in an operative condition in accordance with NFPA 80. Where allowed by the *fire code official*, the application of field-applied labels associated with the maintenance of opening protectives shall follow the requirements of the *approved* third-party certification organization accredited for *listing* the opening protective. Fire doors and *smoke barrier* doors shall not be blocked or obstructed, or otherwise made inoperable. Fusible links shall be replaced promptly whenever fused or damaged. Fire door assemblies shall not be modified.

703.2.1 Signs. Where required by the *fire code official*, a sign shall be permanently displayed on or near each fire door in letters not less than 1 inch (25 mm) high to read as follows:

1. For doors designed to be kept normally open: FIRE DOOR—DO NOT BLOCK.
2. For doors designed to be kept normally closed: FIRE DOOR—KEEP CLOSED.

703.2.2 Hold-open devices and closers. Hold-open devices and automatic door closers, where provided, shall be maintained. During the period that such device is out of service for repairs, the door it operates shall remain in the closed position.

703.2.3 Door operation. Swinging fire doors shall close from the full-open position and latch automatically. The door closer shall exert enough force to close and latch the door from any partially open position.

703.3 Ceilings. The hanging and displaying of salable goods and other decorative materials from acoustical ceiling systems that are part of a fire-resistance-rated horizontal assembly, shall be prohibited.

703.4 Testing. Horizontal and vertical sliding and rolling fire doors shall be inspected and tested annually to confirm proper operation and full closure. Records of inspections and testing shall be maintained.

SECTION 704
FLOOR OPENINGS AND SHAFTS

704.1 Enclosure. Interior vertical shafts including, but not limited to, *stairways*, elevator hoistways, service and utility shafts, that connect two or more stories of a building shall be enclosed or protected as required in Chapter 11. New floor openings in existing buildings shall comply with the *International Building Code*.

704.2 Opening protectives. Where openings are required to be protected, opening protectives shall be maintained self-closing or automatic-closing by smoke detection. Existing fusible-link-type automatic door-closing devices are permitted if the fusible link rating does not exceed 135°F (57°C).

CHAPTER 8

INTERIOR FINISH, DECORATIVE MATERIALS AND FURNISHINGS

SECTION 801 GENERAL

801.1 Scope. The provisions of this chapter shall govern interior finish, interior trim, furniture, furnishings, decorative materials and decorative vegetation in buildings. Existing buildings shall comply with Sections 803 through 808. New buildings shall comply with Sections 804 through 808, and Section 803 of the *International Building Code*.

SECTION 802 DEFINITIONS

802.1 Definitions. The following terms are defined in Chapter 2:

FLAME SPREAD.

FLAME SPREAD INDEX.

INTERIOR FLOOR-WALL BASE.

SITE-FABRICATED STRETCH SYSTEM.

SMOKE-DEVELOPED INDEX.

SECTION 803 INTERIOR WALL AND CEILING FINISH AND TRIM IN EXISTING BUILDINGS

803.1 General. The provisions of this section shall limit the allowable fire performance and smoke development of interior wall and ceiling finishes and interior wall and ceiling trim in existing buildings based on location and occupancy classification. Interior wall and ceiling finishes shall be classified in accordance with Section 803 of the *International Building Code*. Such materials shall be grouped in accordance with ASTM E 84, as indicated in Section 803.1.1, or in accordance with NFPA 286, as indicated in Section 803.1.2.

Exceptions:

1. Materials having a thickness less than 0.036 inch (0.9 mm) applied directly to the surface of walls and ceilings.
2. Exposed portions of structural members complying with the requirements of buildings of Type IV construction in accordance with the *International Building Code* shall not be subject to interior finish requirements.

803.1.1 Classification in accordance with ASTM E 84. Interior finish materials shall be grouped in the following classes in accordance with their flame spread and smoke-developed index where tested in accordance with ASTM E 84.

Class A: flame spread index 0–25; smoke-developed index 0–450.

Class B: flame spread index 26–75; smoke-developed index 0–450.

Class C: flame spread index 76–200; smoke-developed index 0–450.

803.1.2 Classification in accordance with NFPA 286. Interior wall or ceiling finishes shall be allowed to be tested in accordance with NFPA 286. Finishes tested in accordance with NFPA 286 shall comply with Section 803.1.2.1. Interior wall and ceiling finish materials tested in accordance with NFPA 286 and meeting the acceptance criteria of Section 803.1.2.1 shall be allowed to be used where a Class A classification in accordance with ASTM E 84 is required.

803.1.2.1 Acceptance criteria for NFPA 286. The interior finish shall comply with the following:

1. During the 40 kW exposure, flames shall not spread to the ceiling.
2. The flame shall not spread to the outer extremity of the sample on any wall or ceiling.
3. Flashover, as defined in NFPA 286, shall not occur.
4. The peak heat release rate throughout the test shall not exceed 800 kW.
5. The total smoke released throughout the test shall not exceed 1,000 m^2.

803.2 Stability. Interior finish materials regulated by this chapter shall be applied or otherwise fastened in such a manner that such materials will not readily become detached where subjected to room temperatures of 200°F (93°C) for not less than 30 minutes.

803.3 Interior finish requirements based on occupancy. Interior wall and ceiling finish shall have a flame spread index not greater than that specified in Table 803.3 for the group and location designated.

803.4 Fire-retardant coatings. The required flame spread or smoke-developed index of surfaces in existing buildings shall be allowed to be achieved by application of *approved* fire-retardant coatings, paints or solutions to surfaces having a flame spread index exceeding that allowed. Such applications shall comply with NFPA 703 and the required fire-retardant properties shall be maintained or renewed in accordance with the manufacturer's instructions.

803.5 Textiles. Where used as interior wall or ceiling finish materials, textiles, including materials having woven or nonwoven, napped, tufted, looped or similar surface, shall comply with the requirements of this section.

803.5.1 Textile wall or ceiling coverings. Textile wall or ceiling coverings shall comply with one of the following:

1. The wall or ceiling covering shall have a Class A flame spread index in accordance with ASTM E 84

or UL 723, and be protected by automatic sprinklers installed in accordance with Section 903.3.1.1 or 903.3.1.2.

2. The wall covering shall meet the criteria of Section 803.5.1.1 when tested in the manner intended for use in accordance with NFPA 265 using the product-mounting system, including adhesive, of actual use.
3. The wall or ceiling covering shall meet the criteria of Section 803.1.2.1 when tested in accordance with NFPA 286 using the product-mounting system, including adhesive, of actual use.

803.5.1.1 Method B test protocol. During the Method B protocol, the textile wall covering or expanded vinyl wall covering shall comply with the following:

1. During the 40-kW exposure, flames shall not spread to the ceiling.
2. The flame shall not spread to the outer extremities of the samples on the 8-foot by 12-foot (203 by 305 mm) walls.
3. Flashover, as defined in NFPA 265, shall not occur.
4. For newly introduced wall and ceiling coverings, the total smoke released throughout the test shall not exceed 1,000 m^2.

803.5.2 Newly introduced textile wall and ceiling coverings. Newly introduced textile wall and ceiling coverings shall comply with one of the following:

1. The wall or ceiling covering shall have a Class A flame spread index in accordance with ASTM E 84 or UL 723, and be protected by automatic sprinklers installed in accordance with Section 903.3.1.1 or 903.3.1.2. Test specimen preparation and mounting shall be in accordance with ASTM E 2404.

TABLE 803.3
INTERIOR WALL AND CEILING FINISH REQUIREMENTS BY OCCUPANCY[k]

GROUP	SPRINKLERED[l]			NONSPRINKLERED		
	Interior exit stairways and interior exit ramps and exit passageways[a, b]	Corridors and enclosure for exit access stairways and exit access ramps	Rooms and enclosed spaces[c]	Interior exit stairways and interior exit ramps and exit passageways[a, b]	Corridors and enclosure for exit access stairways and exit access ramps	Rooms and enclosed spaces[c]
A-1 & A-2	B	B	C	A	A[d]	B[e]
A-3[f], A-4, A-5	B	B	C	A	A[d]	C
B, E, M, R-1, R-4	B	C	C	A	B	C
F	C	C	C	B	C	C
H	B	B	C[g]	A	A	B
I-1	B	C	C	A	B	B
I-2	B	B	B[h, i]	A	A	B
I-3	A	A[j]	C	A	A	B
I-4	B	B	B[h, i]	A	A	B
R-2	C	C	C	B	B	C
R-3	C	C	C	C	C	C
S	C	C	C	B	B	C
U	No Restrictions			No Restrictions		

For SI: 1 inch = 25.4 mm, 1 square foot = 0.0929 m^2.

a. Class C interior finish materials shall be allowed for wainscoting or paneling of not more than 1,000 square feet of applied surface area in the grade lobby where applied directly to a noncombustible base or over furring strips applied to a noncombustible base and fireblocked as required by Section 803.11 of the *International Building Code*.

b. In exit enclosures of buildings less than three stories in height of other than Group I-3, Class B interior finish for nonsprinklered buildings and Class C for sprinklered buildings shall be permitted.

c. Requirements for rooms and enclosed spaces shall be based upon spaces enclosed by partitions. Where a fire-resistance rating is required for structural elements, the enclosing partitions shall extend from the floor to the ceiling. Partitions that do not comply with this shall be considered as enclosing spaces and the rooms or spaces on both sides shall be considered as one. In determining the applicable requirements for rooms and enclosed spaces, the specific occupancy thereof shall be the governing factor regardless of the group classification of the building or structure.

d. Lobby areas in Group A-1, A-2 and A-3 occupancies shall not be less than Class B materials.

e. Class C interior finish materials shall be allowed in Group A occupancies with an occupant load of 300 persons or less.

f. In places of religious worship, wood used for ornamental purposes, trusses, paneling or chancel furnishing shall be allowed.

g. Class B material is required where the building exceeds two stories.

h. Class C interior finish materials shall be allowed in administrative spaces.

i. Class C interior finish materials shall be allowed in rooms with a capacity of four persons or less.

j. Class B materials shall be allowed as wainscoting extending not more than 48 inches above the finished floor in corridors.

k. Finish materials as provided for in other sections of this code.

l. Applies when the vertical exits, exit passageways, corridors or rooms and spaces are protected by an approved automatic sprinkler system installed in accordance with Section 903.3.1.1 or 903.3.1.2.

2. The wall covering shall meet the criteria of Section 803.5.1.1 when tested in the manner intended for use in accordance with NFPA 265 using the product-mounting system (including adhesive) of actual use.
3. The wall or ceiling covering shall meet the criteria of Section 803.1.2.1 when tested in accordance with NFPA 286 using the product-mounting system (including adhesive) of actual use.

803.6 Expanded vinyl wall or ceiling coverings. Expanded vinyl wall or ceiling coverings shall comply with one of the following:

1. The wall or ceiling covering shall have a Class A flame spread index in accordance with ASTM E 84 or UL 723, and be protected by automatic sprinklers installed in accordance with Section 903.3.1.1 or 903.3.1.2. Test specimen preparation and mounting shall be in accordance with ASTM E 2404.
2. The wall covering shall meet the criteria of Section 803.5.1.1 when tested in the manner intended for use in accordance with NFPA 265 using the product-mounting system (including adhesive) of actual use.
3. The wall or ceiling covering shall meet the criteria of Section 803.1.2.1 when tested in accordance with NFPA 286 using the product-mounting system (including adhesive) of actual use.

803.7 Facings or wood veneers intended to be applied on site over a wood substrate. Facings or veneers intended to be applied on site over a wood substrate shall comply with one of the following:

1. The facing or veneer shall have a Class A, B or C flame spread index and smoke-developed index, based on the requirements of Table 803.3, in accordance with ASTM E 84 or UL 723. Test specimen preparation and mounting shall be in accordance with ASTM E 2404.
2. The facing or veneer shall meet the criteria of Section 803.1.2.1 when tested in accordance with NFPA 286 using the product-mounting system, including adhesive, described in Section 5.8.9 of NFPA 286.

803.8 Foam plastic materials. Foam plastic materials shall not be used as interior wall and ceiling finish unless specifically allowed by Section 803.8.1 or 803.8.2. Foam plastic materials shall not be used as interior trim unless specifically allowed by Section 803.8.3.

803.8.1 Combustibility characteristics. Foam plastic materials shall be allowed on the basis of fire tests that substantiate their combustibility characteristics for the use intended under actual fire conditions, as indicated in Section 2603.9 of the *International Building Code*. This section shall apply both to exposed foam plastics and to foam plastics used in conjunction with a textile or vinyl facing or cover.

803.8.2 Thermal barrier. Foam plastic material shall be allowed if it is separated from the interior of the building by a thermal barrier in accordance with Section 2603.4 of the *International Building Code*.

803.8.3 Trim. Foam plastic shall be allowed for trim in accordance with Section 804.2.

[BF] 803.9 High-density polyethylene (HDPE) and polypropylene (PP). Where high-density polyethylene or polypropylene is used as an interior finish it shall comply with Section 803.1.2.

[BF] 803.10 Site-fabricated stretch systems. Where used as newly installed interior wall or interior ceiling finish materials, site-fabricated stretch systems containing all three components described in the definition in Chapter 2 shall be tested in the manner intended for use, and shall comply with the requirements of Section 803.1.1 or 803.1.2. If the materials are tested in accordance with ASTM E 84 or UL 723, specimen preparation and mounting shall be in accordance with ASTM E 2573.

SECTION 804
INTERIOR WALL AND CEILING TRIM AND INTERIOR FLOOR FINISH IN NEW AND EXISTING BUILDINGS

804.1 Interior trim. Material, other than foam plastic, used as interior trim in new and existing buildings shall have minimum Class C flame spread and smoke-developed indices, when tested in accordance with ASTM E 84 or UL 723, as described in Section 803.1.1. Combustible trim, excluding handrails and guardrails, shall not exceed 10 percent of the specific wall or ceiling areas to which it is attached.

804.1.1 Alternative testing. When the interior trim material has been tested as an interior finish in accordance with NFPA 286 and complies with the acceptance criteria in Section 803.1.2.1, it shall not be required to be tested for flame spread index and smoke-developed index in accordance with ASTM E 84.

804.2 Foam plastic. Foam plastic used as interior trim shall comply with Sections 804.2.1 through 804.2.4.

804.2.1 Density. The minimum density of the interior trim shall be 20 pounds per cubic foot (320 kg/m^3).

804.2.2 Thickness. The maximum thickness of the interior trim shall be $^1/_2$ inch (12.7 mm) and the maximum width shall be 8 inches (203 mm).

804.2.3 Area limitation. The interior trim shall not constitute more than 10 percent of the specific wall or ceiling area to which it is attached.

804.2.4 Flame spread. The flame spread index shall not exceed 75 where tested in accordance with ASTM E 84 or UL 723. The smoke-developed index shall not be limited.

> **Exception:** When the interior trim material has been tested as an interior finish in accordance with NFPA 286 and complies with the acceptance criteria in Section 803.1.2.1, it shall not be required to be tested for flame spread index in accordance with ASTM E 84 or UL 723.

804.3 New interior floor finish. New interior floor finish and floor covering materials in new and existing buildings shall comply with Sections 804.3.1 through 804.3.3.2.

Exception: Floor finishes and coverings of a traditional type, such as wood, vinyl, linoleum or terrazzo, and resilient floor covering materials that are not composed of fibers.

804.3.1 Classification. Interior floor finish and floor covering materials required by Section 804.3.3.2 to be of Class I or II materials shall be classified in accordance with NFPA 253. The classification referred to herein corresponds to the classifications determined by NFPA 253 as follows: Class I, 0.45 watts/cm^2 or greater; Class II, 0.22 watts/cm^2 or greater.

804.3.2 Testing and identification. Interior floor finish and floor covering materials shall be tested by an *approved* agency in accordance with NFPA 253 and identified by a hang tag or other suitable method so as to identify the manufacturer or supplier and style, and shall indicate the interior floor finish or floor covering classification in accordance with Section 804.3.1. Carpet-type floor coverings shall be tested as proposed for use, including underlayment. Test reports confirming the information provided in the manufacturer's product identification shall be furnished to the *fire code official* upon request.

804.3.3 Interior floor finish requirements. New interior floor coverings materials shall comply with Sections 804.3.3.1 and 804.3.3.2, and interior floor finish materials shall comply with Section 804.3.1.

804.3.3.1 Pill test. In all occupancies, new floor covering materials shall comply with the requirements of the DOC FF-1 "pill test" (CPSC 16 CFR Part 1630) or of ASTM D 2859.

804.3.3.2 Minimum critical radiant flux. In all occupancies, new interior floor finish and floor covering materials in enclosures for *stairways* and *ramps*, *exit passageways*, *corridors* and rooms or spaces not separated from *corridors* by full-height partitions extending from the floor to the underside of the ceiling shall withstand a minimum critical radiant flux. The minimum critical radiant flux shall be not less than Class I in Groups I-1, I-2 and I-3 and not less than Class II in Groups A, B, E, H, I-4, M, R-1, R-2 and S.

Exception: Where a building is equipped throughout with an *automatic sprinkler system* in accordance with Section 903.3.1.1 or 903.3.1.2, Class II materials shall be permitted in any area where Class I materials are required and materials complying with DOC FF-1 "pill test" (CPSC 16 CFR Part 1630) or with ASTM D 2859 shall be permitted in any area where Class II materials are required.

804.4 Interior floor-wall base. Interior floor-wall base that is 6 inches (152 mm) or less in height shall be tested in accordance with NFPA 253 and shall be not less than Class II. Where a Class I floor finish is required, the floor-wall base shall be Class I. The classification referred to herein corresponds to the classifications determined by NFPA 253 as follows: Class I, 0.45 watt/cm^2 or greater; Class II, 0.22 watts/cm^2 or greater.

Exception: Interior trim materials that comply with Section 804.1.

SECTION 805
UPHOLSTERED FURNITURE AND MATTRESSES IN NEW AND EXISTING BUILDINGS

805.1 Group I-1, Condition 2. The requirements in Sections 805.1.1 through 805.1.2 shall apply to facilities in Group I-1, Condition 2.

805.1.1 Upholstered furniture. Newly introduced upholstered furniture shall meet the requirements of Sections 805.1.1.1 through 805.1.1.3.

805.1.1.1 Ignition by cigarettes. Newly introduced upholstered furniture shall be shown to resist ignition by cigarettes as determined by tests conducted in accordance with one of the following:

1. Mocked-up composites of the upholstered furniture shall have a char length not exceeding 1.5 inches (38 mm) when tested in accordance with NFPA 261.
2. The components of the upholstered furniture shall meet the requirements for Class I when tested in accordance with NFPA 260.

805.1.1.2 Heat release rate. Newly introduced upholstered furniture shall have limited rates of heat release when tested in accordance with ASTM E 1537 or California Technical Bulletin 133, as follows:

1. The peak rate of heat release for the single upholstered furniture item shall not exceed 80 kW.

   Exception: Upholstered furniture in rooms or spaces protected by an *approved automatic sprinkler system* installed in accordance with Section 903.3.1.1.

2. The total energy released by the single upholstered furniture item during the first 10 minutes of the test shall not exceed 25 megajoules (MJ).

   Exception: Upholstered furniture in rooms or spaces protected by an *approved automatic sprinkler system* installed in accordance with Section 903.3.1.1.

805.1.1.3 Identification. Upholstered furniture shall bear the label of an *approved* agency, confirming compliance with the requirements of Sections 805.1.1.1 and 805.1.1.2.

805.1.2 Mattresses. Newly introduced mattresses shall meet the requirements of Sections 805.1.2.1 through 805.1.2.3.

805.1.2.1 Ignition by cigarettes. Newly introduced mattresses shall be shown to resist ignition by cigarettes as determined by tests conducted in accordance with DOC 16 CFR Part 1632 and shall have a char length not exceeding 2 inches (51 mm).

805.1.2.2 Heat release rate. Newly introduced mattresses shall have limited rates of heat release when tested in accordance with ASTM E 1590 or California Technical Bulletin 129, as follows:

1. The peak rate of heat release for the single mattress shall not exceed 100 kW.

   Exception: Mattresses in rooms or spaces protected by an *approved automatic sprinkler system* installed in accordance with Section 903.3.1.1.

2. The total energy released by the single mattress during the first 10 minutes of the test shall not exceed 25 MJ.

   Exception: Mattresses in rooms or spaces protected by an *approved automatic sprinkler system* installed in accordance with Section 903.3.1.1.

805.1.2.3 Identification. Mattresses shall bear the label of an *approved* agency, confirming compliance with the requirements of Sections 805.2.2.1 and 805.2.2.2.

805.2 Group I-2, nursing homes and hospitals. The requirements in Sections 805.2.1 through 805.2.2 shall apply to nursing homes and hospitals classified in Group I-2.

805.2.1 Upholstered furniture. Newly introduced upholstered furniture shall meet the requirements of Sections 805.2.1.1 through 805.2.1.3.

805.2.1.1 Ignition by cigarettes. Newly introduced upholstered furniture shall be shown to resist ignition by cigarettes as determined by tests conducted in accordance with one of the following: (a) mocked-up composites of the upholstered furniture shall have a char length not exceeding 1.5 inches (38 mm) when tested in accordance with NFPA 261 or (b) the components of the upholstered furniture shall meet the requirements for Class I when tested in accordance with NFPA 260.

Exception: Upholstered furniture belonging to the patients in sleeping rooms of nursing homes (Group I-2), provided that a smoke detector is installed in such rooms. Battery-powered, single-station smoke alarms shall be allowed.

805.2.1.2 Heat release rate. Newly introduced upholstered furniture shall have limited rates of heat release when tested in accordance with ASTM E 1537 or California Technical Bulletin 133, as follows:

1. The peak rate of heat release for the single upholstered furniture item shall not exceed 80 kW.

   Exception: Upholstered furniture in rooms or spaces protected by an *approved automatic sprinkler system* installed in accordance with Section 903.3.1.1.

2. The total energy released by the single upholstered furniture item during the first 10 minutes of the test shall not exceed 25 MJ.

   Exception: Upholstered furniture in rooms or spaces protected by an *approved automatic sprinkler system* installed in accordance with Section 903.3.1.1.

805.2.1.3 Identification. Upholstered furniture shall bear the label of an *approved* agency, confirming compliance with the requirements of Sections 805.2.1.1 and 805.2.1.2.

805.2.2 Mattresses. Newly introduced mattresses shall meet the requirements of Sections 805.2.2.1 through 805.2.2.3.

805.2.2.1 Ignition by cigarettes. Newly introduced mattresses shall be shown to resist ignition by cigarettes as determined by tests conducted in accordance with DOC 16 CFR Part 1632 and shall have a char length not exceeding 2 inches (51 mm).

805.2.2.2 Heat release rate. Newly introduced mattresses shall have limited rates of heat release when tested in accordance with ASTM E 1590 or California Technical Bulletin 129, as follows:

1. The peak rate of heat release for the single mattress shall not exceed 100 kW.

   Exception: Mattresses in rooms or spaces protected by an *approved automatic sprinkler system* installed in accordance with Section 903.3.1.1.

2. The total energy released by the single mattress during the first 10 minutes of the test shall not exceed 25 MJ.

   Exception: Mattresses in rooms or spaces protected by an *approved automatic sprinkler system* installed in accordance with Section 903.3.1.1.

805.2.2.3 Identification. Mattresses shall bear the label of an *approved* agency, confirming compliance with the requirements of Sections 805.2.2.1 and 805.2.2.2.

805.3 Group I-3, detention and correction facilities. The requirements in Sections 805.3.1 through 805.3.2 shall apply to detention and correction facilities classified in Group I-3.

805.3.1 Upholstered furniture. Newly introduced upholstered furniture shall meet the requirements of Sections 805.3.1.1 through 805.3.1.3

805.3.1.1 Ignition by cigarettes. Newly introduced upholstered furniture shall be shown to resist ignition by cigarettes as determined by tests conducted in accordance with one of the following:

1. Mocked-up composites of the upholstered furniture shall have a char length not exceeding 1.5 inches (38 mm) when tested in accordance with NFPA 261.
2. The components of the upholstered furniture shall meet the requirements for Class I when tested in accordance with NFPA 260.

805.3.1.2 Heat release rate. Newly introduced upholstered furniture shall have limited rates of heat release when tested in accordance with ASTM E 1537, as follows:

1. The peak rate of heat release for the single upholstered furniture item shall not exceed 80 kW.
2. The total energy released by the single upholstered furniture item during the first 10 minutes of the test shall not exceed 25 MJ.

805.3.1.3 Identification. Upholstered furniture shall bear the label of an *approved* agency, confirming compliance with the requirements of Sections 805.3.1.1 and 805.3.1.2.

805.3.2 Mattresses. Newly introduced mattresses shall meet the requirements of Sections 805.3.2.1 through 805.3.2.3.

805.3.2.1 Ignition by cigarettes. Newly introduced mattresses shall be shown to resist ignition by cigarettes as determined by tests conducted in accordance with DOC 16 CFR Part 1632 and shall have a char length not exceeding 2 inches (51 mm).

805.3.2.2 Fire performance tests. Newly introduced mattresses shall be tested in accordance with Section 805.3.2.2.1 or 805.3.2.2.2.

805.3.2.2.1 Heat release rate. Newly introduced mattresses shall have limited rates of heat release when tested in accordance with ASTM E 1590 or California Technical Bulletin 129, as follows:

1. The peak rate of heat release for the single mattress shall not exceed 100 kW.
2. The total energy released by the single mattress during the first 10 minutes of the test shall not exceed 25 MJ.

805.3.2.2.2 Mass loss test. Newly introduced mattresses shall have a mass loss not exceeding 15 percent of the initial mass of the mattress where tested in accordance with the test in Annex A of ASTM F 1085.

805.3.2.3 Identification. Mattresses shall bear the label of an *approved* agency, confirming compliance with the requirements of Sections 805.3.2.1 and 805.3.2.2.

805.4 Group R-2 college and university dormitories. The requirements of Sections 805.4.1 through 805.4.2.3 shall apply to college and university dormitories classified in Group R-2, including decks, porches and balconies.

805.4.1 Upholstered furniture. Newly introduced upholstered furniture shall meet the requirements of Sections 805.4.1.1 through 805.4.1.3

805.4.1.1 Ignition by cigarettes. Newly introduced upholstered furniture shall be shown to resist ignition by cigarettes as determined by tests conducted in accordance with one of the following:

1. Mocked-up composites of the upholstered furniture shall have a char length not exceeding $1^1/_2$ inches (38 mm) when tested in accordance with NFPA 261.
2. The components of the upholstered furniture shall meet the requirements for Class I when tested in accordance with NFPA 260.

805.4.1.2 Heat release rate. Newly introduced upholstered furniture shall have limited rates of heat release when tested in accordance with ASTM E 1537 or California Technical Bulletin 133, as follows:

1. The peak rate of heat release for the single upholstered furniture item shall not exceed 80 kW.

   Exception: Upholstered furniture in rooms or spaces protected by an *approved automatic sprinkler system* installed in accordance with Section 903.3.1.1.

2. The total energy released by the single upholstered furniture item during the first 10 minutes of the test shall not exceed 25 MJ.

   Exception: Upholstered furniture in rooms or spaces protected by an *approved automatic sprinkler system* installed in accordance with Section 903.3.1.1.

805.4.1.3 Identification. Upholstered furniture shall bear the label of an *approved* agency, confirming compliance with the requirements of Sections 805.4.1.1 and 805.4.1.2.

805.4.2 Mattresses. Newly introduced mattresses shall meet the requirements of Sections 805.4.2.1 through 805.4.2.3.

805.4.2.1 Ignition by cigarettes. Newly introduced mattresses shall be shown to resist ignition by cigarettes as determined by tests conducted in accordance with DOC 16 CFR Part 1632 and shall have a char length not exceeding 2 inches (51 mm).

805.4.2.2 Heat release rate. Newly introduced mattresses shall have limited rates of heat release when tested in accordance with ASTM E 1590 or California Technical Bulletin 129, as follows:

1. The peak rate of heat release for the single mattress shall not exceed 100 kW.

   Exception: Mattresses in rooms or spaces protected by an *approved automatic sprinkler system* installed in accordance with Section 903.3.1.1.

2. The total energy released by the single mattress during the first 10 minutes of the test shall not exceed 25 MJ.

   Exception: Mattresses in rooms or spaces protected by an *approved automatic sprinkler system* installed in accordance with Section 903.3.1.1.

805.4.2.3 Identification. Mattresses shall bear the label of an *approved* agency, confirming compliance with the requirements of Sections 805.4.2.1 and 805.4.2.2.

SECTION 806 DECORATIVE VEGETATION IN NEW AND EXISTING BUILDINGS

806.1 Natural cut trees. Natural cut trees, where allowed by this section, shall have the trunk bottoms cut off not less than 0.5 inch (12.7 mm) above the original cut and shall be placed in a support device complying with Section 806.1.2.

806.1.1 Restricted occupancies. Natural cut trees shall be prohibited within ambulatory care facilities and Group A, E, I-1, I-2, I-3, I-4, M, R-1, R-2 and R-4 occupancies.

Exceptions:

1. Trees located in areas protected by an *approved automatic sprinkler system* installed in accordance with Section 903.3.1.1 or 903.3.1.2 shall not be prohibited in Groups A, E, M, R-1 and R-2.
2. Trees shall be allowed within *dwelling units* in Group R-2 occupancies.

806.1.2 Support devices. The support device that holds the tree in an upright position shall be of a type that is stable and that meets all of the following criteria:

1. The device shall hold the tree securely and be of adequate size to avoid tipping over of the tree.
2. The device shall be capable of containing a minimum two-day supply of water.
3. The water level, when full, shall cover the tree stem not less than 2 inches (51 mm). The water level shall be maintained above the fresh cut and checked not less than once daily.

806.1.3 Dryness. The tree shall be removed from the building whenever the needles or leaves fall off readily when a tree branch is shaken or if the needles are brittle and break when bent between the thumb and index finger. The tree shall be checked daily for dryness.

806.2 Artificial vegetation. Artificial decorative vegetation shall meet the flame propagation performance criteria of Test Method 1 or Test Method 2, as appropriate, of NFPA 701. Meeting the flame propagation performance criteria of Test Method 1 or Test Method 2, as appropriate, of NFPA 701 shall be documented and certified by the manufacturer in an *approved* manner. Alternatively, the artificial decorative vegetation item shall be tested in accordance with NFPA 289, using the 20 kW ignition source, and shall have a maximum heat release rate of 100 kW.

806.3 Obstruction of means of egress. The required width of any portion of a *means of egress* shall not be obstructed by decorative vegetation. Natural cut trees shall not be located within an exit, corridor, or a lobby or vestibule.

806.4 Open flame. Candles and open flames shall not be used on or near decorative vegetation. Natural cut trees shall be kept a distance from heat vents and any open flame or heat-producing devices at least equal to the height of the tree.

806.5 Electrical fixtures and wiring. The use of unlisted electrical wiring and lighting on natural cut trees and artificial decorative vegetation shall be prohibited. The use of electrical wiring and lighting on artificial trees constructed entirely of metal shall be prohibited.

SECTION 807 DECORATIVE MATERIALS OTHER THAN DECORATIVE VEGETATION IN NEW AND EXISTING BUILDINGS

807.1 General. Combustible decorative materials, other than decorative vegetation, shall comply with Sections 807.2 through 807.5.6. *

807.2 Limitations. The following requirements shall apply to all occupancies:

1. Furnishings or decorative materials of an explosive or highly flammable character shall not be used.
2. Fire-retardant coatings in existing buildings shall be maintained so as to retain the effectiveness of the treatment under service conditions encountered in actual use.
3. Furnishings or other objects shall not be placed to obstruct exits, access thereto, egress therefrom or visibility thereof.
4. The permissible amount of noncombustible decorative materials shall not be limited.

807.3 Combustible decorative materials. In other than Group I-3, curtains, draperies, fabric hangings and other similar combustible decorative materials suspended from walls or ceilings shall comply with Section 807.4 and shall not exceed 10 percent of the specific wall or ceiling area to which they are attached.

Fixed or movable walls and partitions, paneling, wall pads and crash pads applied structurally or for decoration, acoustical correction, surface insulation or other purposes shall be considered *interior finish*, shall comply with Section 803 and shall not be considered *decorative materials* or furnishings.

Exceptions:

1. In auditoriums in Group A, the permissible amount of curtains, draperies, fabric hangings and other similar combustible decorative material suspended from walls or ceilings shall not exceed 75 percent of the aggregate wall area where the building is equipped throughout with an *approved automatic sprinkler system* in accordance with Section 903.3.1.1, and where the material is installed in accordance with Section 803.11 of the *International Building Code*.
2. In Group R-2 dormitories, within sleeping units and dwelling units, the permissible amount of curtains, draperies, fabric hangings and other similar decorative materials suspended from walls or ceilings shall not exceed 50 percent of the aggregate wall areas where the building is equipped throughout with an *approved automatic sprinkler system* installed in accordance with Section 903.3.1. **

3. In Group B and M occupancies, the amount of combustible fabric partitions suspended from the ceiling and not supported by the floor shall comply with Section 807.4 and shall not be limited.

807.4 Acceptance criteria and reports. Where required to exhibit improved fire performance, curtains, draperies, fabric hangings and other similar combustible decorative materials suspended from walls or ceilings shall be tested by an *approved* agency and meet the flame propagation performance criteria of Test Method 1 or Test Method 2, as appropriate, of NFPA 701 or exhibit a maximum rate of heat release of 100 kW when tested in accordance with NFPA 289, using the 20 kW ignition source. Reports of test results shall be prepared in accordance with the test method used and furnished to the *fire code official* upon request.

807.5 Occupancy-based requirements. In occupancies specified, combustible decorative materials not complying with Section 807.3 shall comply with Sections 807.5.1 through 807.5.6.

807.5.1 Group A. In Group A occupancies, the requirements in Sections 807.5.1.1 through 807.5.1.4 shall apply.

807.5.1.1 Foam plastics. Exposed foam plastic materials and unprotected materials containing foam plastic used for decorative purposes or stage scenery or exhibit booths shall have a maximum heat release rate of 100 kW when tested in accordance with UL 1975, or when tested in accordance with NFPA 289 using the 20 kW ignition source.

Exceptions:

1. Individual foam plastic items or items containing foam plastic where the foam plastic does not exceed 1 pound (0.45 kg) in weight.
2. Cellular or foam plastic shall be allowed for trim in accordance with Section 804.2.

807.5.1.2 Motion picture screens. The screens upon which motion pictures are projected in new and existing buildings of Group A shall either meet the flame propagation performance criteria of Test Method 1 or Test Method 2, as appropriate, of NFPA 701 or shall comply with the requirements for a Class B interior finish in accordance with Section 803 of the *International Building Code*.

807.5.1.3 Wood use in places of religious worship. In places of religious worship, wood used for ornamental purposes, trusses, paneling or chancel furnishing shall not be limited.

807.5.1.4 Pyroxylin plastic. Imitation leather or other material consisting of or coated with a pyroxylin or similarly hazardous base shall not be used.

807.5.2 Group E. Group E occupancies shall comply with Sections 807.5.2.1 through 807.5.2.3.

807.5.2.1 Storage in corridors and lobbies. Clothing and personal effects shall not be stored in *corridors* and lobbies.

Exceptions:

1. *Corridors* protected by an *approved automatic sprinkler system* installed in accordance with Section 903.3.1.1.
2. *Corridors* protected by an *approved* fire alarm system installed in accordance with Section 907.
3. Storage in metal lockers, provided the minimum required egress width is maintained.

807.5.2.2 Artwork in corridors. Artwork and teaching materials shall be limited on the walls of *corridors* to not more than 20 percent of the wall area.

807.5.2.3 Artwork in classrooms. Artwork and teaching materials shall be limited on walls of classrooms to not more than 50 percent of the specific wall area to which they are attached.

807.5.3 Groups I-1 and I-2. In Group I-1 and I-2 occupancies, combustible *decorative materials* shall comply with Sections 807.5.3.1 through 807.5.3.4.

807.5.3.1 Group I-1 and I-2 Condition 1 within units. In Group I-1 and Group I-2 Condition 1 occupancies, equipped throughout with an *approved automatic sprinkler system* installed in accordance with Section 903.3.1.1, within sleeping units and dwelling units, combustible decorative materials placed on walls shall be limited to not more than 50 percent of the wall area to which they are attached.

807.5.3.2 In Group I-1 and I-2 Condition 1 for areas other than within units. In Group I-1 and Group I-2 Condition 1 occupancies, equipped throughout with an *approved automatic sprinkler system* installed in accordance with Section 903.3.1.1, combustible decorative materials placed on walls in areas other than within dwelling and sleeping units shall be limited to not more than 30 percent of the wall area to which they are attached.

807.5.3.3 In Group I-2 Condition 2. In Group I-2 Condition 2 occupancies, equipped throughout with an *approved automatic sprinkler system* installed in accordance with Section 903.3.1.1, combustible decorative materials placed on walls shall be limited to not more than 30 percent of the wall area to which they are attached.

807.5.3.4 Other areas in Groups I-1 and I-2. In Group I-1 and I-2 occupancies, in areas not equipped throughout with an *approved automatic sprinkler system*, combustible decorative materials shall be of such limited quantities that a hazard of fire development or spread is not present.

807.5.4 Group I-3. In Group I-3, combustible *decorative materials* are prohibited.

807.5.5 Group I-4. Group I-4 occupancies shall comply with the requirements in Sections 807.5.5.1 through 807.5.5.3.

807.5.5.1 Storage in corridors and lobbies. Clothing and personal effects shall not be stored in *corridors* and lobbies.

Exceptions:

1. *Corridors* protected by an *approved automatic sprinkler system* installed in accordance with Section 903.3.1.1.
2. *Corridors* protected by an *approved* fire alarm system installed in accordance with Section 907.
3. Storage in metal lockers, provided the minimum required egress width is maintained.

807.5.5.2 Artwork in corridors. Artwork and teaching materials shall be limited on walls of *corridors* to not more than 20 percent of the wall area.

807.5.5.3 Artwork in classrooms. Artwork and teaching materials shall be limited on walls of classrooms to not more than 50 percent of the specific wall area to which they are attached.

** **807.5.6 Dormitories in Group R-2.** In Group R-2 dormitories, within sleeping units and dwelling units, the combustible decorative materials shall be of limited quantities such that a hazard of fire development or spread is not present.

SECTION 808
FURNISHINGS OTHER THAN UPHOLSTERED FURNITURE AND MATTRESSES OR DECORATIVE MATERIALS IN NEW AND EXISTING BUILDINGS

808.1 Wastebaskets and linen containers in Group I-1, I-2 and I-3 occupancies. Wastebaskets, linen containers and other waste containers, including their lids, located in Group I-1, I-2 and I-3 occupancies shall be constructed of noncombustible materials or of materials that meet a peak rate of heat release not exceeding 300 kW/m^2 when tested in accordance with ASTM E 1354 at an incident heat flux of 50 kW/m^2 in the horizontal orientation. Metal wastebaskets and other metal waste containers with a capacity of 20 gallons (75.7 L) or more shall be *listed* in accordance with UL 1315 and shall be provided with a noncombustible lid. Portable containers exceeding 32 gallons (121 L) shall be stored in an area classified as a waste and linen collection room and constructed in accordance with Table 509 of the *International Building Code*.

808.2 Waste containers with a capacity of 20 gallons or more in Group R-2 college and university dormitories. Waste containers, including their lids, located in Group R-2 college and university dormitories, and with a capacity of 20 gallons (75.7 L) or more, shall be constructed of noncombustible materials or of materials that meet a peak rate of heat release not exceeding 300 kW/m^2 when tested in accordance with ASTM E 1354 at an incident heat flux of 50 kW/m^2 in the horizontal orientation. Metal wastebaskets and other metal waste containers with a capacity of 20 gallons (75.7 L) or more shall be *listed* in accordance with UL 1315 and shall be provided with a noncombustible lid. Portable containers exceeding 32 gallons (121 L) shall be stored in an area classified as a waste and linen collection room constructed in accordance with Table 509 of the *International Building Code*.

808.3 Signs. Foam plastic signs that are not affixed to interior building surfaces shall have a maximum heat release rate of 150 kW when tested in accordance with UL 1975, or when tested in accordance with NFPA 289 using the 20-kW ignition source.

Exception: Where the aggregate area of foam plastic signs is less than 10 percent of the floor area or wall area of the room or space in which the signs are located, whichever is less, subject to the approval of the *fire code official*.

808.4 Combustible lockers. Where lockers constructed of combustible materials are used, the lockers shall be considered interior finish and shall comply with Section 803.

Exception: Lockers constructed entirely of wood and noncombustible materials shall be permitted to be used wherever interior finish materials are required to meet a Class C classification in accordance with Section 803.1.1.

CHAPTER 9

FIRE PROTECTION SYSTEMS

SECTION 901
GENERAL

901.1 Scope. The provisions of this chapter shall specify where *fire protection systems* are required and shall apply to the design, installation, inspection, operation, testing and maintenance of all *fire protection systems*.

901.2 Construction documents. The *fire code official* shall have the authority to require *construction documents* and calculations for all *fire protection systems* and to require permits be issued for the installation, rehabilitation or modification of any *fire protection system*. *Construction documents* for *fire protection systems* shall be submitted for review and approval prior to system installation.

901.2.1 Statement of compliance. Before requesting final approval of the installation, where required by the *fire code official*, the installing contractor shall furnish a written statement to the *fire code official* that the subject *fire protection system* has been installed in accordance with *approved* plans and has been tested in accordance with the manufacturer's specifications and the appropriate installation standard. Any deviations from the design standards shall be noted and copies of the approvals for such deviations shall be attached to the written statement.

901.3 Permits. Permits shall be required as set forth in Sections 105.6 and 105.7.

901.4 Installation. *Fire protection systems* shall be maintained in accordance with the original installation standards for that system. Required systems shall be extended, altered or augmented as necessary to maintain and continue protection where the building is altered, remodeled or added to. *Alterations* to *fire protection systems* shall be done in accordance with applicable standards.

901.4.1 Required fire protection systems. *Fire protection systems* required by this code or the *International Building Code* shall be installed, repaired, operated, tested and maintained in accordance with this code. A *fire protection system* for which a design option, exception or reduction to the provisions of this code or the *International Building Code* has been granted shall be considered to be a required system.

901.4.2 Nonrequired fire protection systems. A *fire protection system* or portion thereof not required by this code or the *International Building Code* shall be allowed to be furnished for partial or complete protection provided such installed system meets the applicable requirements of this code and the *International Building Code*.

901.4.3 Fire areas. Where buildings, or portions thereof, are divided into *fire areas* so as not to exceed the limits established for requiring a *fire protection system* in accordance with this chapter, such *fire areas* shall be separated by *fire barriers* constructed in accordance with Section 707 of the *International Building Code* or *horizontal assemblies* constructed in accordance with Section 711 of the *International Building Code*, or both, having a fire-resistance rating of not less than that determined in accordance with Section 707.3.10 of the *International Building Code*.

901.4.4 Additional fire protection systems. In occupancies of a hazardous nature, where special hazards exist in addition to the normal hazards of the occupancy, or where the *fire code official* determines that access for fire apparatus is unduly difficult, the *fire code official* shall have the authority to require additional safeguards. Such safeguards include, but shall not be limited to, the following: automatic fire detection systems, fire alarm systems, automatic fire-extinguishing systems, standpipe systems, or portable or fixed extinguishers. Fire protection equipment required under this section shall be installed in accordance with this code and the applicable referenced standards.

901.4.5 Appearance of equipment. Any device that has the physical appearance of life safety or fire protection equipment but that does not perform that life safety or fire protection function shall be prohibited.

901.4.6 Pump and riser room size. Where provided, fire pump rooms and *automatic sprinkler system* riser rooms shall be designed with adequate space for all equipment necessary for the installation, as defined by the manufacturer, with sufficient working space around the stationary equipment. Clearances around equipment to elements of permanent construction, including other installed equipment and appliances, shall be sufficient to allow inspection, service, repair or replacement without removing such elements of permanent construction or disabling the function of a required fire-resistance-rated assembly. Fire pump and *automatic sprinkler system* riser rooms shall be provided with a door(s) and an unobstructed passageway large enough to allow removal of the largest piece of equipment.

901.5 Installation acceptance testing. Fire detection and alarm systems, fire-extinguishing systems, fire hydrant systems, fire standpipe systems, fire pump systems, private fire service mains and all other *fire protection systems* and appurtenances thereto shall be subject to acceptance tests as contained in the installation standards and as *approved* by the *fire code official*. The *fire code official* shall be notified before any required acceptance testing.

901.5.1 Occupancy. It shall be unlawful to occupy any portion of a building or structure until the required fire detection, alarm and suppression systems have been tested and *approved*.

901.6 Inspection, testing and maintenance. Fire detection, alarm, and extinguishing systems, mechanical smoke exhaust systems, and smoke and heat vents shall be maintained in an operative condition at all times, and shall be replaced or repaired where defective. Nonrequired *fire protection systems* and equipment shall be inspected, tested and maintained or removed.

901.6.1 Standards. *Fire protection systems* shall be inspected, tested and maintained in accordance with the referenced standards *listed* in Table 901.6.1.

TABLE 901.6.1
FIRE PROTECTION SYSTEM MAINTENANCE STANDARDS

SYSTEM	STANDARD
Portable fire extinguishers	NFPA 10
Carbon dioxide fire-extinguishing system	NFPA 12
Halon 1301 fire-extinguishing systems	NFPA 12A
Dry-chemical extinguishing systems	NFPA 17
Wet-chemical extinguishing systems	NFPA 17A
Water-based fire protection systems	NFPA 25
Fire alarm systems	NFPA 72
Smoke and heat vents	NFPA 204
Water-mist systems	NFPA 750
Clean-agent extinguishing systems	NFPA 2001

901.6.2 Records. Records of all system inspections, tests and maintenance required by the referenced standards shall be maintained.

901.6.2.1 Records information. Initial records shall include the name of the installation contractor, type of components installed, manufacturer of the components, location and number of components installed per floor. Records shall also include the manufacturers' operation and maintenance instruction manuals. Such records shall be maintained for the life of the installation.

901.7 Systems out of service. Where a required *fire protection system* is out of service, the fire department and the *fire code official* shall be notified immediately and, where required by the *fire code official*, the building shall be either evacuated or an *approved* fire watch shall be provided for all occupants left unprotected by the shutdown until the *fire protection system* has been returned to service.

Where utilized, fire watches shall be provided with not less than one *approved* means for notification of the fire department and their only duty shall be to perform constant patrols of the protected premises and keep watch for fires.

901.7.1 Impairment coordinator. The building *owner* shall assign an impairment coordinator to comply with the requirements of this section. In the absence of a specific designee, the *owner* shall be considered the impairment coordinator.

901.7.2 Tag required. A tag shall be used to indicate that a system, or portion thereof, has been removed from service.

901.7.3 Placement of tag. The tag shall be posted at each fire department connection, system control valve, fire alarm control unit, fire alarm annunciator and *fire command center*, indicating which system, or part thereof, has been removed from service. The *fire code official* shall specify where the tag is to be placed.

901.7.4 Preplanned impairment programs. Preplanned impairments shall be authorized by the impairment coordinator. Before authorization is given, a designated individual shall be responsible for verifying that all of the following procedures have been implemented:

1. The extent and expected duration of the impairment have been determined.
2. The areas or buildings involved have been inspected and the increased risks determined.
3. Recommendations have been submitted to management or the building *owner*/manager.
4. The fire department has been notified.
5. The insurance carrier, the alarm company, the building *owner*/manager and other authorities having jurisdiction have been notified.
6. The supervisors in the areas to be affected have been notified.
7. A tag impairment system has been implemented.
8. Necessary tools and materials have been assembled on the impairment site.

901.7.5 Emergency impairments. Where unplanned impairments occur, appropriate emergency action shall be taken to minimize potential injury and damage. The impairment coordinator shall implement the steps outlined in Section 901.7.4.

901.7.6 Restoring systems to service. When impaired equipment is restored to normal working order, the impairment coordinator shall verify that all of the following procedures have been implemented:

1. Necessary inspections and tests have been conducted to verify that affected systems are operational.
2. Supervisors have been advised that protection is restored.
3. The fire department has been advised that protection is restored.
4. The building *owner*/manager, insurance carrier, alarm company and other involved parties have been advised that protection is restored.
5. The impairment tag has been removed.

901.8 Removal of or tampering with equipment. It shall be unlawful for any person to remove, tamper with or otherwise disturb any fire hydrant, fire detection and alarm system, fire suppression system or other fire appliance required by this code except for the purpose of extinguishing fire, training purposes, recharging or making necessary repairs or where *approved* by the *fire code official.*

901.8.1 Removal of or tampering with appurtenances. Locks, gates, doors, barricades, chains, enclosures, signs, tags or seals that have been installed by or at the direction of the *fire code official* shall not be removed, unlocked, destroyed, tampered with or otherwise vandalized in any manner.

901.8.2 Removal of existing occupant-use hose lines. The *fire code official* is authorized to permit the removal of existing occupant-use hose lines where all of the following conditions exist:

1. Installation is not required by this code or the *International Building Code*.
2. The hose line would not be utilized by trained personnel or the fire department.
3. The remaining outlets are compatible with local fire department fittings.

901.9 Termination of monitoring service. For fire alarm systems required to be monitored by this code, notice shall be made to the *fire code official* whenever alarm monitoring services are terminated. Notice shall be made in writing, to the *fire code official* by the monitoring service provider being terminated.

901.10 Recall of fire protection components. Any *fire protection system* component regulated by this code that is the subject of a voluntary or mandatory recall under federal law shall be replaced with *approved, listed* components in compliance with the referenced standards of this code. The *fire code official* shall be notified in writing by the building *owner* when the recalled component parts have been replaced.

SECTION 902 DEFINITIONS

902.1 Definitions. The following terms are defined in Chapter 2:

ALARM NOTIFICATION APPLIANCE.

ALARM SIGNAL.

ALARM VERIFICATION FEATURE.

ANNUNCIATOR.

AUDIBLE ALARM NOTIFICATION APPLIANCE.

AUTOMATIC.

AUTOMATIC FIRE-EXTINGUISHING SYSTEM.

AUTOMATIC SMOKE DETECTION SYSTEM.

AUTOMATIC SPRINKLER SYSTEM.

AUTOMATIC WATER MIST SYSTEM.

AVERAGE AMBIENT SOUND LEVEL.

CARBON DIOXIDE EXTINGUISHING SYSTEM.

CLEAN AGENT.

COMMERCIAL MOTOR VEHICLE.

CONSTANTLY ATTENDED LOCATION.

DELUGE SYSTEM.

DETECTOR, HEAT.

DRY-CHEMICAL EXTINGUISHING AGENT.

ELEVATOR GROUP.

EMERGENCY ALARM SYSTEM.

EMERGENCY VOICE/ALARM COMMUNICATIONS.

FIRE ALARM BOX, MANUAL.

FIRE ALARM CONTROL UNIT.

FIRE ALARM SIGNAL.

FIRE ALARM SYSTEM.

FIRE AREA.

FIRE DETECTOR, AUTOMATIC.

FIRE PROTECTION SYSTEM.

FIRE SAFETY FUNCTIONS.

FIXED BASE OPERATOR (FBO).

FOAM-EXTINGUISHING SYSTEM.

HALOGENATED EXTINGUISHING SYSTEM.

IMPAIRMENT COORDINATOR.

INITIATING DEVICE.

MANUAL FIRE ALARM BOX.

MULTIPLE-STATION ALARM DEVICE.

MULTIPLE-STATION SMOKE ALARM.

NOTIFICATION ZONE.

NUISANCE ALARM.

PRIVATE GARAGE.

RECORD DRAWINGS.

SINGLE-STATION SMOKE ALARM.

SLEEPING UNIT.

SMOKE ALARM.

SMOKE DETECTOR.

STANDPIPE SYSTEM, CLASSES OF.

- **Class I system.**
- **Class II system.**
- **Class III system.**

STANDPIPE, TYPES OF.

- **Automatic dry.**
- **Automatic wet.**
- **Manual dry.**
- **Manual wet.**
- **Semiautomatic dry.**

SUPERVISING STATION.

SUPERVISORY SERVICE.

SUPERVISORY SIGNAL.

SUPERVISORY SIGNAL-INITIATING DEVICE.

TIRES, BULK STORAGE OF.

TRANSIENT AIRCRAFT.

TROUBLE SIGNAL.

VISIBLE ALARM NOTIFICATION APPLIANCE.

WET-CHEMICAL EXTINGUISHING AGENT.

WIRELESS PROTECTION SYSTEM.

ZONE.

ZONE, NOTIFICATION.

SECTION 903
AUTOMATIC SPRINKLER SYSTEMS

903.1 General. *Automatic sprinkler systems* shall comply with this section.

903.1.1 Alternative protection. Alternative automatic fire-extinguishing systems complying with Section 904 shall be permitted instead of automatic sprinkler protection where recognized by the applicable standard and *approved* by the *fire code official.*

903.2 Where required. *Approved automatic sprinkler systems* in new buildings and structures shall be provided in the locations described in Sections 903.2.1 through 903.2.12.

Exception: Spaces or areas in telecommunications buildings used exclusively for telecommunications equipment, associated electrical power distribution equipment, batteries and standby engines, provided those spaces or areas are equipped throughout with an automatic smoke detection system in accordance with Section 907.2 and are separated from the remainder of the building by not less than 1-hour *fire barriers* constructed in accordance with Section 707 of the *International Building Code* or not less than 2-hour *horizontal assemblies* constructed in accordance with Section 711 of the *International Building Code,* or both.

903.2.1 Group A. An *automatic sprinkler system* shall be provided throughout buildings and portions thereof used as Group A occupancies as provided in this section. For Group A-1, A-2, A-3 and A-4 occupancies, the *automatic sprinkler system* shall be provided throughout the story where the *fire area* containing the Group A-1, A-2, A-3 or A-4 occupancy is located, and throughout all stories from the Group A occupancy to, and including, the *levels of exit discharge* serving the Group A occupancy. For Group A-5 occupancies, the *automatic sprinkler system* shall be provided in the spaces indicated in Section 903.2.1.5.

903.2.1.1 Group A-1. An *automatic sprinkler system* shall be provided for *fire areas* containing Group A-1 occupancies and intervening floors of the building where one of the following conditions exists:

1. The *fire area* exceeds 12,000 square feet (1115 m^2).
2. The *fire area* has an *occupant load* of 300 or more.
3. The *fire area* is located on a floor other than a *level of exit discharge* serving such occupancies.
4. The *fire area* contains a multitheater complex.

903.2.1.2 Group A-2. An *automatic sprinkler system* shall be provided for *fire areas* containing Group A-2 occupancies and intervening floors of the building where one of the following conditions exists:

1. The *fire area* exceeds 5,000 square feet (464 m^2).
2. The *fire area* has an *occupant load* of 100 or more.
3. The *fire area* is located on a floor other than a *level of exit discharge* serving such occupancies.

903.2.1.3 Group A-3. An *automatic sprinkler system* shall be provided for *fire areas* containing Group A-3 occupancies and intervening floors of the building where one of the following conditions exists:

1. The *fire area* exceeds 12,000 square feet (1115 m^2).
2. The *fire area* has an *occupant load* of 300 or more.
3. The *fire area* is located on a floor other than a *level of exit discharge* serving such occupancies.

903.2.1.4 Group A-4. An *automatic sprinkler system* shall be provided for *fire areas* containing Group A-4 occupancies and intervening floors of the building where one of the following conditions exists:

1. The *fire area* exceeds 12,000 square feet (1115 m^2).
2. The *fire area* has an *occupant load* of 300 or more.
3. The *fire area* is located on a floor other than a *level of exit discharge* serving such occupancies.

903.2.1.5 Group A-5. An *automatic sprinkler system* shall be provided for Group A-5 occupancies in the following areas: concession stands, retail areas, press boxes and other accessory use areas in excess of 1,000 square feet (93 m^2).

903.2.1.6 Assembly occupancies on roofs. Where an occupied roof has an assembly occupancy with an *occupant load* exceeding 100 for Group A-2 and 300 for other Group A occupancies, all floors between the occupied roof and the *level of exit discharge* shall be equipped with an *automatic sprinkler system* in accordance with Section 903.3.1.1 or 903.3.1.2.

Exception: Open parking garages of Type I or Type II construction.

903.2.1.7 Multiple fire areas. An *automatic sprinkler system* shall be provided where multiple fire areas of Group A-1, A-2, A-3 or A-4 occupancies share exit or *exit access* components and the combined *occupant load* of these fire areas is 300 or more.

903.2.2 Ambulatory care facilities. An *automatic sprinkler system* shall be installed throughout the entire floor containing an ambulatory care facility where either of the following conditions exist at any time:

1. Four or more care recipients are incapable of self-preservation, whether rendered incapable by staff or

staff has accepted responsibility for care recipients already incapable.

2. One or more care recipients that are incapable of self-preservation are located at other than the level of exit discharge serving such a facility.

In buildings where ambulatory care is provided on levels other than the *level of exit discharge*, an *automatic sprinkler system* shall be installed throughout the entire floor where such care is provided as well as all floors below, and all floors between the level of ambulatory care and the nearest *level of exit discharge*, including the *level of exit discharge*.

903.2.3 Group E. An *automatic sprinkler system* shall be provided for Group E occupancies as follows:

1. Throughout all Group E *fire areas* greater than 12,000 square feet (1115 m^2) in area.
2. Throughout every portion of educational buildings below the lowest *level of exit discharge* serving that portion of the building.

Exception: An *automatic sprinkler system* is not required in any area below the lowest *level of exit discharge* serving that area where every classroom throughout the building has not fewer than one exterior *exit* door at ground level.

903.2.4 Group F-1. An *automatic sprinkler system* shall be provided throughout all buildings containing a Group F-1 occupancy where one of the following conditions exists:

1. A Group F-1 *fire area* exceeds 12,000 square feet (1115 m^2).
2. A Group F-1 *fire area* is located more than three stories above grade plane.
3. The combined area of all Group F-1 *fire areas* on all floors, including any mezzanines, exceeds 24,000 square feet (2230 m^2).
4. A Group F-1 occupancy used for the manufacture of upholstered furniture or mattresses exceeds 2,500 square feet (232 m^2).

903.2.4.1 Woodworking operations. An *automatic sprinkler system* shall be provided throughout all Group F-1 occupancy *fire areas* that contain woodworking operations in excess of 2,500 square feet in area (232 m^2) that generate finely divided combustible waste or use finely divided combustible materials.

903.2.5 Group H. *Automatic sprinkler systems* shall be provided in high-hazard occupancies as required in Sections 903.2.5.1 through 903.2.5.3.

903.2.5.1 General. An *automatic sprinkler system* shall be installed in Group H occupancies.

903.2.5.2 Group H-5 occupancies. An *automatic sprinkler system* shall be installed throughout buildings containing Group H-5 occupancies. The design of the sprinkler system shall be not less than that required under the *International Building Code* for the occupancy hazard classifications in accordance with Table 903.2.5.2.

Where the design area of the sprinkler system consists of a *corridor* protected by one row of sprinklers, the maximum number of sprinklers required to be calculated is 13.

TABLE 903.2.5.2
GROUP H-5 SPRINKLER DESIGN CRITERIA

LOCATION	OCCUPANCY HAZARD CLASSIFICATION
Fabrication areas	Ordinary Hazard Group 2
Service corridors	Ordinary Hazard Group 2
Storage rooms without dispensing	Ordinary Hazard Group 2
Storage rooms with dispensing	Extra Hazard Group 2
Corridors	Ordinary Hazard Group 2

903.2.5.3 Pyroxylin plastics. An *automatic sprinkler system* shall be provided in buildings, or portions thereof, where cellulose nitrate film or pyroxylin plastics are manufactured, stored or handled in quantities exceeding 100 pounds (45 kg).

903.2.6 Group I. An *automatic sprinkler system* shall be provided throughout buildings with a Group I *fire area.*

Exceptions:

1. An *automatic sprinkler system* installed in accordance with Section 903.3.1.2 shall be permitted in Group I-1 Condition 1 facilities.
2. An *automatic sprinkler system* is not required where Group I-4 day care facilities are at the *level of exit discharge* and where every room where care is provided has not fewer than one exterior *exit* door.
3. In buildings where Group I-4 day care is provided on levels other than the *level of exit discharge*, an *automatic sprinkler system* in accordance with Section 903.3.1.1 shall be installed on the entire floor where care is provided, all floors between the level of care and the *level of exit discharge* and all floors below the level of exit discharge other than areas classified as an open parking garage.

903.2.7 Group M. An *automatic sprinkler system* shall be provided throughout buildings containing a Group M occupancy where one of the following conditions exists:

1. A Group M *fire area* exceeds 12,000 square feet (1115 m^2).
2. A Group M *fire area* is located more than three stories above grade plane.
3. The combined area of all Group M *fire areas* on all floors, including any mezzanines, exceeds 24,000 square feet (2230 m^2).
4. A Group M occupancy used for the display and sale of upholstered furniture or mattresses exceeds 5,000 square feet (464 m^2).

903.2.7.1 High-piled storage. An *automatic sprinkler system* shall be provided as required in Chapter 32 in all buildings of Group M where storage of merchandise is in high-piled or rack storage arrays.

903.2.8 Group R. An *automatic sprinkler system* installed in accordance with Section 903.3 shall be provided throughout all buildings with a Group R *fire area.*

903.2.8.1 Group R-3. An *automatic sprinkler system* installed in accordance with Section 903.3.1.3 shall be permitted in Group R-3 occupancies.

[F] 903.2.8.2 Group R-4 Condition 1. An *automatic sprinkler system* installed in accordance with Section 903.3.1.3 shall be permitted in Group R-4 Condition 1 occupancies.

[F] 903.2.8.3 Group R-4 Condition 2. An *automatic sprinkler system* installed in accordance with Section 903.3.1.2 shall be permitted in Group R-4 Condition 2 occupancies. Attics shall be protected in accordance with Section 903.2.8.3.1 or 903.2.8.3.2.

[F] 903.2.8.3.1 Attics used for living purposes, storage or fuel-fired equipment. Attics used for living purposes, storage or fuel-fired equipment shall be protected throughout with an *automatic sprinkler system* installed in accordance with Section 903.3.1.2.

[F] 903.2.8.3.2 Attics not used for living purposes, storage or fuel-fired equipment. Attics not used for living purposes, storage or fuel-fired equipment shall be protected in accordance with one of the following:

1. Attics protected throughout by a heat detector system arranged to activate the building fire alarm system in accordance with Section 907.2.10.
2. Attics constructed of noncombustible materials.
3. Attics constructed of fire-retardant-treated wood framing complying with Section 2303.2 of the *International Building Code.*
4. The *automatic sprinkler system* shall be extended to provide protection throughout the attic space.

[F] 903.2.8.4 Care facilities. An *automatic sprinkler system* installed in accordance with Section 903.3.1.3 shall be permitted in care facilities with five or fewer individuals in a single-family dwelling.

903.2.9 Group S-1. An *automatic sprinkler system* shall be provided throughout all buildings containing a Group S-1 occupancy where one of the following conditions exists:

1. A Group S-1 *fire area* exceeds 12,000 square feet (1115 m^2).
2. A Group S-1 *fire area* is located more than three stories above grade plane.
3. The combined area of all Group S-1 *fire areas* on all floors, including any mezzanines, exceeds 24,000 square feet (2230 m^2).
4. A Group S-1 *fire area* used for the storage of commercial motor vehicles where the *fire area* exceeds 5,000 square feet (464 m^2).
5. A Group S-1 occupancy used for the storage of upholstered furniture or mattresses exceeds 2,500 square feet (232 m^2).

903.2.9.1 Repair garages. An *automatic sprinkler system* shall be provided throughout all buildings used as repair garages in accordance with Section 406.8 of the *International Building Code*, as shown:

1. Buildings having two or more stories above grade plane, including *basements*, with a *fire area* containing a repair garage exceeding 10,000 square feet (929 m^2).
2. Buildings not more than one story above grade plane, with a *fire area* containing a repair garage exceeding 12,000 square feet (1115 m^2).
3. Buildings with repair garages servicing vehicles parked in *basements.*
4. A Group S-1 *fire area* used for the repair of commercial motor vehicles where the *fire area* exceeds 5,000 square feet (464 m^2).

903.2.9.2 Bulk storage of tires. Buildings and structures where the area for the storage of tires exceeds 20,000 cubic feet (566 m^3) shall be equipped throughout with an *automatic sprinkler system* in accordance with Section 903.3.1.1.

903.2.10 Group S-2 enclosed parking garages. An *automatic sprinkler system* shall be provided throughout buildings classified as enclosed parking garages in accordance with Section 406.6 of the *International Building Code* where either of the following conditions exists:

1. Where the *fire area* of the enclosed parking garage exceeds 12,000 square feet (1115 m^2).
2. Where the enclosed parking garage is located beneath other groups.

Exception: Enclosed parking garages located beneath Group R-3 occupancies.

903.2.10.1 Commercial parking garages. An *automatic sprinkler system* shall be provided throughout buildings used for storage of commercial motor vehicles where the *fire area* exceeds 5,000 square feet (464 m^2).

903.2.11 Specific buildings areas and hazards. In all occupancies other than Group U, an *automatic sprinkler system* shall be installed for building design or hazards in the locations set forth in Sections 903.2.11.1 through 903.2.11.6.

903.2.11.1 Stories without openings. An *automatic sprinkler system* shall be installed throughout all stories, including *basements*, of all buildings where the floor area exceeds 1,500 square feet (139.4 m²) and where there is not provided not fewer than one of the following types of *exterior wall* openings:

1. Openings below grade that lead directly to ground level by an exterior *stairway* complying with Section 1009 or an outside ramp complying with Section 1010. Openings shall be located in each 50 linear feet (15 240 mm), or fraction thereof, of *exterior wall* in the story on at least one side. The required openings shall be distributed such that the lineal distance between adjacent openings does not exceed 50 feet (15 240 mm).
2. Openings entirely above the adjoining ground level totaling not less than 20 square feet (1.86 m²) in each 50 linear feet (15 240 mm), or fraction thereof, of *exterior wall* in the story on at least one side. The required openings shall be distributed such that the lineal distance between adjacent openings does not exceed 50 feet (15 240 mm). The height of the bottom of the clear opening shall not exceed 44 inches (1118 mm) measured from the floor.

903.2.11.1.1 Opening dimensions and access. Openings shall have a minimum dimension of not less than 30 inches (762 mm). Such openings shall be accessible to the fire department from the exterior and shall not be obstructed in a manner such that fire fighting or rescue cannot be accomplished from the exterior.

903.2.11.1.2 Openings on one side only. Where openings in a story are provided on only one side and the opposite wall of such story is more than 75 feet (22 860 mm) from such openings, the story shall be equipped throughout with an *approved automatic sprinkler system* or openings as specified above shall be provided on not fewer than two sides of the story.

903.2.11.1.3 Basements. Where any portion of a *basement* is located more than 75 feet (22 860 mm) from openings required by Section 903.2.11.1, or where walls, partitions or other obstructions are installed that restrict the application of water from hose streams, the *basement* shall be equipped throughout with an *approved automatic sprinkler system.*

903.2.11.2 Rubbish and linen chutes. An *automatic sprinkler system* shall be installed at the top of rubbish and linen chutes and in their terminal rooms. Chutes shall have additional sprinkler heads installed at alternate floors and at the lowest intake. Where a rubbish chute extends through a building more than one floor below the lowest intake, the extension shall have sprinklers installed that are recessed from the drop area of the chute and protected from freezing in accordance with Section 903.3.1.1. Such sprinklers shall be installed at alternate floors beginning with the second level below the last intake and ending with the floor above the discharge. Chute sprinklers shall be accessible for servicing.

903.2.11.3 Buildings 55 feet or more in height. An *automatic sprinkler system* shall be installed throughout buildings that have one or more stories with an *occupant load* of 30 or more located 55 feet (16 764 mm) or more above the lowest level of fire department vehicle access, measured to the finished floor.

Exceptions:

1. Open parking structures.
2. Occupancies in Group F-2.

903.2.11.4 Ducts conveying hazardous exhausts. Where required by the *International Mechanical Code*, automatic sprinklers shall be provided in ducts conveying hazardous exhaust or flammable or combustible materials.

Exception: Ducts where the largest cross-sectional diameter of the duct is less than 10 inches (254 mm).

903.2.11.5 Commercial cooking operations. An *automatic sprinkler system* shall be installed in commercial kitchen exhaust hood and duct systems where an *automatic sprinkler system* is used to comply with Section 904.

903.2.11.6 Other required suppression systems. In addition to the requirements of Section 903.2, the provisions indicated in Table 903.2.11.6 require the installation of a fire suppression system for certain buildings and areas.

903.2.12 During construction. *Automatic sprinkler systems* required during construction, *alteration* and demolition operations shall be provided in accordance with Section 3313.

903.3 Installation requirements. *Automatic sprinkler systems* shall be designed and installed in accordance with Sections 903.3.1 through 903.3.8.

903.3.1 Standards. Sprinkler systems shall be designed and installed in accordance with Section 903.3.1.1, unless otherwise permitted by Sections 903.3.1.2 and 903.3.1.3 and other chapters of this code, as applicable.

903.3.1.1 NFPA 13 sprinkler systems. Where the provisions of this code require that a building or portion thereof be equipped throughout with an *automatic sprinkler system* in accordance with this section, sprinklers shall be installed throughout in accordance with NFPA 13 except as provided in Sections 903.3.1.1.1 and 903.3.1.1.2.

TABLE 903.2.11.6
ADDITIONAL REQUIRED FIRE SUPPRESSION SYSTEMS

SECTION	SUBJECT
914.2.1	Covered and open mall buildings
914.3.1	High-rise buildings
914.4.1	Atriums
914.5.1	Underground structures
914.6.1	Stages
914.7.1	Special amusement buildings
914.8.2	Airport traffic control towers
914.8.3, 914.8.6	Aircraft hangars
914.9	Flammable finishes
914.10	Drying rooms
914.11.1	Ambulatory care facilities
1029.6.2.3	Smoke-protected assembly seating
1103.5.1	Pyroxylin plastic storage in existing buildings
1103.5.2	Existing Group I-2 occupancies
1103.5.3	Existing Group I-2 Condition 2 occupancies
1103.5.4	Pyroxylin plastics
2108.2	Dry cleaning plants
2108.3	Dry cleaning machines
2309.3.2.6.2	Hydrogen motor fuel-dispensing area canopies
2404.2	Spray finishing in Group A, E, I or R
2404.4	Spray booths and spray rooms
2405.2	Dip-tank rooms in Group A, I or R
2405.4.1	Dip tanks
2405.9.4	Hardening and tempering tanks
2703.10	HPM facilities
2703.10.1.1	HPM work station exhaust
2703.10.2	HPM gas cabinets and exhausted enclosures
2703.10.3	HPM exit access corridor
2703.10.4	HPM exhaust ducts
2703.10.4.1	HPM noncombustible ducts
2703.10.4.2	HPM combustible ducts
2807.3	Lumber production conveyor enclosures
2808.7	Recycling facility conveyor enclosures
3006.1	Class A and B ovens
3006.2	Class C and D ovens
Table 3206.2	Storage fire protection
3206.4	Storage
5003.8.4.1	Gas rooms
5003.8.5.3	Exhausted enclosures
5004.5	Indoor storage of hazardous materials
5005.1.8	Indoor dispensing of hazardous materials
5104.4.1	Aerosol warehouses

(continued)

TABLE 903.2.11.6—continued
ADDITIONAL REQUIRED FIRE SUPPRESSION SYSTEMS

SECTION	SUBJECT
5106.3.2	Aerosol display and merchandising areas
5204.5	Storage of more than 1,000 cubic feet of loose combustible fibers
5306.2.1	Exterior medical gas storage room
5306.2.2	Interior medical gas storage room
5306.2.3	Medical gas storage cabinet
5606.5.2.1	Storage of smokeless propellant
5606.5.2.3	Storage of small arms primers
5704.3.7.5.1	Flammable and combustible liquid storage rooms
5704.3.8.4	Flammable and combustible liquid storage warehouses
5705.3.7.3	Flammable and combustible liquid Group H-2 or H-3 areas
6004.1.2	Gas cabinets for highly toxic and toxic gas
6004.1.3	Exhausted enclosures for highly toxic and toxic gas
6004.2.2.6	Gas rooms for highly toxic and toxic gas
6004.3.3	Outdoor storage for highly toxic and toxic gas
6504.1.1	Pyroxylin plastic storage cabinets
6504.1.3	Pyroxylin plastic storage vaults
6504.2	Pyroxylin plastic storage and manufacturing

For SI: 1 cubic foot = 0.023 m^3.

903.3.1.1.1 Exempt locations. Automatic sprinklers shall not be required in the following rooms or areas where such rooms or areas are protected with an *approved* automatic fire detection system in accordance with Section 907.2 that will respond to visible or invisible particles of combustion. Sprinklers shall not be omitted from a room merely because it is damp, of fire-resistance-rated construction or contains electrical equipment.

1. A room where the application of water, or flame and water, constitutes a serious life or fire hazard.
2. A room or space where sprinklers are considered undesirable because of the nature of the contents, where *approved* by the *fire code official*.
3. Generator and transformer rooms separated from the remainder of the building by walls and floor/ceiling or roof/ceiling assemblies having a *fire-resistance rating* of not less than 2 hours.
4. Rooms or areas that are of noncombustible construction with wholly noncombustible contents.
5. Fire service access elevator machine rooms and machinery spaces.

6. Machine rooms, machinery spaces, control rooms and control spaces associated with occupant evacuation elevators designed in accordance with Section 3008 of the *International Building Code*.

903.3.1.1.2 Bathrooms. In Group R occupancies, other than Group R-4 occupancies, sprinklers shall not be required in bathrooms that do not exceed 55 square feet (5 m^2) in area and are located within individual *dwelling units* or *sleeping units*, provided that walls and ceilings, including the walls and ceilings behind a shower enclosure or tub, are of noncombustible or limited-combustible materials with a 15-minute thermal barrier rating.

903.3.1.2 NFPA 13R sprinkler systems. *Automatic sprinkler systems* in Group R occupancies up to and including four stories in height in buildings not exceeding 60 feet (18 288 mm) in height above grade plane shall be permitted to be installed throughout in accordance with NFPA 13R.

The number of stories of Group R occupancies constructed in accordance with Sections 510.2 and 510.4 of the *International Building Code* shall be measured from the horizontal assembly creating separate buildings.

903.3.1.2.1 Balconies and decks. Sprinkler protection shall be provided for exterior balconies, decks and ground floor patios of *dwelling units* and *sleeping units* where the building is of Type V construction, provided there is a roof or deck above. Sidewall sprinklers that are used to protect such areas shall be permitted to be located such that their deflectors are within 1 inch (25 mm) to 6 inches (152 mm) below the structural members and a maximum distance of 14 inches (356 mm) below the deck of the exterior balconies and decks that are constructed of open wood joist construction.

903.3.1.2.2 Open-ended corridors. Sprinkler protection shall be provided in *open-ended corridors* and associated *exterior stairways* and *ramps* as specified in Section 1027.6, Exception 3.

903.3.1.3 NFPA 13D sprinkler systems. *Automatic sprinkler systems* installed in one- and two-family *dwellings*; Group R-3; Group R-4 Condition 1 and *townhouses* shall be permitted to be installed throughout in accordance with NFPA 13D.

903.3.2 Quick-response and residential sprinklers. Where *automatic sprinkler systems* are required by this code, quick-response or residential automatic sprinklers shall be installed in all of the following areas in accordance with Section 903.3.1 and their listings:

1. Throughout all spaces within a smoke compartment containing care recipient *sleeping units* in Group I-2 in accordance with the *International Building Code.*
2. Throughout all spaces within a smoke compartment containing treatment rooms in ambulatory care facilities.
3. *Dwelling units* and *sleeping units* in Group I-1 and R occupancies.
4. Light-hazard occupancies as defined in NFPA 13.

903.3.3 Obstructed locations. Automatic sprinklers shall be installed with due regard to obstructions that will delay activation or obstruct the water distribution pattern. Automatic sprinklers shall be installed in or under covered kiosks, displays, booths, concession stands or equipment that exceeds 4 feet (1219 mm) in width. Not less than a 3-foot (914 mm) clearance shall be maintained between automatic sprinklers and the top of piles of *combustible fibers.*

Exception: Kitchen equipment under exhaust hoods protected with a fire-extinguishing system in accordance with Section 904.

903.3.4 Actuation. *Automatic sprinkler systems* shall be automatically actuated unless specifically provided for in this code.

903.3.5 Water supplies. Water supplies for *automatic sprinkler systems* shall comply with this section and the standards referenced in Section 903.3.1. The potable water supply shall be protected against backflow in accordance with the requirements of this section and the *International Plumbing Code*. For connections to public waterworks systems, the water supply test used for design of fire protection systems shall be adjusted to account for seasonal and daily pressure fluctuations based on information from the water supply authority and as approved by the *fire code official*.

903.3.5.1 Domestic services. Where the domestic service provides the water supply for the *automatic sprinkler system*, the supply shall be in accordance with this section.

903.3.5.2 Residential combination services. A single combination water supply shall be allowed provided that the domestic demand is added to the sprinkler demand as required by NFPA 13R. *

903.3.6 Hose threads. Fire hose threads and fittings used in connection with *automatic sprinkler systems* shall be as prescribed by the *fire code official*.

903.3.7 Fire department connections. Fire department connections for *automatic sprinkler systems* shall be installed in accordance with Section 912.

903.3.8 Limited area sprinkler systems. Limited area sprinkler systems shall be in accordance with the standards listed in Section 903.3.1 except as provided in Sections 903.3.8.1 through 903.3.8.5.

903.3.8.1 Number of sprinklers. Limited area sprinkler systems shall not exceed six sprinklers in any single fire area.

903.3.8.2 Occupancy hazard classification. Only areas classified by NFPA 13 as Light Hazard or Ordinary Hazard Group 1 shall be permitted to be protected by limited area sprinkler systems.

903.3.8.3 Piping arrangement. Where a limited area sprinkler system is installed in a building with an automatic wet standpipe system, sprinklers shall be supplied by the standpipe system. Where a limited area sprinkler system is installed in a building without an automatic wet standpipe system, water shall be permitted to be supplied by the plumbing system provided that the plumbing system is capable of simultaneously supplying domestic and sprinkler demands.

903.3.8.4 Supervision. Control valves shall not be installed between the water supply and sprinklers unless the valves are of an *approved* indicating type that are supervised or secured in the open position.

903.3.8.5 Calculations. Hydraulic calculations in accordance with NFPA 13 shall be provided to demonstrate that the available water flow and pressure are adequate to supply all sprinklers installed in any single *fire area* with discharge densities corresponding to the hazard classification.

903.4 Sprinkler system supervision and alarms. Valves controlling the water supply for *automatic sprinkler systems*, pumps, tanks, water levels and temperatures, critical air pressures and waterflow switches on all sprinkler systems shall be electrically supervised by a *listed* fire alarm control unit.

Exceptions:

1. *Automatic sprinkler systems* protecting one- and two-family *dwellings*.
2. Limited area sprinkler systems in accordance with Section 903.3.8.
3. *Automatic sprinkler systems* installed in accordance with NFPA 13R where a common supply main is used to supply both domestic water and the *automatic sprinkler system*, and a separate shutoff valve for the *automatic sprinkler system* is not provided.
4. Jockey pump control valves that are sealed or locked in the open position.
5. Control valves to commercial kitchen hoods, paint spray booths or dip tanks that are sealed or locked in the open position.
6. Valves controlling the fuel supply to fire pump engines that are sealed or locked in the open position.
7. Trim valves to pressure switches in dry, preaction and deluge sprinkler systems that are sealed or locked in the open position.

903.4.1 Monitoring. Alarm, supervisory and trouble signals shall be distinctly different and shall be automatically transmitted to an *approved* supervising station or, where *approved* by the *fire code official*, shall sound an audible signal at a constantly attended location.

Exceptions:

1. Underground key or hub valves in roadway boxes provided by the municipality or public utility are not required to be monitored.
2. Backflow prevention device test valves located in limited area sprinkler system supply piping shall be locked in the open position. In occupancies required to be equipped with a fire alarm system, the backflow preventer valves shall be electrically supervised by a tamper switch installed in accordance with NFPA 72 and separately annunciated.

903.4.2 Alarms. An approved audible device, located on the exterior of the building in an *approved* location, shall be connected to each *automatic sprinkler system*. Such sprinkler waterflow alarm devices shall be activated by water flow equivalent to the flow of a single sprinkler of the smallest orifice size installed in the system. Where a fire alarm system is installed, actuation of the *automatic sprinkler system* shall actuate the building fire alarm system.

903.4.3 Floor control valves. *Approved* supervised indicating control valves shall be provided at the point of connection to the riser on each floor in high-rise buildings.

903.5 Testing and maintenance. Sprinkler systems shall be tested and maintained in accordance with Section 901.

903.6 Where required in existing buildings and structures. An *automatic sprinkler system* shall be provided in existing buildings and structures where required in Chapter 11.

SECTION 904 ALTERNATIVE AUTOMATIC FIRE-EXTINGUISHING SYSTEMS

904.1 General. Automatic fire-extinguishing systems, other than *automatic sprinkler systems*, shall be designed, installed, inspected, tested and maintained in accordance with the provisions of this section and the applicable referenced standards.

904.1.1 Certification of service personnel for fire-extinguishing equipment. Service personnel providing or conducting maintenance on automatic fire-extinguishing systems, other than *automatic sprinkler systems*, shall possess a valid certificate issued by an *approved* governmental agency, or other *approved* organization for the type of system and work performed.

904.2 Where permitted. Automatic fire-extinguishing systems installed as an alternative to the required *automatic sprinkler systems* of Section 903 shall be *approved* by the *fire code official.*

904.2.1 Restriction on using automatic sprinkler system exceptions or reductions. Automatic fire-extinguishing systems shall not be considered alternatives for the purposes of exceptions or reductions allowed for *automatic sprinkler systems* or by other requirements of this code.

904.2.2 Commercial hood and duct systems. Each required commercial kitchen exhaust hood and duct system required by Section 609 to have a Type I hood shall be protected with an *approved* automatic fire-extinguishing system installed in accordance with this code.

904.3 Installation. Automatic fire-extinguishing systems shall be installed in accordance with this section.

904.3.1 Electrical wiring. Electrical wiring shall be in accordance with NFPA 70.

904.3.2 Actuation. Automatic fire-extinguishing systems shall be automatically actuated and provided with a manual means of actuation in accordance with Section 904.11.1.

Where more than one hazard could be simultaneously involved in fire due to their proximity, all hazards shall be protected by a single system designed to protect all hazards that could become involved.

Exception: Multiple systems shall be permitted to be installed if they are designed to operate simultaneously.

904.3.3 System interlocking. Automatic equipment interlocks with fuel shutoffs, ventilation controls, door closers, window shutters, conveyor openings, smoke and heat vents and other features necessary for proper operation of the fire-extinguishing system shall be provided as required by the design and installation standard utilized for the hazard.

904.3.4 Alarms and warning signs. Where alarms are required to indicate the operation of automatic fire-extinguishing systems, distinctive audible, visible alarms and warning signs shall be provided to warn of pending agent discharge. Where exposure to automatic-extinguishing agents poses a hazard to persons and a delay is required to ensure the evacuation of occupants before agent discharge, a separate warning signal shall be provided to alert occupants once agent discharge has begun. Audible signals shall be in accordance with Section 907.5.2.

904.3.5 Monitoring. Where a building fire alarm system is installed, automatic fire-extinguishing systems shall be monitored by the building fire alarm system in accordance with NFPA 72.

904.4 Inspection and testing. Automatic fire-extinguishing systems shall be inspected and tested in accordance with the provisions of this section prior to acceptance.

904.4.1 Inspection. Prior to conducting final acceptance tests, all of the following items shall be inspected:

1. Hazard specification for consistency with design hazard.
2. Type, location and spacing of automatic- and manual-initiating devices.
3. Size, placement and position of nozzles or discharge orifices.
4. Location and identification of audible and visible alarm devices.
5. Identification of devices with proper designations.
6. Operating instructions.

904.4.2 Alarm testing. Notification appliances, connections to fire alarm systems and connections to *approved* supervising stations shall be tested in accordance with this section and Section 907 to verify proper operation.

904.4.2.1 Audible and visible signals. The audibility and visibility of notification appliances signaling agent discharge or system operation, where required, shall be verified.

904.4.3 Monitor testing. Connections to protected premises and supervising station fire alarm systems shall be tested to verify proper identification and retransmission of alarms from automatic fire-extinguishing systems.

904.5 Wet-chemical systems. Wet-chemical extinguishing systems shall be installed, maintained, periodically inspected and tested in accordance with NFPA 17A and their listing. Records of inspections and testing shall be maintained.

904.5.1 System test. Systems shall be inspected and tested for proper operation at six-month intervals. Tests shall include a check of the detection system, alarms and releasing devices, including manual stations and other associated equipment. Extinguishing system units shall be weighed and the required amount of agent verified. Stored pressure-type units shall be checked for the required pressure. The cartridge of cartridge-operated units shall be weighed and replaced at intervals indicated by the manufacturer.

904.5.2 Fusible link maintenance. Fixed temperature-sensing elements shall be maintained to ensure proper operation of the system.

904.6 Dry-chemical systems. Dry-chemical extinguishing systems shall be installed, maintained, periodically inspected and tested in accordance with NFPA 17 and their listing. Records of inspections and testing shall be maintained.

904.6.1 System test. Systems shall be inspected and tested for proper operation at six-month intervals. Tests shall include a check of the detection system, alarms and releasing devices, including manual stations and other associated equipment. Extinguishing system units shall be weighed, and the required amount of agent verified. Stored pressure-type units shall be checked for the required pressure. The cartridge of cartridge-operated units shall be weighed and replaced at intervals indicated by the manufacturer.

904.6.2 Fusible link maintenance. Fixed temperature-sensing elements shall be maintained to ensure proper operation of the system.

904.7 Foam systems. Foam-extinguishing systems shall be installed, maintained, periodically inspected and tested in accordance with NFPA 11 and NFPA 16 and their listing. Records of inspections and testing shall be maintained.

904.7.1 System test. Foam-extinguishing systems shall be inspected and tested at intervals in accordance with NFPA 25.

904.8 Carbon dioxide systems. Carbon dioxide extinguishing systems shall be installed, maintained, periodically inspected and tested in accordance with NFPA 12 and their listing. Records of inspections and testing shall be maintained.

904.8.1 System test. Systems shall be inspected and tested for proper operation at 12-month intervals.

904.8.2 High-pressure cylinders. High-pressure cylinders shall be weighed and the date of the last hydrostatic test shall be verified at six-month intervals. Where a container shows a loss in original content of more than 10 percent, the cylinder shall be refilled or replaced.

904.8.3 Low-pressure containers. The liquid-level gauges of low-pressure containers shall be observed at one-week intervals. Where a container shows a content loss of more than 10 percent, the container shall be refilled to maintain the minimum gas requirements.

904.8.4 System hoses. System hoses shall be examined at 12-month intervals for damage. Damaged hoses shall be replaced or tested. At five-year intervals, all hoses shall be tested.

904.8.4.1 Test procedure. Hoses shall be tested at not less than 2,500 pounds per square inch (psi) (17 238 kPa) for high-pressure systems and at not less than 900 psi (6206 kPa) for low-pressure systems.

904.8.5 Auxiliary equipment. Auxiliary and supplementary components, such as switches, door and window releases, interconnected valves, damper releases and supplementary alarms, shall be manually operated at 12-month intervals to ensure that such components are in proper operating condition.

904.9 Halon systems. Halogenated extinguishing systems shall be installed, maintained, periodically inspected and tested in accordance with NFPA 12A and their listing. Records of inspections and testing shall be maintained.

904.9.1 System test. Systems shall be inspected and tested for proper operation at 12-month intervals.

904.9.2 Containers. The extinguishing agent quantity and pressure of containers shall be checked at six-month intervals. Where a container shows a loss in original weight of more than 5 percent or a loss in original pressure (adjusted for temperature) of more than 10 percent, the container shall be refilled or replaced. The weight and pressure of the container shall be recorded on a tag attached to the container.

904.9.3 System hoses. System hoses shall be examined at 12-month intervals for damage. Damaged hoses shall be replaced or tested. At five-year intervals, all hoses shall be tested.

904.9.3.1 Test procedure. For Halon 1301 systems, hoses shall be tested at not less than 1,500 psi (10 343 kPa) for 600 psi (4137 kPa) charging pressure systems and not less than 900 psi (6206 kPa) for 360 psi (2482 kPa) charging pressure systems. For Halon 1211 hand-hose line systems, hoses shall be tested at 2,500 psi (17 238 kPa) for high-pressure systems and 900 psi (6206 kPa) for low-pressure systems.

904.9.4 Auxiliary equipment. Auxiliary and supplementary components, such as switches, door and window releases, interconnected valves, damper releases and supplementary alarms, shall be manually operated at 12-month intervals to ensure such components are in proper operating condition.

904.10 Clean-agent systems. Clean-agent fire-extinguishing systems shall be installed, maintained, periodically inspected and tested in accordance with NFPA 2001 and their listing. Records of inspections and testing shall be maintained.

904.10.1 System test. Systems shall be inspected and tested for proper operation at 12-month intervals.

904.10.2 Containers. The extinguishing agent quantity and pressure of the containers shall be checked at six-month intervals. Where a container shows a loss in original weight of more than 5 percent or a loss in original pressure, adjusted for temperature, of more than 10 percent, the container shall be refilled or replaced. The weight and pressure of the container shall be recorded on a tag attached to the container.

904.10.3 System hoses. System hoses shall be examined at 12-month intervals for damage. Damaged hoses shall be replaced or tested. All hoses shall be tested at five-year intervals.

904.11 Automatic water mist systems. *Automatic water mist systems* shall be permitted in applications that are consistent with the applicable listing or approvals and shall comply with Sections 904.11.1 through 904.11.3.

904.11.1 Design and installation requirements. *Automatic water mist systems* shall be designed and installed in accordance with Sections 904.11.1.1 through 904.11.1.4.

904.11.1.1 General. *Automatic water mist systems* shall be designed and installed in accordance with NFPA 750 and the manufacturer's instructions.

904.11.1.2 Actuation. *Automatic water mist systems* shall be automatically actuated.

904.11.1.3 Water supply protection. Connections to a potable water supply shall be protected against backflow in accordance with the *International Plumbing Code*.

904.11.1.4 Secondary water supply. Where a secondary water supply is required for an *automatic sprinkler system*, an *automatic water mist system* shall be provided with an *approved* secondary water supply.

904.11.2 Water mist system supervision and alarms. Supervision and alarms shall be provided as required for *automatic sprinkler systems* in accordance with Section 903.4.

904.11.2.1 Monitoring. Monitoring shall be provided as required for *automatic sprinkler systems* in accordance with Section 903.4.1.

904.11.2.2 Alarms. Alarms shall be provided as required for *automatic sprinkler systems* in accordance with Section 903.4.2.

904.11.2.3 Floor control valves. Floor control valves shall be provided as required for *automatic sprinkler systems* in accordance with Section 903.4.3.

904.11.3 Testing and maintenance. *Automatic water mist systems* shall be tested and maintained in accordance with Section 901.6.

904.12 Commercial cooking systems. The automatic fire-extinguishing system for commercial cooking systems shall be of a type recognized for protection of commercial cooking equipment and exhaust systems of the type and arrangement protected. Preengineered automatic dry- and wet-chemical extinguishing systems shall be tested in accordance with UL 300 and *listed* and *labeled* for the intended application. Other types of automatic fire-extinguishing systems shall be *listed* and *labeled* for specific use as protection for commercial cooking operations. The system shall be installed in accordance with this code, its listing and the manufacturer's installation instructions. Automatic fire-extinguishing systems of the following types shall be installed in accordance with the referenced standard indicated, as follows:

1. Carbon dioxide extinguishing systems, NFPA 12.
2. *Automatic sprinkler systems*, NFPA 13.
3. Foam-water sprinkler system or foam-water spray systems, NFPA 16.
4. Dry-chemical extinguishing systems, NFPA 17.
5. Wet-chemical extinguishing systems, NFPA 17A.

Exception: Factory-built commercial cooking recirculating systems that are tested in accordance with UL 710B and *listed*, *labeled* and installed in accordance with Section 304.1 of the *International Mechanical Code*.

904.12.1 Manual system operation. A manual actuation device shall be located at or near a *means of egress* from the cooking area not less than 10 feet (3048 mm) and not more than 20 feet (6096 mm) from the kitchen exhaust system. The manual actuation device shall be installed not more than 48 inches (1200 mm) nor less than 42 inches (1067 mm) above the floor and shall clearly identify the hazard protected. The manual actuation shall require a maximum force of 40 pounds (178 N) and a maximum movement of 14 inches (356 mm) to actuate the fire suppression system.

Exception: *Automatic sprinkler systems* shall not be required to be equipped with manual actuation means.

904.12.2 System interconnection. The actuation of the fire extinguishing system shall automatically shut down the fuel or electrical power supply to the cooking equipment. The fuel and electrical supply reset shall be manual.

904.12.3 Carbon dioxide systems. Where carbon dioxide systems are used, there shall be a nozzle at the top of the ventilating duct. Additional nozzles that are symmetrically arranged to give uniform distribution shall be installed within vertical ducts exceeding 20 feet (6096 mm) and horizontal ducts exceeding 50 feet (15 240 mm). Dampers shall be installed at either the top or the bottom of the duct and shall be arranged to operate automatically upon activation of the fire-extinguishing system. Where the damper is installed at the top of the duct, the top nozzle shall be immediately below the damper. Automatic carbon dioxide fire-extinguishing systems shall be sufficiently sized to protect all hazards venting through a common duct simultaneously.

904.12.3.1 Ventilation system. Commercial-type cooking equipment protected by an automatic carbon dioxide extinguishing system shall be arranged to shut off the ventilation system upon activation.

904.12.4 Special provisions for automatic sprinkler systems. *Automatic sprinkler systems* protecting commercial-type cooking equipment shall be supplied from a separate, readily accessible, indicating-type control valve that is identified.

904.12.4.1 Listed sprinklers. Sprinklers used for the protection of fryers shall be tested in accordance with UL 199E, *listed* for that application and installed in accordance with their listing.

904.12.5 Portable fire extinguishers for commercial cooking equipment. Portable fire extinguishers shall be provided within a 30-foot (9144 mm) distance of travel from commercial-type cooking equipment. Cooking equipment involving solid fuels or vegetable or animal oils and fats shall be protected by a Class K rated portable extinguisher in accordance with Section 904.12.5.1 or 904.12.5.2, as applicable.

904.12.5.1 Portable fire extinguishers for solid fuel cooking appliances. Solid fuel cooking appliances, whether or not under a hood, with fireboxes 5 cubic feet (0.14 m^3) or less in volume shall have a minimum 2.5-gallon (9 L) or two 1.5-gallon (6 L) Class K wet-chemical portable fire extinguishers located in accordance with Section 904.12.5.

904.12.5.2 Class K portable fire extinguishers for deep fat fryers. Where hazard areas include deep fat fryers, listed Class K portable fire extinguishers shall be provided as follows:

1. For up to four fryers having a maximum cooking medium capacity of 80 pounds (36.3 kg) each: one Class K portable fire extinguisher of a minimum 1.5-gallon (6 L) capacity.
2. For every additional group of four fryers having a maximum cooking medium capacity of 80 pounds (36.3 kg) each: one additional Class K portable fire extinguisher of a minimum 1.5-gallon (6 L) capacity shall be provided.
3. For individual fryers exceeding 6 square feet (0.55 m^2) in surface area: Class K portable fire extinguishers shall be installed in accordance with the extinguisher manufacturer's recommendations.

904.12.6 Operations and maintenance. Automatic fire-extinguishing systems protecting commercial cooking systems shall be maintained in accordance with Sections 904.12.6.1 through 904.12.6.3.

904.12.6.1 Existing automatic fire-extinguishing systems. Where changes in the cooking media, positioning of cooking equipment or replacement of cooking equipment occur in existing commercial cooking systems, the automatic fire-extinguishing system shall be required to comply with the applicable provisions of Sections 904.12 through 904.12.4.

904.12.6.2 Extinguishing system service. Automatic fire-extinguishing systems shall be serviced at least every six months and after activation of the system. Inspection shall be by qualified individuals, and a certificate of inspection shall be forwarded to the *fire code official* upon completion.

904.12.6.3 Fusible link and sprinkler head replacement. Fusible links and automatic sprinkler heads shall be replaced at least annually, and other protection devices shall be serviced or replaced in accordance with the manufacturer's instructions.

Exception: Frangible bulbs are not required to be replaced annually.

904.13 Domestic cooking systems in Group I-2 Condition 1. In Group I-2 Condition 1 occupancies where cooking facilities are installed in accordance with Section 407.2.6 of the *International Building Code*, the domestic cooking hood provided over the cooktop or range shall be equipped with an automatic fire-extinguishing system of a type recognized for protection of domestic cooking equipment. Preengineered automatic extinguishing systems shall be tested in accordance with UL 300A and listed and labeled for the intended application. The system shall be installed in accordance with this code, its listing and the manufacturer's instructions.

904.13.1 Manual system operation and interconnection. Manual actuation and system interconnection for the hood suppression system shall be in accordance with Sections 904.12.1 and 904.12.2, respectively.

904.13.2 Portable fire extinguishers for domestic cooking equipment in Group I-2 Condition 1. A portable fire extinguisher complying with Section 906 shall be installed within a 30-foot (9144 mm) distance of travel from domestic cooking appliances.

SECTION 905 STANDPIPE SYSTEMS

905.1 General. Standpipe systems shall be provided in new buildings and structures in accordance with Sections 905.2 through 905.10. In buildings used for *high-piled combustible storage*, fire protection shall be in accordance with Chapter 32.

905.2 Installation standard. Standpipe systems shall be installed in accordance with this section and NFPA 14. Fire department connections for standpipe systems shall be in accordance with Section 912.

905.3 Required Installations. Standpipe systems shall be installed where required by Sections 905.3.1 through 905.3.8. Standpipe systems are allowed to be combined with *automatic sprinkler systems*.

Exception: Standpipe systems are not required in Group R-3 occupancies.

905.3.1 Height. Class III standpipe systems shall be installed throughout buildings where the floor level of the highest story is located more than 30 feet (9144 mm) above the lowest level of the fire department vehicle access, or where the floor level of the lowest story is located more than 30 feet (9144 mm) below the highest level of fire department vehicle access.

Exceptions:

1. Class I standpipes are allowed in buildings equipped throughout with an *automatic sprinkler system* in accordance with Section 903.3.1.1 or 903.3.1.2.
2. Class I manual standpipes are allowed in open parking garages where the highest floor is located not more than 150 feet (45 720 mm) above the lowest level of fire department vehicle access.
3. Class I manual dry standpipes are allowed in open parking garages that are subject to freezing temperatures, provided that the hose connections are located as required for Class II standpipes in accordance with Section 905.5.
4. Class I standpipes are allowed in *basements* equipped throughout with an *automatic sprinkler system*.
5. In determining the lowest level of fire department vehicle access, it shall not be required to consider either of the following:
   5.1. Recessed loading docks for four vehicles or less.
   5.2. Conditions where topography makes access from the fire department vehicle to the building impractical or impossible.

905.3.2 Group A. Class I automatic wet standpipes shall be provided in nonsprinklered Group A buildings having an *occupant load* exceeding 1,000 persons.

Exceptions:

1. Open-air-seating spaces without enclosed spaces.
2. Class I automatic dry and semiautomatic dry standpipes or manual wet standpipes are allowed in buildings that are not high-rise buildings.

905.3.3 Covered and open mall buildings. Covered mall and open mall buildings shall be equipped throughout with a standpipe system where required by Section 905.3.1. Mall buildings not required to be equipped with a standpipe system by Section 905.3.1 shall be equipped with Class I hose connections connected to the *automatic sprinkler system* sized to deliver water at 250 gallons per minute (946.4 L/min) at the most hydraulically remote hose connection while concurrently supplying the automatic sprinkler system demand. The standpipe system shall be designed not to exceed a 50 pounds per square inch (psi) (345 kPa) residual pressure loss with a flow of 250 gallons per minute (946.4 L/min) from the fire department connection to the hydraulically most remote hose connection. Hose connections shall be provided at each of the following locations:

1. Within the mall at the entrance to each exit passageway or corridor.
2. At each floor-level landing within *interior exit stairways* opening directly on the mall.
3. At exterior public entrances to the mall of a covered mall building
4. At public entrances at the perimeter line of an open mall building.

5. At other locations as necessary so that the distance to reach all portions of a tenant space does not exceed 200 feet (60 960 mm) from a hose connection.

905.3.4 Stages. Stages greater than 1,000 square feet (93 m^2) in area shall be equipped with a Class III wet standpipe system with $1^1/_2$-inch and $2^1/_2$-inch (38 mm and 64 mm) hose connections on each side of the stage.

Exception: Where the building or area is equipped throughout with an *automatic sprinkler system*, a $1^1/_2$-inch (38 mm) hose connection shall be installed in accordance with NFPA 13 or in accordance with NFPA 14 for Class II or III standpipes.

905.3.4.1 Hose and cabinet. The $1^1/_2$-inch (38 mm) hose connections shall be equipped with sufficient lengths of $1^1/_2$-inch (38 mm) hose to provide fire protection for the stage area. Hose connections shall be equipped with an *approved* adjustable fog nozzle and be mounted in a cabinet or on a rack.

905.3.5 Underground buildings. Underground buildings shall be equipped throughout with a Class I automatic wet or manual wet standpipe system.

905.3.6 Helistops and heliports. Buildings with a rooftop *helistop* or *heliport* shall be equipped with a Class I or III standpipe system extended to the roof level on which the *helistop* or *heliport* is located in accordance with Section 2007.5.

905.3.7 Marinas and boatyards. Standpipes in marinas and boatyards shall comply with Chapter 36.

905.3.8 Rooftop gardens and landscaped roofs. Buildings or structures that have rooftop gardens or landscaped roofs and that are equipped with a standpipe system shall have the standpipe system extended to the roof level on which the rooftop garden or landscaped roof is located.

905.4 Location of Class I standpipe hose connections. Class I standpipe hose connections shall be provided in all of the following locations:

1. In every required *interior exit stairway*, a hose connection shall be provided for each story above and below grade plane. Hose connections shall be located at an intermediate landing between stories, unless otherwise *approved* by the *fire code official*.
2. On each side of the wall adjacent to the *exit* opening of a horizontal *exit*.

   Exception: Where floor areas adjacent to a horizontal *exit* are reachable from an *interior exit stairway* hose connection by a 30-foot (9144 mm) hose stream from a nozzle attached to 100 feet (30 480 mm) of hose, a hose connection shall not be required at the horizontal *exit*.
3. In every *exit* passageway, at the entrance from the exit passageway to other areas of a building.

   Exception: Where floor areas adjacent to an exit passageway are reachable from an *interior exit stairway* hose connection by a 30-foot (9144 mm) hose stream from a nozzle attached to 100 feet (30 480 mm) of hose, a hose connection shall not be required at the entrance from the exit passageway to other areas of the building.
4. In covered mall buildings, adjacent to each exterior public entrance to the mall and adjacent to each entrance from an *exit* passageway or *exit corridor* to the mall. In open mall buildings, adjacent to each public entrance to the mall at the perimeter line and adjacent to each entrance from an exit passageway or exit corridor to the mall.
5. Where the roof has a slope less than four units vertical in 12 units horizontal (33.3-percent slope), a hose connection shall be located to serve the roof or at the highest landing of an *interior exit stairway* with access to the roof provided in accordance with Section 1011.12.
6. Where the most remote portion of a nonsprinklered floor or story is more than 150 feet (45 720 mm) from a hose connection or the most remote portion of a sprinklered floor or story is more than 200 feet (60 960 mm) from a hose connection, the *fire code official* is authorized to require that additional hose connections be provided in *approved* locations.

905.4.1 Protection. Risers and laterals of Class I standpipe systems not located within an *interior exit stairway* shall be protected by a degree of *fire resistance* equal to that required for vertical enclosures in the building in which they are located.

Exception: In buildings equipped throughout with an *approved automatic sprinkler system*, laterals that are not located within an *interior exit stairway* are not required to be enclosed within fire-resistance-rated construction.

905.4.2 Interconnection. In buildings where more than one standpipe is provided, the standpipes shall be interconnected in accordance with NFPA 14.

905.5 Location of Class II standpipe hose connections. Class II standpipe hose connections shall be accessible and shall be located so that all portions of the building are within 30 feet (9144 mm) of a nozzle attached to 100 feet (30 480 mm) of hose.

905.5.1 Groups A-1 and A-2. In Group A-1 and A-2 occupancies with *occupant loads* of more than 1,000, hose connections shall be located on each side of any stage, on each side of the rear of the auditorium, on each side of the balcony and on each tier of dressing rooms.

905.5.2 Protection. Fire-resistance-rated protection of risers and laterals of Class II standpipe systems is not required.

905.5.3 Class II system 1-inch hose. A minimum 1-inch (25 mm) hose shall be allowed to be used for hose stations in light-hazard occupancies where investigated and *listed* for this service and where *approved* by the *fire code official*.

905.6 Location of Class III standpipe hose connections. Class III standpipe systems shall have hose connections located as required for Class I standpipes in Section 905.4 and shall have Class II hose connections as required in Section 905.5.

905.6.1 Protection. Risers and laterals of Class III standpipe systems shall be protected as required for Class I systems in accordance with Section 905.4.1.

905.6.2 Interconnection. In buildings where more than one Class III standpipe is provided, the standpipes shall be interconnected in accordance with NFPA 14.

905.7 Cabinets. Cabinets containing fire-fighting equipment, such as standpipes, fire hose, fire extinguishers or fire department valves, shall not be blocked from use or obscured from view.

905.7.1 Cabinet equipment identification. Cabinets shall be identified in an *approved* manner by a permanently attached sign with letters not less than 2 inches (51 mm) high in a color that contrasts with the background color, indicating the equipment contained therein.

Exceptions:

1. Doors not large enough to accommodate a written sign shall be marked with a permanently attached pictogram of the equipment contained therein.
2. Doors that have either an *approved* visual identification clear glass panel or a complete glass door panel are not required to be marked.

905.7.2 Locking cabinet doors. Cabinets shall be unlocked.

Exceptions:

1. Visual identification panels of glass or other *approved* transparent frangible material that is easily broken and allows access.
2. *Approved* locking arrangements.
3. Group I-3 occupancies.

905.8 Dry standpipes. Dry standpipes shall not be installed.

Exception: Where subject to freezing and in accordance with NFPA 14.

905.9 Valve supervision. Valves controlling water supplies shall be supervised in the open position so that a change in the normal position of the valve will generate a supervisory signal at the supervising station required by Section 903.4. Where a fire alarm system is provided, a signal shall be transmitted to the control unit.

Exceptions:

1. Valves to underground key or hub valves in roadway boxes provided by the municipality or public utility do not require supervision.
2. Valves locked in the normal position and inspected as provided in this code in buildings not equipped with a fire alarm system.

905.10 During construction. Standpipe systems required during construction and demolition operations shall be provided in accordance with Section 3313.

905.11 Existing buildings. Where required in Chapter 11, existing structures shall be equipped with standpipes installed in accordance with Section 905.

SECTION 906
PORTABLE FIRE EXTINGUISHERS

906.1 Where required. Portable fire extinguishers shall be installed in all of the following locations:

1. In new and existing Group A, B, E, F, H, I, M, R-1, R-2, R-4 and S occupancies.

   Exception: In Group R-2 occupancies, portable fire extinguishers shall be required only in locations specified in Items 2 through 6 where each *dwelling unit* is provided with a portable fire extinguisher having a minimum rating of 1-A:10-B:C.
2. Within 30 feet (9144 mm) of commercial cooking equipment.
3. In areas where flammable or *combustible liquids* are stored, used or dispensed.
4. On each floor of structures under construction, except Group R-3 occupancies, in accordance with Section 3315.1.
5. Where required by the sections indicated in Table 906.1.
6. Special-hazard areas, including but not limited to laboratories, computer rooms and generator rooms, where required by the *fire code official.*

906.2 General requirements. Portable fire extinguishers shall be selected, installed and maintained in accordance with this section and NFPA 10.

Exceptions:

1. The distance of travel to reach an extinguisher shall not apply to the spectator seating portions of Group A-5 occupancies.
2. Thirty-day inspections shall not be required and maintenance shall be allowed to be once every 3 years for dry-chemical or halogenated agent portable fire extinguishers that are supervised by a *listed* and *approved* electronic monitoring device, provided that all of the following conditions are met:
   - 2.1. Electronic monitoring shall confirm that extinguishers are properly positioned, properly charged and unobstructed.
   - 2.2. Loss of power or circuit continuity to the electronic monitoring device shall initiate a trouble signal.

**TABLE 906.1
ADDITIONAL REQUIRED PORTABLE FIRE EXTINGUISHERS**

SECTION	SUBJECT
303.5	Asphalt kettles
307.5	Open burning
308.1.3	Open flames—torches
309.4	Powered industrial trucks
2005.2	Aircraft towing vehicles
2005.3	Aircraft welding apparatus
2005.4	Aircraft fuel-servicing tank vehicles
2005.5	Aircraft hydrant fuel-servicing vehicles
2005.6	Aircraft fuel-dispensing stations
2007.7	Heliports and helistops
2108.4	Dry cleaning plants
2305.5	Motor fuel-dispensing facilities
2310.6.4	Marine motor fuel-dispensing facilities
2311.6	Repair garages
2404.4.1	Spray-finishing operations
2405.4.2	Dip-tank operations
2406.4.2	Powder-coating areas
2804.3	Lumberyards/woodworking facilities
2808.8	Recycling facilities
2809.5	Exterior lumber storage
2903.5	Organic-coating areas
3006.3	Industrial ovens
3104.12	Tents and membrane structures
3206.10	High-piled storage
3315.1	Buildings under construction or demolition
3317.3	Roofing operations
3408.2	Tire rebuilding/storage
3504.2.6	Welding and other hot work
3604.4	Marinas
3703.6	Combustible fibers
5703.2.1	Flammable and combustible liquids, general
5704.3.3.1	Indoor storage of flammable and combustible liquids
5704.3.7.5.2	Liquid storage rooms for flammable and combustible liquids
5705.4.9	Solvent distillation units
5706.2.7	Farms and construction sites—flammable and combustible liquids storage
5706.4.10.1	Bulk plants and terminals for flammable and combustible liquids
5706.5.4.5	Commercial, industrial, governmental or manufacturing establishments—fuel dispensing
5706.6.4	Tank vehicles for flammable and combustible liquids
5906.5.7	Flammable solids
6108.2	LP-gas

2.3. The extinguishers shall be installed inside of a building or cabinet in a noncorrosive environment.

2.4. Electronic monitoring devices and supervisory circuits shall be tested every 3 years when extinguisher maintenance is performed.

2.5. A written log of required hydrostatic test dates for extinguishers shall be maintained by the *owner* to verify that hydrostatic tests are conducted at the frequency required by NFPA 10.

3. In Group I-3, portable fire extinguishers shall be permitted to be located at staff locations.

906.2.1 Certification of service personnel for portable fire extinguishers. Service personnel providing or conducting maintenance on portable fire extinguishers shall possess a valid certificate issued by an *approved* governmental agency, or other *approved* organization for the type of work performed.

906.3 Size and distribution. The size and distribution of portable fire extinguishers shall be in accordance with Sections 906.3.1 through 906.3.4.

906.3.1 Class A fire hazards. Portable fire extinguishers for occupancies that involve primarily Class A fire hazards, the minimum sizes and distribution shall comply with Table 906.3(1).

**TABLE 906.3(1)
FIRE EXTINGUISHERS FOR CLASS A FIRE HAZARDS**

	LIGHT (Low) HAZARD OCCUPANCY	ORDINARY (Moderate) HAZARD OCCUPANCY	EXTRA (High) HAZARD OCCUPANCY
Minimum rated single extinguisher	2-A[c]	2-A	4-A[a]
Maximum floor area per unit of A	3,000 square feet	1,500 square feet	1,000 square feet
Maximum floor area for extinguisher[b]	11,250 square feet	11,250 square feet	11,250 square feet
Maximum distance of travel to extinguisher	75 feet	75 feet	75 feet

For SI: 1 foot = 304.8 mm, 1 square foot = 0.0929 m^2, 1 gallon = 3.785 L.

a. Two $2^1/_2$-gallon water-type extinguishers shall be deemed the equivalent of one 4-A rated extinguisher.

b. Annex E.3.3 of NFPA 10 provides more details concerning application of the maximum floor area criteria.

c. Two water-type extinguishers each with a 1-A rating shall be deemed the equivalent of one 2-A rated extinguisher for Light (Low) Hazard Occupancies.

906.3.2 Class B fire hazards. Portable fire extinguishers for occupancies involving flammable or *combustible liquids* with depths of less than or equal to 0.25-inch (6.4 mm) shall be selected and placed in accordance with Table 906.3(2).

Portable fire extinguishers for occupancies involving flammable or *combustible liquids* with a depth of greater than 0.25-inch (6.4 mm) shall be selected and placed in accordance with NFPA 10.

**TABLE 906.3(2)
FLAMMABLE OR COMBUSTIBLE LIQUIDS WITH DEPTHS OF LESS THAN OR EQUAL TO 0.25-INCH[a]**

TYPE OF HAZARD	BASIC MINIMUM EXTINGUISHER RATING	MAXIMUM DISTANCE OF TRAVEL TO EXTINGUISHERS (feet)
Light (Low)	5-B 10-B	30 50
Ordinary (Moderate)	10-B 20-B	30 50
Extra (High)	40-B 80-B	30 50

For SI: 1 inch = 25.4 mm, 1 foot = 304.8 mm.

a. For requirements on water-soluble flammable liquids and alternative sizing criteria, see Section 5.5 of NFPA 10.

906.3.3 Class C fire hazards. Portable fire extinguishers for Class C fire hazards shall be selected and placed on the basis of the anticipated Class A or B hazard.

906.3.4 Class D fire hazards. Portable fire extinguishers for occupancies involving combustible metals shall be selected and placed in accordance with NFPA 10.

906.4 Cooking grease fires. Fire extinguishers provided for the protection of cooking grease fires shall be of an *approved* type compatible with the automatic fire-extinguishing system agent and in accordance with Section 904.12.5.

906.5 Conspicuous location. Portable fire extinguishers shall be located in conspicuous locations where they will be readily accessible and immediately available for use. These locations shall be along normal paths of travel, unless the *fire code official* determines that the hazard posed indicates the need for placement away from normal paths of travel.

906.6 Unobstructed and unobscured. Portable fire extinguishers shall not be obstructed or obscured from view. In rooms or areas in which visual obstruction cannot be completely avoided, means shall be provided to indicate the locations of extinguishers.

906.7 Hangers and brackets. Hand-held portable fire extinguishers, not housed in cabinets, shall be installed on the hangers or brackets supplied. Hangers or brackets shall be securely anchored to the mounting surface in accordance with the manufacturer's installation instructions.

906.8 Cabinets. Cabinets used to house portable fire extinguishers shall not be locked.

Exceptions:

1. Where portable fire extinguishers subject to malicious use or damage are provided with a means of ready access.
2. In Group I-3 occupancies and in mental health areas in Group I-2 occupancies, access to portable fire extinguishers shall be permitted to be locked or to be located in staff locations provided the staff has keys.

906.9 Extinguisher installation. The installation of portable fire extinguishers shall be in accordance with Sections 906.9.1 through 906.9.3.

906.9.1 Extinguishers weighing 40 pounds or less. Portable fire extinguishers having a gross weight not exceeding 40 pounds (18 kg) shall be installed so that their tops are not more than 5 feet (1524 mm) above the floor.

906.9.2 Extinguishers weighing more than 40 pounds. Hand-held portable fire extinguishers having a gross weight exceeding 40 pounds (18 kg) shall be installed so that their tops are not more than 3.5 feet (1067 mm) above the floor.

906.9.3 Floor clearance. The clearance between the floor and the bottom of installed hand-held portable fire extinguishers shall be not less than 4 inches (102 mm).

906.10 Wheeled units. Wheeled fire extinguishers shall be conspicuously located in a designated location.

SECTION 907
FIRE ALARM AND DETECTION SYSTEMS

907.1 General. This section covers the application, installation, performance and maintenance of fire alarm systems and their components in new and existing buildings and structures. The requirements of Section 907.2 are applicable to new buildings and structures. The requirements of Section 907.9 are applicable to existing buildings and structures.

907.1.1 Construction documents. *Construction documents* for fire alarm systems shall be of sufficient clarity to indicate the location, nature and extent of the work proposed and show in detail that it will conform to the provisions of this code, the *International Building Code* and relevant laws, ordinances, rules and regulations, as determined by the *fire code official.*

907.1.2 Fire alarm shop drawings. Shop drawings for fire alarm systems shall be submitted for review and approval prior to system installation, and shall include, but not be limited to, all of the following where applicable to the system being installed:

1. A floor plan that indicates the use of all rooms.
2. Locations of alarm-initiating devices.
3. Locations of alarm notification appliances, including candela ratings for visible alarm notification appliances.
4. Design minimum audibility level for occupant notification.
5. Location of fire alarm control unit, transponders and notification power supplies.
6. Annunciators.
7. Power connection.
8. Battery calculations.
9. Conductor type and sizes.
10. Voltage drop calculations.
11. Manufacturers' data sheets indicating model numbers and listing information for equipment, devices and materials.

12. Details of ceiling height and construction.
13. The interface of fire safety control functions.
14. Classification of the supervising station.

907.1.3 Equipment. Systems and components shall be *listed* and *approved* for the purpose for which they are installed.

907.2 Where required—new buildings and structures. An *approved* fire alarm system installed in accordance with the provisions of this code and NFPA 72 shall be provided in new buildings and structures in accordance with Sections 907.2.1 through 907.2.23 and provide occupant notification in accordance with Section 907.5, unless other requirements are provided by another section of this code.

Not fewer than one manual fire alarm box shall be provided in an *approved* location to initiate a fire alarm signal for fire alarm systems employing automatic fire detectors or waterflow detection devices. Where other sections of this code allow elimination of fire alarm boxes due to sprinklers, a single fire alarm box shall be installed.

Exceptions:

1. The manual fire alarm box is not required for fire alarm systems dedicated to elevator recall control and supervisory service.
2. The manual fire alarm box is not required for Group R-2 occupancies unless required by the *fire code official* to provide a means for fire watch personnel to initiate an alarm during a sprinkler system impairment event. Where provided, the manual fire alarm box shall not be located in an area that is accessible to the public.

907.2.1 Group A. A manual fire alarm system that activates the occupant notification system in accordance with Section 907.5 shall be installed in Group A occupancies where the occupant load due to the assembly occupancy is 300 or more. Group A occupancies not separated from one another in accordance with Section 707.3.10 of the *International Building Code* shall be considered as a single occupancy for the purposes of applying this section. Portions of Group E occupancies occupied for assembly purposes shall be provided with a fire alarm system as required for the Group E occupancy.

Exception: Manual fire alarm boxes are not required where the building is equipped throughout with an automatic sprinkler system installed in accordance with Section 903.3.1.1 and the occupant notification appliances will activate throughout the notification zones upon sprinkler water flow.

907.2.1.1 System initiation in Group A occupancies with an occupant load of 1,000 or more. Activation of the fire alarm in Group A occupancies with an *occupant load* of 1,000 or more shall initiate a signal using an emergency voice/alarm communications system in accordance with Section 907.5.2.2.

Exception: Where *approved*, the prerecorded announcement is allowed to be manually deactivated for a period of time, not to exceed 3 minutes, for the sole purpose of allowing a live voice announcement from an *approved*, constantly attended location.

907.2.1.2 Emergency voice/alarm communication system captions. Stadiums, arenas and grandstands required to caption audible public announcements shall be in accordance with Section 907.5.2.2.4.

907.2.2 Group B. A manual fire alarm system shall be installed in Group B occupancies where one of the following conditions exists:

1. The combined Group B *occupant load* of all floors is 500 or more.
2. The Group B *occupant load* is more than 100 persons above or below the lowest *level of exit discharge*.
3. The *fire area* contains an ambulatory care facility.

Exception: Manual fire alarm boxes are not required where the building is equipped throughout with an *automatic sprinkler system* installed in accordance with Section 903.3.1.1 and the occupant notification appliances will activate throughout the notification zones upon sprinkler water flow.

907.2.2.1 Ambulatory care facilities. *Fire areas* containing ambulatory care facilities shall be provided with an electronically supervised automatic smoke detection system installed within the ambulatory care facility and in public use areas outside of tenant spaces, including public *corridors* and elevator lobbies.

Exception: Buildings equipped throughout with an *automatic sprinkler system* in accordance with Section 903.3.1.1 provided the occupant notification appliances will activate throughout the notification zones upon sprinkler water flow.

907.2.3 Group E. A manual fire alarm system that initiates the occupant notification signal utilizing an emergency voice/alarm communication system meeting the requirements of Section 907.5.2.2 and installed in accordance with Section 907.6 shall be installed in Group E occupancies. When *automatic sprinkler systems* or smoke detectors are installed, such systems or detectors shall be connected to the building fire alarm system.

Exceptions:

1. A manual fire alarm system is not required in Group E occupancies with an *occupant load* of 50 or less.
2. Emergency voice/alarm communication systems meeting the requirements of Section 907.5.2.2 and installed in accordance with Section 907.6 shall not be required in Group E occupancies with occupant loads of 100 or less, provided that activation of the manual fire alarm system initiates an *approved* occupant notification signal in accordance with Section 907.5.

3. Manual fire alarm boxes are not required in Group E occupancies where all of the following apply:
   - 3.1. Interior *corridors* are protected by smoke detectors.
   - 3.2. Auditoriums, cafeterias, gymnasiums and similar areas are protected by *heat detectors* or other *approved* detection devices.
   - 3.3. Shops and laboratories involving dusts or vapors are protected by *heat detectors* or other *approved* detection devices.
4. Manual fire alarm boxes shall not be required in Group E occupancies where all of the following apply:
   - 4.1. The building is equipped throughout with an *approved automatic sprinkler system* installed in accordance with Section 903.3.1.1.
   - 4.2. The emergency voice/alarm communication system will activate on sprinkler water flow.
   - 4.3. Manual activation is provided from a normally occupied location.

907.2.4 Group F. A manual fire alarm system that activates the occupant notification system in accordance with Section 907.5 shall be installed in Group F occupancies where both of the following conditions exist:

1. The Group F occupancy is two or more stories in height.
2. The Group F occupancy has a combined *occupant load* of 500 or more above or below the lowest *level of exit discharge*.

Exception: Manual fire alarm boxes are not required where the building is equipped throughout with an *automatic sprinkler system* installed in accordance with Section 903.3.1.1 and the occupant notification appliances will activate throughout the notification zones upon sprinkler water flow.

907.2.5 Group H. A manual fire alarm system that activates the occupant notification system in accordance with Section 907.5 shall be installed in Group H-5 occupancies and in occupancies used for the manufacture of organic coatings. An automatic smoke detection system shall be installed for highly toxic gases, organic peroxides and oxidizers in accordance with Chapters 60, 62 and 63, respectively.

907.2.6 Group I. A manual fire alarm system that activates the occupant notification system in accordance with Section 907.5 shall be installed in Group I occupancies. An automatic smoke detection system that activates the occupant notification system in accordance with Section 907.5 shall be provided in accordance with Sections 907.2.6.1, 907.2.6.2 and 907.2.6.3.3.

Exceptions:

1. Manual fire alarm boxes in *sleeping units* of Group I-1 and I-2 occupancies shall not be required at *exits* if located at all care providers' control stations or other constantly attended staff locations, provided such stations are visible and continuously accessible and that the distances of travel required in Section 907.4.2.1 are not exceeded.
2. Occupant notification systems are not required to be activated where private mode signaling installed in accordance with NFPA 72 is *approved* by the *fire code official* and staff evacuation responsibilities are included in the fire safety and evacuation plan required by Section 404.

907.2.6.1 Group I-1. An automatic smoke detection system shall be installed in *corridors*, waiting areas open to *corridors* and *habitable spaces* other than *sleeping units* and kitchens. The system shall be activated in accordance with Section 907.5.

Exceptions:

1. For Group I-1 Condition 1 occupancies, smoke detection in *habitable spaces* is not required where the facility is equipped throughout with an *automatic sprinkler system* installed in accordance with Section 903.3.1.1.
2. Smoke detection is not required for exterior balconies.

907.2.6.1.1 Smoke alarms. Single- and multiple-station smoke alarms shall be installed in accordance with Section 907.2.11.

907.2.6.2 Group I-2. An automatic smoke detection system shall be installed in *corridors* in Group I-2 Condition 1 facilities and spaces permitted to be open to the *corridors* by Section 407.2 of the *International Building Code*. The system shall be activated in accordance with Section 907.4. Group I-2 Condition 2 occupancies shall be equipped with an automatic smoke detection system as required in Section 407 of the *International Building Code*.

Exceptions:

1. *Corridor* smoke detection is not required in smoke compartments that contain *sleeping units* where such units are provided with smoke detectors that comply with UL 268. Such detectors shall provide a visual display on the *corridor* side of each *sleeping unit* and shall provide an audible and visual alarm at the care providers' station attending each unit.

2. *Corridor* smoke detection is not required in smoke compartments that contain *sleeping units* where *sleeping unit* doors are equipped with automatic door-closing devices with integral smoke detectors on the unit sides installed in accordance with their listing, provided that the integral detectors perform the required alerting function.

907.2.6.3 Group I-3 occupancies. Group I-3 occupancies shall be equipped with a manual fire alarm system and automatic smoke detection system installed for alerting staff.

907.2.6.3.1 System initiation. Actuation of an automatic fire-extinguishing system, *automatic sprinkler system*, a manual fire alarm box or a fire detector shall initiate an approved fire alarm signal that automatically notifies staff.

907.2.6.3.2 Manual fire alarm boxes. Manual fire alarm boxes are not required to be located in accordance with Section 907.4.2 where the fire alarm boxes are provided at staff-attended locations having direct supervision over areas where manual fire alarm boxes have been omitted.

907.2.6.3.2.1 Manual fire alarms boxes in detainee areas. Manual fire alarm boxes are allowed to be locked in areas occupied by detainees, provided that staff members are present within the subject area and have keys readily available to operate the manual fire alarm boxes.

907.2.6.3.3 Automatic smoke detection system. An automatic smoke detection system shall be installed throughout resident housing areas, including *sleeping units* and contiguous day rooms, group activity spaces and other common spaces normally accessible to residents.

Exceptions:

1. Other *approved* smoke detection arrangements providing equivalent protection, including, but not limited to, placing detectors in exhaust ducts from cells or behind protective guards *listed* for the purpose, are allowed when necessary to prevent damage or tampering.
2. *Sleeping units* in Use Conditions 2 and 3 as described in Section 308 of the *International Building Code*.
3. Smoke detectors are not required in *sleeping units* with four or fewer occupants in smoke compartments that are equipped throughout with an *automatic sprinkler system* installed in accordance with Section 903.3.1.1.

907.2.7 Group M. A manual fire alarm system that activates the occupant notification system in accordance with Section 907.5 shall be installed in Group M occupancies where one of the following conditions exists:

1. The combined Group M *occupant load* of all floors is 500 or more persons.
2. The Group M *occupant load* is more than 100 persons above or below the lowest *level of exit discharge*.

Exceptions:

1. A manual fire alarm system is not required in covered or open mall buildings complying with Section 402 of the *International Building Code*.
2. Manual fire alarm boxes are not required where the building is equipped throughout with an *automatic sprinkler system* installed in accordance with Section 903.3.1.1 and the occupant notification appliances will automatically activate throughout the notification zones upon sprinkler water flow.

907.2.7.1 Occupant notification. During times that the building is occupied, the initiation of a signal from a manual fire alarm box or from a waterflow switch shall not be required to activate the alarm notification appliances when an alarm signal is activated at a constantly attended location from which evacuation instructions shall be initiated over an emergency voice/alarm communication system installed in accordance with Section 907.5.2.2.

907.2.8 Group R-1. Fire alarm systems and smoke alarms shall be installed in Group R-1 occupancies as required in Sections 907.2.8.1 through 907.2.8.3.

907.2.8.1 Manual fire alarm system. A manual fire alarm system that activates the occupant notification system in accordance with Section 907.5 shall be installed in Group R-1 occupancies.

Exceptions:

1. A manual fire alarm system is not required in buildings not more than two stories in height where all individual *sleeping units* and contiguous attic and crawl spaces to those units are separated from each other and public or common areas by not less than 1-hour *fire partitions* and each individual *sleeping unit* has an *exit* directly to a *public way*, *egress court* or yard.
2. Manual fire alarm boxes are not required throughout the building where all of the following conditions are met:
   - 2.1. The building is equipped throughout with an *automatic sprinkler system* installed in accordance with Section 903.3.1.1 or 903.3.1.2.
   - 2.2. The notification appliances will activate upon sprinkler water flow.

2.3. Not fewer than one manual fire alarm box is installed at an *approved* location.

907.2.8.2 Automatic smoke detection system. An automatic smoke detection system that activates the occupant notification system in accordance with Section 907.5 shall be installed throughout all interior *corridors* serving *sleeping units*.

Exception: An automatic smoke detection system is not required in buildings that do not have interior *corridors* serving *sleeping units* and where each *sleeping unit* has a *means of egress* door opening directly to an *exit* or to an exterior *exit access* that leads directly to an *exit*.

907.2.8.3 Smoke alarms. Single- and multiple-station smoke alarms shall be installed in accordance with Section 907.2.11.

907.2.9 Group R-2. Fire alarm systems and smoke alarms shall be installed in Group R-2 occupancies as required in Sections 907.2.9.1 and 907.2.9.3.

907.2.9.1 Manual fire alarm system. A manual fire alarm system that activates the occupant notification system in accordance with Section 907.5 shall be installed in Group R-2 occupancies where any of the following conditions apply:

1. Any *dwelling unit* or *sleeping unit* is located three or more stories above the lowest *level of exit discharge*.
2. Any *dwelling unit* or *sleeping unit* is located more than one story below the highest *level of exit discharge* of *exits* serving the *dwelling unit* or *sleeping unit*.
3. The building contains more than 16 *dwelling units* or *sleeping units*.

Exceptions:

1. A fire alarm system is not required in buildings not more than two stories in height where all *dwelling units* or *sleeping units* and contiguous attic and crawl spaces are separated from each other and public or common areas by not less than 1-hour *fire partitions* and each *dwelling unit* or *sleeping unit* has an *exit* directly to a *public way*, *egress court* or yard.
2. Manual fire alarm boxes are not required where the building is equipped throughout with an *automatic sprinkler system* installed in accordance with Section 903.3.1.1 or 903.3.1.2 and the occupant notification appliances will automatically activate throughout the notification zones upon a sprinkler water flow.
3. A fire alarm system is not required in buildings that do not have interior *corridors* serving *dwelling units* and are protected by an *approved automatic sprinkler system* installed in accordance with Section 903.3.1.1 or 903.3.1.2, provided that *dwelling units* either have a *means of egress* door opening directly to an exterior *exit access* that leads directly to the *exits* or are served by open-ended *corridors* designed in accordance with Section 1027.6, Exception 3.

907.2.9.2 Smoke alarms. Single- and multiple-station smoke alarms shall be installed in accordance with Section 907.2.11.

907.2.9.3 Group R-2 college and university buildings. An automatic smoke detection system that activates the occupant notification system in accordance with Section 907.5 shall be installed in Group R-2 occupancies operated by a college or university for student or staff housing in all of the following locations:

1. Common spaces outside of *dwelling units* and *sleeping units*.
2. Laundry rooms, mechanical equipment rooms and storage rooms.
3. All interior corridors serving *sleeping units* or *dwelling units*.

Exception: An automatic smoke detection system is not required in buildings that do not have interior *corridors* serving *sleeping units* or *dwelling units* and where each *sleeping unit* or *dwelling unit* either has a *means of egress* door opening directly to an exterior *exit access* that leads directly to an *exit* or a *means of egress* door opening directly to an *exit*.

Required smoke alarms in *dwelling units* and *sleeping units* in Group R-2 occupancies operated by a college or university for student or staff housing shall be interconnected with the fire alarm system in accordance with NFPA 72.

907.2.10 Group R-4. Fire alarm systems and smoke alarms shall be installed in Group R-4 occupancies as required in Sections 907.2.10.1 through 907.2.10.3.

907.2.10.1 Manual fire alarm system. A manual fire alarm system that activates the occupant notification system in accordance with Section 907.5 shall be installed in Group R-4 occupancies.

Exceptions:

1. A manual fire alarm system is not required in buildings not more than two stories in height where all individual *sleeping units* and contiguous attic and crawl spaces to those units are separated from each other and public or common areas by not less than 1-hour *fire partitions* and each individual *sleeping unit* has an *exit* directly to a *public way*, *egress court* or yard.

2. Manual fire alarm boxes are not required throughout the building where all of the following conditions are met:

   2.1. The building is equipped throughout with an *automatic sprinkler system* installed in accordance with Section 903.3.1.1 or 903.3.1.2.

   2.2. The notification appliances will activate upon sprinkler water flow.

   2.3. Not fewer than one manual fire alarm box is installed at an *approved* location.

3. Manual fire alarm boxes in resident or patient sleeping areas shall not be required at *exits* where located at all nurses' control stations or other constantly attended staff locations, provided such stations are visible and continuously accessible and that the distances of travel required in Section 907.4.2.1 are not exceeded.

907.2.10.2 Automatic smoke detection system. An automatic smoke detection system that activates the occupant notification system in accordance with Section 907.5 shall be installed in *corridors*, waiting areas open to *corridors* and *habitable spaces* other than *sleeping units* and kitchens.

Exceptions:

1. Smoke detection in *habitable spaces* is not required where the facility is equipped throughout with an *automatic sprinkler system* installed in accordance with Section 903.3.1.1.
2. An automatic smoke detection system is not required in buildings that do not have interior *corridors* serving *sleeping units* and where each *sleeping unit* has a *means of egress* door opening directly to an *exit* or to an exterior *exit access* that leads directly to an exit.

907.2.10.3 Smoke alarms. Single- and multiple-station smoke alarms shall be installed in accordance with Section 907.2.11.

907.2.11 Single- and multiple-station smoke alarms. *Listed* single- and multiple-station smoke alarms complying with UL 217 shall be installed in accordance with Sections 907.2.11.1 through 907.2.11.6 and NFPA 72.

907.2.11.1 Group R-1. Single- or multiple-station smoke alarms shall be installed in all of the following locations in Group R-1:

1. In sleeping areas.
2. In every room in the path of the *means of egress* from the sleeping area to the door leading from the *sleeping unit*.
3. In each story within the *sleeping unit*, including *basements*. For *sleeping units* with split levels and without an intervening door between the adjacent levels, a smoke alarm installed on the upper level shall suffice for the adjacent lower level provided that the lower level is less than one full story below the upper level.

907.2.11.2 Groups R-2, R-3, R-4 and I-1. Single or multiple-station smoke alarms shall be installed and maintained in Groups R-2, R-3, R-4 and I-1 regardless of *occupant load* at all of the following locations:

1. On the ceiling or wall outside of each separate sleeping area in the immediate vicinity of bedrooms.
2. In each room used for sleeping purposes.
3. In each story within a *dwelling unit*, including *basements* but not including crawl spaces and uninhabitable attics. In *dwellings* or *dwelling units* with split levels and without an intervening door between the adjacent levels, a smoke alarm installed on the upper level shall suffice for the adjacent lower level provided that the lower level is less than one full story below the upper level.

907.2.11.3 Installation near cooking appliances. Smoke alarms shall not be installed in the following locations unless this would prevent placement of a smoke alarm in a location required by Section 907.2.11.1 or 907.2.11.2:

1. Ionization smoke alarms shall not be installed less than 20 feet (6096 mm) horizontally from a permanently installed cooking appliance.
2. Ionization smoke alarms with an alarm-silencing switch shall not be installed less than 10 feet (3048 mm) horizontally from a permanently installed cooking appliance.
3. Photoelectric smoke alarms shall not be installed less than 6 feet (1829 mm) horizontally from a permanently installed cooking appliance.

907.2.11.4 Installation near bathrooms. Smoke alarms shall be installed not less than 3 feet (914 mm) horizontally from the door or opening of a bathroom that contains a bathtub or shower unless this would prevent placement of a smoke alarm required by Section 907.2.11.1 or 907.2.11.2.

907.2.11.5 Interconnection. Where more than one smoke alarm is required to be installed within an individual *dwelling unit* or *sleeping unit* in Group R or I-1 occupancies, the smoke alarms shall be interconnected in such a manner that the activation of one alarm will activate all of the alarms in the individual unit. Physical interconnection of smoke alarms shall not be required where listed wireless alarms are installed and all alarms sound upon activation of one alarm. The alarm shall be clearly audible in all bedrooms over background noise levels with all intervening doors closed.

907.2.11.6 Power source. In new construction, required smoke alarms shall receive their primary power from the building wiring where such wiring is served from a commercial source and shall be equipped with a battery backup. Smoke alarms with integral strobes that are not equipped with battery back-up shall be connected to an emergency electrical system in accordance with Section 604. Smoke alarms shall emit a signal when the batteries are low. Wiring shall be permanent and without a disconnecting switch other than as required for overcurrent protection.

Exception: Smoke alarms are not required to be equipped with battery backup where they are connected to an emergency electrical system that complies with Section 604.

907.2.11.7 Smoke detection system. Smoke detectors listed in accordance with UL 268 and provided as part of the building fire alarm system shall be an acceptable alternative to single- and multiple-station *smoke alarms* and shall comply with the following:

1. The fire alarm system shall comply with all applicable requirements in Section 907.
2. Activation of a smoke detector in a *dwelling unit* or *sleeping unit* shall initiate alarm notification in the *dwelling unit* or *sleeping unit* in accordance with Section 907.5.2.
3. Activation of a smoke detector in a *dwelling unit* or *sleeping unit* shall not activate alarm notification appliances outside of the *dwelling unit* or *sleeping unit*, provided that a supervisory signal is generated and monitored in accordance with Section 907.6.6.

907.2.12 Special amusement buildings. An automatic smoke detection system shall be provided in special amusement buildings in accordance with Sections 907.2.12.1 through 907.2.12.3.

907.2.12.1 Alarm. Activation of any single smoke detector, the *automatic sprinkler system* or any other automatic fire detection device shall immediately activate an audible and visible alarm at the building at a constantly attended location from which emergency action can be initiated, including the capability of manual initiation of requirements in Section 907.2.12.2.

907.2.12.2 System response. The activation of two or more smoke detectors, a single smoke detector equipped with an alarm verification feature, the *automatic sprinkler system* or other *approved* fire detection device shall automatically do all of the following:

1. Cause illumination of the *means of egress* with light of not less than 1 footcandle (11 lux) at the walking surface level.
2. Stop any conflicting or confusing sounds and visual distractions.
3. Activate an *approved* directional *exit* marking that will become apparent in an emergency.
4. Activate a prerecorded message, audible throughout the special amusement building, instructing patrons to proceed to the nearest exit. Alarm signals used in conjunction with the prerecorded message shall produce a sound that is distinctive from other sounds used during normal operation.

907.2.12.3 Emergency voice/alarm communication system. An emergency voice/alarm communication system, which is also allowed to serve as a public address system, shall be installed in accordance with Section 907.5.2.2 and be audible throughout the entire special amusement building.

907.2.13 High-rise buildings. High-rise buildings shall be provided with an automatic smoke detection system in accordance with Section 907.2.13.1, a fire department communication system in accordance with Section 907.2.13.2 and an emergency voice/alarm communication system in accordance with Section 907.5.2.2.

Exceptions:

1. Airport traffic control towers in accordance with Section 907.2.22 of this code and Section 412 of the *International Building Code*.
2. Open parking garages in accordance with Section 406.5 of the *International Building Code*.
3. Buildings with an occupancy in Group A-5 in accordance with Section 303.1 of the *International Building Code*.
4. Low-hazard special occupancies in accordance with Section 503.1.1 of the *International Building Code*.
5. Buildings with an occupancy in Group H-1, H-2 or H-3 in accordance with Section 415 of the *International Building Code*.
6. In Group I-1 and I-2 occupancies, the alarm shall sound at a constantly attended location and occupant notification shall be broadcast by the emergency voice/alarm communication system.

907.2.13.1 Automatic smoke detection. Automatic smoke detection in high-rise buildings shall be in accordance with Sections 907.2.13.1.1 and 907.2.13.1.2.

907.2.13.1.1 Area smoke detection. Area smoke detectors shall be provided in accordance with this section. Smoke detectors shall be connected to an automatic fire alarm system. The activation of any detector required by this section shall activate the emergency voice/alarm communication system in accordance with Section 907.5.2.2. In addition to smoke detectors required by Sections 907.2.1 through 907.2.10, smoke detectors shall be located as follows:

1. In each mechanical equipment, electrical, transformer, telephone equipment or similar room that is not provided with sprinkler protection.
2. In each elevator machine room, machinery space, control room and control space and in elevator lobbies.

[M] **907.2.13.1.2 Duct smoke detection.** Duct smoke detectors complying with Section 907.3.1 shall be located as follows:

1. In the main return air and exhaust air plenum of each air-conditioning system having a capacity greater than 2,000 cubic feet per minute (cfm) (0.94 m^3/s). Such detectors shall be located in a serviceable area downstream of the last duct inlet.
2. At each connection to a vertical duct or riser serving two or more stories from a return air duct or plenum of an air-conditioning system. In Group R-1 and R-2 occupancies, a smoke detector is allowed to be used in each return air riser carrying not more than 5,000 cfm (2.4 m^3/s) and serving not more than 10 air-inlet openings.

907.2.13.2 Fire department communication system. Where a wired communication system is *approved* in lieu of an emergency responder radio coverage system in accordance with Section 510, the wired fire department communication system shall be designed and installed in accordance with NFPA 72 and shall operate between a *fire command center* complying with Section 508, elevators, elevator lobbies, emergency and standby power rooms, fire pump rooms, areas of refuge and inside *interior exit stairways*. The fire department communication device shall be provided at each floor level within the *interior exit stairway*.

907.2.14 Atriums connecting more than two stories. A fire alarm system shall be installed in occupancies with an atrium that connects more than two stories, with smoke detection in locations required by a rational analysis in Section 909.4 and in accordance with the system operation requirements in Section 909.17. The system shall be activated in accordance with Section 907.5. Such occupancies in Group A, E or M shall be provided with an emergency voice/alarm communication system complying with the requirements of Section 907.5.2.2.

907.2.15 High-piled combustible storage areas. An automatic smoke detection system shall be installed throughout *high-piled combustible storage* areas where required by Section 3206.5.

907.2.16 Aerosol storage uses. Aerosol storage rooms and general-purpose warehouses containing aerosols shall be provided with an *approved* manual fire alarm system where required by this code.

907.2.17 Lumber, wood structural panel and veneer mills. Lumber, wood structural panel and veneer mills shall be provided with a manual fire alarm system.

907.2.18 Underground buildings with smoke control systems. Where a smoke control system is installed in an underground building in accordance with the *International Building Code*, automatic smoke detectors shall be provided in accordance with Section 907.2.18.1.

907.2.18.1 Smoke detectors. Not fewer than one smoke detector *listed* for the intended purpose shall be installed in all of the following areas:

1. Mechanical equipment, electrical, transformer, telephone equipment, elevator machine or similar rooms.
2. Elevator lobbies.
3. The main return and exhaust air plenum of each air-conditioning system serving more than one story and located in a serviceable area downstream of the last duct inlet.
4. Each connection to a vertical duct or riser serving two or more floors from return air ducts or plenums of heating, ventilating and air-conditioning systems, except that in Group R occupancies, a *listed* smoke detector is allowed to be used in each return air riser carrying not more than 5,000 cfm (2.4 m^3/s) and serving not more than 10 air inlet openings.

907.2.18.2 Alarm required. Activation of the smoke control system shall activate an audible alarm at a constantly attended location.

907.2.19 Deep underground buildings. Where the lowest level of a structure is more than 60 feet (18 288 mm) below the finished floor of the lowest *level of exit discharge*, the structure shall be equipped throughout with a manual fire alarm system, including an emergency voice/alarm communication system installed in accordance with Section 907.5.2.2.

907.2.20 Covered and open mall buildings. Where the total floor area exceeds 50,000 square feet (4645 m^2) within either a covered mall building or within the perimeter line of an open mall building, an emergency voice/alarm communication system shall be provided. Emergency voice/alarm communication systems serving a mall, required or otherwise, shall be accessible to the fire department. The system shall be provided in accordance with Section 907.5.2.2.

907.2.21 Residential aircraft hangars. Not fewer than one single-station smoke alarm shall be installed within a residential aircraft hangar as defined in Chapter 2 of the *International Building Code* and shall be interconnected into the residential smoke alarm or other sounding device to provide an alarm that will be audible in all sleeping areas of the *dwelling*.

907.2.22 Airport traffic control towers. An automatic smoke detection system that activates the occupant notification system in accordance with Section 907.5 shall be provided in airport control towers in accordance with Sections 907.2.22.1 and 907.2.22.2.

Exception: Audible appliances shall not be installed within the control tower cab.

907.2.22.1 Airport traffic control towers with multiple exits and automatic sprinklers. Airport traffic control towers with multiple *exits* and equipped throughout with an *automatic sprinkler system* in accordance with Section 903.3.1.1 shall be provided with smoke detectors in all of the following locations:

1. Airport traffic control cab.
2. Electrical and mechanical equipment rooms.
3. Airport terminal radar and electronics rooms.
4. Outside each opening into *interior exit stairways*.
5. Along the single *means of egress* permitted from observation levels.
6. Outside each opening into the single *means of egress* permitted from observation levels.

907.2.22.2 Other airport traffic control towers. Airport traffic control towers with a single *exit* or where sprinklers are not installed throughout shall be provided with smoke detectors in all of the following locations:

1. Airport traffic control cab.
2. Electrical and mechanical equipment rooms.
3. Airport terminal radar and electronics rooms.
4. Office spaces incidental to the tower operation.
5. Lounges for employees, including sanitary facilities.
6. *Means of egress.*
7. Accessible utility shafts.

907.2.23 Battery rooms. An automatic smoke detection system shall be installed in areas containing stationary storage battery systems with a liquid capacity of more than 50 gallons (189 L).

907.3 Fire safety functions. Automatic fire detectors utilized for the purpose of performing fire safety functions shall be connected to the building's fire alarm control unit where a fire alarm system is required by Section 907.2. Detectors shall, upon actuation, perform the intended function and activate the alarm notification appliances or activate a visible and audible supervisory signal at a constantly attended location. In buildings not equipped with a fire alarm system, the automatic fire detector shall be powered by normal electrical service and, upon actuation, perform the intended function. The detectors shall be located in accordance with NFPA 72.

907.3.1 Duct smoke detectors. Smoke detectors installed in ducts shall be *listed* for the air velocity, temperature and humidity present in the duct. Duct smoke detectors shall be connected to the building's fire alarm control unit when a fire alarm system is required by Section 907.2. Activation of a duct smoke detector shall initiate a visible and audible supervisory signal at a *constantly attended location* and shall perform the intended fire safety function in accordance with this code and the *International Mechanical Code*. In facilities that are required to be monitored by a supervising station, duct smoke detectors shall report only as a supervisory signal and not as a fire alarm. They shall not be used as a substitute for required open area detection.

Exceptions:

1. The supervisory signal at a constantly attended location is not required where duct smoke detectors activate the building's alarm notification appliances.
2. In occupancies not required to be equipped with a fire alarm system, actuation of a smoke detector shall activate a visible and an audible signal in an *approved* location. Smoke detector trouble conditions shall activate a visible or audible signal in an *approved* location and shall be identified as air duct detector trouble.

907.3.2 Delayed egress locks. Where delayed egress locks are installed on *means of egress* doors in accordance with Section 1010.1.9.7, an automatic smoke or heat detection system shall be installed as required by that section.

907.3.3 Elevator emergency operation. Automatic fire detectors installed for elevator emergency operation shall be installed in accordance with the provisions of ASME A17.1 and NFPA 72.

907.3.4 Wiring. The wiring to the auxiliary devices and equipment used to accomplish the fire safety functions shall be monitored for integrity in accordance with NFPA 72.

907.4 Initiating devices. Where manual or automatic alarm initiation is required as part of a fire alarm system, the initiating devices shall be installed in accordance with Sections 907.4.1 through 907.4.3.1.

907.4.1 Protection of fire alarm control unit. In areas that are not continuously occupied, a single smoke detector shall be provided at the location of each fire alarm control unit, notification appliance circuit power extenders and supervising station transmitting equipment.

Exception: Where ambient conditions prohibit installation of smoke detector, a *heat detector* shall be permitted.

907.4.2 Manual fire alarm boxes. Where a manual fire alarm system is required by another section of this code, it shall be activated by fire alarm boxes installed in accordance with Sections 907.4.2.1 through 907.4.2.6.

907.4.2.1 Location. Manual fire alarm boxes shall be located not more than 5 feet (1524 mm) from the entrance to each *exit*. In buildings not protected by an *automatic sprinkler system* in accordance with Section 903.3.1.1 or 903.3.1.2, additional manual fire alarm boxes shall be located so that the *exit access* travel distance to the nearest box does not exceed 200 feet (60 960 mm).

907.4.2.2 Height. The height of the manual fire alarm boxes shall be not less than 42 inches (1067 mm) and not more than 48 inches (1372 mm) measured vertically, from the floor level to the activating handle or lever of the box.

907.4.2.3 Color. Manual fire alarm boxes shall be red in color.

907.4.2.4 Signs. Where fire alarm systems are not monitored by a supervising station, an *approved* permanent sign shall be installed adjacent to each manual fire alarm box that reads: WHEN ALARM SOUNDS—CALL FIRE DEPARTMENT.

Exception: Where the manufacturer has permanently provided this information on the manual fire alarm box.

907.4.2.5 Protective covers. The *fire code official* is authorized to require the installation of *listed* manual fire alarm box protective covers to prevent malicious false alarms or to provide the manual fire alarm box with protection from physical damage. The protective cover shall be transparent or red in color with a transparent face to permit visibility of the manual fire alarm box. Each cover shall include proper operating instructions. A protective cover that emits a local alarm signal shall not be installed unless *approved.* Protective covers shall not project more than that permitted by Section 1003.3.3.

907.4.2.6 Unobstructed and unobscured. Manual fire alarm boxes shall be accessible, unobstructed, unobscured and visible at all times.

907.4.3 Automatic smoke detection. Where an automatic smoke detection system is required it shall utilize smoke detectors unless ambient conditions prohibit such an installation. In spaces where smoke detectors cannot be utilized due to ambient conditions, *approved* automatic *heat detectors* shall be permitted.

907.4.3.1 Automatic sprinkler system. For conditions other than specific fire safety functions noted in Section 907.3, in areas where ambient conditions prohibit the installation of smoke detectors, an *automatic sprinkler system* installed in such areas in accordance with Section 903.3.1.1 or 903.3.1.2 and that is connected to the fire alarm system shall be *approved* as automatic heat detection.

907.5 Occupant notification systems. A fire alarm system shall annunciate at the fire alarm control unit and shall initiate occupant notification upon activation, in accordance with Sections 907.5.1 through 907.5.2.3.3. Where a fire alarm system is required by another section of this code, it shall be activated by:

1. Automatic fire detectors.
2. Automatic sprinkler system waterflow devices.
3. Manual fire alarm boxes.
4. Automatic fire-extinguishing systems.

Exception: Where notification systems are allowed elsewhere in Section 907 to annunciate at a constantly attended location.

907.5.1 Presignal feature. A presignal feature shall not be installed unless *approved* by the *fire code official* and the fire department. Where a presignal feature is provided, a signal shall be annunciated at a constantly attended location *approved* by the fire department, so that occupant notification can be activated in the event of fire or other emergency.

907.5.2 Alarm notification appliances. Alarm notification appliances shall be provided and shall be *listed* for their purpose.

907.5.2.1 Audible alarms. Audible alarm notification appliances shall be provided and emit a distinctive sound that is not to be used for any purpose other than that of a fire alarm.

Exceptions:

1. Audible alarm notification appliances are not required in critical care areas of Group I-2 Condition 2 occupancies that are in compliance with Section 907.2.6, Exception 2.
2. A visible alarm notification appliance installed in a nurses' control station or other continuously attended staff location in a Group I-2 Condition 2 suite shall be an acceptable alternative to the installation of audible alarm notification appliances throughout the suite in Group I-2 Condition 2 occupancies that are in compliance with Section 907.2.6, Exception 2.
3. Where provided, audible notification appliances located in each occupant evacuation elevator lobby in accordance with Section 3008.9.1 of the *International Building Code* shall be connected to a separate notification zone for manual paging only.

907.5.2.1.1 Average sound pressure. The audible alarm notification appliances shall provide a sound pressure level of 15 decibels (dBA) above the average ambient sound level or 5 dBA above the maximum sound level having a duration of not less than 60 seconds, whichever is greater, in every occupiable space within the building.

907.5.2.1.2 Maximum sound pressure. The maximum sound pressure level for audible alarm notification appliances shall be 110 dBA at the minimum hearing distance from the audible appliance. Where the average ambient noise is greater than 95 dBA, visible alarm notification appliances shall be provided in accordance with NFPA 72 and audible alarm notification appliances shall not be required.

907.5.2.2 Emergency voice/alarm communication systems. Emergency voice/alarm communication systems required by this code shall be designed and installed in accordance with NFPA 72. The operation of any automatic fire detector, sprinkler waterflow device or manual fire alarm box shall automatically sound an alert tone followed by voice instructions giving *approved* information and directions for a general or staged evacuation in accordance with the building's fire safety and evacuation plans required by Section 404. In high-rise buildings, the system shall operate on at least the alarming floor, the floor above and the floor below. Speakers shall be provided throughout the building by paging zones. At a minimum, paging zones shall be provided as follows:

1. Elevator groups.
2. *Interior exit stairways*.
3. Each floor.
4. *Areas of refuge* as defined in Chapter 2.

Exception: In Group I-1 and I-2 occupancies, the alarm shall sound in a constantly attended area and a general occupant notification shall be broadcast over the overhead page.

907.5.2.2.1 Manual override. A manual override for emergency voice communication shall be provided on a selective and all-call basis for all paging zones.

907.5.2.2.2 Live voice messages. The emergency voice/alarm communication system shall have the capability to broadcast live voice messages by paging zones on a selective and all-call basis.

907.5.2.2.3 Alternate uses. The emergency voice/alarm communication system shall be allowed to be used for other announcements, provided the manual fire alarm use takes precedence over any other use.

907.5.2.2.4 Emergency voice/alarm communication captions. Where stadiums, arenas and grandstands are required to caption audible public announcements in accordance with Section 1108.2.7.3 of the *International Building Code*, the emergency/voice alarm communication system shall be captioned. Prerecorded or live emergency captions shall be from an *approved* location constantly attended by personnel trained to respond to an emergency.

907.5.2.2.5 Emergency power. Emergency voice/alarm communications systems shall be provided with emergency power in accordance with Section 604. The system shall be capable of powering the required load for a duration of not less than 24 hours, as required in NFPA 72.

907.5.2.3 Visible alarms. Visible alarm notification appliances shall be provided in accordance with Sections 907.5.2.3.1 through 907.5.2.3.3.

Exceptions:

1. Visible alarm notification appliances are not required in *alterations*, except where an existing fire alarm system is upgraded or replaced, or a new fire alarm system is installed.
2. Visible alarm notification appliances shall not be required in *exits* as defined in Chapter 2.
3. Visible alarm notification appliances shall not be required in elevator cars.
4. Visual alarm notification appliances are not required in critical care areas of Group I-2 Condition 2 occupancies that are in compliance with Section 907.2.6, Exception 2.

907.5.2.3.1 Public use areas and common use areas. Visible alarm notification appliances shall be provided in *public use areas* and *common use areas*.

Exception: Where employee work areas have audible alarm coverage, the notification appliance circuits serving the employee work areas shall be initially designed with not less than 20-percent spare capacity to account for the potential of adding visible notification appliances in the future to accommodate hearing-impaired employee(s).

907.5.2.3.2 Groups I-1 and R-1. Group I-1 and R-1 *dwelling units* or *sleeping units* in accordance with Table 907.5.2.3.2 shall be provided with a visible alarm notification appliance, activated by both the in-room smoke alarm and the building fire alarm system.

TABLE 907.5.2.3.2
VISIBLE ALARMS

NUMBER OF SLEEPING UNITS	SLEEPING ACCOMMODATIONS WITH VISIBLE ALARMS
6 to 25	2
26 to 50	4
51 to 75	7
76 to 100	9
101 to 150	12
151 to 200	14
201 to 300	17
301 to 400	20
401 to 500	22
501 to 1,000	5% of total
1,001 and over	50 plus 3 for each 100 over 1,000

907.5.2.3.3 Group R-2. In Group R-2 occupancies required by Section 907 to have a fire alarm system, all *dwelling units* and *sleeping units* shall be provided with the capability to support visible alarm notification appliances in accordance with Chapter 10 of ICC A117.1. Such capability shall be permitted to include the potential for future interconnection of the building fire alarm system with the unit smoke alarms, replacement of audible appliances with combination audible/visible appliances, or future extension of the existing wiring from the unit smoke alarm locations to required locations for visible appliances.

907.6 Installation and monitoring. A fire alarm system shall be installed and monitored in accordance with Sections 907.6.1 through 907.6.6.2 and NFPA 72.

907.6.1 Wiring. Wiring shall comply with the requirements of NFPA 70 and NFPA 72. Wireless protection systems utilizing radio-frequency transmitting devices shall comply with the special requirements for supervision of low-power wireless systems in NFPA 72.

907.6.2 Power supply. The primary and secondary power supply for the fire alarm system shall be provided in accordance with NFPA 72.

Exception: Backup power for single-station and multiple-station smoke alarms as required in Section 907.2.11.6.

907.6.3 Initiating device identification. The fire alarm system shall identify the specific initiating device address, location, device type, floor level where applicable and status including indication of normal, alarm, trouble and supervisory status, as appropriate.

Exceptions:

1. Fire alarm systems in single-story buildings less than 22,500 square feet (2090 m^2) in area.
2. Fire alarm systems that only include manual fire alarm boxes, waterflow initiating devices and not more than 10 additional alarm-initiating devices.
3. Special initiating devices that do not support individual device identification.
4. Fire alarm systems or devices that are replacing existing equipment.

907.6.3.1 Annunciation. The initiating device status shall be annunciated at an *approved* on-site location.

907.6.4 Zones. Each floor shall be zoned separately and a zone shall not exceed 22,500 square feet (2090 m^2). The length of any zone shall not exceed 300 feet (91 440 mm) in any direction.

Exception: *Automatic sprinkler system* zones shall not exceed the area permitted by NFPA 13.

907.6.4.1 Zoning indicator panel. A zoning indicator panel and the associated controls shall be provided in an *approved* location. The visual zone indication shall lock in until the system is reset and shall not be canceled by the operation of an audible alarm-silencing switch.

907.6.4.2 High-rise buildings. In high-rise buildings, a separate zone by floor shall be provided for each of the following types of alarm-initiating devices where provided:

1. Smoke detectors.
2. Sprinkler waterflow devices.
3. Manual fire alarm boxes.
4. Other *approved* types of automatic fire detection-devices or suppression systems.

907.6.5 Access. Access shall be provided to each fire alarm device and notification appliance for periodic inspection, maintenance and testing.

907.6.6 Monitoring. Fire alarm systems required by this chapter or by the *International Building Code* shall be monitored by an *approved* supervising station in accordance with NFPA 72.

Exception: Monitoring by a supervising station is not required for:

1. Single- and multiple-station smoke alarms required by Section 907.2.11.
2. Smoke detectors in Group I-3 occupancies.
3. *Automatic sprinkler systems* in one- and two-family dwellings.

907.6.6.1 Automatic telephone-dialing devices. Automatic telephone-dialing devices used to transmit an emergency alarm shall not be connected to any fire department telephone number unless *approved* by the fire chief.

907.6.6.2 Termination of monitoring service. Termination of fire alarm monitoring services shall be in accordance with Section 901.9.

907.7 Acceptance tests and completion. Upon completion of the installation, the fire alarm system and all fire alarm components shall be tested in accordance with NFPA 72.

907.7.1 Single- and multiple-station alarm devices. When the installation of the alarm devices is complete, each device and interconnecting wiring for multiple-station alarm devices shall be tested in accordance with the smoke alarm provisions of NFPA 72.

907.7.2 Record of completion. A record of completion in accordance with NFPA 72 verifying that the system has been installed and tested in accordance with the *approved* plans and specifications shall be provided.

907.7.3 Instructions. Operating, testing and maintenance instructions and record drawings ("as builts") and equipment specifications shall be provided at an *approved* location.

907.8 Inspection, testing and maintenance. The maintenance and testing schedules and procedures for fire alarm and fire detection systems shall be in accordance with Sections 907.8.1 through 907.8.5 and NFPA 72. Records of inspection, testing and maintenance shall be maintained.

907.8.1 Maintenance required. Where required for compliance with the provisions of this code, devices, equipment, systems, conditions, arrangements, levels of protection or other features shall thereafter be continuously maintained in accordance with applicable NFPA requirements or as directed by the *fire code official.*

907.8.2 Testing. Testing shall be performed in accordance with the schedules in NFPA 72 or more frequently where required by the *fire code official.* Records of testing shall be maintained.

Exception: Devices or equipment that are inaccessible for safety considerations shall be tested during scheduled shutdowns where *approved* by the *fire code official*, but not less than every 18 months.

907.8.3 Smoke detector sensitivity. Smoke detector sensitivity shall be checked within one year after installation and every alternate year thereafter. After the second calibration test, where sensitivity tests indicate that the detector has remained within its *listed* and marked sensitivity range (or 4-percent obscuration light grey smoke, if not marked), the length of time between calibration tests shall be permitted to be extended to not more than 5 years. Where the frequency is extended, records of detector-caused nuisance alarms and subsequent trends of these alarms shall be maintained. In zones or areas where nuisance alarms show any increase over the previous year, calibration tests shall be performed.

907.8.4 Sensitivity test method. To verify that each smoke detector is within its *listed* and marked sensitivity range, it shall be tested using one of the following methods:

1. A calibrated test method.
2. The manufacturer's calibrated sensitivity test instrument.
3. *Listed* control equipment arranged for the purpose.
4. A smoke detector/control unit arrangement whereby the detector causes a signal at the control unit where the detector's sensitivity is outside its acceptable sensitivity range.
5. Another calibrated sensitivity test method acceptable to the *fire code official*.

Detectors found to have a sensitivity outside the *listed* and marked sensitivity range shall be cleaned and recalibrated or replaced.

Exceptions:

1. Detectors *listed* as field adjustable shall be permitted to be either adjusted within the *listed* and marked sensitivity range and cleaned and recalibrated or they shall be replaced.
2. This requirement shall not apply to single-station smoke alarms.

907.8.4.1 Sensitivity testing device. Smoke detector sensitivity shall not be tested or measured using a device that administers an unmeasured concentration of smoke or other aerosol into the detector.

907.8.5 Inspection, testing and maintenance. The building *owner* shall be responsible to maintain the fire and life safety systems in an operable condition at all times. Service personnel shall meet the qualification requirements of NFPA 72 for inspection, testing and maintenance of such systems. Records of inspection, testing and maintenance shall be maintained.

907.9 Where required in existing buildings and structures. An *approved* fire alarm system shall be provided in existing buildings and structures where required in Chapter 11.

SECTION 908
EMERGENCY ALARM SYSTEMS

908.1 Group H occupancies. Emergency alarms for the detection and notification of an emergency condition in Group H occupancies shall be provided as required in Chapter 50.

908.2 Group H-5 occupancy. Emergency alarms for notification of an emergency condition in an HPM facility shall be provided as required in Section 2703.12. A continuous gas detection system shall be provided for HPM gases in accordance with Section 2703.13.

908.3 Highly toxic and toxic materials. Where required by Section 6004.2.2.10, a gas detection system shall be provided for indoor storage and use of highly toxic and toxic *compressed gases*.

908.4 Ozone gas-generator rooms. A gas detection system shall be provided in ozone gas-generator rooms in accordance with Section 6005.3.2.

908.5 Repair garages. A flammable-gas detection system shall be provided in repair garages for vehicles fueled by non-odorized gases in accordance with Section 2311.7.2.

908.6 Refrigeration systems. Refrigeration system machinery rooms shall be provided with a refrigerant detector in accordance with Section 606.9.

908.7 Carbon dioxide (CO_2) systems. Emergency alarm systems in accordance with Section 5307.5.2 shall be provided where required for compliance with Section 5307.5.

SECTION 909
SMOKE CONTROL SYSTEMS

909.1 Scope and purpose. This section applies to mechanical or passive smoke control systems where they are required for new buildings or portions thereof by provisions of the *International Building Code* or this code. The purpose of this section is to establish minimum requirements for the design, installation and acceptance testing of smoke control systems that are intended to provide a tenable environment for the evacuation or relocation of occupants. These provisions are not intended for the preservation of contents, the timely restoration of operations or for assistance in fire suppression or overhaul activities. Smoke control systems regulated by this section serve a different purpose than the smoke- and heat-venting provisions found in Section 910. Mechanical smoke control systems shall not be considered exhaust systems under Chapter 5 of the *International Mechanical Code*.

909.2 General design requirements. Buildings, structures, or parts thereof required by the *International Building Code* or this code to have a smoke control system or systems shall have such systems designed in accordance with the applicable requirements of Section 909 and the generally accepted and well-established principles of engineering relevant to the design. The *construction documents* shall include sufficient information and detail to describe adequately the elements of the design necessary for the proper implementation of the smoke control systems. These documents shall be accompanied with sufficient information and analysis to demonstrate compliance with these provisions.

909.3 Special inspection and test requirements. In addition to the ordinary inspection and test requirements that buildings, structures and parts thereof are required to undergo, smoke control systems subject to the provisions of Section 909 shall undergo special inspections and tests sufficient to verify the proper commissioning of the smoke control design in its final installed condition. The design submission accompanying the *construction documents* shall clearly detail procedures and methods to be used and the items subject to such inspections and tests. Such commissioning shall be in accordance with generally accepted engineering practice and, where possible, based on published standards for the particular testing involved. The special inspections and tests required by this section shall be conducted under the same terms as in Section 1704 of the *International Building Code.*

909.4 Analysis. A rational analysis supporting the types of smoke control systems to be employed, the methods of their operations, the systems supporting them and the methods of construction to be utilized shall accompany the *construction documents* submission and include, but not be limited to, the items indicated in Sections 909.4.1 through 909.4.7.

909.4.1 Stack effect. The system shall be designed such that the maximum probable normal or reverse stack effect will not adversely interfere with the system's capabilities. In determining the maximum probable stack effect, altitude, elevation, weather history and interior temperatures shall be used.

909.4.2 Temperature effect of fire. Buoyancy and expansion caused by the design fire in accordance with Section 909.9 shall be analyzed. The system shall be designed such that these effects do not adversely interfere with the system's capabilities.

909.4.3 Wind effect. The design shall consider the adverse effects of wind. Such consideration shall be consistent with the wind-loading provisions of the *International Building Code.*

909.4.4 Systems. The design shall consider the effects of the heating, ventilating and air-conditioning (HVAC) systems on both smoke and fire transport. The analysis shall include all permutations of systems status. The design shall consider the effects of the fire on the heating, ventilating and air-conditioning systems.

909.4.5 Climate. The design shall consider the effects of low temperatures on systems, property and occupants. Air inlets and exhausts shall be located so as to prevent snow or ice blockage.

909.4.6 Duration of operation. All portions of active or engineered smoke control systems shall be capable of continued operation after detection of the fire event for a period of not less than either 20 minutes or 1.5 times the calculated egress time, whichever is greater.

909.4.7 Smoke control system interaction. The design shall consider the interaction effects of the operation of multiple smoke control systems for all design scenarios.

909.5 Smoke barrier construction. *Smoke barriers* required for passive smoke control and a smoke control system using the pressurization method shall comply with Section 709 of the *International Building Code. Smoke barriers* shall be constructed and sealed to limit leakage areas exclusive of protected openings. The maximum allowable leakage area shall be the aggregate area calculated using the following leakage area ratios:

1. Walls: $A/A_w = 0.00100$
2. Interior *exit stairways* and *ramps* and *exit passageways*: $A/A_w = 0.00035$
3. Enclosed *exit access stairways* and *ramps* and all other shafts: $A/A_w = 0.00150$
4. Floors and roofs: $A/A_F = 0.00050$

where:

A = Total leakage area, square feet (m^2).

A_F = Unit floor or roof area of barrier, square feet (m^2).

A_w = Unit wall area of barrier, square feet (m^2).

The leakage area ratios shown do not include openings due to gaps around doors and operable windows. The total leakage area of the *smoke barrier* shall be determined in accordance with Section 909.5.1 and tested in accordance with Section 909.5.2.

909.5.1 Total leakage area. Total leakage area of the barrier is the product of the *smoke barrier* gross area multiplied by the allowable leakage area ratio, plus the area of other openings such as gaps around doors and operable windows.

909.5.2 Testing of leakage area. Compliance with the maximum total leakage area shall be determined by achieving the minimum air pressure difference across the barrier with the system in the smoke control mode for mechanical smoke control systems utilizing the pressurization method. Compliance with the maximum total leakage area of passive smoke control systems shall be verified through methods such as door fan testing or other methods, as *approved* by the *fire code official.*

909.5.3 Opening protection. Openings in *smoke barriers* shall be protected by automatic-closing devices actuated by the required controls for the mechanical smoke control system. Door openings shall be protected by fire door assemblies complying with Section 716.5.3 of the *International Building Code.*

Exceptions:

1. Passive smoke control systems with automatic-closing devices actuated by spot-type smoke detectors *listed* for releasing service installed in accordance with Section 907.3.
2. Fixed openings between smoke zones that are protected utilizing the airflow method.

3. In Group I-1 Condition 2, Group I-2 and ambulatory care facilities, where a pair of opposite-swinging doors are installed across a corridor in accordance with Section 909.5.3.1, the doors shall not be required to be protected in accordance with Section 716 of the *International Building Code*. The doors shall be close-fitting within operational tolerances and shall not have a center mullion or undercuts in excess of $^3/_4$-inch (19.1 mm) louvers or grilles. The doors shall have head and jamb stops and astragals or rabbets at meeting edges and, where permitted by the door manufacturer's listing, positive-latching devices are not required.
4. In Group I-2 and ambulatory care facilities, where such doors are special-purpose horizontal sliding, accordion or folding door assemblies installed in accordance with Section 1010.1.4.3 and are automatic closing by smoke detection in accordance with Section 716.5.9.3 of the *International Building Code*.
5. Group I-3.
6. Openings between smoke zones with clear ceiling heights of 14 feet (4267 mm) or greater and bank-down capacity of greater than 20 minutes as determined by the design fire size.

909.5.3.1 Group I-1 Condition 2, Group I-2 and ambulatory care facilities. In Group I-1 Condition 2, Group I-2 and *ambulatory care facilities*, where doors are installed across a *corridor*, the doors shall be automatic closing by smoke detection in accordance with Section 716.5.9.3 of the *International Building Code* and shall have a vision panel with fire-protection-rated glazing materials in fire-protection-rated frames, the area of which shall not exceed that tested.

909.5.3.2 Ducts and air transfer openings. Ducts and air transfer openings are required to be protected with a minimum Class II, 250°F (121°C) smoke damper complying with Section 717 of the *International Building Code*.

909.6 Pressurization method. The primary mechanical means of controlling smoke shall be by pressure differences across *smoke barriers*. Maintenance of a tenable environment is not required in the smoke-control zone of fire origin.

909.6.1 Minimum pressure difference. The minimum pressure difference across a *smoke barrier* shall be 0.05-inch water gage (0.0124 kPa) in fully sprinklered buildings.

In buildings allowed to be other than fully sprinklered, the smoke control system shall be designed to achieve pressure differences not less than two times the maximum calculated pressure difference produced by the design fire.

909.6.2 Maximum pressure difference. The maximum air pressure difference across a *smoke barrier* shall be determined by required door-opening or closing forces. The actual force required to open *exit* doors when the system is in the smoke control mode shall be in accordance with Section 1010.1.3. Opening and closing forces for other doors shall be determined by standard engineering methods for the resolution of forces and reactions. The calculated force to set a side-hinged, swinging door in motion shall be determined by:

$$F = F_{dc} + K(WA\Delta P)/2(W - d) \qquad \textbf{(Equation 9-1)}$$

where:

A = Door area, square feet (m^2).

d = Distance from door handle to latch edge of door, feet (m).

F = Total door opening force, pounds (N).

F_{dc} = Force required to overcome closing device, pounds (N).

K = Coefficient 5.2 (1.0).

W = Door width, feet (m).

ΔP = Design pressure difference, inches of water (Pa).

909.6.3 Pressurized stairways and elevator hoistways. Where stairways or elevator hoistways are pressurized, such pressurization systems shall comply with Section 909 as smoke control systems, in addition to the requirements of Section 909.21 of this code and Section 909.20 of the *International Building Code*.

909.7 Airflow design method. Where *approved* by the *fire code official*, smoke migration through openings fixed in a permanently open position, which are located between smoke control zones by the use of the airflow method, shall be permitted. The design airflow shall be in accordance with this section. Airflow shall be directed to limit smoke migration from the fire zone. The geometry of openings shall be considered to prevent flow reversal from turbulent effects. Smoke control systems using the airflow method shall be designed in accordance with NFPA 92.

909.7.1 Prohibited conditions. This method shall not be employed where either the quantity of air or the velocity of the airflow will adversely affect other portions of the smoke control system, unduly intensify the fire, disrupt plume dynamics or interfere with exiting. In no case shall airflow toward the fire exceed 200 feet per minute (1.02 m/s). Where the calculated airflow exceeds this limit, the airflow method shall not be used.

909.8 Exhaust method. Where *approved* by the *fire code official*, mechanical smoke control for large enclosed volumes, such as in atriums or malls, shall be permitted to utilize the exhaust method. Smoke control systems using the exhaust method shall be designed in accordance with NFPA 92.

909.8.1 Smoke layer. The height of the lowest horizontal surface of the smoke layer interface shall be maintained not less than 6 feet (1829 mm) above a walking surface that forms a portion of a required egress system within the smoke zone.

909.9 Design fire. The design fire shall be based on a rational analysis performed by the registered design professional and *approved* by the *fire code official*. The design fire shall be based on the analysis in accordance with Section 909.4 and this section.

909.9.1 Factors considered. The engineering analysis shall include the characteristics of the fuel, fuel load, effects included by the fire and whether the fire is likely to be steady or unsteady.

909.9.2 Design fire fuel. Determination of the design fire shall include consideration of the type of fuel, fuel spacing and configuration.

909.9.3 Heat-release assumptions. The analysis shall make use of best available data from *approved* sources and shall not be based on excessively stringent limitations of combustible material.

909.9.4 Sprinkler effectiveness assumptions. A documented engineering analysis shall be provided for conditions that assume fire growth is halted at the time of sprinkler activation.

909.10 Equipment. Equipment including, but not limited to, fans, ducts, automatic dampers and balance dampers shall be suitable for their intended use, suitable for the probable exposure temperatures that the rational analysis indicates, and as *approved* by the *fire code official*.

909.10.1 Exhaust fans. Components of exhaust fans shall be rated and certified by the manufacturer for the probable temperature rise to which the components will be exposed. This temperature rise shall be computed by:

$$T_s = (Q_c/mc) + (T_a) \quad \textbf{(Equation 9-3)}$$

where:

c = Specific heat of smoke at smoke layer temperature, Btu/lb°F · (kJ/kg · K).

m = Exhaust rate, pounds per second (kg/s).

Q_c = Convective heat output of fire, Btu/s (kW).

T_a = Ambient temperature, °F (K).

T_s = Smoke temperature, °F (K).

Exception: Reduced T_s as calculated based on the assurance of adequate dilution air.

909.10.2 Ducts. Duct materials and joints shall be capable of withstanding the probable temperatures and pressures to which they are exposed as determined in accordance with Section 909.10.1. Ducts shall be constructed and supported in accordance with the *International Mechanical Code*. Ducts shall be leak tested to 1.5 times the maximum design pressure in accordance with nationally accepted practices. Measured leakage shall not exceed 5 percent of design flow. Results of such testing shall be a part of the documentation procedure. Ducts shall be supported directly from fire-resistance-rated structural elements of the building by substantial, noncombustible supports.

Exception: Flexible connections, for the purpose of vibration isolation, complying with the *International Mechanical Code* and that are constructed of *approved* fire-resistance-rated materials.

909.10.3 Equipment, inlets and outlets. Equipment shall be located so as to not expose uninvolved portions of the building to an additional fire hazard. Outside air inlets shall be located so as to minimize the potential for introducing smoke or flame into the building. Exhaust outlets shall be so located as to minimize reintroduction of smoke into the building and to limit exposure of the building or adjacent buildings to an additional fire hazard.

909.10.4 Automatic dampers. Automatic dampers, regardless of the purpose for which they are installed within the smoke control system, shall be *listed* and conform to the requirements of *approved* recognized standards.

909.10.5 Fans. In addition to other requirements, belt-driven fans shall have 1.5 times the number of belts required for the design duty with the minimum number of belts being two. Fans shall be selected for stable performance based on normal temperature and, where applicable, elevated temperature. Calculations and manufacturer's fan curves shall be part of the documentation procedures. Fans shall be supported and restrained by noncombustible devices in accordance with the structural design requirements of Chapter 16 of the *International Building Code*.

Motors driving fans shall not be operated beyond their nameplate horsepower (kilowatts) as determined from measurement of actual current draw and shall have a minimum service factor of 1.15.

909.11 Standby power. Smoke control systems shall be provided with standby power in accordance with Section 604.

909.11.1 Equipment room. The standby power source and its transfer switches shall be in a room separate from the normal power transformers and switch gears and ventilated directly to and from the exterior. The room shall be enclosed with not less than 1-hour fire barriers constructed in accordance with Section 707 of the *International Building Code* or horizontal assemblies constructed in accordance with Section 711 of the *International Building Code*, or both.

909.11.2 Power sources and power surges. Elements of the smoke control system relying on volatile memories or the like shall be supplied with uninterruptable power sources of sufficient duration to span 15-minute primary power interruption. Elements of the smoke control system susceptible to power surges shall be suitably protected by conditioners, suppressors or other *approved* means.

909.12 Detection and control systems. Fire detection systems providing control input or output signals to mechanical smoke control systems or elements thereof shall comply with the requirements of Section 907. Such systems shall be equipped with a control unit complying with UL 864 and *listed* as smoke control equipment.

909.12.1 Verification. Control systems for mechanical smoke control systems shall include provisions for verification. Verification shall include positive confirmation of actuation, testing, manual override and the presence of power downstream of all disconnects. A preprogrammed weekly test sequence shall report abnormal conditions audibly, visually and by printed report. The preprogrammed weekly test shall operate all devices, equipment, and components used for smoke control.

Exception: Where verification of individual components tested through the preprogrammed weekly testing sequence will interfere with, and produce unwanted effects to, normal building operation, such individual components are permitted to be bypassed from the preprogrammed weekly testing, where *approved* by the *fire code official* and in accordance with both of the following:

1. Where the operation of components is bypassed from the preprogrammed weekly test, presence of power downstream of all disconnects shall be verified weekly by a listed control unit.
2. Testing of all components bypassed from the preprogrammed weekly test shall be in accordance with Section 909.20.6.

909.12.2 Wiring. In addition to meeting requirements of NFPA 70, all wiring, regardless of voltage, shall be fully enclosed within continuous raceways.

909.12.3 Activation. Smoke control systems shall be activated in accordance with this section.

909.12.3.1 Pressurization, airflow or exhaust method. Mechanical smoke control systems using the pressurization, airflow or exhaust method shall have completely automatic control.

909.12.3.2 Passive method. Passive smoke control systems actuated by *approved* spot-type detectors *listed* for releasing service shall be permitted.

909.12.4 Automatic control. Where completely automatic control is required or used, the automatic-control sequences shall be initiated from an appropriately zoned *automatic sprinkler system* complying with Section 903.3.1.1, manual controls that are readily accessible to the fire department and any smoke detectors required by the engineering analysis.

909.13 Control air tubing. Control air tubing shall be of sufficient size to meet the required response times. Tubing shall be flushed clean and dry prior to final connections and shall be adequately supported and protected from damage. Tubing passing through concrete or masonry shall be sleeved and protected from abrasion and electrolytic action.

909.13.1 Materials. Control air tubing shall be hard drawn copper, Type L, ACR in accordance with ASTM B 42, ASTM B 43, ASTM B 68, ASTM B 88, ASTM B 251 and ASTM B 280. Fittings shall be wrought copper or brass, solder type, in accordance with ASME B 16.18 or ASME B 16.22. Changes in direction shall be made with appropriate tool bends. Brass compression-type fittings shall be used at final connection to devices; other joints shall be brazed using a BCuP5 brazing alloy with solidus above 1,100°F (593°C) and liquidus below 1,500°F (816°C). Brazing flux shall be used on copper-to-brass joints only.

Exception: Nonmetallic tubing used within control panels and at the final connection to devices, provided all of the following conditions are met:

1. Tubing shall comply with the requirements of Section 602.2.1.3 of the *International Mechanical Code*.
2. Tubing and the connected device shall be completely enclosed within a galvanized or paint-grade steel enclosure having a minimum thickness of 0.0296 inch (0.7534 mm) (No. 22 gage). Entry to the enclosure shall be by copper tubing with a protective grommet of neoprene or Teflon or by suitable brass compression to male-barbed adapter.
3. Tubing shall be identified by appropriately documented coding.
4. Tubing shall be neatly tied and supported within the enclosure. Tubing bridging cabinets and doors or moveable devices shall be of sufficient length to avoid tension and excessive stress. Tubing shall be protected against abrasion. Tubing serving devices on doors shall be fastened along hinges.

909.13.2 Isolation from other functions. Control tubing serving other than smoke control functions shall be isolated by automatic isolation valves or shall be an independent system.

909.13.3 Testing. Control air tubing shall be tested at three times the operating pressure for not less than 30 minutes without any noticeable loss in gauge pressure prior to final connection to devices.

909.14 Marking and identification. The detection and control systems shall be clearly marked at all junctions, accesses and terminations.

909.15 Control diagrams. Identical control diagrams showing all devices in the system and identifying their location and function shall be maintained current and kept on file with the *fire code official*, the fire department and in the *fire command center* in a format and manner *approved* by the fire chief.

909.16 Fire fighter's smoke control panel. A fire fighter's smoke control panel for fire department emergency response purposes only shall be provided and shall include manual control or override of automatic control for mechanical smoke control systems. The panel shall be located in a *fire command center* complying with Section 508 in high-rise buildings or buildings with smoke-protected assembly seating. In all other buildings, the fire fighter's smoke control panel shall be installed in an *approved* location adjacent to the fire alarm control panel. The fire fighter's smoke control panel shall comply with Sections 909.16.1 through 909.16.3.

909.16.1 Smoke control systems. Fans within the building shall be shown on the fire fighter's control panel. A clear indication of the direction of airflow and the relationship of components shall be displayed. Status indicators shall be provided for all smoke control equipment, annunciated by fan and zone and by pilot-lamp-type indicators as follows:

1. Fans, dampers and other operating equipment in their normal status—WHITE.
2. Fans, dampers and other operating equipment in their off or closed status—RED.
3. Fans, dampers and other operating equipment in their on or open status—GREEN.
4. Fans, dampers and other operating equipment in a fault status—YELLOW/AMBER.

909.16.2 Smoke control panel. The fire fighter's control panel shall provide control capability over the complete smoke control system equipment within the building as follows:

1. ON-AUTO-OFF control over each individual piece of operating smoke control equipment that can also be controlled from other sources within the building. This includes *stairway* pressurization fans; smoke exhaust fans; supply, return and exhaust fans; elevator shaft fans; and other operating equipment used or intended for smoke control purposes.
2. OPEN-AUTO-CLOSE control over individual dampers relating to smoke control and that are also controlled from other sources within the building.
3. ON-OFF or OPEN-CLOSE control over smoke control and other critical equipment associated with a fire or smoke emergency and that can only be controlled from the fire fighter's control panel.

Exceptions:

1. Complex systems, where *approved*, where the controls and indicators are combined to control and indicate all elements of a single smoke zone as a unit.
2. Complex systems, where *approved*, where the control is accomplished by computer interface using *approved*, plain English commands.

909.16.3 Control action and priorities. The fire fighter's control panel actions shall be as follows:

1. ON-OFF and OPEN-CLOSE control actions shall have the highest priority of any control point within the building. Once issued from the fire fighter's control panel, automatic or manual control from any other control point within the building shall not contradict the control action. Where automatic means are provided to interrupt normal, nonemergency equipment operation or produce a specific result to safeguard the building or equipment including, but not limited to, duct freezestats, duct smoke detectors, high-temperature cutouts, temperature-actuated linkage and similar devices, such means shall be capable of being overridden by the fire fighter's control panel. The last control action as indicated by each fire fighter's control panel switch position shall prevail. Control actions shall not require the smoke control system to assume more than one configuration at any one time.

   Exception: Power disconnects required by NFPA 70.

2. Only the AUTO position of each three-position fire-fighter's control panel switch shall allow automatic or manual control action from other control points within the building. The AUTO position shall be the NORMAL, nonemergency, building control position. Where a fire fighter's control panel is in the AUTO position, the actual status of the device (on, off, open, closed) shall continue to be indicated by the status indicator described in Section 909.16.1. Where directed by an automatic signal to assume an emergency condition, the NORMAL position shall become the emergency condition for that device or group of devices within the zone. Control actions shall not require the smoke control system to assume more than one configuration at any one time.

909.17 System response time. Smoke-control system activation shall be initiated immediately after receipt of an appropriate automatic or manual activation command. Smoke control systems shall activate individual components (such as dampers and fans) in the sequence necessary to prevent physical damage to the fans, dampers, ducts and other equipment. For purposes of smoke control, the fire fighter's control panel response time shall be the same for automatic or manual smoke control action initiated from any other building control point. The total response time, including that necessary for detection, shutdown of operating equipment and smoke control system startup, shall allow for full operational mode to be achieved before the conditions in the space exceed the design smoke condition. The system response time for each component and their sequential relationships shall be detailed in the required rational analysis and verification of their installed condition reported in the required final report.

909.18 Acceptance testing. Devices, equipment, components and sequences shall be individually tested. These tests, in addition to those required by other provisions of this code, shall consist of determination of function, sequence and, where applicable, capacity of their installed condition.

909.18.1 Detection devices. Smoke or fire detectors that are a part of a smoke control system shall be tested in accordance with Chapter 9 in their installed condition. Where applicable, this testing shall include verification of airflow in both minimum and maximum conditions.

909.18.2 Ducts. Ducts that are part of a smoke control system shall be traversed using generally accepted practices to determine actual air quantities.

909.18.3 Dampers. Dampers shall be tested for function in their installed condition.

909.18.4 Inlets and outlets. Inlets and outlets shall be read using generally accepted practices to determine air quantities.

909.18.5 Fans. Fans shall be examined for correct rotation. Measurements of voltage, amperage, revolutions per minute and belt tension shall be made.

909.18.6 Smoke barriers. Measurements using inclined manometers or other *approved* calibrated measuring devices shall be made of the pressure differences across *smoke barriers*. Such measurements shall be conducted for each possible smoke control condition.

909.18.7 Controls. Each smoke zone equipped with an automatic-initiation device shall be put into operation by the actuation of one such device. Each additional device within the zone shall be verified to cause the same sequence without requiring the operation of fan motors in order to prevent damage. Control sequences shall be verified throughout the system, including verification of override from the fire fighter's control panel and simulation of standby power conditions.

909.18.8 Testing for smoke control. Smoke control systems shall be tested by a special inspector in accordance with Section 1705.18 of the *International Building Code*.

909.18.8.1 Scope of testing. Testing shall be conducted in accordance with the following:

1. During erection of ductwork and prior to concealment for the purposes of leakage testing and recording of device location.
2. Prior to occupancy and after sufficient completion for the purposes of pressure-difference testing, flow measurements, and detection and control verification.

909.18.8.2 Qualifications. *Approved* agencies for smoke control testing shall have expertise in fire protection engineering, mechanical engineering and certification as air balancers.

909.18.8.3 Reports. A complete report of testing shall be prepared by the *approved* agency. The report shall include identification of all devices by manufacturer, nameplate data, design values, measured values and identification tag or mark. The report shall be reviewed by the responsible registered design professional and, when satisfied that the design intent has been achieved, the responsible registered design professional shall sign, seal and date the report.

909.18.8.3.1 Report filing. A copy of the final report shall be filed with the *fire code official* and an identical copy shall be maintained in an *approved* location at the building.

909.18.9 Identification and documentation. Charts, drawings and other documents identifying and locating each component of the smoke control system, and describing their proper function and maintenance requirements, shall be maintained on file at the building as an attachment to the report required by Section 909.18.8.3. Devices shall have an *approved* identifying tag or mark on them consistent with the other required documentation and shall be dated indicating the last time they were successfully tested and by whom.

909.19 System acceptance. Buildings, or portions thereof, required by this code to comply with this section shall not be issued a certificate of occupancy until such time that the *fire code official* determines that the provisions of this section have been fully complied with and that the fire department has received satisfactory instruction on the operation, both automatic and manual, of the system and a written maintenance program complying with the requirements of Section 909.20.1 has been submitted and *approved* by the *fire code official*.

Exception: In buildings of phased construction, a temporary certificate of occupancy, as *approved* by the *fire code official*, shall be allowed, provided that those portions of the building to be occupied meet the requirements of this section and that the remainder does not pose a significant hazard to the safety of the proposed occupants or adjacent buildings.

909.20 Maintenance. Smoke control systems shall be maintained to ensure to a reasonable degree that the system is capable of controlling smoke for the duration required. The system shall be maintained in accordance with the manufacturer's instructions and Sections 909.20.1 through 909.20.6.

909.20.1 Schedule. A routine maintenance and operational testing program shall be initiated immediately after the smoke control system has passed the acceptance tests. A written schedule for routine maintenance and operational testing shall be established.

909.20.2 Records. Records of smoke control system testing and maintenance shall be maintained. The record shall include the date of the maintenance, identification of the servicing personnel and notification of any unsatisfactory condition and the corrective action taken, including parts replaced.

909.20.3 Testing. Operational testing of the smoke control system shall include all equipment such as initiating devices, fans, dampers, controls, doors and windows.

909.20.4 Dedicated smoke control systems. Dedicated smoke control systems shall be operated for each control sequence semiannually. The system shall be tested under standby power conditions.

909.20.5 Nondedicated smoke control systems. Nondedicated smoke control systems shall be operated for each control sequence annually. The system shall be tested under standby power conditions.

909.20.6 Components bypassing weekly test. Where components of the smoke control system are bypassed by the preprogrammed weekly test required by Section 909.12.1, such components shall be tested semiannually. The system shall be tested under standby power conditions.

[BF] 909.21 Elevator hoistway pressurization alternative. Where elevator hoistway pressurization is provided in lieu of required enclosed elevator lobbies, the pressurization system shall comply with Sections 909.21.1 through 909.21.11.

[BF] 909.21.1 Pressurization requirements. Elevator hoistways shall be pressurized to maintain a minimum positive pressure of 0.10 inch of water (25 Pa) and a maximum positive pressure of 0.25 inch of water (67 Pa) with respect to adjacent occupied space on all floors. This pressure shall be measured at the midpoint of each hoistway door, with all elevator cars at the floor of recall and all hoistway doors on the floor of recall open and all other hoistway doors closed. The pressure differential shall be measured between the hoistway and the adjacent elevator landing. The opening and closing of hoistway doors at each level must be demonstrated during this test. The supply air intake shall be from an outside, uncontaminated source located a minimum distance of 20 feet (6096 mm) from any air exhaust system or outlet.

Exceptions:

1. On floors containing only Group R occupancies, the pressure differential is permitted to be measured between the hoistway and a *dwelling unit* or *sleeping unit.*
2. Where an elevator opens into a lobby enclosed in accordance with Section 3007.6 or 3008.6 of the *International Building Code*, the pressure differential is permitted to be measured between the hoistway and the space immediately outside the door(s) from the floor to the enclosed lobby.
3. The pressure differential is permitted to be measured relative to the outdoor atmosphere on floors other than the following:
   - 3.1. The fire floor.
   - 3.2. The two floors immediately below the fire floor.
   - 3.3. The floor immediately above the fire floor.
4. The minimum positive pressure of 0.10 inch of water (25 Pa) and a maximum positive pressure of 0.25 inch of water (67 Pa) with respect to occupied floors is not required at the floor of recall with the doors open.

[BF] 909.21.1.1 Use of ventilation systems. Ventilation systems, other than hoistway supply air systems, are permitted to be used to exhaust air from adjacent spaces on the fire floor, two floors immediately below and one floor immediately above the fire floor to the building's exterior where necessary to maintain positive pressure relationships as required in Section 909.21.1 during operation of the elevator shaft pressurization system.

[BF] 909.21.2 Rational analysis. A rational analysis complying with Section 909.4 shall be submitted with the *construction documents*.

[BF] 909.21.3 Ducts for system. Any duct system that is part of the pressurization system shall be protected with the same *fire-resistance rating* as required for the elevator shaft enclosure.

[BF] 909.21.4 Fan system. The fan system provided for the pressurization system shall be as required by Sections 909.21.4.1 through 909.21.4.4.

[BF] 909.21.4.1 Fire resistance. Where located within the building, the fan system that provides the pressurization shall be protected with the same *fire-resistance rating* required for the elevator shaft enclosure.

[BF] 909.21.4.2 Smoke detection. The fan system shall be equipped with a smoke detector that will automatically shut down the fan system when smoke is detected within the system.

[BF] 909.21.4.3 Separate systems. A separate fan system shall be used for each elevator hoistway.

[BF] 909.21.4.4 Fan capacity. The supply fan shall be either adjustable with a capacity of not less than 1,000 cfm (0.4719 m^3/s) per door, or that specified by a *registered design professional* to meet the requirements of a designed pressurization system.

[BF] 909.21.5 Standby power. The pressurization system shall be provided with standby power in accordance with Section 604.

[BF] 909.21.6 Activation of pressurization system. The elevator pressurization system shall be activated upon activation of either the building fire alarm system or the elevator lobby smoke detectors. Where both a building fire alarm system and elevator lobby smoke detectors are present, each shall be independently capable of activating the pressurization system.

[BF] 909.21.7 Testing. Testing for performance shall be required in accordance with Section 909.18.8. System acceptance shall be in accordance with Section 909.19.

[BF] 909.21.8 Marking and identification. Detection and control systems shall be marked in accordance with Section 909.14.

[BF] 909.21.9 Control diagrams. Control diagrams shall be provided in accordance with Section 909.15.

[BF] 909.21.10 Control panel. A control panel complying with Section 909.16 shall be provided.

[BF] 909.21.11 System response time. Hoistway pressurization systems shall comply with the requirements for smoke control system response time in Section 909.17.

SECTION 910
SMOKE AND HEAT REMOVAL

910.1 General. Where required by this code, smoke and heat vents or mechanical smoke removal systems shall conform to the requirements of this section.

*

910.2 Where required. Smoke and heat vents or a mechanical smoke removal system shall be installed as required by Sections 910.2.1 and 910.2.2.

**

Exceptions:

1. Frozen food warehouses used solely for storage of Class I and II commodities where protected by an *approved automatic sprinkler system.*

2. Smoke and heat removal shall not be required in areas of buildings equipped with early suppression fast-response (ESFR) sprinklers.

3. Smoke and heat removal shall not be required in areas of buildings equipped with control mode special application sprinklers with a response time index of 50 $(m \cdot S)^{1/2}$ or less that are listed to control a fire in stored commodities with 12 or fewer sprinklers.

910.2.1 Group F-1 or S-1. Smoke and heat vents installed in accordance with Section 910.3 or a mechanical smoke removal system installed in accordance with Section 910.4 shall be installed in buildings and portions thereof used as a Group F-1 or S-1 occupancy having more than 50,000 square feet (4645 m^2) of undivided area. In occupied portions of a building equipped throughout with an *automatic sprinkler system* in accordance with Section 903.3.1.1, where the upper surface of the story is not a roof assembly, a mechanical smoke removal system in accordance with Section 910.4 shall be installed.

Exception: Group S-1 aircraft repair hangars.

910.2.2 High-piled combustible storage. Smoke and heat removal required by Table 3206.2 for buildings and portions thereof containing high-piled combustible storage shall be installed in accordance with Section 910.3 in unsprinklered buildings. In buildings and portions thereof containing high-piled combustible storage equipped throughout with an *automatic sprinkler system* in accordance with Section 903.3.1.1, a smoke and heat removal system shall be installed in accordance with Section 910.3 or 910.4. In occupied portions of a building equipped throughout with an *automatic sprinkler system* in accordance with Section 903.3.1.1 where the upper surface of the story is not a roof assembly, a mechanical smoke removal system in accordance with Section 910.4 shall be installed.

910.3 Smoke and heat vents. The design and installation of smoke and heat vents shall be in accordance with Sections 910.3.1 through 910.3.3.

910.3.1 Listing and labeling. Smoke and heat vents shall be *listed* and labeled to indicate compliance with UL 793 or FM 4430.

910.3.2 Smoke and heat vent locations. Smoke and heat vents shall be located 20 feet (6096 mm) or more from adjacent *lot lines* and *fire walls* and 10 feet (3048 mm) or more from *fire barriers.* Vents shall be uniformly located within the roof in the areas of the building where the vents are required to be installed by Section 910.2, with consideration given to roof pitch, sprinkler location and structural members.

910.3.3 Smoke and heat vents area. The required aggregate area of smoke and heat vents shall be calculated as follows:

For buildings equipped throughout with an *automatic sprinkler system* in accordance with Section 903.3.1.1:

$A_{VR} = V/9000$ **(Equation 9-4)**

where:

A_{VR} = The required aggregate vent area (ft^2).

V = Volume (ft^3) of the area that requires smoke removal.

For unsprinklered buildings:

$A_{VR} = A_{FA}/50$ **(Equation 9-5)**

where:

A_{VR} = The required aggregate vent area (ft^2).

A_{FA} = The area of the floor in the area that requires smoke removal.

910.4 Mechanical smoke removal systems. Mechanical smoke removal systems shall be designed and installed in accordance with Sections 910.4.1 through 910.4.7.

910.4.1 Automatic sprinklers required. The building shall be equipped throughout with an approved *automatic sprinkler system* in accordance with Section 903.3.1.1.

910.4.2 Exhaust fan construction. Exhaust fans that are part of a mechanical smoke removal system shall be rated for operation at 221°F (105°C). Exhaust fan motors shall be located outside of the exhaust fan air stream.

910.4.3 System design criteria. The mechanical smoke removal system shall be sized to exhaust the building at a minimum rate of two air changes per hour based upon the volume of the building or portion thereof without contents. The capacity of each exhaust fan shall not exceed 30,000 cubic feet per minute (14.2 m^3/sec).

910.4.3.1 Makeup air. Makeup air openings shall be provided within 6 feet (1829 mm) of the floor level. Operation of makeup air openings shall be manual or automatic. The minimum gross area of makeup air inlets shall be 8 square feet per 1,000 cubic feet per minute (0.74 m^2 per 0.4719 m^3/s) of smoke exhaust.

910.4.4 Activation. The mechanical smoke removal system shall be activated by manual controls only.

910.4.5 Manual control location. Manual controls shall be located so as to be accessible to the fire service from an exterior door of the building and protected against interior fire exposure by not less than 1-hour *fire barriers* constructed in accordance with Section 707 of the *International Building Code* or *horizontal assemblies* constructed in accordance with Section 711 of the *International Building Code*, or both.

910.4.6 Control wiring. Wiring for operation and control of mechanical smoke removal systems shall be connected ahead of the main disconnect in accordance with Section 701.12E of NFPA 70 and be protected against interior fire exposure to temperatures in excess of 1,000°F (538°C) for a period of not less than 15 minutes.

910.4.7 Controls. Where building air-handling and mechanical smoke removal systems are combined or where independent building air-handling systems are provided, fans shall automatically shut down in accordance with the *International Mechanical Code*. The manual controls provided for the smoke removal system shall have the capability to override the automatic shutdown of fans that are part of the smoke removal system.

910.5 Maintenance. Smoke and heat vents and mechanical smoke removal systems shall be maintained in an operative condition in accordance with Section 910.5.1 or 910.5.2, respectively.

910.5.1 Smoke and heat vents. Smoke and heat vents shall be maintained in an operative condition in accordance with NFPA 204 and Section 910.5.1.1

910.5.1.1 Fusible links. Fusible links for smoke and heat vents shall be replaced whenever fused, damaged or painted.

910.5.2 Mechanical smoke removal systems. Mechanical smoke removal systems shall be maintained in accordance with the equipment manufacturer's maintenance instructions and Sections 910.5.2.1 through 910.5.2.4.

910.5.2.1 Frequency. Systems shall be operationally tested not less than once per year. Testing shall include the operation of all system components, including control elements.

910.5.2.2 Testing. Operational testing of the mechanical smoke removal system shall include all equipment such as fans, controls and make-up air openings.

910.5.2.3 Schedule. A routine maintenance and operational testing program shall be initiated and a written schedule for routine maintenance and operational testing shall be established.

910.5.2.4 Written record. A written record of mechanical smoke exhaust system testing and maintenance shall be maintained on the premises. The written record shall include the date of the maintenance, identification of the servicing personnel and notification of any unsatisfactory condition and the corrective action taken, including parts replaced.

SECTION 911
EXPLOSION CONTROL

911.1 General. Explosion control shall be provided in the following locations:

1. Where a structure, room or space is occupied for purposes involving explosion hazards as identified in Table 911.1.
2. Where quantities of hazardous materials specified in Table 911.1 exceed the maximum allowable quantities in Table 5003.1.1(1).

Such areas shall be provided with explosion (*deflagration*) venting, explosion (*deflagration*) prevention systems or *barricades* in accordance with this section and NFPA 69, or NFPA 495 as applicable. *Deflagration* venting shall not be utilized as a means to protect buildings from *detonation* hazards.

911.2 Required deflagration venting. Areas that are required to be provided with *deflagration* venting shall comply with the following:

1. Walls, ceilings and roofs exposing surrounding areas shall be designed to resist a minimum internal pressure of 100 pounds per square foot (psf) (4788 Pa). The minimum internal design pressure shall be not less than five times the maximum internal relief pressure specified in Item 5 of this section.
2. *Deflagration* venting shall be provided only in exterior walls and roofs.

   Exception: Where sufficient *exterior wall* and roof venting cannot be provided because of inadequate exterior wall or roof area, *deflagration* venting shall be allowed by specially designed shafts vented to the exterior of the building.
3. *Deflagration* venting shall be designed to prevent unacceptable structural damage. Where relieving a *deflagration*, vent closures shall not produce projectiles of sufficient velocity and mass to cause life threatening injuries to the occupants or other persons on the property or adjacent *public ways*.
4. The aggregate clear area of vents and venting devices shall be governed by the pressure resistance of the construction assemblies specified in Item 1 of this section and the maximum internal pressure allowed by Item 5 of this section.
5. Vents shall be designed to withstand loads in accordance with the *International Building Code*. Vents shall consist of any one or any combination of the following to relieve at a maximum internal pressure of 20 pounds per square foot (958 Pa), but not less than the loads required by the *International Building Code*:
   5.1. *Exterior walls* designed to release outward.
   5.2. Hatch covers.
   5.3. Outward swinging doors.
   5.4. Roofs designed to uplift.
   5.5. Venting devices *listed* for the purpose.
6. Vents designed to release from the *exterior walls* or roofs of the building when venting a *deflagration* shall discharge directly to the exterior of the building where an unoccupied space not less than 50 feet (15 240 mm) in width is provided between the exterior walls of the building and the lot line.

   Exception: Vents complying with Item 7 of this section.
7. Vents designed to remain attached to the building when venting a *deflagration* shall be so located that the discharge opening shall be not less than 10 feet (3048 mm) vertically from window openings and *exits* in the building and 20 feet (6096 mm) horizontally from *exits* in the building, from window openings and *exits* in adjacent buildings on the same lot and from the lot line.

8. Discharge from vents shall not be into the interior of the building.

911.3 Explosion prevention systems. Explosion prevention systems shall be of an *approved* type and installed in accordance with the provisions of this code and NFPA 69.

911.4 Barricades. *Barricades* shall be designed and installed in accordance with NFPA 495.

SECTION 912 FIRE DEPARTMENT CONNECTIONS

912.1 Installation. Fire department connections shall be installed in accordance with the NFPA standard applicable to the system design and shall comply with Sections 912.2 through 912.7.

912.2 Location. With respect to hydrants, driveways, buildings and landscaping, fire department connections shall be so located that fire apparatus and hose connected to supply the system will not obstruct access to the buildings for other fire apparatus. The location of fire department connections shall be *approved* by the fire chief.

912.2.1 Visible location. Fire department connections shall be located on the street side of buildings, fully visible and recognizable from the street or nearest point of fire department vehicle access or as otherwise *approved* by the fire chief.

TABLE 911.1
EXPLOSION CONTROL REQUIREMENTS[f]

MATERIAL	CLASS	EXPLOSION CONTROL METHODS	
		Barricade construction	Explosion (deflagration) venting or explosion (deflagration) prevention systems
Hazard Category			
Combustible dusts[a]	—	Not required	Required
Cryogenic fluids	Flammable	Not required	Required
Explosives	Division 1.1 Division 1.2 Division 1.3 Division 1.4 Division 1.5 Division 1.6	Required Required Not required Not required Required Required	Not required Not required Required Required Not required Not required
Flammable gas	Gaseous Liquefied	Not required Not required	Required Required
Flammable liquids	IA[b] IB[c]	Not required Not required	Required Required
Organic peroxides	Unclassified detonable I	Required Required	Not permitted Not permitted
Oxidizer liquids and solids	4	Required	Not permitted
Pyrophoric	Gases	Not required	Required
Unstable (reactive)	4 3 detonable 3 nondetonable	Required Required Not required	Not permitted Not permitted Required
Water-reactive liquids and solids	3 2[e]	Not required Not required	Required Required
Special Uses			
Acetylene generator rooms	—	Not required	Required
Grain processing	—	Not required	Required
Liquefied petroleum gas distribution facilities	—	Not required	Required
Where explosion hazards exist[d]	Detonation Deflagration	Required Not required	Not permitted Required

a. Combustible dusts that are generated during manufacturing or processing. See definition of "Combustible dust" in Chapter 2.

b. Storage or use.

c. In open use or dispensing.

d. Rooms containing dispensing and use of hazardous materials where an explosive environment can occur because of the characteristics or nature of the hazardous materials or as a result of the dispensing or use process.

e. A method of explosion control shall be provided where Class 2 water-reactive materials can form potentially explosive mixtures.

f. Explosion venting is not required for Group H-5 Fabrication Areas complying with Chapter 27 and the *International Building Code*.

912.2.2 Existing buildings. On existing buildings, wherever the fire department connection is not visible to approaching fire apparatus, the fire department connection shall be indicated by an *approved* sign mounted on the street front or on the side of the building. Such sign shall have the letters "FDC" not less than 6 inches (152 mm) high and words in letters not less than 2 inches (51 mm) high or an arrow to indicate the location. Such signs shall be subject to the approval of the *fire code official*.

912.3 Fire hose threads. Fire hose threads used in connection with standpipe systems shall be *approved* and shall be compatible with fire department hose threads.

912.4 Access. Immediate access to fire department connections shall be maintained at all times and without obstruction by fences, bushes, trees, walls or any other fixed or moveable object. Access to fire department connections shall be *approved* by the fire chief.

Exception: Fences, where provided with an access gate equipped with a sign complying with the legend requirements of Section 912.5 and a means of emergency operation. The gate and the means of emergency operation shall be *approved* by the fire chief and maintained operational at all times.

912.4.1 Locking fire department connection caps. The *fire code official* is authorized to require locking caps on fire department connections for water-based *fire protection systems* where the responding fire department carries appropriate key wrenches for removal.

912.4.2 Clear space around connections. A working space of not less than 36 inches (914 mm) in width, 36 inches (914 mm) in depth and 78 inches (1981 mm) in height shall be provided and maintained in front of and to the sides of wall-mounted fire department connections and around the circumference of free-standing fire department connections, except as otherwise required or *approved* by the fire chief.

912.4.3 Physical protection. Where fire department connections are subject to impact by a motor vehicle, vehicle impact protection shall be provided in accordance with Section 312.

912.5 Signs. A metal sign with raised letters not less than 1 inch (25 mm) in size shall be mounted on all fire department connections serving automatic sprinklers, standpipes or fire pump connections. Such signs shall read: AUTOMATIC SPRINKLERS or STANDPIPES or TEST CONNECTION or a combination thereof as applicable. Where the fire department connection does not serve the entire building, a sign shall be provided indicating the portions of the building served.

912.6 Backflow protection. The potable water supply to automatic sprinkler and standpipe systems shall be protected against backflow as required by the *International Plumbing Code*.

912.7 Inspection, testing and maintenance. Fire department connections shall be periodically inspected, tested and maintained in accordance with NFPA 25. Records of inspection, testing and maintenance shall be maintained.

SECTION 913
FIRE PUMPS

913.1 General. Where provided, fire pumps shall be installed in accordance with this section and NFPA 20.

913.2 Protection against interruption of service. The fire pump, driver and controller shall be protected in accordance with NFPA 20 against possible interruption of service through damage caused by explosion, fire, flood, earthquake, rodents, insects, windstorm, freezing, vandalism and other adverse conditions.

913.2.1 Protection of fire pump rooms. Rooms where fire pumps are located shall be separated from all other areas of the building in accordance with Section 913.2.1 of the *International Building Code*.

913.2.2 Circuits supplying fire pumps. Cables used for survivability of circuits supplying fire pumps shall be *listed* in accordance with UL 2196. Electrical circuit protective systems shall be installed in accordance with their listing requirements.

913.3 Temperature of pump room. Suitable means shall be provided for maintaining the temperature of a pump room or pump house, where required, above 40°F (5°C).

913.3.1 Engine manufacturer's recommendation. Temperature of the pump room, pump house or area where engines are installed shall never be less than the minimum recommended by the engine manufacturer. The engine manufacturer's recommendations for oil heaters shall be followed.

913.4 Valve supervision. Where provided, the fire pump suction, discharge and bypass valves, and isolation valves on the backflow prevention device or assembly shall be supervised open by one of the following methods:

1. Central-station, proprietary or remote-station signaling service.
2. Local signaling service that will cause the sounding of an audible signal at a constantly attended location.
3. Locking valves open.
4. Sealing of valves and *approved* weekly recorded inspection where valves are located within fenced enclosures under the control of the *owner*.

913.4.1 Test outlet valve supervision. Fire pump test outlet valves shall be supervised in the closed position.

913.5 Testing and maintenance. Fire pumps shall be inspected, tested and maintained in accordance with the requirements of this section and NFPA 25. Records of inspection, testing and maintenance shall be maintained.

913.5.1 Acceptance test. Acceptance testing shall be done in accordance with the requirements of NFPA 20.

913.5.2 Generator sets. Engine generator sets supplying emergency or standby power to fire pump assemblies shall be periodically tested in accordance with NFPA 110. Records of testing shall be maintained.

913.5.3 Transfer switches. Automatic transfer switches shall be periodically tested in accordance with NFPA 110. Records of testing shall be maintained.

913.5.4 Pump room environmental conditions. Tests of pump room environmental conditions, including heating, ventilation and illumination, shall be made to ensure proper manual or automatic operation of the associated equipment.

SECTION 914
FIRE PROTECTION BASED ON SPECIAL DETAILED REQUIREMENTS OF USE AND OCCUPANCY

914.1 General. This section shall specify where *fire protection systems* are required based on the detailed requirements of use and occupancy of the *International Building Code*.

914.2 Covered and open mall buildings. Covered and open mall buildings shall comply with Sections 914.2.1 through 914.2.4.

914.2.1 Automatic sprinkler system. Covered and open mall buildings and buildings connected shall be equipped throughout with an automatic sprinkler system in accordance with Section 903.3.1.1, which shall comply with the all of the following:

1. The automatic sprinkler system shall be complete and operative throughout occupied space in the mall building prior to occupancy of any of the tenant spaces. Unoccupied tenant spaces shall be similarly protected unless provided with approved alternative protection.
2. Sprinkler protection for the mall of a covered mall building shall be independent from that provided for tenant spaces or anchor buildings.
3. Sprinkler protection for the tenant spaces of an open mall building shall be independent from that provided for anchor buildings.
4. Sprinkler protection shall be provided beneath exterior circulation balconies located adjacent to an open mall.
5. Where tenant spaces are supplied by the same system, they shall be independently controlled.

Exception: An *automatic sprinkler system* shall not be required in spaces or areas of open parking garages separated from the covered or open mall in accordance with Section 402.4.2.3 of the *International Building Code* and constructed in accordance with Section 406.5 of the *International Building Code*.

914.2.2 Standpipe system. The covered and open mall building shall be equipped throughout with a standpipe system as required by Section 905.3.3.

914.2.3 Emergency voice/alarm communication system. Where the total floor area exceeds 50,000 square feet (4645 m2) within either a covered mall building or within the perimeter line of an open mall building, an emergency voice/alarm communication system shall be provided. Emergency voice/alarm communication systems serving a mall, required or otherwise, shall be accessible to the fire department. The system shall be provided in accordance with Section 907.5.2.2.

914.2.4 Fire department access to equipment. Rooms or areas containing controls for air-conditioning systems, automatic fire-extinguishing systems, *automatic sprinkler systems* or other detection, suppression or control elements shall be identified for use by the fire department.

914.3 High-rise buildings. High-rise buildings shall comply with Sections 914.3.1 through 914.3.7.

914.3.1 Automatic sprinkler system. Buildings and structures shall be equipped throughout with an *automatic sprinkler system* in accordance with Section 903.3.1.1 and a secondary water supply where required by Section 914.3.3.

Exception: An *automatic sprinkler system* shall not be required in spaces or areas of:

1. Open parking garages in accordance with Section 406.5 of the *International Building Code.*
2. Telecommunications equipment buildings used exclusively for telecommunications equipment, associated electrical power distribution equipment, batteries and standby engines, provided that those spaces or areas are equipped throughout with an automatic fire detection system in accordance with Section 907.2 and are separated from the remainder of the building by not less than 1-hour *fire barriers* constructed in accordance with Section 707 of the *International Building Code* or not less than 2-hour *horizontal assemblies* constructed in accordance with Section 711 of the *International Building Code*, or both.

914.3.1.1 Number of sprinkler risers and system design. Each sprinkler system zone in buildings that are more than 420 feet (128 m) in height shall be supplied by no fewer than two risers. Each riser shall supply sprinklers on alternate floors. If more than two risers are provided for a zone, sprinklers on adjacent floors shall not be supplied from the same riser.

914.3.1.1.1 Riser location. Sprinkler risers shall be placed in interior exit stairways and ramps that are remotely located in accordance with Section 1015.2.

914.3.1.2 Water supply to required fire pumps. In buildings that are more than 420 feet (128 m) in *building height*, required fire pumps shall be supplied by connections to no fewer than two water mains located in different streets. Separate supply piping shall be provided between each connection to the water main and the pumps. Each connection and the supply piping between the connection and the pumps shall be sized to supply the flow and pressure required for the pumps to operate.

Exception: Two connections to the same main shall be permitted provided the main is valved such that an interruption can be isolated so that the water supply will continue without interruption through no fewer than one of the connections.

** **914.3.2 Secondary water supply.** An automatic secondary on-site water supply having a capacity not less than the hydraulically calculated sprinkler demand, including the hose stream requirement, shall be provided for high-rise buildings assigned to Seismic Design Category C, D, E or F as determined by the *International Building Code*. An additional fire pump shall not be required for the secondary water supply unless needed to provide the minimum design intake pressure at the suction side of the fire pump supplying the *automatic sprinkler system*. The secondary water supply shall have a duration of not less than 30 minutes as determined by the occupancy hazard classification in accordance with NFPA 13.

Exception: Existing buildings.

914.3.3 Fire alarm system. A fire alarm system shall be provided in accordance with Section 907.2.13.

914.3.4 Automatic smoke detection. Smoke detection shall be provided in accordance with Section 907.2.13.1.

914.3.5 Emergency voice/alarm communication system. An emergency voice/alarm communication system shall be provided in accordance with Section 907.5.2.2.

914.3.6 Emergency responder radio coverage. Emergency responder radio coverage shall be provided in accordance with Section 510.

914.3.7 Fire command. A *fire command center* complying with Section 508 shall be provided in a location *approved* by the fire department.

914.4 Atriums. Atriums shall comply with Sections 914.4.1 and 914.4.2.

914.4.1 Automatic sprinkler system. An *approved automatic sprinkler system* shall be installed throughout the entire building.

Exceptions:

1. That area of a building adjacent to or above the atrium need not be sprinklered, provided that portion of the building is separated from the atrium portion by not less than a 2-hour *fire barrier* constructed in accordance with Section 707 of the *International Building Code* or *horizontal assemblies* constructed in accordance with Section 711 of the *International Building Code*, or both.
2. Where the ceiling of the atrium is more than 55 feet (16 764 mm) above the floor, sprinkler protection at the ceiling of the atrium is not required.

914.4.2 Fire alarm system. A fire alarm system shall be provided where required by Section 907.2.14.

914.5 Underground buildings. Underground buildings shall comply with Sections 914.5.1 through 914.5.5.

914.5.1 Automatic sprinkler system. The highest *level of exit discharge* serving the underground portions of the building and all levels below shall be equipped with an *automatic sprinkler system* installed in accordance with Section 903.3.1.1. Water-flow switches and control valves shall be supervised in accordance with Section 903.4.

914.5.2 Smoke control system. A smoke control system is required to control the migration of products of combustion in accordance with Section 909 and provisions of this section. Smoke control shall restrict movement of smoke to the general area of fire origin and maintain *means of egress* in a usable condition.

914.5.3 Compartment smoke control system. Where compartmentation is required by Section 405.4 of the *International Building Code*, each compartment shall have an independent smoke control system. The system shall be automatically activated and capable of manual operation in accordance with Section 907.2.18.

914.5.4 Fire alarm system. A fire alarm system shall be provided where required by Sections 907.2.18 and 907.2.19.

914.5.5 Standpipe system. The underground building shall be provided throughout with a standpipe system in accordance with Section 905.

914.6 Stages. Stages shall comply with Sections 914.6.1 and 914.6.2.

914.6.1 Automatic sprinkler system. Stages shall be equipped with an *automatic sprinkler system* in accordance with Section 903.3.1.1. Sprinklers shall be installed under the roof and gridiron and under all catwalks and galleries over the stage. Sprinklers shall be installed in dressing rooms, performer lounges, shops and storerooms accessory to such stages.

Exceptions:

1. Sprinklers are not required under stage areas less than 4 feet (1219 mm) in clear height utilized exclusively for storage of tables and chairs, provided the concealed space is separated from the adjacent spaces by Type X gypsum board not less than $^{5}/_{8}$ inch (15.9 mm) in thickness.
2. Sprinklers are not required for stages 1,000 square feet (93 m^2) or less in area and 50 feet (15 240 mm) or less in height where curtains, scenery or other combustible hangings are not retractable vertically. Combustible hangings shall be limited to a single main curtain, borders, legs and a single backdrop.
3. Sprinklers are not required within portable orchestra enclosures on stages.

914.6.2 Standpipe system. Standpipe systems shall be provided in accordance with Section 905.

914.7 Special amusement buildings. Special amusement buildings shall comply with Sections 914.7.1 and 914.7.2.

914.7.1 Automatic sprinkler system. Special amusement buildings shall be equipped throughout with an *automatic sprinkler system* in accordance with Section 903.3.1.1.

Where the special amusement building is temporary, the sprinkler water supply shall be of an *approved* temporary means.

Exception: Automatic sprinklers are not required where the total floor area of a temporary special amusement building is less than 1,000 square feet (93 m^2) and the *exit access* travel distance from any point to an *exit* is less than 50 feet (15 240 mm).

914.7.2 Automatic smoke detection. Special amusement buildings shall be equipped with an automatic smoke detection system in accordance with Section 907.2.12.

914.8 Aircraft-related occupancies. Aircraft-related occupancies shall comply with Sections 914.8.1 through 914.8.6.

914.8.1 Automatic smoke detection systems. Airport traffic control towers shall be provided with an automatic smoke detection system installed in accordance with Section 907.2.22.

914.8.2 Automatic sprinkler system for new airport traffic control towers. Where an occupied floor is located more than 35 feet (10 668 mm) above the lowest level of fire department vehicle access, new airport traffic control towers shall be equipped with an *automatic sprinkler system* in accordance with Section 903.3.1.1.

914.8.3 Fire suppression for aircraft hangars. Aircraft hangars shall be provided with a fire suppression system designed in accordance with NFPA 409, based upon the classification for the hangar given in Table 914.8.3.

Exception: Where a fixed base operator has separate repair facilities on site, Group II hangars operated by a fixed base operator used for storage of transient aircraft only shall have a fire suppression system, but the system shall be exempt from foam requirements.

914.8.3.1 Hazardous operations. Any Group III aircraft hangar in accordance with Table 914.8.3 that contains hazardous operations including, but not limited to, the following shall be provided with a Group I or II fire suppression system in accordance with NFPA 409 as applicable:

1. Doping.
2. Hot work including, but not limited to, welding, torch cutting and torch soldering.
3. Fuel transfer.
4. Fuel tank repair or maintenance not including defueled tanks in accordance with NFPA 409, inerted tanks or tanks that have never been fueled.
5. Spray finishing operations.
6. Total fuel capacity of all aircraft within the unsprinklered single *fire area* in excess of 1,600 gallons (6057 L).
7. Total fuel capacity of all aircraft within the maximum single *fire area* in excess of 7,500 gallons (28 390 L) for a hangar equipped throughout with an *automatic sprinkler system* installed in accordance with Section 903.3.1.1.

914.8.3.2 Separation of maximum single fire areas. Maximum single *fire areas* established in accordance with hangar classification and construction type in Table 914.8.3 shall be separated by 2-hour *fire walls* constructed in accordance with Section 706 of the *International Building Code*. In determining the maximum single fire area as set forth in Table 914.8.3, ancillary uses that are separated from aircraft servicing areas by not less than a 1-hour *fire barrier* constructed in accordance with Section 707 of the *International Building Code* shall not be included in the area.

914.8.4 Finishing. The process of "doping," involving the use of a volatile flammable solvent, or of painting shall be carried on in a separate detached building equipped with automatic fire-extinguishing equipment in accordance with Section 903.

TABLE 914.8.3
HANGAR FIRE SUPPRESSION REQUIREMENTS[a,b,c]

MAXIMUM SINGLE FIRE AREA (square feet)	INTERNATIONAL BUILDING CODE TYPE OF CONSTRUCTION								
	IA	IB	IIA	IIB	IIIA	IIIB	IV	VA	VB
> 40,001	Group I	Group I	Group I	Group I	Group I	Group I	Group I	Group I	Group I
40,000	Group II	Group II	Group II	Group II	Group II	Group II	Group II	Group II	Group II
30,000	Group III	Group II	Group II	Group II	Group II	Group II	Group II	Group II	Group II
20,000	Group III	Group III	Group II	Group II	Group II	Group II	Group II	Group II	Group II
15,000	Group III	Group III	Group III	Group II	Group III	Group II	Group III	Group II	Group II
12,000	Group III	Group III	Group III	Group III	Group III	Group III	Group III	Group II	Group II
8,000	Group III	Group III	Group III	Group III	Group III	Group III	Group III	Group III	Group II
5,000	Group III	Group III	Group III	Group III	Group III	Group III	Group III	Group III	Group III

For SI: 1 square foot = 0.0929 m^2, 1 foot = 304.8 mm.

a. Aircraft hangars with a door height greater than 28 feet shall be provided with fire suppression for a Group I hangar regardless of maximum fire area.

b. Groups shall be as classified in accordance with NFPA 409.

c. Membrane structures complying with Section 3102 of the *International Building Code* shall be classified as a Group IV hangar.

914.8.5 Residential aircraft hangar smoke alarms. Smoke alarms shall be provided within residential aircraft hangars in accordance with Section 907.2.21.

914.8.6 Aircraft paint hangar fire suppression. Aircraft paint hangars shall be provided with fire suppression as required by NFPA 409.

914.9 Application of flammable finishes. An *automatic sprinkler system* or fire-extinguishing system shall be provided in all spray, dip and immersing spaces and storage rooms, and shall be installed in accordance with Chapter 9.

914.10 Drying rooms. Drying rooms designed for high-hazard materials and processes, including special occupancies as provided for in Chapter 4 of the *International Building Code*, shall be protected by an *approved* automatic fire-extinguishing system complying with the provisions of Chapter 9.

914.11 Ambulatory care facilities. Occupancies classified as ambulatory care facilities shall comply with Sections 914.11.1 through 914.11.3.

914.11.1 Automatic sprinkler systems. An *automatic sprinkler system* shall be provided for ambulatory care facilities in accordance with Section 903.2.2.

914.11.2 Manual fire alarm systems. A manual fire alarm system shall be provided for ambulatory care facilities in accordance with Section 907.2.2.

914.11.3 Fire alarm systems. An automatic smoke detection system shall be provided for ambulatory care facilities in accordance with Section 907.2.2.1.

**

SECTION 915 CARBON MONOXIDE DETECTION

915.1 General. Carbon monoxide detection shall be installed in new buildings in accordance with Sections 915.1.1 through 915.6. Carbon monoxide detection shall be installed in existing buildings in accordance with Section 1103.9.

915.1.1 Where required. Carbon monoxide detection shall be provided in Group I-1, I-2, I-4 and R occupancies and in classrooms in Group E occupancies in the locations specified in Section 915.2 where any of the conditions in Sections 915.1.2 through 915.1.6 exist.

915.1.2 Fuel-burning appliances and fuel-burning fireplaces. Carbon monoxide detection shall be provided in *dwelling units*, *sleeping units* and classrooms that contain a fuel-burning appliance or a fuel-burning fireplace.

915.1.3 Forced-air furnaces. Carbon monoxide detection shall be provided in *dwelling units*, *sleeping units* and classrooms served by a fuel-burning, forced-air furnace.

Exception: Carbon monoxide detection shall not be required in *dwelling units*, *sleeping units* and classrooms where carbon monoxide detection is provided in the first room or area served by each main duct leaving the furnace, and the carbon monoxide alarm signals are automatically transmitted to an approved location.

915.1.4 Fuel-burning appliances outside of dwelling units, sleeping units and classrooms. Carbon monoxide detection shall be provided in *dwelling units*, *sleeping units* and classrooms located in buildings that contain fuel-burning appliances or fuel-burning fireplaces.

Exceptions:

1. Carbon monoxide detection shall not be required in *dwelling units*, *sleeping units* and classrooms where there are no communicating openings between the fuel-burning appliance or fuel-burning fireplace and the *dwelling unit*, *sleeping unit* or classroom.
2. Carbon monoxide detection shall not be required in *dwelling units*, *sleeping units* and classrooms where carbon monoxide detection is provided in one of the following locations:
   - 2.1. In an approved location between the fuel-burning appliance or fuel-burning fireplace and the *dwelling unit*, *sleeping unit* or classroom.
   - 2.2. On the ceiling of the room containing the fuel-burning appliance or fuel-burning fireplace.

915.1.5 Private garages. Carbon monoxide detection shall be provided in *dwelling units*, *sleeping units* and classrooms in buildings with attached private garages.

Exceptions:

1. Carbon monoxide detection shall not be required where there are no communicating openings between the private garage and the *dwelling unit*, *sleeping unit* or classroom.
2. Carbon monoxide detection shall not be required in *dwelling units*, *sleeping units* and classrooms located more than one story above or below a private garage.
3. Carbon monoxide detection shall not be required where the private garage connects to the building through an open-ended corridor.
4. Where carbon monoxide detection is provided in an approved location between openings to a private garage and *dwelling units*, *sleeping units* or classrooms, carbon monoxide detection shall not be required in the *dwelling units*, *sleeping units* or classrooms.

915.1.6 Exempt garages. For determining compliance with Section 915.1.5, an open parking garage complying with Section 406.5 of the *International Building Code* or an enclosed parking garage complying with Section 406.6 of the *International Building Code* shall not be considered a private garage.

915.2 Locations. Where required by Section 915.1.1, carbon monoxide detection shall be installed in the locations specified in Sections 915.2.1 through 915.2.3.

915.2.1 Dwelling units. Carbon monoxide detection shall be installed in *dwelling units* outside of each separate sleeping area in the immediate vicinity of the bedrooms.

Where a fuel-burning appliance is located within a bedroom or its attached bathroom, carbon monoxide detection shall be installed within the bedroom.

915.2.2 Sleeping units. Carbon monoxide detection shall be installed in *sleeping units.*

Exception: Carbon monoxide detection shall be allowed to be installed outside of each separate sleeping area in the immediate vicinity of the *sleeping unit* where the *sleeping unit* or its attached bathroom does not contain a fuel-burning appliance and is not served by a forced air furnace.

915.2.3 Group E occupancies. Carbon monoxide detection shall be installed in classrooms in Group E occupancies. Carbon monoxide alarm signals shall be automatically transmitted to an on-site location that is staffed by school personnel.

Exception: Carbon monoxide alarm signals shall not be required to be automatically transmitted to an on-site location that is staffed by school personnel in Group E occupancies with an occupant load of 30 or less.

915.3 Detection equipment. Carbon monoxide detection required by Sections 915.1 through 915.2.3 shall be provided by carbon monoxide alarms complying with Section 915.4 or carbon monoxide detection systems complying with Section 915.5.

915.4 Carbon monoxide alarms. Carbon monoxide alarms shall comply with Sections 915.4.1 through 915.4.3.

915.4.1 Power source. Carbon monoxide alarms shall receive their primary power from the building wiring where such wiring is served from a commercial source, and when primary power is interrupted, shall receive power from a battery. Wiring shall be permanent and without a disconnecting switch other than that required for overcurrent protection.

Exception: Where installed in buildings without commercial power, battery-powered carbon monoxide alarms shall be an acceptable alternative.

915.4.2 Listings. Carbon monoxide alarms shall be listed in accordance with UL 2034.

915.4.3 Combination alarms. Combination carbon monoxide/smoke alarms shall be an acceptable alternative to carbon monoxide alarms. Combination carbon monoxide/smoke alarms shall be listed in accordance with UL 2034 and UL 217.

915.5 Carbon monoxide detection systems. Carbon monoxide detection systems shall be an acceptable alternative to carbon monoxide alarms and shall comply with Sections 915.5.1 through 915.5.3.

915.5.1 General. Carbon monoxide detection systems shall comply with NFPA 720. Carbon monoxide detectors shall be listed in accordance with UL 2075.

915.5.2 Locations. Carbon monoxide detectors shall be installed in the locations specified in Section 915.2. These locations supersede the locations specified in NFPA 720.

915.5.3 Combination detectors. Combination carbon monoxide/smoke detectors installed in carbon monoxide detection systems shall be an acceptable alternative to carbon monoxide detectors, provided they are listed in accordance with UL 2075 and UL 268.

915.6 Maintenance. Carbon monoxide alarms and carbon monoxide detection systems shall be maintained in accordance with NFPA 720. Carbon monoxide alarms and carbon monoxide detectors that become inoperable or begin producing end-of-life signals shall be replaced.

CHAPTER 10

MEANS OF EGRESS

SECTION 1001 ADMINISTRATION

1001.1 General. Buildings or portions thereof shall be provided with a *means of egress* system as required by this chapter. The provisions of this chapter shall control the design, construction and arrangement of *means of egress* components required to provide an *approved means of egress* from structures and portions thereof. Sections 1003 through 1030 shall apply to new construction. Section 1031 shall apply to existing buildings.

> **Exception:** Detached one- and two-family dwellings and multiple single-family dwellings (townhouses) not more than three stories above grade plane in height with a separate means of egress and their accessory structures shall comply with the *International Residential Code*.

1001.2 Minimum requirements. It shall be unlawful to alter a building or structure in a manner that will reduce the number of *exits* or the capacity of the *means of egress* to less than required by this code.

SECTION 1002 DEFINITIONS

[BE] 1002.1 Definitions. The following terms are defined in Chapter 2:

ACCESSIBLE MEANS OF EGRESS.

AISLE.

AISLE ACCESSWAY.

ALTERNATING TREAD DEVICE.

AREA OF REFUGE.

BLEACHERS.

BREAKOUT.

COMMON PATH OF EGRESS TRAVEL.

CORRIDOR.

DOOR, BALANCED.

EGRESS COURT.

EMERGENCY ESCAPE AND RESCUE OPENING.

EXIT.

EXIT ACCESS.

EXIT ACCESS DOORWAY.

EXIT ACCESS RAMP.

EXIT ACCESS STAIRWAY.

EXIT DISCHARGE.

EXIT DISCHARGE, LEVEL OF.

EXIT, HORIZONTAL.

EXIT PASSAGEWAY.

EXTERIOR EXIT RAMP.

EXTERIOR EXIT STAIRWAY.

FIRE EXIT HARDWARE.

FIXED SEATING.

FLIGHT.

FLOOR AREA, GROSS.

FLOOR AREA, NET.

FOLDING AND TELESCOPIC SEATING.

GRANDSTAND.

GUARD.

HANDRAIL.

INTERIOR EXIT RAMP.

INTERIOR EXIT STAIRWAY.

LOW ENERGY POWER-OPERATED DOOR.

MEANS OF EGRESS.

MERCHANDISE PAD.

NOSING.

OCCUPANT LOAD.

OPEN-ENDED CORRIDOR.

PANIC HARDWARE.

PHOTOLUMINESCENT.

POWER-ASSISTED DOOR.

POWER-OPERATED DOOR.

PUBLIC WAY.

RAMP.

SCISSOR STAIRWAY.

SELF-LUMINOUS.

SMOKE-PROTECTED ASSEMBLY SEATING.

STAIR.

STAIRWAY.

STAIRWAY, INTERIOR.

STAIRWAY, SPIRAL.

WINDER.

SECTION 1003 GENERAL MEANS OF EGRESS

[BE] 1003.1 Applicability. The general requirements specified in Sections 1003 through 1015 shall apply to all three elements of the *means of egress* system, in addition to those specific requirements for the *exit access*, the *exit* and the *exit discharge* detailed elsewhere in this chapter.

[BE] 1003.2 Ceiling height. The *means of egress* shall have a ceiling height of not less than 7 feet 6 inches (2286 mm).

Exceptions:

1. Sloped ceilings in accordance with Section 1208.2.
2. Ceilings of *dwelling units* and *sleeping units* within residential occupancies in accordance with Section 1208.2 of the *International Building Code.*
3. Allowable projections in accordance with Section 1003.3.
4. *Stair* headroom in accordance with Section 1011.3.
5. Door height in accordance with Section 1010.1.1.
6. *Ramp* headroom in accordance with Section 1012.5.2.
7. The clear height of floor levels in vehicular and pedestrian traffic areas of public and private parking garages in accordance with Section 406.4.1 of the *International Building Code.*
8. Areas above and below *mezzanine* floors in accordance with Section 505.2 of the *International Building Code.*

[BE] 1003.3 Protruding objects. Protruding objects on circulation paths shall comply with the requirements of Sections 1003.3.1 through 1003.3.4.

[BE] 1003.3.1 Headroom. Protruding objects are permitted to extend below the minimum ceiling height required by Section 1003.2 where a minimum headroom of 80 inches (2032 mm) is provided over any walking surface, including walks, *corridors*, *aisles* and passageways. Not more than 50 percent of the ceiling area of a *means of egress* shall be reduced in height by protruding objects.

Exception: Door closers and stops shall not reduce headroom to less than 78 inches (1981 mm).

A barrier shall be provided where the vertical clearance is less than 80 inches (2032 mm) high. The leading edge of such a barrier shall be located 27 inches (686 mm) maximum above the floor.

[BE] 1003.3.2 Post-mounted objects. A free-standing object mounted on a post or pylon shall not overhang that post or pylon more than 4 inches (102 mm) where the lowest point of the leading edge is more than 27 inches (686 mm) and less than 80 inches (2032 mm) above the walking surface. Where a sign or other obstruction is mounted between posts or pylons and the clear distance between the posts or pylons is greater than 12 inches (305 mm), the lowest edge of such sign or obstruction shall be 27 inches (686 mm) maximum or 80 inches (2032 mm) minimum above the finished floor or ground.

Exception: These requirements shall not apply to sloping portions of *handrails* between the top and bottom riser of *stairs* and above the *ramp* run.

[BE] 1003.3.3 Horizontal projections. Objects with leading edges more than 27 inches (685 mm) and not more than 80 inches (2030 mm) above the floor shall not project horizontally more than 4 inches (102 mm) into the circulation path.

Exception: *Handrails* are permitted to protrude $4^1/_2$ inches (114 mm) from the wall.

[BE] 1003.3.4 Clear width. Protruding objects shall not reduce the minimum clear width of *accessible routes.*

[BE] 1003.4 Floor surface. Walking surfaces of the *means of egress* shall have a slip-resistant surface and be securely attached.

[BE] 1003.5 Elevation change. Where changes in elevation of less than 12 inches (305 mm) exist in the *means of egress*, sloped surfaces shall be used. Where the slope is greater than one unit vertical in 20 units horizontal (5-percent slope), *ramps* complying with Section 1012 shall be used. Where the difference in elevation is 6 inches (152 mm) or less, the *ramp* shall be equipped with either *handrails* or floor finish materials that contrast with adjacent floor finish materials.

Exceptions:

1. A single step with a maximum riser height of 7 inches (178 mm) is permitted for buildings with occupancies in Groups F, H, R-2, R-3, S and U at exterior doors not required to be accessible by Chapter 11 of the *International Building Code.*
2. A *stair* with a single riser or with two risers and a tread is permitted at locations not required to be *accessible* by Chapter 11 of the *International Building Code*, where the risers and treads comply with Section 1011.5, the minimum depth of the tread is 13 inches (330 mm) and not less than one *handrail* complying with Section 1014 is provided within 30 inches (762 mm) of the centerline of the normal path of egress travel on the *stair.*
3. A step is permitted in *aisles* serving seating that has a difference in elevation less than 12 inches (305 mm) at locations not required to be *accessible* by Chapter 11 of the *International Building Code*, provided that the risers and treads comply with Section 1029.13 and the *aisle* is provided with a *handrail* complying with Section 1029.15.

Throughout a story in a Group I-2 occupancy, any change in elevation in portions of the *means of egress* that serve nonambulatory persons shall be by means of a *ramp* or sloped walkway.

[BE] 1003.6 Means of egress continuity. The path of egress travel along a *means of egress* shall not be interrupted by a building element other than a *means of egress* component as specified in this chapter. Obstructions shall not be placed in the minimum width or required capacity of a *means of egress* component except projections permitted by this chapter. The minimum width or required capacity of a *means of egress* system shall not be diminished along the path of egress travel.

[BE] 1003.7 Elevators, escalators and moving walks. Elevators, escalators and moving walks shall not be used as a component of a required *means of egress* from any other part of the building.

Exception: Elevators used as an *accessible means of egress* in accordance with Section 1009.4.

SECTION 1004
OCCUPANT LOAD

[BE] 1004.1 Design occupant load. In determining *means of egress* requirements, the number of occupants for whom *means of egress* facilities are provided shall be determined in accordance with this section.

[BE] 1004.1.1 Cumulative occupant loads. Where the path of egress travel includes intervening rooms, areas or spaces, cumulative *occupant loads* shall be determined in accordance with this section.

[BE] 1004.1.1.1 Intervening spaces or accessory areas. Where occupants egress from one or more rooms, areas or spaces through others, the design *occupant load* shall be the combined *occupant load* of interconnected accessory or intervening spaces. Design of egress path capacity shall be based on the cumulative portion of *occupant loads* of all rooms, areas or spaces to that point along the path of egress travel.

[BE] 1004.1.1.2 Adjacent levels for mezzanines. That portion of the *occupant load* of a *mezzanine* with required egress through a room, area or space on an adjacent level shall be added to the *occupant load* of that room, area or space.

[BE] 1004.1.1.3 Adjacent stories. Other than for the egress components designed for convergence in accordance with Section 1005.6, the *occupant load* from separate stories shall not be added.

[BE] 1004.1.2 Areas without fixed seating. The number of occupants shall be computed at the rate of one occupant per unit of area as prescribed in Table 1004.1.2. For areas without *fixed seating*, the occupant load shall be not less than that number determined by dividing the floor area under consideration by the *occupant load* factor assigned to the function of the space as set forth in Table 1004.1.2. Where an intended function is not listed in Table 1004.1.2, the *fire code official* shall establish a function based on a listed function that most nearly resembles the intended function.

Exception: Where *approved* by the *fire code official*, the actual number of occupants for whom each occupied space, floor or building is designed, although less than those determined by calculation, shall be permitted to be used in the determination of the design *occupant load.*

[BE] 1004.2 Increased occupant load. The *occupant load* permitted in any building, or portion thereof, is permitted to be increased from that number established for the occupancies in Table 1004.1.2, provided that all other requirements of the code are met based on such modified number and the *occupant load* does not exceed one occupant per 7 square feet (0.65 m^2) of occupiable floor space. Where required by the

[BE] TABLE 1004.1.2
MAXIMUM FLOOR AREA ALLOWANCES PER OCCUPANT

FUNCTION OF SPACE	OCCUPANT LOAD FACTOR[a]
Accessory storage areas, mechanical equipment room	300 gross
Agricultural building	300 gross
Aircraft hangars	500 gross
Airport terminal Baggage claim Baggage handling Concourse Waiting areas	 20 gross 300 gross 100 gross 15 gross
Assembly Gaming floors (keno, slots, etc.) Exhibit gallery and museum	 11 gross 30 net
Assembly with fixed seats	See Section 1004.4
Assembly without fixed seats Concentrated (chairs only – not fixed) Standing space Unconcentrated (tables and chairs)	 7 net 5 net 15 net
Bowling centers, allow 5 persons for each lane including 15 feet of runway, and for additional areas	7 net
Business areas	100 gross
Courtrooms – other than fixed seating areas	40 net
Day care	35 net
Dormitories	50 gross
Educational Classroom area Shops and other vocational room areas	 20 net 50 net
Exercise rooms	50 gross
Group H-5 Fabrication and manufacturing areas	200 gross
Industrial areas	100 gross
Institutional areas Inpatient treatment areas Outpatient areas Sleeping areas	 240 gross 100 gross 120 gross
Kitchens, commercial	200 gross
Library Reading rooms Stack area	 50 net 100 gross
Locker rooms	50 gross
Mall buildings – covered and open	See Section 402.8.2 of the *International Building Code*
Mercantile Storage, stock, shipping areas	60 gross 300 gross
Parking garages	200 gross
Residential	200 gross
Skating rinks, swimming pools Rink and pool Decks	 50 gross 15 gross
Stages and platforms	15 net
Warehouses	500 gross

For SI: 1 square foot = 0.0929 m^2, 1 foot = 304.8 mm.

a. Floor area in square feet per occupant.

fire code official, an *approved aisle*, seating or fixed equipment diagram substantiating any increase in *occupant load* shall be submitted. Where required by the *fire code official*, such diagram shall be posted.

[BE] 1004.3 Posting of occupant load. Every room or space that is an assembly occupancy shall have the *occupant load* of the room or space posted in a conspicuous place, near the main *exit* or *exit access* doorway from the room or space. Posted signs shall be of an *approved* legible permanent design and shall be maintained by the owner or the owner's authorized agent.

[BE] 1004.4 Fixed seating. For areas having fixed seats and *aisles*, the *occupant load* shall be determined by the number of fixed seats installed therein. The *occupant load* for areas in which *fixed seating* is not installed, such as waiting spaces, shall be determined in accordance with Section 1004.1.2 and added to the number of fixed seats.

The *occupant load* of wheelchair spaces and the associated companion seat shall be based on one occupant for each wheelchair space and one occupant for the associated companion seat provided in accordance with Section 1108.2.3 of the *International Building Code*.

For areas having *fixed seating* without dividing arms, the *occupant load* shall be not less than the number of seats based on one person for each 18 inches (457 mm) of seating length.

The *occupant load* of seating booths shall be based on one person for each 24 inches (610 mm) of booth seat length measured at the backrest of the seating booth.

[BE] 1004.5 Outdoor areas. *Yards*, patios, *courts* and similar outdoor areas accessible to and usable by the building occupants shall be provided with *means of egress* as required by this chapter. The *occupant load* of such outdoor areas shall be assigned by the *fire code official* in accordance with the anticipated use. Where outdoor areas are to be used by persons in addition to the occupants of the building, and the path of egress travel from the outdoor areas passes through the building, *means of egress* requirements for the building shall be based on the sum of the *occupant loads* of the building plus the outdoor areas.

Exceptions:

1. Outdoor areas used exclusively for service of the building need only have one *means of egress*.
2. Both outdoor areas associated with Group R-3 and individual dwelling units of Group R-2.

[BE] 1004.6 Multiple occupancies. Where a building contains two or more occupancies, the *means of egress* requirements shall apply to each portion of the building based on the occupancy of that space. Where two or more occupancies utilize portions of the same *means of egress* system, those egress components shall meet the more stringent requirements of all occupancies that are served.

SECTION 1005 MEANS OF EGRESS SIZING

[BE] 1005.1 General. All portions of the *means of egress* system shall be sized in accordance with this section.

Exception: *Aisles* and *aisle accessways* in rooms or spaces used for assembly purposes complying with Section 1029.

[BE] 1005.2 Minimum width based on component. The minimum width, in inches (mm), of any *means of egress* components shall be not less than that specified for such component, elsewhere in this code.

[BE] 1005.3 Required capacity based on occupant load. The required capacity, in inches (mm), of the *means of egress* for any room, area, space or story shall be not less than that determined in accordance with Sections 1005.3.1 and 1005.3.2:

[BE] 1005.3.1 Stairways. The capacity, in inches, of *means of egress stairways* shall be calculated by multiplying the *occupant load* served by such *stairways* by a means of *egress* capacity factor of 0.3 inch (7.6 mm) per occupant. Where *stairways* serve more than one story, only the *occupant load* of each story considered individually shall be used in calculating the required capacity of the *stairways* serving that story.

Exceptions:

1. For other than Group H and I-2 occupancies, the capacity, in inches, of *means of egress stairways* shall be calculated by multiplying the *occupant load* served by such *stairways* by a *means of egress* capacity factor of 0.2 inches (5.1 mm) per occupant in buildings equipped throughout with an *automatic sprinkler system* installed in accordance with Section 903.3.1.1 or 903.3.1.2 and an emergency voice/alarm communication system in accordance with Section 907.5.2.2.
2. Facilities with *smoke-protected assembly seating* shall be permitted to use the capacity factors in Table 1029.6.2 indicated for stepped *aisles* for *exit access* or *exit stairways* where the entire path for *means of egress* from the seating to the *exit discharge* is provided with a smoke control system complying with Section 909.
3. Facilities with outdoor *smoke-protected assembly seating* shall be permitted to the capacity factors in Section 1029.6.3 indicated for stepped *aisles* for *exit access* or *exit stairways* where the entire path for *means of egress* from the seating to the *exit discharge* is open to the outdoors.

[BE] 1005.3.2 Other egress components. The capacity, in inches, of *means of egress* components other than *stairways* shall be calculated by multiplying the *occupant load* served by such component by a *means of egress* capacity factor of 0.2 inches (5.1 mm) per occupant.

Exceptions:

1. For other than Group H and I-2 occupancies, the capacity, in inches, of *means of egress* components other than *stairways* shall be calculated by multiplying the *occupant load* served by such component by a *means of egress* capacity factor of 0.15 inches (3.8 mm) per occupant in buildings equipped throughout with an *automatic sprinkler system* installed in accordance with Section 903.3.1.1 or 903.3.1.2 and an emergency voice/alarm communication system in accordance with Section 907.5.2.2.
2. Facilities with *smoke-protected assembly seating* shall be permitted to use the capacity factors in Table 1029.6.2 indicated for level or ramped aisles for *means of egress* components other than *stairways* where the entire path for *means of egress* from the seating to the *exit discharge* is provided with a smoke control system complying with Section 909.
3. Facilities with outdoor *smoke-protected assembly seating* shall be permitted to the capacity factors in Section 1029.6.3 indicated for level or ramped aisles for *means of egress* components other than *stairways* where the entire path for *means of egress* from the seating to the *exit discharge* is open to the outdoors.

[BE] 1005.4 Continuity. The minimum width or required capacity of the *means of egress* required from any story of a building shall not be reduced along the path of egress travel until arrival at the *public way*.

[BE] 1005.5 Distribution of minimum width and required capacity. Where more than one *exit*, or access to more than one *exit*, is required, the *means of egress* shall be configured such that the loss of any one *exit*, or access to one *exit*, shall not reduce the available capacity or width to less than 50 percent of the required capacity or width.

[BE] 1005.6 Egress convergence. Where the *means of egress* from stories above and below converge at an intermediate level, the capacity of the *means of egress* from the point of convergence shall be not less than the largest minimum width or the sum of the required capacities for the *stairways* or *ramps* serving the two adjacent stories, whichever is larger.

[BE] 1005.7 Encroachment. Encroachments into the required *means of egress* width shall be in accordance with the provisions of this section.

[BE] 1005.7.1 Doors. Doors, when fully opened, shall not reduce the required width by more than 7 inches (178 mm). Doors in any position shall not reduce the required width by more than one-half.

Exceptions:

1. Surface-mounted latch release hardware shall be exempt from inclusion in the 7-inch maximum (178 mm) encroachment where both of the following conditions exists:
   1.1. The hardware is mounted to the side of the door facing away from the adjacent wall where the door is in the open position.
   1.2. The hardware is mounted not less than 34 inches (865 mm) nor more than 48 inches (1219 mm) above the finished floor.
2. The restrictions on door swing shall not apply to doors within individual *dwelling units* and *sleeping units* of Group R-2 occupancies and *dwelling units* of Group R-3 occupancies.

[BE] 1005.7.2 Other projections. *Handrail* projections shall be in accordance with the provisions of Section 1014.8. Other nonstructural projections such as trim and similar decorative features shall be permitted to project into the required width not more than $1^1/_2$ inches (38 mm) on each side.

Exception: Projections are permitted in corridors within Group I-2 Condition 1 in accordance with Section 407.4.3 of the *International Building Code*.

[BE] 1005.7.3 Protruding objects. Protruding objects shall comply with the applicable requirements of Section 1003.3.

SECTION 1006
NUMBERS OF EXITS AND EXIT ACCESS DOORWAYS

[BE] 1006.1 General. The number of *exits* or *exit access doorways* required within the *means of egress* system shall comply with the provisions of Section 1006.2 for spaces, including *mezzanines*, and Section 1006.3 for stories.

[BE] 1006.2 Egress from spaces. Rooms, areas or spaces, including *mezzanines*, within a story or basement shall be provided with the number of *exits* or access to *exits* in accordance with this section.

[BE] 1006.2.1 Egress based on occupant load and common path of egress travel distance. Two *exits* or *exit access doorways* from any space shall be provided where the design *occupant load* or the *common path of egress travel* distance exceeds the values listed in Table 1006.2.1.

Exceptions:

1. In Group R-2 and R-3 occupancies, one *means of egress* is permitted within and from individual *dwelling units* with a maximum *occupant load* of 20 where the *dwelling unit* is equipped throughout with an *automatic sprinkler system* in accordance with Section 903.3.1.1 or 903.3.1.2 and the *common path of egress travel* does not exceed 125 feet (38 100 mm).
2. Care suites in Group I-2 occupancies complying with Section 407.4 of the *International Building Code*.

** **[BE] 1006.2.1.1 Three or more exits or exit access doorways.** Three *exits* or *exit access doorways* shall be provided from any space with an *occupant load* of 501 to 1,000. Four *exits* or *exit access doorways* shall be provided from any space with an *occupant load* greater than 1,000.

[BE] 1006.2.2 Egress based on use. The numbers of *exits* or access to *exits* shall be provided in the uses described in Sections 1006.2.2.1 through 1006.2.2.5.

** **[BE] 1006.2.2.1 Boiler, incinerator and furnace rooms.** Two *exit access doorways* are required in boiler, incinerator and furnace rooms where the area is over 500 square feet (46 m^2) and any fuel-fired equipment exceeds 400,000 British thermal units (Btu) (422 000 KJ) input capacity. Where two *exit access doorways* are required, one is permitted to be a fixed ladder or an *alternating tread device*. *Exit access doorways* shall be separated by a horizontal distance equal to one-half the length of the maximum overall diagonal dimension of the room.

[BE] 1006.2.2.2 Refrigeration machinery rooms. ** Machinery rooms larger than 1,000 square feet (93 m^2) shall have not less than two *exits* or *exit access doorways*. Where two *exit access doorways* are required, one such doorway is permitted to be served by a fixed ladder or an *alternating tread device*. *Exit access* doorways shall be separated by a horizontal distance equal to one-half the maximum horizontal dimension of the room.

All portions of machinery rooms shall be within 150 feet (45 720 mm) of an *exit* or *exit access doorway*. An increase in *exit access* travel distance is permitted in accordance with Section 1017.1.

Doors shall swing in the direction of egress travel, regardless of the *occupant load* served. Doors shall be tight fitting and self-closing.

[BE] 1006.2.2.3 Refrigerated rooms or spaces. Rooms or spaces having a floor area larger than 1,000 ** square feet (93 m^2), containing a refrigerant evaporator and maintained at a temperature below 68°F (20°C), shall have access to not less than two *exits* or *exit access doorways*.

Exit access travel distance shall be determined as specified in Section 1017.1, but all portions of a refrigerated room or space shall be within 150 feet (45 720 mm) of an *exit* or *exit* access *doorway* where such

[BE] TABLE 1006.2.1
SPACES WITH ONE EXIT OR EXIT ACCESS DOORWAY

OCCUPANCY	MAXIMUM OCCUPANT LOAD OF SPACE	MAXIMUM COMMON PATH OF EGRESS TRAVEL DISTANCE (feet)		
		Without Sprinkler System (feet)		With Sprinkler System (feet)
		Occupant Load		
		OL ≤ 30	OL > 30	
A[c], E, M	49	75	75	75[a]
B	49	100	75	100[a]
F	49	75	75	100[a]
H-1, H-2, H-3	3	NP	NP	25[b]
H-4, H-5	10	NP	NP	75[b]
I-1, I-2[d], I-4	10	NP	NP	75[a]
I-3	10	NP	NP	100[a]
R-1	10	NP	NP	75[a]
R-2	10	NP	NP	125[a]
R-3[e]	10	NP	NP	125[a]
R-4[e]	10	75	75	125[a]
S[f]	29	100	75	100[a]
U	49	100	75	75[a]

For SI: 1 foot = 304.8 mm.

NP = Not Permitted.

a. Buildings equipped throughout with an *automatic sprinkler system* in accordance with Section 903.3.1.1 or 903.3.1.2. See Section 903 for occupancies where *automatic sprinkler systems* are permitted in accordance with Section 903.3.1.2

b. Group H occupancies equipped throughout with an *automatic sprinkler system* in accordance with Section 903.2.5.

c. For a room or space used for assembly purposes having *fixed seating*, see Section 1029.8.

d. For the travel distance limitations in Group I-2, see Section 407.4 of the *International Building Code*.

e. The length of common path of egress travel distance in a Group R-3 occupancy located in a mixed occupancy building or within a Group R-3 or R-4 congregate living facility.

f. The length of *common path of egress travel* distance in a Group S-2 open parking garage shall be not more than 100 feet.

rooms are not protected by an *approved automatic sprinkler system*. Egress is allowed through adjoining refrigerated rooms or spaces.

Exception: Where using refrigerants in quantities limited to the amounts based on the volume set forth in the *International Mechanical Code*.

** **[BE] 1006.2.2.4 Day care means of egress.** Day care facilities, rooms or spaces where care is provided for more than 10 children that are $2^1/_2$ years of age or less, shall have access to not less than two *exits* or *exit access doorways*.

** **[BE] 1006.2.2.5 Vehicular ramps.** Vehicular ramps shall not be considered as an *exit access ramp* unless pedestrian facilities are provided.

** **[BE] 1006.3 Egress from stories or occupied roofs.** The *means of egress* system serving any story or occupied roof shall be provided with the number of *exits* or access to *exits* based on the aggregate *occupant load* served in accordance with this section. The path of egress travel to an *exit* shall not pass through more than one adjacent story.

Each story above the second story of a building shall have not less than one *interior* or *exterior exit stairway*, or interior or *exterior exit ramp*. Where nothree or more *exits* or access to *exits* are required, not less than 50 percent of the required *exits* shall be *interior* or *exterior exit stairways* or *ramps*.

Exceptions:

1. *Interior exit stairways* and *interior exit ramps* are not required in open parking garages where the *means of egress* serves only the open parking garage.
2. *Interior exit stairways* and *interior exit ramps* are not required in outdoor facilities where all portions of the *means of egress* are essentially open to the outside.

** **[BE] 1006.3.1 Egress based on occupant load.** Each story and occupied roof shall have the minimum number of *exits*, or access to *exits*, as specified in Table 1006.3.1. A single *exit* or access to a single *exit* shall be permitted in accordance with Section 1006.3.2. The required number of *exits*, or *exit access stairways* or *ramps* providing access to *exits*, from any story or occupied roof shall be maintained until arrival at the *exit discharge* or a *public way*.

**[BE] TABLE 1006.3.1
MINIMUM NUMBER OF EXITS OR ACCESS TO EXITS PER STORY**

OCCUPANT LOAD PER STORY	MINIMUM NUMBER OF EXITS OR ACCESS TO EXITS FROM STORY
1-500	2
501-1,000	3
More than 1,000	4

[BE] 1006.3.2 Single exits. A single *exit* or access to a single *exit* shall be permitted from any story or occupied roof, where one of the following conditions exists: **

1. The *occupant load*, number of *dwelling units* and *exit access* travel distance do not exceed the values in Table 1006.3.2(1) or 1006.3.2(2).
2. Rooms, areas and spaces complying with Section 1006.2.1 with *exits* that discharge directly to the exterior at the *level of exit discharge*, are permitted to have one *exit* or access to a single *exit*.
3. Parking garages where vehicles are mechanically parked shall be permitted to have one *exit* or access to a single *exit*.
4. Group R-3 and R-4 occupancies shall be permitted to have one *exit* or access to a single *exit*.
5. Individual single-story or multistory *dwelling units* shall be permitted to have a single *exit* or access to a single *exit* from the *dwelling unit* provided that both of the following criteria are met:
   5.1. The *dwelling unit* complies with Section 1006.2.1 as a space with one means of egress.
   5.2. Either the *exit* from the *dwelling unit* discharges directly to the exterior at the *level of exit discharge*, or the *exit* access outside the *dwelling unit*'s entrance door provides access to not less than two approved independent *exits*.

[BE] 1006.3.2.1 Mixed occupancies. Where one *exit*, or *exit access stairway* or *ramp* providing access to exits at other stories, is permitted to serve individual stories, mixed occupancies shall be permitted to be served by single *exits* provided each individual occupancy complies with the applicable requirements of Table 1006.3.2(1) or 1006.3.2(2) for that occupancy. **

**[BE] TABLE 1006.3.2(1)
STORIES WITH ONE EXIT OR ACCESS TO ONE EXIT FOR R-2 OCCUPANCIES** **

STORY	OCCUPANCY	MAXIMUM NUMBER OF DWELLING UNITS	MAXIMUM COMMON PATH OF EGRESS TRAVEL DISTANCE
Basement, first, second or third story above grade plane	R-2[a, b]	4 dwelling units	125 feet
Fourth story above grade plane and higher	NP	NA	NA

For SI: 1 foot = 3048 mm.

NP – Not Permitted

NA – Not Applicable

a. Buildings classified as Group R-2 equipped throughout with an *automatic sprinkler system* in accordance with Section 903.3.1.1 or 903.3.1.2 and provided with emergency escape and rescue openings in accordance with Section 1030.

b. This Table is used for R-2 occupancies consisting of *dwelling units*. For R-2 occupancies consisting of *sleeping units*, use Table 1006.3.2(2).

Where applicable, cumulative *occupant loads* from adjacent occupancies shall be considered in accordance with the provisions of Section 1004.1. In each story of a mixed occupancy building, the maximum number of occupants served by a single *exit* shall be such that the sum of the ratios of the calculated number of occupants of the space divided by the allowable number of occupants indicated in Table 1006.3.2(2) for each occupancy does not exceed one. Where *dwelling units* are located on a story with other occupancies, the actual number of *dwelling units* divided by four plus the ratio from the other occupancy does not exceed one.

** **[BE] 1006.3.2.2 Basements.** A basement provided with one *exit* shall not be located more than one story below grade plane.

SECTION 1007 EXIT AND EXIT ACCESS DOORWAY CONFIGURATION

[BE] 1007.1 General. *Exits*, *exit access doorways*, and *exit access stairways* and *ramps* serving spaces, including individual building stories, shall be separated in accordance with the provisions of this section.

** **[BE] 1007.1.1 Two exits or exit access doorways.** Where two *exits*, *exit access doorways*, *exit access stairways* or *ramps*, or any combination thereof, are required from any portion of the *exit access*, they shall be placed a distance apart equal to not less than one-half of the length of the maximum overall diagonal dimension of the building or area to be served measured in a straight line between them. Interlocking or *scissor stairways* shall be counted as one exit *stairway*.

Exceptions:

1. Where *interior exit stairways* or *ramps* are interconnected by a 1-hour fire-resistance-rated corridor conforming to the requirements of Section 1020, the required *exit* separation shall be measured along the shortest direct line of travel within the *corridor*.
2. Where a building is equipped throughout with an *automatic sprinkler system* in accordance with Section 903.3.1.1 or 903.3.1.2, the separation distance shall be not less than one-third of the length of the maximum overall diagonal dimension of the area served.

[BE] 1007.1.1.1 Measurement point. The separation distance required in Section 1007.1.1 shall be measured in accordance with the following:

1. The separation distance to *exit* or *exit access doorways* shall be measured to any point along the width of the doorway.
2. The separation distance to *exit access stairways* shall be measured to the closest riser.
3. The separation distance to *exit access ramps* shall be measured to the start of the ramp run.

[BE] 1007.1.2 Three or more exits or exit access doorways. Where access to three or more exits is required, not less than two exit or *exit access doorways* shall be arranged in accordance with the provisions of Section 1007.1.1. Additional required exit or *exit access doorways* shall be arranged a reasonable distance apart so that if one becomes blocked, the others will be available. **

[BE] 1007.1.3 Remoteness of exit access stairways or ramps. Where two *exit* access *stairways* or *ramps* provide the required *means of egress* to *exits* at another story, the required separation distance shall be maintained for all portions of such *exit access stairways* or *ramps*.

[BE] 1007.1.3.1 Three or more exit access stairways or ramps. Where more than two *exit access stairways* or *ramps* provide the required *means of egress*, not less than two shall be arranged in accordance with Section 1007.1.3.

** **[BE] 1006.3.2(2)**
STORIES WITH ONE EXIT OR ACCESS TO ONE EXIT FOR OTHER OCCUPANCIES

STORY	OCCUPANCY	MAXIMUM OCCUPANT LOAD PER STORY	MAXIMUM COMMON PATH OF EGRESS TRAVEL DISTANCE (feet)
First story above or below grade plane	A, B[b], E F[b], M, U	49	75
	H-2, H-3	3	25
	H-4, H-5, I, R-1, R-2[a, c], R-4	10	75
	S[b,d]	29	75
Second story above grade plane	B, F, M, S[d]	29	75
Third story above grade plane and higher	NP	NA	NA

For SI: 1 foot = 304.8 mm.
NP = Not Permitted.
NA = Not Applicable.

a. Buildings classified as Group R-2 equipped throughout with an *automatic sprinkler system* in accordance with Section 903.3.1.1 or 903.3.1.2 and provided with emergency escape and rescue openings in accordance with Section 1030.

b. Group B, F and S occupancies in buildings equipped throughout with an *automatic sprinkler system* in accordance with Section 903.3.1.1 shall have a maximum *exit access* travel distance of 100 feet.

c. This table is used for R-2 occupancies consisting of *sleeping units*. For R-2 occupancies consisting of *dwelling units*, use Table 1006.3.2(1).

d. The length of *exit access* travel distance in a Group S-2 open parking garage shall be not more than 100 feet.

SECTION 1008
MEANS OF EGRESS ILLUMINATION

[BE] 1008.1 Means of egress illumination. Illumination shall be provided in the *means of egress* in accordance with Section 1008.2. Under emergency power, *means of egress* illumination shall comply with Section 1008.3.

[BE] 1008.2 Illumination required. The *means of egress* serving a room or space shall be illuminated at all times that the room or space is occupied.

Exceptions:

1. Occupancies in Group U.
2. *Aisle accessways* in Group A.
3. *Dwelling units* and *sleeping units* in Groups R-1, R-2 and R-3.
4. *Sleeping units* of Group I occupancies.

[BE] 1008.2.1 Illumination level under normal power. The *means of egress* illumination level shall be not less than 1 footcandle (11 lux) at the walking surface.

Exception: For auditoriums, theaters, concert or opera halls and similar assembly occupancies, the illumination at the walking surface is permitted to be reduced during performances by one of the following methods provided that the required illumination is automatically restored upon activation of a premises' fire alarm system:

1. Externally illuminated walking surfaces shall be permitted to be illuminated to not less than 0.2 footcandle (2.15 lux),
2. Steps, landings and the sides of *ramps* shall be permitted to be marked with self-luminous materials in accordance with Sections 1025.2.1, 1025.2.2 and 1025.2.4 by systems *listed* in accordance with UL 1994.

[BE] 1008.2.2 Exit discharge. In Group I-2 occupancies where two or more *exits* are required, on the exterior landings required by Section 1010.6.1, means of egress illumination levels for the *exit discharge* shall be provided such that failure of any single lighting unit shall not reduce the illumination level at the landing to less than 1 footcandle (11 lux).

[BE] 1008.3 Emergency power for illumination. The power supply for *means of egress* illumination shall normally be provided by the premises' electrical supply.

[BE] 1008.3.1 General. In the event of power supply failure in rooms and spaces that require two or more *means of egress* an emergency electrical system shall automatically illuminate all of the following areas:

1. *Aisles.*
2. *Corridors.*
3. *Exit access stairways* and *ramps.*

[BE] 1008.3.2 Buildings. In the event of power supply failure, in buildings that require two or more *means of egress*, an emergency electrical system shall automatically illuminate all of the following areas:

1. Interior *exit access stairways* and *ramps*
2. *Interior* and *exterior exit stairways* and *ramps.*
3. *Exit passageways.*
4. Vestibules and areas on the *level of discharge* used for *exit discharge* in accordance with Section 1028.1.
5. Exterior landings as required by Section 1010.1.6 for exit doorways that lead directly to the *exit discharge.*

[BE] 1008.3.3 Rooms and spaces. In the event of power supply failure, an emergency electrical system shall automatically illuminate all of the following areas:

1. Electrical equipment rooms.
2. Fire command centers.
3. Fire pump rooms.
4. Generator rooms.
5. Public restrooms with an area greater than 300 square feet (27.87 m^2).

[BE] 1008.3.4 Duration. The emergency power system shall provide power for a duration of not less than 90 minutes and shall consist of storage batteries, unit equipment or an on-site generator. The installation of the emergency power system shall be in accordance with Section 604.

[BE] 1008.3.5 Illumination level under emergency power. Emergency lighting facilities shall be arranged to provide initial illumination that is not less than an average of 1 footcandle (11 lux) and a minimum at any point of 0.1 footcandle (1 lux) measured along the path of egress at floor level. Illumination levels shall be permitted to decline to 0.6 footcandle (6 lux) average and a minimum at any point of 0.06 footcandle (0.6 lux) at the end of the emergency lighting time duration. A maximum-to-minimum illumination uniformity ratio of 40 to 1 shall not be exceeded. In Group I-2 occupancies, failure of any single lighting unit shall not reduce the illumination level to less than 0.2 foot-candle (2.2 lux).

SECTION 1009
ACCESSIBLE MEANS OF EGRESS

[BE] 1009.1 Accessible means of egress required. *Accessible means of egress* shall comply with this section. Accessible spaces shall be provided with not less than one *accessible means of egress.* Where more than one *means of egress* is required by Section 1006.2 or 1006.3 from an accessible space, each accessible portion of the space shall be served by not less than two *accessible means of egress.*

Exceptions:

1. *Accessible means of egress* are not required to be provided in existing buildings.

2. One *accessible means of egress* is required from an accessible *mezzanine* level in accordance with Section 1009.3, 1009.4 or 1009.5.

3. In assembly areas with ramped *aisles* or stepped *aisles*, one *accessible means of egress* is permitted where the common path of travel is accessible and meets the requirements in Section 1029.8.

[BE] 1009.2 Continuity and components. Each required *accessible means of egress* shall be continuous to a public way and shall consist of one or more of the following components:

1. *Accessible routes* complying with Section 1104 of the *International Building Code.*
2. *Interior exit stairways* complying with Sections 1009.3 and 1023.
3. *Exit access stairways* complying with Sections 1009.3 and 1019.3 or 1019.4.
4. *Exterior exit stairways* complying with Sections 1009.3 and 1027 and serving levels other than the *level of exit discharge.*
5. Elevators complying with Section 1009.4.
6. Platform lifts complying with Section 1009.5.
7. *Horizontal exits* complying with Section 1026.
8. *Ramps* complying with Section 1012.
9. *Areas of refuge* complying with Section 1009.6.
10. Exterior areas for assisted rescue complying with Section 1009.7 serving *exits* at the *level of exit discharge.*

[BE] 1009.2.1 Elevators required. In buildings where a required accessible floor is four or more stories above or below a *level of exit discharge*, not less than one required *accessible means of egress* shall be an elevator complying with Section 1009.4.

Exceptions:

1. In buildings equipped throughout with an *automatic sprinkler system* installed in accordance with Section 903.3.1.1 or 903.3.1.2, the elevator shall not be required on floors provided with a *horizontal exit* and located at or above the *levels of exit discharge.*
2. In buildings equipped throughout with an *automatic sprinkler system* installed in accordance with Section 903.3.1.1 or 903.3.1.2, the elevator shall not be required on floors provided with a *ramp* conforming to the provisions of Section 1012.

[BE] 1009.3 Stairways. In order to be considered part of an *accessible means of egress*, a *stairway* between stories shall have a clear width of 48 inches (1219 mm) minimum between *handrails* and shall either incorporate an *area of refuge* within an enlarged floor-level landing or shall be accessed from an *area of refuge* complying with Section 1009.6. *Exit access stairways* that connect levels in the same story are not permitted as part of an *accessible means of egress.*

Exceptions:

1. *Exit access stairways* providing *means of egress* from *mezzanines* are permitted as part of an *accessible means of egress.*
2. The clear width of 48 inches (1219 mm) between *handrails* is not required in buildings equipped throughout with an *automatic sprinkler system* installed in accordance with Section 903.3.1.1 or 903.3.1.2.
3. The clear width of 48 inches (1219 mm) between *handrails* is not required for *stairways* accessed from a refuge area in conjunction with a *horizontal exit.*
4. *Areas of refuge* are not required at *exit access stairways* where a two-way communication is provided at the elevator landing in accordance with Section 1009.8.
5. *Areas of refuge* are not required at *stairways* in buildings equipped throughout with an *automatic sprinkler system* installed in accordance with Section 903.3.1.1 or 903.3.1.2.
6. *Areas of refuge* are not required at *stairways* serving open parking garages.
7. Areas of *refuge* are not required for *smoke protected assembly seating* areas complying with Section 1029.6.2.
8. *Areas of refuge* are not required at *stairways* in Group R-2 occupancies.
9. *Areas of refuge* are not required for *stairways* accessed from a refuge area in conjunction with a *horizontal exit.*

[BE] 1009.4 Elevators. In order to be considered part of an *accessible means of egress*, an elevator shall comply with the emergency operation and signaling device requirements of Section 2.27 of ASME A17.1. Standby power shall be provided in accordance with Section 604 of this code and Section 3003 of the *International Building Code.* The elevator shall be accessed from an *area of refuge* complying with Section 1009.6.

Exceptions:

1. *Areas of refuge* are not required at the elevator in open parking garages.
2. *Areas of refuge* are not required in buildings and facilities equipped throughout with an *automatic sprinkler system* installed in accordance with Section 903.3.1.1 or 903.3.1.2.
3. *Areas of refuge* are not required at elevators not required to be located in a shaft in accordance with Section 712 of the *International Building Code.*
4. *Areas of refuge* are not required at elevators serving *smoke protected assembly seating* areas complying with Section 1029.6.2.

5. *Areas of refuge* are not required for elevators accessed from a refuge area in conjunction with a *horizontal exit*.

[BE] 1009.5 Platform lifts. Platform lifts shall be permitted to serve as part of an *accessible means of egress* where allowed as part of a required *accessible route* in Section 1109.8 of the *International Building Code* except for Item 10. Standby power for the platform lift shall be provided in accordance with Section 604.

[BE] 1009.6 Areas of refuge. Every required *area of refuge* shall be accessible from the space it serves by an *accessible means of egress*.

[BE] 1009.6.1 Travel distance. The maximum travel distance from any accessible space to an *area of refuge* shall not exceed the *exit access* travel distance permitted for the occupancy in accordance with Section 1017.1.

[BE] 1009.6.2 Stairway or elevator access. Every required *area of refuge* shall have direct access to a *stairway* complying with Sections 1009.3 and 1023 or an elevator complying with Section 1009.4.

[BE] 1009.6.3 Size. Each *area of refuge* shall be sized to accommodate one wheelchair space of 30 inches by 48 inches (762 mm by 1219 mm) for each 200 occupants or portion thereof, based on the *occupant load* of the *area of refuge* and areas served by the *area of refuge*. Such wheelchair spaces shall not reduce the *means of egress* minimum width or required capacity. Access to any of the required wheelchair spaces in an *area of refuge* shall not be obstructed by more than one adjoining wheelchair space.

[BE] 1009.6.4 Separation. Each *area of refuge* shall be separated from the remainder of the story by a *smoke barrier* complying with Section 709 of the *International Building Code* or a *horizontal* exit complying with Section 1026. Each *area of refuge* shall be designed to minimize the intrusion of smoke.

Exceptions:

1. *Areas of refuge* located within an enclosure for *interior exit stairways* complying with Section 1023.
2. *Areas of refuge* in outdoor facilities where *exit access* is essentially open to the outside.

[BE] 1009.6.5 Two-way communication. *Areas of refuge* shall be provided with a two-way communication system complying with Sections 1009.8.1 and 1009.8.2.

[BE] 1009.7 Exterior areas for assisted rescue. Exterior areas for assisted rescue shall be accessed by an *accessible route* from the area served.

Where the *exit discharge* does not include an *accessible route* from an *exit* located on the *level of exit discharge* to a *public way*, an exterior area of assisted rescue shall be provided on the exterior landing in accordance with Sections 1009.7.1 through 1009.7.4.

[BE] 1009.7.1 Size. Each exterior area for assisted rescue shall be sized to accommodate wheelchair spaces in accordance with Section 1009.6.3.

[BE] 1009.7.2 Separation. *Exterior walls* separating the exterior area of assisted rescue from the interior of the building shall have a minimum fire-resistance rating of 1 hour, rated for exposure to fire from the inside. The fire-resistance-rated *exterior wall* construction shall extend horizontally 10 feet (3048 mm) beyond the landing on either side of the landing or equivalent fire-resistance-rated construction is permitted to extend out perpendicular to the *exterior wall* 4 feet (1220 mm) minimum on the side of the landing. The fire-resistance-rated construction shall extend vertically from the ground to a point 10 feet (3048 mm) above the floor level of the area for assisted rescue or to the roof line, whichever is lower. Openings within such fire-resistance-rated *exterior walls* shall be protected in accordance with Section 716 of the *International Building Code*.

[BE] 1009.7.3 Openness. The exterior area for assisted rescue shall be open to the outside air. The sides other than the separation walls shall be not less than 50 percent open, and the open area shall be distributed so as to minimize the accumulation of smoke or toxic gases.

[BE] 1009.7.4 Stairways. *Stairways* that are part of the *means of egress* for the exterior area for assisted rescue shall provide a clear width of 48 inches (1220 mm) between *handrails*.

Exception: The clear width of 48 inches (1220 mm) between *handrails* is not required at *stairways* serving buildings equipped throughout with an *automatic sprinkler system* installed in accordance with Section 903.3.1.1 or 903.3.1.2.

[BE] 1009.8 Two-way communication. A two-way communication system complying with Sections 1009.8.1 and 1009.8.2 shall be provided at the landing serving each elevator or bank of elevators on each accessible floor that is one or more stories above or below the *level of exit discharge*.

Exceptions:

1. Two-way communication systems are not required at the landing serving each elevator or bank of elevators where the two-way communication system is provided within *areas of refuge* in accordance with Section 1009.6.5.
2. Two-way communication systems are not required on floors provided with *ramps* conforming to the provisions of Section 1012.
3. Two-way communication systems are not required at the landings serving only service elevators that are not designated as part of the *accessible means of egress* or serve as part of the required *accessible route* into a facility.
4. Two-way communication systems are not required at the landings serving only freight elevators.
5. Two-way communication systems are not required at the landing serving a private residence elevator.

[BE] 1009.8.1 System requirements. Two-way communication systems shall provide communication between each required location and the fire command center or a central control point location *approved* by the fire department. Where the central control point is not constantly attended, a two-way communication system shall have a timed automatic telephone dial-out capability to a monitoring location or 9-1-1. The two-way communication system shall include both audible and visible signals.

[BE] 1009.8.2 Directions. Directions for the use of the two-way communication system, instructions for summoning assistance via the two-way communication system and written identification of the location shall be posted adjacent to the two-way communication system. Signage shall comply with the ICC A117.1 requirements for visual characters.

[BE] 1009.9 Signage. Signage indicating special accessibility provisions shall be provided as shown:

1. Each door providing access to an *area of refuge* from an adjacent floor area shall be identified by a sign stating: AREA OF REFUGE.
2. Each door providing access to an exterior area for assisted rescue shall be identified by a sign stating: EXTERIOR AREA FOR ASSISTED RESCUE.

Signage shall comply with the ICC A117.1 requirements for visual characters and include the International Symbol of Accessibility. Where exit sign illumination is required by Section 1013.3, the signs shall be illuminated. Additionally, visual characters, raised character and braille signage complying with ICC A117.1 shall be located at each door to an *area of refuge* and exterior area for assisted rescue in accordance with Section 1013.4.

[BE] 1009.10 Directional signage. Directional signage indicating the location of all other means of egress and which of those are *accessible means of egress* shall be provided at the following:

1. At *exits* serving a required accessible space but not providing an *approved accessible means of egress*.
2. At elevator landings.
3. Within *areas of refuge*.

[BE] 1009.11 Instructions. In *areas of refuge* and exterior areas for assisted rescue, instructions on the use of the area under emergency conditions shall be posted. Signage shall comply with the ICC A117.1 requirements for visual characters. The instructions shall include all of the following:

1. Persons able to use the *exit stairway* do so as soon as possible, unless they are assisting others.
2. Information on planned availability of assistance in the use of *stairs* or supervised operation of elevators and how to summon such assistance.
3. Directions for use of the two-way communication system where provided.

SECTION 1010
DOORS, GATES AND TURNSTILES

[BE] 1010.1 Doors. *Means of egress* doors shall meet the requirements of this section. Doors serving a *means of egress* system shall meet the requirements of this section and Section 1022.2. Doors provided for egress purposes in numbers greater than required by this code shall meet the requirements of this section.

Means of egress doors shall be readily distinguishable from the adjacent construction and finishes such that the doors are easily recognizable as doors. Mirrors or similar reflecting materials shall not be used on *means of egress* doors. *Means of egress* doors shall not be concealed by curtains, drapes, decorations or similar materials.

[BE] 1010.1.1 Size of doors. The required capacity of each door opening shall be sufficient for the *occupant load* thereof and shall provide a minimum clear width of 32 inches (813 mm). Clear openings of doorways with swinging doors shall be measured between the face of the door and the stop, with the door open 90 degrees (1.57 rad). Where this section requires a minimum clear width of 32 inches (813 mm) and a door opening includes two door leaves without a mullion, one leaf shall provide a clear opening width of 32 inches (813 mm). The maximum width of a swinging door leaf shall be 48 inches (1219 mm) nominal. *Means of egress* doors in a Group I-2 occupancy used for the movement of beds shall provide a clear width not less than $41^1/_2$ inches (1054 mm). The height of door openings shall be not less than 80 inches (2032 mm).

Exceptions:

1. The minimum and maximum width shall not apply to door openings that are not part of the required *means of egress* in Group R-2 and R-3 occupancies.
2. Door openings to resident *sleeping units* in Group I-3 occupancies shall have a clear width of not less than 28 inches (711 mm).
3. Door openings to storage closets less than 10 square feet (0.93 m^2) in area shall not be limited by the minimum width.
4. Width of door leaves in revolving doors that comply with Section 1010.1.4.1 shall not be limited.
5. Door openings within a *dwelling unit* or *sleeping unit* shall be not less than 78 inches (1981 mm) in height.
6. Exterior door openings in *dwelling units* and *sleeping units*, other than the required *exit* door, shall be not less than 76 inches (1930 mm) in height.
7. In other than Group R-1 occupancies, the minimum widths shall not apply to interior egress doors within a *dwelling unit* or *sleeping unit* that is not required to be an Accessible unit, Type A unit or Type B unit.
8. Door openings required to be *accessible* within Type B units shall have a minimum clear width of 31.75 inches (806 mm).
9. Doors to walk-in freezers and coolers less than 1,000 square feet (93 m^2) in area shall have a maximum width of 60 inches (1524 mm).

10. In Group R-1 *dwelling units* or *sleeping units* not required to be Accessible units, the minimum width shall not apply to doors for showers or saunas.

[BE] 1010.1.1.1 Projections into clear width. There shall not be projections into the required clear width lower than 34 inches (864 mm) above the floor or ground. Projections into the clear opening width between 34 inches (864 mm) and 80 inches (2032 mm) above the floor or ground shall not exceed 4 inches (102 mm).

Exception: Door closers and door stops shall be permitted to be 78 inches (1980 mm) minimum above the floor.

[BE] 1010.1.2 Door swing. Egress doors shall be of the pivoted or side-hinged swinging type.

Exceptions:

1. Private garages, office areas, factory and storage areas with an *occupant load* of 10 or less.
2. Group I-3 occupancies used as a place of detention.
3. Critical or intensive care patient rooms within suites of health care facilities.
4. Doors within or serving a single *dwelling unit* in Groups R-2 and R-3.
5. In other than Group H occupancies, revolving doors complying with Section 1010.1.4.1.
6. In other than Group H occupancies, special purpose horizontal sliding, accordion or folding door assemblies complying with Section 1010.1.4.3.
7. Power-operated doors in accordance with Section 1010.1.4.2.
8. Doors serving a bathroom within an individual *sleeping unit* in Group R-1.
9. In other than Group H occupancies, manually operated horizontal sliding doors are permitted in a *means of egress* from spaces with an *occupant load* of 10 or less.

[BE] 1010.1.2.1 Direction of swing. Pivot or side-hinged swinging doors shall swing in the direction of egress travel where serving a room or area containing an occupant load of 50 or more persons or a Group H occupancy.

[BE] 1010.1.3 Door opening force. The force for pushing or pulling open interior swinging egress doors, other than fire doors, shall not exceed 5 pounds (22 N). These forces do not apply to the force required to retract latch bolts or disengage other devices that hold the door in a closed position. For other swinging doors, as well as sliding and folding doors, the door latch shall release when subjected to a 15-pound (67 N) force. The door shall be set in motion when subjected to a 30-pound (133 N) force. The door shall swing to a full-open position when subjected to a 15-pound (67 N) force.

[BE] 1010.1.3.1 Location of applied forces. Forces shall be applied to the latch side of the door.

[BE] 1010.1.4 Special doors. Special doors and security grilles shall comply with the requirements of Sections 1010.1.4.1 through 1010.1.4.4.

[BE] 1010.1.4.1 Revolving doors. Revolving doors shall comply with the following:

1. Revolving doors shall comply with BHMA A156.27 and shall be installed in accordance with the manufacturer's instructions.
2. Each revolving door shall be capable of *breakout* in accordance with BHMA A156.27 and shall provide an aggregate width of not less than 36 inches (914 mm).
3. A revolving door shall not be located within 10 feet (3048 mm) of the foot or top of stairways or escalators. A dispersal area shall be provided between the stairways or escalators and the revolving doors.
4. The revolutions per minute (rpm) for a revolving door shall not exceed the maximum rpm as specified in BHMA A156.27. Manual revolving doors shall comply with Table 1010.1.4.1(1). Automatic or power-operated revolving doors shall comply with Table 1010.1.4.1(2).
5. An emergency stop switch shall be provided near each entry point of a revolving door within 48 inches (1220 mm) of the door and between 24 inches (610 mm) and 48 inches (1220 mm) above the floor. The activation area of the emergency stop switch button shall be not less than 1 inch (25 mm) in diameter and shall be red.
6. Each revolving door shall have a side-hinged swinging door that complies with Section 1010.1 in the same wall and within 10 feet (3048 mm) of the revolving door.
7. Revolving doors shall not be part of an *accessible route* required by Section 1009 of this code and Chapter 11 of the *International Building Code*.

**[BE] TABLE 1010.1.4.1(1)
MAXIMUM DOOR SPEED MANUAL REVOLVING DOORS**

REVOLVING DOOR MAXIMUM NOMINAL DIAMETER (FT-IN)	MAXIMUM ALLOWABLE REVOLVING DOOR SPEED (RPM)
6-0	12
7-0	11
8-0	10
9-0	9
10-0	8

For SI: 1 inch = 25.4 mm, 1 foot = 304.8 mm.

[BE] TABLE 1010.1.4(2)
MAXIMUM DOOR SPEED AUTOMATIC OR POWER-OPERATED REVOLVING DOORS

REVOLVING DOOR MAXIMUM NOMINAL DIAMETER (FT-IN)	MAXIMUM ALLOWABLE REVOLVING DOOR SPEED (RPM)
8-0	7.2
9-0	6.4
10-0	5.7
11-0	5.2
12-0	4.8
12-6	4.6
14-0	4.1
16-0	3.6
17-0	3.4
18-0	3.2
20-0	2.9
24-0	2.4

For SI: 1 inch = 25.4 mm, 1 foot = 304.8 mm.

[BE] 1010.1.4.1.1 Egress component. A revolving door used as a component of a *means of egress* shall comply with Section 1010.1.4.1 and the following three conditions:

1. Revolving doors shall not be given credit for more than 50 percent of the minimum width or required capacity.
2. Each revolving door shall be credited with a capacity based on not more than a 50-person occupant load.
3. Each revolving door shall provide for egress in accordance with BHMA A156.27 with a *breakout* force of not more than 130 pounds (578 N).

[BE] 1010.1.4.1.2 Other than egress component. A revolving door used as other than a component of a *means of egress* shall comply with Section 1010.1.4.1. The *breakout* force of a revolving door not used as a component of a *means of egress* shall not be more than 180 pounds (801 N).

Exception: A *breakout* force in excess of 180 pounds (801 N) is permitted if the collapsing force is reduced to not more than 130 pounds (578 N) when not less than one of the following conditions is satisfied:

1. There is a power failure or power is removed to the device holding the door wings in position.
2. There is an actuation of the *automatic sprinkler system* where such system is provided.
3. There is an actuation of a smoke detection system that is installed in accordance with Section 907 to provide coverage in areas within the building that are within 75 feet (22 860 mm) of the revolving doors.
4. There is an actuation of a manual control switch, in an *approved* location and clearly identified, that reduces the *breakout* force to not more than 130 pounds (578 N).

[BE] 1010.1.4.2 Power-operated doors. Where *means of egress* doors are operated or assisted by power, the design shall be such that in the event of power failure, the door is capable of being opened manually to permit *means of egress* travel or closed where necessary to safeguard *means of egress*. The forces required to open these doors manually shall not exceed those specified in Section 1010.1.3, except that the force to set the door in motion shall not exceed 50 pounds (220 N). The door shall be capable of swinging open from any position to the full width of the opening in which such door is installed when a force is applied to the door on the side from which egress is made. Power-operated swinging doors, power-operated sliding doors and power-operated folding doors shall comply with BHMA A156.10. Power-assisted swinging doors and low energy power-operated swinging doors shall comply with BHMA A156.19.

Exceptions:

1. Occupancies in Group I-3.
2. Horizontal sliding doors complying with Section 1010.1.4.3.
3. For a biparting door in the emergency *breakout* mode, a door leaf located within a multiple-leaf opening shall be exempt from the minimum 32-inch (813 mm) single-leaf requirement of Section 1010.1.1, provided a minimum 32-inch (813 mm) clear opening is provided when the two biparting leaves meeting in the center are broken out.

[BE] 1010.1.4.3 Special purpose horizontal sliding, accordion or folding doors. In other than Group H occupancies, special purpose horizontal sliding, accordion, or folding door assemblies permitted to be a component of a *means of egress* in accordance with Exception 6 to Section 1010.1.2 shall comply with all of the following criteria:

1. The doors shall be power operated and shall be capable of being operated manually in the event of power failure.
2. The doors shall be openable by a simple method from both sides without special knowledge or effort.
3. The force required to operate the door shall not exceed 30 pounds (133 N) to set the door in motion and 15 pounds (67 N) to close or open the door to the minimum required width.
4. The door shall be openable with a force not to exceed 15 pounds (67 N) when a force of 250 pounds (1100 N) is applied perpendicular to the door adjacent to the operating device.

5. The door assembly shall comply with the applicable *fire protection rating* and, where rated, shall be self-closing or automatic closing by smoke detection in accordance with Section 716.5.9.3 of the *International Building Code*, shall be installed in accordance with NFPA 80 and shall comply with Section 716 of the *International Building Code*.
6. The door assembly shall have an integrated standby power supply.
7. The door assembly power supply shall be electrically supervised.
8. The door shall open to the minimum required width within 10 seconds after activation of the operating device.

[BE] 1010.1.4.4 Security grilles. In Groups B, F, M and S, horizontal sliding or vertical security grilles are permitted at the main *exit* and shall be openable from the inside without the use of a key or special knowledge or effort during periods that the space is occupied. The grilles shall remain secured in the full-open position during the period of occupancy by the general public. Where two or more *means of egress* are required, not more than one-half of the *exits* or *exit access doorways* shall be equipped with horizontal sliding or vertical security grilles.

[BE] 1010.1.5 Floor elevation. There shall be a floor or landing on each side of a door. Such floor or landing shall be at the same elevation on each side of the door. Landings shall be level except for exterior landings, which are permitted to have a slope not to exceed 0.25 unit vertical in 12 units horizontal (2-percent slope).

Exceptions:

1. Doors serving individual *dwelling units* in Groups R-2 and R-3 where the following apply:
   1.1. A door is permitted to open at the top step of an interior *flight* of *stairs*, provided the door does not swing over the top step.
   1.2. Screen doors and storm doors are permitted to swing over *stairs* or landings.
2. Exterior doors as provided for in Section 1003.5, Exception 1, and Section 1022.2, which are not on an *accessible route*.
3. In Group R-3 occupancies not required to be Accessible units, Type A units or Type B units, the landing at an exterior doorway shall be not more than 7$^{3}/_{4}$ inches (197 mm) below the top of the threshold, provided the door, other than an exterior storm or screen door, does not swing over the landing.
4. Variations in elevation due to differences in finish materials, but not more than $^{1}/_{2}$ inch (12.7 mm).
5. Exterior decks, patios or balconies that are part of Type B *dwelling units*, have impervious surfaces and that are not more than 4 inches (102 mm) below the finished floor level of the adjacent interior space of the dwelling unit.
6. Doors serving equipment spaces not required to be accessible in accordance with Section 1103.2.9 of the *International Building Code* and serving an *occupant load* of five or less shall be permitted to have a landing on one side to be not more than 7 inches (178 mm) above or below the landing on the egress side of the door.

[BE] 1010.1.6 Landings at doors. Landings shall have a width not less than the width of the *stairway* or the door, whichever is greater. Doors in the fully open position shall not reduce a required dimension by more than 7 inches (178 mm). Where a landing serves an *occupant load* of 50 or more, doors in any position shall not reduce the landing to less than one-half its required width. Landings shall have a length measured in the direction of travel of not less than 44 inches (1118 mm).

Exception: Landing length in the direction of travel in Groups R-3 and U and within individual units of Group R-2 need not exceed 36 inches (914 mm).

[BE] 1010.1.7 Thresholds. Thresholds at doorways shall not exceed $^{3}/_{4}$ inch (19.1 mm) in height above the finished floor or landing for sliding doors serving *dwelling units* or $^{1}/_{2}$ inch (12.7 mm) above the finished floor or landing for other doors. Raised thresholds and floor level changes greater than $^{1}/_{4}$ inch (6.4 mm) at doorways shall be beveled with a slope not greater than one unit vertical in two units horizontal (50-percent slope).

Exceptions:

1. In occupancy Group R-2 or R-3, threshold heights for sliding and side-hinged exterior doors shall be permitted to be up to 7$^{3}/_{4}$ inches (197 mm) in height if all of the following apply:
   1.1. The door is not part of the required *means of egress*.
   1.2. The door is not part of an *accessible route* as required by Chapter 11 of the *International Building Code*.
   1.3. The door is not part of an accessible unit, Type A unit or Type B unit.
2. In Type B units, where Exception 5 to Section 1010.1.5 permits a 4-inch (102 mm) elevation change at the door, the threshold height on the exterior side of the door shall not exceed 4$^{3}/_{4}$ inches (120 mm) in height above the exterior deck, patio or balcony for sliding doors or 4$^{1}/_{2}$ inches (114 mm) above the exterior deck, patio or balcony for other doors.

[BE] 1010.1.8 Door arrangement. Space between two doors in a series shall be 48 inches (1219 mm) minimum plus the width of a door swinging into the space. Doors in a series shall swing either in the same direction or away from the space between the doors.

Exceptions:

1. The minimum distance between horizontal sliding power-operated doors in a series shall be 48 inches (1219 mm).
2. Storm and screen doors serving individual *dwelling units* in Groups R-2 and R-3 need not be spaced 48 inches (1219 mm) from the other door.
3. Doors within individual *dwelling units* in Groups R-2 and R-3 other than within Type A dwelling units.

[BE] 1010.1.9 Door operations. Except as specifically permitted by this section, egress doors shall be readily openable from the egress side without the use of a key or special knowledge or effort.

[BE] 1010.1.9.1 Hardware. Door handles, pulls, latches, locks and other operating devices on doors required to be accessible by Chapter 11 of the *International Building Code* shall not require tight grasping, tight pinching or twisting of the wrist to operate.

[BE] 1010.1.9.2 Hardware height. Door handles, pulls, latches, locks and other operating devices shall be installed 34 inches (864 mm) minimum and 48 inches (1219 mm) maximum above the finished floor. Locks used only for security purposes and not used for normal operation are permitted at any height.

Exception: Access doors or gates in barrier walls and fences protecting pools, spas and hot tubs shall be permitted to have operable parts of the release of latch on self-latching devices at 54 inches (1370 mm) maximum above the finished floor or ground, provided the self-latching devices are not also self-locking devices operated by means of a key, electronic opener or integral combination lock.

[BE] 1010.1.9.3 Locks and latches. Locks and latches shall be permitted to prevent operation of doors where any of the following exist:

1. Places of detention or restraint.
2. In buildings in occupancy Group A having an *occupant load* of 300 or less, Groups B, F, M and S, and in places of religious worship, the main door or doors are permitted to be equipped with key-operated locking devices from the egress side provided:
   - 2.1. The locking device is readily distinguishable as locked.
   - 2.2. A readily visible durable sign is posted on the egress side on or adjacent to the door stating: THIS DOOR TO REMAIN UNLOCKED WHEN THIS SPACE IS OCCUPIED. The sign shall be in letters 1 inch (25 mm) high on a contrasting background.
   - 2.3. The use of the key-operated locking device is revokable by the *fire code official* for due cause.
3. Where egress doors are used in pairs, *approved* automatic flush bolts shall be permitted to be used, provided that the door leaf having the automatic flush bolts does not have a doorknob or surface-mounted hardware.
4. Doors from individual *dwelling* or *sleeping units* of Group R occupancies having an *occupant load* of 10 or less are permitted to be equipped with a night latch, dead bolt or security chain, provided such devices are openable from the inside without the use of a key or tool.
5. Fire doors after the minimum elevated temperature has disabled the unlatching mechanism in accordance with *listed* fire door test procedures.

[BE] 1010.1.9.4 Bolt locks. Manually operated flush bolts or surface bolts are not permitted.

Exceptions:

1. On doors not required for egress in individual *dwelling units* or *sleeping units.*
2. Where a pair of doors serves a storage or equipment room, manually operated edge- or surface-mounted bolts are permitted on the inactive leaf.
3. Where a pair of doors serves an *occupant load* of less than 50 persons in a Group B, F or S occupancy, manually operated edge- or surface-mounted bolts are permitted on the inactive leaf. The inactive leaf shall not contain doorknobs, panic bars or similar operating hardware.
4. Where a pair of doors serves a Group B, F or S occupancy, manually operated edge- or surface-mounted bolts are permitted on the inactive leaf provided such inactive leaf is not needed to meet egress capacity requirements and the building is equipped throughout with an *automatic sprinkler system* in accordance with Section 903.3.1.1. The inactive leaf shall not contain doorknobs, panic bars or similar operating hardware.
5. Where a pair of doors serves patient care rooms in Group I-2 occupancies, self-latching edge- or surface-mounted bolts are permitted on the inactive leaf provided that the inactive leaf is not needed to meet egress capacity requirements and the inactive leaf shall not contain doorknobs, panic bars or similar operating hardware.

[BE] 1010.1.9.5 Unlatching. The unlatching of any door or leaf shall not require more than one operation.

Exceptions:

1. Places of detention or restraint.
2. Where manually operated bolt locks are permitted by Section 1010.1.9.4.
3. Doors with automatic flush bolts as permitted by Section 1010.1.9.3, Item 3.
4. Doors from individual *dwelling units* and *sleeping units* of Group R occupancies as permitted by Section 1010.1.9.3, Item 4.

[BE] 1010.1.9.5.1 Closet and bathroom doors in Group R-4 occupancies. In Group R-4 occupancies, closet doors that latch in the closed position shall be openable from inside the closet, and bathroom doors that latch in the closed position shall be capable of being unlocked from the ingress side.

[BE] 1010.1.9.6 Controlled egress doors in Groups I-1 and I-2. Electric locking systems, including electro-mechanical locking systems and electromagnetic locking systems, shall be permitted to be locked in the means of egress in Group I-1 or I-2 occupancies where the clinical needs of persons receiving care require their containment. Controlled egress doors shall be permitted in such occupancies where the building is equipped throughout with an *automatic sprinkler system* in accordance with Section 903.3.1.1 or an approved automatic smoke or heat detection system installed in accordance with Section 907, provided that the doors are installed and operate in accordance with all of the following:

1. The door locks shall unlock upon actuation of the *automatic sprinkler system* or automatic fire detection system.
2. The door locks shall unlock upon loss of power controlling the lock or lock mechanism.
3. The door locking system shall be installed to have the capability of being unlocked by a switch located at the fire command center, a nursing station or other approved location. The switch shall directly break power to the lock.
4. A building occupant shall not be required to pass through more than one door equipped with a controlled egress locking system before entering an exit.
5. The procedures for unlocking the doors shall be described and approved as part of the emergency planning and preparedness required by Chapter 4.
6. All clinical staff shall have the keys, codes or other means necessary to operate the locking systems.
7. Emergency lighting shall be provided at the door.
8. The door locking system units shall be *listed* in accordance with UL 294.

Exceptions:

1. Items 1 through 4 shall not apply to doors to areas occupied by persons who, because of clinical needs, require restraint or containment as part of the function of a psychiatric treatment area.
2. Items 1 through 4 shall not apply to doors to areas where a *listed* egress control system is utilized to reduce the risk of child abduction from nursery and obstetric areas of a Group I-2 hospital.

[BE] 1010.1.9.7 Delayed egress. Delayed egress locking systems, shall be permitted to be installed on doors serving any occupancy except Group A, E and H in buildings that are equipped throughout with an *automatic sprinkler system* in accordance with Section 903.3.1.1 or an *approved* automatic smoke or heat detection system installed in accordance with Section 907. The locking system shall be installed and operated in accordance with all of the following:

1. The delay electronics of the delayed egress locking system shall deactivate upon actuation of the *automatic sprinkler system* or automatic fire detection system, allowing immediate, free egress.
2. The delay electronics of the delayed egress locking system shall deactivate upon loss of power controlling the lock or lock mechanism, allowing immediate free egress.
3. The delayed egress locking system shall have the capability of being deactivated at the fire command center and other approved locations.
4. An attempt to egress shall initiate an irreversible process that shall allow such egress in not more than 15 seconds when a physical effort to exit is applied to the egress side door hardware for not more than 3 seconds. Initiation of the irreversible process shall activate an audible signal in the vicinity of the door. Once the delay electronics have been deactivated, rearming the delay electronics shall be by manual means only.

   Exception: Where *approved*, a delay of not more than 30 seconds is permitted on a delayed egress door.

5. The egress path from any point shall not pass through more than one delayed egress locking system.

   Exception: In Group I-2 or I-3 occupancies, the egress path from any point in the building shall not pass through more than two delayed egress locking systems provided the combined delay does not exceed 30 seconds.

6. A sign shall be provided on the door and shall be located above and within 12 inches (305 mm) of the door exit hardware:
   6.1. For doors that swing in the direction of egress, the sign shall read: PUSH UNTIL ALARM SOUNDS. DOOR CAN BE OPENED IN 15 [30] SECONDS.
   6.2. For doors that swing in the opposite direction of egress, the sign shall read: PULL UNTIL ALARM SOUNDS. DOOR CAN BE OPENED IN 15 [30] SECONDS.
   6.3 The sign shall comply with the visual character requirements in ICC A117.1.

   Exception: Where *approved*, in Group I occupancies, the installation of a sign is not required where care recipients who, because of clinical needs, require restraint or containment as part of the function of the treatment area.
7. Emergency lighting shall be provided on the egress side of the door.
8. The delayed egress locking system units shall be *listed* in accordance with UL 294.

[BE] 1010.1.9.8 Sensor release of electrically locked egress doors. The electric locks on sensor-released doors located in a *means of egress* in buildings with an occupancy in Groups A, B, E, I-1, I-2, I-4, M, R-1 or R-2 and entrance doors to tenant spaces in occupancies in Groups A, B, E, I-1, I-2, I-4, M, R-1 or R-2 are permitted where installed and operated in accordance with all of the following criteria:

1. The sensor shall be installed on the egress side, arranged to detect an occupant approaching the doors. The doors shall be arranged to unlock by a signal from or loss of power to the sensor.
2. Loss of power to the lock or locking system shall automatically unlock the doors.
3. The doors shall be arranged to unlock from a manual unlocking device located 40 inches to 48 inches (1016 mm to 1219 mm) vertically above the floor and within 5 feet (1524 mm) of the secured doors. Ready access shall be provided to the manual unlocking device and the device shall be clearly identified by a sign that reads "PUSH TO EXIT." When operated, the manual unlocking device shall result in direct interruption of power to the lock—independent of other electronics—and the doors shall remain unlocked for not less than 30 seconds.
4. Activation of the building fire alarm system, where provided, shall automatically unlock the doors, and the doors shall remain unlocked until the fire alarm system has been reset.
5. Activation of the building *automatic sprinkler system* or fire detection system, where provided, shall automatically unlock the doors. The doors shall remain unlocked until the fire alarm system has been reset.
6. The door locking system units shall be listed in accordance with UL 294.

[BE] 1010.1.9.9 Electromagnetically locked egress doors. Doors in the *means of egress* in buildings with an occupancy in Group A, B, E, I-1, I-2, I-4, M, R-1 or R-2 and doors to tenant spaces in Group A, B, E, I-1, I-2, I-4, M, R-1 or R-2 shall be permitted to be locked with an electromagnetic locking system where equipped with hardware that incorporates a built-in switch and where installed and operated in accordance with all of the following:

1. The hardware that is affixed to the door leaf has an obvious method of operation that is readily operated under all lighting conditions.
2. The hardware is capable of being operated with one hand.
3. Operation of the hardware directly interrupts the power to the electromagnetic lock and unlocks the door immediately.
4. Loss of power to the locking system automatically unlocks the door.
5. Where *panic* or *fire exit hardware* is required by Section 1010.1.10, operation of the *panic* or *fire exit hardware* also releases the electromagnetic lock.
6. The locking system units shall be *listed* in accordance with UL 294.

[BE] 1010.1.9.10 Locking arrangements in correctional facilities. In occupancies in Groups A-2, A-3, A-4, B, E, F, I-2, I-3, M and S within correctional and detention facilities, doors in *means of egress* serving rooms or spaces occupied by persons whose movements are controlled for security reasons shall be permitted to be locked where equipped with egress control devices that shall unlock manually and by not less than one of the following means:

1. Activation of an *automatic sprinkler system* installed in accordance with Section 903.3.1.1.
2. Activation of an *approved* manual fire alarm box.
3. A signal from a constantly attended location.

[BE] 1010.1.9.11 Stairway doors. Interior *stairway means of egress* doors shall be openable from both sides without the use of a key or special knowledge or effort.

Exceptions:

1. *Stairway* discharge doors shall be openable from the egress side and shall only be locked from the opposite side.
2. This section shall not apply to doors arranged in accordance with Section 403.5.3 of the *International Building Code*.

3. In *stairways* serving not more than four stories, doors are permitted to be locked from the side opposite the egress side, provided they are openable from the egress side and capable of being unlocked simultaneously without unlatching upon a signal from the fire command center, if present, or a signal by emergency personnel from a single location inside the main entrance to the building.
4. *Stairway* exit doors shall be openable from the egress side and shall only be locked from the opposite side in Group B, F, M and S occupancies where the only interior access to the tenant space is from a single *exit stairway* where permitted in Section 1006.3.2.
5. *Stairway* exit doors shall be openable from the egress side and shall only be locked from the opposite side in Group R-2 occupancies where the only interior access to the *dwelling unit* is from a single exit stairway where permitted in Section 1006.3.2.

[BE] 1010.1.10 Panic and fire exit hardware. Doors serving a Group H occupancy and doors serving rooms or spaces with an *occupant load* of 50 or more in a Group A or E occupancy shall not be provided with a latch or lock other than *panic hardware* or *fire exit hardware*.

Exceptions:

1. A main *exit* of a Group A occupancy shall be permitted to be locking in accordance with Section 1010.1.9.3, Item 2.
2. Doors serving a Group A or E occupancy shall be permitted to be electromagnetically locked in accordance with Section 1010.1.9.9.

Electrical rooms with equipment rated 1,200 amperes or more and over 6 feet (1829 mm) wide, and that contain over-current devices, switching devices or control devices with exit or exit access doors, shall be equipped with *panic hardware* or *fire exit hardware*. The doors shall swing in the direction of egress travel.

[BE] 1010.1.10.1 Installation. Where *panic* or *fire exit hardware* is installed, it shall comply with the following:

1. *Panic hardware* shall be *listed* in accordance with UL 305.
2. *Fire exit hardware* shall be *listed* in accordance with UL 10C and UL 305.
3. The actuating portion of the releasing device shall extend not less than one-half of the door leaf width.
4. The maximum unlatching force shall not exceed 15 pounds (67 N).

[BE] 1010.1.10.2 Balanced doors. If *balanced doors* are used and *panic hardware* is required, the *panic hardware* shall be the push-pad type and the pad shall not extend more than one-half the width of the door measured from the latch side.

[BE] 1010.2 Gates. Gates serving the *means of egress* system shall comply with the requirements of this section. Gates used as a component in a *means of egress* shall conform to the applicable requirements for doors.

Exception: Horizontal sliding or swinging gates exceeding the 4-foot (1219 mm) maximum leaf width limitation are permitted in fences and walls surrounding a stadium.

[BE] 1010.2.1 Stadiums. *Panic hardware* is not required on gates surrounding stadiums where such gates are under constant immediate supervision while the public is present, and where safe dispersal areas based on 3 square feet (0.28 m^2) per occupant are located between the fence and enclosed space. Such required safe dispersal areas shall not be located less than 50 feet (15 240 mm) from the enclosed space. See Section 1028.5 for *means of egress* from safe dispersal areas.

[BE] 1010.3 Turnstiles. Turnstiles or similar devices that restrict travel to one direction shall not be placed so as to obstruct any required *means of egress*.

Exception: Each turnstile or similar device shall be credited with a capacity based on not more than a 50-person *occupant load* where all of the following provisions are met:

1. Each device shall turn free in the direction of egress travel when primary power is lost and on the manual release by an employee in the area.
2. Such devices are not given credit for more than 50 percent of the required egress capacity or width.
3. Each device is not more than 39 inches (991 mm) high.
4. Each device has not less than $16^1/_2$ inches (419 mm) clear width at and below a height of 39 inches (991 mm) and not less than 22 inches (559 mm) clear width at heights above 39 inches (991 mm).

Where located as part of an *accessible route*, turnstiles shall have not less than 36 inches (914 mm) clear at and below a height of 34 inches (864 mm), not less than 32 inches (813 mm) clear width between 34 inches (864 mm) and 80 inches (2032 mm) and shall consist of a mechanism other than a revolving device.

[BE] 1010.3.1 High turnstile. Turnstiles more than 39 inches (991 mm) high shall meet the requirements for revolving doors.

[BE] 1010.3.2 Additional door. Where serving an *occupant load* greater than 300, each turnstile that is not portable shall have a side-hinged swinging door that conforms to Section 1010.1 within 50 feet (15 240 mm).

SECTION 1011
STAIRWAYS

[BE] 1011.1 General. *Stairways* serving occupied portions of a building shall comply with the requirements of Sections 1011.2 through 1011.13. Alternating tread devices shall comply with Section 1011.14. Ships ladders shall comply with Section 1011.15. Ladders shall comply with Section 1011.16.

Exception: Within rooms or spaces used for assembly purposes, stepped *aisles* shall comply with Section 1029.

* **[BE] 1011.2 Width and capacity.** The required capacity of *stairways* shall be determined as specified in Section 1005.1, but the minimum width shall be not less than 44 inches (1118 mm). See Section 1009.3 for *accessible means of egress stairways*.

Exceptions:

1. *Stairways* serving an *occupant load* of less than 50 shall have a width of not less than 36 inches (914 mm).
2. *Spiral stairways* as provided for in Section 1011.10.
3. Where an incline platform lift or *stairway* chairlift is installed on *stairways* serving occupancies in Group R-3, or within *dwelling units* in occupancies in Group R-2, a clear passage width not less than 20 inches (508 mm) shall be provided. Where the seat and platform can be folded when not in use, the distance shall be measured from the folded position.

[BE] 1011.3 Headroom. *Stairways* shall have a headroom clearance of not less than 80 inches (2032 mm) measured vertically from a line connecting the edge of the *nosings*. Such headroom shall be continuous above the *stairway* to the point where the line intersects the landing below, one tread depth beyond the bottom riser. The minimum clearance shall be maintained the full width of the *stairway* and landing.

Exceptions:

1. *Spiral stairways* complying with Section 1011.10 are permitted a 78-inch (1981 mm) headroom clearance.
2. In Group R-3 occupancies; within *dwelling units* in Group R-2 occupancies; and in Group U occupancies that are accessory to a Group R-3 occupancy or accessory to individual *dwelling units* in Group R-2 occupancies; where the *nosings* of treads at the side of a *flight* extend under the edge of a floor opening through which the *stair* passes, the floor opening shall be allowed to project horizontally into the required headroom not more than $4^3/_4$ inches (121 mm).

[BE] 1011.4 Walkline. The walkline across *winder* treads shall be concentric to the direction of travel through the turn and located 12 inches (305 mm) from the side where the *winders* are narrower. The 12-inch (305 mm) dimension shall be measured from the widest point of the clear *stair* width at the walking surface of the *winder*. Where *winders* are adjacent within the *flight*, the point of the widest clear *stair* width of the adjacent *winders* shall be used.

[BE] 1011.5 Stair treads and risers. *Stair* treads and risers shall comply with Sections 1011.5.1 through 1011.5.5.3.

[BE] 1011.5.1 Dimension reference surfaces. For the purpose of this section, all dimensions are exclusive of carpets, rugs or runners.

[BE] 1011.5.2 Riser height and tread depth. *Stair* riser heights shall be 7 inches (178 mm) maximum and 4 inches (102 mm) minimum. The riser height shall be measured vertically between the *nosings* of adjacent treads. Rectangular tread depths shall be 11 inches (279 mm) minimum measured horizontally between the vertical planes of the foremost projection of adjacent treads and at a right angle to the tread's *nosing*. *Winder* treads shall have a minimum tread depth of 11 inches (279 mm) between the vertical planes of the foremost projection of adjacent treads at the intersections with the walkline and a minimum tread depth of 10 inches (254 mm) within the clear width of the *stair*.

Exceptions:

1. *Spiral stairways* in accordance with Section 1011.10.
2. *Stairways* connecting stepped *aisles* to cross aisles or concourses shall be permitted to use the riser/tread dimension in Section 1029.13.2.
3. In Group R-3 occupancies; within *dwelling units* in Group R-2 occupancies; and in Group U occupancies that are accessory to a Group R-3 occupancy or accessory to individual *dwelling units* in Group R-2 occupancies; the maximum riser height shall be $7^3/_4$ inches (197 mm); the minimum tread depth shall be 10 inches (254 mm); the minimum *winder* tread depth at the walkline shall be 10 inches (254 mm); and the minimum *winder* tread depth shall be 6 inches (152 mm). A *nosing* projection not less than ¾ inch (19.1 mm) but not more than $1^1/_4$ inches (32 mm) shall be provided on *stairways* with solid risers where the tread depth is less than 11 inches (279 mm).
4. See Section 403.1 of the *International Existing Building Code* for the replacement of existing *stairways*.
5. In Group I-3 facilities, *stairways* providing access to guard towers, observation stations and control rooms, not more than 250 square feet (23 m^2) in area, shall be permitted to have a maximum riser height of 8 inches (203 mm) and a minimum tread depth of 9 inches (229 mm).

[BE] 1011.5.3 Winder treads. *Winder* treads are not permitted in *means of egress stairways* except within a *dwelling unit*.

Exceptions:

1. Curved *stairways* in accordance with Section 1011.9.
2. *Spiral stairways* in accordance with Section 1011.10.

[BE] 1011.5.4 Dimensional uniformity. *Stair* treads and risers shall be of uniform size and shape. The tolerance between the largest and smallest riser height or between the largest and smallest tread depth shall not exceed $^{3}/_{8}$ inch (9.5 mm) in any *flight* of *stairs*. The greatest *winder* tread depth at the walkline within any *flight* of *stairs* shall not exceed the smallest by more than $^{3}/_{8}$ inch (9.5 mm).

Exceptions:

1. *Stairways* connecting stepped *aisles* to cross *aisles* or concourses shall be permitted to comply with the dimensional nonuniformity in Section 1029.13.2.
2. Consistently shaped *winders*, complying with Section 1011.5, differing from rectangular treads in the same *flight* of *stairs*.
3. Nonuniform riser dimension complying with Section 1011.5.4.1.

[BE] 1011.5.4.1 Nonuniform height risers. Where the bottom or top riser adjoins a sloping *public way*, walkway or driveway having an established grade and serving as a landing, the bottom or top riser is permitted to be reduced along the slope to less than 4 inches (102 mm) in height, with the variation in height of the bottom or top riser not to exceed one unit vertical in 12 units horizontal (8-percent slope) of stair width. The nosings or leading edges of treads at such nonuniform height risers shall have a distinctive marking stripe, different from any other *nosing* marking provided on the *stair flight*. The distinctive marking stripe shall be visible in descent of the *stair* and shall have a slip-resistant surface. Marking stripes shall have a width of not less than 1 inch (25 mm) but not more than 2 inches (51 mm).

[BE] 1011.5.5 Nosing and riser profile. *Nosings* shall have a curvature or bevel of not less than $^{1}/_{16}$ inch (1.6 mm) but not more than $^{9}/_{16}$ inch (14.3 mm) from the foremost projection of the tread. Risers shall be solid and vertical or sloped under the tread above from the underside of the *nosing* above at an angle not more than 30 degrees (0.52 rad) from the vertical.

[BE] 1011.5.5.1 Nosing projection size. The leading edge (*nosings*) of treads shall project not more than $1^{1}/_{4}$ inches (32 mm) beyond the tread below.

[BE] 1011.5.5.2 Nosing projection uniformity. *Nosing* projections of the leading edges shall be of uniform size, including the projections of the *nosing's* leading edge of the floor at the top of a *flight*.

[BE] 1011.5.5.3 Solid risers. Risers shall be solid.

Exceptions:

1. Solid risers are not required for *stairways* that are not required to comply with Section 1009.3, provided that the opening between treads does not permit the passage of a sphere with a diameter of 4 inches (102 mm).
2. Solid risers are not required for occupancies in Group I-3 or in Group F, H and S occupancies other than areas accessible to the public. There are no restrictions on the size of the opening in the riser.
3. Solid risers are not required for *spiral stairways* constructed in accordance with Section 1011.10.

[BE] 1011.6 Stairway landings. There shall be a floor or landing at the top and bottom of each *stairway*. The width of landings shall be not less than the width of *stairways* served. Every landing shall have a minimum width measured perpendicular to the direction of travel equal to the width of the *stairway*. Where the *stairway* has a straight run the depth need not exceed 48 inches (1219 mm). Doors opening onto a landing shall not reduce the landing to less than one-half the required width. When fully open, the door shall not project more than 7 inches (178 mm) into a landing. Where wheelchair spaces are required on the *stairway* landing in accordance with Section 1009.6.3, the wheelchair space shall not be located in the required width of the landing and doors shall not swing over the wheelchair spaces.

Exception: Where *stairways* connect stepped *aisles* to cross *aisles* or concourses, *stairway* landings are not required at the transition between *stairways* and stepped *aisles* constructed in accordance with Section 1029.

[BE] 1011.7 Stairway construction. *Stairways* shall be built of materials consistent with the types permitted for the type of construction of the building, except that wood *handrails* shall be permitted for all types of construction.

[BE] 1011.7.1 Stairway walking surface. The walking surface of treads and landings of a *stairway* shall not be sloped steeper than one unit vertical in 48 units horizontal (2-percent slope) in any direction. *Stairway* treads and landings shall have a solid surface. Finish floor surfaces shall be securely attached.

Exceptions:

1. Openings in stair walking surfaces shall be a size that does not permit the passage of $^{1}/_{2}$-inch-diameter (12.7 mm) sphere. Elongated openings shall be placed so that the long dimension is perpendicular to the direction of travel.
2. In Group F, H and S occupancies, other than areas of parking structures accessible to the public, openings in treads and landings shall not be prohibited provided a sphere with a diameter of $1^{1}/_{8}$ inches (29 mm) cannot pass through the opening.

[BE] 1011.7.2 Outdoor conditions. Outdoor *stairways* and outdoor approaches to *stairways* shall be designed so that water will not accumulate on walking surfaces.

[BE] 1011.7.3 Enclosures under interior stairways. The walls and soffits within enclosed usable spaces under enclosed and unenclosed *stairways* shall be protected by 1-hour fire-resistance- rated construction or the *fire-resis-*

tance rating of the *stairway* enclosure, whichever is greater. Access to the enclosed space shall not be directly from within the *stairway* enclosure.

Exception: Spaces under *stairways* serving and contained within a single residential *dwelling unit* in Group R-2 or R-3 shall be permitted to be protected on the enclosed side with $^1/_2$-inch (12.7 mm) gypsum board.

[BE] 1011.7.4 Enclosures under exterior stairways. There shall not be enclosed usable space under *exterior exit stairways* unless the space is completely enclosed in 1-hour fire-resistance-rated construction. The open space under *exterior stairways* shall not be used for any purpose.

[BE] 1011.8 Vertical rise. A *flight* of *stairs* shall not have a vertical rise greater than 12 feet (3658 mm) between floor levels or landings.

Exception: *Spiral stairways* used as a *means of egress* from technical production areas.

[BE] 1011.9 Curved stairways. Curved *stairways* with *winder* treads shall have treads and risers in accordance with Section 1011.5 and the smallest radius shall be not less than twice the minimum width or required capacity of the *stairway*.

Exception: The radius restriction shall not apply to curved *stairways* in Group R-3 and within individual *dwelling units* in Group R-2.

[BE] 1011.10 Spiral stairways. *Spiral stairways* are permitted to be used as a component in the *means of egress* only within *dwelling units* or from a space not more than 250 square feet (23 m^2) in area and serving not more than five occupants, or from technical production areas in accordance with Section 410.6 of the *International Building Code*.

A *spiral stairway* shall have a $7^1/_2$-inch (191 mm) minimum clear tread depth at a point 12 inches (305 mm) from the narrow edge. The risers shall be sufficient to provide a headroom of 78 inches (1981 mm) minimum, but riser height shall not be more than $9^1/_2$ inches (241 mm). The minimum *stairway* clear width at and below the *handrail* shall be 26 inches (660 mm).

[BE] 1011.11 Handrails. *Stairways* shall have *handrails* on each side and shall comply with Section 1014. Where glass is used to provide the *handrail*, the *handrail* shall also comply with Section 2407 of the *International Building Code*.

Exceptions:

1. *Stairways* within *dwelling units*, and *spiral stairways* are permitted to have a *handrail* on one side only.
2. Decks, patios and walkways that have a single change in elevation where the landing depth on each side of the change of elevation is greater than what is required for a landing do not require *handrails*.
3. In Group R-3 occupancies, a change in elevation consisting of a single riser at an entrance or egress door does not require *handrails*.
4. Changes in room elevations of three or fewer risers within *dwelling units* and *sleeping units* in Group R-2 and R-3 do not require *handrails*.

[BE] 1011.12 Stairway to roof. In buildings four or more stories above grade plane, one *stairway* shall extend to the roof surface, unless the roof has a slope steeper than four units vertical in 12 units horizontal (33-percent slope).

Exception: Other than where required by Section 1011.12.1, in buildings without an occupied roof, access to the roof from the top story shall be permitted to be by an *alternating tread device*, a ships ladder or a permanent ladder.

[BE] 1011.12.1 Stairway to elevator equipment. Roofs and penthouses containing elevator equipment that must be accessed for maintenance are required to be accessed by a *stairway*.

[BE] 1011.12.2 Roof access. Where a *stairway* is provided to a roof, access to the roof shall be provided through a penthouse complying with Section 1510.2 of the *International Building Code*.

Exception: In buildings without an occupied roof, access to the roof shall be permitted to be a roof hatch or trap door not less than 16 square feet (1.5 m^2) in area and having a minimum dimension of 2 feet (610 mm).

[BE] 1011.13 Guards. *Guards* shall be provided along *stairways* and landings where required by Section 1015 and shall be constructed in accordance with Section 1015. Where the roof hatch opening providing the required access is located within 10 feet (3049 mm) of the roof edge, such roof access or roof edge shall be protected by *guards* installed in accordance with Section 1015.

[BE] 1011.14 Alternating tread devices. *Alternating tread devices* are limited to an element of a *means of egress* in buildings of Groups F, H and S from a *mezzanine* not more than 250 square feet (23 m^2) in area and that serves not more than five occupants; in buildings of Group I-3 from a guard tower, observation station or control room not more than 250 square feet (23 m^2) in area and for access to unoccupied roofs. *Alternating tread devices* used as a *means of egress* shall not have a rise greater than 20 feet (6096 mm) between floor levels or landings.

[BE] 1011.14.1 Handrails of alternating tread devices. *Handrails* shall be provided on both sides of *alternating tread devices* and shall comply with Section 1012.

[BE] 1011.14.2 Treads of alternating tread devices. *Alternating tread devices* shall have a minimum tread depth of 5 inches (127 mm), a minimum projected tread depth of $8^1/_2$ inches (216 mm), a minimum tread width of 7 inches (178 mm) and a maximum riser height of $9^1/_2$ inches (241 mm). The tread depth shall be measured horizontally between the vertical planes of the foremost projections of adjacent treads. The riser height shall be measured vertically between the leading edges of adjacent treads. The riser height and tread depth provided shall result in an angle of ascent from the horizontal of between 50 and 70 degrees (0.87 and 1.22 rad). The initial tread of the device shall begin at the same elevation as the platform, landing or floor surface.

Exception: *Alternating tread devices* used as an element of a *means of egress* in buildings from a *mezzanine* area not more than 250 square feet (23 m^2) in area that serves not more than five occupants shall have a minimum tread depth of 3 inches (76 mm) with a minimum projected tread depth of $10^1/_2$ inches (267 mm). The rise to the next alternating tread surface shall not exceed 8 inches (203 mm).

[BE] 1011.15 Ships ladders. Ships ladders are permitted to be used in Group I-3 as a component of a *means of egress* to and from control rooms or elevated facility observation stations not more than 250 square feet (23 m^2) with not more than three occupants and for access to unoccupied roofs. The minimum clear width at and below the *handrails* shall be 20 inches (508 mm).

[BE] 1011.15.1 Handrails of ships ladders. *Handrails* shall be provided on both sides of ships ladders.

[BE] 1011.15.2 Treads of ships ladders. Ships ladders shall have a minimum tread depth of 5 inches (127 mm). The tread shall be projected such that the total of the tread depth plus the *nosing* projection is not less than $8^1/_2$ inches (216 mm). The maximum riser height shall be $9^1/_2$ inches (241 mm).

[BE] 1011.16 Ladders. Permanent ladders shall not serve as a part of the *means of egress* from occupied spaces within a building. Permanent ladders shall be permitted to provide access to the following areas:

1. Spaces frequented only by personnel for maintenance, repair or monitoring of equipment.
2. Nonoccupiable spaces accessed only by catwalks, crawl spaces, freight elevators or very narrow passageways.
3. Raised areas used primarily for purposes of security, life safety or fire safety including, but not limited to, observation galleries, prison guard towers, fire towers or lifeguard stands.
4. Elevated levels in Group U not open to the general public.
5. Nonoccupied roofs that are not required to have *stairway* access in accordance with Section 1011.12.1.
6. Ladders shall be constructed in accordance with Section 306.5 of the *International Mechanical Code*.

SECTION 1012 RAMPS

[BE] 1012.1 Scope. The provisions of this section shall apply to ramps used as a component of a *means of egress*.

Exceptions:

1. Ramped *aisles* within assembly rooms or spaces shall comply with the provisions in Section 1029.
2. Curb *ramps* shall comply with ICC A117.1.
3. Vehicle *ramps* in parking garages for pedestrian *exit access* shall not be required to comply with Sections 1012.3 through 1012.10 where they are not an *accessible route* serving accessible parking spaces, other required accessible elements or part of an *accessible means of egress*.

[BE] 1012.2 Slope. *Ramps* used as part of a *means of egress* shall have a running slope not steeper than one unit vertical in 12 units horizontal (8-percent slope). The slope of other pedestrian *ramps* shall not be steeper than one unit vertical in eight units horizontal (12.5-percent slope).

[BE] 1012.3 Cross slope. The slope measured perpendicular to the direction of travel of a *ramp* shall not be steeper than one unit vertical in 48 units horizontal (2-percent slope).

[BE] 1012.4 Vertical rise. The rise for any *ramp* run shall be 30 inches (762 mm) maximum.

[BE] 1012.5 Minimum dimensions. The minimum dimensions of *means of egress ramps* shall comply with Sections 1012.5.1 through 1012.5.3.

[BE] 1012.5.1 Width and capacity. The minimum width and required capacity of a *means of egress ramp* shall be not less than that required for *corridors* by Section 1020.2. The clear width of a *ramp* between *handrails*, if provided, or other permissible projections shall be 36 inches (914 mm) minimum.

[BE] 1012.5.2 Headroom. The minimum headroom in all parts of the *means of egress ramp* shall be not less than 80 inches (2032 mm).

[BE] 1012.5.3 Restrictions. *Means of egress ramps* shall not reduce in width in the direction of egress travel. Projections into the required *ramp* and landing width are prohibited. Doors opening onto a landing shall not reduce the clear width to less than 42 inches (1067 mm).

[BE] 1012.6 Landings. *Ramps* shall have landings at the bottom and top of each *ramp*, points of turning, entrance, *exits* and at doors. Landings shall comply with Sections 1012.6.1 through 1012.6.5.

[BE] 1012.6.1 Slope. Landings shall have a slope not steeper than one unit vertical in 48 units horizontal (2-percent slope) in any direction. Changes in level are not permitted.

[BE] 1012.6.2 Width. The landing width shall be not less than the width of the widest *ramp* run adjoining the landing.

[BE] 1012.6.3 Length. The landing length shall be 60 inches (1525 mm) minimum.

Exceptions:

1. In Group R-2 and R-3 individual *dwelling* and *sleeping units* that are not required to be Accessible units, Type A units or Type B units in accordance with Section 1107 of the *International Building Code*, landings are permitted to be 36 inches (914 mm) minimum.

2. Where the *ramp* is not a part of an *accessible route*, the length of the landing shall not be required to be more than 48 inches (1220 mm) in the direction of travel.

[BE] 1012.6.4 Change in direction. Where changes in direction of travel occur at landings provided between *ramp* runs, the landing shall be 60 inches by 60 inches (1524 mm by 1524 mm) minimum.

Exception: In Group R-2 and R-3 individual *dwelling* or *sleeping units* that are not required to be Accessible units, Type A units or Type B units in accordance with Section 1107 of the *International Building Code*, landings are permitted to be 36 inches by 36 inches (914 mm by 914 mm) minimum.

[BE] 1012.6.5 Doorways. Where doorways are located adjacent to a *ramp* landing, maneuvering clearances required by ICC A117.1 are permitted to overlap the required landing area.

[BE] 1012.7 Ramp construction. *Ramps* shall be built of materials consistent with the types permitted for the type of construction of the building, except that wood *handrails* shall be permitted for all types of construction.

[BE] 1012.7.1 Ramp surface. The surface of *ramps* shall be of slip-resistant materials that are securely attached.

[BE] 1012.7.2 Outdoor conditions. Outdoor *ramps* and outdoor approaches to *ramps* shall be designed so that water will not accumulate on walking surfaces.

[BE] 1012.8 Handrails. *Ramps* with a rise greater than 6 inches (152 mm) shall have *handrails* on both sides. *Handrails* shall comply with Section 1014.

[BE] 1012.9 Guards. *Guards* shall be provided where required by Section 1015 and shall be constructed in accordance with Section 1015.

[BE] 1012.10 Edge protection. Edge protection complying with Section 1012.10.1 or 1012.10.2 shall be provided on each side of *ramp* runs and at each side of *ramp* landings.

Exceptions:

1. Edge protection is not required on *ramps* that are not required to have *handrails*, provided they have flared sides that comply with the ICC A117.1 curb *ramp* provisions.
2. Edge protection is not required on the sides of *ramp* landings serving an adjoining *ramp* run or *stairway*.
3. Edge protection is not required on the sides of *ramp* landings having a vertical dropoff of not more than $^1/_2$ inch (12.7 mm) within 10 inches (254 mm) horizontally of the required landing area.

[BE] 1012.10.1 Curb, rail, wall or barrier. A curb, rail, wall or barrier shall be provided to serve as edge protection. A curb shall be not less than 4 inches (102 mm) in height. Barriers shall be constructed so that the barrier prevents the passage of a 4-inch-diameter (102 mm) sphere, where any portion of the sphere is within 4 inches (102 mm) of the floor or ground surface.

[BE] 1012.10.2 Extended floor or ground surface. The floor or ground surface of the *ramp* run or landing shall extend 12 inches (305 mm) minimum beyond the inside face of a *handrail* complying with Section 1014.

SECTION 1013
EXIT SIGNS

[BE] 1013.1 Where required. *Exits* and *exit access* doors shall be marked by an *approved* exit sign readily visible from any direction of egress travel. The path of egress travel to *exits* and within *exits* shall be marked by readily visible exit signs to clearly indicate the direction of egress travel in cases where the *exit* or the path of egress travel is not immediately visible to the occupants. Intervening *means of egress* doors within *exits* shall be marked by exit signs. Exit sign placement shall be such that no point in an *exit access corridor* or *exit passageway* is more than 100 feet (30 480 mm) or the *listed* viewing distance for the sign, whichever is less, from the nearest visible *exit* sign.

Exceptions:

1. Exit signs are not required in rooms or areas that require only one *exit* or *exit access*.
2. Main exterior *exit* doors or gates that are obviously and clearly identifiable as *exits* need not have *exit* signs where *approved* by the *fire code official.*
3. Exit signs are not required in occupancies in Group U and individual *sleeping units* or *dwelling units* in Group R-1, R-2 or R-3.
4. Exit signs are not required in dayrooms, sleeping rooms or dormitories in occupancies in Group I-3.
5. In occupancies in Groups A-4 and A-5, exit signs are not required on the seating side of vomitories or openings into seating areas where exit signs are provided in the concourse that are readily apparent from the vomitories. Egress lighting is provided to identify each vomitory or opening within the seating area in an emergency.

[BE] 1013.2 Floor-level exit signs in Group R-1. Where exit signs are required in Group R-1 occupancies by Section 1013.1, additional low-level exit signs shall be provided in all areas serving guest rooms in Group R-1 occupancies and shall comply with Section 1013.5.

The bottom of the sign shall be not less than 10 inches (254 mm) nor more than 12 inches (305 mm) above the floor level. The sign shall be flush mounted to the door or wall. Where mounted on the wall, the edge of the sign shall be within 4 inches (102 mm) of the door frame on the latch side.

[BE] 1013.3 Illumination. Exit signs shall be internally or externally illuminated.

Exception: Tactile signs required by Section 1013.4 need not be provided with illumination.

[BE] 1013.4 Raised character and braille exit signs. A sign stating EXIT in visual characters, raised characters and braille and complying with ICC A117.1 shall be provided adjacent to each door to an area of refuge, an exterior area for assisted rescue, an exit stairway or ramp, an exit passageway and the exit discharge.

[BE] 1013.5 Internally illuminated exit signs. Electrically powered, *self-luminous* and *photoluminescent exit* signs shall be *listed* and labeled in accordance with UL 924 and shall be installed in accordance with the manufacturer's instructions and Section 604. Exit signs shall be illuminated at all times.

[BE] 1013.6 Externally illuminated exit signs. Externally illuminated exit signs shall comply with Sections 1013.6.1 through 1013.6.3.

[BE] 1013.6.1 Graphics. Every exit sign and directional exit sign shall have plainly legible letters not less than 6 inches (152 mm) high with the principal strokes of the letters not less than $^3/_4$ inch (19.1 mm) wide. The word "EXIT" shall have letters having a width not less than 2 inches (51 mm) wide, except the letter "I," and the minimum spacing between letters shall be not less than $^3/_8$ inch (9.5 mm). Signs larger than the minimum established in this section shall have letter widths, strokes and spacing in proportion to their height.

The word "EXIT" shall be in high contrast with the background and shall be clearly discernible when the means of exit sign illumination is or is not energized. If a chevron directional indicator is provided as part of the exit sign, the construction shall be such that the direction of the chevron directional indicator cannot be readily changed.

[BE] 1013.6.2 Exit sign illumination. The face of an exit sign illuminated from an external source shall have an intensity of not less than 5 footcandles (54 lux).

[BE] 1013.6.3 Power source. Exit signs shall be illuminated at all times. To ensure continued illumination for a duration of not less than 90 minutes in case of primary power loss, the sign illumination means shall be connected to an emergency power system provided from storage batteries, unit equipment or an on-site generator. The installation of the emergency power system shall be in accordance with Section 604.

Exceptions:

1. *Approved* exit sign illumination means that provide continuous illumination independent of external power sources for a duration of not less than 90 minutes, in case of primary power loss, are not required to be connected to an emergency electrical system.
2. Group I-2 Condition 2 exit sign illumination shall not be provided by unit equipment battery only.

SECTION 1014 HANDRAILS

[BE] 1014.1 Where required. *Handrails* serving *stairways*, *ramps*, stepped *aisles* and ramped *aisles* shall be adequate in strength and attachment in accordance with Section 1607.8 of the *International Building Code*. *Handrails* required for *stairways* by Section 1011.11 shall comply with Sections 1014.2 through 1014.9. *Handrails* required for *ramps* by Section 1012.8 shall comply with Sections 1014.2 through 1014.8. *Handrails* for stepped *aisles* and ramped *aisles* required by Section 1029.15 shall comply with Sections 1014.2 through 1014.8.

[BE] 1014.2 Height. *Handrail* height, measured above *stair* tread *nosings*, or finish surface of *ramp* slope, shall be uniform, not less than 34 inches (864 mm) and not more than 38 inches (965 mm). *Handrail* height of *alternating tread devices* and ships ladders, measured above tread *nosings*, shall be uniform, not less than 30 inches (762 mm) and not more than 34 inches (864 mm).

Exceptions:

1. Where handrail fittings or bendings are used to provide continuous transition between *flights*, the fittings or bendings shall be permitted to exceed the maximum height.
2. In Group R-3 occupancies; within *dwelling units* in Group R-2 occupancies; and in Group U occupancies that are associated with a Group R-3 occupancy or associated with individual *dwelling units* in Group R-2 occupancies; where handrail fittings or bendings are used to provide continuous transition between *flights*, transition at *winder* treads, transition from *handrail* to *guard*, or where used at the start of a *flight*, the *handrail* height at the fittings or bendings shall be permitted to exceed the maximum height.
3. *Handrails* on top of a *guard* where permitted along stepped *aisles* and ramped *aisles* in accordance with Section 1029.15.

[BE] 1014.3 Handrail graspability. Required *handrails* shall comply with Section 1014.3.1 or shall provide equivalent graspability.

Exception: In Group R-3 occupancies; within *dwelling units* in Group R-2 occupancies; and in Group U occupancies that are accessory to a Group R-3 occupancy or accessory to individual *dwelling units* in Group R-2 occupancies; *handrails* shall be Type I in accordance with Section 1014.3.1, Type II in accordance with Section 1014.3.2 or shall provide equivalent graspability.

[BE] 1014.3.1 Type I. *Handrails* with a circular cross section shall have an outside diameter of not less than $1^1/_4$ inches (32 mm) and not greater than 2 inches (51 mm). Where the *handrail* is not circular, it shall have a perimeter dimension of not less than 4 inches (102 mm) and not greater than $6^1/_4$ inches (160 mm) with a maximum cross-sectional dimension of $2^1/_4$ inches (57 mm) and minimum cross-sectional dimension of 1 inch (25 mm). Edges shall have a minimum radius of 0.01 inch (0.25 mm).

[BE] 1014.3.2 Type II. *Handrails* with a perimeter greater than $6^1/_4$ inches (160 mm) shall provide a graspable finger recess area on both sides of the profile. The finger recess shall begin within a distance of $^3/_4$ inch (19 mm) measured vertically from the tallest portion of the profile and achieve a depth of not less than $^5/_{16}$ inch (8 mm) within $^7/_8$ inch (22 mm) below the widest portion of the profile. This required depth shall continue for not less than $^3/_8$ inch (10 mm) to a level that is not less than $1^3/_4$ inches (45 mm) below the tallest portion of the profile. The minimum width of the *handrail* above the recess shall be not less than $1^1/_4$ inches (32 mm) to a maximum of $2^3/_4$ inches (70 mm). Edges shall have a minimum radius of 0.01 inch (0.25 mm).

[BE] 1014.4 Continuity. *Handrail* gripping surfaces shall be continuous, without interruption by newel posts or other obstructions.

Exceptions:

1. *Handrails* within *dwelling units* are permitted to be interrupted by a newel post at a turn or landing.
2. Within a *dwelling unit*, the use of a volute, turnout, starting easing or starting newel is allowed over the lowest tread.
3. Handrail brackets or balusters attached to the bottom surface of the *handrail* that do not project horizontally beyond the sides of the *handrail* within $1^1/_2$ inches (38 mm) of the bottom of the *handrail* shall not be considered obstructions. For each $^1/_2$ inch (12.7 mm) of additional *handrail* perimeter dimension above 4 inches (102 mm), the vertical clearance dimension of $1^1/_2$ inches (38 mm) shall be permitted to be reduced by $^1/_8$ inch (3.2 mm).
4. Where *handrails* are provided along walking surfaces with slopes not steeper than 1:20, the bottoms of the handrail gripping surfaces shall be permitted to be obstructed along their entire length where they are integral to crash rails or bumper guards.
5. *Handrails* serving stepped *aisles* or ramped *aisles* are permitted to be discontinuous in accordance with Section 1029.15.1.

[BE] 1014.5 Fittings. *Handrails* shall not rotate within their fittings.

[BE] 1014.6 Handrail extensions. *Handrails* shall return to a wall, *guard* or the walking surface or shall be continuous to the *handrail* of an adjacent *flight* of *stairs* or ramp run. Where *handrails* are not continuous between *flights* the *handrails* shall extend horizontally not less than 12 inches (305 mm) beyond the top riser and continue to slope for the depth of one tread beyond the bottom riser. At *ramps* where *handrails* are not continuous between runs, the *handrails* shall extend horizontally above the landing 12 inches (305 mm) minimum beyond the top and bottom of *ramp* runs. The extensions of *handrails* shall be in the same direction of the *flights* of *stairs* at *stairways* and the *ramp* runs at *ramps*.

Exceptions:

1. *Handrails* within a *dwelling unit* that is not required to be accessible need extend only from the top riser to the bottom riser.
2. *Handrails* serving *aisles* in rooms or spaces used for assembly purposes are permitted to comply with the *handrail* extensions in accordance with Section 1029.15.
3. *Handrails* for *alternating tread devices* and ships ladders are permitted to terminate at a location vertically above the top and bottom risers. *Handrails* for *alternating tread devices* are not required to be continuous between *flights* or to extend beyond the top or bottom risers.

[BE] 1014.7 Clearance. Clear space between a *handrail* and a wall or other surface shall be not less than $1^1/_2$ inches (38 mm). A *handrail* and a wall or other surface adjacent to the *handrail* shall be free of any sharp or abrasive elements.

[BE] 1014.8 Projections. On *ramps* and on ramped *aisles* that are part of an *accessible route*, the clear width between *handrails* shall be 36 inches (914 mm) minimum. Projections into the required width of *aisles*, *stairways* and *ramps* at each side shall not exceed $4^1/_2$ inches (114 mm) at or below the *handrail* height. Projections into the required width shall not be limited above the minimum headroom height required in Section 1011.3. Projections due to intermediate *handrails* shall not constitute a reduction in the egress width. Where a pair of intermediate *handrails* are provided within the *stairway* width without a walking surface between the pair of intermediate *handrails* and the distance between the pair of intermediate *handrails* is greater than 6 inches (152 mm), the available egress width shall be reduced by the distance between the closest edges of each such intermediate pair of *handrails* that is greater than 6 inches (152 mm).

[BE] 1014.9 Intermediate handrails. *Stairways* shall have intermediate *handrails* located in such a manner that all portions of the *stairway* minimum width or required capacity are within 30 inches (762 mm) of a *handrail*. On monumental *stairs*, *handrails* shall be located along the most direct path of egress travel.

SECTION 1015 GUARDS

[BE] 1015.1 General. *Guards* shall comply with the provisions of Section 1015.2 through 1015.6. Operable windows with sills located more than 72 inches (1829 mm) above finished grade or other surface below shall comply with Section 1015.7.

[BE] 1015.2 Where required. *Guards* shall be located along open-sided walking surfaces, including *mezzanines*, equipment platforms, *aisles*, *stairs*, *ramps* and landings that are located more than 30 inches (762 mm) measured vertically to the floor or grade below at any point within 36 inches (914 mm) horizontally to the edge of the open side. *Guards* shall be adequate in strength and attachment in accordance with Section 1607.8 of the *International Building Code*.

Exception: *Guards* are not required for the following locations:

1. On the loading side of loading docks or piers.
2. On the audience side of stages and raised platforms, including *stairs* leading up to the stage and raised platforms.
3. On raised stage and platform floor areas, such as runways, *ramps* and side stages used for entertainment or presentations.

4. At vertical openings in the performance area of stages and platforms.
5. At elevated walking surfaces appurtenant to stages and platforms for access to and utilization of special lighting or equipment.
6. Along vehicle service pits not accessible to the public.
7. In assembly seating areas at cross aisles in accordance with Section 1029.16.2.

[BE] 1015.2.1 Glazing. Where glass is used to provide a *guard* or as a portion of the *guard* system, the *guard* shall comply with Section 2407 of the *International Building Code*. Where the glazing provided does not meet the strength and attachment requirements of Section 1607.8 of the *International Building Code*, complying *guards* shall be located along glazed sides of open-sided walking surfaces.

[BE] 1015.3 Height. Required *guards* shall be not less than 42 inches (1067 mm) high, measured vertically as follows:

1. From the adjacent walking surfaces.
2. On *stairways* and stepped *aisles*, from the line connecting the leading edges of the tread *nosings*.
3. On *ramps* and ramped *aisles*, from the *ramp* surface at the *guard*.

Exceptions:

1. For occupancies in Group R-3 not more than three stories above grade in height and within individual *dwelling units* in occupancies in Group R-2 not more than three stories above grade in height with separate *means of egress*, required *guards* shall be not less than 36 inches (914 mm) in height measured vertically above the adjacent walking surfaces or adjacent *fixed seating*.
2. For occupancies in Group R-3, and within individual *dwelling units* in occupancies in Group R-2, *guards* on the open sides of *stairs* shall have a height not less than 34 inches (864 mm) measured vertically from a line connecting the leading edges of the treads.
3. For occupancies in Group R-3, and within individual *dwelling units* in occupancies in Group R-2, where the top of the *guard* also serves as a *handrail* on the open sides of *stairs*, the top of the *guard* shall be not less than 34 inches (864 mm) and not more than 38 inches (965 mm) measured vertically from a line connecting the leading edges of the treads.
4. The *guard* height in assembly seating areas shall comply with Section 1029.16 as applicable.
5. Along *alternating tread devices* and ships ladders, *guards* where the top rail also serves as a *handrail* shall have height not less than 30 inches (762 mm) and not more than 34 inches (864 mm), measured vertically from the leading edge of the device tread *nosing*.

[BE] 1015.4 Opening limitations. Required *guards* shall not have openings that allow passage of a sphere 4 inches (102 mm) in diameter from the walking surface to the required *guard* height.

Exceptions:

1. From a height of 36 inches (914 mm) to 42 inches (1067 mm), *guards* shall not have openings that allow passage of a sphere $4^3/_8$ inches (111 mm) in diameter.
2. The triangular openings at the open sides of a *stair*, formed by the riser, tread and bottom rail shall not allow passage of a sphere 6 inches (152 mm) in diameter.
3. At elevated walking surfaces for access to and use of electrical, mechanical or plumbing systems or equipment, *guards* shall not have openings that allow passage of a sphere 21 inches (533 mm) in diameter.
4. In areas that are not open to the public within occupancies in Group I-3, F, H or S, and for *alternating tread devices* and ships ladders, *guards* shall not have openings that allow passage of a sphere 21 inches (533 mm) in diameter.
5. In assembly seating areas, *guards* required at the end of *aisles* in accordance with Section 1029.16.4 shall not have openings that allow passage of a sphere 4 inches (102 mm) in diameter up to a height of 26 inches (660 mm). From a height of 26 inches (660 mm) to 42 inches (1067 mm) above the adjacent walking surfaces, *guards* shall not have openings that allow passage of a sphere 8 inches (203 mm) in diameter.
6. Within individual *dwelling units* and *sleeping units* in Group R-2 and R-3 occupancies, *guards* on the open sides of *stairs* shall not have openings that allow passage of a sphere $4^3/_8$ (111 mm) inches in diameter.

[BE] 1015.5 Screen porches. Porches and decks that are enclosed with insect screening shall be provided with *guards* where the walking surface is located more than 30 inches (762 mm) above the floor or grade below.

[BE] 1015.6 Mechanical equipment, systems and devices. *Guards* shall be provided where various components that require service are located within 10 feet (3048 mm) of a roof edge or open side of a walking surface and such edge or open side is located more than 30 inches (762 mm) above the floor, roof or grade below. The *guard* shall extend not less than 30 inches (762 mm) beyond each end of such components. The *guard* shall be constructed so as to prevent the passage of a sphere 21 inches (533 mm) in diameter.

Exception: *Guards* are not required where permanent fall arrest/restraint anchorage connector devices that comply with ANSI/ASSE Z 359.1 are affixed for use during the entire roof covering lifetime. The devices shall be re-evaluated for possible replacement when the entire roof covering is replaced. The devices shall be placed not more than

10 feet (3048 mm) on center along hip and ridge lines and placed not less than 10 feet (3048 mm) from the roof edge or open side of the walking surface.

[BE] 1015.7 Roof access. *Guards* shall be provided where the roof hatch opening is located within 10 feet (3048 mm) of a roof edge or open side of a walking surface and such edge or open side is located more than 30 inches (762 mm) above the floor, roof or grade below. The *guard* shall be constructed so as to prevent the passage of a sphere 21 inches (533 mm) in diameter.

Exception: *Guards* are not required where permanent fall arrest/restraint anchorage connector devices that comply with ANSI/ASSE Z 359.1 are affixed for use during the entire roof covering lifetime. The devices shall be reevaluated for possible replacement when the entire roof covering is replaced. The devices shall be placed not more than 10 feet (3048 mm) on center along hip and ridge lines and placed not less than 10 feet (3048 mm) from the roof edge or open side of the walking surface.

[BE] 1015.8 Window openings. Windows in Group R-2 and R-3 buildings including *dwelling units*, where the top of the sill of an operable window opening is located less than 36 inches above the finished floor and more than 72 inches (1829 mm) above the finished grade or other surface below on the exterior of the building, shall comply with one of the following:

1. Operable windows where the top of the sill of the opening is located more than 75 feet (22 860 mm) above the finished grade or other surface below and that are provided with window fall prevention devices that comply with ASTM F 2006.
2. Operable windows where the openings will not allow a 4-inch-diameter (102 mm) sphere to pass through the opening when the window is in its largest opened position.
3. Operable windows where the openings are provided with window fall prevention devices that comply with ASTM F 2090.
4. Operable windows that are provided with window opening control devices that comply with Section 1015.8.1.

[BE] 1015.8.1 Window opening control devices. Window opening control devices shall comply with ASTM F 2090. The window opening control device, after operation to release the control device allowing the window to fully open, shall not reduce the minimum net clear opening area of the window unit to less than the area required by Section 1030.2.

SECTION 1016
EXIT ACCESS

[BE] 1016.1 General. The *exit access* shall comply with the applicable provisions of Sections 1003 through 1015. *Exit access* arrangement shall comply with Sections 1016 through 1021.

[BE] 1016.2 Egress through intervening spaces. Egress through intervening spaces shall comply with this section.

1. Exit access through an enclosed elevator lobby is permitted. Access to not less than one of the required *exits* shall be provided without travel through the enclosed elevator lobbies required by Section 3006.2, 3007 or 3008 of the *International Building Code*. Where the path of *exit access* travel passes through an enclosed elevator lobby the level of protection required for the enclosed elevator lobby is not required to be extended to the *exit* unless direct access to an *exit* is required by other sections of this code.
2. Egress from a room or space shall not pass through adjoining or intervening rooms or areas, except where such adjoining rooms or areas and the area served are accessory to one or the other, are not a Group H occupancy and provide a discernible path of egress travel to an *exit*.

   Exception: *Means of egress* are not prohibited through adjoining or intervening rooms or spaces in a Group H, S or F occupancy where the adjoining or intervening rooms or spaces are the same or a lesser hazard occupancy group.
3. An *exit access* shall not pass through a room that can be locked to prevent egress.
4. *Means of egress* from *dwelling units* or sleeping areas shall not lead through other sleeping areas, toilet rooms or bathrooms.
5. Egress shall not pass through kitchens, storage rooms, closets or spaces used for similar purposes.

   Exceptions:

   1. *Means of egress* are not prohibited through a kitchen area serving adjoining rooms constituting part of the same *dwelling unit* or *sleeping unit.*
   2. *Means of egress* are not prohibited through stockrooms in Group M occupancies where all of the following are met:

      2.1. The stock is of the same hazard classification as that found in the main retail area.

      2.2. Not more than 50 percent of the *exit access* is through the stockroom.

      2.3. The stockroom is not subject to locking from the egress side.

      2.4. There is a demarcated, minimum 44-inch-wide (1118 mm) *aisle* defined by full- or partial-height fixed walls or similar construction that will maintain the required width and lead directly from the retail area to the *exit* without obstructions.

[BE] 1016.2.1 Multiple tenants. Where more than one tenant occupies any one floor of a building or structure, each tenant space, *dwelling unit* and *sleeping unit* shall be provided with access to the required *exits* without passing through adjacent tenant spaces, *dwelling units* and *sleeping units*.

Exception: The *means of egress* from a smaller tenant space shall not be prohibited from passing through a larger adjoining tenant space where such rooms or spaces of the smaller tenant occupy less than 10 percent of the area of the larger tenant space through which they pass; are the same or similar occupancy group; a discernable path of egress travel to an *exit* is provided; and the *means of egress* into the adjoining space is not subject to locking from the egress side. A required *means of egress* serving the larger tenant space shall not pass through the smaller tenant space or spaces.

SECTION 1017 EXIT ACCESS TRAVEL DISTANCE

[BE] 1017.1 General. Travel distance within the *exit access* portion of the *means of egress* system shall be in accordance with this section.

[BE] 1017.2 Limitations. *Exit access* travel distance shall not exceed the values given in Table 1017.2.

[BE] 1017.2.1 Exterior egress balcony increase. *Exit access* travel distances specified in Table 1017.2 shall be increased up to an additional 100 feet (30 480 mm) provided the last portion of the *exit access* leading to the *exit* occurs on an exterior egress balcony constructed in accordance with Section 1021. The length of such balcony shall be not less than the amount of the increase taken.

[BE] 1017.2.2 Group F-1 and S-1 increase. The maximum *exit access* travel distance shall be 400 feet (122 m) in Group F-1 or S-1 occupancies where all of the following conditions are met:

1. The portion of the building classified as Group F-1 or S-1 is limited to one story in height.
2. The minimum height from the finished floor to the bottom of the ceiling or roof slab or deck is 24 feet (7315 mm).
3. The building is equipped throughout with an *automatic sprinkler system* in accordance with Section 903.3.1.1.

[BE] 1017.3 Measurement. *Exit access* travel distance shall be measured from the most remote point within a story along the natural and unobstructed path of horizontal and vertical egress travel to the entrance to an *exit*.

Exception: In open parking garages, *exit access* travel distance is permitted to be measured to the closest riser of an *exit access stairway* or the closest slope of an *exit access ramp*.

[BE] 1017.3.1 Exit access stairways and ramps. Travel distance on *exit access stairways* or *ramps* shall be included in the *exit access* travel distance measurement. The measurement along *stairways* shall be made on a plane parallel and tangent to the *stair* tread *nosings* in the center of the *stair* and landings. The measurement along *ramps* shall be made on the walking surface in the center of the *ramp* and landings.

[BE] TABLE 1017.2 EXIT ACCESS TRAVEL DISTANCE[a]

OCCUPANCY	WITHOUT SPRINKLER SYSTEM (feet)	WITH SPRINKLER SYSTEM (feet)
A, E, F-1, M, R, S-1	200	250[b]
I-1	Not Permitted	250[b]
B	200	300[c]
F-2, S-2, U	300	400[c]
H-1	Not Permitted	75[d]
H-2	Not Permitted	100[d]
H-3	Not Permitted	150[d]
H-4	Not Permitted	175[d]
H-5	Not Permitted	200[c]
I-2, I-3, I-4	Not Permitted	200[c]

For SI: 1 foot = 304.8 mm.

a. See the following sections for modifications to exit access travel distance requirements:

Section 402.8 of the *International Building Code*: For the distance limitation in malls.

Section 404.9 of the *International Building Code*: For the distance limitation through an atrium space.

Section 407.4 of the *International Building Code*: For the distance limitation in Group I-2.

Sections 408.6.1 and 408.8.1 of the *International Building Code*: For the distance limitations in Group I-3.

Section 411.4 of the *International Building Code*: For the distance limitation in special amusement buildings.

Section 412.7 of the *International Building Code*: For the distance limitations in aircraft manufacturing facilities.

Section 1006.2.2.2: For the distance limitation in refrigeration machinery rooms.

Section 1006.2.2.3: For the distance limitation in refrigerated rooms and spaces.

Section 1006.3.2: For buildings with one *exit*.

Section 1017.2.2: For increased distance limitation in Groups F-1 and S-1.

Section 1029.7: For increased limitation in assembly seating.

Section 3103.4 of the *International Building Code*: For temporary structures.

Section 3104.9 of the *International Building Code*: For pedestrian walkways.

b. Buildings equipped throughout with an *automatic sprinkler system* in accordance with Section 903.3.1.1 or 903.3.1.2. See Section 903 for occupancies where *automatic sprinkler systems* are permitted in accordance with Section 903.3.1.2.

c. Buildings equipped throughout with an *automatic sprinkler system* in accordance with Section 903.3.1.1.

d. Group H occupancies equipped throughout with an *automatic sprinkler system* in accordance with Section 903.2.5.1.

SECTION 1018 AISLES

[BE] 1018.1 General. *Aisles* and *aisle accessways* serving as a portion of the *exit access* in the *means of egress* system shall comply with the requirements of this section. *Aisles* or *aisle accessways* shall be provided from all occupied portions of the *exit access* that contain seats, tables, furnishings, displays and similar fixtures or equipment. The minimum width or required capacity of *aisles* shall be unobstructed.

Exception: Encroachments complying with Section 1005.7.

[BE] 1018.2 Aisles in assembly spaces. *Aisles and aisle accessways* serving a room or space used for assembly purposes shall comply with Section 1029.

[BE] 1018.3 Aisles in Groups B and M. In Group B and M occupancies, the minimum clear *aisle* width shall be determined by Section 1005.1 for the *occupant load* served, but shall be not less than that required for *corridors* by Section 1020.2.

Exception: Nonpublic *aisles* serving less than 50 people and not required to be accessible by Chapter 11 of the *International Building Code* need not exceed 28 inches (711 mm) in width.

[BE] 1018.4 Aisle accessways in Group M. An *aisle accessway* shall be provided on not less than one side of each element within the *merchandise pad*. The minimum clear width for an *aisle accessway* not required to be accessible shall be 30 inches (762 mm). The required clear width of the *aisle accessway* shall be measured perpendicular to the elements and merchandise within the *merchandise pad*. The 30-inch (762 mm) minimum clear width shall be maintained to provide a path to an adjacent *aisle* or *aisle accessway*. The *common path of egress travel* shall not exceed 30 feet (9144 mm) from any point in the *merchandise pad*.

Exception: For areas serving not more than 50 occupants, the *common path of egress travel* shall not exceed 75 feet (22 860 mm).

[BE] 1018.5 Aisles in other than assembly spaces and Groups B and M. In other than rooms or spaces used for assembly purposes and Group B and M occupancies, the minimum clear *aisle* capacity shall be determined by Section 1005.1 for the *occupant load* served, but the width shall be not less than that required for *corridors* by Section 1020.2.

Exception: Nonpublic *aisles* serving less than 50 people and not required to be accessible by Chapter 11 of the *International Building Code* need not exceed 28 inches (711 mm) in width.

SECTION 1019 EXIT ACCESS STAIRWAYS AND RAMPS

**

[BE] 1019.1 General. *Exit access stairways* and *ramps* serving as an *exit access* component in a *means of egress* system shall comply with the requirements of this section. The number of stories connected by *exit access stairways* and *ramps* shall include basements, but not *mezzanines*.

[BE] 1019.2 All occupancies. *Exit access stairways* and *ramps* that serve floor levels within a single story are not required to be enclosed.

[BE] 1019.3 Occupancies other than Groups I-2 and I-3. In other than Group I-2 and I-3 occupancies, floor openings containing *exit access stairways* or *ramps* that do not comply with one of the conditions listed in this section shall be enclosed with a shaft enclosure constructed in accordance with Section 713 of the *International Building Code*. *

1. *Exit access stairways* and *ramps* that serve, or atmospherically communicate between, only two stories. Such interconnected stories shall not be open to other stories.
2. In Group R-1, R-2 or R-3 occupancies, *exit access stairways* and *ramps* connecting four stories or less serving and contained within an individual *dwelling unit* or *sleeping unit* or live/work unit.
3. *Exit access stairways* serving and contained within a Group R-3 congregate residence or a Group R-4 facility are not required to be enclosed.
4. *Exit access stairways* and *ramps* in buildings equipped throughout with an *automatic sprinkler system* in accordance with Section 903.3.1.1, where the area of the vertical opening between stories does not exceed twice the horizontal projected area of the *stairway* or *ramp*, and the opening is protected by a draft curtain and closely spaced sprinklers in accordance with NFPA 13. In other than Group B and M occupancies, this provision is limited to openings that do not connect more than four stories.
5. *Exit access stairways* and *ramps* within an atrium complying with the provisions of Section 404 of the *International Building Code*.
6. *Exit access stairways* and *ramps* in open parking garages that serve only the parking garage.
7. *Exit access stairways* and *ramps* serving open-air seating complying with the *exit access* travel distance requirements of Section 1029.7.
8. *Exit access stairways* and *ramps* serving the balcony, gallery or press box and the main assembly floor in occupancies such as theaters, *places of religious worship*, auditoriums and sports facilities.

[BE] 1019.4 Group I-2 and I-3 occupancies. In Group I-2 and I-3 occupancies, floor openings between stories containing *exit access stairways* or *ramps* are required to be enclosed with a shaft enclosure constructed in accordance with Section 713 of the *International Building Code*.

Exception: In Group I-3 occupancies, *exit access stairways* or *ramps* constructed in accordance with Section 408 of the *International Building Code* are not required to be enclosed.

SECTION 1020 CORRIDORS

[BE] 1020.1 Construction. *Corridors* shall be fire-resistance rated in accordance with Table 1020.1. The *corridor* walls required to be fire-resistance rated shall comply with Section 708 of the *International Building Code* for fire partitions.

Exceptions:

1. A fire-resistance rating is not required for *corridors* in an occupancy in Group E where each room that is used for instruction has not less than one door opening directly to the exterior and rooms for assembly purposes have not less than one-half of the required means of egress doors opening directly to the exterior. Exterior doors specified in this exception are required to be at ground level.
2. A fire-resistance rating is not required for corridors contained within a *dwelling unit* or *sleeping unit* in an occupancy in Groups I-1 and R.
3. A fire-resistance rating is not required for *corridors* in open parking garages.
4. A fire-resistance rating is not required for *corridors* in an occupancy in Group B that is a space requiring only a single *means of egress* complying with Section 1006.2.
5. *Corridors* adjacent to the *exterior walls* of buildings shall be permitted to have unprotected openings on unrated *exterior walls* where unrated walls are permitted by Table 602 of the *International Building Code* and unprotected openings are permitted by Table 705.8 of the *International Building Code*.

[BE] TABLE 1020.1
CORRIDOR FIRE-RESISTANCE RATING

OCCUPANCY	OCCUPANT LOAD SERVED BY CORRIDOR	REQUIRED FIRE-RESISTANCE RATING (hours)	
		Without sprinkler system	With sprinkler system[c]
H-1, H-2, H-3	All	Not Permitted	1
H-4, H-5	Greater than 30	Not Permitted	1
A, B, E, F, M, S, U	Greater than 30	1	0
R	Greater than 10	Not Permitted	0.5
I-2[a], I-4	All	Not Permitted	0
I-1, I-3	All	Not Permitted	1[b]

a. For requirements for occupancies in Group I-2, see Sections 407.2 and 407.3 of the *International Building Code*.

b. For a reduction in the fire-resistance rating for occupancies in Group I-3, see Section 408.8 of the *International Building Code*.

c. Buildings equipped throughout with an *automatic sprinkler system* in accordance with Section 903.3.1.1 or 903.3.1.2 where allowed.

[BE] 1020.2 Width and capacity. The required capacity of *corridors* shall be determined as specified in Section 1005.1, but the minimum width shall be not less than that specified in Table 1020.2.

Exception: In Group I-2 occupancies, *corridors* are not required to have a clear width of 96 inches (2438 mm) in areas where there will not be stretcher or bed movement for access to care or as part of the defend-in-place strategy.

[BE] TABLE 1020.2
MINIMUM CORRIDOR WIDTH

OCCUPANCY	MINIMUM WIDTH (inches)
Any facilities not listed below	44
Access to and utilization of mechanical, plumbing or electrical systems or equipment	24
With an occupant load of less than 50	36
Within a dwelling unit	36
In Group E with a corridor having a occupant load of 100 or more	72
In corridors and areas serving stretcher traffic in ambulatory care facilities	72
Group I-2 in areas where required for bed movement	96

For SI: 1 inch = 25.4 mm.

[BE] 1020.3 Obstruction. The minimum width or required capacity of *corridors* shall be unobstructed.

Exception: Encroachments complying with Section 1005.7.

[BE] 1020.4 Dead ends. Where more than one *exit* or exit access doorway is required, the *exit access* shall be arranged such that there are no dead ends in *corridors* more than 20 feet (6096 mm) in length.

Exceptions:

1. In occupancies in Group I-3 of Condition 2, 3 or 4, the dead end in a *corridor* shall not exceed 50 feet (15 240 mm).
2. In occupancies in Groups B, E, F, I-1, M, R-1, R-2, R-4, S and U, where the building is equipped throughout with an *automatic sprinkler system* in accordance with Section 903.3.1.1, the length of the dead-end *corridors* shall not exceed 50 feet (15 240 mm).
3. A dead-end *corridor* shall not be limited in length where the length of the dead-end *corridor* is less than 2.5 times the least width of the dead-end *corridor*.

[BE] 1020.5 Air movement in corridors. *Corridors* shall not serve as supply, return, exhaust, relief or ventilation air ducts.

Exceptions:

1. Use of a *corridor* as a source of makeup air for exhaust systems in rooms that open directly onto such *corridors*, including toilet rooms, bathrooms, dressing rooms, smoking lounges and janitor closets, shall be permitted, provided that each such *corridor* is directly supplied with outdoor air at a rate greater than the rate of makeup air taken from the *corridor*.

2. Where located within a *dwelling unit*, the use of *corridors* for conveying return air shall not be prohibited.

3. Where located within tenant spaces of 1,000 square feet (93 m^2) or less in area, utilization of *corridors* for conveying return air is permitted.

4. Incidental air movement from pressurized rooms within health care facilities, provided that the *corridor* is not the primary source of supply or return to the room.

[BE] 1020.5.1 Corridor ceiling. Use of the space between the *corridor* ceiling and the floor or roof structure above as a return air plenum is permitted for one or more of the following conditions:

1. The *corridor* is not required to be of fire-resistance-rated construction.

2. The *corridor* is separated from the plenum by fire-resistance-rated construction.

3. The air-handling system serving the *corridor* is shut down upon activation of the air-handling unit smoke detectors required by the *International Mechanical Code*.

4. The air-handling system serving the *corridor* is shut down upon detection of sprinkler water flow where the building is equipped throughout with an *automatic sprinkler system*.

5. The space between the *corridor* ceiling and the floor or roof structure above the *corridor* is used as a component of an *approved* engineered smoke control system.

[BE] 1020.6 Corridor continuity. Fire-resistance-rated *corridors* shall be continuous from the point of entry to an *exit*, and shall not be interrupted by intervening rooms. Where the path of egress travel within a fire-resistance-rated *corridor* to the *exit* includes travel along unenclosed *exit access stairways* or *ramps*, the fire-resistance-rating shall be continuous for the length of the *stairway* or *ramp* and for the length of the connecting *corridor* on the adjacent floor leading to the *exit*.

Exceptions:

1. Foyers, lobbies or reception rooms constructed as required for *corridors* shall not be construed as intervening rooms.

2. Enclosed elevator lobbies as permitted by Item 1 of Section 1016.2 shall not be construed as intervening rooms.

SECTION 1021
EGRESS BALCONIES

[BE] 1021.1 General. Balconies used for egress purposes shall conform to the same requirements as *corridors* for minimum width, required capacity, headroom, dead ends and projections.

[BE] 1021.2 Wall separation. Exterior egress balconies shall be separated from the interior of the building by walls and opening protectives as required for *corridors*.

Exception: Separation is not required where the exterior egress balcony is served by not less than two *stairways* and a dead-end travel condition does not require travel past an unprotected opening to reach a *stairway*.

[BE] 1021.3 Openness. The long side of an egress balcony shall be at least 50 percent open, and the open area above the guards shall be so distributed as to minimize the accumulation of smoke or toxic gases.

[BE] 1021.4 Location. Exterior egress balconies shall have a minimum fire separation distance of 10 feet (3048 mm) measured at right angles from the exterior edge of the egress balcony to the following:

1. Adjacent lot lines.

2. Other portions of the building.

3. Other buildings on the same lot unless the adjacent building *exterior walls* and openings are protected in accordance with Section 705 of the *International Building Code* based on fire separation distance.

For the purposes of this section, other portions of the building shall be treated as separate buildings.

SECTION 1022
EXITS

[BE] 1022.1 General. *Exits* shall comply with Sections 1022 through 1027 and the applicable requirements of Sections 1003 through 1015. An *exit* shall not be used for any purpose that interferes with its function as a *means of egress*. Once a given level of *exit* protection is achieved, such level of protection shall not be reduced until arrival at the *exit discharge*. *Exits* shall be continuous from the point of entry into the *exit* to the *exit discharge*.

[BE] 1022.2 Exterior exit doors. Buildings or structures used for human occupancy shall have not less than one exterior door that meets the requirements of Section 1010.1.1.

[BE] 1022.2.1 Detailed requirements. Exterior exit doors shall comply with the applicable requirements of Section 1010.1.

[BE] 1022.2.2 Arrangement. Exterior exit doors shall lead directly to the *exit discharge* or the *public way*.

*

SECTION 1023
INTERIOR EXIT STAIRWAYS AND RAMPS

[BE] 1023.1 General. *Interior exit stairways* and *ramps* serving as an *exit* component in a *means of egress* system shall comply with the requirements of this section. *Interior exit stairways* and *ramps* shall be enclosed and lead directly to the exterior of the building or shall be extended to the exterior of the building with an *exit passageway* conforming to the requirements of Section 1024, except as permitted in Section

1028.1. An *interior exit* stairway or *ramp* shall not be used for any purpose other than as a *means of egress* and a circulation path.

[BE] 1023.2 Construction. Enclosures for *interior exit stairways* and *ramps* shall be constructed as *fire barriers* in accordance with Section 707 of the *International Building Code* or *horizontal assemblies* constructed in accordance with Section 711 of the *International Building Code*, or both. *Interior exit stairway* and *ramp* enclosures shall have a *fire-resistance rating* of not less than 2 hours where connecting four stories or more and not less than 1 hour where connecting less than four stories. The number of stories connected by the *interior exit stairways* or *ramps* shall include any basements, but not any *mezzanines*. *Interior exit stairways* and *ramps* shall have a fire-resistance rating not less than the floor assembly penetrated, but need not exceed 2 hours.

Exceptions:

1. *Interior exit stairways* and *ramps* in Group I-3 occupancies in accordance with the provisions of Section 408.3.8 of the *International Building Code.*
2. *Interior exit stairways* within an atrium enclosed in accordance with Section 404.6 of the *International Building Code.*

[BE] 1023.3 Termination. *Interior exit stairways* and *ramps* shall terminate at an *exit discharge* or a *public way*.

Exception: A combination of *interior exit stairways*, *interior exit ramps* and *exit* passageways, constructed in accordance with Sections 1023.2, 1023.3.1 and 1024, respectively, and forming a continuous protected enclosure, shall be permitted to extend an *interior exit stairway* or *ramp* to the *exit discharge* or a *public way*.

[BE] 1023.3.1 Extension. Where *interior exit stairways* and *ramps* are extended to an *exit discharge* or a *public way* by an *exit passageway*, the *interior exit stairway* and *ramp* shall be separated from the *exit passageway* by a *fire barrier* constructed in accordance with Section 707 of the *International Building Code* or a *horizontal assembly* constructed in accordance with Section 711 of the *International Building Code*, or both. The *fire-resistance rating* shall be not less than that required for the *interior exit stairway* and *ramp*. A *fire door* assembly complying with Section 716.5 of the *International Building Code* shall be installed in the *fire barrier* to provide a *means of egress* from the *interior exit stairway* and *ramp* to the *exit passageway*. Openings in the *fire barrier* other than the *fire door* assembly are prohibited. Penetrations of the *fire barrier* are prohibited.

Exceptions:

1. Penetrations of the *fire barrier* in accordance with Section 1023.5 shall be permitted.
2. Separation between an *interior exit stairway* or *ramp* and the *exit passageway* extension shall not be required where there are no openings into the *exit passageway* extension.

[BE] 1023.4 Openings. *Interior exit stairway* and *ramp* opening protectives shall be in accordance with the requirements of Section 716 of the *International Building Code.*

Openings in *interior exit stairways* and *ramps* other than unprotected exterior openings shall be limited to those necessary for *exit access* to the enclosure from normally occupied spaces and for egress from the enclosure.

Elevators shall not open into *interior exit stairways* and *ramps*.

[BE] 1023.5 Penetrations. Penetrations into or through *interior exit stairways* and *ramps* are prohibited except for equipment and ductwork necessary for independent ventilation or pressurization, sprinkler piping, standpipes, electrical raceway for fire department communication systems and electrical raceway serving the *interior exit stairway* and *ramp* and terminating at a steel box not exceeding 16 square inches (0.010 m^2). Such penetrations shall be protected in accordance with Section 714 of the *International Building Code.* There shall not be penetrations or communication openings, whether protected or not, between adjacent *interior exit stairways* and *ramps*.

Exception: Membrane penetrations shall be permitted on the outside of the *interior exit stairway* and *ramp*. Such penetrations shall be protected in accordance with Section 714.3.2 of the *International Building Code.*

[BE] 1023.6 Ventilation. Equipment and ductwork for *interior exit stairway* and *ramp* ventilation as permitted by Section 1023.5 shall comply with one of the following items:

1. Such equipment and ductwork shall be located exterior to the building and shall be directly connected to the *interior exit stairway* and *ramp* by ductwork enclosed in construction as required for shafts.
2. Where such equipment and ductwork is located within the *interior exit stairway* and *ramp*, the intake air shall be taken directly from the outdoors and the exhaust air shall be discharged directly to the outdoors, or such air shall be conveyed through ducts enclosed in construction as required for shafts.
3. Where located within the building, such equipment and ductwork shall be separated from the remainder of the building, including other mechanical equipment, with construction as required for shafts.

In each case, openings into the fire-resistance-rated construction shall be limited to those needed for maintenance and operation and shall be protected by opening protectives in accordance with Section 716 of the *International Building Code* for shaft enclosures.

The *interior exit stairway* and *ramp* ventilation systems shall be independent of other building ventilation systems.

[BE] 1023.7 Interior exit stairway and ramp exterior walls. Exterior walls of the *interior exit stairway* or *ramp* shall comply with the requirements of Section 705 of the *International Building Code* for *exterior walls*. Where nonrated walls or unprotected openings enclose the exterior of the *stairway* or *ramps* and the walls or openings are exposed by other parts of the building at an angle of less than 180 degrees (3.14 rad), the building *exterior walls* within 10 feet (3048 mm) horizontally of a nonrated wall or unprotected opening shall have a *fire-resistance rating* of not less than 1 hour. Openings within such *exterior walls* shall be protected by opening protectives having a *fire protection rating* of not less than $^3/_4$ hour. This construction shall extend vertically from the ground to a point 10 feet (3048 mm) above the topmost landing of the *stairway* or *ramp*, or to the roof line, whichever is lower.

[BE] 1023.8 Discharge identification. An *interior exit stairway* and *ramp* shall not continue below its *level of exit discharge* unless an *approved* barrier is provided at the *level of exit discharge* to prevent persons from unintentionally continuing into levels below. Directional exit signs shall be provided as specified in Section 1013.

[BE] 1023.9 Stairway identification signs. A sign shall be provided at each floor landing in an *interior exit stairway* and *ramp* connecting more than three stories designating the floor level, the terminus of the top and bottom of the *interior exit stairway* and *ramp* and the identification of the *stairway* or *ramp*. The signage shall also state the story of, and the direction to, the *exit discharge* and the availability of roof access from the *interior exit stairway* and *ramp* for the fire department. The sign shall be located 5 feet (1524 mm) above the floor landing in a position that is readily visible when the doors are in the open and closed positions. In addition to the *stairway* identification sign, a floor-level sign in visual characters, raised characters and braille complying with ICC A117.1 shall be located at each floor-level landing adjacent to the door leading from the *interior exit stairway* and *ramp* into the *corridor* to identify the floor level.

[BE] 1023.9.1 Signage requirements. *Stairway* identification signs shall comply with all of the following requirements:

1. The signs shall be a minimum size of 18 inches (457 mm) by 12 inches (305 mm).
2. The letters designating the identification of the *interior exit stairway* and *ramp* shall be not less than $1^1/_2$ inches (38 mm) in height.
3. The number designating the floor level shall be not less than of 5 inches (127 mm) in height and located in the center of the sign.
4. Other lettering and numbers shall be not less than 1 inch (25 mm) in height.
5. Characters and their background shall have a nonglare finish. Characters shall contrast with their background, with either light characters on a dark background or dark characters on a light background.
6. Where signs required by Section 1023.9 are installed in the *interior exit stairways* and *ramps* of buildings subject to Section 1025, the signs shall be made of the same materials as required by Section 1025.4.

[BE] 1023.10 Elevator lobby identification signs. At landings in *interior exit stairways* where two or more doors lead to the floor level, any door with direct access to an enclosed elevator lobby shall be identified by signage located on the door or directly adjacent to the door stating "Elevator Lobby." Signage shall be in accordance with Section 1023.9.1, Items 4, 5 and 6.

[BE] 1023.11 Smokeproof enclosures. Where required by Section 403.5.4 or 405.7.2 of the *International Building Code*, *interior exit stairways* and *ramps* shall be *smokeproof enclosures* in accordance with Section 909.20.

[BE] 1023.11.1 Termination and extension. A *smokeproof enclosure* shall terminate at an *exit discharge* or a *public way*. The *smokeproof enclosure* shall be permitted to be extended by an *exit passageway* in accordance with Section 1023.3. The *exit passageway* shall be without openings other than the *fire door assembly* required by Section 1023.3.1 and those necessary for egress from the *exit passageway*. The *exit passageway* shall be separated from the remainder of the building by 2-hour *fire barriers* constructed in accordance with Section 707 of the *International Building Code* or *horizontal assemblies* constructed in accordance with Section 711 of the *International Building Code*, or both.

Exceptions:

1. Openings in the *exit passageway* serving a *smokeproof enclosure* are permitted where the *exit passageway* is protected and pressurized in the same manner as the *smokeproof enclosure*, and openings are protected as required for access from other floors.
2. The *fire barrier* separating the *smokeproof enclosure* from the *exit passageway* is not required, provided the *exit passageway* is protected and pressurized in the same manner as the *smokeproof enclosure*.
3. A *smokeproof enclosure* shall be permitted to egress through areas on the *level of exit discharge* or vestibules as permitted by Section 1028.

[BE] 1023.11.2 Enclosure access. Access to the *stairway* or *ramp* within a *smokeproof enclosure* shall be by way of a vestibule or an open exterior balcony.

Exception: Access is not required by way of a vestibule or exterior balcony for *stairways* and *ramps* using the pressurization alternative complying with Section 909.20.5 of the *International Building Code*.

SECTION 1024
EXIT PASSAGEWAYS

[BE] 1024.1 Exit passageways. *Exit passageways* serving as an exit component in a *means of egress* system shall comply with the requirements of this section. An *exit passageway* shall not be used for any purpose other than as a *means of egress* and a circulation path.

[BE] 1024.2 Width. The required capacity of *exit passageways* shall be determined as specified in Section 1005.1 but the minimum width shall be not less than 44 inches (1118 mm), except that *exit passageways* serving an *occupant load* of less than 50 shall be not less than 36 inches (914 mm) in width. The minimum width or required capacity of *exit passageways* shall be unobstructed.

Exception: Encroachments complying with Section 1005.7.

[BE] 1024.3 Construction. *Exit passageway* enclosures shall have walls, floors and ceilings of not less than a 1-hour *fire-resistance rating*, and not less than that required for any connecting *interior exit stairway* or *ramp*. *Exit passageways* shall be constructed as *fire barriers* in accordance with Section 707 of the *International Building Code* or *horizontal assemblies* constructed in accordance with Section 711 of the *International Building Code*, or both.

[BE] 1024.4 Termination. *Exit passageways* on the *level of exit discharge* shall terminate at an *exit discharge*. *Exit passageways* on other levels shall terminate at an *exit*.

[BE] 1024.5 Openings. *Exit passageway* opening protectives shall be in accordance with the requirements of Section 716 of the *International Building Code*.

Except as permitted in Section 402.8.7 of the *International Building Code*, openings in *exit passageways* other than unprotected exterior openings shall be limited to those necessary for *exit access* to the *exit passageway* from normally occupied spaces and for egress from the *exit passageway*.

Where an *interior exit stairway* or *ramp* is extended to an *exit discharge* or a *public way* by an *exit passageway*, the *exit passageway* shall comply with Section 1023.3.1.

Elevators shall not open into an *exit passageway*.

[BE] 1024.6 Penetrations. Penetrations into or through an *exit passageway* are prohibited except for equipment and ductwork necessary for independent pressurization, sprinkler piping, standpipes, electrical raceway for fire department communication and electrical raceway serving the *exit passageway* and terminating at a steel box not exceeding 16 square inches (0.010 m^2). Such penetrations shall be protected in accordance with Section 714 of the *International Building Code*. There shall not be penetrations or communicating openings, whether protected or not, between adjacent *exit passageways*.

Exception: Membrane penetrations shall be permitted on the outside of the *exit passageway*. Such penetrations shall be protected in accordance with Section 714.3.2 of the *International Building Code*.

[BE] 1024.7 Ventilation. Equipment and ductwork for *exit passageway* ventilation as permitted by Section 1024.6 shall comply with one of the following:

1. The equipment and ductwork shall be located exterior to the building and shall be directly connected to the *exit passageway* by ductwork enclosed in construction as required for shafts.
2. Where the equipment and ductwork is located within the *exit passageway*, the intake air shall be taken directly from the outdoors and the exhaust air shall be discharged directly to the outdoors, or the air shall be conveyed through ducts enclosed in construction as required for shafts.
3. Where located within the building, the equipment and ductwork shall be separated from the remainder of the building, including other mechanical equipment, with construction as required for shafts.

In each case, openings into the fire-resistance-rated construction shall be limited to those needed for maintenance and operation and shall be protected by opening protectives in accordance with Section 716 of the *International Building Code* for shaft enclosures.

Exit passageway ventilation systems shall be independent of other building ventilation systems.

SECTION 1025
LUMINOUS EGRESS PATH MARKINGS

[BE] 1025.1 General. *Approved* luminous egress path markings delineating the exit path shall be provided in high-rise buildings of Group A, B, E, I, M, and R-1 occupancies in accordance with Sections 1025.1 through 1025.5.

Exception: Luminous egress path markings shall not be required on the *level of exit discharge* in lobbies that serve as part of the exit path in accordance with Section 1028.1, Exception 1.

[BE] 1025.2 Markings within exit components. Egress path markings shall be provided in *interior exit stairways*, *interior exit ramps* and *exit passageways*, in accordance with Sections 1025.2.1 through 1025.2.6.

[BE] 1025.2.1 Steps. A solid and continuous stripe shall be applied to the horizontal leading edge of each step and shall extend for the full length of the step. Outlining stripes shall have a minimum horizontal width of 1 inch (25 mm) and a maximum width of 2 inches (51 mm). The leading edge of the stripe shall be placed not more than $^1/_2$ inch (12.7 mm) from the leading edge of the step and the stripe shall not overlap the leading edge of the step by not more than $^1/_2$ inch (12.7 mm) down the vertical face of the step.

Exception: The minimum width of 1 inch (25 mm) shall not apply to outlining stripes listed in accordance with UL 1994.

[BE] 1025.2.2 Landings. The leading edge of landings shall be marked with a stripe consistent with the dimensional requirements for steps.

[BE] 1025.2.3 Handrails. *Handrails* and handrail extensions shall be marked with a solid and continuous stripe having a minimum width of 1 inch (25 mm). The stripe shall be placed on the top surface of the *handrail* for the entire length of the *handrail*, including extensions and newel post caps. Where *handrails* or handrail extensions bend or turn corners, the stripe shall not have a gap of more than 4 inches (102 mm).

Exception: The minimum width of 1 inch (25 mm) shall not apply to outlining stripes listed in accordance with UL 1994.

[BE] 1025.2.4 Perimeter demarcation lines. Stair landings and other floor areas within *interior exit stairways*, *interior exit ramps* and *exit passageways*, with the exception of the sides of steps, shall be provided with solid and continuous demarcation lines on the floor or on the walls or a combination of both. The stripes shall be 1 to 2 inches (25 mm to 51 mm) wide with interruptions not exceeding 4 inches (102 mm).

Exception: The minimum width of 1 inch (25 mm) shall not apply to outlining stripes *listed* in accordance with UL 1994.

[BE] 1025.2.4.1 Floor-mounted demarcation lines. Perimeter demarcation lines shall be placed within 4 inches (102 mm) of the wall and shall extend to within 2 inches (51 mm) of the markings on the leading edge of landings. The demarcation lines shall continue across the floor in front of all doors.

Exception: Demarcation lines shall not extend in front of *exit discharge* doors that lead out of an *exit* and through which occupants must travel to complete the exit path.

[BE] 1025.2.4.2 Wall-mounted demarcation lines. Perimeter demarcation lines shall be placed on the wall with the bottom edge of the stripe not more than 4 inches (102 mm) above the finished floor. At the top or bottom of the *stairs*, demarcation lines shall drop vertically to the floor within 2 inches (51 mm) of the step or landing edge. Demarcation lines on walls shall transition vertically to the floor and then extend across the floor where a line on the floor is the only practical method of outlining the path. Where the wall line is broken by a door, demarcation lines on walls shall continue across the face of the door or transition to the floor and extend across the floor in front of such door.

Exception: Demarcation lines shall not extend in front of *exit discharge* doors that lead out of an *exit* and through which occupants must travel to complete the exit path.

[BE] 1025.2.4.3 Transition. Where a wall-mounted demarcation line transitions to a floor-mounted demarcation line, or vice-versa, the wall-mounted demarcation line shall drop vertically to the floor to meet a complimentary extension of the floor-mounted demarcation line, thus forming a continuous marking.

[BE] 1025.2.5 Obstacles. Obstacles at or below 6 feet 6 inches (1981 mm) in height and projecting more than 4 inches (102 mm) into the egress path shall be outlined with markings not less than 1 inch (25 mm) in width comprised of a pattern of alternating equal bands, of luminous material and black, with the alternating bands not more than 2 inches (51 mm) thick and angled at 45 degrees (0.79 rad). Obstacles shall include, but are not limited to, standpipes, hose cabinets, wall projections, and restricted height areas. However, such markings shall not conceal any required information or indicators including but not limited to instructions to occupants for the use of standpipes.

[BE] 1025.2.6 Doors within the exit path. Doors through which occupants must pass in order to complete the exit path shall be provided with markings complying with Sections 1025.2.6.1 through 1025.2.6.3.

[BE] 1025.2.6.1 Emergency exit symbol. The doors shall be identified by a low-location luminous emergency exit symbol complying with NFPA 170. The exit symbol shall be not less than 4 inches (102 mm) in height and shall be mounted on the door, centered horizontally, with the top of the symbol not higher than 18 inches (457 mm) above the finished floor.

[BE] 1025.2.6.2 Door hardware markings. Door hardware shall be marked with not less than 16 square inches (406 mm^2) of luminous material. This marking shall be located behind, immediately adjacent to, or on the door handle or escutcheon. Where a panic bar is installed, such material shall be not less than 1 inch (25 mm) wide for the entire length of the actuating bar or touchpad.

[BE] 1025.2.6.3 Door frame markings. The top and sides of the door frame shall be marked with a solid and continuous 1-inch- to 2-inch-wide (25 mm to 51 mm) stripe. Where the door molding does not provide sufficient flat surface on which to locate the stripe, the stripe shall be permitted to be located on the wall surrounding the frame.

[BE] 1025.3 Uniformity. Placement and dimensions of markings shall be consistent and uniform throughout the same enclosure.

[BE] 1025.4 Self-luminous and photoluminescent. Luminous egress path markings shall be permitted to be made of any material, including paint, provided that an electrical charge is not required to maintain the required luminance. Such materials shall include, but not be limited to, *self-luminous* materials and *photoluminescent* materials. Materials shall comply with either of the following standards:

1. UL 1994.
2. ASTM E 2072, except that the charging source shall be 1 footcandle (11 lux) of fluorescent illumination for 60 minutes, and the minimum luminance shall be 30 milicandelas per square meter at 10 minutes and 5 milicandelas per square meter after 90 minutes.

[BE] 1025.5 Illumination. Where *photoluminescent* exit path markings are installed, they shall be provided with not less than 1 footcandle (11 lux) of illumination for not less than 60 minutes prior to periods when the building is occupied and continuously during the building occupancy.

SECTION 1026
HORIZONTAL EXITS

[BE] 1026.1 Horizontal exits. *Horizontal exits* serving as an *exit* in a *means of egress* system shall comply with the requirements of this section. A *horizontal exit* shall not serve as the only *exit* from a portion of a building, and where two or more *exits* are required, not more than one-half of the total number of *exits* or total *exit* minimum width or required capacity shall be *horizontal exits*.

Exceptions:

1. *Horizontal exits* are permitted to comprise two-thirds of the required *exits* from any building or floor area for occupancies in Group I-2.

2. *Horizontal exits* are permitted to comprise 100 percent of the *exits* required for occupancies in Group I-3. Not less than 6 square feet (0.6 m^2) of accessible space per occupant shall be provided on each side of the *horizontal exit* for the total number of people in adjoining compartments.

[BE] 1026.2 Separation. The separation between buildings or refuge areas connected by a *horizontal exit* shall be provided by a *fire wall* complying with Section 706 of the *International Building Code*; or by a *fire barrier* complying with Section 707 of the *International Building Code* or a *horizontal assembly* complying with Section 711 of the *International Building Code*, or both. The minimum *fire-resistance rating* of the separation shall be 2 hours. Opening protectives in *horizontal exits* shall also comply with Section 716 of the *International Building Code*. Duct and air transfer openings in a *fire wall* or *fire barrier* that serves as a *horizontal exit* shall also comply with Section 717 of the *International Building Code*. The *horizontal exit* separation shall extend vertically through all levels of the building unless floor assemblies have a *fire-resistance rating* of not less than 2 hours with no unprotected openings.

Exception: A *fire-resistance rating* is not required at *horizontal exits* between a building area and an above-grade pedestrian walkway constructed in accordance with Section 3104 of the *International Building Code*, provided that the distance between connected buildings is more than 20 feet (6096 mm).

Horizontal exits constructed as *fire barriers* shall be continuous from *exterior wall* to *exterior wall* so as to divide completely the floor served by the *horizontal exit*.

[BE] 1026.3 Opening protectives. *Fire doors* in *horizontal exits* shall be self-closing or automatic-closing when activated by a *smoke detector* in accordance with Section 716.5.9.3 of the *International Building Code*. Doors, where located in a cross-corridor condition, shall be automatic-closing by activation of a *smoke detector* installed in accordance with Section 716.5.9.3 of the *International Building Code*.

[BE] 1026.4 Refuge area. The refuge area of a *horizontal exit* shall be a space occupied by the same tenant or a public area and each such refuge area shall be adequate to accommodate the original *occupant load* of the refuge area plus the *occupant load* anticipated from the adjoining compartment. The anticipated *occupant load* from the adjoining compartment shall be based on the capacity of the *horizontal exit* doors entering the refuge area.

[BE] 1026.4.1 Capacity. The capacity of the refuge area shall be computed based on a net floor area allowance of 3 square feet (0.2787 m^2) for each occupant to be accommodated therein.

Exceptions: The net floor area allowable per occupant shall be as follows for the indicated occupancies:

1. Six square feet (0.6 m^2) per occupant for occupancies in Group I-3.

2. Fifteen square feet (1.4 m^2) per occupant for ambulatory occupancies in Group I-2.

3. Thirty square feet (2.8 m^2) per occupant for non-ambulatory occupancies in Group I-2.

[BE] 1026.4.2 Number of exits. The refuge area into which a *horizontal exit* leads shall be provided with *exits* adequate to meet the occupant requirements of this chapter, but not including the added *occupant load* imposed by persons entering it through *horizontal exits* from other areas. Not less than one refuge area *exit* shall lead directly to the exterior or to an *interior exit stairway* or *ramp*.

Exception: The adjoining compartment shall not be required to have a *stairway* or door leading directly outside, provided the refuge area into which a *horizontal exit* leads has *stairways* or doors leading directly outside and are so arranged that egress shall not require the occupants to return through the compartment from which egress originates.

SECTION 1027
EXTERIOR EXIT STAIRWAYS AND RAMPS

[BE] 1027.1 Exterior exit stairways and ramps. *Exterior exit stairways* and *ramps* serving as an element of a required *means of egress* shall comply with this section.

[BE] 1027.2 Use in a means of egress. *Exterior exit stairways* shall not be used as an element of a required *means of egress* for Group I-2 occupancies. For occupancies in other than Group I-2, *exterior exit stairways* and *ramps* shall be permitted as an element of a required *means of egress* for buildings not exceeding six stories above *grade plane* or that are not high-rise buildings.

[BE] 1027.3 Open side. *Exterior exit stairways* and *ramps* serving as an element of a required *means of egress* shall be open on not less than one side, except for required structural columns, beams, handrails and guards. An open side shall have not less than 35 square feet (3.3 m^2) of aggregate open area adjacent to each floor level and the level of each intermediate landing. The required open area shall be located not less than 42 inches (1067 mm) above the adjacent floor or landing level.

[BE] 1027.4 Side yards. The open areas adjoining *exterior exit stairways* or *ramps* shall be either *yards*, *courts* or *public ways*; the remaining sides are permitted to be enclosed by the *exterior walls* of the building.

[BE] 1027.5 Location. *Exterior exit stairways* and *ramps* shall have a minimum fire separation distance of 10 feet (3048 mm) measured at right angles from the exterior edge of the stairway or ramps, including landings, to:

1. Adjacent lot lines.
2. Other portions of the building.
3. Other buildings on the same lot unless the adjacent building *exterior walls* and openings are protected in accordance with Section 705 of the *International Building Code* based on fire separation distance.

For the purposes of this section, other portions of the building shall be treated as separate buildings.

[BE] 1027.6 Exterior exit stairway and ramp protection. *Exterior exit stairways* and *ramps* shall be separated from the interior of the building as required in Section 1023.2. Openings shall be limited to those necessary for egress from normally occupied spaces. Where a vertical plane projecting from the edge of an *exterior exit stairway* or *ramp* and landings is exposed by other parts of the building at an angle of less than 180 degrees (3.14 rad), the *exterior wall* shall be rated in accordance with Section 1023.7.

Exceptions:

1. Separation from the interior of the building is not required for occupancies, other than those in Group R-1 or R-2, in buildings that are not more than two stories above grade plane where a *level of exit discharge* serving such occupancies is the first story above grade plane.
2. Separation from the interior of the building is not required where the *exterior exit stairway* or *ramp* is served by an *exterior exit ramp* or balcony that connects two remote *exterior exit stairways* or other approved *exits*, with a perimeter that is not less than 50 percent open. To be considered open, the opening shall be not less than 50 percent of the height of the enclosing wall, with the top of the openings not less than 7 feet (2134 mm) above the top of the balcony.
3. Separation from the *open-ended corridor* of the building is not required for *exterior exit stairways* or *ramps*, provided that Items 3.1 through 3.5 are met:
   - 3.1. The building, including *open-ended corridors*, and *stairways* and *ramps*, shall be equipped throughout with an *automatic sprinkler system* in accordance with Section 903.3.1.1 or 903.3.1.2.
   - 3.2. The *open-ended corridors* comply with Section 1020.
   - 3.3. The *open-ended corridors* are connected on each end to an *exterior exit stairway* or *ramp* complying with Section 1027.
   - 3.4. The *exterior walls* and openings adjacent to the *exterior exit stairway* or *ramp* comply with Section 1023.7.
   - 3.5. At any location in an *open-ended corridor* where a change of direction exceeding 45 degrees (0.79 rad) occurs, a clear opening of not less than 35 square feet (3.3 m^2) or an exterior *stairway* or *ramp* shall be provided. Where clear openings are provided, they shall be located so as to minimize the accumulation of smoke or toxic gases.

SECTION 1028
EXIT DISCHARGE

[BE] 1028.1 General. *Exits* shall discharge directly to the exterior of the building. The *exit discharge* shall be at grade or shall provide a direct path of egress travel to grade. The *exit discharge* shall not reenter a building. The combined use of Exceptions 1 and 2 shall not exceed 50 percent of the number and minimum width or required capacity of the required *exits*.

Exceptions:

1. Not more than 50 percent of the number and minimum width or required capacity of *interior exit stairways* and *ramps* is permitted to egress through areas on the *level of discharge* provided all of the following conditions are met:
   - 1.1. Discharge of *interior exit stairways* and *ramps* shall be provided with a free and unobstructed path of travel to an exterior exit door and such *exit* is readily visible and identifiable from the point of termination of the enclosure.
   - 1.2. The entire area of the *level of exit discharge* is separated from areas below by construction conforming to the *fire-resistance rating* for the enclosure.
   - 1.3. The egress path from the *interior exit stairway* and *ramp* on the *level of exit discharge* is protected throughout by an *approved automatic sprinkler system.* Portions of the *level of exit discharge* with access to the egress path shall either be equipped throughout with an *automatic sprinkler system* installed in accordance with Section 903.3.1.1 or 903.3.1.2, or separated from the egress path in accordance with the requirements for the enclosure of *interior exit stairways* or *ramps*.
   - 1.4. Where a required *interior exit stairway* or *ramp* and an *exit access stairway* or *ramp* serve the same floor level and terminate at the same *level of exit discharge*, the termination of the *exit access stairway* or *ramp* and the exit discharge door of the *interior exit stairway* or *ramp* shall be separated by a distance of not less than 30 feet (9144 mm) or not less than one-fourth the length of the maximum overall diagonal dimension of the building, whichever is less. The distance shall be measured in a straight line between the exit discharge door from the

interior exit stairway or ramp and the last tread of the *exit access stairway* or termination of slope of the *exit access ramp*.

2. Not more than 50 percent of the number and minimum width or required capacity of the interior *exit stairways* and *ramps* is permitted to egress through a vestibule provided all of the following conditions are met:

    2.1. The entire area of the vestibule is separated from areas below by construction conforming to the *fire-resistance rating* of the *interior exit stairway* or *ramp* enclosure.

    2.2. The depth from the exterior of the building is not greater than 10 feet (3048 mm) and the length is not greater than 30 feet (9144 mm).

    2.3 The area is separated from the remainder of the *level of exit discharge* by a *fire partition* constructed in accordance with Section 708 of the *International* Building Code.

    Exception: The maximum transmitted temperature rise is not required.

    2.4. The area is used only for *means of egress* and *exits* directly to the outside.

3. *Horizontal exits* complying with Section 1026 shall not be required to discharge directly to the exterior of the building.

[BE] 1028.2 Exit discharge width or capacity. The minimum width or required capacity of the *exit discharge* shall be not less than the minimum width or required capacity of the *exits* being served.

[BE] 1028.3 Exit discharge components. *Exit discharge* components shall be sufficiently open to the exterior so as to minimize the accumulation of smoke and toxic gases.

[BE] 1028.4 Egress courts. *Egress courts* serving as a portion of the *exit discharge* in the *means of egress* system shall comply with the requirements of Sections 1028.4.1 and 1028.4.2.

[BE] 1028.4.1 Width or capacity. The required capacity of *egress courts* shall be determined as specified in Section 1005.1, but the minimum width shall be not less than 44 inches (1118 mm), except as specified herein. *Egress courts* serving Group R-3 and U occupancies shall be not less than 36 inches (914 mm) in width. The required capacity and width of *egress courts* shall be unobstructed to a height of 7 feet (2134 mm).

Exception: Encroachments complying with Section 1005.7.

Where an *egress court* exceeds the minimum required width and the width of such *egress court* is then reduced along the path of exit travel, the reduction in width shall be gradual. The transition in width shall be affected by a guard not less than 36 inches (914 mm) in height and shall not create an angle of more than 30 degrees (0.52 rad) with respect to the axis of the *egress court* along the path of egress travel. The width of the *egress court* shall not be less than the required capacity.

[BE] 1028.4.2 Construction and openings. Where an *egress court* serving a building or portion thereof is less than 10 feet (3048 mm) in width, the *egress court* walls shall have not less than 1-hour fire-resistance-rated construction for a distance of 10 feet (3048 mm) above the floor of the *egress court*. Openings within such walls shall be protected by opening protectives having a fire protection rating of not less than $^3/_4$ hour.

Exceptions:

1. *Egress courts* serving an *occupant load* of less than 10.
2. *Egress courts* serving Group R-3.

[BE] 1028.5 Access to a public way. The *exit discharge* shall provide a direct and unobstructed access to a *public way*.

Exception: Where access to a *public way* cannot be provided, a safe dispersal area shall be provided where all of the following are met:

1. The area shall be of a size to accommodate *not less than* 5 square feet (0.46 m^2) for each person.
2. The area shall be located on the same lot not less than 50 feet (15 240 mm) away from the building requiring egress.
3. The area shall be permanently maintained and identified as a safe dispersal area.
4. The area shall be provided with a safe and unobstructed path of travel from the building.

[BE] TABLE 1029.6.2
CAPACITY FOR AISLES FOR SMOKE-PROTECTED ASSEMBLY

TOTAL NUMBER OF SEATS IN THE SMOKE-PROTECTED ASSEMBLY SEATING	INCHES OF CAPACITY PER SEAT SERVED			
	Stepped aisles with handrails within 30 inches	Stepped aisles without handrails within 30 inches	Level aisles or ramped aisles not steeper than 1 in 10 in slope	Ramped aisles steeper than 1 in 10 in slope
Equal to or less than 5,000	0.200	0.250	0.150	0.165
10,000	0.130	0.163	0.100	0.110
15,000	0.096	0.120	0.070	0.077
20,000	0.076	0.095	0.056	0.062
Equal to or greater than 25,000	0.060	0.075	0.044	0.048

For SI: 1 inch = 25.4 mm.

SECTION 1029
ASSEMBLY

[BE] 1029.1 General. A room or space used for assembly purposes that contains seats, tables, displays, equipment or other material shall comply with this section.

[BE] 1029.1.1 Bleachers. *Bleachers*, *grandstands* and *folding and telescopic seating*, that are not building elements, shall comply with ICC 300.

[BE] 1029.1.1.1 Spaces under grandstands and bleachers. Where spaces under *grandstands* or *bleachers* are used for purposes other than ticket booths less than 100 square feet (9.29 m^2) and toilet rooms, such spaces shall be separated by *fire barriers* complying with Section 707 of the *International Building Code* and *horizontal assemblies* complying with Section 711 of the *International Building Code* with not less than 1-hour fire-resistance-rated construction.

[BE] 1029.2 Assembly main exit. A building, room or space used for assembly purposes that has an *occupant load* of greater than 300 and is provided with a main *exit*, that main *exit* shall be of sufficient capacity to accommodate not less than one-half of the *occupant load*, but such capacity shall be not less than the total required capacity of all *means of egress* leading to the *exit*. Where the building is classified as a Group A occupancy, the main *exit* shall front on not less than one street or an unoccupied space of not less than 10 feet (3048 mm) in width that adjoins a street or *public way*. In a building, room or space used for assembly purposes where there is not a well-defined main *exit* or where multiple main *exits* are provided, *exits* shall be permitted to be distributed around the perimeter of the building provided that the total capacity of egress is not less than 100 percent of the required capacity

[BE] 1029.3 Assembly other exits. In addition to having access to a main *exit*, each level in a building used for assembly purposes having an *occupant load* greater than 300 and provided with a main *exit*, shall be provided with additional *means of egress* that shall provide an egress capacity for not less than one-half of the total *occupant load* served by that level and shall comply with Section 1007.1. In a building used for assembly purposes where there is not a well-defined main *exit* or where multiple main *exits* are provided, *exits* for each level shall be permitted to be distributed around the perimeter of the building, provided that the total width of egress is not less than 100 percent of the required width.

[BE] 1029.4 Foyers and lobbies. In Group A-1 occupancies, where persons are admitted to the building at times when seats are not available, such persons shall be allowed to wait in a lobby or similar space, provided such lobby or similar space shall not encroach upon the minimum width or required capacity of the *means of egress*. Such foyer, if not directly connected to a public street by all the main entrances or *exits*, shall have a straight and unobstructed *corridor* or path of travel to every such main entrance or *exit*.

[BE] 1029.5 Interior balcony and gallery means of egress. For balconies, galleries or press boxes having a seating capacity of 50 or more located in a building, room or space used for assembly purposes, not less than two *means of egress* shall be provided, with one from each side of every balcony, gallery or press box.

[BE] 1029.6 Capacity of aisle for assembly. The required capacity of *aisles* shall be not less than that determined in accordance with Section 1029.6.1 where *smoke-protected assembly seating* is not provided and with Section 1029.6.2 or 1029.6.3 where *smoke-protected assembly seating* is provided.

[BE] 1029.6.1 Without smoke protection. The required capacity in inches (mm) of the *aisles* for assembly seating without smoke protection shall be not less than the *occupant load* served by the egress element in accordance with all of the following, as applicable:

1. Not less than 0.3 inch (7.6 mm) of *aisle* capacity for each occupant served shall be provided on stepped *aisles* having riser heights 7 inches (178 mm) or less and tread depths 11 inches (279 mm) or greater, measured horizontally between tread *nosings*.
2. Not less than 0.005 inch (0.127 mm) of additional *aisle* capacity for each occupant shall be provided for each 0.10 inch (2.5mm) of riser height above 7 inches (178 mm).
3. Where egress requires stepped *aisle* descent, not less than 0.075 inch (1.9 mm) of additional *aisle* capacity for each occupant shall be provided on those portions of *aisle* capacity having no *handrail* within a horizontal distance of 30 inches (762 mm).
4. Ramped *aisles*, where slopes are steeper than one unit vertical in 12 units horizontal (8-percent slope), shall have not less than 0.22 inch (5.6 mm) of clear *aisle* capacity for each occupant served. Level or ramped aisles, where slopes are not steeper than one unit vertical in 12 units horizontal (8-percent slope), shall have not less than 0.20 inch (5.1 mm) of clear *aisle* capacity for each occupant served.

[BE] 1029.6.2 Smoke-protected assembly seating. The required capacity in inches (mm) of the *aisle* for *smoke-protected assembly seating* shall be not less than the *occupant load* served by the egress element multiplied by the appropriate factor in Table 1029.6.2. The total number of seats specified shall be those within the space exposed to the same smoke-protected environment. Interpolation is permitted between the specific values shown. A life safety evaluation, complying with NFPA 101, shall be done for a facility utilizing the reduced width requirements of Table 1029.6.2 for *smoke-protected assembly seating*.

Exception: For outdoor *smoke-protected assembly seating* with an *occupant load* not greater than 18,000, the required capacity in inches (mm) shall be determined using the factors in Section 1029.6.3.

[BE] 1029.6.2.1 Smoke control. *Aisles* and *aisle accessways* serving a *smoke-protected assembly seating* area shall be provided with a smoke control system complying with Section 909 or natural ventilation designed to maintain the smoke level not less than 6 feet (1829 mm) above the floor of the *means of egress*.

[BE] 1029.6.2.2 Roof height. A *smoke-protected assembly seating area* with a roof shall have the lowest portion of the roof deck not less than 15 feet (4572 mm) above the highest *aisle* or *aisle accessway*.

Exception: A roof canopy in an outdoor stadium shall be permitted to be less than 15 feet (4572 mm) above the highest *aisle* or *aisle accessway* provided that there are no objects less than 80 inches (2032 mm) above the highest *aisle* or *aisle accessway*.

[BE] 1029.6.2.3 Automatic sprinklers. Enclosed areas with walls and ceilings in buildings or structures containing *smoke-protected assembly seating* shall be protected with an *approved automatic sprinkler system* in accordance with Section 903.3.1.1.

Exceptions:

1. The floor area used for contests, performances or entertainment provided the roof construction is more than 50 feet (15 240 mm) above the floor level and the use is restricted to low fire hazard uses.
2. Press boxes and storage facilities less than 1,000 square feet (93 m^2) in area.
3. Outdoor seating facilities where seating and the *means of egress* in the seating area are essentially open to the outside.

[BE] 1029.6.3 Outdoor smoke-protected assembly seating. The required capacity in inches (mm) of *aisles* shall be not less than the total *occupant load* served by the egress element multiplied by 0.08 (2.0 mm) where egress is by stepped *aisle* and multiplied by 0.06 (1.52 mm) where egress is by level *aisles* and ramped *aisles*.

Exception: The required capacity in inches (mm) of *aisles* shall be permitted to comply with Section 1029.6.2 for the number of seats in the outdoor *smoke-protected assembly seating* where Section 1029.6.2 permits less capacity.

[BE] 1029.7 Travel distance. *Exits* and *aisles* shall be so located that the travel distance to an exit door shall be not greater than 200 feet (60 960 mm) measured along the line of travel in nonsprinklered buildings. Travel distance shall be not more than 250 feet (76 200 mm) in sprinklered buildings. Where *aisles* are provided for seating, the distance shall be measured along the *aisles* and *aisle accessways* without travel over or on the seats.

Exceptions:

1. *Smoke-protected assembly seating*: The travel distance from each seat to the nearest entrance to a vomitory or concourse shall not exceed 200 feet (60 960 mm). The travel distance from the entrance to the vomitory or concourse to a *stairway*, *ramp* or walk on the exterior of the building shall not exceed 200 feet (60 960 mm).
2. Open-air seating: The travel distance from each seat to the building exterior shall not exceed 400 feet (122 m). The travel distance shall not be limited in facilities of Type I or II construction.

[BE] 1029.8 Common path of egress travel. The *common path of egress travel* shall not exceed 30 feet (9144 mm) from any seat to a point where an occupant has a choice of two paths of egress travel to two *exits*.

Exceptions:

1. For areas serving less than 50 occupants, the *common path of egress travel* shall not exceed 75 feet (22 860 mm).
2. For *smoke-protected assembly seating*, the *common path of egress travel* shall not exceed 50 feet (15 240 mm).

[BE] 1029.8.1 Path through adjacent row. Where one of the two paths of travel is across the *aisle* through a row of seats to another *aisle*, there shall be not more than 24 seats between the two *aisles*, and the minimum clear width between rows for the row between the two *aisles* shall be 12 inches (305 mm) plus 0.6 inch (15.2 mm) for each additional seat above seven in the row between *aisles*.

Exception: For *smoke-protected assembly seating* there shall be not more than 40 seats between the two *aisles* and the minimum clear width shall be 12 inches (305 mm) plus 0.3 inch (7.6 mm) for each additional seat.

[BE] 1029.9 Assembly aisles are required. Every occupied portion of any building, room or space used for assembly purposes that contains seats, tables, displays, similar fixtures or equipment shall be provided with *aisles* leading to *exits* or *exit access doorways* in accordance with this section.

[BE] 1029.9.1 Minimum aisle width. The minimum clear width for *aisles* shall comply with one of the following:

1. Forty-eight inches (1219 mm) for stepped *aisles* having seating on each side.

   Exception: Thirty-six inches (914 mm) where the stepped *aisles* serve less than 50 seats.

2. Thirty-six inches (914 mm) for stepped *aisles* having seating on only one side.

   Exception: Twenty-three inches (584 mm) between an aisle stair *handrail* and seating where a stepped *aisle* does not serve more than five rows on one side.

3. Twenty-three inches (584 mm) between a stepped aisle *handrail* or *guard* and seating where the stepped aisle is subdivided by a mid-aisle *handrail*.
4. Forty-two inches (1067 mm) for level or ramped *aisles* having seating on both sides.

   Exceptions:

   1. Thirty-six inches (914 mm) where the *aisle* serves less than 50 seats.
   2. Thirty inches (762 mm) where the *aisle* does not serve more than 14 seats.

5. Thirty-six inches (914 mm) for level or ramped *aisles* having seating on only one side.

   Exception: For other than ramped *aisles* that serve as part of an accessible route, 30 inches (762 mm) where the ramped *aisle* does not serve more than 14 seats.

[BE] 1029.9.2 Aisle catchment area. The *aisle* shall provide sufficient capacity for the number of persons accommodated by the catchment area served by the *aisle*. The catchment area served by an *aisle* is that portion of the total space served by that section of the *aisle*. In establishing catchment areas, the assumption shall be made that there is a balanced use of all *means of egress*, with the number of persons in proportion to egress capacity.

[BE] 1029.9.3 Converging aisles. Where *aisles* converge to form a single path of egress travel, the required capacity of that path shall be not less than the combined required capacity of the converging *aisles*.

[BE] 1029.9.4 Uniform width and capacity. Those portions of *aisles*, where egress is possible in either of two directions, shall be uniform in minimum width or required capacity.

[BE] 1029.9.5 Dead end aisles. Each end of an *aisle* shall be continuous to a cross *aisle*, foyer, doorway, vomitory, concourse or *stairway* in accordance with Section 1029.9.7 having access to an *exit*.

Exceptions:

1. Dead-end *aisles* shall not be greater than 20 feet (6096 mm) in length.
2. Dead-end *aisles* longer than 16 rows are permitted where seats beyond the 16th row dead-end *aisle* are not more than 24 seats from another *aisle*, measured along a row of seats having a minimum clear width of 12 inches (305 mm) plus 0.6 inch (15.2 mm) for each additional seat above seven in the row where seats have backrests or beyond 10 where seats are without backrests in the row.
3. For *smoke-protected assembly seating*, the dead end *aisle* length of vertical *aisles* shall not exceed a distance of 21 rows.
4. For *smoke-protected assembly seating*, a longer dead-end *aisle* is permitted where seats beyond the 21-row dead-end *aisle* are not more than 40 seats from another *aisle*, measured along a row of seats having an *aisle accessway* with a minimum clear width of 12 inches (305 mm) plus 0.3 inch (7.6 mm) for each additional seat above seven in the row where seats have backrests or beyond 10 where seats are without backrests in the row.

[BE] 1029.9.6 Aisle measurement. The clear width for *aisles* shall be measured to walls, edges of seating and tread edges except for permitted projections.

Exception: The clear width of *aisles* adjacent to seating at tables shall be permitted to be measured in accordance with Section 1029.12.1.

[BE] 1029.9.6.1 Assembly aisle obstructions. There shall not be obstructions in the minimum width or required capacity of *aisles*.

Exception: *Handrails* are permitted to project into the required width of stepped *aisles* and ramped aisles in accordance with Section 1014.8.

[BE] 1029.9.7 Stairways connecting to stepped aisles. A *stairway* that connects a stepped *aisle* to a cross *aisle* or concourse shall be permitted to comply with the assembly *aisle* walking surface requirements of Section 1029.12. Transitions between *stairways* and stepped *aisles* shall comply with Section 1029.10.

[BE] 1029.9.8 Stairways connecting to vomitories. A stairway that connects a vomitory to a cross aisle or concourse shall be permitted to comply with the assembly *aisle* walking surface requirements of Section 1029.12. Transitions between *stairways* and stepped *aisles* shall comply with Section 1029.10.

[BE] 1029.10 Transitions. Transitions between *stairways* and stepped *aisles* shall comply with either Section 1029.10.1 or 1029.10.2.

[BE] 1029.10.1 Transitions and stairways that maintain stepped aisle riser and tread dimensions. Stepped *aisles*, transitions and *stairways* that maintain riser and tread dimensions shall comply with Section 1029.12 as one *exit access* component.

[BE] 1029.10.2 Transitions to stairways that do not maintain stepped aisle riser and tread dimensions. Transitions between stepped *aisles* with riser and tread dimensions that differ from the *stairways* shall comply with Sections 1029.10.2.1 and 1029.10.3.

[BE] 1029.10.2.1 Stairways and stepped aisles in a straight run. Transitions where the *stairway* is a straight run from the stepped *aisle* shall have a minimum depth of 22 inches (559 mm) where the treads on the descending side of the transition have greater depth and 30 inches (762 mm) where the treads on the descending side of the transition have lesser depth.

[BE] 1029.10.2.2 Stairways and stepped aisles that change direction. Transitions where the *stairway* changes direction from the stepped *aisle* shall have a minimum depth of 11 inches (280 mm) or the stepped *aisle* tread depth, whichever is greater, between the stepped *aisle* and *stairway*.

[BE] 1029.10.3 Transition marking. A distinctive marking stripe shall be provided at each *nosing* or leading edge adjacent to the transition. Such stripe shall be not less than 1 inch (25 mm), and not more than 2 inches (51 mm), wide. The edge marking stripe shall be distinctively different from the stepped *aisle* contrasting marking stripe.

[BE] 1029.11 Construction. *Aisles*, stepped *aisles* and ramped *aisles* shall be built of materials consistent with the types permitted for the type of construction of the building.

Exception: Wood *handrails* shall be permitted for all types of construction.

[BE] 1029.11.1 Walking surface. The surface of *aisles*, stepped *aisles* and ramped *aisles* shall be of slip-resistant materials that are securely attached. The surface for stepped *aisles* shall comply with Section 1011.7.1.

[BE] 1029.11.2 Outdoor conditions. Outdoor *aisles*, stepped *aisles* and ramped *aisles* and outdoor approaches to *aisles*, stepped *aisles* and ramped *aisles* shall be designed so that water will not accumulate on the walking surface.

[BE] 1029.12 Aisle accessways. *Aisle accessways* for seating at tables shall comply with Section 1029.12.1. *Aisle accessways* for seating in rows shall comply with Section 1029.12.2.

[BE] 1029.12.1 Seating at tables. Where seating is located at a table or counter and is adjacent to an *aisle* or *aisle accessway*, the measurement of required clear width of the *aisle* or *aisle accessway* shall be made to a line 19 inches (483 mm) away from and parallel to the edge of the table or counter. The 19-inch (483 mm) distance shall be measured perpendicular to the side of the table or counter. In the case of other side boundaries for *aisles* or *aisle accessways*, the clear width shall be measured to walls, edges of seating and tread edges.

Exception: Where tables or counters are served by fixed seats, the width of the *aisle* or *aisle accessway* shall be measured from the back of the seat.

[BE] 1029.12.1.1 Aisle accessway capacity and width for seating at tables. *Aisle accessways* serving arrangements of seating at tables or counters shall comply with the capacity requirements of Section 1005.1 but shall not have less than 12 inches (305 mm) of width plus $^1/_2$ inch (12.7 mm) of width for each additional 1 foot (305 mm), or fraction thereof, beyond 12 feet (3658 mm) of *aisle accessway* length measured from the center of the seat farthest from an *aisle*.

Exception: Portions of an *aisle accessway* having a length not exceeding 6 feet (1829 mm) and used by a total of not more than four persons.

[BE] 1029.12.1.2 Seating at table aisle accessway length. The length of travel along the *aisle accessway* shall not exceed 30 feet (9144 mm) from any seat to the point where a person has a choice of two or more paths of egress travel to separate *exits*.

[BE] 1029.12.2 Clear width of aisle accessways serving seating in rows. Where seating rows have 14 or fewer seats, the minimum clear *aisle accessway* width shall be not less than 12 inches (305 mm) measured as the clear horizontal distance from the back of the row ahead and the nearest projection of the row behind. Where chairs have automatic or self-rising seats, the measurement shall be made with seats in the raised position. Where any chair in the row does not have an automatic or self-rising seat, the measurements shall be made with the seat in the down position. For seats with folding tablet arms, row spacing shall be determined with the tablet arm in the used position.

Exception: For seats with folding tablet arms, row spacing is permitted to be determined with the tablet arm in the stored position where the tablet arm when raised manually to vertical position in one motion automatically returns to the stored position by force of gravity.

[BE] 1029.12.2.1 Dual access. For rows of seating served by *aisles* or doorways at both ends, there shall be not more than 100 seats per row. The minimum clear width of 12 inches (305 mm) between rows shall be increased by 0.3 inch (7.6 mm) for every additional seat beyond 14 seats where seats have backrests or beyond 21 where seats are without backrests. The minimum clear width is not required to exceed 22 inches (559 mm).

Exception: For *smoke-protected assembly seating*, the row length limits for a 12-inch-wide (305 mm) *aisle accessway*, beyond which the *aisle accessway* minimum clear width shall be increased, are in Table 1029.12.2.1.

[BE] 1029.12.2.2 Single access. For rows of seating served by an *aisle* or doorway at only one end of the row, the minimum clear width of 12 inches (305 mm) between rows shall be increased by 0.6 inch (15.2 mm) for every additional seat beyond seven seats where

[BE] TABLE 1029.12.2.1
SMOKE-PROTECTED ASSEMBLY AISLE ACCESSWAYS

TOTAL NUMBER OF SEATS IN THE SMOKE-PROTECTED ASSEMBLY SEATING	MAXIMUM NUMBER OF SEATS PER ROW PERMITTED TO HAVE A MINIMUM 12-INCH CLEAR WIDTH AISLE ACCESSWAY			
	Aisle or doorway at both ends of row		Aisle or doorway at one end of row only	
	Seats with backrests	Seats without backrests	Seats with backrests	Seats without backrests
Less than 4,000	14	21	7	10
4,000	15	22	7	10
7,000	16	23	8	11
10,000	17	24	8	11
13,000	18	25	9	12
16,000	19	26	9	12
19,000	20	27	10	13
22,000 and greater	21	28	11	14

For SI: 1 inch = 25.4 mm.

seats have backrests or beyond 10 where seats are without backrests. The minimum clear width is not required to exceed 22 inches (559 mm).

Exception: For *smoke-protected assembly seating*, the row length limits for a 12-inch-wide (305 mm) *aisle accessway*, beyond which the *aisle accessway* minimum clear width shall be increased, are in Table 1029.12.2.1.

[BE] 1029.13 Assembly aisle walking surfaces. Ramped aisles shall comply with Sections 1029.13.1 through 1029.13.1.3. Stepped *aisles* shall comply with Sections 1029.13.2 through 1029.13.2.4.

[BE] 1029.13.1 Ramped aisles. *Aisles* that are sloped more than one unit vertical in 20 units horizontal (5-percent slope) shall be considered a ramped *aisle*. Ramped *aisles* that serve as part of an accessible route in accordance with Sections 1009 of this code and Section 1108.2 of the *International Building Code* shall have a maximum slope of one unit vertical in 12 units horizontal (8-percent slope). The slope of other ramped *aisles* shall not exceed one unit vertical in 8 units horizontal (12.5-percent slope).

[BE] 1029.13.1.1 Cross slope. The slope measured perpendicular to the direction of travel of a ramped *aisle* shall not be steeper than one unit vertical in 48 units horizontal (2-percent slope).

[BE] 1029.13.1.2 Landings. Ramped *aisles* shall have landings in accordance with Sections 1012.6 through 1012.6.5. Landings for ramped *aisles* shall be permitted to overlap required *aisles* or cross *aisles*.

[BE] 1029.13.1.3 Edge protection. Ramped *aisles* shall have edge protection in accordance with Section 1012.11.

Exception: In assembly spaces with *fixed seating*, edge protection is not required on the sides of ramped *aisles* where the ramped *aisles* provide access to the adjacent seating and *aisle accessways*.

[BE] 1029.13.2 Stepped aisles. *Aisles* with a slope exceeding one unit vertical in eight units horizontal (12.5-percent slope) shall consist of a series of risers and treads that extends across the full width of *aisles* and complies with Sections 1029.13.2.1 through 1029.13.2.4.

[BE] 1029.13.2.1 Treads. Tread depths shall be not less than 11 inches (279 mm) and shall have dimensional uniformity.

Exception: The tolerance between adjacent treads shall not exceed $^3/_{16}$ inch (4.8 mm).

[BE] 1029.13.2.2 Risers. Where the gradient of stepped *aisles* is to be the same as the gradient of adjoining seating areas, the riser height shall be not less than 4 inches (102 mm) nor more than 8 inches (203 mm) and shall be uniform within each flight.

Exceptions:

1. Riser height nonuniformity shall be limited to the extent necessitated by changes in the gradient of the adjoining seating area to maintain adequate sightlines. Where nonuniformities exceed $^3/_{16}$ inch (4.8 mm) between adjacent risers, the exact location of such nonuniformities shall be indicated with a distinctive marking stripe on each tread at the *nosing* or leading edge adjacent to the nonuniform risers. Such stripe shall be not less than 1 inch (25 mm), and not more than 2 inches (51 mm), wide. The edge marking stripe shall be distinctively different from the contrasting marking stripe.
2. Riser heights not exceeding 9 inches (229 mm) shall be permitted where they are necessitated by the slope of the adjacent seating areas to maintain sightlines.

[BE] 1029.13.2.2.1 Construction Tolerances. The tolerance between adjacent risers on a stepped *aisle* that were designed to be equal height shall not exceed $^3/_{16}$ inch (4.8 mm). Where the stepped *aisle* is designed in accordance with Exception 1 of Section 1029.3.2.2, the stepped *aisle* shall be constructed so that each riser of unequal height, determined in the direction of descent, is not more than $^3/_8$ inch (9.5 mm) in height different from adjacent risers where stepped *aisle* treads are less than 22 inches (560 mm) in depth and $^3/_4$ inch (19.1 mm) in height different from adjacent risers where stepped *aisle* treads are 22 inches (560 mm) or greater in depth.

[BE] 1029.13.2.3 Tread contrasting marking stripe. A contrasting marking stripe shall be provided on each tread at the *nosing* or leading edge such that the location of each tread is readily apparent when viewed in descent. Such stripe shall be not less than 1 inch (25 mm), and not more than 2 inches (51 mm), wide.

Exception: The contrasting marking stripe is permitted to be omitted where tread surfaces are such that the location of each tread is readily apparent when viewed in descent.

[BE] 1029.13.2.4 Nosing and profile. *Nosing* and riser profile shall comply with Sections 1011.5.5 through 1011.5.5.3.

[BE] 1029.14 Seat stability. In a building, room or space used for assembly purposes, the seats shall be securely fastened to the floor.

Exceptions:

1. In a building, room or space used for assembly purposes or portions thereof without ramped or tiered floors for seating and with 200 or fewer seats, the seats shall not be required to be fastened to the floor.
2. In a building, room or space used for assembly purposes or portions thereof without ramped or tiered floors for seating, the seats shall not be required to be fastened to the floor.
3. In a building, room or space used for assembly purposes or portions thereof without ramped or tiered floors for seating and with greater than 200 seats, the

seats shall be fastened together in groups of not less than three or the seats shall be securely fastened to the floor.

4. In a building, room or space used for assembly purposes where flexibility of the seating arrangement is an integral part of the design and function of the space and seating is on tiered levels, not more than 200 seats shall not be required to be fastened to the floor. Plans showing seating, tiers and *aisles* shall be submitted for approval.
5. Groups of seats within a building, room or space used for assembly purposes separated from other seating by railings, *guards*, partial height walls or similar barriers with level floors and having not more than 14 seats per group shall not be required to be fastened to the floor.
6. Seats intended for musicians or other performers and separated by railings, *guards*, partial height walls or similar barriers shall not be required to be fastened to the floor.

[BE] 1029.15 Handrails. Ramped *aisles* having a slope exceeding one unit vertical in 15 units horizontal (6.7-percent slope) and stepped *aisles* shall be provided with *handrails* in compliance with Section 1014 located either at one or both sides of the *aisle* or within the *aisle* width.

Exceptions:

1. *Handrails* are not required for ramped *aisles* with seating on both sides.
2. *Handrails* are not required where, at the side of the *aisle*, there is a *guard* with a top surface that complies with the graspability requirements of *handrails* in accordance with Section 1014.3.
3. *Handrail* extensions are not required at the top and bottom of stepped *aisles* and ramped *aisles* to permit crossovers within the *aisles*.

[BE] 1029.15.1 Discontinuous handrails. Where there is seating on both sides of the *aisle*, the mid-aisle *handrails* shall be discontinuous with gaps or breaks at intervals not exceeding five rows to facilitate access to seating and to permit crossing from one side of the *aisle* to the other. These gaps or breaks shall have a clear width of not less than 22 inches (559 mm) and not greater than 36 inches (914 mm), measured horizontally, and the mid-aisle *handrail* shall have rounded terminations or bends.

[BE] 1029.15.2 Handrail termination. *Handrails* located on the side of stepped *aisles* shall return to a wall, *guard* or the walking surfaces or shall be continuous to the *handrail* of an adjacent stepped *aisle* flight.

[BE] 1029.15.3 Mid-aisle termination. Mid-aisle *handrails* shall not extend beyond the lowest riser and shall terminate within 18 inches (381 mm), measured horizontally, from the lowest riser. *Handrail* extensions are not required.

Exception: Mid-aisle *handrails* shall be permitted to extend beyond the lowest riser where the *handrail* extensions do not obstruct the width of the cross *aisle*.

[BE] 1029.15.4 Rails. Where mid-aisle *handrails* are provided in stepped *aisles*, there shall be an additional rail located approximately 12 inches (305 mm) below the *handrail*. The rail shall be adequate in strength and attachment in accordance with Section 1607.8.1.2 of the *International Building Code*.

[BE] 1029.16 Assembly guards. *Guards* adjacent to seating in a building, room or space used for assembly purposes shall be provided where required by Section 1015 and shall be constructed in accordance with Section 1015 except where provided in accordance with Sections 1029.16.1 through 1029.16.4. At *bleachers, grandstands and folding and telescopic seating*, *guards* must be provided where required by ICC 300 and Section 1029.16.1.

[BE] 1029.16.1 Perimeter guards. Perimeter *guards* shall be provided where the footboards or walking surface of seating facilities are more than 30 inches (762 mm) above the floor or grade below. Where the seatboards are adjacent to the perimeter, *guard* height shall be 42 inches (1067 mm) high minimum, measured from the seatboard. Where the seats are self-rising, *guard* height shall be 42 inches (1067 mm) high minimum, measured from the floor surface. Where there is an *aisle* between the seating and the perimeter, the *guard* height shall be measured in accordance with Section 1015.2.

Exceptions:

1. *Guards* that impact sightlines shall be permitted to comply with Section 1029.16.3.
2. *Bleachers, grandstands and folding and telescopic seating* shall not be required to have perimeter *guards* where the seating is located adjacent to a wall and the space between the wall and the seating is less than 4 inches (102 mm).

[BE] 1029.16.2 Cross aisles. Cross *aisles* located more than 30 inches (762 mm) above the floor or grade below shall have *guards* in accordance with Section 1015.

Where an elevation change of 30 inches (762 mm) or less occurs between a cross *aisle* and the adjacent floor or grade below, *guards* not less than 26 inches (660 mm) above the *aisle* floor shall be provided.

Exception: Where the backs of seats on the front of the cross *aisle* project 24 inches (610 mm) or more above the adjacent floor of the *aisle*, a *guard* need not be provided.

[BE] 1029.16.3 Sightline-constrained guard heights. Unless subject to the requirements of Section 1029.16.4, a fascia or railing system in accordance with the *guard* requirements of Section 1015 and having a minimum height of 26 inches (660 mm) shall be provided where the floor or footboard elevation is more than 30 inches (762 mm) above the floor or grade below and the fascia or railing would otherwise interfere with the sightlines of immediately adjacent seating.

[BE] 1029.16.4 Guards at the end of aisles. A fascia or railing system complying with the *guard* requirements of Section 1015 shall be provided for the full width of the *aisle* where the foot of the *aisle* is more than 30 inches

(762 mm) above the floor or grade below. The fascia or railing shall be a minimum of 36 inches (914 mm) high and shall provide a minimum 42 inches (1067 mm) measured diagonally between the top of the rail and the *nosing* of the nearest tread.

SECTION 1030 EMERGENCY ESCAPE AND RESCUE

[BE] 1030.1 General. In addition to the *means of egress* required by this chapter, provisions shall be made for *emergency escape and rescue openings* in Group R-2 occupancies in accordance with Tables 1006.3.2(1) and 1006.3.2(2) and Group R-3 occupancies. Basements and sleeping rooms below the fourth story above *grade plane* shall have at least one exterior *emergency escape and rescue opening* in accordance with this section. Where basements contain one or more sleeping rooms, *emergency escape and rescue openings* shall be required in each sleeping room, but shall not be required in adjoining areas of the basement. Such openings shall open directly into a *public way* or to a *yard* or *court* that opens to a *public way*.

Exceptions:

1. Basements with a ceiling height of less than 80 inches (2032 mm) shall not be required to have *emergency escape and rescue openings.*
2. *Emergency escape and rescue openings* are not required from basements or sleeping rooms that have an *exit* door or *exit access* door that opens directly into a *public way* or to a *yard, court* or exterior exit balcony that opens to a *public way.*
3. Basements without *habitable spaces* and having not more than 200 square feet (18.6 m^2) in floor area shall not be required to have *emergency escape and rescue openings.*

[BE] 1030.2 Minimum size. *Emergency escape and rescue openings* shall have a minimum net clear opening of 5.7 square feet (0.53 m^2).

Exception: The minimum net clear opening for grade-floor *emergency escape and rescue openings* shall be 5 square feet (0.46 m^2).

[BE] 1030.2.1 Minimum dimensions. The minimum net clear opening height dimension shall be 24 inches (610 mm). The minimum net clear opening width dimension shall be 20 inches (508 mm). The net clear opening dimensions shall be the result of normal operation of the opening.

[BE] 1030.3 Maximum height from floor. *Emergency escape and rescue openings* shall have the bottom of the clear opening not greater than 44 inches (1118 mm) measured from the floor.

[BE] 1030.4 Operational constraints. *Emergency escape and rescue openings* shall be operational from the inside of the room without the use of keys or tools. Bars, grilles, grates or similar devices are permitted to be placed over *emergency escape and rescue openings* provided the minimum net clear opening size complies with Section 1030.2 and such devices shall be releasable or removable from the inside without the use of a key, tool or force greater than that which is required for normal operation of the escape and rescue opening. Where such bars, grilles, grates or similar devices are installed in existing buildings, *smoke alarms* shall be installed in accordance with Section 907.2.11 regardless of the valuation of the *alteration.*

[BE] 1030.5 Window wells. An *emergency escape and rescue opening* with a finished sill height below the adjacent ground level shall be provided with a window well in accordance with Sections 1030.5.1 and 1030.5.2.

[BE] 1030.5.1 Minimum size. The minimum horizontal area of the window well shall be 9 square feet (0.84 m^2), with a minimum dimension of 36 inches (914 mm). The area of the window well shall allow the *emergency escape and rescue opening* to be fully opened.

[BE] 1030.5.2 Ladders or steps. Window wells with a vertical depth of more than 44 inches (1118 mm) shall be equipped with an *approved* permanently affixed ladder or steps. Ladders or rungs shall have an inside width of at least 12 inches (305 mm), shall project at least 3 inches (76 mm) from the wall and shall be spaced not more than 18 inches (457 mm) on center (o.c.) vertically for the full height of the window well. The ladder or steps shall not encroach into the required dimensions of the window well by more than 6 inches (152 mm). The ladder or steps shall not be obstructed by the *emergency escape and rescue opening.* Ladders or steps required by this section are exempt from the *stairway* requirements of Section 1011.

SECTION 1031 MAINTENANCE OF THE MEANS OF EGRESS

1031.1 General. The *means of egress* for buildings or portions thereof shall be maintained in accordance with this section.

1031.2 Reliability. Required *exit accesses, exits* and *exit discharges* shall be continuously maintained free from obstructions or impediments to full instant use in the case of fire or other emergency where the building area served by the *means of egress* is occupied. An *exit* or *exit passageway* shall not be used for any purpose that interferes with a *means of egress.*

1031.2.1 Security devices and egress locks. Security devices affecting *means of egress* shall be subject to approval of the *fire code official.* Security devices and locking arrangements in the *means of egress* that restrict, control, or delay egress shall be installed and maintained as required by this chapter.

1031.3 Obstructions. A *means of egress* shall be free from obstructions that would prevent its use, including the accumulation of snow and ice.

1031.3.1 Group I-2. In Group I-2, the required clear width for *aisles, corridors* and *ramps* that are part of the required *means of egress* shall comply with Section 1020.2. The facility shall have a plan to maintain the required clear width during emergency situations.

Exception: In areas required for bed movement, equipment shall be permitted in the required width where all the following provisions are met:

1. The equipment is low hazard and wheeled.
2. The equipment does not reduce the effective clear width for the *means of egress* to less than 5 feet (1525 mm).
3. The equipment is limited to:
    3.1 Equipment and carts in use.
    3.2 Medical emergency equipment.
    3.3 Infection control carts.
    3.4 Patient lift and transportation equipment.
4. Medical emergency equipment and patient lift and transportation equipment, when not in use, is required to be located on one side of the corridor.
5. The equipment is limited in number to a maximum of one per patient sleeping room or patient care room within each smoke compartment.

[BE] 1031.4 Exit signs. Exit signs shall be installed and maintained in accordance with Section 1013. Decorations, furnishings, equipment or adjacent signage that impairs the visibility of exit signs, creates confusion or prevents identification of the *exit* shall not be allowed.

1031.5 Nonexit identification. Where a door is adjacent to, constructed similar to and can be confused with a *means of egress* door, that door shall be identified with an *approved* sign that identifies the room name or use of the room.

1031.6 Finishes, furnishings and decorations. Means of egress doors shall be maintained in such a manner as to be distinguishable from the adjacent construction and finishes such that the doors are easily recognizable as doors. Furnishings, decorations or other objects shall not be placed so as to obstruct *exits*, access thereto, egress therefrom, or visibility thereof. Hangings and draperies shall not be placed over exit doors or otherwise be located to conceal or obstruct an *exit*. Mirrors shall not be placed on *exit* doors. Mirrors shall not be placed in or adjacent to any *exit* in such a manner as to confuse the direction of exit.

1031.7 Emergency escape and rescue openings. Required *emergency escape and rescue openings* shall be maintained in accordance with the code in effect at the time of construction, and the following: Required *emergency escape and rescue openings* shall be operational from the inside of the room without the use of keys or tools. Bars, grilles, grates or similar devices are allowed to be placed over *emergency escape and rescue openings* provided the minimum net clear opening size complies with the code that was in effect at the time of construction and such devices shall be releasable or removable from the inside without the use of a key, tool or force greater than that which is required for normal operation of the *emergency escape and rescue opening*.

1031.8 Inspection, testing and maintenance. All two-way communication systems for *areas of refuge* shall be inspected and tested on a yearly basis to verify that all components are operational. Where required, the tests shall be conducted in the presence of the *fire code official*. Records of inspection, testing and maintenance shall be maintained.

1031.9 Floor identification signs. The floor identification signs required by Sections 1023.9 and 1104.24 shall be maintained in an *approved* manner.

CHAPTER 11

CONSTRUCTION REQUIREMENTS FOR EXISTING BUILDINGS

SECTION 1101 GENERAL

1101.1 Scope. The provisions of this chapter shall apply to existing buildings constructed prior to the adoption of this code.

1101.2 Intent. The intent of this chapter is to provide a minimum degree of fire and life safety to persons occupying existing buildings by providing minimum construction requirements where such existing buildings do not comply with the minimum requirements of the *International Building Code.*

1101.3 Permits. Permits shall be required as set forth in Sections 105.6 and 105.7 and the *International Building Code.*

1101.4 Owner notification. When a building is found to be in noncompliance with this chapter, the *fire code official* shall duly notify the *owner* of the building. Upon receipt of such notice, the *owner* shall, subject to the following time limits, take necessary actions to comply with the provisions of this chapter.

1101.4.1 Construction documents. *Construction documents* necessary to comply with this chapter shall be completed and submitted within a time schedule *approved* by the *fire code official.*

1101.4.2 Completion of work. Work necessary to comply with this chapter shall be completed within a time schedule *approved* by the *fire code official.*

1101.4.3 Extension of time. The *fire code official* is authorized to grant necessary extensions of time when it can be shown that the specified time periods are not physically practical or pose an undue hardship. The granting of an extension of time for compliance shall be based on the showing of good cause and subject to the filing of an acceptable systematic plan of correction with the *fire code official.*

SECTION 1102 DEFINITIONS

1102.1 Definitions. The following terms are defined in Chapter 2:

DUTCH DOOR.

EXISTING.

SECTION 1103 FIRE SAFETY REQUIREMENTS FOR EXISTING BUILDINGS

1103.1 Required construction. Existing buildings shall comply with not less than the minimum provisions specified in Table 1103.1 and as further enumerated in Sections 1103.2 through 1103.10.

The provisions of this chapter shall not be construed to allow the elimination of *fire protection systems* or a reduction in the level of fire safety provided in buildings constructed in accordance with previously adopted codes.

Exceptions:

1. Where a change in fire-resistance rating has been approved in accordance with Section 803.6 of the *International Existing Building Code.*
2. Group U occupancies.

TABLE 1103.1 OCCUPANCY AND USE REQUIREMENTS[a]

SECTION	USE			OCCUPANCY CLASSIFICATION																		
	High-rise	Atrium or covered mall	Under-ground building	A	B	E	F	H-1	H-2	H-3	H-4	H-5	I-1	I-2	I-3	I-4	M	R-1	R-2	R-3	R-4	S
1103.2	R	R	R	R	R	R	R	R	R	R	R	R	R	R	R	R	R	R	R	R	R	R
1103.3	R	—	R	R	R	R	R	R	R	R	R	R	R	R	R	R	R	R	R	R	R	R
1103.4.1	R	—	R	—	—	—	—	—	—	—	—	—	—	R	R	—	—	—	—	—	—	—
1103.4.2	R	—	R	R	R	R	R	R	R	R	R	R	R	—	—	R	R	R	R	—	R	R
1103.4.3	R	—	R	R	R	R	R	R	R	R	R	R	R	—	—	R	R	R	R	—	R	R
1103.4.4	—	R	—	—	—	—	—	—	—	—	—	—	—	—	—	—	—	—	—	—	—	—
1103.4.5	—	—	—	—	R	—	—	—	—	—	—	—	—	—	—	—	R	—	—	—	—	—
1103.4.6	—	—	—	R	—	R	R	R	R	R	R	R	R	R	R	R	—	R	R	R	R	R
1103.4.7	—	—	—	R	—	R	R	R	R	R	R	R	R	R	R	R	—	R	R	R	R	R
1103.4.8	R	—	R	R	R	R	R	R	R	R	R	R	R	—	—	R	R	R	R	R	R	R
1103.4.9	R	—	—	—	—	—	—	—	—	—	—	—	—	R	—	—	—	—	—	—	—	—

(continued)

TABLE 1103.1
OCCUPANCY AND USE REQUIREMENTS[a]—continued

SECTION	USE			OCCUPANCY CLASSIFICATION																		
	High-rise	Atrium or covered mall	Under-ground building	A	B	E	F	H-1	H-2	H-3	H-4	H-5	I-1	I-2	I-3	I-4	M	R-1	R-2	R-3	R-4	S
1103.5.1	—	—	—	R[c]	—	—	—	—	—	—	—	—	—	—	—	—	—	—	—	—	—	—
1103.5.2, 1103.5.3[b]	—	—	—	—	—	—	—	—	—	—	—	—	—	R	—	—	—	—	—	—	—	—
1103.5.4	—	—	—	R	R	R	R	R	R	R	R	R	R	R	R	R	R	R	R	R	R	R
1103.6.1	R	—	R	R	R	R	R	R	R	R	R	R	R	R	R	R	R	R	R	—	R	R
1103.6.2	R	—	R	R	R	R	R	R	R	R	R	R	R	R	R	R	R	R	R	—	R	R
1103.7.1	—	—	—	—	—	R	—	—	—	—	—	—	—	—	—	—	—	—	—	—	—	—
1103.7.2	—	—	—	—	—	—	—	—	—	—	—	—	R	—	—	—	—	—	—	—	—	—
1103.7.3	—	—	—	—	—	—	—	—	—	—	—	—	—	R	—	—	—	—	—	—	—	—
1103.7.4	—	—	—	—	—	—	—	—	—	—	—	—	—	—	R	—	—	—	—	—	—	—
1103.7.5	—	—	—	—	—	—	—	—	—	—	—	—	—	—	—	—	—	R	—	—	—	—
1103.7.6	—	—	—	—	—	—	—	—	—	—	—	—	—	—	—	—	—	—	R	—	—	—
1103.7.7	—	—	—	—	—	—	—	—	—	—	—	—	—	—	—	—	—	—	—	—	R	—
1103.8	—	—	—	—	—	—	—	—	—	—	—	—	R	—	—	—	—	R	R	R	R	—
1103.9	R	—	—	—	—	—	—	—	—	—	—	—	R	R	—	R	—	R	R	R	R	—
1104	R	R	R	R	R	R	R	R	R	R	R	R	R	R	R	R	R	R	R	R	R	R
1105	—	—	—	—	—	—	—	—	—	—	—	—	—	R	—	—	—	—	—	—	—	—
1106	—	—	—	—	—	—	—	—	—	—	—	—	—	R	—	—	—	—	—	—	—	—

a. Existing buildings shall comply with the sections identified as "Required" (R) based on occupancy classification or use, or both, whichever is applicable.
b. Only applies to Group I-2 Condition 2 as established by the adopting ordinance.
c. Only applies to Group A-2 occupancies.
R = The building is required to comply.

1103.1.1 Historic buildings. Facilities designated as historic buildings shall develop a fire protection plan in accordance with NFPA 914. The fire protection plans shall comply with the maintenance and availability provisions in Sections 404.3 and 404.4.

1103.2 Emergency responder radio coverage in existing buildings. Existing buildings that do not have approved radio coverage for emergency responders within the building, based upon the existing coverage levels of the public safety communication systems of the jurisdiction at the exterior of the building, shall be equipped with such coverage according to one of the following:

1. Where an existing wired communication system cannot be repaired or is being replaced, or where not approved in accordance with Section 510.1, Exception 1.
2. Within a time frame established by the adopting authority.

Exception: Where it is determined by the fire code official that the radio coverage system is not needed.

1103.3 Existing elevators. Existing elevators, escalators and moving walks shall comply with the requirements of Sections 1103.3.1 and 1103.3.2.

1103.3.1 Elevators, escalators and moving walks. Existing elevators, escalators and moving walks in Group I-2 Condition 2 occupancies shall comply with ASME A17.3.

1103.3.2 Elevator emergency operation. Existing elevators with a travel distance of 25 feet (7620 mm) or more above or below the main floor or other level of a building and intended to serve the needs of emergency personnel for fire-fighting or rescue purposes shall be provided with emergency operation in accordance with ASME A17.3.

Exceptions:

1. Buildings without occupied floors located more than 55 feet (16 764 mm) above or 25 feet (7620 mm) below the lowest level of fire department vehicle access where protected at the elevator shaft openings with additional fire doors in accordance with Section 716.5 of the *International*

Building Code and where all of the following conditions are met:

1.1 The doors shall be provided with vision panels of approved fire protection-rated glazing so located as to furnish clear vision of the approach to the elevator. Such glazing shall not exceed 100 square inches (0.065 m^2) in area.

1.2 The doors shall be held open but be automatic-closing by activation of a fire alarm initiating device installed in accordance with the requirements of NFPA 72 as for Phase I Emergency Recall Operation, and shall be located at each floor served by the elevator; in the associated elevator machine room, control space, or control room; and in the elevator hoistway, where sprinklers are located in those hoistways.

1.3 The doors, when closed, shall have signs visible from the approach area stating: WHEN THESE DOORS ARE CLOSED OR IN FIRE EMERGENCY, DO NOT USE ELEVATOR. USE EXIT STAIRWAYS.

2. Buildings without occupied floors located more than 55 feet (16 764 mm) above or 25 feet (7620 mm) below the lowest level of fire department vehicle access where provided with *automatic sprinkler systems* installed in accordance with Section 903.3.1.1 or 903.3.1.2.

3. Freight elevators in buildings provided with both *automatic sprinkler systems* installed in accordance with Section 903.3.1.1 or 903.3.1.2 and not less than one ASME 17.3-compliant elevator serving the same floors.

Elimination of previously installed Phase I emergency recall or Phase II emergency in-car systems shall not be permitted.

1103.4 Vertical openings. Interior vertical openings, including but not limited to *stairways*, elevator hoistways, service and utility shafts, that connect two or more stories of a building, shall be enclosed or protected as specified in Sections 1103.4.1 through 1103.4.10.

1103.4.1 Group I-2 and I-3 occupancies. In Group I-2 and I-3 occupancies, interior vertical openings connecting two or more stories shall be protected with 1-hour fire-resistance-rated construction.

Exceptions:

1. In Group I-2, unenclosed vertical openings not exceeding two connected stories and not concealed within the building construction shall be permitted as follows:

   1.1. The unenclosed vertical openings shall be separated from other unenclosed vertical openings serving other floors by a smoke barrier.

   1.2. The unenclosed vertical openings shall be separated from corridors by smoke partitions.

   1.3. The unenclosed vertical openings shall be separated from other fire or smoke compartments on the same floors by a smoke barrier.

   1.4. On other than the lowest level, the unenclosed vertical openings shall not serve as a required means of egress.

2. In Group I-2, atriums connecting three or more stories shall not require 1-hour fire-resistance-rated construction where the building is equipped throughout with an *automatic sprinkler system* installed in accordance with Section 903.3, and all of the following conditions are met:

   2.1. For other than existing approved atriums with a smoke control system, where the atrium was constructed and is maintained in accordance with the code in effect at the time the atrium was created, the atrium shall have a smoke control system that is in compliance with Section 909.

   2.2. Glass walls forming a smoke partition or a glass-block wall assembly shall be permitted when in compliance with Condition 2.2.1 or 2.2.2.

      2.2.1. Glass walls forming a smoke partition shall be permitted where all of the following conditions are met:

         2.2.1.1. Automatic sprinklers are provided along both sides of the separation wall and doors, or on the room side only if there is not a walkway or occupied space on the atrium side.

         2.2.1.2. The sprinklers shall be not more than 12 inches (305 mm) away from the face of the glass and at intervals along the glass of not greater than 72 inches (1829 mm).

         2.2.1.3. Windows in the glass wall shall be non-operating type.

         2.2.1.4. The glass wall and windows shall be installed in a gasketed frame in a manner that the framing system deflects without breaking (loading) the glass before the sprinkler system operates.

2.2.1.5. The sprinkler system shall be designed so that the entire surface of the glass is wet upon activation of the sprinkler system without obstruction.

2.2.2. A fire barrier is not required where a glass-block wall assembly complying with Section 2110 of the *International Building Code* and having a $^{3}/_{4}$-hour fire protection rating is provided.

2.3. Where doors are provided in the glass wall, they shall be either self-closing or automatic-closing and shall be constructed to resist the passage of smoke.

3. In Group I-3 occupancies, exit *stairways* or ramps and *exit access stairways* or *ramps* constructed in accordance with Section 408 in the *International Building Code.*

1103.4.2 Three to five stories. In other than Group I-2 and I-3 occupancies, interior vertical openings connecting three to five stories shall be protected by either 1-hour fire-resistance-rated construction or an *automatic sprinkler system* shall be installed throughout the building in accordance with Section 903.3.1.1 or 903.3.1.2.

Exceptions:

1. Vertical opening protection is not required for Group R-3 occupancies.
2. Vertical opening protection is not required for open parking garages.
3. Vertical opening protection for escalators shall be in accordance with Section 1103.4.5, 1103.4.6 or 1103.4.7.
4. *Exit access stairways* and *ramps* shall be in accordance with Section 1103.4.8.

1103.4.3 More than five stories. In other than Group I-2 and I-3 occupancies, interior vertical openings connecting more than five stories shall be protected by 1-hour fire-resistance-rated construction.

Exceptions:

1. Vertical opening protection is not required for Group R-3 occupancies.
2. Vertical opening protection is not required for open parking garages.
3. Vertical opening protection for escalators shall be in accordance with Section 1103.4.5, 1103.4.6 or 1103.4.7.
4. *Exit access stairways* and *ramps* shall be in accordance with Section 1103.4.8.

1103.4.4 Atriums and covered malls. In other than Group I-2 and I-3 occupancies, interior vertical openings in a covered mall building or a building with an atrium shall be protected by either 1-hour fire-resistance-rated construction or an *automatic sprinkler system* shall be installed throughout the building in accordance with Section 903.3.1.1 or 903.3.1.2.

Exceptions:

1. Vertical opening protection is not required for Group R-3 occupancies.
2. Vertical opening protection is not required for open parking garages.
3. *Exit access stairways* and *ramps* shall be in accordance with Section 1103.4.8.

1103.4.5 Escalators in Group B and M occupancies. In Group B and M occupancies, escalators creating vertical openings connecting any number of stories shall be protected by either 1-hour fire-resistance-rated construction or an *automatic sprinkler system* in accordance with Section 903.3.1.1 installed throughout the building, with a draft curtain and closely spaced sprinklers around the escalator opening.

1103.4.6 Escalators connecting four or fewer stories. In other than Group B and M occupancies, escalators creating vertical openings connecting four or fewer stories shall be protected by either 1-hour fire-resistance-rated construction or an *automatic sprinkler system* in accordance with Section 903.3.1.1 or 903.3.1.2 shall be installed throughout the building, and a draft curtain with closely spaced sprinklers shall be installed around the escalator opening.

1103.4.7 Escalators connecting more than four stories. In other than Group B and M occupancies, escalators creating vertical openings connecting five or more stories shall be protected by 1-hour fire-resistance-rated construction.

1103.4.8 Occupancies other than Group I-2 and I-3. In other than Group I-2 and I-3 occupancies, floor openings containing *exit access stairways* or *ramps* that do not comply with one of the conditions listed in this section shall be protected by 1-hour fire-resistance-rated construction.

1. *Exit access stairways* and *ramps* that serve, or atmospherically communicate between, only two stories. Such interconnected stories shall not be open to other stories.
2. In Group R-1, R-2 or R-3 occupancies, *exit access stairways* and *ramps* connecting four stories or less serving and contained within an individual *dwelling unit* or *sleeping unit* or live/work unit.
3. *Exit access stairways* and *ramps* in buildings equipped throughout with an *automatic sprinkler system* in accordance with Section 903.3.1.1, where the area of the vertical opening between stories does

not exceed twice the horizontal projected area of the *stairway* or *ramp*, and the opening is protected by a draft curtain and closely spaced sprinklers in accordance with NFPA 13. In other than Group B and M occupancies, this provision is limited to openings that do not connect more than four stories.

4. *Exit access stairways* and *ramps* within an atrium complying with the provisions of Section 404 of the *International Building Code*.
5. *Exit access stairways* and *ramps* in open parking garages that serve only the parking garage.
6. *Exit access stairways* and *ramps* serving open-air seating complying with the exit access travel distance requirements of Section 1029.7 of the *International Building Code*.
7. *Exit access stairways* and *ramps* serving the balcony, gallery or press box and the main assembly floor in occupancies such as theaters, places of religious worship, auditoriums and sports facilities.

1103.4.9 Waste and linen chutes. In Group I-2 occupancies, existing waste and linen chutes shall comply with Sections 1103.4.9.1 through 1103.4.9.5.

1103.4.9.1 Enclosure. Chutes shall be enclosed with 1-hour fire-resistance-rated construction. Opening protectives shall be in accordance with Section 716 of the *International Building Code* and have a fire protection rating of not less than1 hour.

1103.4.9.2 Chute intakes. Chute intakes shall comply with Section 1103.4.9.2.1 or 1103.4.9.2.2.

1103.4.9.2.1 Chute intake direct from corridor. Where intake to chutes is direct from a *corridor*, the intake opening shall be equipped with a chute-intake door in accordance with Section 716 of the *International Building Code* and having a fire protection rating of not less than 1 hour.

1103.4.9.2.2 Chute intake via a chute-intake room. Where the intake to chutes is accessed through a chute-intake room, the room shall be enclosed with 1-hour fire-resistance-rated construction. Opening protectives for the intake room shall be in accordance with Section 716 of the *International Building Code* and have a fire protection rating of not less than $^3/_4$ hour. Opening protective for the chute enclosure shall be in accordance with Section 1103.4.9.1.

1103.4.9.3 Automatic sprinkler system. Chutes shall be equipped with an *approved automatic sprinkler system* in accordance with Section 903.2.11.2.

1103.4.9.4 Chute discharge rooms. Chutes shall terminate in a dedicated chute discharge room. Such rooms shall be separated from the remainder of the building by not less than 1-hour fire-resistance-rated construction. Opening protectives shall be in accordance with Section 716 of the *International Building Code* and have a fire protection rating of not less than 1 hour.

1103.4.9.5 Chute discharge protection. Chute discharges shall be equipped with a self-closing or automatic-closing opening protective in accordance with Section 716 of the *International Building Code* and having a fire protection rating of not less than 1 hour.

1103.4.10 Flue-fed incinerators. Existing flue-fed incinerator rooms and associated flue shafts shall be protected with 1-hour fire-resistance-rated construction and shall not have other vertical openings connected with the space other than the associated flue. Opening protectives shall be in accordance with Section 716 of the *International Building Code* and have a fire protection rating of not less than 1 hour.

1103.5 Sprinkler systems. An *automatic sprinkler system* shall be provided in existing buildings in accordance with Sections 1103.5.1 through 1103.5.4.

1103.5.1 Group A-2. An *automatic sprinkler system* shall be installed in accordance with Section 903.3.1.1 throughout existing buildings or portions thereof used as Group A-2 occupancies with an occupant load of 300 or more.

1103.5.2 Group I-2. In Group I-2, an *automatic sprinkler system* shall be provided in accordance with Section 1105.8.

1103.5.3 Group I-2 Condition 2. In addition to the requirements of Section 1103.5.2, existing buildings of Group I-2 Condition 2 occupancy shall be equipped throughout with an *approved automatic sprinkler system* in accordance with Section 903.3.1.1. The *automatic sprinkler system* shall be installed as established by the adopting ordinance.

1103.5.4 Pyroxylin plastics. An *automatic sprinkler system* shall be provided throughout existing buildings where cellulose nitrate film or pyroxylin plastics are manufactured, stored or handled in quantities exceeding 100 pounds (45 kg). Vaults located within buildings for the storage of raw pyroxylin shall be protected with an *approved automatic sprinkler system* capable of discharging 1.66 gallons per minute per square foot (68 L/min/m^2) over the area of the vault.

1103.6 Standpipes. Existing structures shall be equipped with standpipes installed in accordance with Section 905 where required in Sections 1103.6.1 and 1103.6.2. The *fire code official* is authorized to approve the installation of manual standpipe systems to achieve compliance with this section where the responding fire department is capable of providing the required hose flow at the highest standpipe outlet.

1103.6.1 Existing multiple-story buildings. Existing buildings with occupied floors located more than 50 feet (15 240 mm) above the lowest level of fire department access or more than 50 feet (15 240 mm) below the highest level of fire department access shall be equipped with standpipes.

1103.6.2 Existing helistops and heliports. Existing buildings with a rooftop helistop or heliport located more than 30 feet (9144 mm) above the lowest level of fire department access to the roof level on which the helistop or heliport is located shall be equipped with standpipes in accordance with Section 2007.5.

1103.7 Fire alarm systems. An *approved* fire alarm system shall be installed in existing buildings and structures in accordance with Sections 1103.7.1 through 1103.7.7 and provide occupant notification in accordance with Section 907.5 unless other requirements are provided by other sections of this code.

Exception: Occupancies with an existing, previously *approved* fire alarm system.

1103.7.1 Group E. A fire alarm system shall be installed in existing Group E occupancies in accordance with Section 907.2.3.

Exceptions:

1. A manual fire alarm system is not required in a building with a maximum area of 1,000 square feet (93 m^2) that contains a single classroom and is located not closer than 50 feet (15 240 mm) from another building.
2. A manual fire alarm system is not required in Group E occupancies with an *occupant load* less than 50.

1103.7.2 Group I-1. An automatic fire alarm system shall be installed in existing Group I-1 facilities in accordance with Section 907.2.6.1.

Exception: Where each sleeping room has a *means of egress* door opening directly to an exterior egress balcony that leads directly to the *exits* in accordance with Section 1021, and the building is not more than three stories in height.

1103.7.3 Group I-2. In Group I-2, an automatic fire alarm system shall be installed in accordance with Section 1105.9.

1103.7.4 Group I-3. An automatic and manual fire alarm system shall be installed in existing Group I-3 occupancies in accordance with Section 907.2.6.3.

1103.7.5 Group R-1. A fire alarm system and smoke alarms shall be installed in existing Group R-1 occupancies in accordance with Sections 1103.7.5.1 through 1103.7.5.2.1.

1103.7.5.1 Group R-1 hotel and motel manual fire alarm system. A manual fire alarm system that activates the occupant notification system in accordance with Section 907.5 shall be installed in existing Group R-1 hotels and motels more than three stories or with more than 20 *sleeping units.*

Exceptions:

1. Buildings less than two stories in height where all *sleeping units,* attics and crawl spaces are separated by 1-hour fire-resistance-rated construction and each *sleeping unit* has direct access to a *public way, egress court* or yard.
2. Manual fire alarm boxes are not required throughout the building where the following conditions are met:
   - 2.1. The building is equipped throughout with an *automatic sprinkler system* installed in accordance with Section 903.3.1.1 or 903.3.1.2.
   - 2.2. The notification appliances will activate upon sprinkler water flow.
   - 2.3. Not less than one manual fire alarm box is installed at an *approved* location.

1103.7.5.1.1 Group R-1 hotel and motel automatic smoke detection system. An automatic smoke detection system that activates the occupant notification system in accordance with Section 907.5 shall be installed in existing Group R-1 hotels and motels throughout all interior *corridors* serving sleeping rooms not equipped with an *approved,* supervised sprinkler system installed in accordance with Section 903.

Exception: An automatic smoke detection system is not required in buildings that do not have interior *corridors* serving *sleeping units* and where each sleeping unit has a *means of egress* door opening directly to an *exit* or to an exterior *exit access* that leads directly to an *exit.*

1103.7.5.2 Group R-1 boarding and rooming houses manual fire alarm system. A manual fire alarm system that activates the occupant notification system in accordance with Section 907.5 shall be installed in existing Group R-1 boarding and rooming houses.

Exception: Buildings less than two stories in height where all *sleeping units,* attics and crawl spaces are separated by 1-hour fire-resistance-rated construction and each *sleeping unit* has direct access to a *public way, egress court* or yard.

1103.7.5.2.1 Group R-1 boarding and rooming houses automatic smoke detection system. An automatic smoke detection system that activates the occupant notification system in accordance with Section 907.5 shall be installed in existing Group R-1 boarding and rooming houses throughout all interior *corridors* serving *sleeping units* not equipped with an *approved,* supervised sprinkler system installed in accordance with Section 903.

Exception: Buildings equipped with single-station smoke alarms meeting or exceeding the requirements of Section 907.2.11.1 and where the fire alarm system includes not less than one manual fire alarm box per floor arranged to initiate the alarm.

1103.7.6 Group R-2. A manual fire alarm system that activates the occupant notification system in accordance with Section 907.5 shall be installed in existing Group R-2 occupancies more than three stories in height or with more than 16 *dwelling* or *sleeping units*.

Exceptions:

1. Where each living unit is separated from other contiguous living units by *fire barriers* having a *fire-resistance rating* of not less than $^3/_4$ hour, and where each living unit has either its own independent *exit* or its own independent stairway or ramp discharging at grade.
2. A separate fire alarm system is not required in buildings that are equipped throughout with an *approved* supervised *automatic sprinkler system* installed in accordance with Section 903.3.1.1 or 903.3.1.2 and having a local alarm to notify all occupants.
3. A fire alarm system is not required in buildings that do not have interior *corridors* serving *dwelling units* and are protected by an *approved automatic sprinkler system* installed in accordance with Section 903.3.1.1 or 903.3.1.2, provided that *dwelling units* either have a *means of egress* door opening directly to an exterior *exit access* that leads directly to the *exits* or are served by open-ended *corridors* designed in accordance with Section 1027.6, Exception 3.
4. A fire alarm system is not required in buildings that do not have interior *corridors* serving *dwelling units*, do not exceed three stories in height and comply with both of the following:
   4.1. Each *dwelling unit* is separated from other contiguous *dwelling units* by *fire barriers* having a *fire-resistance rating* of not less than $^3/_4$ hour.
   4.2. Each *dwelling unit* is provided with hard-wired, interconnected smoke alarms as required for new construction in Section 907.2.11.

1103.7.7 Group R-4. A manual fire alarm system that activates the occupant notification system in accordance with Section 907.5 shall be installed in existing Group R-4 residential care/assisted living facilities in accordance with Section 907.2.10.1.

Exceptions:

1. Where there are interconnected smoke alarms meeting the requirements of Section 907.2.11 and there is not less than one manual fire alarm box per floor arranged to continuously sound the smoke alarms.
2. Other manually activated, continuously sounding alarms *approved* by the *fire code official*.

1103.8 Single- and multiple-station smoke alarms. Single- and multiple-station smoke alarms shall be installed in existing Group I-1 and R occupancies in accordance with Sections 1103.8.1 through 1103.8.3.

1103.8.1 Where required. Existing Group I-1 and R occupancies shall be provided with single-station smoke alarms in accordance with Section 907.2.11. Interconnection and power sources shall be in accordance with Sections 1103.8.2 and 1103.8.3, respectively.

Exceptions:

1. Where the code that was in effect at the time of construction required smoke alarms and smoke alarms complying with those requirements are already provided.
2. Where smoke alarms have been installed in occupancies and dwellings that were not required to have them at the time of construction, additional smoke alarms shall not be required provided that the existing smoke alarms comply with requirements that were in effect at the time of installation.
3. Where smoke detectors connected to a fire alarm system have been installed as a substitute for smoke alarms.

1103.8.2 Interconnection. Where more than one smoke alarm is required to be installed within an individual *dwelling* or *sleeping unit*, the smoke alarms shall be interconnected in such a manner that the activation of one alarm will activate all of the alarms in the individual unit. Physical interconnection of smoke alarms shall not be required where listed wireless alarms are installed and all alarms sound upon activation of one alarm. The alarm shall be clearly audible in all bedrooms over background noise levels with all intervening doors closed.

Exceptions:

1. Interconnection is not required in buildings that are not undergoing *alterations*, repairs or construction of any kind.
2. Smoke alarms in existing areas are not required to be interconnected where *alterations* or repairs do not result in the removal of interior wall or ceiling finishes exposing the structure, unless there is an attic, crawl space or *basement* available that could provide access for interconnection without the removal of interior finishes.

1103.8.3 Power source. Single-station smoke alarms shall receive their primary power from the building wiring provided that such wiring is served from a commercial source and shall be equipped with a battery backup. Smoke alarms with integral strobes that are not equipped with battery backup shall be connected to an emergency electrical system. Smoke alarms shall emit a signal when the batteries are low. Wiring shall be permanent and without a disconnecting switch other than as required for overcurrent protection.

Exceptions:

1. Smoke alarms are permitted to be solely battery operated in existing buildings where construction is not taking place.

2. Smoke alarms are permitted to be solely battery operated in buildings that are not served from a commercial power source.
3. Smoke alarms are permitted to be solely battery operated in existing areas of buildings undergoing *alterations* or repairs that do not result in the removal of interior walls or ceiling finishes exposing the structure, unless there is an attic, crawl space or *basement* available that could provide access for building wiring without the removal of interior finishes.

1103.9 Carbon monoxide alarms. Existing Group I-1, I-2, I-4 and R occupancies shall be equipped with carbon monoxide alarms in accordance with Section 915, except that the carbon monoxide alarms shall be allowed to be solely battery operated.

1103.10 Medical gases. Medical gases stored and transferred in health-care-related facilities shall be in accordance with Chapter 53.

SECTION 1104 MEANS OF EGRESS FOR EXISTING BUILDINGS

1104.1 General. *Means of egress* in existing buildings shall comply with the minimum egress requirements where specified in Table 1103.1 as further enumerated in Sections 1104.2 through 1104.25, and the building code that applied at the time of construction. Where the provisions of this chapter conflict with the building code that applied at the time of construction, the most restrictive provision shall apply. Existing buildings that were not required to comply with a building code at the time of construction shall comply with the minimum egress requirements where specified in Table 1103.1 as further enumerated in Sections 1104.2 through 1104.25.

1104.2 Elevators, escalators and moving walks. Elevators, escalators and moving walks shall not be used as a component of a required *means of egress.*

Exceptions:

1. Elevators used as an *accessible means of egress* where allowed by Section 1009.4.
2. Previously *approved* elevators, escalators and moving walks in existing buildings.

1104.3 Exit sign illumination. Exit signs shall be internally or externally illuminated. The face of an exit sign illuminated from an external source shall have an intensity of not less than 5 footcandles (54 lux). Internally illuminated signs shall provide equivalent luminance and be *listed* for the purpose.

Exception: *Approved* self-luminous signs that provide evenly illuminated letters shall have a minimum luminance of 0.06 foot-lamberts (0.21 cd/m^2).

1104.4 Power source. Where emergency illumination is required in Section 1104.5, exit signs shall be visible under emergency illumination conditions.

Exception: *Approved* signs that provide continuous illumination independent of external power sources are not required to be connected to an emergency electrical system.

1104.5 Illumination emergency power. Where *means of egress* illumination is provided, the power supply for *means of egress* illumination shall normally be provided by the premises' electrical supply. In the event of power supply failure, illumination shall be automatically provided from an emergency system for the following occupancies where such occupancies require two or more *means of egress*:

1. Group A having 50 or more occupants.

   Exception: Assembly occupancies used exclusively as a place of worship and having an *occupant load* of less than 300.

2. Group B buildings three or more stories in height, buildings with 100 or more occupants above or below a *level of exit discharge* serving the occupants or buildings with 1,000 or more total occupants.
3. Group E in interior *exit access* and *exit stairways* and *ramps*, *corridors*, windowless areas with student occupancy, shops and laboratories.
4. Group F having more than 100 occupants.

   Exception: Buildings used only during daylight hours and that are provided with windows for natural light in accordance with the *International Building Code.*

5. Group I.
6. Group M.

   Exception: Buildings less than 3,000 square feet (279 m^2) in gross sales area on one story only, excluding mezzanines.

7. Group R-1.

   Exception: Where each *sleeping unit* has direct access to the outside of the building at grade.

8. Group R-2.

   Exception: Where each *dwelling unit* or *sleeping unit* has direct access to the outside of the building at grade.

9. Group R-4.

   Exception: Where each *sleeping unit* has direct access to the outside of the building at ground level.

1104.5.1 Emergency power duration and installation. Emergency power for *means of egress* illumination shall be provided in accordance with Section 604. In other than

Group I-2, emergency power shall be provided for not less than 60 minutes for systems requiring emergency power. In Group I-2, essential electrical systems shall comply with Sections 1105.5.1 and 1105.5.2.

1104.6 Guards. Guards complying with this section shall be provided at the open sides of *means of egress* that are more than 30 inches (762 mm) above the floor or grade below.

1104.6.1 Height of guards. Guards shall form a protective barrier not less than 42 inches (1067 mm) high.

Exceptions:

1. Existing guards on the open side of exit access and exit *stairways* and *ramps* shall be not less than 30 inches (760 mm) high.
2. Existing *guards* within *dwelling units* shall be not less than 36 inches (910 mm) high.
3. Existing *guards* in assembly seating areas.

1104.6.2 Opening limitations. Open *guards* shall have balusters or ornamental patterns such that a 6-inch-diameter (152 mm) sphere cannot pass through any opening up to a height of 34 inches (864 mm).

Exceptions:

1. At elevated walking surfaces for access to, and use of, electrical, mechanical or plumbing systems or equipment, guards shall have balusters or be of solid materials such that a sphere with a diameter of 21 inches (533 mm) cannot pass through any opening.
2. In occupancies in Group I-3, F, H or S, the clear distance between intermediate rails measured at right angles to the rails shall not exceed 21 inches (533 mm).
3. *Approved* existing open guards.

1104.7 Size of doors. The minimum width of each door opening shall be sufficient for the *occupant load* thereof and shall provide a clear width of not less than 28 inches (711 mm). Where this section requires a minimum clear width of 28 inches (711 mm) and a door opening includes two door leaves without a mullion, one leaf shall provide a clear opening width of 28 inches (711 mm). In ambulatory care facilities, doors serving as *means of egress* from patient treatment rooms or patient sleeping rooms shall provide a clear width of not less than 32 inches (813 mm). In Group I-2, *means of egress* doors where used for the movement of beds shall provide a clear width not less than $41^1/_2$ inches (1054 mm). The maximum width of a swinging door leaf shall be 48 inches (1219 mm) nominal. The height of door openings shall be not less than 80 inches (2032 mm).

Exceptions:

1. The minimum and maximum width shall not apply to door openings that are not part of the required *means of egress* in occupancies in Groups R-2 and R-3.
2. Door openings to storage closets less than 10 square feet (0.93 m^2) in area shall not be limited by the minimum width.
3. Width of door leafs in revolving doors that comply with Section 1010.1.1 shall not be limited.
4. Door openings within a *dwelling unit* shall be not less than 78 inches (1981 mm) in height.
5. Exterior door openings in *dwelling units*, other than the required *exit* door, shall be not less than 76 inches (1930 mm) in height.
6. *Exit access* doors serving a room not larger than 70 square feet (6.5 m^2) shall be not less than 24 inches (610 mm) in door width.
7. Door closers and door stops shall be permitted to be 78 inches (1980 mm) minimum above the floor.

1104.8 Opening force for doors. The opening force for interior side-swinging doors without closers shall not exceed a 5-pound (22 N) force. The opening forces do not apply to the force required to retract latch bolts or disengage other devices that hold the door in a closed position. For other side-swinging, sliding and folding doors, the door latch shall release when subjected to a force of not more than 15 pounds (66 N). The door shall be set in motion when subjected to a force not exceeding 30 pounds (133 N). The door shall swing to a full-open position when subjected to a force of not more than 50 pounds (222 N). Forces shall be applied to the latch side.

1104.9 Revolving doors. Revolving doors shall comply with the following:

1. A revolving door shall not be located within 10 feet (3048 mm) of the foot or top of *stairways* or escalators. A dispersal area shall be provided between the *stairways* or escalators and the revolving doors.
2. The revolutions per minute for a revolving door shall not exceed those shown in Table 1104.9.
3. Each revolving door shall have a conforming side-hinged swinging door in the same wall as the revolving door and within 10 feet (3048 mm).

Exceptions:

1. A revolving door is permitted to be used without an adjacent swinging door for street-floor elevator lobbies provided a stairway, escalator or door from other parts of the building does not discharge through the lobby and the lobby does not have any occupancy or use other than as a means of travel between elevators and a street.
2. Existing revolving doors where the number of revolving doors does not exceed the number of swinging doors within 20 feet (6096 mm).

TABLE 1104.9
REVOLVING DOOR SPEEDS

INSIDE DIAMETER (feet-inches)	POWER-DRIVEN-TYPE SPEED CONTROL (rpm)	MANUAL-TYPE SPEED CONTROL (rpm)
6-6	11	12
7-0	10	11
7-6	9	11
8-0	9	10
8-6	8	9
9-0	8	9
9-6	7	8
10-0	7	8

For SI: 1 inch = 25.4 mm, 1 foot = 304.8 mm.

1104.9.1 Egress component. A revolving door used as a component of a *means of egress* shall comply with Section 1104.9 and all of the following conditions:

1. Revolving doors shall not be given credit for more than 50 percent of the required egress capacity.
2. Each revolving door shall be credited with not more than a 50-person capacity.
3. Revolving doors shall be capable of being collapsed when a force of not more than 130 pounds (578 N) is applied within 3 inches (76 mm) of the outer edge of a wing.

1104.10 Stair dimensions for existing stairways. Existing *stairways* in buildings shall be permitted to remain if the rise does not exceed $8^1/_4$ inches (210 mm) and the run is not less than 9 inches (229 mm). Existing *stairways* can be rebuilt.

Exception: Other *stairways approved* by the *fire code official.*

1104.10.1 Dimensions for replacement stairways. The replacement of an existing *stairway* in a structure shall not be required to comply with the new *stairway* requirements of Section 1009 where the existing space and construction will not allow a reduction in pitch or slope.

1104.11 Winders. Existing winders shall be allowed to remain in use if they have a minimum tread depth of 6 inches (152 mm) and a minimum tread depth of 9 inches (229 mm) at a point 12 inches (305 mm) from the narrowest edge.

1104.12 Curved stairways. Existing curved *stairways* shall be allowed to continue in use, provided the minimum depth of tread is 10 inches (254 mm) and the smallest radius shall be not less than twice the width of the *stairway.*

1104.13 Stairway handrails. *Stairways* shall have *handrails* on at least one side. *Handrails* shall be located so that all portions of the *stairway* width required for egress capacity are within 44 inches (1118 mm) of a *handrail.*

Exception: *Aisle stairs* provided with a center *handrail* are not required to have additional *handrails.*

1104.13.1 Height. *Handrail* height, measured above *stair* tread nosings, shall be uniform, not less than 30 inches (762 mm) and not more than 42 inches (1067 mm).

1104.14 Slope of ramps. *Ramp* runs utilized as part of a *means of egress* shall have a running slope not steeper than one unit vertical in 10 units horizontal (10-percent slope). The slope of other *ramps* shall not be steeper than one unit vertical in eight units horizontal (12.5-percent slope).

1104.15 Width of ramps. Existing *ramps* are permitted to have a minimum width of 30 inches (762 mm) but not less than the width required for the number of occupants served as determined by Section 1005.1. In Group I-2, *ramps* serving as a *means of egress* and used for the movement of patients in beds shall comply with Section 1105.5.4.

1104.16 Fire escape stairways. Fire escape *stairways* shall comply with Sections 1104.16.1 through 1104.16.7.

1104.16.1 Existing means of egress. Fire escape *stairways* shall be permitted in existing buildings but shall not constitute more than 50 percent of the required *exit* capacity.

1104.16.2 Protection of openings. Openings within 10 feet (3048 mm) of fire escape *stairways* shall be protected by opening protectives having a minimum $^3/_4$-hour *fire protection rating.*

Exception: In buildings equipped throughout with an *approved automatic sprinkler system,* opening protection is not required.

1104.16.3 Dimensions. Fire escape *stairways* shall meet the minimum width, capacity, riser height and tread depth as specified in Section 1104.10.

1104.16.4 Access. Access to a fire escape *stairway* from a *corridor* shall not be through an intervening room. Access to a fire escape *stairway* shall be from a door or window meeting the criteria of Section 1005.1. Access to a fire escape *stairway* shall be directly to a balcony, landing or platform. These shall not be higher than the floor or window sill level and not lower than 8 inches (203 mm) below the floor level or 18 inches (457 mm) below the window sill.

1104.16.5 Materials and strength. Components of fire escape *stairways* shall be constructed of noncombustible materials. Fire escape *stairways* and balconies shall support the dead load plus a live load of not less than 100 pounds per square foot (4.78 kN/m^2). Fire escape *stairways* and balconies shall be provided with a top and intermediate *handrail* on each side.

1104.16.5.1 Examination. Fire escape *stairways* and balconies shall be examined for structural adequacy and safety in accordance with Section 1104.16.5 by a registered design professional or others acceptable to the *fire code official* every 5 years, or as required by the *fire code official.* An inspection report shall be submitted to the *fire code official* after such examination.

1104.16.6 Termination. The lowest balcony shall not be more than 18 feet (5486 mm) from the ground. Fire escape *stairways* shall extend to the ground or be provided with counterbalanced *stairs* reaching the ground.

Exception: For fire escape *stairways* serving 10 or fewer occupants, an *approved* fire escape ladder is allowed to serve as the termination.

1104.16.7 Maintenance. Fire escape *stairways* shall be kept clear and unobstructed at all times and shall be maintained in good working order.

1104.17 Corridor construction. Corridors serving an occupant load greater than 30 and the openings therein shall provide an effective barrier to resist the movement of smoke. Transoms, louvers, doors and other openings shall be kept closed or be self-closing. In Group I-2, corridors in areas housing patient sleeping or care rooms shall comply with Section 1105.4.

Exceptions:

1. *Corridors* in occupancies other than in Group H, that are equipped throughout with an *approved automatic sprinkler system.*
2. *Corridors* in occupancies in Group E where each room utilized for instruction or assembly has not less than one-half of the required *means of egress* doors opening directly to the exterior of the building at ground level.
3. *Corridors* that are in accordance with the *International Building Code.*

1104.17.1 Corridor openings. Openings in *corridor* walls shall comply with the requirements of the *International Building Code.*

Exceptions:

1. Where 20-minute fire door assemblies are required, solid wood doors not less than 1.75 inches (44 mm) thick or insulated steel doors are allowed.
2. Openings protected with fixed wire glass set in steel frames.
3. Openings covered with 0.5-inch (12.7 mm) gypsum wallboard or 0.75-inch (19.1 mm) plywood on the room side.
4. Opening protection is not required where the building is equipped throughout with an *approved automatic sprinkler system.*

1104.18 Dead end corridors. Where more than one exit or exit access doorway is required, the *exit access* shall be arranged such that dead ends do not exceed the limits specified in Table 1104.18. In Group I-2, in smoke compartments containing patient sleeping rooms and treatment rooms, dead end *corridors* shall be in accordance with Section 1105.5.6.

Exception: A dead-end passageway or *corridor* shall not be limited in length where the length of the dead-end passageway or *corridor* is less than 2.5 times the least width of the dead-end passageway or *corridor.*

1104.19 Exit access travel distance. *Exits* shall be located so that the maximum length of exit access travel, measured from the most remote point to an *approved exit* along the natural and unobstructed path of egress travel, does not exceed the distances given in Table 1104.18.

1104.20 Common path of egress travel. The *common path of egress travel* shall not exceed the distances given in Table 1104.18.

1104.21 Stairway discharge identification. An interior *exit stairway* or *ramp* that continues below its *level of exit discharge* shall be arranged and marked to make the direction of egress to a *public way* readily identifiable.

Exception: *Stairways* that continue one-half story beyond their *levels of exit discharge* need not be provided with barriers where the *exit discharge* is obvious.

1104.22 Exterior stairway protection. *Exterior exit stairways* shall be separated from the interior of the building as required in Section 1027.6. Openings shall be limited to those necessary for egress from normally occupied spaces.

Exceptions:

1. Separation from the interior of the building is not required for buildings that are two stories or less above grade where the *level of exit discharge* serving such occupancies is the first story above grade.
2. Separation from the interior of the building is not required where the exterior *stairway* is served by an exterior balcony that connects two remote exterior *stairways* or other *approved exits*, with a perimeter that is not less than 50 percent open. To be considered open, the opening shall be not less than 50 percent of the height of the enclosing wall, with the top of the opening not less than 7 feet (2134 mm) above the top of the balcony.
3. Separation from the interior of the building is not required for an exterior *stairway* located in a building or structure that is permitted to have unenclosed interior *stairways* in accordance with Section 1023.
4. Separation from the open-ended corridors of the building is not required for exterior *stairways* provided that:
    - 4.1. The open-ended *corridors* comply with Section 1020.
    - 4.2. The open-ended *corridors* are connected on each end to an *exterior exit stairway* complying with Section 1027.
    - 4.3. At any location in an open-ended *corridor* where a change of direction exceeding 45 degrees (0.79 rad) occurs, a clear opening of not less than 35 square feet (3 m^2) or an exterior *stairway* shall be provided. Where clear openings are provided, they shall be located so as to minimize the accumulation of smoke or toxic gases.

1104.23 Minimum aisle width. The minimum clear width of *aisles* shall be:

1. Forty-two inches (1067 mm) for aisle stairs having seating on each side.

   Exception: Thirty-six inches (914 mm) where the *aisle* serves less than 50 seats.

2. Thirty-six inches (914 mm) for stepped *aisles* having seating on only one side.

   Exceptions:

   1. Thirty inches (760 mm) for catchment areas serving not more than 60 seats.
   2. Twenty-three inches (584 mm) between a stepped aisle *handrail* and seating where an *aisle* does not serve more than five rows on one side.

3. Twenty inches (508 mm) between a stepped *aisle handrail* or *guard* and seating where the *aisle* is subdivided by the *handrail*.

4. Forty-two inches (1067 mm) for level or ramped *aisles* having seating on both sides.

   Exception: Thirty-six inches (914 mm) where the *aisle* serves less than 50 seats.

5. Thirty-six inches (914 mm) for level or ramped *aisles* having seating on only one side.

   Exception: Thirty inches (760 mm) for catchment areas serving not more than 60 seats.

6. In Group I-2, where *aisles* are used for movement of patients in beds, *aisles* shall comply with Section 1105.5.8.

TABLE 1104.18
COMMON PATH, DEAD-END AND TRAVEL DISTANCE LIMITS (by occupancy)

OCCUPANCY	COMMON PATH LIMIT		DEAD-END LIMIT		TRAVEL DISTANCE LIMIT	
	Unsprinklered (feet)	Sprinklered (feet)	Unsprinklered (feet)	Sprinklered (feet)	Unsprinklered (feet)	Sprinklered (feet)
Group A	20/75[a]	20/75[a]	20[b]	20[b]	200	250
Group B[h]	75	100	50	50	200	300
Group E	75	75	20	50	200	250
Group F-1, S-1[d, h]	75	100	50	50	200	250
Group F-2, S-2[d, h]	75	100	50	50	300	400
Group H-1	25	25	0	0	75	75
Group H-2	50	100	0	0	75	100
Group H-3	50	100	20	20	100	150
Group H-4	75	75	20	20	150	175
Group H-5	75	75	20	50	150	200
Group I-1	75	75	20	50	200	250
Group I-2	Notes e, g	Notes e, g	Note f	Note f	150	200[c]
Group I-3	100	100	NR	NR	150[c]	200[c]
Group I-4 (Day care centers)	NR	NR	20	20	200	250
Group M (Covered or open mall)	75	100	50	50	200	400
Group M (Mercantile)	75	100	50	50	200	250
Group R-1 (Hotels)	75	75	50	50	200	250
Group R-2 (Apartments)	75	125	50	50	200	250
Group R-3 (One- and two-family)	NR	NR	NR	NR	NR	NR
Group R-4 (Residential care/assisted living)	NR	NR	NR	NR	NR	NR
Group U[h]	75	100	20	50	300	400

NR = No requirements.

For SI: 1 foot = 304.8 mm, 1 square foot = 0.0929 m^2.

a. 20 feet for common path serving 50 or more persons; 75 feet for common path serving less than 50 persons.

b. See Section 1029.9.5 for dead-end aisles in Group A occupancies.

c. This dimension is for the total travel distance, assuming incremental portions have fully utilized their allowable maximums. For travel distance within the room, and from the room exit access door to the exit, see the appropriate occupancy chapter.

d. See the *International Building Code* for special requirements on spacing of doors in aircraft hangars.

e. In Group I-2, separation of exit access doors within a care recipient sleeping room, or any suite that includes care recipient sleeping rooms, shall comply with Section 1105.5.7.

f. In Group I-2, in smoke compartments containing care recipient sleeping rooms and treatment rooms, dead-end corridors shall comply with Section 1105.5.6.

g. In Group I-2 Condition 2, care recipient sleeping rooms, or any suite that includes care recipient sleeping rooms, shall comply with Section 1105.6.

h. Where a tenant space in Group B, S and U occupancies has an occupant load of not more than 30, the length of a common path of egress travel shall not be more than 100 feet.

1104.24 Stairway floor number signs. Existing *stairways* shall be marked in accordance with Section 1023.9.

1104.25 Egress path markings. Existing high-rise buildings of Group A, B, E, I, M and R-1 occupancies shall be provided with luminous *egress* path markings in accordance with Section 1025.

Exception: Open, unenclosed stairwells in historic buildings designated as historic under a state or local historic preservation program.

SECTION 1105 CONSTRUCTION REQUIREMENTS FOR EXISTING GROUP I-2

1105.1 General. Existing Group I-2 shall meet all of the following requirements:

1. The minimum fire safety requirements in Section 1103.
2. The minimum mean of egress requirements in Section 1104.
3. The additional egress and construction requirements in Section 1105.

Where the provisions of this chapter conflict with the construction requirements that applied at the time of construction, the most restrictive provision shall apply.

1105.2 Construction. Group I-2 Condition 2 shall not be located on a floor level higher than the floor level limitation in Table 1105.2 based on the type of construction.

1105.3 Incidental uses in existing Group I-2. Incidental uses associated with and located within existing single-occupancy or mixed-occupancy Group I-2 buildings and that generally pose a greater level of risk to such occupancies shall comply with the provisions of Sections 1105.3.1 through 1105.3.3.2.1. Incidental uses in Group I-2 occupancies are limited to those listed in Table 1105.3.

1105.3.1 Occupancy classification. Incidental uses shall not be individually classified in accordance with Section 302.1 of the *International Building Code*. Incidental uses shall be included in the building occupancies within which they are located.

1105.3.2 Area limitations. Incidental uses shall not occupy more than 10 percent of the building area of the story in which they are located.

1105.3.3 Separation and protection. The incidental uses listed in Table 1105.3 shall be separated from the remainder of the building or equipped with an *automatic sprinkler system*, or both, in accordance with the provisions of that table.

1105.3.3.1 Separation. Where Table 1105.3 specifies a fire-resistance-rated separation, the incidental uses shall be separated from the remainder of the building in accordance with Section 509.4.1 of the *International Building Code*.

1105.3.3.2 Protection. Where Table 1105.3 permits an *automatic sprinkler system* without a fire-resistance-

TABLE 1105.2
FLOOR LEVEL LIMITATIONS FOR GROUP I-2 CONDITION 2

CONSTRUCTION TYPE	AUTOMATIC SPRINKLER SYSTEM	ALLOWABLE FLOOR LEVEL[a]			
		1	2	3	4 or more
IA	Note b	P	P	P	P
	Note c	P	P	P	P
IB	Note b	P	P	P	P
	Note c	P	P	P	P
IIA	Note b	P	P	P	NP
	Note c	P	NP	NP	NP
IIB	Note b	P	P	NP	NP
	Note c	NP	NP	NP	NP
IIIA	Note b	P	P	NP	NP
	Note c	P	NP	NP	NP
IIIB	Note b	P	NP	NP	NP
	Note c	NP	NP	NP	NP
IV	Note b	P	P	NP	NP
	Note c	NP	NP	NP	NP
VA	Note b	P	P	NP	NP
	Note c	NP	NP	NP	NP
VB	Note b	P	NP	NP	NP
	Note c	NP	NP	NP	NP

P = Permitted; NP = Not permitted.

a. Floor level shall be counted based on the number of stories above grade.

b. The building is equipped throughout with an automatic sprinkler system in accordance with Section 903.3.1.1.

c. The building is equipped with an automatic sprinkler system in accordance with Section 1105.8.

rated separation, the incidental uses shall be separated from the remainder of the building by construction capable of resisting the passage of smoke in accordance with Section 509.4.2 of the *International Building Code*.

1105.3.3.2.1 Protection limitation. Except as otherwise specified in Table 1105.2 for certain incidental uses, where an *automatic sprinkler system* is provided in accordance with Table 1105.3, only the space occupied by the incidental use need be equipped with such a system.

1105.4 Corridor construction. In Group I-2, in areas housing patient sleeping or care rooms, *corridor* walls and the opening protectives therein shall provide a barrier designed to resist the passage of smoke in accordance with Sections 1105.4.1 through 1105.4.7.

1105.4.1 Materials. The walls shall be of materials permitted by the building type of construction.

1105.4.2 Fire-resistance rating. Unless required elsewhere in this code, corridor walls are not required to have a fire-resistance rating.

1105.4.3 Corridor wall continuity. *Corridor* walls shall extend from the top of the foundation or floor below to one of the following:

1. The underside of the floor or roof sheathing, deck or slab above.
2. The underside of a ceiling above where the ceiling membrane is constructed to limit the passage of smoke.
3. The underside of a lay-in ceiling system where the ceiling system is constructed to limit the passage of smoke and where the ceiling tiles weigh not less than 1 pound per square foot (4.88 kg/m^2)of tile.

1105.4.4 Openings in corridor walls. Openings in *corridor* walls shall provide protection in accordance with 1105.4.4.1 through 1105.4.4.3.

1105.4.4.1 Windows. Windows in *corridor* walls shall be sealed to limit the passage of smoke, or the window shall be automatic-closing upon detection of smoke, or the window opening shall be protected by an automatic closing device that closes upon detection of smoke.

Exception: In smoke compartments not containing patient sleeping rooms, pass-through windows or similar openings shall be permitted in accordance with Section 1105.4.4.3.

1105.4.4.2 Doors. Doors in *corridor* walls shall comply with Sections 1105.4.4.2.1 through 1105.4.4.2.3.

1105.4.4.2.1 Louvers. Doors in *corridor* walls shall not include louvers, transfer grills or similar openings.

Exception: Doors shall be permitted to have louvers, transfer grills or similar openings at toilet rooms or bathrooms; storage rooms that do not contain storage of flammable or combustible material; and storage rooms that are not required to be separated as incidental uses.

TABLE 1105.3
INCIDENTAL USES IN EXISTING GROUP I-2 OCCUPANCIES

ROOM OR AREA	SEPARATION AND/OR PROTECTION
Furnace room where any piece of equipment is over 400,000 Btu per hour input	1 hour or provide automatic sprinkler system
Rooms with boilers where the largest piece of equipment is over 15 psi and 10 horsepower	1 hour or provide automatic sprinkler system
Refrigerant machinery room	1 hour or provide automatic sprinkler system
Hydrogen fuel gas rooms, not classified as Group H	2 hours
Incinerator rooms	2 hours and provide automatic sprinkler system
Paint shops not classified as Group H	2 hours; or 1 hour and provide automatic sprinkler system
Laboratories and vocational shops, not classified as Group H	1 hour or provide automatic sprinkler system
Laundry rooms over 100 square feet	1 hour or provide automatic sprinkler system
Patient rooms equipped with padded surfaces	1 hour or provide automatic sprinkler system
Physical plant maintenance shops	1 hour or provide automatic sprinkler system
Waste and linen collection rooms with containers with total volume of 10 cubic feet or greater	1 hour or provide automatic sprinkler system
Storage rooms greater than 100 square feet	1 hour or provide automatic sprinkler system
Stationary storage battery systems having a liquid electrolyte capacity of more than 50 gallons for flooded lead-acid, nickel cadmium or VRLA, or more than 1,000 pounds for lithium-ion and lithium metal polymer used for facility standby power, emergency power or uninterruptable power supplies	2 hours

For SI: 1 square foot = 0.0929 m^2, 1 pound per square inch (psi) = 6.9 kPa, 1 British thermal unit (Btu) per hour = 0.293 watts, 1 horsepower = 746 watts, 1 gallon = 3.785 L.

1105.4.4.2.2 Corridor doors. Doors in *corridor* walls shall limit the transfer of smoke by complying with the following:

1. Doors shall be constructed of not less than $1^3/_4$ inch-thick (44 mm) solid bonded-core wood or capable of resisting fire not less than $^1/_3$ hour.

   Exception: Corridor doors in buildings equipped throughout with an automatic sprinkler system.

2. Frames for side-hinged swinging doors shall have stops on the sides and top to limit transfer of smoke.
3. Where provided, vision panels in doors shall be a fixed glass window assembly installed to limit the passage of smoke. Existing wired glass panels with steel frames shall be permitted to remain in place.
4. Door undercuts shall not exceed 1 inch (25 mm).
5. Doors shall be positive latching with devices that resist not less than 5 pounds (22.2 N). Roller latches are prohibited.
6. Mail slots or similar openings shall be permitted in accordance with Section 1105.4.4.3.

1105.4.4.2.3 Dutch doors. Where provided, dutch doors shall comply with Section 1105.4.4.2.2. In addition, dutch doors shall be equipped with latching devices on either the top or bottom leaf to allow leaves to latch together. The space between the leaves shall be protected with devices such as astragals to limit the passage of smoke.

1105.4.4.2.4 Self- or automatic-closing doors. Where self- or automatic-closing doors are required, closers shall be maintained in operational condition.

1105.4.4.3 Openings in corridor walls and doors. In other than smoke compartments containing patient sleeping rooms, mail slots, pass-through windows or similar openings shall not be required to be protected where the aggregate area of the openings between the *corridor* and a room are not greater than 80 square inches (51 613 mm^2) and are located with the top edge of any opening not higher than 48 inches above the floor.

1105.4.5 Penetrations. The space around penetrating items shall be filled with an *approved* material to limit the passage of smoke.

1105.4.6 Joints. Joints shall be filled with an *approved* material to limit the passage of smoke.

1105.4.7 Ducts and air transfer openings. The space around a duct penetrating a smoke partition shall be filled with an *approved* material to limit the passage of smoke. Air transfer openings in smoke partitions shall be provided with a smoke damper complying with Section 717.3.2.2 of the *International Building Code*.

Exception: Where the installation of a smoke damper will interfere with the operation of a required smoke control system in accordance with Section 909, approved alternative protection shall be utilized.

1105.5 Means of egress. In addition to the *means of egress* requirements in Section 1104, Group I-2 facilities shall meet the *means of egress* requirements in Section 1105.5.1 through 1105.5.8.

1105.5.1 Exit signs and emergency illumination. The power system for exit signs and emergency illumination for the *means of egress* shall provide power for not less than 90 minutes and consist of storage batteries, unit equipment or an on-site generator.

1105.5.2 Emergency power for operational needs. The essential electrical system shall be capable of supplying services in accordance with NFPA 99.

1105.5.3 Size of door. Means of egress doors used for the movement of patients in beds shall provide a minimum clear width of $41^1/_2$ inches (1054 mm). The height of the door opening shall be not less than 80 inches (2032 mm).

Exceptions:

1. Door closers and door stops shall be permitted to be 78 inches (1981 mm) minimum above the floor.
2. In Group I-2 Condition 1, existing means of egress doors used for the movement of patients in beds that provide a minimum clear width of 32 inches (813 mm) shall be permitted to remain.

1105.5.4 Ramps. In areas where *ramps* are used for movement of patients in beds, the clear width of the *ramp* shall be not less than 48 inches (1219 mm).

1105.5.5 Corridor width. In areas where *corridors* are used for movement of patients in beds, the clear width of the *corridor* shall be not less than 48 inches (1219 mm).

1105.5.6 Dead-end corridors. In smoke compartments containing patient sleeping rooms and treatment rooms, dead-end *corridors* shall not exceed 30 feet (9144 mm) unless approved by the *fire code official*.

1105.5.7 Separation of exit access doors. Patient sleeping rooms, or any suite that includes patient sleeping rooms, of more than 1,000 square feet (92.9 m^2) shall have not less than two exit access doors placed a distance apart equal to not less than one-third of the length of the maximum overall diagonal dimension of the patient sleeping room or suite to be served, measured in a straight line between exit access doors.

1105.5.8 Aisles. In areas where *aisles* are used for movement of patients in beds, the clear width of the *aisle* shall be not less than 48 inches (1219 mm).

1105.6 Smoke compartments. Smoke compartments shall be provided in existing Group I-2 Condition 2, in accordance with Sections 1105.6.1 through 1105.6.4.

1105.6.1 Design. Smoke barriers shall be provided to subdivide each story used for patients sleeping with an occupant load of more than 30 patients into not fewer than two smoke compartments.

1105.6.1.1 Refuge areas. Refuge areas shall be provided within each smoke compartment. The size of the refuge area shall accommodate the occupants and care recipients from the adjoining smoke compartment. Where a smoke compartment is adjoined by two or more smoke compartments, the minimum area of the refuge area shall accommodate the largest occupant load of the adjoining compartments.

The size of the refuge area shall provide the following:

1. Not less than 30 net square feet (2.8 m^2) for each care recipient confined to a bed or stretcher.
2. Not less than 15 square feet (1.4 m^2) for each resident in a Group I-2 using mobility assistance devices.
3. Not less than 6 square feet (0.56 m^2) for each occupant not addressed in Items 1 and 2.

Areas of spaces permitted to be included in the calculation of the refuge area are *corridors*, sleeping areas, treatment rooms, lounge or dining areas and other low-hazard areas.

1105.6.2 Smoke barriers. Smoke barriers shall be constructed in accordance with Section 709 of the *International Building Code*.

Exceptions:

1. Existing smoke barriers are permitted to remain where the existing smoke barrier has a minimum fire-resistance rating of $^1/_2$ hour.
2. Smoke barriers shall be permitted to terminate at an atrium enclosure in accordance with Section 404.6 of the *International Building Code*.

1105.6.3 Opening protectives. Openings in smoke barriers shall be protected in accordance with Section 716 of the *International Building Code*. Opening protectives shall have a minimum fire-protection-rating of $^1/_3$ hour.

Exception: Existing wired glass vision panels in doors shall be permitted to remain.

1105.6.4 Penetrations. Penetrations of smoke barriers shall comply with the *International Building Code*.

Exception: Approved existing materials and methods of construction.

1105.6.5 Joints. Joints made in or between smoke barriers shall comply with the *International Building Code*.

Exception: Approved existing materials and methods of construction.

1105.6.6 Duct and air transfer openings. Penetrations in a smoke barrier by duct and air transfer openings shall comply with Section 717 of the *International Building Code*.

Exception: Where existing duct and air transfer openings in smoke barriers exist without smoke dampers, they shall be permitted to remain. Any changes to existing smoke dampers shall be submitted for review and approved in accordance with Section 717 of the *International Building Code*.

1105.7 Group I-2 care suites. Care suites in existing Group I-2 Condition 2 occupancies shall comply with Sections 407.4.3 through 407.4.3.6.2 of the *International Building Code*.

1105.8 Group I-2 automatic sprinkler system. An *automatic sprinkler system* installed in accordance with Section 903.3.1.1 shall be provided throughout existing Group I-2 fire areas. The sprinkler system shall be provided throughout the floor where the Group I-2 occupancy is located, and in all floors between the Group I-2 occupancy and the *level of exit discharge*.

1105.9 Group I-2 automatic fire alarm system. An automatic fire alarm system shall be installed in existing Group I-2 occupancies in accordance with Section 907.2.6.2.

Exception: Manual fire alarm boxes in patient sleeping areas shall not be required at *exits* if located at all nurses' control stations or other constantly attended staff locations, provided such stations are visible and continuously accessible and that travel distances required in Section 907.5.2.1 are not exceeded.

1105.10 Essential electrical systems. Essential electrical systems in Group I-2 Condition 2 occupancies shall be in accordance with Sections 1105.10.1 and 1105.10.2.

1105.10.1 Where required. In Group I-2 Condition 2 occupancies where life support is being provided, an essential electrical system shall be provided in accordance with NFPA 99.

1105.10.2 Installation and duration. In Group I-2 Condition 2 occupancies, the installation and duration of operation of existing essential electrical systems shall be based on a hazard vulnerability analysis conducted in accordance with NFPA 99.

SECTION 1106 REQUIREMENTS FOR OUTDOOR OPERATIONS

1106.1 Tire storage yards. Existing tire storage yards shall be provided with fire apparatus access roads in accordance with Sections 1106.1.1 and 1106.1.2.

1106.1.1 Access to piles. Access roadways shall be within 150 feet (45 720 mm) of any point in the storage yard where storage piles are located not less than 20 feet (6096 mm) from any storage pile.

1106.1.2 Location within piles. Fire apparatus access roads shall be located within all pile clearances identified in Section 3405.4 and within all fire breaks required in Section 3405.5.

CHAPTERS 12 through 19

RESERVED

Part IV—Special Occupancies and Operations

CHAPTER 20

AVIATION FACILITIES

SECTION 2001 GENERAL

2001.1 Scope. Airports, heliports, helistops and aircraft hangars shall be in accordance with this chapter.

2001.2 Regulations not covered. Regulations not specifically contained herein pertaining to airports, aircraft maintenance, aircraft hangars and appurtenant operations shall be in accordance with nationally recognized standards.

2001.3 Permits. For permits to operate aircraft-refueling vehicles, application of flammable or combustible finishes and hot work, see Section 105.6.

SECTION 2002 DEFINITIONS

2002.1 Definitions. The following terms are defined in Chapter 2:

AIRCRAFT OPERATION AREA (AOA).

AIRPORT.

HELIPORT.

HELISTOP.

SECTION 2003 GENERAL PRECAUTIONS

2003.1 Sources of ignition. Open flames, flame-producing devices and other sources of ignition shall not be permitted in a hangar, except in *approved* locations or in any location within 50 feet (15 240 mm) of an aircraft-fueling operation.

2003.2 Smoking. Smoking shall be prohibited in aircraft-refueling vehicles, aircraft hangars and aircraft operation areas used for cleaning, paint removal, painting operations or fueling. "No Smoking" signs shall be provided in accordance with Section 310.

Exception: Designated and *approved* smoking areas.

2003.3 Housekeeping. The aircraft operation area (AOA) and related areas shall be kept free from combustible debris at all times.

2003.4 Fire department access. Fire apparatus access roads shall be provided and maintained in accordance with Chapter 5. Fire apparatus access roads and aircraft parking positions shall be designed in a manner so as to preclude the possibility of fire vehicles traveling under any portion of a parked aircraft.

2003.5 Dispensing of flammable and combustible liquids. The dispensing, transferring and storage of flammable and *combustible liquids* shall be in accordance with this chapter and Chapter 57. Aircraft motor vehicle fuel-dispensing facilities shall be in accordance with Chapter 23.

2003.6 Combustible storage. Combustible materials stored in aircraft hangars shall be stored in *approved* locations and containers.

2003.7 Hazardous material storage. Hazardous materials shall be stored in accordance with Chapter 50.

SECTION 2004 AIRCRAFT MAINTENANCE

2004.1 Transferring flammable and combustible liquids. Flammable and *combustible liquids* shall not be dispensed into or removed from a container, tank, vehicle or aircraft except in *approved* locations.

2004.2 Application of flammable and combustible liquid finishes. The application of flammable or Class II *combustible liquid* finishes is prohibited unless both of the following conditions are met:

1. The application of the liquid finish is accomplished in an *approved* location.
2. The application methods and procedures are in accordance with Chapter 24.

2004.3 Cleaning parts. Class IA flammable liquids shall not be used to clean aircraft, aircraft parts or aircraft engines. Cleaning with other flammable and *combustible liquids* shall be in accordance with Section 5705.3.6.

2004.4 Spills. Sections 2004.4.1 through 2004.4.3 shall apply to spills of flammable and *combustible liquids* and other hazardous materials. Fuel spill control shall also comply with Section 2006.11.

2004.4.1 Cessation of work. Activities in the affected area not related to the mitigation of the spill shall cease until the spilled material has been removed or the hazard has been mitigated.

2004.4.2 Vehicle movement. Aircraft or other vehicles shall not be moved through the spill area until the spilled material has been removed or the hazard has been mitigated.

2004.4.3 Mitigation. Spills shall be reported, documented and mitigated in accordance with the provisions of this chapter and Section 5003.3.

2004.5 Running engines. Aircraft engines shall not be run in aircraft hangars except in *approved* engine test areas.

2004.6 Open flame. Repairing of aircraft requiring the use of open flames, spark-producing devices or the heating of parts above 500°F (260°C) shall only be done outdoors or in an area complying with the provisions of the *International Building Code* for a Group F-1 occupancy.

2004.7 Other aircraft maintenance. Maintenance, repairs, modifications, or construction performed upon aircraft not addressed elsewhere in this code shall be conducted in accordance with NFPA 410.

SECTION 2005 PORTABLE FIRE EXTINGUISHERS

2005.1 General. Portable fire extinguishers suitable for flammable or *combustible liquid* and electrical-type fires shall be provided as specified in Sections 2005.2 through 2005.6 and Section 906. Extinguishers required by this section shall be inspected and maintained in accordance with Section 906.

2005.2 On towing vehicles. Vehicles used for towing aircraft shall be equipped with not less than one *listed* portable fire extinguisher complying with Section 906 and having a minimum rating of 20-B:C.

2005.3 On welding apparatus. Welding apparatus shall be equipped with not less than one *listed* portable fire extinguisher complying with Section 906 and having a minimum rating of 2-A:20-B:C.

2005.4 On aircraft fuel-servicing tank vehicles. Aircraft fuel-servicing tank vehicles shall be equipped with not less than two *listed* portable fire extinguishers complying with Section 906, each having a minimum rating of 20-B:C. A portable fire extinguisher shall be readily accessible from either side of the vehicle.

2005.5 On hydrant fuel-servicing vehicles. Hydrant fuel-servicing vehicles shall be equipped with not less than one *listed* portable fire extinguisher complying with Section 906, and having a minimum rating of 20-B:C.

2005.6 At fuel-dispensing stations. Portable fire extinguishers at fuel-dispensing stations shall be located such that pumps or dispensers are not more than 75 feet (22 860 mm) from one such extinguisher. Fire extinguishers shall be provided as follows:

1. Where the open-hose discharge capacity of the fueling system is not more than 200 gallons per minute (13 L/s), not less than two *listed* portable fire extinguishers complying with Section 906 and having a minimum rating of 20-B:C shall be provided.
2. Where the open-hose discharge capacity of the fueling system is more than 200 gallons per minute (13 L/s) but not more than 350 gallons per minute (22 L/s), not less than one *listed* wheeled extinguisher complying with Section 906 and having a minimum extinguishing rating of 80-B:C, and a minimum agent capacity of 125 pounds (57 kg), shall be provided.
3. Where the open-hose discharge capacity of the fueling system is more than 350 gallons per minute (22 L/s), not less than two *listed* wheeled extinguishers complying with Section 906 and having a minimum rating of 80-B:C each, and a minimum capacity agent of 125 pounds (57 kg) of each, shall be provided.

2005.7 Fire extinguisher access. Portable fire extinguishers required by this chapter shall be accessible at all times. Where necessary, provisions shall be made to clear accumulations of snow, ice and other forms of weather-induced obstructions.

2005.7.1 Cabinets. Cabinets and enclosed compartments used to house portable fire extinguishers shall be clearly marked with the words FIRE EXTINGUISHER in letters not less than 2 inches (51 mm) high. Cabinets and compartments shall be readily accessible at all times.

2005.8 Reporting use. Use of a fire extinguisher under any circumstances shall be reported to the manager of the airport and the *fire code official* immediately after use.

SECTION 2006 AIRCRAFT FUELING

2006.1 Aircraft motor vehicle fuel-dispensing facilities. Aircraft motor vehicle fuel-dispensing facilities shall be in accordance with Chapter 23.

2006.2 Airport fuel systems. Airport fuel systems shall be designed and constructed in accordance with NFPA 407.

2006.3 Construction of aircraft-fueling vehicles and accessories. Aircraft-fueling vehicles shall comply with this section and shall be designed and constructed in accordance with NFPA 407.

2006.3.1 Transfer apparatus. Aircraft-fueling vehicles shall be equipped and maintained with an *approved* transfer apparatus.

2006.3.1.1 Internal combustion type. Where such transfer apparatus is operated by an individual unit of the internal-combustion-motor type, such power unit shall be located as remotely as practicable from pumps, piping, meters, air eliminators, water separators, hose reels and similar equipment, and shall be housed in a separate compartment from any of the aforementioned items. The fuel tank in connection therewith shall be suitably designed and installed, and the maximum fuel capacity shall not exceed 5 gallons (19 L) where the tank is installed on the engine. The exhaust pipe, muffler and tail pipe shall be shielded.

2006.3.1.2 Gear operated. Where operated by gears or chains, the gears, chains, shafts, bearings, housing and all parts thereof shall be of an *approved* design and shall be installed and maintained in an *approved* manner.

2006.3.1.3 Vibration isolation. Flexible connections for the purpose of eliminating vibration are allowed if the material used therein is designed, installed and maintained in an *approved* manner, provided such connections do not exceed 24 inches (610 mm) in length.

2006.3.2 Pumps. Pumps of a positive-displacement type shall be provided with a bypass relief valve set at a pressure of not more than 35 percent in excess of the normal working pressure of such unit. Such units shall be equipped and maintained with a pressure gauge on the discharge side of the pump.

2006.3.3 Dispensing hoses and nozzles. Hoses shall be designed for the transferring of hydrocarbon liquids and shall not be any longer than necessary to provide efficient fuel transfer operations. Hoses shall be equipped with an *approved* shutoff nozzle. Fuel-transfer nozzles shall be self-closing and designed to be actuated by hand pressure only. Notches and other devices shall not be used for holding a nozzle valve handle in the open position. Nozzles shall be equipped with a bonding cable complete with proper attachment for aircraft to be serviced.

2006.3.4 Protection of electrical equipment. Electric wiring, switches, lights and other sources of ignition, where located in a compartment housing piping, pumps, air eliminators, water separators, hose reels or similar equipment, shall be enclosed in a vapor-tight housing. Electrical motors located in such a compartment shall be of a type *approved* for use as specified in NFPA 70.

2006.3.5 Venting of equipment compartments. Compartments housing piping, pumps, air eliminators, water separators, hose reels and similar equipment shall be adequately ventilated at floor level or within the floor itself.

2006.3.6 Accessory equipment. Ladders, hose reels and similar accessory equipment shall be of an *approved* type and constructed substantially as follows:

1. Ladders constructed of noncombustible material are allowed to be used with or attached to aircraft-fueling vehicles, provided the manner of attachment or use of such ladders is *approved* and does not constitute an additional fire or accident hazard in the operation of such fueling vehicles.
2. Hose reels used in connection with fueling vehicles shall be constructed of noncombustible materials and shall be provided with a packing gland or other device that will preclude fuel leakage between reels and fuel manifolds.

2006.3.7 Electrical bonding provisions. Transfer apparatus shall be metallically interconnected with tanks, chassis, axles and springs of aircraft-fueling vehicles.

2006.3.7.1 Bonding cables. Aircraft-fueling vehicles shall be provided and maintained with a substantial heavy-duty electrical cable of sufficient length to be bonded to the aircraft to be serviced. Such cable shall be metallically connected to the transfer apparatus or chassis of the aircraft-fueling vehicle on one end and shall be provided with a suitable metal clamp on the other end, to be fixed to the aircraft.

2006.3.7.2 Bonding cable protection. The bonding cable shall be bare or have a transparent protective sleeve and be stored on a reel or in a compartment provided for no other purpose. It shall be carried in such a manner that it will not be subjected to sharp kinks or accidental breakage under conditions of general use.

2006.3.8 Smoking. Smoking in aircraft-fueling vehicles is prohibited. Signs to this effect shall be conspicuously posted in the driver's compartment of all fueling vehicles.

2006.3.9 Smoking equipment. Smoking equipment such as cigarette lighters and ash trays shall not be provided in aircraft-fueling vehicles.

2006.4 Operation, maintenance and use of aircraft-fueling vehicles. The operation, maintenance and use of aircraft-fueling vehicles shall be in accordance with Sections 2006.4.1 through 2006.4.4 and other applicable provisions of this chapter.

2006.4.1 Proper maintenance. Aircraft-fueling vehicles and all related equipment shall be properly maintained and kept in good repair. Accumulations of oil, grease, fuel and other flammable or combustible materials is prohibited. Maintenance and servicing of such equipment shall be accomplished in *approved* areas.

2006.4.2 Vehicle integrity. Tanks, pipes, hoses, valves and other fuel delivery equipment shall be maintained leak free at all times.

2006.4.3 Removal from service. Aircraft-fueling vehicles and related equipment that are in violation of Section 2006.4.1 or 2006.4.2 shall be immediately defueled and removed from service and shall not be returned to service until proper repairs have been made.

2006.4.4 Operators. Aircraft-fueling vehicles that are operated by a person, firm or corporation other than the permittee or the permittee's authorized employee shall be provided with a legible sign visible from outside the vehicle showing the name of the person, firm or corporation operating such unit.

2006.5 Fueling and defueling. Aircraft-fueling and defueling operations shall be in accordance with Sections 2006.5.1 through 2006.5.5.

2006.5.1 Positioning of aircraft-fueling vehicles. Aircraft-fueling vehicles shall not be located, parked or permitted to stand in a position where such unit would obstruct egress from an aircraft should a fire occur during fuel-transfer operations. Aircraft-fueling vehicles shall not be located, parked or permitted to stand under any portion of an aircraft.

Exception: Aircraft-fueling vehicles shall be allowed to be located under aircraft wings during underwing fueling of turbine-engine powered aircraft.

2006.5.1.1 Fueling vehicle egress. A clear path shall be maintained for aircraft-fueling vehicles to provide for prompt and timely egress from the fueling area.

2006.5.1.2 Aircraft vent openings. A clear space of not less than 10 feet (3048 mm) shall be maintained between aircraft fuel-system vent openings and any part or portion of an aircraft-fueling vehicle.

2006.5.1.3 Parking. Prior to leaving the cab, the aircraft-fueling vehicle operator shall ensure that the parking brake has been set. Not less than two chock blocks not less than 5 inches by 5 inches by 12 inches (127 mm by 127 mm by 305 mm) in size and dished to fit the contour of the tires shall be utilized and positioned in such a manner as to preclude movement of the vehicle in any direction.

2006.5.2 Electrical bonding. Aircraft-fueling vehicles shall be electrically bonded to the aircraft being fueled or defueled. Bonding connections shall be made prior to making fueling connections and shall not be disconnected until the fuel-transfer operations are completed and the fueling connections have been removed.

Where a hydrant service vehicle or cart is used for fueling, the hydrant coupler shall be connected to the hydrant system prior to bonding the fueling equipment to the aircraft.

2006.5.2.1 Conductive hose. In addition to the bonding cable required by Section 2006.5.2, conductive hose shall be used for all fueling operations.

2006.5.2.2 Bonding conductors on transfer nozzles. Transfer nozzles shall be equipped with *approved* bonding conductors that shall be clipped or otherwise positively engaged with the bonding attachment provided on the aircraft adjacent to the fuel tank cap prior to removal of the cap.

Exception: In the case of overwing fueling where no appropriate bonding attachment adjacent to the fuel fill port has been provided on the aircraft, the fueling operator shall touch the fuel tank cap with the nozzle spout prior to removal of the cap. The nozzle shall be kept in contact with the fill port until fueling is completed.

2006.5.2.3 Funnels. Where required, metal funnels are allowed to be used during fueling operations. Direct contact between the fueling receptacle, the funnel and the fueling nozzle shall be maintained during the fueling operation.

2006.5.3 Training. Aircraft-fueling vehicles shall be attended and operated only by persons instructed in methods of proper use and operation and who are qualified to use such fueling vehicles in accordance with minimum safety requirements.

2006.5.3.1 Fueling hazards. Fuel-servicing personnel shall know and understand the hazards associated with each type of fuel dispensed by the airport fueling-system operator.

2006.5.3.2 Fire safety training. Employees of fuel agents who fuel aircraft, accept fuel shipments or otherwise handle fuel shall receive *approved* fire safety training.

2006.5.3.2.1 Fire extinguisher training. Fuel-servicing personnel shall receive *approved* training in the operation of fire-extinguishing equipment.

2006.5.3.2.2 Records. The airport fueling-system operator shall maintain records of all training administered to its employees.

2006.5.4 Transfer personnel. During fuel-transfer operations, a qualified person shall be in control of each transfer nozzle and another qualified person shall be in immediate control of the fuel-pumping equipment to shut off or otherwise control the flow of fuel from the time fueling operations are begun until they are completed.

Exceptions:

1. For underwing refueling, the person stationed at the point of fuel intake is not required.
2. For overwing refueling, the person stationed at the fuel pumping equipment shall not be required where the person at the fuel dispensing device is within 75 feet (22 800 mm) of the emergency shutoff device; is not on the wing of the aircraft and has a clear and unencumbered path to the fuel pumping equipment; and the fuel dispensing line does not exceed 50 feet (15 240 mm) in length.

The fueling operator shall monitor the panel of the fueling equipment and the aircraft control panel during pressure fueling or shall monitor the fill port during overwing fueling.

2006.5.5 Fuel flow control. Fuel flow-control valves shall be operable only by the direct hand pressure of the operator. Removal of the operator's hand pressure shall cause an immediate cessation of the flow of fuel.

2006.6 Emergency fuel shutoff. Emergency fuel shutoff controls and procedures shall comply with Sections 2006.6.1 through 2006.6.4.

2006.6.1 Accessibility. Emergency fuel shutoff controls shall be readily accessible at all times when the fueling system is being operated.

2006.6.2 Notification of the fire department. The fueling-system operator shall establish a procedure by which the fire department will be notified in the event of an activation of an emergency fuel shutoff control.

2006.6.3 Determining cause. Prior to reestablishment of normal fuel flow, the cause of fuel shutoff conditions shall be determined and corrected.

2006.6.4 Testing. Emergency fuel shutoff devices shall be operationally tested at intervals not exceeding three months. The fueling-system operator shall maintain testing records.

2006.7 Protection of hoses. Before an aircraft-fueling vehicle is moved, fuel transfer hoses shall be properly placed on the *approved* reel or in the compartment provided, or stored on the top decking of the fueling vehicle if proper height rail is provided for security and protection of such equipment. Fuel-transfer hose shall not be looped or draped over any part of the fueling vehicle, except as herein provided. Fuel-transfer hose shall not be dragged when such fueling vehicle is moved from one fueling position to another.

2006.8 Loading and unloading. Aircraft-fueling vehicles shall be loaded only at an *approved* loading rack. Such loading racks shall be in accordance with Section 5706.5.1.12.

Exceptions:

1. Aircraft-refueling units are allowed to be loaded from the fuel tanks of an aircraft during defueling operations.
2. Fuel transfer between tank vehicles is allowed to be performed in accordance with Section 5706.6 when the operation is not less than 200 feet (60 960 mm) from an aircraft.

The fuel cargo of such units shall be unloaded only by *approved* transfer apparatus into the fuel tanks of aircraft, underground storage tanks or *approved* gravity storage tanks.

2006.9 Passengers. Passenger traffic is allowed during the time fuel transfer operations are in progress, provided the following provisions are strictly enforced by the *owner* of the aircraft or the *owner*'s authorized employee:

1. Smoking and producing an open flame in the cabin of the aircraft or the outside thereof within 50 feet (15 240 mm) of such aircraft shall be prohibited.

   A qualified employee of the aircraft *owner* shall be responsible for seeing that the passengers are not allowed to smoke when remaining aboard the aircraft or while going across the ramp from the gate to such aircraft, or vice versa.
2. Passengers shall not be permitted to linger about the plane, but shall proceed directly between the loading gate and the aircraft.
3. Passenger loading stands or walkways shall be left in loading position until all fuel transfer operations are completed.
4. Fuel transfer operations shall not be performed on the main *exit* side of any aircraft containing passengers except when the *owner* of such aircraft or a capable and qualified employee of such *owner* remains inside the aircraft to direct and assist the escape of such passengers through regular and emergency *exits* in the event fire should occur during fuel transfer operations.

2006.10 Sources of ignition. Smoking and producing open flames within 50 feet (15 240 mm) of a point where fuel is being transferred shall be prohibited. Electrical and motor-driven devices shall not be connected to or disconnected from an aircraft at any time fueling operations are in progress on such aircraft.

2006.11 Fuel spill prevention and procedures. Fuel spill prevention and the procedures for handling spills shall comply with Sections 2006.11.1 through 2006.11.7.

2006.11.1 Fuel-service equipment maintenance. Aircraft fuel-servicing equipment shall be maintained and kept free from leaks. Fuel-servicing equipment that malfunctions or leaks shall not be continued in service.

2006.11.2 Transporting fuel nozzles. Fuel nozzles shall be carried utilizing appropriate handles. Dragging fuel nozzles along the ground shall be prohibited.

2006.11.3 Drum fueling. Fueling from drums or other containers having a capacity greater than 5 gallons (19 L) shall be accomplished with the use of an *approved* pump.

2006.11.4 Fuel spill procedures. The fueling-system operator shall establish procedures to follow in the event of a fuel spill. These procedures shall be comprehensive and shall provide for all of the following:

1. Upon observation of a fuel spill, the aircraft-fueling operator shall immediately stop the delivery of fuel by releasing hand pressure from the fuel flow-control valve.
2. Failure of the fuel control valve to stop the continued spillage of fuel shall be cause for the activation of the appropriate emergency fuel shutoff device.
3. A supervisor for the fueling-system operator shall respond to the fuel spill area immediately.

2006.11.5 Notification of the fire department. The fire department shall be notified of any fuel spill that is considered a hazard to people or property or which meets one or more of the following criteria:

1. Any dimension of the spill is greater than 10 feet (3048 mm).
2. The spill area is greater than 50 square feet (4.65 m^2).
3. The fuel flow is continuous in nature.

2006.11.6 Investigation required. An investigation shall be conducted by the fueling-system operator of all spills requiring notification of the fire department. The investigation shall provide conclusive proof of the cause and verification of the appropriate use of emergency procedures. Where it is determined that corrective measures are necessary to prevent future incidents of the same nature, they shall be implemented immediately.

2006.11.7 Multiple fuel delivery vehicles. Simultaneous delivery of fuel from more than one aircraft-fueling vehicle to a single aircraft-fueling manifold is prohibited unless proper backflow prevention devices are installed to prevent fuel flow into the tank vehicles.

2006.12 Aircraft engines and heaters. Operation of aircraft onboard engines and combustion heaters shall be terminated prior to commencing fuel service operations and shall remain off until the fuel-servicing operation is completed.

Exception: In an emergency, a single jet engine is allowed to be operated during fuel servicing where all of the following conditions are met:

1. The emergency shall have resulted from an onboard failure of the aircraft's auxiliary power unit.
2. Restoration of auxiliary power to the aircraft by ground support services is not available.

3. The engine to be operated is either at the rear of the aircraft or on the opposite side of the aircraft from the fuel service operation.
4. The emergency operation is in accordance with a written procedure *approved* by the *fire code official.*

2006.13 Vehicle and equipment restrictions. During aircraft-fueling operations, only the equipment actively involved in the fueling operation is allowed within 50 feet (15 240 mm) of the aircraft being fueled. Other equipment shall be prohibited in this area until the fueling operation is complete.

Exception: Aircraft-fueling operations utilizing single-point refueling with a sealed, mechanically locked fuel line connection and the fuel is not a Class I flammable liquid.

A clear space of not less than 10 feet (3048 mm) shall be maintained between aircraft fuel-system vent openings and any part or portion of aircraft-servicing vehicles or equipment.

2006.13.1 Overwing fueling. Vehicles or equipment shall not be allowed beneath the trailing edge of the wing when aircraft fueling takes place over the wing and the aircraft fuel-system vents are located on the upper surface of the wing.

2006.14 Electrical equipment. Electrical equipment, including but not limited to, battery chargers, ground or auxiliary power units, fans, compressors or tools, shall not be operated, nor shall they be connected or disconnected from their power source, during fuel service operations.

2006.14.1 Other equipment. Electrical or other spark-producing equipment shall not be used within 10 feet (3048 mm) of fueling equipment, aircraft fill or vent points, or spill areas unless that equipment is intrinsically safe and *approved* for use in an explosive atmosphere.

2006.15 Open flames. Open flames and open-flame devices are prohibited within 50 feet (15 240 mm) of any aircraft fuel-servicing operation or fueling equipment.

2006.15.1 Other areas. The *fire code official* is authorized to establish other locations where open flames and open-flame devices are prohibited.

2006.15.2 Matches and lighters. Personnel assigned to and engaged in fuel-servicing operations shall not carry matches or lighters on or about their person. Matches or lighters shall be prohibited in, on or about aircraft-fueling equipment.

2006.16 Lightning procedures. The *fire code official* is authorized to require the airport authority and the fueling-system operator to establish written procedures to follow when lightning flashes are detected on or near the airport. These procedures shall establish criteria for the suspension and resumption of aircraft-fueling operations.

2006.17 Fuel-transfer locations. Aircraft fuel-transfer operations shall be prohibited indoors.

Exception: In aircraft hangars built in accordance with the provisions of the *International Building Code* for Group F-1 occupancies, aircraft fuel-transfer operations are allowed where either of the following conditions exist:

1. Necessary to accomplish aircraft fuel-system maintenance operations. Such operations shall be performed in accordance with nationally recognized standards.
2. The fuel being used has a *flash point* greater than 100°F (37.8°C).

2006.17.1 Position of aircraft. Aircraft being fueled shall be positioned such that any fuel system vents and other fuel tank openings are not less than:

1. Twenty-five feet (7620 mm) from buildings or structures other than jet bridges; and
2. Fifty feet (15 240 mm) from air intake vents for boiler, heater or incinerator rooms.

2006.17.2 Fire equipment access. Access for fire service equipment to aircraft shall be maintained during fuel-servicing operations.

2006.18 Defueling operations. The requirements for fueling operations contained in this section shall also apply to aircraft defueling operations. Additional procedures shall be established by the fueling-system operator to prevent overfilling of the tank vehicle used in the defueling operation.

2006.19 Maintenance of aircraft-fueling hose. Aircraft-fueling hoses shall be maintained in accordance with Sections 2006.19.1 through 2006.19.4.

2006.19.1 Inspections. Hoses used to fuel or defuel aircraft shall be inspected periodically to ensure their serviceability and suitability for continued service. The fuel-service operator shall maintain records of all tests and inspections performed on fueling hoses. Hoses found to be defective or otherwise damaged shall be immediately removed from service.

2006.19.1.1 Daily inspection. Each hose shall be inspected daily. This inspection shall include a complete visual scan of the exterior for evidence of damage, blistering or leakage. Each coupling shall be inspected for evidence of leaks, slippage or misalignment.

2006.19.1.2 Monthly inspection. A more thorough inspection, including pressure testing, shall be accomplished for each hose on a monthly basis. This inspection shall include examination of the fuel delivery inlet screen for rubber particles, which indicates problems with the hose lining.

2006.19.2 Damaged hose. Hose that has been subjected to severe abuse shall be immediately removed from service. Such hoses shall be hydrostatically tested prior to being returned to service.

2006.19.3 Repairing hose. Hoses are allowed to be repaired by removing the damaged portion and recoupling the undamaged end. When recoupling hoses, only couplings designed and *approved* for the size and type of hose in question shall be used. Hoses repaired in this manner shall be visually inspected and hydrostatically tested prior to being placed back in service.

2006.19.4 New hose. New hose shall be visually inspected prior to being placed into service.

2006.20 Aircraft fuel-servicing vehicles parking. Unattended aircraft fuel-servicing vehicles shall be parked in areas that provide for both the unencumbered dispersal of vehicles in the event of an emergency and the control of leakage such that adjacent buildings and storm drains are not contaminated by leaking fuel.

2006.20.1 Parking area design. Parking areas for tank vehicles shall be designed and utilized such that a clearance of 10 feet (3048 mm) is maintained between each parked vehicle for fire department access. In addition, a minimum clearance of 50 feet (15 240 mm) shall be maintained between tank vehicles and parked aircraft and structures other than those used for the maintenance and/or garaging of aircraft fuel-servicing vehicles.

2006.21 Radar equipment. Aircraft fuel-servicing operations shall be prohibited while the weather-mapping radar of that aircraft is operating.

Aircraft fuel-servicing or other operations in which flammable liquids, vapors or mists could be present shall not be conducted within 300 feet (91 440 mm) of an operating aircraft surveillance radar.

Aircraft fuel-servicing operations shall not be conducted within 300 feet (91 440 mm) of airport flight traffic surveillance radar equipment.

Aircraft fuel-servicing or other operations in which flammable liquids, vapors or mists could be present shall not be conducted within 100 feet (30 480 mm) of airport ground traffic surveillance radar equipment.

2006.21.1 Direction of radar beams. The beam from ground radar equipment shall not be directed toward fuel storage or loading racks.

Exceptions:

1. Fuel storage and loading racks in excess of 300 feet (91 440 mm) from airport flight traffic surveillance equipment.
2. Fuel storage and loading racks in excess of 100 feet (30 480 mm) from airport ground traffic surveillance equipment.

SECTION 2007
HELISTOPS AND HELIPORTS

2007.1 General. Helistops and heliports shall be maintained in accordance with Sections 2007.2 through 2007.8. Helistops and heliports on buildings shall be constructed in accordance with the *International Building Code.*

2007.2 Clearances. The touchdown area shall be surrounded on all sides by a clear area having minimum average width at roof level of 15 feet (4572 mm) but no width less than 5 feet (1524 mm). The clear area shall be maintained.

2007.3 Flammable and Class II combustible liquid spillage. Landing areas on structures shall be maintained so as to confine flammable or Class II *combustible liquid* spillage to the landing area itself, and provisions shall be made to drain such spillage away from *exits* or *stairways* serving the helicopter landing area or from a structure housing such *exit* or *stairway.*

2007.4 Exits. *Exits* and *stairways* shall be maintained in accordance with Section 412.7 of the *International Building Code.*

2007.5 Standpipe systems. A building with a rooftop helistop or heliport shall be provided with a Class I or III standpipe system extended to the roof level on which the helistop or heliport is located. All portions of the helistop and heliport area shall be within 150 feet (45 720 mm) of a $2^1/_2$-inch (63.5 mm) outlet on the standpipe system.

2007.6 Foam protection. Foam fire-protection capabilities shall be provided for rooftop heliports. Such systems shall be designed, installed and maintained in accordance with the applicable provisions of Sections 903, 904 and 905.

2007.7 Fire extinguishers. Not less than one portable fire extinguisher having a minimum 80-B:C rating shall be provided for each permanent takeoff and landing area and for the aircraft parking areas. Installation, inspection and maintenance of these extinguishers shall be in accordance with Section 906.

2007.8 Federal approval. Before operating helicopters from helistops and heliports, approval shall be obtained from the Federal Aviation Administration.

CHAPTER 21

DRY CLEANING

SECTION 2101 GENERAL

2101.1 Scope. Dry cleaning plants and their operations shall comply with the requirements of this chapter.

2101.2 Permit required. Permits shall be required as set forth in Section 105.6.

SECTION 2102 DEFINITIONS

2102.1 Definitions. The following terms are defined in Chapter 2:

DRY CLEANING.

DRY CLEANING PLANT.

DRY CLEANING ROOM.

DRY CLEANING SYSTEM.

SOLVENT OR LIQUID CLASSIFICATIONS.

Class I solvents.

Class II solvents.

Class IIIA solvents.

Class IIIB solvents.

Class IV solvents.

SECTION 2103 CLASSIFICATIONS

2103.1 Solvent classification. Dry cleaning solvents shall be classified according to their *flash points* as follows:

1. Class I solvents are liquids having a *flash point* below 100°F (38°C).
2. Class II solvents are liquids having a *flash point* at or above 100°F (38°C) and below 140°F (60°C).
3. Class IIIA solvents are liquids having a *flash point* at or above 140°F (60°C) and below 200°F (93°C).
4. Class IIIB solvents are liquids having a *flash point* at or above 200°F (93°C).
5. Class IV solvents are liquids classified as nonflammable.

2103.2 Classification of dry cleaning plants and systems. Dry cleaning plants and systems shall be classified based on the solvents used as follows:

1. Type I—systems using Class I solvents.
2. Type II—systems using Class II solvents.
3. Type III-A—systems using Class IIIA solvents.
4. Type III-B—systems using Class IIIB solvents.
5. Type IV—systems using Class IV solvents in which dry cleaning is not conducted by the public.
6. Type V—systems using Class IV solvents in which dry cleaning is conducted by the public.

Spotting and pretreating operations conducted in accordance with Section 2106 shall not change the type of the dry cleaning plant.

2103.2.1 Multiple solvents. Dry cleaning plants using more than one class of solvent for dry cleaning shall be classified based on the numerically lowest solvent class.

2103.3 Design. The occupancy classification, design and construction of dry cleaning plants shall comply with the applicable requirements of the *International Building Code*.

SECTION 2104 GENERAL REQUIREMENTS

2104.1 Prohibited use. Type I dry cleaning plants shall be prohibited. Limited quantities of Class I solvents stored and used in accordance with this section shall not be prohibited in dry cleaning plants.

2104.2 Building services. Building services and systems shall be designed, installed and maintained in accordance with this section and Chapter 6.

2104.2.1 Ventilation. Ventilation shall be provided in accordance with Section 502 of the *International Mechanical Code* and DOL 29 CFR Part 1910.1000, where applicable.

2104.2.2 Heating. In Type II dry cleaning plants, heating shall be by indirect means using steam, hot water or hot oil only.

2104.2.3 Electrical wiring and equipment. Electrical wiring and equipment in dry cleaning rooms or other locations subject to flammable vapors shall be installed in accordance with NFPA 70.

2104.2.4 Bonding and grounding. Storage tanks, treatment tanks, filters, pumps, piping, ducts, dry cleaning units, stills, tumblers, drying cabinets and other such equipment, where not inherently electrically conductive, shall be bonded together and grounded. Isolated equipment shall be grounded.

SECTION 2105 OPERATING REQUIREMENTS

2105.1 General. The operation of dry cleaning systems shall comply with the requirements of Sections 2105.1.1 through 2105.3.

2105.1.1 Written instructions. Written instructions covering the proper installation and safe operation and use of equipment and solvent shall be given to the buyer.

2105.1.1.1 Type II, III-A, III-B and IV systems. In Type II, III-A, III-B and IV dry cleaning systems, machines shall be operated in accordance with the operating instructions furnished by the machinery manufacturer. Employees shall be instructed as to the hazards involved in their departments and in the work they perform.

2105.1.1.2 Type V systems. Operating instructions for customer use of Type V dry cleaning systems shall be conspicuously posted in a location near the dry cleaning unit. A telephone number shall be provided for emergency assistance.

2105.1.2 Equipment identification. The manufacturer shall provide nameplates on dry cleaning machines indicating the class of solvent for which each machine is designed.

2105.1.3 Open systems prohibited. Dry cleaning by immersion and agitation in open vessels shall be prohibited.

2105.1.4 Prohibited use of solvent. The use of solvents with a *flash point* below that for which a machine is designed or *listed* shall be prohibited.

2105.1.5 Equipment maintenance and housekeeping. Proper maintenance and operating practices shall be observed in order to prevent the leakage of solvent or the accumulation of lint. The handling of waste material generated by dry cleaning operations and the maintenance of facilities shall comply with the provisions of this section.

2105.1.5.1 Floors. Class I and II liquids shall not be used for cleaning floors.

2105.1.5.2 Filters. Filter residue and other residues containing solvent shall be handled and disposed of in covered metal containers.

2105.1.5.3 Lint. Lint and refuse shall be removed from traps daily, deposited in *approved* waste cans, removed from the premises, and disposed of safely. At all other times, traps shall be held securely in place.

2105.1.5.4 Customer areas. In Type V dry cleaning systems, customer areas shall be kept clean.

2105.2 Type II systems. Special operating requirements for Type II dry cleaning systems shall comply with the provisions of Sections 2105.2.1 through 2105.2.3.

2105.2.1 Inspection of materials. Materials to be dry cleaned shall be searched thoroughly and foreign materials, including matches and metallic substances, shall be removed.

2105.2.2 Material transfer. In removing materials from the washer, provisions shall be made for minimizing the dripping of solvent on the floor. Where materials are transferred from a washer to a drain tub, a nonferrous metal drip apron shall be placed so that the apron rests on the drain tub and the cylinder of the washer.

2105.2.3 Ventilation. A mechanical ventilation system which is designed to exhaust 1 cubic foot of air per minute for each square foot of floor area [0.0058 $m^3/(s \cdot m^2)$] shall be installed in dry cleaning rooms and in drying rooms. The ventilation system shall operate automatically when the dry cleaning equipment is in operation and shall have manual controls at an *approved* location.

2105.3 Type IV and V systems. Type IV and V dry cleaning systems shall be provided with an automatically activated exhaust ventilation system to maintain a minimum of 100 feet per minute (0.51 m/s) air velocity through the loading door when the door is opened. Such systems for dry cleaning equipment shall comply with the *International Mechanical Code*.

Exception: Dry cleaning units are not required to be provided with exhaust ventilation where an exhaust hood is installed immediately outside of and above the loading door which operates at an airflow rate as follows:

$$Q = 100 \times A_{LD} \quad \textbf{(Equation 21-1)}$$

where:

Q = flow rate exhausted through the hood, cubic feet per minute (m^3/s).

A_{LD} = area of the loading door, square feet (m^2).

SECTION 2106
SPOTTING AND PRETREATING

2106.1 General. Spotting and pretreating operations and equipment shall comply with the provisions of Sections 2106.2 through 2106.5.

2106.2 Class I solvents. The maximum quantity of Class I solvents permitted at any work station shall be 1 gallon (4 L). Spotting or prespotting shall be permitted to be conducted with Class I solvents where they are stored in and dispensed from *approved* safety cans or in sealed DOT-approved metal shipping containers of not more than 1-gallon (4 L) capacity.

2106.2.1 Spotting and prespotting. Spotting and prespotting shall be permitted to be conducted with Class I solvents where dispensed from plastic containers of not more than 1 pint (0.5 L) capacity.

2106.3 Class II and III solvents. Scouring, brushing, and spotting and pretreating shall be permitted to be conducted with Class II or III solvents. The maximum quantity of Class II or III solvents permitted at any work station shall be 1 gallon (4 L). In other than Group H-2 occupancy, the aggregate quantities of solvents shall not exceed the maximum allowable quantity per control area for use-open system.

2106.3.1 Spotting tables. Scouring, brushing or spotting tables on which articles are soaked in solvent shall have a liquid-tight top with a curb on all sides not less than 1 inch (25 mm) high. The top of the table shall be pitched to ensure thorough draining to a $1^1/_2$-inch (38 mm) drain connected to an *approved* container.

2106.3.2 Special handling. Where *approved*, articles that cannot be washed in the usual washing machines are allowed to be cleaned in scrubbing tubs. Scrubbing tubs shall comply with the following:

1. Only Class II or III liquids shall be used.

2. The total amount of solvent used in such open containers shall not exceed 3 gallons (11 L).
3. Scrubbing tubs shall be secured to the floor.
4. Scrubbing tubs shall be provided with permanent $1^1/_2$-inch (38 mm) drains. Such drain shall be provided with a trap and shall be connected to an *approved* container.

2106.3.3 Ventilation. Scrubbing tubs, scouring, brushing or spotting operations shall be located such that solvent vapors are captured and exhausted by the ventilating system.

2106.3.4 Bonding and grounding. Metal scouring, brushing and spotting tables and scrubbing tubs shall be permanently and effectively bonded and grounded.

2106.4 Type IV systems. Flammable and combustible liquids used for spotting operations shall be stored in *approved* safety cans or in sealed DOTn-approved shipping containers of not more than 1 gallon (4 L) in capacity. Aggregate amounts shall not exceed 10 gallons (38 L).

2106.5 Type V systems. Spotting operations using flammable or *combustible liquids* are prohibited in Type V dry cleaning systems.

SECTION 2107 DRY CLEANING SYSTEMS

2107.1 General equipment requirements. Dry cleaning systems, including dry cleaning units, washing machines, stills, drying cabinets, tumblers and their appurtenances, including pumps, piping, valves, filters and solvent coolers, shall be installed and maintained in accordance with NFPA 32. The construction of buildings in which such systems are located shall comply with the requirements of this section and the *International Building Code*.

2107.2 Type II systems. Type II dry cleaning and solvent tank storage rooms shall not be located below grade or above the lowest floor level of the building and shall comply with Sections 2107.2.1 through 2107.2.3.

Exception: Solvent storage tanks installed underground, in vaults or in special enclosures in accordance with Chapter 57.

2107.2.1 Fire-fighting access. Type II dry cleaning plants shall be located so that access is provided and maintained from one side for fire-fighting and fire control purposes in accordance with Section 503.

2107.2.2 Number of means of egress. Type II dry cleaning rooms shall have not less than two *means of egress* doors located at opposite ends of the room, not less than one of which shall lead directly to the outside.

2107.2.3 Spill control and secondary containment. Curbs, drains or other provisions for spill control and secondary containment shall be provided in accordance with Section 5004.2 to collect solvent leakage and fire protection water and direct it to a safe location.

2107.3 Solvent storage tanks. Solvent storage tanks for Class II, IIIA and IIIB liquids shall conform to the requirements of Chapter 57 and be located underground or outside, above ground.

Exception: As provided in NFPA 32 for inside storage or treatment tanks.

SECTION 2108 FIRE PROTECTION

2108.1 General. Where required by this section, *fire protection systems*, devices and equipment shall be installed, inspected, tested and maintained in accordance with Chapter 9.

2108.2 Automatic sprinkler system. An *automatic sprinkler system* shall be installed in accordance with Section 903.3.1.1 throughout dry cleaning plants containing Type II, Type III-A or Type III-B dry cleaning systems.

Exceptions:

1. An *automatic sprinkler system* shall not be required in Type III-A dry cleaning plants where the aggregate quantity of Class III-A solvent in dry cleaning machines and storage does not exceed 330 gallons (1250 L) and dry cleaning machines are equipped with a feature that will accomplish any one of the following:
   1.1. Prevent oxygen concentrations from reaching 8 percent or more by volume.
   1.2. Keep the temperature of the solvent not less than 30°F (16.7°C) below the flash point.
   1.3. Maintain the solvent vapor concentration at a level lower than 25 percent of the lower explosive limit (LEL).
   1.4. Utilize equipment *approved* for use in Class I, Division 2 hazardous locations in accordance with NFPA 70.
   1.5. Utilize an integrated dry-chemical, clean agent or water-mist automatic fire-extinguishing system designed in accordance with Chapter 9.
2. An *automatic sprinkler system* shall not be required in Type III-B dry cleaning plants where the aggregate quantity of Class III-B solvent in dry cleaning machines and storage does not exceed 3,300 gallons (12 490 L).

2108.3 Automatic fire-extinguishing systems. Type II dry cleaning units, washer-extractors, and drying tumblers in Type II dry cleaning plants shall be provided with an *approved* automatic fire-extinguishing system installed and maintained in accordance with Chapter 9.

Exception: Where *approved*, a manual steam jet not less than $^3/_4$ inch (19 mm) with a continuously available steam supply at a pressure not less than 15 pounds per square inch gauge (psig) (103 kPa) is allowed to be substituted for the automatic fire-extinguishing system.

2108.4 Portable fire extinguishers. Portable fire extinguishers shall be selected, installed and maintained in accordance with this section and Section 906. A minimum of two,

2-A:10-B:C portable fire extinguishers shall be provided near the doors inside dry cleaning rooms containing Type II, Type III-A and Type III-B dry cleaning systems.

CHAPTER 22

COMBUSTIBLE DUST-PRODUCING OPERATIONS

SECTION 2201
GENERAL

2201.1 Scope. The equipment, processes and operations involving dust explosion hazards shall comply with the provisions of this chapter.

2201.2 Permits. Permits shall be required for *combustible dust*-producing operations as set forth in Section 105.6.

SECTION 2202
DEFINITION

2202.1 Definition. The following term is defined in Chapter 2:

COMBUSTIBLE DUST.

SECTION 2203
PRECAUTIONS

2203.1 Sources of ignition. Smoking or the use of heating or other devices employing an open flame, or the use of spark-producing equipment is prohibited in areas where *combustible dust* is generated, stored, manufactured, processed or handled.

2203.2 Housekeeping. Accumulation of *combustible dust* shall be kept to a minimum in the interior of buildings. Accumulated *combustible dust* shall be collected by vacuum cleaning or other means that will not place *combustible dust* into suspension in air. Forced air or similar methods shall not be used to remove dust from surfaces.

SECTION 2204
EXPLOSION PROTECTION

2204.1 Standards. The *fire code official* is authorized to enforce applicable provisions of the codes and standards listed in Table 2204.1 to prevent and control dust explosions.

TABLE 2204.1
EXPLOSION PROTECTION STANDARDS

STANDARD	SUBJECT
NFPA 61	Standard for the Prevention of Fires and Dust Explosions in Agricultural and Food Processing Facilities
NFPA 69	Standard on Explosion Prevention Systems
NFPA 70	National Electrical Code
NFPA 85	Boiler and Combustion System Hazards Code
NFPA 120	Standard for Fire Prevention and Control in Coal Mines
NFPA 484	Standard for Combustible Metals
NFPA 654	Standard for Prevention of Fire and Dust Explosions from the Manufacturing, Processing and Handling of Combustible Particulate Solids
NFPA 655	Standard for the Prevention of Sulfur Fires and Explosions
NFPA 664	Standard for the Prevention of Fires and Explosions in Wood Processing and Woodworking Facilities

CHAPTER 23

MOTOR FUEL-DISPENSING FACILITIES AND REPAIR GARAGES

SECTION 2301 GENERAL

2301.1 Scope. Automotive motor fuel-dispensing facilities, marine motor fuel-dispensing facilities, fleet vehicle motor fuel-dispensing facilities, aircraft motor-vehicle fuel-dispensing facilities and repair garages shall be in accordance with this chapter and the *International Building Code, International Fuel Gas Code* and *International Mechanical Code.* Such operations shall include both those that are accessible to the public and private operations.

2301.2 Permits. Permits shall be required as set forth in Section 105.6.

2301.3 Construction documents. *Construction documents* shall be submitted for review and approval prior to the installation or construction of automotive, marine or fleet vehicle motor fuel-dispensing facilities and repair garages in accordance with Section 105.4.

2301.4 Indoor motor fuel-dispensing facilities. Motor fuel-dispensing facilities located inside buildings shall comply with the *International Building Code* and NFPA 30A.

2301.4.1 Protection of floor openings in indoor motor fuel-dispensing facilities. Where motor fuel-dispensing facilities are located inside buildings and the dispensers are located above spaces within the building, openings beneath dispensers shall be sealed to prevent the flow of leaked fuel to lower building spaces.

2301.5 Electrical. Electrical wiring and equipment shall be suitable for the locations in which they are installed and shall comply with Section 605, NFPA 30A and NFPA 70.

2301.6 Heat-producing appliances. Heat-producing appliances shall be suitable for the locations in which they are installed and shall comply with NFPA 30A and the *International Fuel Gas Code* or the *International Mechanical Code.*

SECTION 2302 DEFINITIONS

2302.1 Definitions. The following terms are defined in Chapter 2:

AIRCRAFT MOTOR-VEHICLE FUEL-DISPENSING FACILITY.

ALCOHOL-BLENDED FUELS.

AUTOMOTIVE MOTOR FUEL-DISPENSING FACILITY.

DISPENSING DEVICE, OVERHEAD TYPE.

FLEET VEHICLE MOTOR FUEL-DISPENSING FACILITY.

LIQUEFIED NATURAL GAS (LNG).

MARINE MOTOR FUEL-DISPENSING FACILITY.

REPAIR GARAGE.

SELF-SERVICE MOTOR FUEL-DISPENSING FACILITY.

SECTION 2303 LOCATION OF DISPENSING DEVICES

2303.1 Location of dispensing devices. Dispensing devices shall be located as follows:

1. Ten feet (3048 mm) or more from *lot lines.*
2. Ten feet (3048 mm) or more from buildings having combustible exterior wall surfaces or buildings having noncombustible exterior wall surfaces that are not part of a 1-hour fire-resistance-rated assembly or buildings having combustible overhangs.

   Exception: Canopies constructed in accordance with the *International Building Code* providing weather protection for the fuel islands.
3. Such that all portions of the vehicle being fueled will be on the premises of the motor fuel-dispensing facility.
4. Such that the nozzle, when the hose is fully extended, will not reach within 5 feet (1524 mm) of building openings.
5. Twenty feet (6096 mm) or more from fixed sources of ignition.

2303.2 Emergency disconnect switches. An *approved,* clearly identified and readily accessible emergency disconnect switch shall be provided at an *approved* location to stop the transfer of fuel to the fuel dispensers in the event of a fuel spill or other emergency. The emergency disconnect switch for exterior fuel dispensers shall be located within 100 feet (30 480 mm) of, but not less than 20 feet (6096 mm) from, the fuel dispensers. For interior fuel-dispensing operations, the emergency disconnect switch shall be installed at an *approved* location. Such devices shall be distinctly *labeled* as: EMERGENCY FUEL SHUTOFF. Signs shall be provided in *approved* locations.

SECTION 2304 DISPENSING OPERATIONS

2304.1 Supervision of dispensing. The dispensing of fuel at motor fuel-dispensing facilities shall be conducted by a qualified attendant or shall be under the supervision of a qualified attendant at all times or shall be in accordance with Section 2304.3.

2304.2 Attended self-service motor fuel-dispensing facilities. Attended self-service motor fuel-dispensing facilities shall comply with Sections 2304.2.1 through 2304.2.5. Attended self-service motor fuel-dispensing facilities shall have not less than one qualified attendant on duty while the facility is open for business. The attendant's primary function shall be to supervise, observe and control the dispensing of fuel. The attendant shall prevent the dispensing of fuel into containers that do not comply with Section 2304.4.1, control sources of ignition, give immediate attention to accidental spills or releases, and be prepared to use fire extinguishers.

2304.2.1 Special-type dispensers. *Approved* special-dispensing devices and systems such as, but not limited to, card- or coin-operated and remote-preset types, are allowed at motor fuel-dispensing facilities provided there is not less than one qualified attendant on duty while the facility is open to the public. Remote preset-type devices shall be set in the "off" position while not in use so that the dispenser cannot be activated without the knowledge of the attendant.

2304.2.2 Emergency controls. *Approved* emergency controls shall be provided in accordance with Section 2303.2.

2304.2.3 Operating instructions. Dispenser operating instructions shall be conspicuously posted in *approved* locations on every dispenser.

2304.2.4 Obstructions to view. Dispensing devices shall be in clear view of the attendant at all times. Obstructions shall not be placed between the dispensing area and the attendant.

2304.2.5 Communications. The attendant shall be able to communicate with persons in the dispensing area at all times. An *approved* method of communicating with the fire department shall be provided for the attendant.

2304.3 Unattended self-service motor fuel-dispensing facilities. Unattended self-service motor fuel-dispensing facilities shall comply with Sections 2304.3.1 through 2304.3.7.

2304.3.1 General. Where *approved,* unattended self-service motor fuel-dispensing facilities are allowed. As a condition of approval, the *owner* or operator shall provide, and be accountable for, daily site visits, regular equipment inspection and maintenance.

2304.3.2 Dispensers. Dispensing devices shall comply with Section 2306.7. Dispensing devices operated by the insertion of coins or currency shall not be used unless *approved.*

2304.3.3 Emergency controls. *Approved* emergency controls shall be provided in accordance with Section 2303.2. Emergency controls shall be of a type that is only manually resettable.

2304.3.4 Operating instructions. Dispenser operating instructions shall be conspicuously posted in *approved* locations on every dispenser and shall indicate the location of the emergency controls required by Section 2304.3.3.

2304.3.5 Emergency procedures. An *approved* emergency procedures sign, in addition to the signs required by Section 2305.6, shall be posted in a conspicuous location and shall read:

IN CASE OF FIRE, SPILL OR RELEASE

1. USE EMERGENCY PUMP SHUTOFF

2. REPORT THE ACCIDENT!

FIRE DEPARTMENT TELEPHONE NO.______

FACILITY ADDRESS ____________________

2304.3.6 Communications. A telephone not requiring a coin to operate or other *approved,* clearly identified means to notify the fire department shall be provided on the site in a location *approved* by the *fire code official.*

2304.3.7 Quantity limits. Dispensing equipment used at unsupervised locations shall comply with one of the following:

1. Dispensing devices shall be programmed or set to limit uninterrupted fuel delivery to 25 gallons (95 L) and require a manual action to resume delivery.
2. The amount of fuel being dispensed shall be limited in quantity by a preprogrammed card as *approved.*

2304.4 Dispensing into portable containers. The dispensing of flammable or *combustible liquids* into portable *approved* containers shall comply with Sections 2304.4.1 through 2304.4.3.

2304.4.1 Approved containers required. Class I, II and IIIA liquids shall not be dispensed into a portable container unless such container does not exceed a 6-gallon (22.7 L) capacity, is *listed* or of *approved* material and construction, and has a tight closure with a screwed or spring-loaded cover so designed that the contents can be dispensed without spilling. Liquids shall not be dispensed into portable or cargo tanks.

2304.4.2 Nozzle operation. A hose nozzle valve used for dispensing Class I liquids into a portable container shall be in compliance with Section 2306.7.6 and be manually held open during the dispensing operation.

2304.4.3 Location of containers being filled. Portable containers shall not be filled while located inside the trunk, passenger compartment or truck bed of a vehicle.

SECTION 2305 OPERATIONAL REQUIREMENTS

2305.1 Tank filling operations for Class I, II or III liquids. Delivery operations to tanks for Class I, II or III liquids shall comply with Sections 2305.1.1 through 2305.1.3 and the applicable requirements of Chapter 57.

2305.1.1 Delivery vehicle location. Where liquid delivery to above-ground storage tanks is accomplished by positive-pressure operation, tank vehicles shall be positioned not less than 25 feet (7620 mm) from tanks receiving Class I liquids and 15 feet (4572 mm) from tanks receiving Class II and IIIA liquids.

2305.1.2 Tank capacity calculation. The driver, operator or attendant of a tank vehicle shall, before making delivery to a tank, determine the unfilled, available capacity of such tank by an *approved* gauging device.

2305.1.3 Tank fill connections. Delivery of flammable liquids to tanks more than 1,000 gallons (3785 L) in capacity shall be made by means of *approved* liquid- and vapor-tight connections between the delivery hose and tank fill pipe. Where tanks are equipped with any type of vapor recovery system, all connections required to be made for the safe and proper functioning of the particular vapor recovery process shall be made. Such connections shall be made liquid and vapor tight and remain connected throughout the unloading process. Vapors shall not be discharged at grade level during delivery.

2305.2 Equipment maintenance and inspection. Motor fuel-dispensing facility equipment shall be maintained in proper working order at all times in accordance with Sections 2305.2.1 through 2305.2.5.

2305.2.1 Inspections. Flammable and *combustible liquid* fuel-dispensing and containment equipment shall be periodically inspected where required by the *fire code official* to verify that the equipment is in proper working order and not subject to leakage. Records of inspections shall be maintained.

2305.2.2 Repairs and service. The *fire code official* is authorized to require damaged or unsafe containment and dispensing equipment to be repaired or serviced in an *approved* manner.

2305.2.3 Dispensing devices. Where maintenance to Class I liquid dispensing devices becomes necessary and such maintenance could allow the accidental release or ignition of liquid, the following precautions shall be taken before such maintenance is begun:

1. Only persons knowledgeable in performing the required maintenance shall perform the work.
2. Electrical power to the dispensing device and pump serving the dispenser shall be shut off at the main electrical disconnect panel.
3. The emergency shutoff valve at the dispenser, where installed, shall be closed.
4. Vehicle traffic and unauthorized persons shall be prevented from coming within 12 feet (3658 mm) of the dispensing device.

2305.2.4 Emergency shutoff valves. Automatic emergency shutoff valves required by Section 2306.7.4 shall be checked not less than once per year by manually tripping the hold-open linkage.

2305.2.5 Leak detectors. Leak detection devices required by Section 2306.7.7.1 shall be checked and tested not less than annually in accordance with the manufacturer's specifications to ensure proper installation and operation.

2305.3 Spill control. Provisions shall be made to prevent liquids spilled during dispensing operations from flowing into buildings. Acceptable methods include, but shall not be limited to, grading driveways, raising doorsills or other *approved* means.

2305.4 Sources of ignition. Smoking and open flames shall be prohibited in areas where fuel is dispensed. The engines of vehicles being fueled shall be shut off during fueling. Electrical equipment shall be in accordance with NFPA 70.

2305.5 Fire extinguishers. *Approved* portable fire extinguishers complying with Section 906 with a minimum rating of 2-A:20-B:C shall be provided and located such that an extinguisher is not more than 75 feet (22 860 mm) from pumps, dispensers or storage tank fill-pipe openings.

2305.6 Warning signs. Warning signs shall be conspicuously posted within sight of each dispenser in the fuel-dispensing area and shall state the following:

1. No smoking.
2. Shut off motor.
3. Discharge your static electricity before fueling by touching a metal surface away from the nozzle.
4. To prevent static charge, do not reenter your vehicle while gasoline is pumping.
5. If a fire starts, do not remove nozzle—back away immediately.
6. It is unlawful and dangerous to dispense gasoline into unapproved containers.
7. No filling of portable containers in or on a motor vehicle. Place container on ground before filling.

2305.7 Control of brush and debris. Fenced and diked areas surrounding above-ground tanks shall be kept free from vegetation, debris and other material that is not necessary to the proper operation of the tank and piping system.

Weeds, grass, brush, trash and other combustible materials shall be kept not less than 10 feet (3048 mm) from fuel-handling equipment.

SECTION 2306 FLAMMABLE AND COMBUSTIBLE LIQUID MOTOR FUEL-DISPENSING FACILITIES

2306.1 General. Storage of flammable and *combustible liquids* shall be in accordance with Chapter 57 and Sections 2306.2 through 2306.6.3.

2306.2 Method of storage. *Approved* methods of storage for Class I, II and III liquid fuels at motor fuel-dispensing facilities shall be in accordance with Sections 2306.2.1 through 2306.2.6.

2306.2.1 Underground tanks. Underground tanks for the storage of Class I, II and IIIA liquid fuels shall comply with Chapter 57.

2306.2.1.1 Inventory control for underground tanks. Accurate daily inventory records shall be maintained and reconciled on underground fuel storage tanks for indication of possible leakage from tanks and piping. The records shall include records for each product showing daily reconciliation between sales, use, receipts and inventory on hand. Where there is more than one system consisting of tanks serving separate pumps or dispensers for a product, the reconciliation shall be ascertained separately for each tank system. A consistent or accidental loss of product shall be immediately reported to the *fire code official*.

2306.2.2 Above-ground tanks located inside buildings. Above-ground tanks for the storage of Class I, II and IIIA liquid fuels are allowed to be located in buildings. Such tanks shall be located in special enclosures complying with Section 2306.2.6, in a liquid storage room or a liquid storage warehouse complying with Chapter 57, or shall be *listed* and *labeled* as protected above-ground tanks in accordance with UL 2085.

2306.2.3 Above-ground tanks located outside, above grade. Above-ground tanks shall not be used for the storage of Class I, II or III liquid motor fuels, except as provided by this section.

1. Above-ground tanks used for outside, above-grade storage of Class I liquids shall be *listed* and *labeled* as protected above-ground tanks in accordance with UL 2085 and shall be in accordance with Chapter 57. Such tanks shall be located in accordance with Table 2306.2.3.
2. Above-ground tanks used for outside, above-grade storage of Class II or IIIA liquids shall be *listed* and *labeled* as protected above-ground tanks in accordance with UL 2085 and shall be installed in accordance with Chapter 57. Tank locations shall be in accordance with Table 2306.2.3.

   Exception: Other above-ground tanks that comply with Chapter 57 where *approved* by the *fire code official.*
3. Tanks containing fuels shall not exceed 12,000 gallons (45 420 L) in individual capacity or 48,000 gallons (181 680 L) in aggregate capacity. Installations with the maximum allowable aggregate capacity shall be separated from other such installations by not less than 100 feet (30 480 mm).
4. Tanks located at farms, construction projects, or rural areas shall comply with Section 5706.2.
5. Above-ground tanks used for outside above-grade storage of Class IIIB liquid motor fuel shall be *listed* and *labeled* in accordance with UL 142 or *listed* and *labeled* as protected above-ground tanks in accordance with UL 2085 and shall be installed in accordance with Chapter 57. Tank locations shall be in accordance with Table 2306.2.3.

2306.2.4 Above-ground tanks located in above-grade vaults or below-grade vaults. Above-ground tanks used for storage of Class I, II or IIIA liquid motor fuels are allowed to be installed in vaults located above grade or below grade in accordance with Section 5704.2.8 and shall comply with Sections 2306.2.4.1 and 2306.2.4.2. Tanks in above-grade vaults shall also comply with Table 2306.2.3.

2306.2.4.1 Tank capacity limits. Tanks storing Class I and Class II liquids at an individual site shall be limited to a maximum individual capacity of 15,000 gallons (56 775 L) and an aggregate capacity of 48,000 gallons (181 680 L).

2306.2.4.2 Fleet vehicle motor fuel-dispensing facilities. Tanks storing Class II and Class IIIA liquids at a fleet vehicle motor fuel-dispensing facility shall be limited to a maximum individual capacity of 20,000 gallons (75 700 L) and an aggregate capacity of 80,000 gallons (302 800 L).

2306.2.5 Portable tanks. Where approved by the fire code official, portable tanks are allowed to be temporarily used in conjunction with the dispensing of Class I, II or III liquids into the fuel tanks of motor vehicles or motorized equipment on premises not normally accessible to the public. The approval shall include a definite time limit.

2306.2.6 Special enclosures. Where installation of tanks in accordance with Section 5704.2.11 is impractical, or

TABLE 2306.2.3
MINIMUM SEPARATION REQUIREMENTS FOR ABOVE-GROUND TANKS

CLASS OF LIQUID AND TANK TYPE	INDIVIDUAL TANK CAPACITY (gallons)	MINIMUM DISTANCE FROM NEAREST IMPORTANT BUILDING ON SAME PROPERTY (feet)	MINIMUM DISTANCE FROM NEAREST FUEL DISPENSER (feet)	MINIMUM DISTANCE FROM LOT LINE THAT IS OR CAN BE BUILT UPON, INCLUDING THE OPPOSITE SIDE OF A PUBLIC WAY (feet)	MINIMUM DISTANCE FROM NEAREST SIDE OF ANY PUBLIC WAY (feet)	MINIMUM DISTANCE BETWEEN TANKS (feet)
Class I protected above-ground tanks	Less than or equal to 6,000	5	25[a]	15	5	3
	Greater than 6,000	15	25[a]	25	15	3
Class II and III protected above-ground tanks	Same as Class I	Same as Class I	Same as Class I[c]	Same as Class I	Same as Class I	Same as Class I
Tanks in vaults	0–20,000	0[b]	0	0[b]	0	Separate compartment required for each tank
Other tanks	All	50	50	100	50	3

For SI: 1 foot = 304.8 mm, 1 gallon = 3.785 L.

a. At fleet vehicle motor fuel-dispensing facilities, a minimum separation distance is not required.

b. Underground vaults shall be located such that they will not be subject to loading from nearby structures, or they shall be designed to accommodate applied loads from existing or future structures that can be built nearby.

c. For Class IIIB liquids in protected above-ground tanks, a minimum separation distance is not required.

because of property or building limitations, tanks for liquid motor fuels are allowed to be installed in buildings in special enclosures in accordance with all of the following:

1. The special enclosure shall be liquid tight and vapor tight.
2. The special enclosure shall not contain backfill.
3. Sides, top and bottom of the special enclosure shall be of reinforced concrete not less than 6 inches (152 mm) thick, with openings for inspection through the top only.
4. Tank connections shall be piped or closed such that neither vapors nor liquid can escape into the enclosed space between the special enclosure and any tanks inside the special enclosure.
5. Means shall be provided whereby portable equipment can be employed to discharge to the outside any vapors that might accumulate inside the special enclosure should leakage occur.
6. Tanks containing Class I, II or IIIA liquids inside a special enclosure shall not exceed 6,000 gallons (22 710 L) in individual capacity or 18,000 gallons (68 130 L) in aggregate capacity.
7. Each tank within special enclosures shall be surrounded by a clear space of not less than 3 feet (910 mm) to allow for maintenance and inspection.

2306.3 Security. Above-ground tanks for the storage of liquid motor fuels shall be safeguarded from public access or unauthorized entry in an *approved* manner.

2306.4 Physical protection. Guard posts complying with Section 312 or other *approved* means shall be provided to protect above-ground tanks against impact by a motor vehicle unless the tank is *listed* as a protected above-ground tank with vehicle impact protection.

2306.5 Secondary containment. Above-ground tanks shall be provided with drainage control or diking in accordance with Chapter 57. Drainage control and diking is not required for *listed* secondary containment tanks. Secondary containment systems shall be monitored either visually or automatically. Enclosed secondary containment systems shall be provided with emergency venting in accordance with Section 2306.6.2.5.

2306.6 Piping, valves, fittings and ancillary equipment for use with flammable or combustible liquids. The design, fabrication, assembly, testing and inspection of piping, valves, fittings and ancillary equipment for use with flammable or *combustible liquids* shall be in accordance with Chapter 57 and Sections 2306.6.1 through 2306.6.3.

2306.6.1 Protection from damage. Piping shall be located such that it is protected from physical damage.

2306.6.2 Piping, valves, fittings and ancillary equipment for above-ground tanks for Class I, II and III liquids. Piping, valves, fittings and ancillary equipment for above-ground tanks storing Class I, II and III liquids shall comply with Sections 2306.6.2.1 through 2306.6.2.6.

2306.6.2.1 Tank openings. Tank openings for above-ground tanks shall be through the top only.

2306.6.2.2 Fill-pipe connections. The fill pipe for above-ground tanks shall be provided with a means for making a direct connection to the tank vehicle's fuel-delivery hose so that the delivery of fuel is not exposed to the open air during the filling operation. Where any portion of the fill pipe exterior to the tank extends below the level of the top of the tank, a check valve shall be installed in the fill pipe not more than 12 inches (305 mm) from the fill-hose connection.

2306.6.2.3 Overfill protection. Overfill protection shall be provided for above-ground flammable and *combustible liquid* storage tanks in accordance with Sections 5704.2.7.5.8 and 5704.2.9.7.6.

2306.6.2.4 Siphon prevention. An *approved* antisiphon method shall be provided in the piping system to prevent flow of liquid by siphon action.

2306.6.2.5 Emergency relief venting. Above-ground storage tanks, tank compartments and enclosed secondary containment spaces shall be provided with emergency relief venting in accordance with Chapter 57.

2306.6.2.6 Spill containers. A spill container having a capacity of not less than 5 gallons (19 L) shall be provided for each fill connection. For tanks with a top fill connection, spill containers shall be noncombustible and shall be fixed to the tank and equipped with a manual drain valve that drains into the primary tank. For tanks with a remote fill connection, a portable spill container is allowed.

2306.6.3 Piping, valves, fittings and ancillary equipment for underground tanks. Piping, valves, fittings and ancillary equipment for underground tanks shall comply with Chapter 57 and NFPA 30A.

2306.7 Fuel-dispensing systems for flammable or combustible liquids. The design, fabrication and installation of fuel-dispensing systems for flammable or *combustible liquid* fuels shall be in accordance with Sections 2306.7.1 through 2306.7.9.2.4. Alcohol-blended fuel-dispensing systems shall also comply with Section 2306.8.

2306.7.1 Listed equipment. Electrical equipment, dispensers, hose, nozzles and submersible or subsurface pumps used in fuel-dispensing systems shall be *listed*.

2306.7.2 Fixed pumps required. Class I and II liquids shall be transferred from tanks by means of fixed pumps designed and equipped to allow control of the flow and prevent leakage or accidental discharge.

2306.7.3 Mounting of dispensers. Dispensing devices, except those installed on top of a protected above-ground tank that qualifies as vehicle-impact resistant, shall be protected against physical damage by mounting on a concrete island 6 inches (152 mm) or more in height, or shall be protected in accordance with Section 312. Dispensing devices shall be installed and securely fastened to their mounting surface in accordance with the dispenser manufacturer's instructions. Dispensing devices installed indoors shall be located in an *approved* position where they cannot be struck by an out-of-control vehicle descending a ramp or other slope.

2306.7.4 Dispenser emergency shutoff valve. An *approved* automatic emergency shutoff valve designed to close in the event of a fire or impact shall be properly installed in the liquid supply line at the base of each dispenser supplied by a remote pump. The valve shall be installed so that the shear groove is flush with or within $^1/_2$ inch (12.7 mm) of the top of the concrete dispenser island and there is clearance provided for maintenance purposes around the valve body and operating parts. The valve shall be installed at the liquid supply line inlet of each overhead-type dispenser. Where installed, a vapor return line located inside the dispenser housing shall have a shear section or *approved* flexible connector for the liquid supply line emergency shutoff valve to function. Emergency shutoff valves shall be installed and maintained in accordance with the manufacturer's instructions, tested at the time of initial installation and not less than yearly thereafter in accordance with Section 2305.2.4.

2306.7.5 Dispenser hose. Dispenser hoses shall be not more than 18 feet (5486 mm) in length unless otherwise *approved*. Dispenser hoses shall be *listed* and *approved*. When not in use, hoses shall be reeled, racked or otherwise protected from damage.

2306.7.5.1 Emergency breakaway devices. Dispenser hoses for Class I and II liquids shall be equipped with a *listed* emergency breakaway device designed to retain liquid on both sides of a breakaway point. Such devices shall be installed and maintained in accordance with the manufacturer's instructions. Where hoses are attached to hose-retrieving mechanisms, the emergency breakaway device shall be located between the hose nozzle and the point of attachment of the hose-retrieval mechanism to the hose.

2306.7.6 Fuel delivery nozzles. A listed automatic-closing-type hose nozzle valve with or without a latch-open device shall be provided on island-type dispensers used for dispensing Class I, II or III liquids.

Overhead-type dispensing units shall be provided with a *listed* automatic-closing-type hose nozzle valve without a latch-open device.

Exception: A *listed* automatic-closing-type hose nozzle valve with latch-open device is allowed to be used on overhead-type dispensing units where the design of the system is such that the hose nozzle valve will close automatically in the event the valve is released from a fill opening or upon impact with a driveway.

2306.7.6.1 Special requirements for nozzles. Where dispensing of Class I, II or III liquids is performed, a listed automatic-closing-type hose nozzle valve shall be used incorporating all of the following features:

1. The hose nozzle valve shall be equipped with an integral latch-open device.
2. Where the flow of product is normally controlled by devices or equipment other than the hose nozzle valve, the hose nozzle valve shall not be capable of being opened unless the delivery hose is pressurized. If pressure to the hose is lost, the nozzle shall close automatically.

   Exception: Vapor recovery nozzles incorporating insertion interlock devices designed to achieve shutoff on disconnect from the vehicle fill pipe.

3. The hose nozzle shall be designed such that the nozzle is retained in the fill pipe during the filling operation.
4. The system shall include *listed* equipment with a feature that causes or requires the closing of the hose nozzle valve before the product flow can be resumed or before the hose nozzle valve can be replaced in its normal position in the dispenser.

2306.7.7 Remote pumping systems. Remote pumping systems for liquid fuels shall comply with Sections 2306.7.7.1 and 2306.7.7.2.

2306.7.7.1 Leak detection. Where remote pumps are used to supply fuel dispensers, each pump shall have installed on the discharge side a *listed* leak detection device that will detect a leak in the piping and dispensers and provide an indication. A leak detection device is not required if the piping from the pump discharge to under the dispenser is above ground and visible.

2306.7.7.2 Location. Remote pumps installed above grade, outside of buildings, shall be located not less than 10 feet (3048 mm) from lines of adjoining property that can be built upon and not less than 5 feet (1524 mm) from any building opening. Where an outside pump location is impractical, pumps are permitted to be installed inside buildings as provided for dispensers in Section 2301.4 and Chapter 57. Pumps shall be substantially anchored and protected against physical damage.

2306.7.8 Gravity and pressure dispensing. Flammable liquids shall not be dispensed by gravity from tanks, drums, barrels or similar containers. Flammable or *combustible liquids* shall not be dispensed by a device operating through pressure within a storage tank, drum or container.

2306.7.9 Vapor-recovery and vapor-processing systems. Vapor-recovery and vapor-processing systems shall be in accordance with Sections 2306.7.9.1 through 2306.7.9.2.4.

2306.7.9.1 Vapor-balance systems. Vapor-balance systems shall comply with Sections 2306.7.9.1.1 through 2306.7.9.1.5.

2306.7.9.1.1 Dispensing devices. Dispensing devices incorporating provisions for vapor recovery shall be *listed* and *labeled*. Where existing *listed* or *labeled* dispensing devices are modified for vapor recovery, such modifications shall be *listed* by report by a nationally recognized testing laboratory. The listing by report shall contain a description of the component parts used in the modification and recommended method of installation on specific dispensers. Such report shall be made available on request of the *fire code official*.

Means shall be provided to shut down fuel dispensing in the event the vapor return line becomes blocked.

2306.7.9.1.2 Vapor-return line closeoff. An acceptable method shall be provided to close off the vapor return line from dispensers when the product is not being dispensed.

2306.7.9.1.3 Piping. Piping in vapor-balance systems shall be in accordance with Sections 5703.6, 5704.2.9 and 5704.2.11. Nonmetallic piping shall be installed in accordance with the manufacturer's instructions.

Existing and new vent piping shall be in accordance with Sections 5703.6 and 5704.2. Vapor return piping shall be installed in a manner that drains back to the tank, without sags or traps in which liquid can become trapped. If necessary, because of grade, condensate tanks are allowed in vapor return piping. Condensate tanks shall be designed and installed so that they can be drained without opening.

2306.7.9.1.4 Flexible joints and shear joints. Flexible joints shall be installed in accordance with Section 5703.6.9.

An *approved* shear joint shall be rigidly mounted and connected by a union in the vapor return piping at the base of each dispensing device. The shear joint shall be mounted flush with the top of the surface on which the dispenser is mounted.

2306.7.9.1.5 Testing. Vapor return lines and vent piping shall be tested in accordance with Section 5703.6.3.

2306.7.9.2 Vapor-processing systems. Vapor-processing systems shall comply with Sections 2306.7.9.2.1 through 2306.7.9.2.4.

2306.7.9.2.1 Equipment. Equipment in vapor-processing systems, including hose nozzle valves, vapor pumps, flame arresters, fire checks or systems for prevention of flame propagation, controls and vapor-processing equipment, shall be individually *listed* for the intended use in a specified manner.

Vapor-processing systems that introduce air into the underground piping or storage tanks shall be provided with equipment for prevention of flame propagation that has been tested and *listed* as suitable for the intended use.

2306.7.9.2.2 Location. Vapor-processing equipment shall be located at or above grade. Sources of ignition shall be located not less than 50 feet (15 240 mm) from fuel-transfer areas and not less than 18 inches (457 mm) above tank fill openings and tops of dispenser islands. Vapor-processing units shall be located not less than 10 feet (3048 mm) from the nearest building or *lot line* of a property that can be built upon.

Exception: Where the required distances to buildings, *lot lines* or fuel-transfer areas cannot be obtained, means shall be provided to protect equipment against fire exposure. Acceptable means shall include but not be limited to either of the following:

1. *Approved* protective enclosures, which extend not less than 18 inches (457 mm) above the equipment, constructed of fire-resistant or noncombustible materials.
2. Fire protection using an *approved* water-spray system.

2306.7.9.2.2.1 Distance from dispensing devices. Vapor-processing equipment shall be located not less than 20 feet (6096 mm) from dispensing devices.

2306.7.9.2.2.2 Physical protection. Vapor-processing equipment shall be protected against physical damage by guardrails, curbs, protective enclosures or fencing. Where *approved* protective enclosures are used, *approved* means shall be provided to ventilate the volume within the enclosure to prevent pocketing of flammable vapors.

2306.7.9.2.2.3 Downslopes. Where a downslope exists toward the location of the vapor-processing unit from a fuel-transfer area, the *fire code official* is authorized to require additional separation by distance and height.

2306.7.9.2.3 Installation. Vapor-processing units shall be securely mounted on concrete, masonry or structural steel supports on concrete or other noncombustible foundations. Vapor-recovery and vapor-processing equipment is allowed to be installed on roofs where *approved*.

2306.7.9.2.4 Piping. Piping in a mechanical-assist system shall be in accordance with Section 5703.6.

2306.8 Alcohol-blended fuel-dispensing operations. The design, fabrication and installation of alcohol-blended fuel-dispensing systems shall be in accordance with Section 2306.7 and Sections 2306.8.1 through 2306.8.5.

2306.8.1 Listed equipment. Dispensers shall be *listed* in accordance with UL 87A. Hoses, nozzles, breakaway fittings, swivels, flexible connectors or dispenser emergency shutoff valves, vapor recovery systems, leak detection devices and pumps used in alcohol-blended fuel-dispensing systems shall be *listed* for the specific purpose.

2306.8.2 Compatibility. Dispensers shall be used only with the fuels for which they have been *listed* and which are marked on the product. Field-installed components including hose assemblies, breakaway fittings, swivel connectors and hose nozzle valves shall be provided in accordance with the listing and the marking on the unit.

2306.8.3 Facility identification. Facilities dispensing alcohol-blended fuels shall be identified by an *approved* means.

2306.8.4 Marking. Dispensers shall be marked in an *approved* manner to identify the types of alcohol-blended fuels to be dispensed.

2306.8.5 Maintenance and inspection. Equipment shall be maintained and inspected in accordance with Section 2305.2.

SECTION 2307
LIQUEFIED PETROLEUM GAS MOTOR FUEL-DISPENSING FACILITIES

2307.1 General. Motor fuel-dispensing facilities for liquefied petroleum gas (LP-gas) fuel shall be in accordance with this section and Chapter 61.

2307.2 Approvals. Storage vessels and equipment used for the storage or dispensing of LP-gas shall be *approved* or *listed* in accordance with Sections 2307.2.1 and 2307.2.2.

2307.2.1 Approved equipment. Containers, pressure relief devices (including pressure relief valves), pressure regulators and piping for LP-gas shall be *approved.*

2307.2.2 Listed equipment. Hoses, hose connections, vehicle fuel connections, dispensers, LP-gas pumps and electrical equipment used for LP-gas shall be *listed.*

2307.3 Attendants. Motor fuel-dispensing operations for LP-gas shall be conducted by qualified attendants or in accordance with Section 2307.6 by persons trained in the proper handling of LP-gas.

2307.4 Location of dispensing operations and equipment. The point of transfer for LP-gas dispensing operations shall be separated from buildings and other exposures in accordance with the following:

1. Not less than 25 feet (7620 mm) from buildings where the *exterior wall* is not part of a fire-resistance-rated assembly having a rating of 1 hour or greater.
2. Not less than 25 feet (7620 mm) from combustible overhangs on buildings, measured from a vertical line dropped from the face of the overhang at a point nearest the point of transfer.
3. Not less than 25 feet (7620 mm) from the lot line of property that can be built upon.
4. Not less than 25 feet (7620 mm) from the centerline of the nearest mainline railroad track.
5. Not less than 10 feet (3048 mm) from public streets, highways, thoroughfares, sidewalks and driveways.
6. Not less than 10 feet (3048 mm) from buildings where the *exterior wall* is part of a fire-resistance-rated assembly having a rating of 1 hour or greater.

Exception: The point of transfer for LP-gas dispensing operations need not be separated from canopies that are constructed in accordance with the *International Building Code* and that provide weather protection for the dispensing equipment.

LP-gas containers shall be located in accordance with Chapter 61. LP-gas storage and dispensing equipment shall be located outdoors.

2307.5 Additional requirements for LP-gas dispensers and equipment. LP-gas dispensers and related equipment shall comply with the following provisions.

1. Pumps shall be fixed in place and shall be designed to allow control of the flow and to prevent leakage and accidental discharge.
2. Dispensing devices installed within 10 feet (3048 mm) of where vehicle traffic occurs shall be protected against physical damage by mounting on a concrete island 6 inches (152 mm) or more in height, or shall be protected in accordance with Section 312.
3. Dispensing devices shall be securely fastened to their mounting surface in accordance with the dispenser manufacturer's instructions.

2307.6 Installation of LP-gas dispensing devices and equipment. The installation and operation of LP-gas dispensing systems shall be in accordance with Sections 2307.6.1 through 2307.6.4 and Chapter 61. LP-gas dispensers and dispensing stations shall be installed in accordance with the manufacturer's specifications and their listing.

2307.6.1 Product control valves. The dispenser system piping shall be protected from uncontrolled discharge in accordance with the following:

1. Where mounted on a concrete base, a means shall be provided and installed within $^1/_2$ inch (12.7 mm) of the top of the concrete base that will prevent flow from the supply piping in the event that the dispenser is displaced from its mounting.
2. A manual shutoff valve and an excess flow-control check valve shall be located in the liquid line between the pump and the dispenser inlet where the dispensing device is installed at a remote location and is not part of a complete storage and dispensing unit mounted on a common base.
3. An excess flow-control check valve or an emergency shutoff valve shall be installed in or on the dispenser at the point at which the dispenser hose is connected to the liquid piping.
4. A *listed* automatic-closing type hose nozzle valve with or without a latch-open device shall be provided on island-type dispensers.

2307.6.2 Hoses. Hoses and piping for the dispensing of LP-gas shall be provided with hydrostatic relief valves. The hose length shall not exceed 18 feet (5486 mm). An *approved* method shall be provided to protect the hose against mechanical damage.

2307.6.3 Emergency breakaway devices. Dispenser hoses shall be equipped with a *listed* emergency breakaway device designed to retain liquid on both sides of the breakaway point. Where hoses are attached to hose-retrieving mechanisms, the emergency breakaway device shall be located such that the breakaway device activates to protect the dispenser from being displaced.

2307.6.4 Vehicle impact protection. Where installed within 10 feet of vehicle traffic, LP-gas storage containers, pumps and dispensers shall be protected in accordance with Section 2307.5, Item 2.

2307.7 Public fueling of motor vehicles. Self-service LP-gas dispensing systems, including key, code and card lock dispensing systems, shall be limited to the filling of permanently mounted containers providing fuel to the LP-gas powered vehicle.

The requirements for self-service LP-gas dispensing systems shall be in accordance with the following:

1. The arrangement and operation of the transfer of product into a vehicle shall be in accordance with this section and Chapter 61.
2. The system shall be provided with an emergency shutoff switch located within 100 feet (30 480 mm) of, but not less than 20 feet (6096 mm) from, dispensers.
3. The *owner* of the LP-gas motor fuel-dispensing facility or the owner's designee shall provide for the safe operation of the system and the training of users.
4. The dispenser and hose-end valve shall release not more than $^{1}/_{8}$ fluid ounce (4 cc) of liquid to the atmosphere upon breaking the connection with the fill valve on the vehicle.
5. Portable fire extinguishers shall be provided in accordance with Section 2305.5.
6. Warning signs shall be provided in accordance with Section 2305.6.
7. The area around the dispenser shall be maintained in accordance with Section 2305.7.

2307.8 Overfilling. LP-gas containers shall not be filled with LP-gas in excess of the volume determined using the fixed maximum liquid level gauge installed on the container, the volume determined by the overfilling prevention device installed on the container or the weight determined by the required percentage of the water capacity marked on the container.

SECTION 2308
COMPRESSED NATURAL GAS MOTOR FUEL-DISPENSING FACILITIES

2308.1 General. Motor fuel-dispensing facilities for compressed natural gas (CNG) fuel shall be in accordance with this section and Chapter 53.

2308.2 Approvals. Storage vessels and equipment used for the storage, compression or dispensing of CNG shall be *approved* or *listed* in accordance with Sections 2308.2.1 and 2308.2.2.

2308.2.1 Approved equipment. Containers, compressors, pressure relief devices (including pressure relief valves), and pressure regulators and piping used for CNG shall be *approved.*

2308.2.2 Listed equipment. Hoses, hose connections, dispensers, gas detection systems and electrical equipment used for CNG shall be *listed.* Vehicle-fueling connections shall be *listed* and *labeled.*

2308.3 Location of dispensing operations and equipment. Compression, storage and dispensing equipment shall be located above ground, outside.

Exceptions:

1. Compression, storage or dispensing equipment shall be allowed in buildings of noncombustible construction, as set forth in the *International Building Code*, that are unenclosed for three-quarters or more of the perimeter.
2. Compression, storage and dispensing equipment shall be allowed indoors or in vaults in accordance with Chapter 53.

2308.3.1 Location on property. In addition to the requirements of Section 2303.1, compression, storage and dispensing equipment not located in vaults complying with Chapter 53 shall be installed as follows:

1. Not beneath power lines.
2. Ten feet (3048 mm) or more from the nearest building or *lot line* that could be built on, public street, sidewalk or source of ignition.

   Exception: Dispensing equipment need not be separated from canopies that are constructed in accordance with the *International Building Code* and that provide weather protection for the dispensing equipment.
3. Twenty-five feet (7620 mm) or more from the nearest rail of any railroad track and 50 feet (15 240 mm) or more from the nearest rail of any railroad main track or any railroad or transit line where power for train propulsion is provided by an outside electrical source, such as third rail or overhead catenary.

4. Fifty feet (15 240 mm) or more from the vertical plane below the nearest overhead wire of a trolley bus line.

2308.4 Private fueling of motor vehicles. Self-service CNG-dispensing systems, including key, code and card lock dispensing systems, shall be limited to the filling of permanently mounted fuel containers on CNG-powered vehicles.

In addition to the requirements in Section 2305, the *owner* of a self-service CNG motor fuel-dispensing facility shall ensure the safe operation of the system and the training of users.

2308.5 Pressure regulators. Pressure regulators shall be designed and installed or protected so that their operation will not be affected by the elements (freezing rain, sleet, snow or ice), mud or debris. The protection is allowed to be an integral part of the regulator.

2308.6 Valves. Gas piping to equipment shall be provided with a remote, readily accessible manual shutoff valve.

2308.7 Emergency shutdown control. An emergency shutdown control shall be located within 75 feet (22 860 mm) of, but not less than 25 feet (7620 mm) from, dispensers and shall also be provided in the compressor area. Upon activation, the emergency shutdown system shall automatically shut off the power supply to the compressor and close valves between the main gas supply and the compressor and between the storage containers and dispensers.

2308.8 Discharge of CNG from motor vehicle fuel storage containers. The discharge of CNG from motor vehicle fuel cylinders for the purposes of maintenance, cylinder certification, calibration of dispensers or other activities shall be in accordance with Sections 2308.8.1 through 2308.8.1.2.6.

2308.8.1 Methods of discharge. The discharge of CNG from motor vehicle fuel cylinders shall be accomplished through a closed transfer system in accordance with Section 2308.8.1.1 or an *approved* method of atmospheric venting in accordance with Section 2308.8.1.2.

2308.8.1.1 Closed transfer system. A documented procedure that explains the logical sequence for discharging the cylinder shall be provided to the *fire code official* for review and approval. The procedure shall include what actions the operator will take in the event of a low-pressure or high-pressure natural gas release during the discharging activity. A drawing illustrating the arrangement of piping, regulators and equipment settings shall be provided to the *fire code official* for review and approval. The drawing shall illustrate the piping and regulator arrangement and shall be shown in spatial relation to the location of the compressor, storage vessels and emergency shutdown devices.

2308.8.1.2 Atmospheric venting. Atmospheric venting of CNG shall comply with Sections 2308.8.1.2.1 through 2308.8.1.2.6.

2308.8.1.2.1 Plans and specifications. A drawing illustrating the location of the vessel support, piping, the method of grounding and bonding, and other requirements specified herein shall be provided to the *fire code official* for review and approval.

2308.8.1.2.2 Cylinder stability. A method of rigidly supporting the vessel during the venting of CNG shall be provided. The selected method shall provide not less than two points of support and shall prevent the horizontal and lateral movement of the vessel. The system shall be designed to prevent the movement of the vessel based on the highest gas-release velocity through valve orifices at the vessel's rated pressure and volume. The structure or appurtenance shall be constructed of noncombustible materials.

2308.8.1.2.3 Separation. The structure or appurtenance used for stabilizing the cylinder shall be separated from the site equipment, features and exposures and shall be located in accordance with Table 2308.8.1.2.3.

TABLE 2308.8.1.2.3
SEPARATION DISTANCE FOR ATMOSPHERIC VENTING OF CNG

EQUIPMENT OR FEATURE	MINIMUM SEPARATION (feet)
Buildings	25
Building openings	25
CNG compressor and storage vessels	25
CNG dispensers	25
Lot lines	15
Public ways	15
Vehicles	25

For SI: 1 foot = 304.8 mm.

2308.8.1.2.4 Grounding and bonding. The structure or appurtenance used for supporting the cylinder shall be grounded in accordance with NFPA 70. The cylinder valve shall be bonded prior to the commencement of venting operations.

2308.8.1.2.5 Vent tube. A vent tube that will divert the gas flow to atmosphere shall be installed on the cylinder prior to commencement of the venting and purging operation. The vent tube shall be constructed of pipe or tubing materials *approved* for use with CNG in accordance with Chapter 53.

The vent tube shall be capable of dispersing the gas not less than 10 feet (3048 mm) above grade level. The vent tube shall not be provided with a rain cap or other feature that would limit or obstruct the gas flow.

At the connection fitting of the vent tube and the CNG cylinder, a *listed* bidirectional *detonation* flame arrester shall be provided.

2308.8.1.2.6 Signage. *Approved* "No Smoking" signs complying with Section 310 shall be posted within 10 feet (3048 mm) of the cylinder support structure or appurtenance. *Approved* CYLINDER SHALL BE BONDED signs shall be posted on the cylinder support structure or appurtenance.

SECTION 2309 HYDROGEN MOTOR FUEL-DISPENSING AND GENERATION FACILITIES

2309.1 General. Hydrogen motor fuel-dispensing and generation facilities shall be in accordance with this section and Chapter 58. Where a fuel-dispensing facility includes a repair garage, the repair operation shall comply with Section 2311.

2309.2 Equipment. Equipment used for the generation, compression, storage or dispensing of hydrogen shall be designed for the specific application in accordance with Sections 2309.2.1 through 2309.2.3.

2309.2.1 Approved equipment. Cylinders, containers and tanks; pressure relief devices, including pressure valves; hydrogen vaporizers; pressure regulators; and piping used for gaseous hydrogen systems shall be designed and constructed in accordance with Chapters 53, 55 and 58.

2309.2.2 Listed or approved equipment. Hoses, hose connections, compressors, hydrogen generators, dispensers, detection systems and electrical equipment used for hydrogen shall be *listed* or *approved* for use with hydrogen. Hydrogen motor-fueling connections shall be *listed* and *labeled* or *approved* for use with hydrogen.

2309.2.3 Electrical equipment. Electrical installations shall be in accordance with NFPA 70.

2309.3 Location on property. In addition to the requirements of Section 2303.1, dispensing equipment shall be located in accordance with Sections 2309.3.1 through Section 2309.3.2.

2309.3.1 Location of operations and equipment. Generation, compression, storage and dispensing equipment shall be located in accordance with Sections 2309.3.1.1 through 2309.3.1.5.5.

2309.3.1.1 Outdoors. Generation, compression, or storage equipment shall be allowed outdoors in accordance with Chapter 58 and NFPA 2.

2309.3.1.2 Indoors. Generation, compression, storage and dispensing equipment shall be located in indoor rooms or areas constructed in accordance with the requirements of the *International Building Code*, the *International Fuel Gas Code*, the *International Mechanical Code* and NFPA 2.

2309.3.1.2.1 Maintenance. Gaseous hydrogen systems and detection devices shall be maintained in accordance with the manufacturer's instructions.

2309.3.1.2.2 Smoking. Smoking shall be prohibited in hydrogen cutoff rooms. "No Smoking" signs shall be provided at all entrances to hydrogen fuel gas rooms.

2309.3.1.2.3 Ignition source control. Open flames, flame-producing devices and other sources of ignition shall be controlled in accordance with Chapter 58.

2309.3.1.2.4 Housekeeping. Hydrogen fuel gas rooms shall be kept free from combustible debris and storage.

2309.3.1.3 Gaseous hydrogen storage. Storage of gaseous hydrogen shall be in accordance with Chapters 53 and 58.

2309.3.1.4 Liquefied hydrogen storage. Storage of liquefied hydrogen shall be in accordance with Chapters 55 and 58.

2309.3.1.5 Canopy tops. Gaseous hydrogen compression and storage equipment located on top of motor fuel-dispensing facility canopies shall be in accordance with Sections 2309.3.1.5.1 through 2309.3.1.5.5, Chapters 53 and 58 and the *International Fuel Gas Code*.

2309.3.1.5.1 Construction. Canopies shall be constructed in accordance with the motor fuel-dispensing facility canopy requirements of Section 406.7 of the *International Building Code*.

2309.3.1.5.2 Fire-extinguishing systems. Fuel-dispensing areas under canopies shall be equipped throughout with an *approved automatic sprinkler system* in accordance with Section 903.3.1.1. The design of the sprinkler system shall be not less than that required for Extra Hazard Group 2 occupancies. Operation of the sprinkler system shall activate the emergency functions of Sections 2309.3.1.5.3 and 2309.3.1.5.4.

2309.3.1.5.3 Emergency discharge. Operation of the *automatic sprinkler system* shall activate an automatic emergency discharge system, which will discharge the hydrogen gas from the equipment on the canopy top through the vent pipe system.

2309.3.1.5.4 Emergency shutdown control. Operation of the *automatic sprinkler system* shall activate the emergency shutdown control required by Section 2309.5.3.

2309.3.1.5.5 Signage. *Approved* signage having 2-inch (51 mm) block letters shall be affixed at *approved* locations on the exterior of the canopy structure stating: CANOPY TOP HYDROGEN STORAGE.

2309.3.2 Canopies. Dispensing equipment need not be separated from canopies of Type I or II construction that are constructed in a manner that prevents the accumulation of hydrogen gas and in accordance with Section 406.7 of the *International Building Code*.

2309.4 Dispensing into motor vehicles at self-service hydrogen motor fuel-dispensing facilities. Self-service hydrogen motor fuel-dispensing systems, including key, code and card lock dispensing systems, shall be limited to the filling of permanently mounted fuel containers on hydrogen-powered vehicles.

In addition to the requirements in Section 2311, the *owner* of a self-service hydrogen motor fuel-dispensing facility shall provide for the safe operation of the system through the institution of a fire safety plan submitted in accordance with Section 404, the training of employees and operators who use and maintain the system in accordance with Section 406, and provisions for hazard communication in accordance with Section 407.

2309.4.1 Dispensing systems. Dispensing systems shall be equipped with an overpressure protection device set at not greater than 140 percent of the service pressure of the fueling nozzle it supplies.

2309.5 Safety precautions. Safety precautions at hydrogen motor fuel-dispensing and generation facilities shall be in accordance with Sections 2309.5.1 through 2309.5.3.1.

2309.5.1 Protection from vehicles. Guard posts or other *approved* means shall be provided to protect hydrogen storage systems and use areas subject to vehicular damage in accordance with Section 312.

2309.5.1.1 Vehicle fueling pad. The vehicle shall be fueled on noncoated concrete or other *approved* paving material having a resistance not exceeding 1 megohm as determined by the methodology specified in EN 1081.

2309.5.2 Emergency shutoff valves. A manual emergency shutoff valve shall be provided to shut down the flow of gas from the hydrogen supply to the piping system.

2309.5.2.1 Identification. Manual emergency shutoff valves shall be identified and the location shall be clearly visible, accessible and indicated by means of a sign.

2309.5.3 Emergency shutdown controls. In addition to the manual emergency shutoff valve required by Section 2309.5.2, a remotely located, manually activated emergency shutdown control shall be provided. An emergency shutdown control shall be located within 75 feet (22 860 mm) of, but not less than 25 feet (7620 mm) from, dispensers and hydrogen generators.

2309.5.3.1 System requirements. Activation of the emergency shutdown control shall automatically shut off the power supply to all hydrogen storage, compression and dispensing equipment; shut off natural gas or other fuel supply to the hydrogen generator; and close valves between the main supply and the compressor and between the storage containers and dispensing equipment.

** **2309.6 Defueling of hydrogen from fuel storage containers.** The discharge or defueling of hydrogen from fuel storage tanks for the purpose of maintenance, cylinder certification, calibration of dispensers or other activities shall be in accordance with Sections 2309.6.1 through 2309.6.1.2.4.

2309.6.1 Methods of discharge. The discharge of hydrogen from fuel storage tanks shall be accomplished through a closed transfer system in accordance with Section 2309.6.1.1 or an *approved* method of atmospheric venting in accordance with Section 2309.6.1.2.

2309.6.1.1 Closed transfer system. A documented procedure that explains the logic sequence for discharging the storage tank shall be provided to the *fire code official* for review and approval. The procedure shall include what actions the operator is required to take in the event of a low-pressure or high-pressure hydrogen release during discharging activity. Schematic design documents shall be provided illustrating the arrangement of piping, regulators and equipment settings. The *construction documents* shall illustrate the piping and regulator arrangement and shall be shown in spatial relation to the location of the compressor, storage vessels and emergency shutdown devices.

2309.6.1.2 Atmospheric venting of hydrogen from fuel storage containers. Where atmospheric venting is used for the discharge of hydrogen from fuel storage tanks, such venting shall be in accordance with Sections 2309.6.1.2.1 through 2309.6.1.2.1.4.

2309.6.1.2.1 Defueling equipment. Equipment used for defueling shall be listed and labeled or *approved* for the intended use.

2309.6.1.2.1.1 Manufacturer's equipment required. Equipment supplied by the manufacturer shall be used to connect the storage tanks to be defueled to the vent pipe system.

2309.6.1.2.1.2 Vent pipe maximum diameter. Defueling vent pipes shall have a maximum inside diameter of 1 inch (25 mm).

2309.6.1.2.1.3 Maximum flow rate. The maximum rate of hydrogen flow through the vent pipe system shall not exceed 1,000 cfm at NTP (0.47 m^3/s) and shall be controlled by means of the manufacturer's equipment, at low pressure and without adjustment.

2309.6.1.2.1.4 Isolated use. The vent pipe used for defueling shall not be connected to another venting system used for any other purpose.

2309.6.1.2.2 Construction documents. *Construction documents* shall be provided illustrating the defueling system to be utilized. Plan details shall be of sufficient detail and clarity to allow for evaluation of the piping and control systems to be utilized and include the method of support for cylinders, containers or tanks to be used as part of a closed transfer system, the method of grounding and bonding and other requirements specified herein.

2309.6.1.2.3 Stability of cylinders, containers and tanks. A method of rigidly supporting cylinders, containers or tanks used during the closed transfer system discharge or defueling of hydrogen shall be provided. The method shall provide not less than two points of support and shall be designed to resist lateral movement of the receiving cylinder, container or tank. The system shall be designed to resist movement of the receiver based on the highest gas-release velocity through valve orifices at the receiver's rated service pressure and volume. Supporting structures or appurtenances used to support receivers shall be constructed of noncombustible materials in accordance with the *International Building Code*.

2309.6.1.2.4 Grounding and bonding. Cylinders, containers or tanks and piping systems used for defueling shall be bonded and grounded. Structures or appurtenances used for supporting the cylinders, containers or tanks shall be grounded in accordance with NFPA 70. The valve of the vehicle storage tank shall be bonded with the defueling system prior to the commencement of discharge or defueling operations.

2309.6.2 Repair of hydrogen piping. Piping systems containing hydrogen shall not be opened to the atmosphere for repair without first purging the piping with an inert gas to achieve 1-percent hydrogen or less by volume. Defueling operations and exiting purge flow shall be vented in accordance with Section 2309.6.1.2.

2309.6.3 Purging. Each individual manufactured component of a hydrogen generating, compression, storage or dispensing system shall have a label affixed as well as a description in the installation and owner's manuals describing the procedure for purging air from the system during startup, regular maintenance and for purging hydrogen from the system prior to disassembly (to admit air).

For the interconnecting piping between the individual manufactured components, the pressure rating must be not less than 20 times the absolute pressure present in the piping when any hydrogen meets any air.

2309.6.3.1 System purge required. After installation, repair or maintenance, the hydrogen piping system shall be purged of air in accordance with the manufacturer's procedure for purging air from the system.

SECTION 2310
MARINE MOTOR FUEL-DISPENSING FACILITIES

2310.1 General. The construction of marine motor fuel-dispensing facilities shall be in accordance with the *International Building Code* and NFPA 30A. The storage of Class I, II or IIIA liquids at marine motor fuel-dispensing facilities shall be in accordance with this chapter and Chapter 57.

2310.2 Storage and handling. The storage and handling of Class I, II or IIIA liquids at marine motor fuel-dispensing facilities shall be in accordance with Sections 2310.2.1 through 2310.2.3.

2310.2.1 Class I, II or IIIA liquid storage. Class I, II or IIIA liquids stored inside of buildings used for marine motor fuel-dispensing facilities shall be stored in *approved* containers or portable tanks. Storage of Class I liquids shall not exceed 10 gallons (38 L).

Exception: Storage in liquid storage rooms in accordance with Section 5704.3.7.

2310.2.2 Class II or IIIA liquid storage and dispensing. Class II or IIIA liquids stored or dispensed inside of buildings used for marine motor fuel-dispensing facilities shall be stored in and dispensed from *approved* containers or portable tanks. Storage of Class II and IIIA liquids shall not exceed 120 gallons (454 L).

2310.2.3 Heating equipment. Heating equipment installed in Class I, II or IIIA liquid storage or dispensing areas shall comply with Section 2301.6.

2310.3 Dispensing. The dispensing of liquid fuels at marine motor fuel-dispensing facilities shall comply with Sections 2310.3.1 through 2310.3.5.

2310.3.1 General. Wharves, piers or floats at marine motor fuel-dispensing facilities shall be used exclusively for the dispensing or transfer of petroleum products to or from marine craft, except that transfer of essential ship stores is allowed.

2310.3.2 Supervision. Marine motor fuel-dispensing facilities shall have an attendant or supervisor who is fully aware of the operation, mechanics and hazards inherent to fueling of boats on duty whenever the facility is open for business. The attendant's primary function shall be to supervise, observe and control the dispensing of Class I, II or IIIA liquids or flammable gases.

2310.3.3 Hoses and nozzles. Dispensing of Class I, II or IIIA liquids into the fuel tanks of marine craft shall be by means of an *approved*-type hose equipped with a *listed* automatic-closing nozzle without a latch-open device.

Hoses used for dispensing or transferring Class I, II or IIIA liquids, when not in use, shall be reeled, racked or otherwise protected from mechanical damage.

2310.3.4 Portable containers. Dispensing of Class I, II or IIIA liquids into containers, other than fuel tanks, shall be in accordance with Section 2304.4.1.

2310.3.5 Liquefied petroleum gas. Liquefied petroleum gas cylinders shall not be filled at marine motor fuel-dispensing facilities unless *approved*. *Approved* storage facilities for LP-gas cylinders shall be provided. See also Section 2307.

2310.4 Fueling of marine vehicles at other than approved marine motor fuel-dispensing facilities. Fueling of floating marine craft at other than a marine motor fuel-dispensing facility shall comply with Sections 2310.4.1 and 2310.4.2.

2310.4.1 Class I liquid fuels. Fueling of floating marine craft with Class I fuels at other than a marine motor fuel-dispensing facility is prohibited.

2310.4.2 Class II or III liquid fuels. Fueling of floating marine craft with Class II or III fuels at other than a marine motor fuel-dispensing facility shall be in accordance with all of the following:

1. The premises and operations shall be *approved* by the *fire code official*.
2. Tank vehicles and fueling operations shall comply with Section 5706.6.
3. The dispensing nozzle shall be of the *listed* automatic-closing type without a latch-open device.
4. Nighttime deliveries shall only be made in lighted areas.
5. The tank vehicle flasher lights shall be in operation while dispensing.
6. Fuel expansion space shall be left in each fuel tank to prevent overflow in the event of temperature increase.

2310.5 Fire prevention regulations. General fire safety regulations for marine motor fuel-dispensing facilities shall comply with Sections 2310.5.1 through 2310.5.7.

2310.5.1 Housekeeping. Marine motor fuel-dispensing facilities shall be maintained in a neat and orderly manner. Accumulations of rubbish or waste oils in excessive amounts shall be prohibited.

2310.5.2 Spills. Spills of Class I, II or IIIA liquids at or on the water shall be reported immediately to the fire department and jurisdictional authorities.

2310.5.3 Rubbish containers. Containers with tight-fitting or self-closing lids shall be provided for temporary storage of combustible debris, rubbish and waste material. The rubbish containers shall be constructed entirely of materials that comply with any one of the following:

1. Noncombustible materials.
2. Materials that meet a peak rate of heat release not exceeding 300 kW/m^2 when tested in accordance with ASTM E 1354 at an incident heat flux of 50 kW/m^2 in the horizontal orientation.

2310.5.4 Marine vessels and craft. Vessels or craft shall not be made fast to fuel docks serving other vessels or craft occupying a berth at a marine motor fuel-dispensing facility.

2310.5.5 Sources of ignition. Construction, maintenance, repair and reconditioning work involving the use of open flames, arcs or spark-producing devices shall not be performed at marine motor fuel-dispensing facilities or within 50 feet (15 240 mm) of the dispensing facilities, including piers, wharves or floats, except for emergency repair work *approved* in writing by the *fire code official*. Fueling shall not be conducted at the pier, wharf or float during the course of such emergency repairs.

2310.5.5.1 Smoking. Smoking or open flames shall be prohibited within 50 feet (15 240 mm) of fueling operations. "No Smoking" signs complying with Section 310 shall be posted conspicuously about the premises. Such signs shall have letters not less than 4 inches (102 mm) in height on a background of contrasting color.

2310.5.6 Preparation of tanks for fueling. Boat *owners* and operators shall not offer their craft for fueling unless the tanks being filled are properly vented to dissipate fumes to the outside atmosphere.

2310.5.7 Warning signs. Warning signs shall be prominently displayed at the face of each wharf, pier or float at such elevation as to be clearly visible from the decks of marine craft being fueled. Such signs shall have letters not less than 3 inches (76 mm) in height on a background of contrasting color bearing the following or *approved* equivalent wording:

WARNING

NO SMOKING—STOP ENGINE WHILE FUELING, SHUT OFF ELECTRICITY

DO NOT START ENGINE UNTIL AFTER BELOW DECK SPACES ARE VENTILATED.

2310.6 Fire protection. Fire protection features for marine motor fuel-dispensing facilities shall comply with Sections 2310.6.1 through 2310.6.4.

2310.6.1 Standpipe hose stations. Fire hose, where provided, shall be enclosed within a cabinet, and hose stations shall be labeled: FIRE HOSE—EMERGENCY USE ONLY.

2310.6.2 Obstruction of fire protection equipment. Materials shall not be placed on a pier in such a manner as to obstruct access to fire-fighting equipment or piping system control valves.

2310.6.3 Access. Where the pier is accessible to vehicular traffic, an unobstructed roadway to the shore end of the wharf shall be maintained for access by fire apparatus.

2310.6.4 Portable fire extinguishers. Portable fire extinguishers in accordance with Section 906, each having a minimum rating of 20-B:C, shall be provided as follows:

1. One on each float.
2. One on the pier or wharf within 25 feet (7620 mm) of the head of the gangway to the float, unless the office is within 25 feet (7620 mm) of the gangway or is on the float and an extinguisher is provided thereon.

SECTION 2311 REPAIR GARAGES

2311.1 General. Repair garages shall comply with this section and the *International Building Code*. Repair garages for vehicles that use more than one type of fuel shall comply with the applicable provisions of this section for each type of fuel used.

Where a repair garage includes a motor fuel-dispensing facility, the fuel-dispensing operation shall comply with the requirements of this chapter for motor fuel-dispensing facilities.

2311.2 Storage and use of flammable and combustible liquids. The storage and use of flammable and *combustible liquids* in repair garages shall comply with Chapter 57 and Sections 2311.2.1 through 2311.2.4.

2311.2.1 Cleaning of parts. Cleaning of parts shall be conducted in *listed* and *approved* parts-cleaning machines in accordance with Chapter 57.

2311.2.2 Waste oil, motor oil and other Class IIIB liquids. Waste oil, motor oil and other Class IIIB liquids shall be stored in *approved* tanks or containers, which are allowed to be stored and dispensed from inside repair garages.

2311.2.2.1 Tank location. Tanks storing Class IIIB liquids in repair garages are allowed to be located at, below or above grade, provided that adequate drainage or containment is provided.

2311.2.2.2 Liquid classification. Crankcase drainings shall be classified as Class IIIB liquids unless otherwise determined by testing.

2311.2.3 Drainage and disposal of liquids and oil-soaked waste. Garage floor drains, where provided, shall drain to *approved* oil separators or traps discharging to a sewer in accordance with the *International Plumbing Code*. Contents of oil separators, traps and floor drainage systems shall be collected at sufficiently frequent intervals and removed from the premises to prevent oil from being carried into the sewers.

2311.2.3.1 Disposal of liquids. Crankcase drainings and liquids shall not be dumped into sewers, streams or on the ground, but shall be stored in *approved* tanks or containers in accordance with Chapter 57 until removed from the premises.

2311.2.3.2 Disposal of oily waste. Self-closing metal cans shall be used for oily waste.

2311.2.4 Spray finishing. Spray finishing with flammable or *combustible liquids* shall comply with Chapter 24.

2311.3 Sources of ignition. Sources of ignition shall not be located within 18 inches (457 mm) of the floor and shall comply with Chapters 3 and 35.

2311.3.1 Equipment. Appliances and equipment installed in a repair garage shall comply with the provisions of the *International Building Code*, the *International Mechanical Code* and NFPA 70.

2311.3.2 Smoking. Smoking shall not be allowed in repair garages except in *approved* locations.

2311.4 Below-grade areas. Pits and below-grade work areas in repair garages shall comply with Sections 2311.4.1 through 2311.4.3.

2311.4.1 Construction. Pits and below-grade work areas shall be constructed in accordance with the *International Building Code*.

2311.4.2 Means of egress. Pits and below-grade work areas shall be provided with *means of egress* in accordance with Chapter 10.

2311.4.3 Ventilation. Where Class I liquids or LP-gas are stored or used within a building having a *basement* or pit wherein flammable vapors could accumulate, the *basement* or pit shall be provided with mechanical ventilation in accordance with the *International Mechanical Code*, at a minimum rate of $1^1/_2$ cubic feet per minute per square foot (cfm/ft^2) [0.008 m^3/(s · m^2)] to prevent the accumulation of flammable vapors.

2311.5 Preparation of vehicles for repair. For vehicles powered by gaseous fuels, the fuel shutoff valves shall be closed prior to repairing any portion of the vehicle fuel system.

Vehicles powered by gaseous fuels in which the fuel system has been damaged shall be inspected and evaluated for fuel system integrity prior to being brought into the repair garage. The inspection shall include testing of the entire fuel delivery system for leakage.

2311.6 Fire extinguishers. Fire extinguishers shall be provided in accordance with Section 906.

2311.7 Repair garages for vehicles fueled by lighter-than-air fuels. Repair garages for the conversion and repair of vehicles that use CNG, liquefied natural gas (LNG), hydrogen or other lighter-than-air motor fuels shall be in accordance with Sections 2311.7 through 2311.7.2.3 in addition to the other requirements of Section 2311.

Exceptions:

1. Repair garages where work is not performed on the fuel system and is limited to exchange of parts and maintenance not requiring open flame or welding on the CNG-, LNG-, hydrogen- or other lighter-than-air-fueled motor vehicle.
2. Repair garages for hydrogen-fueled vehicles where work is not performed on the hydrogen storage tank and is limited to the exchange of parts and maintenance not requiring open flame or welding on the hydrogen-fueled vehicle. During the work, the entire hydrogen fuel system shall contain a quantity that is less than 200 cubic feet (5.6 m^3) of hydrogen.

2311.7.1 Ventilation. Repair garages used for the repair of natural gas- or hydrogen-fueled vehicles shall be provided with an *approved* mechanical ventilation system. The mechanical ventilation system shall be in accordance with the *International Mechanical Code* and Sections 2311.7.1.1 and 2311.7.1.2.

Exception: Repair garages with natural ventilation when *approved*.

2311.7.1.1 Design. Indoor locations shall be ventilated utilizing air supply inlets and exhaust outlets arranged to provide uniform air movement to the extent practical. Inlets shall be uniformly arranged on exterior walls near floor level. Outlets shall be located at the high point of the room in exterior walls or the roof.

Ventilation shall be by a continuous mechanical ventilation system or by a mechanical ventilation system activated by a continuously monitoring natural gas detection system or, for hydrogen, a continuously monitoring flammable gas detection system, each activating at a gas concentration of not more than 25 percent of the lower flammable limit (LFL). In all cases, the system shall shut down the fueling system in the event of failure of the ventilation system.

The ventilation rate shall be not less than 1 cubic foot per minute per 12 cubic feet [0.00139 m^3 × (s · m^3)] of room volume.

2311.7.1.2 Operation. The mechanical ventilation system shall operate continuously.

Exceptions:

1. Mechanical ventilation systems that are interlocked with a gas detection system designed in accordance with Sections 2311.7.2 through 2311.7.2.3.
2. Mechanical ventilation systems in repair garages that are used only for repair of vehi-

cles fueled by liquid fuels or odorized gases, such as CNG, where the ventilation system is electrically interlocked with the lighting circuit.

2311.7.2 Gas detection system. Repair garages used for repair of vehicles fueled by nonodorized gases including, but not limited to, hydrogen and nonodorized LNG, shall be provided with a flammable gas detection system.

2311.7.2.1 System design. The flammable gas detection system shall be *listed* or *approved* and shall be calibrated to the types of fuels or gases used by vehicles to be repaired. The gas detection system shall be designed to activate when the level of flammable gas exceeds 25 percent of the lower flammable limit (LFL). Gas detection shall be provided in lubrication or chassis service pits of repair garages used for repairing nonodorized LNG-fueled vehicles.

2311.7.2.1.1 Gas detection system components. Gas detection system control units shall be *listed* and *labeled* in accordance with UL 864 or UL 2017. Gas detectors shall be *listed* and *labeled* in accordance with UL 2075 for use with the gases and vapors being detected.

2311.7.2.2 Operation. Activation of the gas detection system shall result in all the following:

1. Initiation of distinct audible and visual alarm signals in the repair garage.
2. Deactivation of all heating systems located in the repair garage.
3. Activation of the mechanical ventilation system, where the system is interlocked with gas detection.

2311.7.2.3 Failure of the gas detection system. Failure of the gas detection system shall result in the deactivation of the heating system, activation of the mechanical ventilation system where the system is interlocked with the gas detection system and cause a trouble signal to sound in an *approved* location.

2311.8 Defueling equipment required at vehicle maintenance and repair facilities. Facilities for repairing hydrogen fuel systems on hydrogen-fueled vehicles shall have equipment to defuel vehicle storage tanks. Where work must be performed on a vehicle's fuel storage tank for the purpose of maintenance, repair or cylinder certification, defueling and purging shall be conducted in accordance with Section 2309.6.

*

CHAPTER 24
FLAMMABLE FINISHES

SECTION 2401
GENERAL

2401.1 Scope. This chapter shall apply to locations or areas where any of the following activities are conducted:

1. The application of flammable finishes to articles or materials by means of spray apparatus.
2. The application of flammable finishes by dipping or immersing articles or materials into the contents of tanks, vats or containers of flammable or *combustible liquids* for coating, finishing, treatment or similar processes.
3. The application of flammable finishes by applying combustible powders to articles or materials utilizing powder spray guns, electrostatic powder spray guns, fluidized beds or electrostatic fluidized beds.
4. Floor surfacing or finishing operations using Class I or II liquids in areas exceeding 350 square feet (32.5 m^2).
5. The application of flammable finishes consisting of dual-component coatings or Class I or II liquids when applied by brush or roller in quantities exceeding 1 gallon (4 L).

2401.2 Nonapplicability. This chapter shall not apply to spray finishing utilizing flammable or *combustible liquids* that do not sustain combustion, including:

1. Liquids that have no fire point when tested in accordance with ASTM D 92.
2. Liquids with a flashpoint greater than 95°F (35°C) in a water-miscible solution or dispersion with a water and inert (noncombustible) solids content of more than 80 percent by weight.

2401.3 Permits. Permits shall be required as set forth in Sections 105.6 and 105.7.

SECTION 2402
DEFINITIONS

2402.1 Definitions. The following terms are defined in Chapter 2:

DETEARING.

DIP TANK.

ELECTROSTATIC FLUIDIZED BED.

FLAMMABLE FINISHES.

FLAMMABLE VAPOR AREA.

FLUIDIZED BED.

LIMITED SPRAYING SPACE.

RESIN APPLICATION AREA.

ROLL COATING.

SPRAY BOOTH.

SPRAY ROOM.

SPRAYING SPACE.

SECTION 2403
PROTECTION OF OPERATIONS

2403.1 General. Operations covered by this chapter shall be protected as required by Sections 2403.2 through 2403.4.4.

2403.2 Sources of ignition. Protection against sources of ignition shall be provided in accordance with Sections 2403.2.1 through 2403.2.8.

2403.2.1 Electrical wiring and equipment. Electrical wiring and equipment shall comply with this chapter and NFPA 70.

2403.2.1.1 Flammable vapor areas. Electrical wiring and equipment in flammable vapor areas shall be of an explosionproof type *approved* for use in such hazardous locations. Such areas shall be considered to be Class I, Division 1 or Class II, Division 1 hazardous locations in accordance with NFPA 70.

2403.2.1.2 Areas subject to deposits of residues. Electrical equipment, flammable vapor areas or drying operations that are subject to splashing or dripping of liquids shall be specifically *approved* for locations containing deposits of readily ignitable residue and explosive vapors.

Exceptions:

1. This provision shall not apply to wiring in rigid conduit, threaded boxes or fittings not containing taps, splices or terminal connections.
2. This provision shall not apply to electrostatic equipment allowed by Section 2407.

In resin application areas, electrical wiring and equipment that is subject to deposits of combustible residues shall be *listed* for such exposure and shall be installed as required for hazardous (classified) locations. Electrical wiring and equipment not subject to deposits of combustible residues shall be installed as required for ordinary hazard locations.

2403.2.1.3 Areas adjacent to spray booths. Electrical wiring and equipment located outside of, but within 5 feet (1524 mm) horizontally and 3 feet (914 mm) vertically of openings in a spray booth or a spray room, shall be *approved* for Class I, Division 2 or Class II, Division 2 hazardous locations, whichever is applicable.

2403.2.1.4 Areas subject to overspray deposits. Electrical equipment in flammable vapor areas located such that deposits of combustible residues could readily accumulate thereon shall be specifically *approved* for locations containing deposits of readily ignitable residue and *explosive* vapors in accordance with NFPA 70.

Exceptions:

1. Wiring in rigid conduit.
2. Boxes or fittings not containing taps, splices or terminal connections.
3. Equipment allowed by Sections 2404 and 2407 and Chapter 30.

2403.2.2 Open flames and sparks. Open flames and spark-producing devices shall not be located in flammable vapor areas and shall not be located within 20 feet (6096 mm) of such areas unless separated by a permanent partition.

Exception: Drying and baking apparatus complying with Section 2404.6.1.2.

2403.2.3 Hot surfaces. Heated surfaces having a temperature sufficient to ignite vapors shall not be located in flammable vapor areas. Space-heating appliances, steam pipes or hot surfaces in a flammable vapor area shall be located such that they are not subject to accumulation of deposits of combustible residues.

Exception: Drying apparatus complying with Section 2404.6.1.2.

2403.2.4 Equipment enclosures. Equipment or apparatus that is capable of producing sparks or particles of hot metal that would fall into a flammable vapor area shall be totally enclosed.

2403.2.5 Grounding. Metal parts of spray booths, exhaust ducts and piping systems conveying Class I or II liquids shall be electrically grounded in accordance with NFPA 70. Metallic parts located in resin application areas, including but not limited to exhaust ducts, ventilation fans, spray application equipment, workpieces and piping, shall be electrically grounded.

2403.2.6 Smoking prohibited. Smoking shall be prohibited in flammable vapor areas and hazardous materials storage rooms associated with flammable finish processes. "No Smoking" signs complying with Section 310 shall be conspicuously posted in such areas.

2403.2.7 Welding warning signs. Welding, cutting and similar spark-producing operations shall not be conducted in or adjacent to flammable vapor areas or dipping or coating operations unless precautions have been taken to provide safety. Conspicuous signs with the following warning shall be posted in the vicinity of flammable vapor areas, dipping operations and paint storage rooms:

NO WELDING
THE USE OF WELDING OR CUTTING
EQUIPMENT IN OR NEAR THIS AREA
IS DANGEROUS BECAUSE OF FIRE
AND EXPLOSION HAZARDS. WELDING
AND CUTTING SHALL BE DONE ONLY
UNDER THE SUPERVISION OF THE
PERSON IN CHARGE.

2403.2.8 Powered industrial trucks. Powered industrial trucks used in electrically classified areas shall be *listed* for such use.

2403.3 Storage, use and handling of flammable and combustible liquids. The storage, use and handling of flammable and *combustible liquids* shall be in accordance with this section and Chapter 57.

2403.3.1 Use. Containers supplying spray nozzles shall be of a closed type or provided with metal covers that are kept closed. Containers not resting on floors shall be on noncombustible supports or suspended by wire cables. Containers supplying spray nozzles by gravity flow shall not exceed 10 gallons (37.9 L) in capacity.

2403.3.2 Valves. Containers and piping to which a hose or flexible connection is attached shall be provided with a shutoff valve at the connection. Such valves shall be kept shut when hoses are not in use.

2403.3.3 Pumped liquid supplies. Where flammable or *combustible liquids* are supplied to spray nozzles by positive displacement pumps, pump discharge lines shall be provided with an *approved* relief valve discharging to pump suction or a safe detached location.

2403.3.4 Liquid transfer. Where a flammable mixture is transferred from one portable container to another, a bond shall be provided between the two containers. Not less than one container shall be grounded. Piping systems for Class I and II liquids shall be permanently grounded.

2403.3.5 Class I liquids as solvents. Class I liquids used as solvents shall be used in spray gun and equipment cleaning machines that have been *listed* and *approved* for such purpose or shall be used in spray booths or spray rooms in accordance with Sections 2403.3.5.1 and 2403.3.5.2.

2403.3.5.1 Listed devices. Cleaning machines for spray guns and equipment shall not be located in areas open to the public and shall be separated from ignition sources in accordance with their listings or by a distance of 3 feet (914 mm), whichever is greater. The quantity of solvent used in a machine shall not exceed the design capacity of the machine.

2403.3.5.2 Within spray booths and spray rooms. When solvents are used for cleaning spray nozzles and auxiliary equipment within spray booths and spray rooms, the ventilating equipment shall be operated during cleaning.

2403.3.6 Class II and III liquids. Solvents used outside of spray booths, spray rooms or *listed* and *approved* spray gun and equipment cleaning machines shall be restricted to Class II and III liquids.

2403.4 Operations and maintenance. Flammable vapor areas, exhaust fan blades and exhaust ducts shall be kept free from the accumulation of deposits of combustible residues. Where excessive residue accumulates in such areas, spraying operations shall be discontinued until conditions are corrected.

2403.4.1 Tools. Scrapers, spuds and other tools used for cleaning purposes shall be constructed of nonsparking materials.

2403.4.2 Residue. Residues removed during cleaning and debris contaminated with residue shall be immediately removed from the premises and properly disposed.

2403.4.3 Waste cans. *Approved* metal waste cans equipped with self-closing lids shall be provided wherever rags or waste are impregnated with finishing material. Such rags and waste shall be deposited therein immediately after being utilized. The contents of waste cans shall be properly disposed of not less than once daily and at the end of each shift.

2403.4.4 Solvent recycling. Solvent distillation equipment used to recycle and clean dirty solvents shall comply with Section 5705.4.

SECTION 2404
SPRAY FINISHING

2404.1 General. The application of flammable or *combustible liquids* by means of spray apparatus in continuous or intermittent processes shall be in accordance with the requirements of Sections 2403 and 2404.2 through 2404.9.4.

2404.2 Location of spray-finishing operations. Spray-finishing operations conducted in buildings used for Group A, E, I or R occupancies shall be located in a spray room protected with an *approved automatic sprinkler system* installed in accordance with Section 903.3.1.1 and separated vertically and horizontally from other areas in accordance with the *International Building Code*. In other occupancies, spray-finishing operations shall be conducted in a spray room, spray booth or spraying space *approved* for such use.

Exceptions:

1. Automobile undercoating spray operations and spray-on automotive lining operations conducted in areas with *approved* natural or mechanical ventilation shall be exempt from the provisions of Section 2404 when *approved* and where utilizing Class IIIA or IIIB *combustible liquids*.
2. In buildings other than Group A, E, I or R occupancies, *approved* limited spraying space in accordance with Section 2404.9.
3. Resin application areas used for manufacturing of reinforced plastics complying with Section 2409 shall not be required to be located in a spray room, spray booth or spraying space.

2404.3 Design and construction. Design and construction of spray rooms, spray booths and spray spaces shall be in accordance with Sections 2404.3.1 through 2404.3.3.1.

2404.3.1 Spray rooms. Spray rooms shall be constructed and designed in accordance with Section 2404.3.1.1 and the *International Building Code*, and shall comply with Sections 2404.4 through 2404.8.

2404.3.1.1 Floor. Combustible floor construction in spray rooms shall be covered by *approved*, noncombustible, nonsparking material, except where combustible coverings, including but not limited to thin paper or plastic and strippable coatings, are utilized over noncombustible materials to facilitate cleaning operations in spray rooms.

2404.3.2 Spray booths. The design and construction of spray booths shall be in accordance with Sections 2404.3.2.1 through 2404.3.2.6, Sections 2404.4 through 2404.8 and NFPA 33.

2404.3.2.1 Construction. Spray booths shall be constructed of *approved* noncombustible materials. Aluminum shall not be used. Where walls or ceiling assemblies are constructed of sheet metal, single-skin assemblies shall be no thinner than 0.0478 inch (18 gage) (1.2 mm) and each sheet of double-skin assemblies shall be no thinner than 0.0359 inch (20 gage) (0.9 mm). Structural sections of spray booths are allowed to be sealed with latex-based or similar caulks and sealants.

2404.3.2.2 Surfaces. The interior surfaces of spray booths shall be smooth; shall be constructed so as to permit the free passage of exhaust air from all parts of the interior, and to facilitate washing and cleaning; and shall be designed to confine residues within the booth. Aluminum shall not be used.

2404.3.2.3 Floor. Combustible floor construction in spray booths shall be covered by *approved*, noncombustible, nonsparking material, except where combustible coverings, including but not limited to thin paper or plastic and strippable coatings, are utilized over noncombustible materials to facilitate cleaning operations in spray booths.

2404.3.2.4 Means of egress. *Means of egress* shall be provided in accordance with Chapter 10.

Exception: *Means of egress* doors from premanufactured spray booths shall be not less than 30 inches (762 mm) in width by 80 inches (2032 mm) in height.

2404.3.2.5 Clear space. Spray booths shall be installed so that all parts of the booth are readily accessible for cleaning. A clear space of not less than 3 feet (914 mm) shall be maintained on all sides of the spray booth. This clear space shall be kept free of any storage or combustible construction.

Exceptions:

1. This requirement shall not prohibit locating a spray booth closer than 3 feet (914 mm) to or directly against an interior partition, wall or floor/ceiling assembly that has a *fire-resistance rating* of not less than 1 hour, provided the spray booth can be adequately maintained and cleaned.
2. This requirement shall not prohibit locating a spray booth closer than 3 feet (914 mm) to an exterior wall or a roof assembly, provided the wall or roof is constructed of noncombustible

material and the spray booth can be adequately maintained and cleaned.

2404.3.2.6 Size. The aggregate area of spray booths in a building shall not exceed the lesser of 10 percent of the area of any floor of a building or the basic area allowed for a Group H-2 occupancy without area increases, as set forth in the *International Building Code*. The area of an individual spray booth in a building shall not exceed the lesser of the aggregate size limit or 1,500 square feet (139 m^2).

Exception: One individual booth not exceeding 500 square feet (46 m^2).

2404.3.3 Spraying spaces. Spraying spaces shall be designed and constructed in accordance with the *International Building Code*, and Section 2404.3.3.1 and Sections 2404.4 through 2404.8 of this code.

2404.3.3.1 Floor. Combustible floor construction in spraying spaces shall be covered by *approved*, noncombustible, nonsparking material, except where combustible coverings, such as thin paper or plastic and strippable coatings, are utilized over noncombustible materials to facilitate cleaning operations in spraying spaces.

2404.4 Fire protection. Spray booths and spray rooms shall be protected by an *approved* automatic fire-extinguishing system complying with Chapter 9. Protection shall also extend to exhaust plenums, exhaust ducts and both sides of dry filters when such filters are used.

2404.4.1 Fire extinguishers. Portable fire extinguishers complying with Section 906 shall be provided for spraying areas in accordance with the requirements for an extra (high) hazard occupancy.

2404.5 Housekeeping, maintenance and storage of hazardous materials. Housekeeping, maintenance, storage and use of hazardous materials shall be in accordance with Sections 2403.3, 2403.4, 2404.5.1 and 2404.5.2.

2404.5.1 Different coatings. Spray booths, spray rooms and spraying spaces shall not be alternately utilized for different types of coating materials where the combination of materials is conducive to spontaneous ignition, unless all deposits of one material are removed from the booth, room or space and exhaust ducts prior to spraying with a different material.

2404.5.2 Protection of sprinklers. Automatic sprinklers installed in flammable vapor areas shall be protected from the accumulation of residue from spraying operations in an *approved* manner. Bags used as a protective covering shall be 0.003-inch-thick (0.076 mm) polyethylene or cellophane or shall be thin paper. Automatic sprinklers contaminated by overspray particles shall be replaced with new automatic sprinklers.

2404.6 Sources of ignition. Control of sources of ignition shall be in accordance with Section 2403.2 and Sections 2404.6.1 through 2404.6.2.4.

2404.6.1 Drying operations. Spray booths and spray rooms shall not be alternately used for the purpose of drying by arrangements or methods that could cause an increase in the surface temperature of the spray booth or spray room except in accordance with Sections 2404.6.1.1 and 2404.6.1.2. Except as specifically provided in this section, drying or baking units utilizing a heating system having open flames or that are capable of producing sparks shall not be installed in a flammable vapor areas.

2404.6.1.1 Spraying procedure. The spraying procedure shall use low-volume spray application.

2404.6.1.2 Drying apparatus. Fixed drying apparatus shall comply with this chapter and the applicable provisions of Chapter 30. When recirculation ventilation is provided in accordance with Section 2404.7.2, the heating system shall not be within the recirculation air path.

2404.6.1.2.1 Interlocks. The spraying apparatus, drying apparatus and ventilating system for the spray booth or spray room shall be equipped with interlocks arranged to accomplish all of the following:

1. Prevent operation of the spraying apparatus while drying operations are in progress.
2. Where the drying apparatus is located in the spray booth or spray room, prevent operation of the drying apparatus until a timed purge of spray vapors from the spray booth or spray room is complete. This purge time shall be based upon completing at least four air changes of spray booth or spray room volume or for a period of not less than 3 minutes, whichever is greater.
3. Have the ventilating system maintain a safe atmosphere within the spray booth or spray room during the drying process and automatically shut off drying apparatus in the event of a failure of the ventilating system.
4. Shut off the drying apparatus automatically if the air temperature within the booth exceeds 200°F (93°C).

2404.6.1.2.2 Portable infrared apparatus. Where a portable infrared drying apparatus is used, electrical wiring and portable infrared drying equipment shall comply with NFPA 70. Electrical equipment located within 18 inches (457 mm) of floor level shall be *approved* for Class I, Division 2 hazardous locations. Metallic parts of drying apparatus shall be electrically bonded and grounded. During spraying operations, portable drying apparatus and electrical connections and wiring thereto shall not be located within spray booths, spray rooms or other areas where spray residue would be deposited thereon.

2404.6.2 Illumination. Where spraying spaces, spray rooms or spray booths are illuminated through glass panels or other transparent materials, only fixed luminaires shall be utilized as a source of illumination.

2404.6.2.1 Glass panels. Panels for luminaires or for observation shall be of heat-treated glass, wired glass or hammered wire glass and shall be sealed to confine vapors, mists, residues, dusts and deposits to the flammable vapor area. Panels for luminaires shall be separated from the luminaire to prevent the surface temperature of the panel from exceeding 200°F (93°C).

2404.6.2.2 Exterior luminaires. Luminaires attached to the walls or ceilings of a flammable vapor area, but outside of any classified area and separated from the flammable vapor areas by vapor-tight glass panels, shall be suitable for use in ordinary hazard locations. Such luminaires shall be serviced from outside the flammable vapor areas.

2404.6.2.3 Integral luminaires. Luminaires that are an integral part of the walls or ceiling of a flammable vapor area are allowed to be separated from the flammable vapor area by glass panels that are an integral part of the luminaire. Such luminaires shall be *listed* for use in Class I, Division 2 or Class II, Division 2 locations, whichever is applicable, and also shall be suitable for accumulations of deposits of combustible residues. Such luminaires are allowed to be serviced from inside the flammable vapor area.

2404.6.2.4 Portable electric lamps. Portable electric lamps shall not be used in flammable vapor areas during spraying operations. Portable electric lamps used during cleaning or repairing operations shall be of a type *approved* for hazardous locations.

2404.7 Ventilation. Mechanical ventilation of flammable vapor areas shall be provided in accordance with Section 502.7 of the *International Mechanical Code*.

2404.7.1 Operation. Mechanical ventilation shall be kept in operation at all times while spraying operations are being conducted and for a sufficient time thereafter to allow vapors from drying coated articles and finishing material residue to be exhausted. Spraying equipment shall be interlocked with the ventilation of the flammable vapor areas such that spraying operations cannot be conducted unless the ventilation system is in operation.

2404.7.2 Recirculation. Air exhausted from spraying operations shall not be recirculated.

Exceptions:

1. Air exhausted from spraying operations is allowed to be recirculated as makeup air for unmanned spray operations, provided that all of the following conditions exist:
   - 1.1. The solid particulate has been removed.
   - 1.2. The vapor concentration is less than 25 percent of the LFL.
   - 1.3. *Approved* equipment is used to monitor the vapor concentration.
   - 1.4. When the vapor concentration exceeds 25 percent of the LFL, both of the following shall occur:
     - a. An alarm shall sound.
     - b. Spray operations shall automatically shut down.
   - 1.5. In the event of shutdown of the vapor concentration monitor, 100 percent of the air volume specified in Section 510 of the *International Mechanical Code* is automatically exhausted.
2. Air exhausted from spraying operations is allowed to be recirculated as makeup air to manned spraying operations where all of the conditions provided in Exception 1 are included in the installation and documents have been prepared to show that the installation does not pose a life safety hazard to personnel inside the spray booth, spraying space or spray room.

2404.7.3 Air velocity. The ventilation system shall be designed, installed and maintained so that the flammable contaminants are diluted in noncontaminated air to maintain concentrations in the exhaust airflow below 25 percent of the contaminant's lower flammable limit (LFL). In addition, the spray booth shall be provided with mechanical ventilation so that the average air velocity through openings is in accordance with Sections 2404.7.3.1 and 2404.7.3.2.

2404.7.3.1 Open-face or open-front spray booth. For spray application operations conducted in an open-face or open-front spray booth, the ventilation system shall be designed, installed and maintained so that the average air velocity into the spray booth through all openings is not less than 100 feet per minute (0.51 m/s).

Exception: For fixed or automated electrostatic spray application equipment, the average air velocity into the spray booth through all openings shall be not less than 50 feet per minute (0.25 m/s).

2404.7.3.2 Enclosed spray booth or spray room with openings for product conveyance. For spray application operations conducted in an enclosed spray booth or spray room with openings for product conveyance, the ventilation system shall be designed, installed and maintained so that the average air velocity into the spray booth through openings is not less than 100 feet per minute (0.51 m/s).

Exceptions:

1. For fixed or automated electrostatic spray application equipment, the average air velocity into the spray booth through all openings shall be not less than 50 feet per minute (0.25 m/s).
2. Where methods are used to reduce cross drafts that can draw vapors and overspray through openings from the spray booth or spray room, the average air velocity into the spray booth or spray room shall be that necessary to capture and confine vapors and overspray to the spray booth or spray room.

2404.7.4 Ventilation obstruction. Articles being sprayed shall be positioned in a manner that does not obstruct collection of overspray.

2404.7.5 Independent ducts. Each spray booth and spray room shall have an independent exhaust duct system discharging to the outside.

Exceptions:

1. Multiple spray booths having a combined frontal area of 18 square feet (1.67 m^2) or less are allowed to have a common exhaust when identical spray finishing material is used in each booth. If more than one fan serves one booth, fans shall be interconnected such that all fans will operate simultaneously.
2. Where treatment of exhaust is necessary for air pollution control or for energy conservation, ducts shall be allowed to be manifolded if all of the following conditions are met:
   - 2.1. The sprayed materials used are compatible and will not react or cause ignition of the residue in the ducts.
   - 2.2. Nitrocellulose-based finishing material shall not be used.
   - 2.3. A filtering system shall be provided to reduce the amount of overspray carried into the duct manifold.
   - 2.4. Automatic sprinkler protection shall be provided at the junction of each booth exhaust with the manifold, in addition to the protection required by this chapter.

2404.7.6 Termination point. The termination point for exhaust ducts discharging to the atmosphere shall be not less than the following distances:

1. Ducts conveying explosive or flammable vapors, fumes or dusts: 30 feet (9144 mm) from the lot line; 10 feet (3048 mm) from openings into the building; 6 feet (1829 mm) from exterior walls and roofs; 30 feet (9144 mm) from combustible walls or openings into the building that are in the direction of the exhaust discharge; 10 feet (3048 mm) above adjoining grade.
2. Other product-conveying outlets: 10 feet (3048 mm) from the lot line; 3 feet (914 mm) from exterior walls and roofs; 10 feet (3048 mm) from openings into the building; 10 feet (3048 mm) above adjoining grade.

2404.7.7 Fan motors and belts. Electric motors driving exhaust fans shall not be placed inside booths or ducts. Fan rotating elements shall be nonferrous or nonsparking or the casing shall consist of, or be lined with, such material. Belts shall not enter the duct or booth unless the belt and pulley within the duct are tightly enclosed.

2404.7.8 Filters. Air intake filters that are part of a wall or ceiling assembly shall be *listed* as Class I or II in accordance with UL 900. Exhaust filters shall be required.

2404.7.8.1 Supports. Supports and holders for filters shall be constructed of noncombustible materials.

2404.7.8.2 Attachment. Overspray collection filters shall be readily removable and accessible for cleaning or replacement.

2404.7.8.3 Maintaining air velocity. Visible gauges, audible alarms or pressure-activated devices shall be installed to indicate or ensure that the required air velocity is maintained.

2404.7.8.4 Filter rolls. Spray booths equipped with a filter roll that is automatically advanced when the air velocity is reduced to less than 100 feet per minute (0.51 m/s) shall be arranged to shut down the spraying operation if the filter roll fails to advance automatically.

2404.7.8.5 Filter disposal. Discarded filter pads shall be immediately removed to a safe, detached location or placed in a noncombustible container with a tight-fitting lid and disposed of properly.

2404.7.8.6 Spontaneous ignition. Spray booths using dry filters shall not be used for spraying materials that are highly susceptible to spontaneous heating and ignition. Filters shall be changed prior to spraying materials that could react with other materials previously collected. An example of a potentially reactive combination includes lacquer when combined with varnishes, stains or primers.

2404.7.8.7 Waterwash spray booths. Waterwash spray booths shall be of an *approved* design so as to prevent excessive accumulation of deposits in ducts and residue at duct outlets. Such booths shall be arranged so that air and overspray are drawn through a continuously flowing water curtain before entering an exhaust duct to the building exterior.

2404.8 Interlocks. Interlocks for spray application finishes shall be in accordance with Sections 2404.8.1 through 2404.8.2.

2404.8.1 Automated spray application operations. Where protecting automated spray application operations, automatic fire-extinguishing systems shall be equipped with an *approved* interlock feature that will, upon discharge of the system, automatically stop the spraying operations and workpiece conveyors into and out of the flammable vapor areas. Where the building is equipped with a fire alarm system, discharge of the automatic fire-extinguishing system shall also activate the building alarm notification appliances.

2404.8.1.1 Alarm station. A manual fire alarm and emergency system shutdown station shall be installed to serve each flammable vapor area. When activated, the station shall accomplish the functions indicated in Section 2404.8.1.

2404.8.1.2 Alarm station location. Not less than one manual fire alarm and emergency system shutdown station shall be readily accessible to operating personnel. Where access to this station is likely to involve exposure to danger, an additional station shall be located adjacent to an *exit* from the area.

2404.8.2 Ventilation interlock prohibited. Air makeup and flammable vapor area exhaust systems shall not be interlocked with the fire alarm system and shall remain in operation during a fire alarm condition.

Exception: Where the type of fire-extinguishing system used requires such ventilation to be discontinued, air makeup and exhaust systems shall shut down and dampers shall close.

2404.9 Limited spraying spaces. Limited spraying spaces shall comply with Sections 2404.9.1 through 2404.9.4.

2404.9.1 Job size. The aggregate surface area to be sprayed shall not exceed 9 square feet (0.84 m^2).

2404.9.2 Frequency. Spraying operations shall not be of a continuous nature.

2404.9.3 Ventilation. Positive mechanical ventilation providing a minimum of six complete air changes per hour shall be installed. Such system shall meet the requirements of this code for handling flammable vapor areas. Explosion venting is not required.

2404.9.4 Electrical wiring. Electrical wiring within 10 feet (3048 mm) of the floor and 20 feet (6096 mm) horizontally of the limited spraying space shall be designed for Class I, Division 2 locations in accordance with NFPA 70.

SECTION 2405 DIPPING OPERATIONS

2405.1 General. Dip-tank operations shall comply with the requirements of Section 2403 and Sections 2405.2 through 2405.11.

2405.2 Location of dip-tank operations. Dip-tank operations conducted in buildings used for Group A, I or R occupancies shall be located in a room designed for that purpose, equipped with an *approved automatic sprinkler system* and separated vertically and horizontally from other areas in accordance with the *International Building Code*.

2405.3 Construction of dip tanks. Dip tanks shall be constructed in accordance with Sections 2405.3.1 through 2405.3.4.3 and NFPA 34. Dip tanks, including drain boards, shall be constructed of noncombustible material and their supports shall be of heavy metal, reinforced concrete or masonry.

2405.3.1 Overflow. Dip tanks greater than 150 gallons (568 L) in capacity or 10 square feet (0.93 m^2) in liquid surface area shall be equipped with a trapped overflow pipe leading to an *approved* location outside the building. The bottom of the overflow connection shall be not less than 6 inches (152 mm) below the top of the tank.

2405.3.2 Bottom drains. Dip tanks greater than 500 gallons (1893 L) in liquid capacity shall be equipped with bottom drains that are arranged to automatically and manually drain the tank quickly in the event of a fire unless the viscosity of the liquid at normal atmospheric temperature makes this impractical. Manual operation shall be from a safe, accessible location. Where gravity flow is not practicable, automatic pumps shall be provided. Such drains shall be trapped and discharged to a closed, vented salvage tank or to an *approved* outside location.

Exception: Dip tanks containing Class IIIB *combustible liquids* where the liquids are not heated above room temperature and the process area is protected by automatic sprinklers.

2405.3.3 Dipping liquid temperature control. Protection against the accumulation of vapors, self-ignition and excessively high temperatures shall be provided for dipping liquids that are heated directly or heated by the surfaces of the object being dipped.

2405.3.4 Dip-tank covers. Dip-tank covers allowed by Section 2405.4.1 shall be capable of manual operation and shall be automatic closing by *approved* automatic-closing devices designed to operate in the event of a fire.

2405.3.4.1 Construction. Covers shall be constructed of noncombustible material or be of a tin-clad type with enclosing metal applied with locked joints.

2405.3.4.2 Supports. Chain or wire rope shall be utilized for cover supports or operating mechanisms.

2405.3.4.3 Closed covers. Covers shall be kept closed when tanks are not in use.

2405.4 Fire protection. Dip-tank operations shall be protected in accordance with Sections 2405.4.1 through 2405.4.2.

2405.4.1 Fixed fire-extinguishing equipment. An *approved* automatic fire-extinguishing system or dip-tank cover in accordance with Section 2405.3.4 shall be provided for the following dip tanks:

1. Dip tanks less than 150 gallons (568 L) in capacity or 10 square feet (0.93 m^2) in liquid surface area.
2. Dip tanks containing a liquid with a *flash point* below 110°F (43°C) used in such manner that the liquid temperature could equal or be greater than its *flash point* from artificial or natural causes, and having both a capacity of more than 10 gallons (37.9 L) and a liquid surface area of more than 4 square feet (0.37 m^2).

2405.4.1.1 Fire-extinguishing system. An *approved* automatic fire-extinguishing system shall be provided for dip tanks with a 150-gallon (568 L) or more capacity or 10 square feet (0.93 m^2) or larger in a liquid surface area. Fire-extinguishing system design shall be in accordance with NFPA 34.

2405.4.2 Portable fire extinguishers. Areas in the vicinity of dip tanks shall be provided with portable fire extinguishers complying with Section 906 and suitable for flammable and combustible liquid fires as specified for extra (high) hazard occupancies.

2405.5 Housekeeping, maintenance and storage of hazardous materials. Housekeeping, maintenance, storage and use of hazardous materials shall be in accordance with Sections 2403.3 and 2403.4.

2405.6 Sources of ignition. Control of sources of ignition shall be in accordance with Section 2403.2.

2405.7 Ventilation of flammable vapor areas. Flammable vapor areas shall be provided with mechanical ventilation adequate to prevent the dangerous accumulation of vapors. Required ventilation systems shall be arranged such that the failure of any ventilating fan shall automatically stop the dipping conveyor system.

2405.8 Conveyor interlock. Dip tanks utilizing a conveyor system shall be arranged such that in the event of a fire, the conveyor system shall automatically cease motion and the required tank bottom drains shall open.

2405.9 Hardening and tempering tanks. Hardening and tempering tanks shall comply with Sections 2405.3 through 2405.3.3, 2405.4.2 and 2405.8, but shall be exempt from other provisions of Section 2405.

2405.9.1 Location. Tanks shall be located as far as practical from furnaces and shall not be located on or near combustible floors.

2405.9.2 Hoods. Tanks shall be provided with a noncombustible hood and vent or other *approved* venting means, terminating outside of the structure to serve as a vent in case of a fire. Such vent ducts shall be treated as flues and proper clearances shall be maintained from combustible materials.

2405.9.3 Alarms. Tanks shall be equipped with a high-temperature limit switch arranged to sound an alarm when the temperature of the quenching medium reaches 50°F (10°C) below the *flash point*.

2405.9.4 Fire protection. Hardening and tempering tanks greater than 500 gallons (1893 L) in capacity or 25 square feet (2.3 m^2) in liquid surface area shall be protected by an *approved* automatic fire-extinguishing system complying with Chapter 9.

2405.9.5 Use of air pressure. Air under pressure shall not be used to fill or agitate oil in tanks.

2405.10 Flow-coating operations. Flow-coating operations shall comply with the requirements for dip tanks. The area of the sump and any areas on which paint flows shall be considered to be the area of a dip tank.

2405.10.1 Paint supply. Paint shall be supplied by a gravity tank not exceeding 10 gallons (38 L) in capacity or by direct low-pressure pumps arranged to shut down automatically in case of a fire by means of *approved* heat-actuated devices.

2405.11 Roll-coating operations. Roll-coating operations shall comply with Section 2405.10. In roll-coating operations utilizing flammable or *combustible liquids*, sparks from static electricity shall be prevented by electrically bonding and grounding all metallic rotating and other parts of machinery and equipment and by the installation of static collectors, or by maintaining a conductive atmosphere such as a high relative humidity.

SECTION 2406
POWDER COATING

2406.1 General. Operations using finely ground particles of protective finishing material applied in dry powder form by a fluidized bed, an electrostatic fluidized bed, powder spray guns or electrostatic powder spray guns shall comply with Sections 2406.2 through 2406.7. In addition, Section 2407 shall apply to fixed electrostatic equipment used in powder coating operations.

2406.2 Location. Powder coating operations shall be conducted in enclosed powder coating rooms, enclosed powder coating facilities that are ventilated or ventilated spray booths.

2406.3 Construction of powder coating rooms and booths. Powder coating rooms shall be constructed of noncombustible materials. Spray booths shall be constructed in accordance with Section 2404.3.2.

Exception: *Listed* spray-booth assemblies that are constructed of other materials shall be allowed.

2406.4 Fire protection. Areas used for powder coating shall be protected by an *approved* automatic fire-extinguishing system complying with Chapter 9.

2406.4.1 Additional protection for fixed systems. Automated powder application equipment shall be protected by the installation of an *approved*, supervised flame detection apparatus that shall react to the presence of flame within 0.5 second and shall accomplish all of the following:

1. Shutting down of energy supplies (electrical and compressed air) to conveyor, ventilation, application, transfer and powder collection equipment.
2. Closing of segregation dampers in associated ductwork to interrupt airflow from application equipment to powder collectors.
3. Activation of an alarm that is audible throughout the powder coating room or booth.

2406.4.2 Fire extinguishers. Portable fire extinguishers complying with Section 906 shall be provided for areas used for powder coating in accordance with the requirements for an extra-hazard occupancy.

2406.5 Operation and maintenance. Powder coating areas shall be kept free from the accumulation of powder coating dusts, including horizontal surfaces such as ledges, beams, pipes, hoods, booths and floors.

2406.5.1 Cleaning. Surfaces shall be cleaned in such a manner so as to avoid scattering dusts to other places or creating dust clouds. Vacuum sweeping equipment shall be of a type *approved* for use in hazardous locations.

2406.6 Sources of ignition. Control of sources of ignition shall be in accordance with Section 2403.2 and Sections 2406.6.1 through 2406.6.4.

2406.6.1 Drying, curing and fusion equipment. Drying, curing and fusion equipment shall comply with Chapter 30.

2406.6.2 Spark-producing metals. Iron or spark-producing metals shall be prevented from being introduced into the powders being applied by magnetic separators, filter-type separators or by other *approved* means.

2406.6.3 Preheated parts. When parts are heated prior to coating, the temperature of the parts shall not exceed the ignition temperature of the powder to be used.

2406.6.4 Grounding and bonding. Precautions shall be taken to minimize the possibility of ignition by static electrical sparks through static bonding and grounding, where possible, of powder transport, application and recovery equipment.

2406.7 Ventilation. Exhaust ventilation shall be sufficient to maintain the atmosphere below one-half the minimum *explosive* concentration for the material being applied. Nondeposited, air-suspended powders shall be removed through exhaust ducts to the powder recovery system.

SECTION 2407
ELECTROSTATIC APPARATUS

2407.1 General. Electrostatic apparatus and devices used in connection with paint-spraying and paint-detearing operations shall be of an *approved* type.

2407.2 Location and clear space. A space of not less than twice the sparking distance shall be maintained between goods being painted or deteared and electrodes, electrostatic atomizing heads or conductors. A sign stating the sparking distance shall be conspicuously posted near the assembly.

Exception: Portable electrostatic paint-spraying apparatus *listed* for use in Class I, Division 1, locations.

2407.3 Construction of equipment. Electrodes and electrostatic atomizing heads shall be of *approved* construction, rigidly supported in permanent locations and effectively insulated from ground. Insulators shall be nonporous and noncombustible.

Exception: Portable electrostatic paint-spraying apparatus *listed* for use in Class I, Division 1, locations.

2407.3.1 Barriers. Booths, fencing, railings or guards shall be placed about the equipment such that either by their location or character, or both, isolation of the process is maintained from plant storage and personnel. Railings, fencing and guards shall be of conductive material, adequately grounded, and not less than 5 feet (1524 mm) from processing equipment.

Exception: Portable electrostatic paint-spraying apparatus *listed* for use in Class I, Division 1, locations.

2407.4 Fire protection. Areas used for electrostatic spray finishing with fixed equipment shall be protected with an *approved* automatic fire-extinguishing system complying with Chapter 9 and Section 2407.4.1.

2407.4.1 Protection for automated liquid electrostatic spray application equipment. Automated liquid electrostatic spray application equipment shall be protected by the installation of an *approved*, supervised flame detection apparatus that shall, in the event of ignition, react to the presence of flame within 0.5 second and shall accomplish all of the following:

1. Activation of a local alarm in the vicinity of the spraying operation and activation of the building alarm system, if such a system is provided.
2. Shutting down of the coating material delivery system.
3. Termination of all spray application operations.
4. Stopping of conveyors into and out of the flammable vapor areas.
5. Disconnection of power to the high-voltage elements in the flammable vapor areas and disconnection of power to the system.

2407.5 Housekeeping, maintenance and storage of hazardous materials. Housekeeping, maintenance, storage and use of hazardous materials shall be in accordance with Sections 2403.3, 2403.4 and Sections 2407.5.1 and 2407.5.2.

2407.5.1 Maintenance. Insulators shall be kept clean and dry. Drip plates and screens subject to paint deposits shall be removable and taken to a safe place for cleaning. Grounds and bonding means for the paint-spraying apparatus and all associated equipment shall be periodically cleaned and maintained free of overspray.

2407.5.2 Signs. Signs shall be posted to provide the following information:

1. Designate the process zone as dangerous with respect to fire and accident.
2. Identify the grounding requirements for all electrically conductive objects in the flammable vapor area, including persons.
3. Restrict access to qualified personnel only.

2407.6 Sources of ignition. Transformers, power packs, control apparatus and all other electrical portions of the equipment, except high-voltage grids and electrostatic atomizing heads and connections, shall be located outside of the flammable vapor areas or shall comply with Section 2403.2.

2407.7 Ventilation. The flammable vapor area shall be ventilated in accordance with Section 2404.7.

2407.8 Emergency shutdown. Electrostatic apparatus shall be equipped with automatic controls operating without time delay to disconnect the power supply to the high-voltage transformer and signal the operator under any of the following conditions:

1. Stoppage of ventilating fans or failure of ventilating equipment from any cause.
2. Stoppage of the conveyor carrying articles past the high-voltage grid.
3. Occurrence of a ground or an imminent ground at any point of the high-voltage system.
4. Reduction of clearance below that required in Section 2407.2.

2407.9 Ventilation interlock. Hand electrostatic equipment shall be interlocked with the ventilation system for the spraying area so that the equipment cannot be operated unless the ventilating system is in operation.

SECTION 2408
ORGANIC PEROXIDES AND DUAL-COMPONENT COATINGS

2408.1 General. Spraying operations involving the use of organic peroxides and other dual-component coatings shall be in accordance with the requirements of Section 2403 and Sections 2408.2 through 2408.5.

2408.2 Use of organic peroxide coatings. Spraying operations involving the use of organic peroxides and other dual-component coatings shall be conducted in *approved* sprinklered spray booths complying with Section 2404.3.2.

2408.3 Equipment. Spray guns and related handling equipment used with organic peroxides shall be of a type manufactured for such use.

2408.3.1 Pressure tanks. Separate pressure vessels and inserts specifically for the application shall be used for the resin and for the organic peroxide, and shall not be interchanged. Organic peroxide pressure tank inserts shall be constructed of stainless steel or polyethylene.

2408.4 Housekeeping, maintenance, storage and use of hazardous materials. Housekeeping, maintenance, storage and use of hazardous materials shall be in accordance with Sections 2403.3 and 2403.4 and Sections 2408.4.1 through 2408.4.7.

2408.4.1 Contamination prevention. Organic peroxide initiators shall not be contaminated with foreign substances.

2408.4.2 Spilled material. Spilled organic peroxides shall be promptly removed so there are no residues. Spilled material absorbed by using a noncombustible absorbent shall be promptly disposed of in accordance with the manufacturer's recommendation.

2408.4.3 Residue control. Materials shall not be contaminated by dusts and overspray residues resulting from the sanding or spraying of finishing materials containing organic peroxides.

2408.4.4 Handling. Handling of organic peroxides shall be conducted in a manner that avoids shock and friction that produces decomposition and violent reaction hazards.

2408.4.5 Mixing. Organic peroxides shall not be mixed directly with accelerators or promoters.

2408.4.6 Personnel qualifications. Personnel working with organic peroxides and dual-component coatings shall be specifically trained to work with these materials.

2408.4.7 Storage. The storage of organic peroxides shall comply with Chapter 62.

2408.5 Sources of ignition. Only nonsparking tools shall be used in areas where organic peroxides are stored, mixed or applied.

SECTION 2409
INDOOR MANUFACTURING OF REINFORCED PLASTICS

2409.1 General. Indoor manufacturing processes involving spray or hand application of reinforced plastics and using more than 5 gallons (19 L) of resin in a 24-hour period shall be in accordance with Sections 2409.2 through 2409.6.1.

2409.2 Resin application equipment. Equipment used for spray application of resin shall be installed and used in accordance with Section 2408 and Sections 2409.3 through 2409.6.1.

2409.3 Fire protection. Resin application areas shall be protected by an *automatic sprinkler system*. The sprinkler system design shall be not less than that required for Ordinary Hazard, Group 2, with a minimum design area of 3,000 square feet (279 m^2). Where the materials or storage arrangements are required by other regulations to be provided with a higher level of sprinkler system protection, the higher level of sprinkler system protection shall be provided.

2409.4 Housekeeping, maintenance, storage and use of hazardous materials. Housekeeping, maintenance, storage and use of hazardous materials shall be in accordance with Sections 2403.3 and 2403.4 and Sections 2409.4.1 through 2409.4.3.

2409.4.1 Handling of excess catalyzed resin. A noncombustible, open-top container shall be provided for disposal of excess catalyzed resin. Excess catalyzed resin shall be drained into the container while still in the liquid state. Enough water shall be provided in the container to maintain a minimum 2-inch (51 mm) water layer over the contained resin.

2409.4.2 Control of overchop. In areas where chopper guns are used, exposed wall and floor surfaces shall be covered with paper, polyethylene film or other *approved* material to allow for removal of overchop. Overchop shall be allowed to cure for not less than 4 hours prior to removal.

2409.4.2.1 Disposal. Following removal, used wall and floor covering materials required by Section 2409.4.2 shall be placed in a noncombustible container and removed from the facility.

2409.4.3 Storage and use of hazardous materials. Storage and use of organic peroxides shall be in accordance with Section 2408 and Chapter 62. Storage and use of flammable and *combustible liquids* shall be in accordance with Chapter 57. Storage and use of unstable (reactive) materials shall be in accordance with Chapter 66.

2409.5 Sources of ignition in resin application areas. Sources of ignition in resin application areas shall comply with Section 2403.2.

2409.6 Ventilation. Mechanical ventilation shall be provided throughout resin application areas in accordance with Section 2404.7. The ventilation rate shall be adequate to maintain the concentration of flammable vapors in the resin application area at or below 25 percent of the LFL.

Exception: Mechanical ventilation is not required for buildings that have 75 percent of the perimeter unenclosed.

2409.6.1 Local ventilation. Local ventilation shall be provided inside of workpieces where personnel will be under or inside of the workpiece.

SECTION 2410
FLOOR SURFACING AND FINISHING OPERATIONS

2410.1 Scope. Floor surfacing and finishing operations exceeding 350 square feet (33 m^2) and using Class I or II liquids shall comply with Sections 2410.2 through 2410.5.

2410.2 Mechanical system operation. Heating, ventilation and air-conditioning systems shall not be operated during resurfacing or refinishing operations or within 4 hours of the application of flammable or *combustible liquids*.

2410.3 Business operation. Floor surfacing and finishing operations shall not be conducted while an establishment is open to the public.

2410.4 Ignition sources. The power shall be shut down to all electrical sources of ignition within the flammable vapor area, unless those devices are classified for use in Class I, Division 1 hazardous locations.

2410.5 Ventilation. To prevent the accumulation of flammable vapors, mechanical ventilation at a minimum rate of 1 cubic foot per minute per square foot [0.00508 $m^3/(s \cdot m^2)$] of area being finished shall be provided. Such exhaust shall be by *approved* temporary or portable means. Vapors shall be exhausted to the exterior of the building.

CHAPTER 25

FRUIT AND CROP RIPENING

SECTION 2501 GENERAL

2501.1 Scope. Ripening processes where ethylene gas is introduced into a room to promote the ripening of fruits, vegetables and other crops shall comply with this chapter.

Exception: Mixtures of ethylene and one or more inert gases in concentrations that prevent the gas from reaching greater than 25 percent of the lower explosive limit (LEL) when released to the atmosphere.

2501.2 Permits. Permits shall be required as set forth in Section 105.6.

2501.3 Ethylene generators. *Approved* ethylene generators shall be operated and maintained in accordance with Section 2506.

SECTION 2502 DEFINITIONS

2502.1 Terms defined in Chapter 2. Words and terms used in this chapter and defined in Chapter 2 shall have the meanings ascribed to them as defined therein.

SECTION 2503 ETHYLENE GAS

2503.1 Location. Ethylene gas shall be discharged only into *approved* rooms or enclosures designed and constructed for this purpose.

2503.2 Dispensing. Valves controlling discharge of ethylene shall provide positive and fail-closed control of flow and shall be set to limit the concentration of gas in air below 1,000 parts per million (ppm).

SECTION 2504 SOURCES OF IGNITION

2504.1 Ignition prevention. Sources of ignition shall be controlled or protected in accordance with this section and Chapter 3.

2504.2 Electrical wiring and equipment. Electrical wiring and equipment, including luminaires, shall be *approved* for use in Class I, Division 2, Group C hazardous (classified) locations.

2504.3 Static electricity. Containers, piping and equipment used to dispense ethylene shall be bonded and grounded to prevent the discharge of static sparks or arcs.

2504.4 Lighting. Lighting shall be by *approved* electric lamps or luminaires only.

2504.5 Heating. Heating shall be by indirect means utilizing low-pressure steam, hot water or warm air.

Exception: Electric or fuel-fired heaters *approved* for use in hazardous (classified) locations and that are installed and operated in accordance with the applicable provisions of NFPA 70, the *International Mechanical Code* or the *International Fuel Gas Code.*

SECTION 2505 COMBUSTIBLE WASTE

2505.1 Housekeeping. Empty boxes, cartons, pallets and other combustible waste shall be removed from ripening rooms or enclosures and disposed of at regular intervals in accordance with Chapter 3.

SECTION 2506 ETHYLENE GENERATORS

2506.1 Ethylene generators. Ethylene generators shall be *listed* and *labeled* by an *approved* testing laboratory, *approved* by the *fire code official* and used only in *approved* rooms in accordance with the ethylene generator manufacturer's instructions. The listing evaluation shall include documentation that the concentration of ethylene gas does not exceed 25 percent of the lower explosive limit (LEL).

2506.2 Ethylene generator rooms. Ethylene generators shall be used in rooms having a volume of not less than 1,000 cubic feet (28 m^3). Rooms shall have air circulation to ensure even distribution of ethylene gas and shall be free from sparks, open flames or other ignition sources.

SECTION 2507 WARNING SIGNS

2507.1 Where required. *Approved* warning signs indicating the danger involved and necessary precautions shall be posted on all doors and entrances to the premises.

CHAPTER 26

FUMIGATION AND INSECTICIDAL FOGGING

SECTION 2601 GENERAL

2601.1 Scope. Fumigation and insecticidal fogging operations within buildings, structures and spaces shall comply with this chapter.

2601.2 Permits. Permits shall be required as set forth in Section 105.6.

SECTION 2602 DEFINITIONS

2602.1 Definitions. The following terms are defined in Chapter 2:

FUMIGANT.

FUMIGATION.

INSECTICIDAL FOGGING.

SECTION 2603 FIRE SAFETY REQUIREMENTS

2603.1 General. Buildings, structures and spaces in which fumigation and insecticidal fogging operations are conducted shall comply with the fire protection and safety requirements of Sections 2603.2 through 2603.7.

2603.2 Sources of ignition. Fires, open flames and similar sources of ignition shall be eliminated from the space under fumigation or insecticidal fogging. Heating, where needed, shall be of an *approved* type.

2603.2.1 Electricity. Electricity in any part of the building, structure or space where operation of switches or electrical devices, equipment or systems could serve as a source of ignition shall be shut off.

Exception: Circulating fans that have been specifically designed for utilization in hazardous atmospheres and installed in accordance with NFPA 70.

2603.2.2 Electronic devices. Electronic devices, including portable equipment and cellular phones, shall be shut off. Telephone lines shall be disconnected from telephones.

2603.2.3 Duration. Sources of ignition shall be shut off during the fumigation activity and remain shut off until the ventilation required in Section 2603.6 is completed.

2603.3 Notification. The *fire code official* and fire chief shall be notified in writing not less than 48 hours before the building, structure or space is to be closed in connection with the utilization of any toxic or flammable fumigant. Notification shall give the location of the enclosed space to be fumigated or fogged, the occupancy, the fumigants or insecticides to be utilized, the person or persons responsible for the operation, and the date and time at which the operation will begin. Written notice of any fumigation or insecticidal fogging operation shall be given to all affected occupants of the building, structure or space in which such operations are to be conducted with sufficient advance notice to allow the occupants to evacuate the building, structure or space. Such notice shall inform the occupants as to the purposes, anticipated duration and hazards associated with the fumigation or insecticidal fogging operation.

2603.3.1 Warning signs. *Approved* warning signs indicating the danger, type of chemical involved and necessary precautions shall be posted on all doors and entrances to the affected building, structure or space and upon all gangplanks and ladders from the deck, pier or land to a ship. Such notices shall be printed in red ink on a white background. Letters in the headlines shall be not less than 2 inches (51 mm) in height and shall state the date and time of the operation, the name and address of the person, the name of the operator in charge, and a warning stating that the affected building, structure or space shall be vacated not less than 1 hour before the operation begins and shall not be reentered until the danger signs have been removed by the proper authorities.

2603.3.2 Breathing apparatus. Persons engaged in the business of fumigation or insecticidal fogging shall maintain and have available *approved* protective breathing apparatus.

2603.3.3 Watch personnel. During the period fumigation is in progress, except where fumigation is conducted in a gas-tight vault or tank, a responsible watchperson shall remain on duty at the entrance or entrances to the enclosed fumigated space until after the fumigation is completed and the building, structure or space is properly ventilated and safe for occupancy. Sufficient watchers shall be provided to prevent persons from entering the enclosed space under fumigation without being observed.

2603.3.4 Evacuation during fumigation. Occupants of the building, structure or space to be fumigated, except the personnel conducting the fumigation, shall be evacuated from such building, structure or space prior to commencing fumigation operations.

2603.3.5 Evacuation during insecticidal fogging operations. Occupants in the building, structure or space to be fogged, except the personnel conducting the insecticidal fogging operations, shall be evacuated from such building, structure or space prior to commencing fogging operations.

2603.4 Insecticidal fogging liquids. Insecticidal fogging liquids with a *flash point* below 100°F (38°C) shall not be utilized.

2603.5 Sealing of buildings, structures and spaces. Paper and other similar materials that do not meet the flame propagation performance criteria of Test Method 1 or Test Method

2, as appropriate, of NFPA 701 shall not be used to wrap or cover a building, structure or space in excess of that required for the sealing of cracks, casements and similar openings.

2603.5.1 Maintenance of openings. All openings to the building, structure or space to be fumigated or fogged shall be kept securely closed during such operation.

2603.6 Venting and cleanup. At the end of the exposure period, fumigators shall safely and properly ventilate the premises and contents; properly dispose of fumigant containers, residues, debris and other materials used for such fumigation; and clear obstructions from gas-fired appliance vents.

2603.7 Flammable fumigants restricted. The use of carbon disulfide and hydrogen cyanide shall be restricted to agricultural fumigation.

CHAPTER 27

SEMICONDUCTOR FABRICATION FACILITIES

SECTION 2701 GENERAL

2701.1 Scope. Semiconductor fabrication facilities and comparable research and development areas classified as Group H-5 shall comply with this chapter and the *International Building Code*. The use, storage and handling of hazardous materials in Group H-5 shall comply with this chapter, other applicable provisions of this code and the *International Building Code*.

2701.2 Application. The requirements set forth in this chapter are requirements specific only to Group H-5 and shall be applied as exceptions or additions to applicable requirements set forth elsewhere in this code.

2701.3 Multiple hazards. Where a material poses multiple hazards, all hazards shall be addressed in accordance with Section 5001.1.

2701.4 Existing buildings and existing fabrication areas. Existing buildings and existing *fabrication areas* shall comply with this chapter, except that transportation and handling of HPM in *corridors* and enclosures for stairways and ramps shall be allowed where in compliance with Section 2705.3.2 and the *International Building Code*.

2701.5 Permits. Permits shall be required as set forth in Section 105.6.

SECTION 2702 DEFINITIONS

2702.1 Definitions. The following terms are defined in Chapter 2:

CONTINUOUS GAS DETECTION SYSTEM.

EMERGENCY CONTROL STATION.

FABRICATION AREA.

➡ **HAZARDOUS PRODUCTION MATERIAL (HPM).**

HPM ROOM.

PASS-THROUGH.

SEMICONDUCTOR FABRICATION FACILITY.

SERVICE CORRIDOR.

TOOL.

WORKSTATION.

SECTION 2703 GENERAL SAFETY PROVISIONS

2703.1 Emergency control station. An *emergency control station* shall be provided in accordance with Sections 2703.1.1 through 2703.1.3.

2703.1.1 Location. The *emergency control station* shall be located on the premises at an *approved* location outside the fabrication area.

2703.1.2 Staffing. Trained personnel shall continuously staff the *emergency control station.*

2703.1.3 Signals. The *emergency control station* shall receive signals from emergency equipment and alarm and detection systems. Such emergency equipment and alarm and detection systems shall include, but not be limited to, the following where such equipment or systems are required to be provided either in this chapter or elsewhere in this code:

1. *Automatic sprinkler system* alarm and monitoring systems.
2. Manual fire alarm systems.
3. Emergency alarm systems.
4. Continuous gas detection systems.
5. Smoke detection systems.
6. Emergency power system.
7. Automatic detection and alarm systems for pyrophoric liquids and Class 3 water-reactive liquids required by Section 2705.2.3.4.
8. Exhaust ventilation flow alarm devices for pyrophoric liquids and Class 3 water-reactive liquids cabinet exhaust ventilation systems required by Section 2705.2.3.4.

2703.2 Systems, equipment and processes. Systems, equipment and processes shall be in accordance with Sections 2703.2.1 through 2703.2.3.2.

2703.2.1 Application. Systems, equipment and processes shall include, but not be limited to, containers, cylinders, tanks, piping, tubing, valves and fittings.

2703.2.2 General requirements. In addition to the requirements in Section 2703.2, systems, equipment and processes shall also comply with Section 5003.2, other applicable provisions of this code, the *International Building Code* and the *International Mechanical Code.*

2703.2.3 Additional requirements for HPM supply piping. In addition to the requirements in Section 2703.2, HPM supply piping and tubing for HPM gases and liquids shall comply with this section.

2703.2.3.1 General requirements. The requirements set forth in Section 5003.2.2.2 shall apply to supply piping and tubing for HPM gases and liquids.

2703.2.3.2 Health-hazard ranking 3 or 4 HPM. Supply piping and tubing for HPM gases and liquids having a health-hazard ranking of 3 or 4 shall be welded throughout, except for connections located within a ventilation enclosure if the material is a gas, or an *approved* method of drainage or containment provided for connections if the material is a liquid.

2703.3 Construction requirements. Construction of semiconductor fabrication facilities shall be in accordance with Sections 2703.3.1 through 2703.3.9.

2703.3.1 Fabrication areas. Construction and location of *fabrication areas* shall comply with the *International Building Code*.

2703.3.2 Pass-throughs in exit access corridors. Pass-throughs in *exit access corridors* shall be constructed in accordance with the *International Building Code*.

2703.3.3 Liquid storage rooms. Liquid storage rooms shall comply with Chapter 57 and the *International Building Code*.

2703.3.4 HPM rooms. HPM rooms shall comply with the *International Building Code*.

2703.3.5 Gas cabinets. Gas cabinets shall comply with Section 5003.8.6.

2703.3.6 Exhausted enclosures. Exhausted enclosures shall comply with Section 5003.8.5.

2703.3.7 Gas rooms. Gas rooms shall comply with Section 5003.8.4.

2703.3.8 Service corridors. Service corridors shall comply with Section 2705.3 and the *International Building Code*.

2703.3.9 Cabinets containing pyrophoric liquids or water-reactive Class 3 liquids. Cabinets in *fabrication areas* containing pyrophoric liquids or Class 3 water-reactive liquids in containers or in amounts greater than $^1/_2$ gallon (2 L) shall comply with Section 2705.2.3.4.

2703.4 Emergency plan. An emergency plan shall be established as set forth in Section 403.7.1.

2703.5 Maintenance of equipment, machinery and processes. Maintenance of equipment, machinery and processes shall comply with Section 5003.2.6.

2703.6 Security of areas. Areas shall be secured in accordance with Section 5003.9.2.

2703.7 Electrical wiring and equipment. Electrical wiring and equipment in HPM facilities shall comply with Sections 2703.7.1 through 2703.7.3.

2703.7.1 Fabrication areas. Electrical wiring and equipment in *fabrication areas* shall comply with NFPA 70.

2703.7.2 Workstations. Electrical equipment and devices within 5 feet (1524 mm) of workstations in which flammable or pyrophoric gases or flammable liquids are used shall comply with NFPA 70 for Class I, Division 2 hazardous locations. Workstations shall not be energized without adequate exhaust ventilation in accordance with Section 2703.14.

Exception: Class I, Division 2 hazardous electrical equipment is not required where the air removal from the workstation or dilution will prevent the accumulation of flammable vapors and fumes on a continuous basis.

2703.7.3 Hazardous production material (HPM) rooms, gas rooms and liquid storage rooms. Electrical wiring and equipment in HPM rooms, gas rooms and liquid storage rooms shall comply with NFPA 70.

2703.8 Corridors and enclosures for stairways and ramps. Hazardous materials shall not be used or stored in *corridors* or enclosures for stairways and ramps.

2703.9 Service corridors. Hazardous materials shall not be used in an open-system use condition in service corridors.

2703.10 Automatic sprinkler system. An *approved automatic sprinkler system* shall be provided in accordance with Sections 2703.10.1 through 2703.10.5 and Chapter 9.

2703.10.1 Workstations and tools. The design of the sprinkler system in the area shall take into consideration the spray pattern and the effect on the equipment.

2703.10.1.1 Combustible workstations. A sprinkler head shall be installed within each branch exhaust connection or individual plenums of workstations of combustible construction. The sprinkler head in the exhaust connection or plenum shall be located not more than 2 feet (610 mm) from the point of the duct connection or the connection to the plenum. Where necessary to prevent corrosion, the sprinkler head and connecting piping in the duct shall be coated with *approved* or *listed* corrosion-resistant materials. The sprinkler head shall be accessible for periodic inspection.

Exceptions:

1. *Approved* alternative automatic fire-extinguishing systems are allowed. Activation of such systems shall deactivate the related processing equipment.
2. Process equipment that operates at temperatures exceeding 932°F (500°C) and is provided with automatic shutdown capabilities for hazardous materials.
3. Exhaust ducts 10 inches (254 mm) or less in diameter from flammable gas storage cabinets that are part of a workstation.
4. Ducts *listed* or *approved* for use without internal automatic sprinkler protection.

2703.10.1.2 Combustible tools. Where the horizontal surface of a combustible tool is obstructed from ceiling sprinkler discharge, automatic sprinkler protection that covers the horizontal surface of the tool shall be provided.

Exceptions:

1. An automatic gaseous fire-extinguishing local surface application system shall be allowed as an alternative to sprinklers. Gaseous-extinguishing systems shall be actuated by infrared (IR) or ultraviolet/infrared (UV/IR) optical detectors.
2. Tools constructed of materials that are listed as Class 1 or Class 2 in accordance with UL

2360 or *approved* for use without internal fire-extinguishing system protection.

2703.10.2 Gas cabinets and exhausted enclosures. An *approved automatic sprinkler system* shall be provided in gas cabinets and exhausted enclosures containing HPM *compressed gases*.

Exception: Gas cabinets located in an HPM room other than those cabinets containing pyrophoric gases.

2703.10.3 Pass-throughs in existing exit access corridors. Pass-throughs in existing *exit access corridors* shall be protected by an *approved automatic sprinkler system.*

2703.10.4 Exhaust ducts for HPM. An *approved automatic sprinkler system* shall be provided in exhaust ducts conveying gases, vapors, fumes, mists or dusts generated from HPM in accordance with this section and the *International Mechanical Code.*

2703.10.4.1 Metallic and noncombustible nonmetallic exhaust ducts. An *approved automatic sprinkler system* shall be provided in metallic and noncombustible nonmetallic exhaust ducts where all of the following conditions apply:

1. Where the largest cross-sectional diameter is equal to or greater than 10 inches (254 mm).
2. The ducts are within the building.
3. The ducts are conveying flammable gases, vapors or fumes.

2703.10.4.2 Combustible nonmetallic exhaust ducts. An *approved automatic sprinkler system* shall be provided in combustible nonmetallic exhaust ducts where the largest cross-sectional diameter of the duct is equal to or greater than 10 inches (254 mm).

Exceptions:

1. Ducts *listed* or *approved* for applications without *automatic sprinkler system* protection.
2. Ducts not more than 12 feet (3658 mm) in length installed below ceiling level.

2703.10.4.3 Exhaust connections and plenums of combustible workstations. Automatic fire-extinguishing system protection for exhaust connections and plenums of combustible workstations shall comply with Section 2703.10.1.1.

2703.10.4.4 Exhaust duct sprinkler system requirements. Automatic sprinklers installed in exhaust duct systems shall be hydraulically designed to provide 0.5 gallons per minute (gpm) (1.9 L/min) over an area derived by multiplying the distance between the sprinklers in a horizontal duct by the width of the duct. Minimum discharge shall be 20 gpm (76 L/min) per sprinkler from the five hydraulically most remote sprinklers.

2703.10.4.4.1 Sprinkler head locations. Automatic sprinklers shall be installed at 12-foot (3658 mm) intervals in horizontal ducts and at changes in direction. In vertical runs, automatic sprinklers shall be installed at the top and at alternate floor levels.

2703.10.4.4.2 Control valve. A separate indicating control valve shall be provided for sprinklers installed in exhaust ducts.

2703.10.4.4.3 Drainage. Drainage shall be provided to remove sprinkler water discharged in exhaust ducts.

2703.10.4.4.4 Corrosive atmospheres. Where corrosive atmospheres exist, exhaust duct sprinklers and pipe fittings shall be manufactured of corrosion-resistant materials or coated with *approved* materials.

2703.10.4.4.5 Maintenance and inspection. Sprinklers in exhaust ducts shall be accessible for periodic inspection and maintenance.

2703.10.5 Sprinkler alarms and supervision. *Automatic sprinkler systems* shall be electrically supervised and provided with alarms in accordance with Chapter 9. *Automatic sprinkler system* alarm and supervisory signals shall be transmitted to the *emergency control station.*

2703.11 Manual fire alarm system. A manual fire alarm system shall be installed throughout buildings containing a Group H-5 occupancy. Activation of the alarm system shall initiate a local alarm and transmit a signal to the *emergency control station.* Manual fire alarm systems shall be designed and installed in accordance with Section 907.

2703.12 Emergency alarm system. Emergency alarm systems shall be provided in accordance with Sections 2703.12.1 through 2703.12.3, Section 5004.9 and Section 5005.4.4. The *maximum allowable quantity per control area* provisions of Section 5004.1 shall not apply to emergency alarm systems required for HPM.

2703.12.1 Where required. Emergency alarm systems shall be provided in the areas indicated in Sections 2703.12.1.1 through 2703.12.1.3.

2703.12.1.1 Service corridors. An *approved* emergency alarm system shall be provided in service corridors, with not less than one alarm device in the service corridor.

2703.12.1.2 Corridors and interior exit stairways and ramps. Emergency alarms for corridors, interior exit stairways and ramps and exit passageways shall comply with Section 5005.4.4.

2703.12.1.3 Liquid storage rooms, HPM rooms and gas rooms. Emergency alarms for liquid storage rooms, HPM rooms and gas rooms shall comply with Section 5004.9.

2703.12.2 Alarm-initiating devices. An *approved* emergency telephone system, local alarm manual pull stations, or other *approved* alarm-initiating devices are allowed to be used as emergency alarm-initiating devices.

2703.12.3 Alarm signals. Activation of the emergency alarm system shall sound a local alarm and transmit a signal to the *emergency control station.*

2703.13 Continuous gas detection systems. A continuous gas detection system shall be provided for HPM gases where the physiological warning threshold level of the gas is at a higher level than the accepted permissible exposure limit (PEL) for the gas and for flammable gases in accordance with Sections 2703.13.1 through 2703.13.2.2.

2703.13.1 Where required. A continuous gas detection system shall be provided in the areas identified in Sections 2703.13.1.1 through 2703.13.1.4.

2703.13.1.1 Fabrication areas. A continuous gas detection system shall be provided in *fabrication areas* where gas is used in the fabrication area.

2703.13.1.2 HPM rooms. A continuous gas detection system shall be provided in HPM rooms where gas is used in the room.

2703.13.1.3 Gas cabinets, exhausted enclosures and gas rooms. A continuous gas detection system shall be provided in gas cabinets and exhausted enclosures. A continuous gas detection system shall be provided in gas rooms where gases are not located in gas cabinets or exhausted enclosures.

2703.13.1.4 Corridors. Where gases are transported in piping placed within the space defined by the walls of a *corridor* and the floor or roof above the *corridor*, a continuous gas detection system shall be provided where piping is located and in the *corridor*.

Exception: A continuous gas detection system is not required for occasional transverse crossings of the *corridors* by supply piping that is enclosed in a ferrous pipe or tube for the width of the *corridor*.

2703.13.2 Gas detection system operation. The continuous gas detection system shall be capable of monitoring the room, area or equipment in which the gas is located at or below all the following gas concentrations:

1. Immediately dangerous to life and health (IDLH) values where the monitoring point is within an exhausted enclosure, ventilated enclosure or gas cabinet.
2. Permissible exposure limit (PEL) levels where the monitoring point is in an area outside an exhausted enclosure, ventilated enclosure or gas cabinet.
3. For flammable gases, the monitoring detection threshold level shall be vapor concentrations in excess of 25 percent of the lower flammable limit (LFL) where the monitoring is within or outside an exhausted enclosure, ventilated enclosure or gas cabinet.
4. Except as noted in this section, monitoring for highly toxic and toxic gases shall also comply with Chapter 60.

2703.13.2.1 Alarms. The gas detection system shall initiate a local alarm and transmit a signal to the *emergency control station* when a short-term hazard condition is detected. The alarm shall be both visible and audible and shall provide warning both inside and outside the area where the gas is detected. The audible alarm shall be distinct from all other alarms.

2703.13.2.2 Shut off of gas supply. The gas detection system shall automatically close the shutoff valve at the source on gas supply piping and tubing related to the system being monitored for which gas is detected when a short-term hazard condition is detected. Automatic closure of shutoff valves shall comply with the following:

1. Where the gas-detection sampling point initiating the gas detection system alarm is within a gas cabinet or exhausted enclosure, the shutoff valve in the gas cabinet or exhausted enclosure for the specific gas detected shall automatically close.
2. Where the gas-detection sampling point initiating the gas detection system alarm is within a room and *compressed gas* containers are not in gas cabinets or exhausted enclosure, the shutoff valves on all gas lines for the specific gas detected shall automatically close.
3. Where the gas-detection sampling point initiating the gas detection system alarm is within a piping distribution manifold enclosure, the shutoff valve supplying the manifold for the *compressed gas* container of the specific gas detected shall automatically close.

Exception: Where the gas-detection sampling point initiating the gas detection system alarm is at the use location or within a gas valve enclosure of a branch line downstream of a piping distribution manifold, the shutoff valve for the branch line located in the piping distribution manifold enclosure shall automatically close.

2703.14 Exhaust ventilation systems for HPM. Exhaust ventilation systems and materials for exhaust ducts utilized for the exhaust of HPM shall comply with Sections 2703.14.1 through 2703.14.3, other applicable provisions of this code, the *International Building Code* and the *International Mechanical Code*.

2703.14.1 Where required. Exhaust ventilation systems shall be provided in the following locations in accordance with the requirements of this section and the *International Building Code*:

1. *Fabrication areas*: Exhaust ventilation for *fabrication areas* shall comply with the *International Building Code*. The *fire code official* is authorized to require additional manual control switches.
2. Workstations: A ventilation system shall be provided to capture and exhaust gases, fumes and vapors at workstations.
3. Liquid storage rooms: Exhaust ventilation for liquid storage rooms shall comply with Section 5004.3.1 and the *International Building Code*.
4. HPM rooms: Exhaust ventilation for HPM rooms shall comply with Section 5004.3.1 and the *International Building Code*.

5. Gas cabinets: Exhaust ventilation for gas cabinets shall comply with Section 5003.8.6.2. The gas cabinet ventilation system is allowed to connect to a workstation ventilation system. Exhaust ventilation for gas cabinets containing highly toxic or toxic gases shall also comply with Chapter 60.
6. Exhausted enclosures: Exhaust ventilation for exhausted enclosures shall comply with Section 5003.8.5.2. Exhaust ventilation for exhausted enclosures containing highly toxic or toxic gases shall also comply with Chapter 60.
7. Gas rooms: Exhaust ventilation for gas rooms shall comply with Section 5003.8.4.2. Exhaust ventilation for gas rooms containing highly toxic or toxic gases shall also comply with Chapter 60.
8. Cabinets containing pyrophoric liquids or Class 3 water-reactive liquids: Exhaust ventilation for cabinets in *fabrication areas* containing pyrophoric liquids or Class 3 water-reactive liquids shall be as required in Section 2705.2.3.4.

2703.14.2 Penetrations. Exhaust ducts penetrating *fire barriers* constructed in accordance with Section 707 of the *International Building Code* or *horizontal assemblies* constructed in accordance with Section 711 of the *International Building Code* shall be contained in a shaft of equivalent fire-resistance-rated construction. Exhaust ducts shall not penetrate *fire walls*. Fire dampers shall not be installed in exhaust ducts.

2703.14.3 Treatment systems. Treatment systems for highly toxic and toxic gases shall comply with Chapter 60.

2703.15 Emergency power system. An emergency power system shall be provided in Group H-5 occupancies in accordance with Section 604. The emergency power system shall supply power automatically to the electrical systems specified in Section 2703.15.1 when the normal supply system is interrupted.

2703.15.1 Required electrical systems. Emergency power shall be provided for electrically operated equipment and connected control circuits for the following systems:

1. HPM exhaust ventilation systems.
2. HPM gas cabinet ventilation systems.
3. HPM exhausted enclosure ventilation systems.
4. HPM gas room ventilation systems.
5. HPM gas detection systems.
6. Emergency alarm systems.
7. Manual fire alarm systems.
8. *Automatic sprinkler system* monitoring and alarm systems.
9. Automatic alarm and detection systems for pyrophoric liquids and Class 3 water-reactive liquids required in Section 2705.2.3.4.
10. Flow alarm switches for pyrophoric liquids and Class 3 water-reactive liquids cabinet exhaust ventilation systems required in Section 2705.2.3.4.
11. Electrically operated systems required elsewhere in this code or in the *International Building Code* applicable to the use, storage or handling of HPM.

2703.15.2 Exhaust ventilation systems. Exhaust ventilation systems are allowed to be designed to operate at not less than one-half the normal fan speed on the emergency power system where it is demonstrated that the level of exhaust will maintain a safe atmosphere.

2703.16 Sub-atmospheric pressure gas systems. Sub-atmospheric pressure gas systems (SAGS) shall be in accordance with NFPA 318.

SECTION 2704
STORAGE

2704.1 General. Storage of hazardous materials shall comply with Section 2703 and this section and other applicable provisions of this code.

2704.2 Fabrication areas. Hazardous materials storage and the maximum quantities of hazardous materials in use and storage allowed in *fabrication areas* shall be in accordance with Sections 2704.2.1 through 2704.2.2.1.

2704.2.1 Location of HPM storage in fabrication areas. Storage of HPM in *fabrication areas* shall be within *approved* or *listed* storage cabinets, gas cabinets, exhausted enclosures or within a workstation as follows:

1. Flammable and *combustible liquid* storage cabinets shall comply with Section 5704.3.2.
2. Hazardous materials storage cabinets shall comply with Section 5003.8.7.
3. Gas cabinets shall comply with Section 5003.8.6. Gas cabinets for highly toxic or toxic gases shall also comply with Section 6004.1.2.
4. Exhausted enclosures shall comply with Section 5003.8.5. Exhausted enclosures for highly toxic or toxic gases shall also comply with Section 6004.1.3.
5. Workstations shall comply with Section 2705.2.3.

2704.2.2 Maximum aggregate quantities in fabrication areas. The aggregate quantities of hazardous materials stored or used in a single *fabrication area* shall be limited as specified in this section.

Exception: *Fabrication areas* containing quantities of hazardous materials not exceeding the maximum allowable quantities per *control area* established by Sections 5003.1.1, 5704.3.4 and 5704.3.5.

2704.2.2.1 Storage and use in fabrication areas. The maximum quantities of hazardous materials stored or used in a single *fabrication area* shall not exceed the quantities set forth in Table 2704.2.2.1.

TABLE 2704.2.2.1
QUANTITY LIMITS FOR HAZARDOUS MATERIALS IN A SINGLE FABRICATION AREA IN GROUP H-5[a]

HAZARD CATEGORY	SOLIDS (pounds/square foot)	LIQUIDS (gallons/square foot)	GAS (cubic feet @ NTP/square foot)
PHYSICAL-HAZARD MATERIALS			
Combustible dust	Note b	Not Applicable	Not Applicable
Combustible fiber Loose Baled	 Note b Notes b and c	Not Applicable	Not Applicable
Combustible liquid Class II Class IIIA Class IIIB Combination Class I, II and IIIA	Not Applicable	 0.01 0.02 Not Limited 0.04	Not Applicable
Cryogenic gas Flammable Oxidizing	Not Applicable	Not Applicable	 Note d 1.25
Explosives	Note b	Note b	Note b
Flammable gas Gaseous Liquefied	Not Applicable	Not Applicable	 Note d Note d
Flammable liquid Class IA Class IB Class IC Combination Class IA, IB and IC Combination Class I, II and IIIA	Not Applicable	 0.0025 0.025 0.025 0.025 0.04	Not Applicable
Flammable solid	0.001	Not Applicable	Not Applicable
Organic peroxide Unclassified detonable Class I Class II Class III Class IV Class V	 Note b Note b 0.025 0.1 Not Limited Not Limited	Not Applicable	Not Applicable
Oxidizing gas Gaseous Liquefied Combination of Gaseous and Liquefied	Not Applicable	Not Applicable	 1.25 1.25 1.25
Oxidizer Class 4 Class 3 Class 2 Class 1 Combination oxidizer Class 1, 2, 3	 Note b 0.003 0.003 0.003 0.003	 Note b 0.03 0.03 0.03 0.03	Not Applicable
Pyrophoric	0.01	0.00125	Notes d and e
Unstable reactive Class 4 Class 3 Class 2 Class 1	 Note b 0.025 0.1 Not Limited	 Note b 0.0025 0.01 Not Limited	 Note b Note b Note b Not Limited

(continued)

TABLE 2704.2.2.1—continued
QUANTITY LIMITS FOR HAZARDOUS MATERIALS IN A SINGLE FABRICATION AREA IN GROUP H-5

HAZARD CATEGORY	SOLIDS (pounds/square foot)	LIQUIDS (gallons/square foot)	GAS (cubic feet @ NTP/square foot)
PHYSICAL-HAZARD MATERIALS			
Water reactive			
Class 3	Note b	0.00125	Not Applicable
Class 2	0.25	0.025	
Class 1	Not Limited	Not Limited	
HEALTH-HAZARD MATERIALS			
Corrosives	Not Limited	Not Limited	Not Limited
Highly toxics	Not Limited	Not Limited	Note d
Toxics	Not Limited	Not Limited	Note d

For SI: 1 pound per square foot = 4.882 kg/m^2, 1 gallon per square foot = 40.7 L/m^2, 1 cubic foot @ NTP/square foot = 0.305 m^3 @ NTP/m^2, 1 cubic foot = 0.02832 m^3.

a. Hazardous materials within piping shall not be included in the calculated quantities.

b. Quantity of hazardous materials in a single fabrication area shall not exceed the maximum allowable quantities per control area in Tables 5003.1.1(1) and 5003.1.1(2).

c. Densely packed baled cotton that complies with the packing requirements of ISO 8115 shall not be included in this material class.

d. The aggregate quantity of flammable, pyrophoric, toxic and highly toxic gases shall not exceed 9,000 cubic feet at NTP.

e. The aggregate quantity of pyrophoric gases in the building shall not exceed the amounts set forth in Table 5003.8.2.

2704.3 Indoor storage outside of fabrication areas. The indoor storage of hazardous materials outside of *fabrication areas* shall be in accordance with Sections 2704.3.1 through 2704.3.3.

2704.3.1 HPM storage. The indoor storage of HPM in quantities greater than those *listed* in Sections 5003.1.1 and 3404.3.4 shall be in a room complying with the requirements of the *International Building Code* and this code for a liquid storage room, HPM room or gas room as appropriate for the materials stored.

2704.3.2 Other hazardous materials storage. The indoor storage of other hazardous materials shall comply with Sections 5001, 5003 and 5004 and other applicable provisions of this code.

2704.3.3 Separation of incompatible hazardous materials. Incompatible hazardous materials in storage shall be separated from each other in accordance with Section 5003.9.8.

SECTION 2705
USE AND HANDLING

2705.1 General. The use and handling of hazardous materials shall comply with this section, Section 2703 and other applicable provisions of this code.

2705.2 Fabrication areas. The use of hazardous materials in *fabrication areas* shall be in accordance with Sections 2705.2.1 through 2705.2.3.4.

2705.2.1 Location of HPM in use in fabrication areas. Hazardous production materials in use in *fabrication areas* shall be within *approved* or *listed* gas cabinets, exhausted enclosures or a workstation.

2705.2.2 Maximum aggregate quantities in fabrication areas. The aggregate quantities of hazardous materials in a single *fabrication area* shall comply with Section 2704.2.2, and Table 2704.2.2.1. The quantity of HPM in use at a workstation shall not exceed the quantities *listed* in Table 2705.2.2.

2705.2.3 Workstations. Workstations in *fabrication areas* shall be in accordance with Sections 2705.2.3.1 through 2705.2.3.4.

2705.2.3.1 Construction. Workstations in *fabrication areas* shall be constructed of materials compatible with the materials used and stored at the workstation. The portion of the workstation that serves as a cabinet for HPM gases, Class I flammable liquids or Class II or Class IIIA combustible liquids shall be noncombustible and, if of metal, shall be not less than 0.0478-inch (18 gage) (1.2 mm) steel.

2705.2.3.2 Protection of vessels. Vessels containing hazardous materials located in or connected to a workstation shall be protected as follows:

1. HPM: Vessels containing HPM shall be protected from physical damage and shall not project from the workstation.
2. Hazardous *cryogenic fluids*, gases and liquids: Hazardous cryogenic fluid, gas and liquid vessels located within a workstation shall be protected from seismic forces in an *approved* manner in accordance with the *International Building Code*.
3. *Compressed gases*: Protection for *compressed gas* vessels shall also comply with Section 5303.5.
4. *Cryogenic fluids*: Protection for *cryogenic fluid* vessels shall also comply with Section 5503.5.

**TABLE 2705.2.2
MAXIMUM QUANTITIES OF HPM AT A WORKSTATION[d]**

HPM CLASSIFICATION	STATE	MAXIMUM QUANTITY
Flammable, highly toxic, pyrophoric and toxic combined	Gas	Combined aggregate volume of all cylinders at a workstation shall not exceed an internal cylinder volume of 39.6 gallons or 5.29 cubic feet
Flammable	Liquid Solid	15 gallons[a, b] 5 pounds[a, b]
Corrosive	Gas	Combined aggregate volume of all cylinders at a workstation shall not exceed an internal cylinder volume of 39.6 gallons or 5.29 cubic feet
	Liquid	Use-open system: 25 gallons[b] Use-closed system: 150 gallons[b, e]
	Solid	20 pounds[a, b]
Highly toxic	Liquid Solid	15 gallons[a] 5 pounds[a]
Oxidizer	Gas	Combined aggregate volume of all cylinders at a workstation shall not exceed an internal cylinder volume of 39.6 gallons or 5.29 cubic feet
	Liquid	Use-open system: 12 gallons[b] Use-closed system: 60 gallons[b]
	Solid	20 pounds[a, b]
Pyrophoric	Liquid Solid	0.5 gallon[c, f] 4.4 pounds[c, f]
Toxic	Liquid	Use-open system: 15 gallons[b] Use-closed system: 60 gallons[b]
	Solid	5 pounds[a, b]
Unstable reactive Class 3	Liquid Solid	0.5 gallon[a, b] 5 pounds[a, b]
Water-reactive Class 3	Liquid Solid	0.5 gallon[c, f] See Table 2704.2.2.1

For SI: 1 pound = 0.454 kg, 1 gallon = 3.785 L.

a. Maximum allowable quantities shall be increased 100 percent for closed system operations. Where Note b also applies, the increase for both notes shall be allowed.

b. Quantities shall be allowed to be increased 100 percent where workstations are internally protected with an approved automatic fire-extinguishing or suppression system complying with Chapter 9. Where Note b also applies, the increase for both notes shall be allowed. Where Note e also applies, the maximum increase allowed for both Notes b and e shall not exceed 100 percent.

c. Allowed only in workstations that are internally protected with an approved automatic fire-extinguishing or fire protection system complying with Chapter 9 and compatible with the reactivity of materials in use at the workstation.

d. The quantity limits apply only to materials classified as HPM.

e. Quantities shall be allowed to be increased 100 percent for nonflammable, noncombustible corrosive liquids where the materials of construction for workstations are listed or approved for use without internal fire-extinguishing or suppression system protection. Where Note b also applies, the maximum increase allowed for both Notes b and e shall not exceed 100 percent.

f. A maximum quantity of 5.3 gallons of liquids and 44 pounds of total liquids and solids shall be allowed at a workstation where conditions are in accordance with Section 2705.2.3.5.

2705.2.3.3 Drainage and containment for HPM liquids. Each workstation utilizing HPM liquids shall have all of the following:

1. Drainage piping systems connected to a compatible system for disposition of such liquids.
2. The work surface provided with a slope or other means for directing spilled materials to the containment or drainage system.
3. An *approved* means of containing or directing spilled or leaked liquids to the drainage system.

2705.2.3.4 Pyrophoric solids, liquids and Class 3 water-reactive liquids. Pyrophoric liquids and Class 3 water-reactive liquids in containers greater than 0.5-gallon (2 L) but not exceeding 5.3-gallon (20 L) capacity and pyrophoric solids in containers greater than 4.4 pounds (2 kg) but not exceeding 44 pounds (20 kg) shall be allowed at workstations where located inside cabinets and the following conditions are met:

1. Maximum amount per cabinet: The maximum amount per cabinet shall be limited to 5.3 gallons (20 L) of liquids and 44 pounds (20 kg) of total liquids and solids.
2. Cabinet construction: Cabinets shall be constructed in accordance with the following:
   - 2.1. Cabinets shall be constructed of not less than 0.097-inch (2.5 mm) (12 gage) steel.
   - 2.2. Cabinets shall be permitted to have self-closing limited access ports or noncombustible windows that provide access to equipment controls.
   - 2.3. Cabinets shall be provided with self- or manual-closing doors. Manual-closing doors shall be equipped with a door switch that will initiate local audible and visual alarms when the door is in the open position.
3. Cabinet exhaust ventilation system: An exhaust ventilation system shall be provided for cabinets and shall comply with the following:
   - 3.1. The system shall be designed to operate at a negative pressure in relation to the surrounding area.
   - 3.2. The system shall be equipped with monitoring equipment to ensure that required exhaust flow or static pressure is provided.
   - 3.3. Low-flow or static pressure conditions shall send an alarm to the on-site emergency control station. The alarm shall be both visual and audible.
4. Cabinet spill containment: Spill containment shall be provided in each cabinet, with the spill containment capable of holding the contents of the aggregate amount of liquids in containers in each cabinet.

5. Valves: Valves in supply piping between the product containers in the cabinet and the workstation served by the containers shall fail in the closed position upon power failure, loss of exhaust ventilation and upon actuation of the fire control system.
6. Fire detection system: Each cabinet shall be equipped with an automatic fire detection system complying with the following conditions:
    6.1. Automatic detection system: UV/IR, high-sensitivity smoke detection (HSSD) or other *approved* detection systems shall be provided inside each cabinet.
    6.2. Automatic shutoff: Activation of the detection system shall automatically close the shutoff valves at the source on the liquid supply.
    6.3. Alarms and signals: Activation of the detection system shall initiate a local alarm within the *fabrication area* and transmit a signal to the *emergency control station*. The alarms and signals shall be both visual and audible.

2705.3 Transportation and handling. The transportation and handling of hazardous materials shall comply with Sections 2705.3.1 through 2705.3.4.1 and other applicable provisions of this code.

2705.3.1 Corridors and enclosures for stairways and ramps. *Corridors* and enclosures for *exit stairways* and *ramps* in new buildings or serving new fabrication areas shall not contain HPM, except as permitted in corridors by Section 415.11.6.4 of the *International Building Code* and Section 2705.3.2 of this code.

2705.3.2 Transport in corridors and enclosures for stairways and ramps. Transport in *corridors* and enclosures for *stairways* and *ramps* shall be in accordance with Sections 2705.3.2.1 through 2705.3.3.

2705.3.2.1 Fabrication area alterations. Where existing fabrication areas are altered or modified in existing buildings, HPM is allowed to be transported in existing *corridors* where such *corridors* comply with Section 5003.10 of this code and Section 415.11.2 of the *International Building Code*.

2705.3.2.2 HPM transport in corridors and enclosures for stairways and ramps. Nonproduction HPM is allowed to be transported in *corridors* and enclosures for *stairways* and *ramps* where utilized for maintenance, lab work and testing when the transportation is in accordance with Section 5003.10.

2705.3.3 Service corridors. Where a new *fabrication area* is constructed, a service corridor shall be provided where it is necessary to transport HPM from a liquid storage room, HPM room, gas room or from the outside of a building to the perimeter wall of a *fabrication area*. Service corridors shall be designed and constructed in accordance with the *International Building Code*.

2705.3.4 Carts and trucks. Carts and trucks used to transport HPM in *corridors* and enclosures for *stairways* and *ramps* shall comply with Section 5003.10.3.

2705.3.4.1 Identification. Carts and trucks shall be marked to indicate the contents.

CHAPTER 28

LUMBER YARDS AND AGRO-INDUSTRIAL, SOLID BIOMASS AND WOODWORKING FACILITIES

SECTION 2801 GENERAL

2801.1 Scope. The storage, manufacturing and processing of solid biomass feedstock, timber, lumber, plywood, nonmetallic pallets, veneers and agro-industrial byproducts shall be in accordance with this chapter.

2801.2 Permit. Permits shall be required as set forth in Section 105.6.

SECTION 2802 DEFINITIONS

2802.1 Definitions. The following terms are defined in Chapter 2:

AGRO-INDUSTRIAL.

BIOMASS.

COLD DECK.

FINES.

HOGGED MATERIALS.

PLYWOOD AND VENEER MILLS.

RAW PRODUCT.

SOLID BIOFUEL.

SOLID BIOMASS FEEDSTOCK.

STATIC PILES.

TIMBER AND LUMBER PRODUCTION FACILITIES.

SECTION 2803 GENERAL REQUIREMENTS

2803.1 Open yards. Open yards required by the *International Building Code* shall be maintained around structures.

2803.2 Dust control. Equipment or machinery located inside buildings that generates or emits *combustible dust* shall be provided with an *approved* dust collection and exhaust system installed in accordance with Chapter 22 and the *International Mechanical Code*. Equipment or systems that are used to collect, process or convey *combustible dusts* shall be provided with an *approved* explosion control system.

2803.2.1 Explosion venting. Where a dust explosion hazard exists in equipment rooms, buildings or other enclosures, such areas shall be provided with explosion (*deflagration*) venting or an *approved* explosion suppression system complying with Section 911.

2803.3 Waste removal. Sawmills, planning mills and other woodworking plants shall be equipped with a waste removal system that will collect and remove sawdust and shavings. Such systems shall be installed in accordance with Chapter 22 and the *International Mechanical Code*.

Exception: Manual waste removal where *approved*.

2803.3.1 Housekeeping. Provisions shall be made for a systematic and thorough cleaning of the entire plant at sufficient intervals to prevent the accumulations of *combustible dust* and spilled combustible or flammable liquids.

2803.3.2 Metal scrap. Provision shall be made for separately collecting and disposing of any metal scrap so that such scrap will not enter the wood handling or processing equipment.

2803.4 Electrical equipment. Electrical wiring and equipment shall comply with NFPA 70.

2803.5 Control of ignition sources. Protection from ignition sources shall be provided in accordance with Sections 2803.5.1 through 2803.5.3.

2803.5.1 Cutting and welding. Cutting and welding shall comply with Chapter 35.

2803.5.2 Static electricity. Static electricity shall be prevented from accumulating on machines and equipment subject to static electricity buildup by permanent grounding and bonding wires or other *approved* means.

2803.5.3 Smoking. Where smoking constitutes a fire hazard, the *fire code official* is authorized to order the *owner* or occupant to post *approved* "No Smoking" signs complying with Section 310. The *fire code official* is authorized to designate specific locations where smoking is allowed.

2803.6 Fire apparatus access roads. Fire apparatus access roads shall be provided for buildings and facilities in accordance with Section 503.

2803.7 Access plan. Where storage pile configurations could change because of changes in product operations and processing, the access plan shall be submitted for approval when required by the *fire code official*.

SECTION 2804 FIRE PROTECTION

2804.1 General. Fire protection in timber and lumber production mills, plywood and veneer mills and agro-industrial facilities shall comply with Sections 2804.2 through 2804.4.

2804.2 Fire alarms. An *approved* means for transmitting alarms to the fire department shall be provided in timber and lumber production mills and plywood and veneer mills.

2804.2.1 Manual fire alarms. A manual fire alarm system complying with Section 907.2 shall be installed in areas of timber and lumber production mills and for plywood and veneer mills that contain product dryers.

Exception: Where dryers or other sources of ignition are protected by a supervised *automatic sprinkler system* complying with Section 903.

2804.3 Portable fire extinguishers or standpipes and hose. Portable fire extinguishers or standpipes and hose supplied from an *approved* water system shall be provided within a 50-foot (15 240 mm) distance of travel from any machine producing shavings or sawdust. Portable fire extinguishers shall be provided in accordance with Section 906 for extra-high hazards.

2804.4 Automatic sprinkler systems. *Automatic sprinkler systems* shall be installed in accordance with Section 903.3.1.1.

SECTION 2805 PLYWOOD, VENEER AND COMPOSITE BOARD MILLS

2805.1 General. Plant operations of plywood, veneer and composite board mills shall comply with Sections 2805.2 and 2805.3.

2805.2 Dryer protection. Dryers shall be protected throughout by an *approved*, automatic deluge water-spray suppression system complying with Chapter 9. Deluge heads shall be inspected quarterly for pitch buildup. Deluge heads shall be flushed during regular maintenance for functional operation. Manual activation valves shall be located within 75 feet (22 860 mm) of the drying equipment.

2805.3 Thermal oil-heating systems. Facilities that use heat transfer fluids to provide process equipment heat through piped, indirect heating systems shall comply with this code and NFPA 664.

SECTION 2806 LOG STORAGE AREAS

2806.1 General. Log storage areas shall comply with Sections 2806.2 through 2806.3.

2806.2 Cold decks. Cold decks shall not exceed 500 feet (152.4 m) in length, 300 feet (91 440 mm) in width and 20 feet (6096 mm) in height. Cold decks shall be separated from adjacent cold decks or other exposures by not less than 100 feet (30 480 mm).

Exception: The size of cold decks shall be determined by the *fire code official* where the decks are protected by special fire protection including, but not limited to, additional fire flow, portable turrets and deluge sets, and hydrant hose houses equipped with *approved* fire-fighting equipment capable of reaching the entire storage area in accordance with Chapter 9.

2806.3 Pile stability. Log and pole piles shall be stabilized by *approved* means.

SECTION 2807 STORAGE OF WOOD CHIPS AND HOGGED MATERIAL ASSOCIATED WITH TIMBER AND LUMBER PRODUCTION FACILITIES

2807.1 General. The storage of wood chips and hogged materials associated with timber and lumber production facilities shall comply with Sections 2807.2 through 2807.5.

2807.2 Size of piles. Piles shall not exceed 60 feet (18 288 mm) in height, 300 feet (91 440 mm) in width and 500 feet (152 m) in length. Piles shall be separated from adjacent piles or other exposures by *approved* fire apparatus access roads.

Exception: The *fire code official* is authorized to allow the pile size to be increased where additional fire protection is provided in accordance with Chapter 9. The increase shall be based on the capabilities of the system installed.

2807.3 Pile fire protection. Automatic sprinkler protection shall be provided in conveyor tunnels and combustible enclosures that pass under a pile. Combustible or enclosed conveyor systems shall be equipped with an *approved automatic sprinkler system*.

2807.4 Material-handling equipment. *Approved* material-handling equipment shall be readily available for moving wood chips and hogged material.

2807.5 Emergency plan. The *owner* or operator shall develop a plan for monitoring, controlling and extinguishing spot fires. The plan shall be submitted to the *fire code official* for review and approval.

SECTION 2808 STORAGE AND PROCESSING OF WOOD CHIPS, HOGGED MATERIAL, FINES, COMPOST, SOLID BIOMASS FEEDSTOCK AND RAW PRODUCT ASSOCIATED WITH YARD WASTE, AGRO-INDUSTRIAL AND RECYCLING FACILITIES

2808.1 General. The storage and processing of wood chips, hogged materials, fines, compost, solid biomass feedstock and raw product produced from yard waste, debris and agro-industrial and recycling facilities shall comply with Sections 2808.2 through 2808.10.

2808.2 Storage site. Storage sites shall be level and on solid ground, elevated soil lifts or other all-weather surface. Sites shall be thoroughly cleaned before transferring wood products to the site.

2808.3 Size of piles. Piles shall not exceed 25 feet (7620 mm) in height, 150 feet (45 720 mm) in width and 250 feet (76 200 mm) in length.

Exception: The *fire code official* is authorized to allow the pile size to be increased where a fire protection plan is provided for approval that includes, but is not limited to, the following:

1. Storage yard areas and materials-handling equipment selection, design and arrangement shall be

based upon sound fire prevention and protection principles.

2. Factors that lead to spontaneous heating shall be identified in the plan, and control of the various factors shall be identified and implemented, including provisions for monitoring the internal condition of the pile.
3. The plan shall include means for early fire detection and reporting to the public fire department; and facilities needed by the fire department for fire extinguishment including a water supply and fire hydrants.
4. Fire apparatus access roads around the piles and access roads to the top of the piles shall be established, identified and maintained.
5. Regular yard inspections by trained personnel shall be included as part of an effective fire prevention maintenance program.

Additional fire protection called for in the plan shall be provided and shall be installed in accordance with this code. The increase of the pile size shall be based upon the capabilities of the installed fire protection systems and features.

2808.4 Pile separation. Piles shall be separated from adjacent piles by *approved* fire apparatus access roads.

2808.5 Combustible waste. The storage, accumulation and handling of combustible materials and control of vegetation shall comply with Chapter 3.

2808.6 Static pile protection. Static piles shall be monitored by an *approved* means to measure temperatures within the static piles. Internal pile temperatures shall be monitored and recorded weekly. Such records shall be maintained. An operational plan indicating procedures and schedules for the inspection, monitoring and restricting of excessive internal temperatures in static piles shall be submitted to the *fire code official* for review and approval.

2808.7 Pile fire protection. Automatic sprinkler protection shall be provided in conveyor tunnels and combustible enclosures that pass under a pile. Combustible conveyor systems and enclosed conveyor systems shall be equipped with an *approved automatic sprinkler system*.

2808.8 Fire extinguishers. Portable fire extinguishers complying with Section 906 and with a minimum rating of 4-A:60-B:C shall be provided on all vehicles and equipment operating on piles and at all processing equipment.

2808.9 Material-handling equipment. *Approved* material-handling equipment shall be available for moving wood chips, hogged material, wood fines and raw product during fire-fighting operations.

2808.10 Emergency plan. The *owner* or operator shall develop a plan for monitoring, controlling and extinguishing spot fires and submit the plan to the *fire code official* for review and approval.

SECTION 2809 EXTERIOR STORAGE OF FINISHED LUMBER AND SOLID BIOFUEL PRODUCTS

2809.1 General. Exterior storage of finished lumber and solid biofuel products shall comply with Sections 2809.1 through 2809.5.

2809.2 Size of piles. Exterior storage shall be arranged to form stable piles with a maximum height of 20 feet (6096 mm). Piles shall not exceed 150,000 cubic feet (4248 m^3) in volume.

2809.3 Fire apparatus access roads. Fire apparatus access roads in accordance with Section 503 shall be located so that a maximum grid system unit of 50 feet by 150 feet (15 240 mm by 45 720 mm) is established.

2809.4 Security. Permanent storage areas shall be surrounded with an *approved* fence. Fences shall be not less than 6 feet (1829 mm) in height.

Exceptions:

1. Lumber piles inside of buildings and production mills for lumber, plywood and veneer.
2. Solid biofuel piles inside of buildings and agro-industrial processing facilities for solid biomass feedstock.

2809.5 Fire protection. An *approved* hydrant and hose system or portable fire-extinguishing equipment suitable for the fire hazard involved shall be provided for open storage yards. Hydrant and hose systems shall be installed in accordance with NFPA 24. Portable fire extinguishers complying with Section 906 shall be located so that the distance of travel from the nearest unit does not exceed 75 feet (22 860 mm).

CHAPTER 29

MANUFACTURE OF ORGANIC COATINGS

SECTION 2901 GENERAL

2901.1 Scope. Organic coating manufacturing processes shall comply with this chapter, except that this chapter shall not apply to processes manufacturing nonflammable or water-thinned coatings or to operations applying coating materials.

2901.2 Permits. Permits shall be required as set forth in Section 105.6.

2901.3 Maintenance. Structures and their service equipment shall be maintained in accordance with this code and NFPA 35.

SECTION 2902 DEFINITION

2902.1 Definition. The following term is defined in Chapter 2:

ORGANIC COATING.

SECTION 2903 GENERAL PRECAUTIONS

2903.1 Building features. Manufacturing of organic coatings shall be done only in buildings that do not have pits or *basements.*

2903.2 Location. Organic coating manufacturing operations and operations incidental to or connected with organic coating manufacturing shall not be located in buildings having other occupancies.

2903.3 Fire-fighting access. Organic coating manufacturing operations shall be accessible from not less than one side for the purpose of fire control. *Approved aisles* shall be maintained for the unobstructed movement of personnel and fire suppression equipment.

2903.4 Fire protection systems. *Fire protection systems* shall be installed, maintained, periodically inspected and tested in accordance with Chapter 9.

2903.5 Portable fire extinguishers. Not less than one portable fire extinguisher complying with Section 906 for extra hazard shall be provided in organic coating areas.

2903.6 Open flames. Open flames and direct-fired heating devices shall be prohibited in areas where flammable vapor-air mixtures exist.

2903.7 Smoking. Smoking shall be prohibited in accordance with Section 310.

2903.8 Power equipment. Power-operated equipment and industrial trucks shall be of a type *approved* for the location.

2903.9 Tank maintenance. The cleaning of tanks and vessels that have contained flammable or *combustible liquids* shall be performed under the supervision of persons knowledgeable of the fire and explosion potential.

2903.9.1 Repairs. Where necessary to make repairs involving "hot work," the work shall be authorized by the responsible individual before the work begins.

2903.9.2 Empty containers. Empty flammable or *combustible liquid* containers shall be removed to a detached, outside location and, if not cleaned on the premises, the empty containers shall be removed from the plant as soon as practical.

2903.10 Drainage. Drainage facilities shall be provided to direct flammable and *combustible liquid* leakage and fire protection water to an *approved* location away from the building, any other structure, storage area or adjoining premises.

2903.11 Alarm system. An *approved* fire alarm system shall be provided in accordance with Section 907.

SECTION 2904 ELECTRICAL EQUIPMENT AND PROTECTION

2904.1 Wiring and equipment. Electrical wiring and equipment shall comply with this chapter and shall be installed in accordance with NFPA 70.

2904.2 Hazardous locations. Where Class I liquids are exposed to the air, the design of equipment and ventilation of structures shall be such as to limit the Class I, Division 1, locations to the following:

1. Piping trenches.
2. The interior of equipment.
3. The immediate vicinity of pumps or equipment locations, such as dispensing stations, open centrifuges, plate and frame filters, opened vacuum filters, change cans and the surfaces of open equipment. The immediate vicinity shall include a zone extending from the vapor liberation point 5 feet (1524 mm) horizontally in all directions and vertically from the floor to a level 3 feet (914 mm) above the highest point of vapor liberation.

2904.2.1 Other locations. Locations within the confines of the manufacturing room where Class I liquids are handled shall be Class I, Division 2, except locations indicated in Section 2904.2.

2904.2.2 Ordinary equipment. Ordinary electrical equipment, including switchgear, shall be prohibited, except where installed in a room maintained under positive pressure with respect to the hazardous area. The air or other media utilized for pressurization shall be obtained from a source that will not cause any amount or type of flammable vapor to be introduced into the room.

2904.3 Bonding. Equipment including, but not limited to, tanks, machinery and piping shall be bonded and connected to a ground where an ignitable mixture is capable of being present.

2904.3.1 Piping. Electrically isolated sections of metallic piping or equipment shall be grounded or bonded to the other grounded portions of the system.

2904.3.2 Vehicles. Tank vehicles loaded or unloaded through open connections shall be grounded and bonded to the receiving system.

2904.3.3 Containers. Where a flammable mixture is transferred from one portable container to another, a bond shall be provided between the two containers, and one shall be grounded.

2904.4 Ground. Metal framing of buildings shall be grounded with resistance of not more than 5 ohms.

SECTION 2905 PROCESS STRUCTURES

2905.1 Design. Process structures shall be designed and constructed in accordance with the *International Building Code*.

2905.2 Fire apparatus access. Fire apparatus access complying with Section 503 shall be provided for the purpose of fire control to not less than one side of organic coating manufacturing operations.

2905.3 Drainage. Drainage facilities shall be provided in accordance with Section 2903.10 where topographical conditions are such that flammable and *combustible liquids* are capable of flowing from the organic coating manufacturing operation so as to constitute a fire hazard to other premises.

2905.4 Explosion control. Explosion control shall be provided in areas subject to potential *deflagration* hazards as indicated in NFPA 35. Explosion control shall be provided in accordance with Section 911.

2905.5 Ventilation. Enclosed structures in which Class I liquids are processed or handled shall be ventilated at a rate of not less than 1 cubic foot per minute per square foot [0.00508 $m^3/(s \cdot m^2)$] of solid floor area. Ventilation shall be accomplished by exhaust fans that take suction at floor levels and discharge to a safe location outside the structure. Noncontaminated intake air shall be introduced in such a manner that all portions of solid floor areas are provided with continuous uniformly distributed air movement.

2905.6 Heating. Heating provided in hazardous areas shall be by indirect means. Ignition sources such as open flames or electrical heating elements, except as provided for in Section 2904, shall not be permitted within the structure.

SECTION 2906 PROCESS MILLS AND KETTLES

2906.1 Mills. Mills, operating with close clearances, which process flammable and heat-sensitive materials, such as nitrocellulose, shall be located in a detached building or in a noncombustible structure without other occupancies. The amount of nitrocellulose or other flammable material brought into the area shall not be more than the amount required for a batch.

2906.2 Mixers. Mixers shall be of the enclosed type or, where of the open type, shall be provided with properly fitted covers. Where flow is by gravity, a shutoff valve shall be installed as close as practical to the mixer, and a control valve shall be provided near the end of the fill pipe.

2906.3 Open kettles. Open kettles shall be located in an outside area provided with a protective roof; in a separate structure of noncombustible construction; or separated from other areas by a noncombustible wall having a fire-resistance rating of not less than 2 hours.

2906.4 Closed kettles. Contact-heated kettles containing solvents shall be equipped with safety devices that, in case of a fire, will turn off the process heat, turn on the cooling medium and inject inert gas into the kettle.

2906.4.1 Vaporizer location. The vaporizer section of heat-transfer systems that heat closed kettles containing solvents shall be remotely located.

2906.5 Kettle controls. The kettle and thin-down tank shall be instrumented, controlled and interlocked so that any failure of the controls will result in a safe condition. The kettle shall be provided with a pressure-rupture disc in addition to the primary vent. The vent piping from the rupture disc shall be of minimum length and shall discharge to an *approved* location. The thin-down tank shall be adequately vented. Thinning operations shall be provided with an adequate vapor removal system.

SECTION 2907 PROCESS PIPING

2907.1 Design. Piping, valves and fittings shall be designed for the working pressures and structural stresses to which the piping, valves and fittings will be subjected, and shall be of steel or other material *approved* for the service intended.

2907.2 Valves. Valves shall be of an indicating type. Terminal valves on remote pumping systems shall be of the deadman type, shutting off both the pump and the flow of solvent.

2907.3 Support. Piping systems shall be supported adequately and protected against physical damage. Piping shall be pitched to avoid unintentional trapping of liquids, or *approved* drains shall be provided.

2907.4 Connectors. *Approved* flexible connectors shall be installed where vibration exists or frequent movement is necessary. Hose at dispensing stations shall be of an *approved* type.

2907.5 Tests. Before being placed in service, all piping shall be free of leaks when tested for not less than 30 minutes at not less than 1.5 times the working pressure or a minimum of 5 pounds per square inch gauge (psig) (35 kPa) at the highest point in the system.

SECTION 2908
RAW MATERIALS IN PROCESS AREAS

2908.1 Nitrocellulose quantity. The amount of nitrocellulose brought into the operating area shall not exceed the amount required for a work shift. Nitrocellulose spillage shall be promptly swept up and disposed of properly.

2908.2 Organic peroxides quantity. Organic peroxides brought into the operating area shall be in the original shipping container. When in the operating area, the organic peroxide shall not be placed in locations exposed to ignition sources, heat or mechanical shocks.

SECTION 2909
RAW MATERIALS AND FINISHED PRODUCTS

2909.1 General. The storage, handling and use of flammable and *combustible liquids* in process areas shall be in accordance with Chapter 57.

2909.2 Tank storage. Tank storage for flammable and *combustible liquids* located inside of structures shall be limited to storage areas at or above grade which are separated from the processing area in accordance with the *International Building Code*. Processing equipment containing flammable and *combustible liquids* and storage in quantities essential to the continuity of the operations shall not be prohibited in the processing area.

2909.3 Tank vehicle. Tank car and tank vehicle loading and unloading stations for Class I liquids shall be separated from the processing area, other plant structures, nearest lot line of property that can be built upon or public thoroughfare by a minimum clear distance of 25 feet (7620 mm).

2909.3.1 Loading. Loading and unloading structures and platforms for flammable and *combustible liquids* shall be designed and installed in accordance with Chapter 57.

2909.3.2 Safety. Tank cars for flammable liquids shall be unloaded such that the safety to persons and property is ensured. Tank vehicles for flammable and *combustible liquids* shall be loaded and unloaded in accordance with Chapter 57.

2909.4 Nitrocellulose storage. Nitrocellulose storage shall be located on a detached pad or in a separate structure or a room enclosed in accordance with the *International Building Code*. The nitrocellulose storage area shall not be utilized for any other purpose. Electrical wiring and equipment installed in storage areas adjacent to process areas shall comply with Section 2904.2.

2909.4.1 Containers. Nitrocellulose shall be stored in closed containers. Barrels shall be stored on end and not more than two tiers high. Barrels or other containers of nitrocellulose shall not be opened in the main storage structure but at the point of use or other location intended for that purpose.

2909.4.2 Spills. Spilled nitrocellulose shall be promptly wetted with water and disposed of by use or burning in the open at an *approved* detached location.

2909.5 Organic peroxide storage. The storage of organic peroxides shall be in accordance with Chapter 62.

2909.5.1 Size. The size of the package containing organic peroxide shall be selected so that, as nearly as practical, full packages are utilized at one time. Spilled peroxide shall be promptly cleaned up and disposed of as specified by the supplier.

2909.6 Finished products. Finished products that are flammable or *combustible liquids* shall be stored outside of structures, in a separate structure, or in a room separated from the processing area in accordance with the *International Building Code*. The storage of finished products shall be in tanks or closed containers in accordance with Chapter 57.

CHAPTER 30

INDUSTRIAL OVENS

SECTION 3001 GENERAL

3001.1 Scope. This chapter shall apply to the installation and operation of industrial ovens and furnaces. Industrial ovens and furnaces shall comply with the applicable provisions of NFPA 86, the *International Fuel Gas Code*, *International Mechanical Code* and this chapter. The terms "ovens" and "furnaces" are used interchangeably in this chapter.

3001.2 Permits. Permits shall be required as set forth in Sections 105.6 and 105.7.

SECTION 3002 DEFINITIONS

3002.1 Definitions. The following terms are defined in Chapter 2:

FURNACE CLASS A.

FURNACE CLASS B.

FURNACE CLASS C.

FURNACE CLASS D.

SECTION 3003 LOCATION

3003.1 Ventilation. Enclosed rooms or *basements* containing industrial ovens or furnaces shall be provided with combustion air in accordance with the *International Mechanical Code* and the *International Fuel Gas Code*, and with ventilation air in accordance with the *International Mechanical Code.*

3003.2 Exposure. When locating ovens, oven heaters and related equipment, the possibility of fire resulting from overheating or from the escape of fuel gas or fuel oil and the possibility of damage to the building and injury to persons resulting from explosion shall be considered.

3003.3 Ignition source. Industrial ovens and furnaces shall be located so as not to pose an ignition hazard to flammable vapors or mists or *combustible dusts*.

3003.4 Temperatures. Roofs and floors of ovens shall be insulated and ventilated to prevent temperatures at combustible ceilings and floors from exceeding 160°F (71°C).

SECTION 3004 FUEL PIPING

3004.1 Fuel-gas piping. Fuel-gas piping serving industrial ovens shall comply with the *International Fuel Gas Code*. Piping for other fuel sources shall comply with this section.

3004.2 Shutoff valves. Each industrial oven or furnace shall be provided with an *approved* manual fuel shutoff valve in accordance with the *International Mechanical Code* or the *International Fuel Gas Code.*

3004.2.1 Fuel supply lines. Valves for fuel supply lines shall be located within 6 feet (1829 mm) of the appliance served.

Exception: When *approved* and the valve is located in the same general area as the appliance served.

3004.3 Valve position. The design of manual fuel shutoff valves shall incorporate a permanent feature which visually indicates the open or closed position of the valve. Manual fuel shutoff valves shall not be equipped with removable handles or wrenches unless the handle or wrench can only be installed parallel with the fuel line when the valve is in the open position.

SECTION 3005 INTERLOCKS

3005.1 Shut down. Interlocks shall be provided for Class A ovens so that conveyors or sources of flammable or combustible materials shall shut down if either the exhaust or recirculation air supply fails.

SECTION 3006 FIRE PROTECTION

3006.1 Required protection. Class A and B ovens that contain, or are utilized for the processing of, combustible materials shall be protected by an *approved* automatic fire-extinguishing system complying with Chapter 9.

3006.2 Fixed fire-extinguishing systems. Fixed fire-extinguishing systems shall be provided for Class C or D ovens to protect against such hazards as overheating, spillage of molten salts or metals, quench tanks, ignition of hydraulic oil and escape of fuel. It shall be the user's responsibility to consult with the *fire code official* concerning the necessary requirements for such protection.

3006.3 Fire extinguishers. Portable fire extinguishers complying with Section 906 shall be provided not closer than 15 feet (4572 mm) or not more than 50 feet (15 240 mm) or in accordance with NFPA 10. This shall apply to the oven and related equipment.

SECTION 3007 OPERATION AND MAINTENANCE

3007.1 Furnace system information. An *approved*, clearly worded, and prominently displayed safety design data form or manufacturer's nameplate shall be provided stating the safe operating condition for which the furnace system was designed, built, altered or extended.

3007.2 Oven nameplate. Safety data for Class A solvent atmosphere ovens shall be furnished on the manufacturer's nameplate. The nameplate shall provide the following design data:

1. The solvent used.
2. The number of gallons (L) used per batch or per hour of solvent entering the oven.
3. The required purge time.
4. The oven operating temperature.
5. The exhaust blower rating for the number of gallons (L) of solvent per hour or batch at the maximum operating temperature.

Exception: For low-oxygen ovens, the maximum allowable oxygen concentration shall be included in place of the exhaust blower ratings.

3007.3 Training. Operating, maintenance and supervisory personnel shall be thoroughly instructed and trained in the operation of ovens or furnaces.

3007.4 Equipment maintenance. Equipment shall be maintained in accordance with the manufacturer's instructions.

CHAPTER 31

TENTS AND OTHER MEMBRANE STRUCTURES

SECTION 3101 GENERAL

3101.1 Scope. Tents, temporary stage canopies and membrane structures shall comply with this chapter. The provisions of Section 3103 are applicable only to temporary tents and membrane structures. The provisions of Section 3104 are applicable to temporary and permanent tents and membrane structures. Other temporary structures shall comply with the *International Building Code.*

SECTION 3102 DEFINITIONS

3102.1 Definitions. The following terms are defined in Chapter 2:

AIR-INFLATED STRUCTURE.

AIR-SUPPORTED STRUCTURE.

MEMBRANE STRUCTURE.

TEMPORARY STAGE CANOPY.

TENT.

SECTION 3103 TEMPORARY TENTS AND MEMBRANE STRUCTURES

3103.1 General. Tents and membrane structures used for temporary periods shall comply with this section. Other temporary structures erected for a period of 180 days or less shall comply with the *International Building Code.*

3103.2 Approval required. Tents and membrane structures having an area in excess of 400 square feet (37 m^2) shall not be erected, operated or maintained for any purpose without first obtaining a permit and approval from the *fire code official.*

Exceptions:

1. Tents used exclusively for recreational camping purposes.
2. Tents open on all sides that comply with all of the following:
   - 2.1. Individual tents having a maximum size of 700 square feet (65 m^2).
   - 2.2. The aggregate area of multiple tents placed side by side without a fire break clearance of 12 feet (3658 mm), not exceeding 700 square feet (65 m^2) total.
   - 2.3. A minimum clearance of 12 feet (3658 mm) to all structures and other tents.

3103.3 Place of assembly. For the purposes of this chapter, a place of assembly shall include a circus, carnival, tent show, theater, skating rink, dance hall or other place of assembly in or under which persons gather for any purpose.

3103.4 Permits. Permits shall be required as set forth in Sections 105.6 and 105.7.

3103.5 Use period. Temporary tents, air-supported, air-inflated or tensioned membrane structures shall not be erected for a period of more than 180 days within a 12-month period on a single premises.

3103.6 Construction documents. A detailed site and floor plan for tents or membrane structures with an *occupant load* of 50 or more shall be provided with each application for approval. The tent or membrane structure floor plan shall indicate details of the *means of egress* facilities, seating capacity, arrangement of the seating and location and type of heating and electrical equipment.

3103.7 Inspections. The entire tent, air-supported, air-inflated or tensioned membrane structure system shall be inspected at regular intervals, but not less than two times per permit use period, by the permittee, *owner* or agent to determine that the installation is maintained in accordance with this chapter.

Exception: Permit use periods of less than 30 days.

3103.7.1 Inspection report. Where required by the *fire code official,* an inspection report shall be provided and shall consist of maintenance, anchors and fabric inspections.

3103.8 Access, location and parking. Access, location and parking for temporary tents and membrane structures shall be in accordance with this section.

3103.8.1 Access. Fire apparatus access roads shall be provided in accordance with Section 503.

3103.8.2 Location. Tents or membrane structures shall not be located within 20 feet (6096 mm) of *lot lines,* buildings, other tents or membrane structures, parked vehicles or internal combustion engines. For the purpose of determining required distances, support ropes and guy wires shall be considered as part of the temporary membrane structure or tent.

Exceptions:

1. Separation distance between membrane structures and tents not used for cooking is not required where the aggregate floor area does not exceed 15,000 square feet (1394 m^2).
2. Membrane structures or tents need not be separated from buildings when all of the following conditions are met:
   - 2.1. The aggregate floor area of the membrane structure or tent shall not exceed 10,000 square feet (929 m^2).

2.2. The aggregate floor area of the building and membrane structure or tent shall not exceed the allowable floor area including increases as indicated in the *International Building Code*.

2.3. Required *means of egress* are provided for both the building and the membrane structure or tent including travel distances.

2.4. Fire apparatus access roads are provided in accordance with Section 503.

3103.8.3 Location of structures in excess of 15,000 square feet in area. Membrane structures having an area of 15,000 square feet (1394 m^2) or more shall be located not less than 50 feet (15 240 mm) from any other tent or structure as measured from the sidewall of the tent or membrane structure unless joined together by a corridor.

3103.8.4 Membrane structures on buildings. Membrane structures that are erected on buildings, balconies, decks or other structures shall be regulated as permanent membrane structures in accordance with Section 3102 of the *International Building Code*.

3103.8.5 Connecting corridors. Tents or membrane structures are allowed to be joined together by means of corridors. *Exit* doors shall be provided at each end of such corridor. On each side of such corridor and approximately opposite each other, there shall be provided openings not less than 12 feet (3658 mm) wide.

3103.8.6 Fire break. An unobstructed fire break passageway or fire road not less than 12 feet (3658 mm) wide and free from guy ropes or other obstructions shall be maintained on all sides of all tents and membrane structures unless otherwise *approved* by the *fire code official*.

3103.9 Anchorage required. Tents or membrane structures and their appurtenances shall be adequately roped, braced and anchored to withstand the elements of weather and prevent against collapsing. Documentation of structural stability shall be furnished to the *fire code official* on request.

3103.9.1 Tents and membrane structures exceeding one story. Tents and membrane structures exceeding one story shall be designed and constructed to comply with Chapter 16 of the *International Building Code*.

3103.10 Temporary air-supported and air-inflated membrane structures. Temporary air-supported and air-inflated membrane structures shall be in accordance with Sections 3103.10.1 through 3103.10.4.

3103.10.1 Door operation. During high winds exceeding 50 miles per hour (22 m/s) or in snow conditions, the use of doors in air-supported structures shall be controlled to avoid excessive air loss. Doors shall not be left open.

3103.10.2 Fabric envelope design and construction. Air-supported and air-inflated structures shall have the design and construction of the fabric envelope and the method of anchoring in accordance with Architectural Fabric Structures Institute ASI 77.

3103.10.3 Blowers. An air-supported structure used as a place of assembly shall be furnished with not less than two blowers, each of which has adequate capacity to maintain full inflation pressure with normal leakage. The design of the blower shall be so as to provide integral limiting pressure at the design pressure specified by the manufacturer.

3103.10.4 Auxiliary inflation systems. Places of public assembly for more than 200 persons shall be furnished with an auxiliary inflation system capable of powering a blower with the capacity to maintain full inflation pressure with normal leakage in accordance with Section 3103.10.3 for a minimum duration of 4 hours. The auxiliary inflation system shall be either a fully automatic auxiliary engine-generator set or a supplementary blower powered by an internal combustion engine that shall be automatic in operation. The system shall be capable of automatically operating the required blowers at full power within 60 seconds of a commercial power failure.

3103.11 Seating arrangements. Seating in tents or membrane structures shall be in accordance with Chapter 10.

3103.12 Means of egress. *Means of egress* for temporary tents and membrane structures shall be in accordance with Sections 3103.12.1 through 3103.12.8.

3103.12.1 Distribution. *Exits* shall be spaced at approximately equal intervals around the perimeter of the tent or

TABLE 3103.12.2
MINIMUM NUMBER OF MEANS OF EGRESS AND MEANS OF EGRESS WIDTHS FROM TEMPORARY MEMBRANE STRUCTURES AND TENTS

OCCUPANT LOAD	MINIMUM NUMBER OF MEANS OF EGRESS	MINIMUM WIDTH OF EACH MEANS OF EGRESS (inches)	MINIMUM WIDTH OF EACH MEANS OF EGRESS (inches)
		Tent	Membrane Structure
10 to 199	2	72	36
200 to 499	3	72	72
500 to 999	4	96	72
1,000 to 1,999	5	120	96
2,000 to 2,999	6	120	96
Over 3,000[a]	7	120	96

For SI: 1 inch = 25.4 mm.

a. When the occupant load exceeds 3,000, the total width of means of egress (in inches) shall be not less than the total occupant load multiplied by 0.2 inches per person.

membrane structure, and shall be located such that all points are 100 feet (30 480 mm) or less from an *exit*.

3103.12.2 Number. Tents, or membrane structures or a usable portion thereof shall have not less than one *exit* and not less than the number of *exits* required by Table 3103.12.2. The total width of *means of egress* in inches (mm) shall be not less than the total *occupant load* served by a *means of egress* multiplied by 0.2 inches (5 mm) per person.

3103.12.3 Exit openings from tents. *Exit* openings from tents shall remain open unless covered by a flame-resistant curtain. The curtain shall comply with the following requirements:

1. Curtains shall be free sliding on a metal support. The support shall be not less than 80 inches (2032 mm) above the floor level at the *exit*. The curtains shall be so arranged that, when open, no part of the curtain obstructs the *exit*.
2. Curtains shall be of a color, or colors, that contrasts with the color of the tent.

3103.12.4 Doors. *Exit* doors shall swing in the direction of *exit* travel. To avoid hazardous air and pressure loss in air-supported membrane structures, such doors shall be automatic closing against operating pressures. Opening force at the door edge shall not exceed 15 pounds (66 N).

3103.12.5 Aisle. The width of *aisles* without fixed seating shall be in accordance with the following:

1. In areas serving employees only, the minimum *aisle* width shall be 24 inches (610 mm) but not less than the width required by the number of employees served.
2. In public areas, smooth-surfaced, unobstructed *aisles* having a minimum width of not less than 44 inches (1118 mm) shall be provided from seating areas, and *aisles* shall be progressively increased in width to provide, at all points, not less than 1 foot (305 mm) of *aisle* width for each 50 persons served by such *aisle* at that point.

3103.12.5.1 Arrangement and maintenance. The arrangement of *aisles* shall be subject to approval by the *fire code official* and shall be maintained clear at all times during occupancy.

3103.12.6 Exit signs. *Exits* shall be clearly marked. *Exit* signs shall be installed at required *exit* doorways and where otherwise necessary to indicate clearly the direction of egress where the *exit* serves an *occupant load* of 50 or more.

3103.12.6.1 Exit sign illumination. *Exit* signs shall be either *listed* and *labeled* in accordance with UL 924 as the internally illuminated type and used in accordance with the listing or shall be externally illuminated by luminaires supplied in either of the following manners:

1. Two separate circuits, one of which shall be separate from all other circuits, for *occupant loads* of 300 or less.
2. Two separate sources of power, one of which shall be an *approved* emergency system, shall be provided where the *occupant load* exceeds 300. Emergency systems shall be supplied from storage batteries or from the on-site generator set, and the system shall be installed in accordance with NFPA 70. The emergency system provided shall have a minimum duration of 90 minutes when operated at full design demand.

3103.12.7 Means of egress illumination. *Means of egress* shall be illuminated with light having an intensity of not less than 1 footcandle (11 lux) at floor level while the structure is occupied. Fixtures required for *means of egress* illumination shall be supplied from a separate circuit or source of power.

3103.12.8 Maintenance of means of egress. The required width of *exits*, *aisles* and passageways shall be maintained at all times to a *public way*. Guy wires, guy ropes and other support members shall not cross a *means of egress* at a height of less than 8 feet (2438 mm). The surface of *means of egress* shall be maintained in an *approved* manner.

SECTION 3104
TEMPORARY AND PERMANENT TENTS AND MEMBRANE STRUCTURES

3104.1 General. Tents and membrane structures, both temporary and permanent, shall be in accordance with this section. Permanent tents and membrane structures shall also comply with the *International Building Code*.

3104.2 Flame propagation performance treatment. Before a permit is granted, the *owner* or agent shall file with the *fire code official* a certificate executed by an *approved* testing laboratory certifying that the tents and membrane structures and their appurtenances; sidewalls, drops and tarpaulins; floor coverings, bunting and combustible decorative materials and effects, including sawdust where used on floors or passageways, are composed of material meeting the flame propagation performance criteria of Test Method 1 or Test Method 2, as appropriate, of NFPA 701 or shall be treated with a flame retardant in an *approved* manner and meet the flame propagation performance criteria of Test Method 1 or Test Method 2, as appropriate, of NFPA 701, and that such flame propagation performance criteria are effective for the period specified by the permit.

3104.3 Label. Membrane structures or tents shall have a permanently affixed label bearing the identification of size and fabric or material type.

3104.4 Certification. An affidavit or affirmation shall be submitted to the *fire code official* and a copy retained on the premises on which the tent or air-supported structure is located. The affidavit shall attest to all of the following information relative to the flame propagation performance criteria of the fabric:

1. Names and address of the *owners* of the tent or air-supported structure.

2. Date the fabric was last treated with flame-retardant solution.
3. Trade name or kind of chemical used in treatment.
4. Name of person or firm treating the material.
5. Name of testing agency and test standard by which the fabric was tested.

3104.5 Combustible materials. Hay, straw, shavings or similar combustible materials shall not be located within any tent or membrane structure containing an assembly occupancy, except the materials necessary for the daily feeding and care of animals. Sawdust and shavings utilized for a public performance or exhibit shall not be prohibited provided the sawdust and shavings are kept damp. Combustible materials shall not be permitted under stands or seats at any time.

3104.6 Smoking. Smoking shall not be permitted in tents or membrane structures. *Approved* "No Smoking" signs shall be conspicuously posted in accordance with Section 310.

3104.7 Open or exposed flame. Open flame or other devices emitting flame, fire or heat or any flammable or *combustible liquids*, gas, charcoal or other cooking device or any other unapproved devices shall not be permitted inside or located within 20 feet (6096 mm) of the tent or membrane structures while open to the public unless *approved* by the *fire code official.*

3104.8 Fireworks. Fireworks shall not be used within 100 feet (30 480 mm) of tents or membrane structures.

3104.9 Spot lighting. Spot or effect lighting shall only be by electricity, and all combustible construction located within 6 feet (1829 mm) of such equipment shall be protected with *approved* noncombustible insulation not less than $9^1/_4$ inches (235 mm) thick.

3104.10 Safety film. Motion pictures shall not be displayed in tents or membrane structures unless the motion picture film is safety film.

3104.11 Clearance. There shall be a minimum clearance of at least 3 feet (914 mm) between the fabric envelope and all contents located inside membrane structures.

3104.12 Portable fire extinguishers. Portable fire extinguishers shall be provided as required by Section 906.

3104.13 Fire protection equipment. Fire hose lines, water supplies and other auxiliary fire equipment shall be maintained at the site in such numbers and sizes as required by the *fire code official.*

3104.14 Occupant load factors. The *occupant load* allowed in an assembly structure, or portion thereof, shall be determined in accordance with Chapter 10.

3104.15 Heating and cooking equipment. Heating and cooking equipment shall be in accordance with Sections 3104.15.1 through 3104.15.7.

3104.15.1 Installation. Heating or cooking equipment, tanks, piping, hoses, fittings, valves, tubing and other related components shall be installed as specified in the *International Mechanical Code* and the *International Fuel Gas Code,* and shall be *approved* by the *fire code official.*

3104.15.2 Venting. Gas, liquid and solid fuel-burning equipment designed to be vented shall be vented to the outside air as specified in the *International Fuel Gas Code* and the *International Mechanical Code.* Such vents shall be equipped with *approved* spark arresters where required. Where vents or flues are used, all portions of the tent or membrane structure shall be not less than 12 inches (305 mm) from the flue or vent.

3104.15.3 Location. Cooking and heating equipment shall not be located within 10 feet (3048 mm) of *exits* or combustible materials.

3104.15.4 Operations. Operations such as warming of foods, cooking demonstrations and similar operations that use solid flammables, butane or other similar devices that do not pose an ignition hazard, shall be *approved.*

3104.15.5 Cooking tents. Tents with sidewalls or drops where cooking is performed shall be separated from other tents or membrane structures by not less than 20 feet (6096 mm).

3104.15.6 Outdoor cooking. Outdoor cooking that produces sparks or grease-laden vapors shall not be performed within 20 feet (6096 mm) of a tent or membrane structure.

3104.15.7 Electrical heating and cooking equipment. Electrical cooking and heating equipment shall comply with NFPA 70.

3104.16 LP-gas. The storage, handling and use of LP-gas and LP-gas equipment shall be in accordance with Sections 3104.16.1 through 3104.16.3.

3104.16.1 General. LP-gas equipment such as tanks, piping, hoses, fittings, valves, tubing and other related components shall be *approved* and in accordance with Chapter 61 and with the *International Fuel Gas Code.*

3104.16.2 Location of containers. LP-gas containers shall be located outside. Safety release valves shall be pointed away from the tent or membrane structure.

3104.16.2.1 Containers 500 gallons or less. Portable LP-gas containers with a capacity of 500 gallons (1893 L) or less shall have a minimum separation between the container and structure not less than 10 feet (3048 mm).

3104.16.2.2 Containers more than 500 gallons. Portable LP-gas containers with a capacity of more than 500 gallons (1893 L) shall have a minimum separation between the container and structures not less than 25 feet (7620 mm).

3104.16.3 Protection and security. Portable LP-gas containers, piping, valves and fittings that are located outside and are being used to fuel equipment inside a tent or membrane structure shall be adequately protected to prevent tampering, damage by vehicles or other hazards and shall be located in an *approved* location. Portable LP-gas containers shall be securely fastened in place to prevent unauthorized movement.

3104.17 Flammable and combustible liquids. The storage of flammable and *combustible liquids* and the use of flammable-liquid-fueled equipment shall be in accordance with Sections 3104.17.1 through 3104.17.3.

3104.17.1 Use. Flammable-liquid-fueled equipment shall not be used in tents or membrane structures.

3104.17.2 Flammable and combustible liquid storage. Flammable and *combustible liquids* shall be stored outside in an *approved* manner not less than 50 feet (15 240 mm) from tents or membrane structures. Storage shall be in accordance with Chapter 57.

3104.17.3 Refueling. Refueling shall be performed in an *approved* location not less than 20 feet (6096 mm) from tents or membrane structures.

3104.18 Display of motor vehicles. Liquid- and gas-fueled vehicles and equipment used for display within tents or membrane structures shall be in accordance with Sections 3104.18.1 through 3104.18.5.3.

3104.18.1 Batteries. Batteries shall be disconnected in an appropriate manner.

3104.18.2 Fuel. Vehicles or equipment shall not be fueled or defueled within the tent or membrane structure.

3104.18.2.1 Quantity limit. Fuel in the fuel tank shall not exceed one-quarter of the tank capacity or 5 gallons (19 L), whichever is less.

3104.18.2.2 Inspection. Fuel systems shall be inspected for leaks.

3104.18.2.3 Closure. Fuel tank openings shall be locked and sealed to prevent the escape of vapors.

3104.18.3 Location. The location of vehicles or equipment shall not obstruct *means of egress*.

3104.18.4 Places of assembly. When a compressed natural gas (CNG) or liquefied petroleum gas (LP-gas) powered vehicle is parked inside a place of assembly, all the following conditions shall be met:

1. The quarter-turn shutoff valve or other shutoff valve on the outlet of the CNG or LP-gas container shall be closed and the engine shall be operated until it stops. Valves shall remain closed while the vehicle is indoors.
2. The hot lead of the battery shall be disconnected.
3. Dual-fuel vehicles equipped to operate on gasoline and CNG or LP-gas shall comply with this section and Sections 3104.18.1 through 3104.18.5.3 for gasoline-powered vehicles.

3104.18.5 Competitions and demonstrations. Liquid and gas-fueled vehicles and equipment used for competition or demonstration within a tent or membrane structure shall comply with Sections 3104.18.5.1 through 3104.18.5.3.

3104.18.5.1 Fuel storage. Fuel for vehicles or equipment shall be stored in *approved* containers in an *approved* location outside of the structure in accordance with Section 3104.17.2.

3104.18.5.2 Fueling. Refueling shall be performed outside of the structure in accordance with Section 3104.17.3.

3104.18.5.3 Spills. Fuel spills shall be cleaned up immediately.

3104.19 Separation of generators. Generators and other internal combustion power sources shall be separated from tents or membrane structures by not less than 20 feet (6096 mm) and shall be isolated from contact with the public by fencing, enclosure or other *approved* means.

3104.20 Standby personnel. Where, in the opinion of the *fire code official*, it is essential for public safety in a tent or membrane structure used as a place of assembly or any other use where people congregate, because of the number of persons, or the nature of the performance, exhibition, display, contest or activity, the *owner*, agent or lessee shall employ one or more qualified persons, as required and *approved*, to remain on duty during the times such places are open to the public, or when such activity is being conducted.

3104.20.1 Duties. Before each performance or the start of such activity, standby personnel shall keep diligent watch for fires during the time such place is open to the public or such activity is being conducted and take prompt measures for extinguishment of fires that occur and assist in the evacuation of the public from the structure.

3104.20.2 Crowd managers. There shall be trained crowd managers or crowd manager/supervisors at a ratio of one crowd manager/supervisor for every 250 occupants, as *approved*.

3104.21 Combustible vegetation. Combustible vegetation that could create a fire hazard shall be removed from the area occupied by a tent or membrane structure, and from areas within 30 feet (9144 mm) of such structures.

3104.22 Combustible waste material. The floor surface inside tents or membrane structures and the grounds outside and within a 30-foot (9144 mm) perimeter shall be kept free of combustible waste and other combustible materials that could create a fire hazard. Such waste shall be stored in *approved* containers and removed from the premises not less than once a day during the period the structure is occupied by the public.

SECTION 3105
TEMPORARY STAGE CANOPIES

3105.1 General. Temporary stage canopies shall comply with Section 3104, Sections 3105.2 through 3105.8 and ANSI E1.21.

3105.2 Approval. Temporary stage canopies in excess of 400 square feet (37 m^2) shall not be erected, operated or maintained for any purpose without first obtaining approval and a permit from the *fire code official* and the building official.

3105.3 Permits. Permits shall be required as set forth in Sections 105.6 and 105.7.

3105.4 Use period. Temporary stage canopies shall not be erected for a period of more than 45 days.

3105.5 Required documents. The following documents shall be submitted to the *fire code official* and the building official for review before a permit is *approved*:

1. Construction documents: *Construction documents* shall be prepared in accordance with the *International Building Code* by a registered design professional. *Construction* documents shall include:
   - 1.1. A summary sheet showing the building code used, design criteria, loads and support reactions.
   - 1.2. Detailed construction and installation drawings.
   - 1.3. Design calculations.
   - 1.4. Operating limits of the structure explicitly outlined by the registered design professional including environmental conditions and physical forces.
   - 1.5. Effects of additive elements such as video walls, supported scenery, audio equip-ment, vertical and horizontal coverings.
   - 1.6. Means for adequate stability including specific requirements for guying and cross-bracing, ground anchors or ballast for different ground conditions.
2. Designation of responsible party: The *owner* of the temporary stage canopy shall designate in writing a person to have responsibility for the temporary stage canopy on the site. The designated person shall have sufficient knowledge of the construction documents, manufacturer's recommendations and operations plan to make judgments regarding the structure's safety and to coordinate with the *fire code official.*
3. Operations plan: The operations plan shall reflect manufacturer's operational guidelines, procedures for environmental monitoring and actions to be taken under specified conditions consistent with the *construction documents.*

3105.6 Inspections. Inspections shall comply with Section 106 and Sections 3105.6.1 and 3105.6.2.

3105.6.1 Independent inspector. The *owner* of a temporary stage canopy shall employ a qualified, independent approved agency or individual to inspect the installation of a temporary stage canopy.

3105.6.2 Inspection report. The inspecting agency or individual shall furnish an inspection report to the *fire code official.* The inspection report shall indicate that the temporary stage canopy was inspected and was or was not installed in accordance with the approved *construction documents.* Discrepancies shall be brought to the immediate attention of the installer for correction. Where any discrepancy is not corrected, it shall be brought to the attention of the *fire code official* and the designated responsible party.

3105.7 Means of egress. The *means of egress* for temporary stage canopies shall comply with Chapter 10.

3105.8 Location. Temporary stage canopies shall be located a distance from property lines and buildings to accommodate distances indicated in the construction drawings for guy wires, cross-bracing, ground anchors or ballast. Location shall not interfere with egress from a building or encroach on fire apparatus access roads.

CHAPTER 32

HIGH-PILED COMBUSTIBLE STORAGE

SECTION 3201 GENERAL

3201.1 Scope. *High-piled combustible storage* shall be in accordance with this chapter. In addition to the requirements of this chapter, the following material-specific requirements shall apply:

1. Aerosols shall be in accordance with Chapter 51.
2. Flammable and *combustible liquids* shall be in accordance with Chapter 57.
3. Hazardous materials shall be in accordance with Chapter 50.
4. Storage of combustible paper records shall be in accordance with NFPA 13.
5. Storage of *combustible fibers* shall be in accordance with Chapter 37.
6. General storage of combustible material shall be in accordance with Chapter 3.

3201.2 Permits. A permit shall be required as set forth in Section 105.6.

3201.3 Construction documents. At the time of building permit application for new structures designed to accommodate high-piled storage or for requesting a change of occupancy/use, and at the time of application for a storage permit, plans and specifications shall be submitted for review and approval. In addition to the information required by the *International Building Code*, the storage permit submittal shall include the information specified in this section. Following approval of the plans, a copy of the *approved* plans shall be maintained on the premises in an *approved* location. The plans shall include all of the following:

1. Floor plan of the building showing locations and dimensions of *high-piled storage areas*.
2. Usable storage height for each storage area.
3. Number of tiers within each rack, if applicable.
4. Commodity clearance between top of storage and the sprinkler deflector for each storage arrangement.
5. Aisle dimensions between each storage array.
6. Maximum pile volume for each storage array.
7. Location and classification of commodities in accordance with Section 3203.
8. Location of commodities that are banded or encapsulated.
9. Location of required fire department access doors.
10. Type of fire suppression and fire detection systems.
11. Location of valves controlling the water supply of ceiling and in-rack sprinklers.
12. Type, location and specifications of smoke removal and curtain board systems.
13. Dimension and location of transverse and longitudinal flue spaces.
14. Additional information regarding required design features, commodities, storage arrangement and fire protection features within the high-piled storage area shall be provided at the time of permit, when required by the *fire code official*.

3201.4 Evacuation plan. Where required by the *fire code official*, an evacuation plan for public accessible areas and a separate set of plans indicating location and width of *aisles*, location of *exits*, *exit access* doors, *exit* signs, height of storage, and locations of hazardous materials shall be submitted at the time of permit application for review and approval. Following approval of the plans, a copy of the *approved* plans shall be maintained on the premises in an *approved* location.

SECTION 3202 DEFINITIONS

3202.1 Definitions. The following terms are defined in Chapter 2:

ARRAY.

ARRAY, CLOSED.

AUTOMATED RACK STORAGE.

BIN BOX.

COMMODITY.

➡ **EARLY SUPPRESSION FAST-RESPONSE (ESFR) SPRINKLER.**

EXPANDED PLASTIC.

EXTRA-HIGH-RACK COMBUSTIBLE STORAGE.

HIGH-PILED COMBUSTIBLE STORAGE.

HIGH-PILED STORAGE AREA.

LONGITUDINAL FLUE SPACE.

MANUAL STOCKING METHODS.

MECHANICAL STOCKING METHODS.

SHELF STORAGE.

SOLID SHELVING.

TRANSVERSE FLUE SPACE.

SECTION 3203 COMMODITY CLASSIFICATION

3203.1 Classification of commodities. Commodities shall be classified as Class I, II, III, IV or high hazard in accordance with this section. Materials listed within each commodity classification are assumed to be unmodified for improved combustibility characteristics. Use of flame-retarding modifiers or the physical form of the material could change the classification. See Section 3203.7 for classification of Group A, B and C plastics.

3203.2 Class I commodities. Class I commodities are essentially noncombustible products on wooden pallets, in ordinary corrugated cartons with or without single-thickness dividers, or in ordinary paper wrappings with or without pallets. Class I commodities are allowed to contain a limited amount of Group A plastics in accordance with Section 3203.7.4. Examples of Class I commodities include, but are not limited to, the following:

Alcoholic beverages not exceeding 20-percent alcohol
Appliances noncombustible, electrical
Cement in bags
Ceramics
Dairy products in nonwax-coated containers (excluding bottles)
Dry insecticides
Foods in noncombustible containers
Fresh fruits and vegetables in nonplastic trays or containers
Frozen foods
Glass
Glycol in metal cans
Gypsum board
Inert materials, bagged
Insulation, noncombustible
Noncombustible liquids in plastic containers having less than a 5-gallon (19 L) capacity
Noncombustible metal products

3203.3 Class II commodities. Class II commodities are Class I products in slatted wooden crates, solid wooden boxes, multiple-thickness paperboard cartons or equivalent combustible packaging material with or without pallets. Class II commodities are allowed to contain a limited amount of Group A plastics in accordance with Section 3203.7.4. Examples of Class II commodities include, but are not limited to, the following:

Alcoholic beverages not exceeding 20-percent alcohol, in combustible containers
Foods in combustible containers
Incandescent or fluorescent light bulbs in cartons
Thinly coated fine wire on reels or in cartons

3203.4 Class III commodities. Class III commodities are commodities of wood, paper, natural fiber cloth, or Group C plastics or products thereof, with or without pallets. Products are allowed to contain limited amounts of Group A or B plastics, such as metal bicycles with plastic handles, pedals, seats and tires. Group A plastics shall be limited in accordance with Section 3203.7.4. Examples of Class III commodities include, but are not limited to, the following:

Aerosol, Level 1 (see Chapter 51)
Biomass briquettes, bagged, and static piles
Biomass pellets, bagged, and static piles
Charcoal
Combustible fiberboard
Cork, baled
Corn cobs, static piles
Corn stover, baled and chopped
Feed, bagged
Fertilizers, bagged
Firewood
Food in plastic containers
Forest residue, round wood or chipped (branches, bark, cross-cut ends, edgings and treetops)
Furniture: wood, natural fiber, upholstered, nonplastic, wood or metal with plastic-padded and covered armrests
Glycol in combustible containers not exceeding 25 percent
Lubricating or hydraulic fluid in metal cans
Lumber
Mattresses, excluding foam rubber and foam plastics
Noncombustible liquids in plastic containers having a capacity of more than 5 gallons (19 L)
Paints, oil base, in metal cans
Paper, waste, baled
Paper and pulp, horizontal storage, or vertical storage that is banded or protected with approved wrap
Paper in cardboard boxes
Peanut hulls, bagged, and static piles
Pillows, excluding foam rubber and foam plastics
Plastic-coated paper food containers
Plywood
Rags, baled
Recovered construction wood
Rice hulls, bagged, and static piles
Rugs, without foam backing
Seasonal grasses, baled and chopped
Straw, baled
Sugar, bagged
Wood, baled
Wood chips, bagged, and static piles
Woody biomass, round wood or chipped (vase-shaped stubby bushes, bamboo, willows; branches, bark and stem wood)
Wood doors, frames and cabinets
Wood pellets, bagged, and static piles
Yarns of natural fiber and viscose

3203.5 Class IV commodities. Class IV commodities are Class I, II or III products containing Group A plastics in ordinary corrugated cartons and Class I, II and III products with Group A plastic packaging, with or without pallets. Group B plastics and free-flowing Group A plastics are also included in this class. The total amount of nonfree-flowing Group A plastics shall be in accordance with Section 3203.7.4. Examples of Class IV commodities include, but are not limited to, the following:

Aerosol, Level 2 (see Chapter 51)
Alcoholic beverages, exceeding 20-percent but less than 80-percent alcohol, in cans or bottles in cartons
Clothing, synthetic or nonviscose
Combustible metal products (solid)
Furniture, plastic upholstered

Furniture, wood or metal with plastic covering and padding
Glycol in combustible containers (greater than 25 percent and less than 50 percent)
Linoleum products
Paints, oil base in combustible containers
Pharmaceutical, alcoholic elixirs, tonics, etc.
Rugs, foam back
Shingles, asphalt
Thread or yarn, synthetic or nonviscose

3203.6 High-hazard commodities. High-hazard commodities are high-hazard products presenting special fire hazards beyond those of Class I, II, III or IV. Group A plastics not otherwise classified are included in this class. Examples of high-hazard commodities include, but are not limited to, the following:

Aerosol, Level 3 (see Chapter 51)
Alcoholic beverages, exceeding 80-percent alcohol, in bottles or cartons
Commodities of any class in plastic containers in carousel storage
Flammable solids (except solid combustible metals)
Glycol in combustible containers (50 percent or greater)
Lacquers that dry by solvent evaporation, in metal cans or cartons
Lubricating or hydraulic fluid in plastic containers
Mattresses, foam rubber or foam plastics
Pallets and flats that are idle combustible
Paper and pulp, rolled, in vertical storage that is unbanded or not protected with an *approved* wrap
Paper, asphalt, rolled, horizontal storage
Paper, asphalt, rolled, vertical storage
Pillows, foam rubber and foam plastics
Pyroxylin
Rubber tires
Vegetable oil and butter in plastic containers

3203.7 Classification of plastics. Plastics shall be designated as Group A, B or C in accordance with Sections 3203.7.1 through 3203.7.4.

3203.7.1 Group A plastics. Group A plastics are plastic materials having a heat of combustion that is much higher than that of ordinary combustibles, and a burning rate higher than that of Group B plastics. Examples of Group A plastics include, but are not limited to, the following:

ABS (acrylonitrile-butadiene-styrene copolymer)
Acetal (polyformaldehyde)
Acrylic (polymethyl methacrylate)
Butyl rubber
EPDM (ethylene propylene rubber)
FRP (fiberglass-reinforced polyester)
Natural rubber (expanded)
Nitrile rubber (acrylonitrile butadiene rubber)
PET or PETE (polyethylene terephthalate)
Polybutadiene
Polycarbonate
Polyester elastomer
Polyethylene
Polypropylene
Polystyrene (expanded and unexpanded)
Polyurethane (expanded and unexpanded)
PVC (polyvinyl chloride greater than 15-percent plasticized, e.g., coated fabric unsupported film)
SAN (styrene acrylonitrile)
SBR (styrene butadiene rubber)

3203.7.2 Group B plastics. Group B plastics are plastic materials having a heat of combustion and a burning rate higher than that of ordinary combustibles, but not as high as those of Group A plastics. Examples of Group B plastics include, but are not limited to, the following:

Cellulosics (cellulose acetate, cellulose acetate butyrate, ethyl cellulose)
Chloroprene rubber
Fluoroplastics (ECTFE, ethylene-chlorotrifluoroethylene copolymer; ETFE, ethylene-tetrafluoroethylene copolymer; FEP, fluorinated ethylene-propylene copolymer)
Natural rubber (nonexpanded)
Nylon (Nylon 6, Nylon 6/6)
PVC (polyvinyl chloride greater than 5-percent, but not exceeding 15-percent plasticized)
Silicone rubber

3203.7.3 Group C plastics. Group C plastics are plastic materials having a heat of combustion and a burning rate similar to those of ordinary combustibles. Examples of Group C plastics include, but are not limited to, the following:

Fluoroplastics (PCTFE, polychlorotrifluoroethylene; PTFE, polytetrafluoroethylene)
Melamine (melamine formaldehyde)
Phenol
PVC (polyvinyl chloride, rigid or plasticized less than 5 percent, e.g., pipe, pipe fittings)
PVDC (polyvinylidene chloride)
PVDF (polyvinylidene fluoride)
PVF (polyvinyl fluoride)
Urea (urea formaldehyde)

3203.7.4 Limited quantities of Group A plastics in mixed commodities. Figure 3203.7.4 shall be used to determine the quantity of Group A plastics allowed to be stored in a package or carton or on a pallet without increasing the commodity classification.

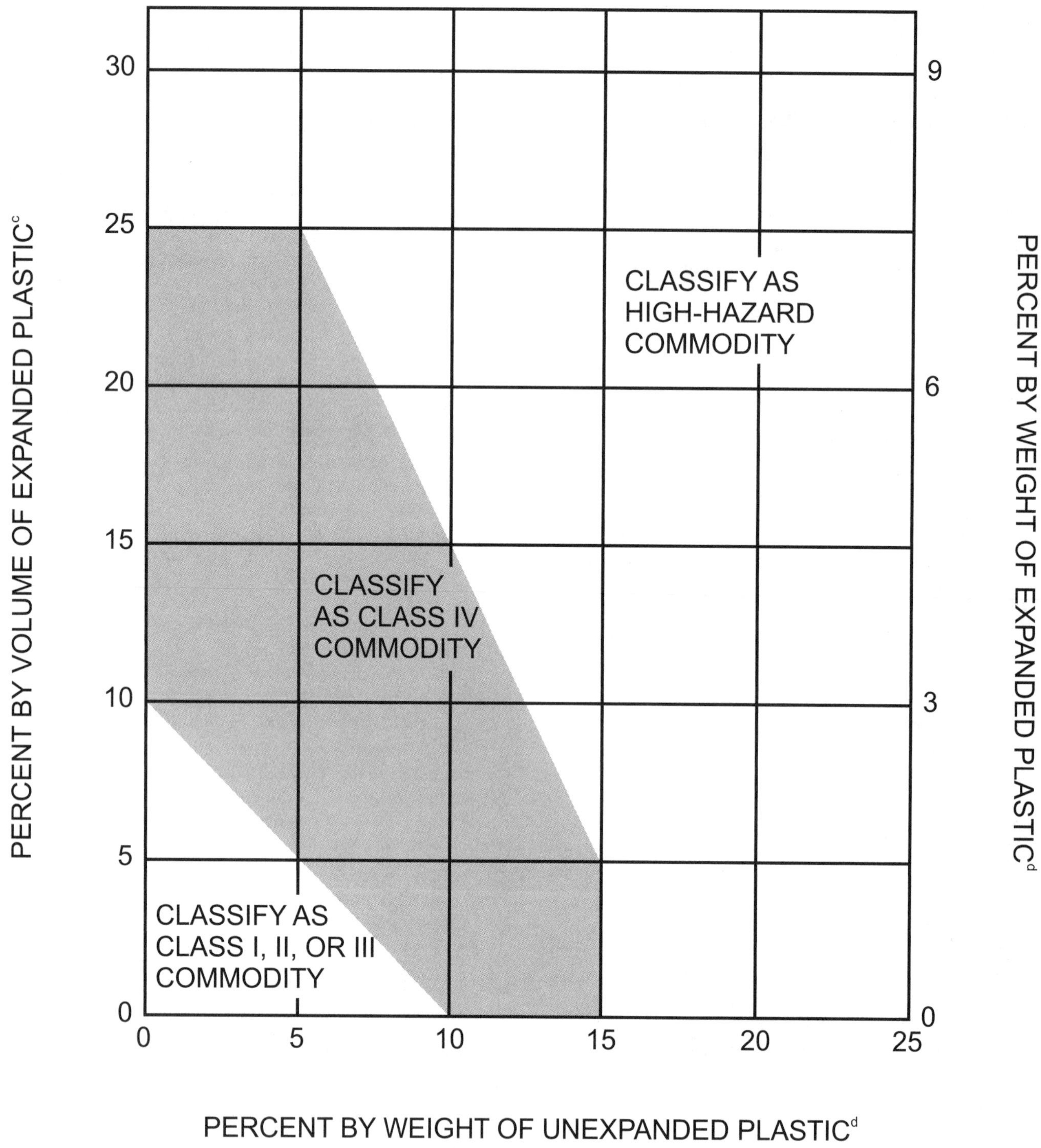

FIGURE 3203.7.4
MIXED COMMODITIES[a, b]

a. This figure is intended to determine the commodity classification of a mixed commodity in a package, carton or on a pallet where plastics are involved.

b. The following is an example of how to apply the figure: A package containing a Class III commodity has 12-percent Group A expanded plastic by volume. The weight of the unexpanded Group A plastic is 10 percent. This commodity is classified as a Class IV commodity. If the weight of the unexpanded plastic is increased to 14 percent, the classification changes to a high-hazard commodity.

c. $\text{Percent by volume} = \dfrac{\text{Volume of plastic in pallet load}}{\text{Total volume of pallet load, including pallet}}$

d. $\text{Percent by weight} = \dfrac{\text{Weight of plastic in pallet load}}{\text{Total weight of pallet load, including pallet}}$

SECTION 3204
DESIGNATION OF HIGH-PILED STORAGE AREAS

3204.1 General. *High-piled storage areas*, and portions of *high-piled storage areas* intended for storage of a different commodity class than adjacent areas, shall be designed and specifically designated to contain Class I, Class II, Class III, Class IV or high-hazard commodities. The designation of a *high-piled combustible storage* area, or portion thereof intended for storage of a different commodity class, shall be based on the highest hazard commodity class stored except as provided in Section 3204.2.

3204.2 Designation based on engineering analysis. The designation of a *high-piled combustible storage* area, or portion thereof, is allowed to be based on a lower hazard class than that of the highest class of commodity stored when a limited quantity of the higher hazard commodity has been demonstrated by engineering analysis to be adequately protected by the *automatic sprinkler system* provided. The engineering analysis shall consider the ability of the sprinkler system to deliver the higher density required by the higher hazard commodity. The higher density shall be based on the actual storage height of the pile or rack and the minimum allowable design area for sprinkler operation as set forth in the density/area figures provided in NFPA 13. The contiguous area occupied by the higher hazard commodity shall not exceed 120 square feet (11 m^2) and additional areas of higher hazard commodity shall be separated from other such areas by 25 feet (7620 mm) or more. The sprinkler system shall be capable of delivering the higher density over a minimum area of 900 square feet (84 m^2) for wet pipe systems and 1,200 square feet (111 m^2) for dry pipe systems. The shape of the design area shall be in accordance with Section 903.

SECTION 3205
HOUSEKEEPING AND MAINTENANCE

3205.1 Rack structures. The structural integrity of racks shall be maintained.

3205.2 Ignition sources. Clearance from ignition sources shall be provided in accordance with Section 305.

3205.3 Smoking. Smoking shall be prohibited. *Approved* "No Smoking" signs shall be conspicuously posted in accordance with Section 310.

3205.4 Aisle maintenance. When restocking is not being conducted, aisles shall be kept clear of storage, waste material and debris. Fire department access doors, aisles and *exit* doors shall not be obstructed. During restocking operations using manual stocking methods, a minimum unobstructed aisle width of 24 inches (610 mm) shall be maintained in 48-inch (1219 mm) or smaller aisles, and a minimum unobstructed aisle width of one-half of the required aisle width shall be maintained in aisles greater than 48 inches (1219 mm). During mechanical stocking operations, a minimum unobstructed aisle width of 44 inches (1118 mm) shall be maintained in accordance with Section 3206.9.

3205.5 Pile dimension and height limitations. Pile dimensions and height limitations shall comply with Section 3207.3.

3205.6 Designation of storage heights. Where required by the *fire code official*, a visual method of indicating the maximum allowable storage height shall be provided.

3205.7 Arrays. Arrays shall comply with Section 3207.4.

3205.8 Flue spaces. Flue spaces shall comply with Section 3208.3.

SECTION 3206
GENERAL FIRE PROTECTION AND LIFE SAFETY FEATURES

3206.1 General. Fire protection and life safety features for *high-piled storage areas* shall be in accordance with Sections 3206.2 through 3206.10.

3206.2 Extent and type of protection. Where required by Table 3206.2, fire detection systems, smoke and heat removal and automatic sprinkler design densities shall extend the lesser of 15 feet (4572 mm) beyond the *high-piled storage area* or to a permanent partition. Where portions of *high-piled storage areas* have different fire protection requirements because of commodity, method of storage or storage height, the fire protection features required by Table 3206.2 within this area shall be based on the most restrictive design requirements.

3206.3 Separation of high-piled storage areas. *High-piled storage areas* shall be separated from other portions of the building where required by Sections 3206.3.1 through 3206.3.2.2.

3206.3.1 Separation from other uses. Mixed occupancies shall be separated in accordance with the *International Building Code.*

3206.3.2 Multiple high-piled storage areas. Multiple *high-piled storage areas* shall be in accordance with Section 3206.3.2.1 or 3206.3.2.2.

3206.3.2.1 Aggregate area. The aggregate of all *high-piled storage areas* within a building shall be used for the application of Table 3206.2 unless such areas are separated from each other by 1-hour *fire barriers* constructed in accordance with Section 707 the *International Building Code*. Openings in such *fire barriers* shall be protected by opening protectives having a 1-hour *fire protection rating*.

3206.3.2.2 Multiclass high-piled storage areas. *High-piled storage areas* classified as Class I through IV not separated from *high-piled storage areas* classified as high hazard shall utilize the aggregate of all *high-piled storage areas* as high hazard for the purposes of the application of Table 3206.2. To be considered as separated, 1-hour *fire barriers* shall be constructed in accordance with Section 707 of the *International Building Code*. Openings in such *fire barriers* shall be protected by opening protectives having a 1-hour *fire protection rating*.

Exception: As provided for in Section 3204.2.

3206.4 Automatic sprinklers. *Automatic sprinkler systems* shall be provided in accordance with Sections 3207, 3208 and 3209.

TABLE 3206.2
GENERAL FIRE PROTECTION AND LIFE SAFETY REQUIREMENTS

COMMODITY CLASS	SIZE OF HIGH-PILED STORAGE AREA[a] (square feet) (see Sections 3206.2 and 3206.4)	ALL STORAGE AREAS (See Sections 3206, 3207 and 3208)[b]				SOLID-PILED STORAGE, SHELF STORAGE AND PALLETIZED STORAGE (see Section 3207.3)		
		Automatic fire-extinguishing system (see Section 3206.4)	Fire detection system (see Section 3206.5)	Building access (see Section 3206.6)	Smoke and heat removal (see Section 3206.7)	Maximum pile dimension[c] (feet)	Maximum permissible storage height[d] (feet)	Maximum pile volume (cubic feet)
I-IV	0-500	Not Required[a]	Not Required	Not Required[e]	Not Required	Not Required	Not Required	Not Required
	501-2,500	Not Required[a]	Yes[i]	Not Required[e]	Not Required	100	40	100,000
	2,501-12,000 Public accessible	Yes	Not Required	Not Required[e]	Not Required	100	40	400,000
	2,501-12,000 Nonpublic accessible (Option 1)	Yes	Not Required	Not Required[e]	Not Required	100	40	400,000
	2,501-12,000 Nonpublic accessible (Option 2)	Not Required[a]	Yes	Yes	Yes[j]	100	30[f]	200,000
	12,001-20,000	Yes	Not Required	Yes	Yes[j]	100	40	400,000
	20,001-500,000	Yes	Not Required	Yes	Yes[j]	100	40	400,000
	Greater than 500,000[g]	Yes	Not Required	Yes	Yes[j]	100	40	400,000
High hazard	0-500	Not Required[a]	Not Required	Not Required[e]	Not Required	50	Not Required	Not Required
	501-2,500 Public accessible	Yes	Not Required	Not Required[e]	Not Required	50	30	75,000
	501-2,500 Nonpublic accessible (Option 1)	Yes	Not Required	Not Required[e]	Not Required	50	30	75,000
	501-2,500 Nonpublic accessible (Option 2)	Not Required[a]	Yes	Yes	Yes[j]	50	20	50,000
	2,501-300,000	Yes	Not Required	Yes	Yes[j]	50	30	75,000
	300,001-500,000[g, h]	Yes	Not Required	Yes	Yes[j]	50	30	75,000

For SI: 1 foot = 304.8 mm, 1 cubic foot = 0.02832 m^3, 1 square foot = 0.0929 m^2.

a. Where automatic sprinklers are required for reasons other than those in Chapter 32, the portion of the sprinkler system protecting the high-piled storage area shall be designed and installed in accordance with Sections 3207 and 3208.

b. For aisles, see Section 3206.9.

c. Piles shall be separated by aisles complying with Section 3206.9.

d. For storage in excess of the height indicated, special fire protection shall be provided in accordance with Note g where required by the fire code official. See Chapters 51 and 57 for special limitations for aerosols and flammable and combustible liquids, respectively.

e. Section 503 shall apply for fire apparatus access.

f. For storage exceeding 30 feet in height, Option 1 shall be used.

g. Special fire protection provisions including, but not limited to, fire protection of exposed steel columns; increased sprinkler density; additional in-rack sprinklers, without associated reductions in ceiling sprinkler density; or additional fire department hose connections shall be provided required by the fire code official.

h. High-piled storage areas shall not exceed 500,000 square feet. A 2-hour fire wall constructed in accordance with Section 706 the *International Building Code* shall be used to divide high-piled storage exceeding 500,000 square feet in area.

i. Not required where an automatic fire-extinguishing system is designed and installed to protect the high-piled storage area in accordance with Sections 3207 and 3208.

j. Not required where storage areas are protected by either early suppression fast response (ESFR) sprinkler systems or control mode special application sprinklers with a response time index of 50 $(m \bullet s)^{1/2}$ or less that are listed to control a fire in the stored commodities with 12 or fewer sprinklers, installed in accordance with NFPA 13.

3206.4.1 Pallets. Automatic sprinkler system requirements based upon the presence of pallets shall be in accordance with NFPA 13.

3206.4.1.1 Plastic pallets. Plastic pallets listed and labeled in accordance with UL 2335 or FM 4996 shall be treated as wood pallets for determining required sprinkler protection.

3206.5 Fire detection. Where fire detection is required by Table 3206.2, an *approved* automatic fire detection system shall be installed throughout the *high-piled storage area*. The system shall be monitored and be in accordance with Section 907.

3206.6 Building access. Where building access is required by Table 3206.2, fire apparatus access roads in accordance with Section 503 shall be provided within 150 feet (45 720 mm) of all portions of the *exterior walls* of buildings used for high-piled storage.

Exception: Where fire apparatus access roads cannot be installed because of topography, railways, waterways, nonnegotiable grades or other similar conditions, the *fire code official* is authorized to require additional fire protection.

3206.6.1 Access doors. Where building access is required by Table 3206.2, fire department access doors shall be provided in accordance with this section. Access doors shall be accessible without the use of a ladder.

3206.6.1.1 Number of doors required. Not less than one access door shall be provided in each 100 linear feet (30 480 mm), or fraction thereof, of the exterior walls that face required fire apparatus access roads. The required access doors shall be distributed such that the lineal distance between adjacent access doors does not exceed 100 feet (30 480 mm).

Exception: The linear distance between adjacent access doors is allowed to exceed 100 feet (30 480 mm) in existing buildings where no change in occupancy is proposed. The number and distribution of access doors in existing buildings shall be approved.

3206.6.1.2 Door size and type. Access doors shall be not less than 3 feet (914 mm) in width and 6 feet 8 inches (2032 mm) in height. Roll-up doors shall not be used unless *approved*.

3206.6.1.3 Locking devices. Only *approved* locking devices shall be used.

3206.7 Smoke and heat removal. Where smoke and heat removal is required by Table 3206.2 it shall be provided in accordance with Section 910.

3206.8 Fire department hose connections. Where *exit* passageways are required by the *International Building Code* for egress, a Class I standpipe system shall be provided in accordance with Section 905.

3206.9 Aisles. Aisles providing access to *exits* and fire department access doors shall be provided in *high-piled storage areas* exceeding 500 square feet (46 m^2), in accordance with Sections 3206.9.1 through 3206.9.3. Aisles separating storage piles or racks shall comply with NFPA 13. Aisles shall also comply with Chapter 10.

Exception: Where aisles are precluded by rack storage systems, alternate methods of access and protection are allowed when *approved*.

3206.9.1 Width. Aisle width shall be in accordance with Sections 3206.9.1.1 and 3206.9.1.2.

Exceptions:

1. Aisles crossing rack structures or storage piles, that are used only for employee access, shall be not less than 24 inches (610 mm) wide.
2. Aisles separating shelves classified as shelf storage shall be not less than 30 inches (762 mm) wide.

3206.9.1.1 Sprinklered buildings. Aisles in sprinklered buildings shall be not less than 44 inches (1118 mm) wide. Aisles shall be not less than 96 inches (2438 mm) wide in *high-piled storage areas* exceeding 2,500 square feet (232 m^2) in area, that are accessible to the public and designated to contain high-hazard commodities.

Exception: Aisles in *high-piled storage areas* exceeding 2,500 square feet (232 m^2) in area, that are accessible to the public and designated to contain high-hazard commodities, are protected by a sprinkler system designed for multiple-row racks of high-hazard commodities shall be not less than 44 inches (1118 mm) wide.

Aisles shall be not less than 96 inches (2438 mm) wide in areas accessible to the public where mechanical stocking methods are used.

3206.9.1.2 Nonsprinklered buildings. Aisles in nonsprinklered buildings shall be not less than 96 inches (2438 mm) wide.

3206.9.2 Clear height. The required aisle width shall extend from floor to ceiling. Rack structural supports and catwalks are allowed to cross aisles at a minimum height of 6 feet 8 inches (2032 mm) above the finished floor level, provided that such supports do not interfere with fire department hose stream trajectory.

3206.9.3 Dead-end aisles. Dead-end aisles shall not exceed 20 feet (6096 mm) in length in Group M occupancies. Dead-end aisles shall not exceed 50 feet (15 240 mm) in length in all other occupancies.

Exception: Dead-end aisles are not limited where the length of the dead-end aisle is less than 2.5 times the least width of the dead-end aisle.

3206.10 Portable fire extinguishers. Portable fire extinguishers shall be provided in accordance with Section 906.

SECTION 3207
SOLID-PILED AND SHELF STORAGE

3207.1 General. Shelf storage and storage in solid piles, solid piles on pallets and bin box storage in bin boxes not exceeding 5 feet (1524 mm) in any dimension, shall be in accordance with Sections 3206 and this section.

3207.2 Fire protection. Where automatic sprinklers are required by Table 3206.2, an *approved automatic sprinkler system* shall be installed throughout the building or to 1-hour *fire barriers* constructed in accordance with Section 707 of the *International Building Code*. Openings in such *fire barriers* shall be protected by opening protectives having a 1-hour *fire protection rating*. The design and installation of the *automatic sprinkler system* and other applicable fire protection shall be in accordance with the *International Building Code* and NFPA 13.

3207.2.1 Shelf storage. Shelf storage greater than 12 feet (3658 mm) but less than 15 feet (4572 mm) in height shall be in accordance with the fire protection requirements set forth in NFPA 13. Shelf storage 15 feet (4572 mm) or more in height shall be protected in an *approved* manner with special fire protection, such as in-rack sprinklers.

3207.3 Pile dimension and height limitations. Pile dimensions, the maximum permissible storage height and pile volume shall be in accordance with Table 3206.2.

3207.4 Array. Where an *automatic sprinkler system* design utilizes protection based on a closed array, array clearances shall be provided and maintained as specified by the standard used.

SECTION 3208 RACK STORAGE

3208.1 General. Rack storage shall be in accordance with Section 3206 and this section. Bin boxes exceeding 5 feet (1524 mm) in any dimension shall be regulated as rack storage.

3208.2 Fire protection. Where automatic sprinklers are required by Table 3206.2, an *approved automatic sprinkler system* shall be installed throughout the building or to 1-hour *fire barriers* constructed in accordance with Section 707 of the *International Building Code*. Openings in such *fire barriers* shall be protected by opening protectives having a 1-hour *fire protection rating*. The design and installation of the *automatic sprinkler system* and other applicable fire protection shall be in accordance with Section 903.3.1.1 and the *International Building Code*.

3208.2.1 Plastic shelves. Storage on plastic shelves shall be protected by *approved* specially engineered *fire protection systems*.

3208.2.2 Racks with solid shelving. Racks with solid shelving having an area greater than 20 square feet (1.9 m^2), measured between *approved* flue spaces at all four edges of the shelf, shall be in accordance with this section.

Exceptions:

1. Racks with mesh, grated, slatted or similar shelves having uniform openings not more than 6 inches (152 mm) apart, comprising not less than 50 percent of the overall shelf area, and with *approved* flue spaces are allowed to be treated as racks without solid shelves.
2. Racks used for the storage of combustible paper records, with solid shelving, shall be in accordance with NFPA 13.

3208.2.2.1 Fire protection. Fire protection for racks with solid shelving shall be in accordance with NFPA 13.

3208.3 Flue spaces. Flue spaces shall be provided in accordance with Table 3208.3. Required flue spaces shall be maintained.

3208.3.1 Flue space protection. Where required by the *fire code official*, flue spaces required by Table 3208.3, in single-, double- or multiple-row rack storage installations shall be equipped with *approved* devices to protect the required flue spaces. Such devices shall not be removed or modified.

3208.4 Column protection. Steel building columns shall be protected in accordance with NFPA 13.

3208.5 Extra-high-rack storage systems. Approval of the *fire code official* shall be obtained prior to installing extra-high-rack combustible storage.

3208.5.1 Fire protection. Buildings with extra-high-rack combustible storage shall be protected with a specially engineered *automatic sprinkler system*. Extra-high-rack combustible storage shall be provided with additional special fire protection, such as separation from other buildings and additional built-in fire protection features and fire department access, where required by the *fire code official*.

SECTION 3209 AUTOMATED STORAGE

3209.1 General. Automated storage shall be in accordance with this section.

3209.2 Automatic sprinklers. Where automatic sprinklers are required by Table 3206.2, the building shall be equipped throughout with an *approved automatic sprinkler system* in accordance with Section 903.3.1.1.

3209.3 Carousel storage. *High-piled storage areas* having greater than 500 square feet (46 m^2) of carousel storage shall be provided with automatic shutdown in accordance with one of the following:

1. An automatic smoke detection system installed in accordance with Section 907, with coverage extending 15 feet (4575 mm) in all directions beyond unenclosed carousel storage systems and that sounds a local alarm at the operator's station and stops the carousel storage system upon the activation of a single detector.
2. An automatic smoke detection system installed in accordance with Section 907 and within enclosed carousel storage systems, that sounds a local alarm at the operator's station and stops the carousel storage system upon the activation of a single detector.
3. A single dead-man-type control switch that allows the operation of the carousel storage system only when the operator is present. The switch shall be in the same

room as the carousel storage system and located to provide for observation of the carousel system.

3209.4 Automated rack storage. *High-piled storage areas* with automated rack storage shall be provided with a manually activated emergency shutdown switch for use by emergency personnel. The switch shall be clearly identified and shall be in a location *approved* by the fire chief.

SECTION 3210 SPECIALTY STORAGE

3210.1 General. Records storage facilities used for the rack or shelf storage of combustible paper records greater than 12 feet (3658 mm) in height shall be in accordance with Sections 3206 and 3208 and NFPA 13. Palletized storage of records shall be in accordance with Section 3207.

TABLE 3208.3
REQUIRED FLUE SPACES FOR RACK STORAGE

RACK CONFIGURATION	AUTOMATIC SPRINKLER PROTECTION		SPRINKLER AT THE CEILING WITH OR WITHOUT MINIMUM IN-RACK SPRINKLERS			IN-RACK SPRINKLERS AT EVERY TIER	NONSPRINKLERED
			≤ 25 feet		> 25 feet	Any height	Any height
	Storage height		Option 1	Option 2			
Single-row rack	Transverse flue space	Size[b]	3 inches	Not Applicable	3 inches	Not Required	Not Required
		Vertically aligned	Not Required	Not Applicable	Yes	Not Applicable	Not Required
	Longitudinal flue space		Not Required	Not Applicable	Not Required	Not Required	Not Required
Double-row rack	Transverse flue space	Size[b]	6 inches[a]	3 inches	3 inches	Not Required	Not Required
		Vertically aligned	Not Required	Not Required	Yes	Not Applicable	Not Required
	Longitudinal flue space		Not Required	6 inches	6 inches	Not Required	Not Required
Multirow rack	Transverse flue space	Size[b]	6 inches	Not Applicable	6 inches	Not Required	Not Required
		Vertically aligned	Not Required	Not Applicable	Yes	Not Applicable	Not Required
	Longitudinal flue space		Not Required	Not Applicable	Not Required	Not Required	Not Required

For SI: 1 inch = 25.4 mm, 1 foot = 304.8 mm.

a. Three-inch transverse flue spaces shall be provided not less than every 10 feet where ESFR sprinkler protection is provided.

b. Random variations are allowed, provided that the configuration does not obstruct water penetration.

CHAPTER 33

FIRE SAFETY DURING CONSTRUCTION AND DEMOLITION

SECTION 3301 GENERAL

3301.1 Scope. This chapter shall apply to structures in the course of construction, *alteration* or demolition, including those in underground locations. Compliance with NFPA 241 is required for items not specifically addressed herein.

3301.2 Purpose. This chapter prescribes minimum safeguards for construction, *alteration* and demolition operations to provide reasonable safety to life and property from fire during such operations.

SECTION 3302 DEFINITIONS

3302.1 Terms defined in Chapter 2. Words and terms used in this chapter and defined in Chapter 2 shall have the meanings ascribed to them as defined therein.

SECTION 3303 TEMPORARY HEATING EQUIPMENT

3303.1 Listed. Temporary heating devices shall be *listed* and *labeled* in accordance with the *International Mechanical Code* or the *International Fuel Gas Code*. Installation, maintenance and use of temporary heating devices shall be in accordance with the terms of the listing.

3303.2 Oil-fired heaters. Oil-fired heaters shall comply with Section 603.

3303.3 LP-gas heaters. Fuel supplies for liquefied-petroleum gas-fired heaters shall comply with Chapter 61 and the *International Fuel Gas Code*.

3303.4 Refueling. Refueling operations for liquid-fueled equipment or appliances shall be conducted in accordance with Section 5705. The equipment or appliance shall be allowed to cool prior to refueling.

3303.5 Installation. Clearance to combustibles from temporary heating devices shall be maintained in accordance with the *labeled* equipment. When in operation, temporary heating devices shall be fixed in place and protected from damage, dislodgement or overturning in accordance with the manufacturer's instructions.

3303.6 Supervision. The use of temporary heating devices shall be supervised and maintained only by competent personnel.

SECTION 3304 PRECAUTIONS AGAINST FIRE

3304.1 Smoking. Smoking shall be prohibited except in *approved* areas. Signs shall be posted in accordance with Section 310. In *approved* areas where smoking is permitted, *approved* ashtrays shall be provided in accordance with Section 310.

3304.2 Combustible debris, rubbish and waste. Combustible debris, rubbish and waste material shall comply with the requirements of Sections 3304.2.1 through 3304.2.4.

3304.2.1 Combustible waste material accumulation. Combustible debris, rubbish and waste material shall not be accumulated within buildings.

3304.2.2 Combustible waste material removal. Combustible debris, rubbish and waste material shall be removed from buildings at the end of each shift of work.

3304.2.3 Rubbish containers. Where rubbish containers with a capacity exceeding 5.33 cubic feet (40 gallons) (0.15 m^3) are used for temporary storage of combustible debris, rubbish and waste material, they shall have tight-fitting or self-closing lids. Such rubbish containers shall be constructed entirely of materials that comply with either of the following:

1. Noncombustible materials.
2. Materials that meet a peak rate of heat release not exceeding 300 kW/m^2 when tested in accordance with ASTM E 1354 at an incident heat flux of 50 kW/m^2 in the horizontal orientation.

3304.2.4 Spontaneous ignition. Materials susceptible to spontaneous ignition, such as oily rags, shall be stored in a listed disposal container.

3304.3 Burning of combustible debris, rubbish and waste. Combustible debris, rubbish and waste material shall not be disposed of by burning on the site unless approved.

3304.4 Open burning. *Open burning* shall comply with Section 307.

3304.5 Fire watch. Where required by the *fire code official* for building demolition, or building construction during working hours that is hazardous in nature, qualified personnel shall be provided to serve as an on-site fire watch. Fire watch personnel shall be provided with not less than one approved means for notification of the fire department and their sole duty shall be to perform constant patrols and watch for the occurrence of fire.

3304.6 Cutting and welding. Operations involving the use of cutting and welding shall be done in accordance with Chapter 35.

3304.7 Electrical. Temporary wiring for electrical power and lighting installations used in connection with the construction, *alteration* or demolition of buildings, structures, equipment or similar activities shall comply with NFPA 70.

SECTION 3305 FLAMMABLE AND COMBUSTIBLE LIQUIDS

3305.1 Storage of flammable and combustible liquids. Storage of flammable and *combustible liquids* shall be in accordance with Section 5704.

3305.2 Class I and Class II liquids. The storage, use and handling of flammable and *combustible liquids* at construction sites shall be in accordance with Section 5706.2. Ventilation shall be provided for operations involving the application of materials containing flammable solvents.

3305.3 Housekeeping. Flammable and combustible liquid storage areas shall be maintained clear of combustible vegetation and waste materials. Such storage areas shall not be used for the storage of combustible materials.

3305.4 Precautions against fire. Sources of ignition and smoking shall be prohibited in flammable and *combustible liquid* storage areas. Signs shall be posted in accordance with Section 310.

3305.5 Handling at point of final use. Class I and II liquids shall be kept in *approved* safety containers.

3305.6 Leakage and spills. Leaking vessels shall be immediately repaired or taken out of service and spills shall be cleaned up and disposed of properly.

SECTION 3306 FLAMMABLE GASES

3306.1 Storage and handling. The storage, use and handling of flammable gases shall comply with Chapter 58.

3306.2 Cleaning with flammable gas. Flammable gases shall not be used to clean or remove debris from piping open to the atmosphere.

3306.2.1 Pipe cleaning and purging. The cleaning and purging of flammable gas piping systems, including cleaning new or existing piping systems, purging piping systems into service and purging piping systems out of service, shall comply with NFPA 56.

Exceptions:

1. Compressed gas piping systems other than fuel gas piping systems where in accordance with Chapter 53.
2. Piping systems regulated by the *International Fuel Gas Code*.
3. Liquefied petroleum gas systems in accordance with Chapter 61.

SECTION 3307 EXPLOSIVE MATERIALS

3307.1 Storage and handling. *Explosive* materials shall be stored, used and handled in accordance with Chapter 56.

3307.2 Supervision. Blasting operations shall be conducted in accordance with Chapter 56.

3307.3 Demolition using explosives. *Approved* fire hoses for use by demolition personnel shall be maintained at the demolition site whenever *explosives* are used for demolition. Such fire hoses shall be connected to an *approved* water supply and shall be capable of being brought to bear on post-*detonation* fires anywhere on the site of the demolition operation.

SECTION 3308 OWNER'S RESPONSIBILITY FOR FIRE PROTECTION

3308.1 Program superintendent. The *owner* shall designate a person to be the fire prevention program superintendent who shall be responsible for the fire prevention program and ensure that it is carried out through completion of the project. The fire prevention program superintendent shall have the authority to enforce the provisions of this chapter and other provisions as necessary to secure the intent of this chapter. Where guard service is provided, the superintendent shall be responsible for the guard service.

3308.2 Prefire plans. The fire prevention program superintendent shall develop and maintain an *approved* prefire plan in cooperation with the fire chief. The fire chief and the *fire code official* shall be notified of changes affecting the utilization of information contained in such prefire plans.

3308.3 Training. Training of responsible personnel in the use of fire protection equipment shall be the responsibility of the fire prevention program superintendent.

3308.4 Fire protection devices. The fire prevention program superintendent shall determine that all fire protection equipment is maintained and serviced in accordance with this code. The quantity and type of fire protection equipment shall be *approved*.

3308.5 Hot work operations. The fire prevention program superintendent shall be responsible for supervising the permit system for hot work operations in accordance with Chapter 35.

3308.6 Impairment of fire protection systems. Impairments to any *fire protection system* shall be in accordance with Section 901.

3308.7 Temporary covering of fire protection devices. Coverings placed on or over fire protection devices to protect them from damage during construction processes shall be immediately removed upon the completion of the construction processes in the room or area in which the devices are installed.

SECTION 3309
FIRE REPORTING

3309.1 Emergency telephone. Readily accessible emergency telephone facilities shall be provided in an *approved* location at the construction site. The street address of the construction site and the emergency telephone number of the fire department shall be posted adjacent to the telephone.

SECTION 3310
ACCESS FOR FIRE FIGHTING

3310.1 Required access. *Approved* vehicle access for fire fighting shall be provided to all construction or demolition sites. Vehicle access shall be provided to within 100 feet (30 480 mm) of temporary or permanent fire department connections. Vehicle access shall be provided by either temporary or permanent roads, capable of supporting vehicle loading under all weather conditions. Vehicle access shall be maintained until permanent fire apparatus access roads are available.

3310.2 Key boxes. Key boxes shall be provided as required by Chapter 5.

SECTION 3311
MEANS OF EGRESS

[B] 3311.1 Stairways required. Where a building has been constructed to a *building height* of 50 feet (15 240 mm) or four stories, or where an existing building exceeding 50 feet (15 240 mm) in *building height* is altered, not less than one temporary lighted *stairway* shall be provided unless one or more of the permanent *stairways* are erected as the construction progresses.

3311.2 Maintenance. Required *means of egress* shall be maintained during construction and demolition, remodeling or *alterations* and additions to any building.

> **Exception:** *Approved* temporary *means of egress* systems and facilities.

SECTION 3312
WATER SUPPLY FOR FIRE PROTECTION

3312.1 When required. An *approved* water supply for fire protection, either temporary or permanent, shall be made available as soon as combustible material arrives on the site.

SECTION 3313
STANDPIPES

3313.1 Where required. In buildings required to have standpipes by Section 905.3.1, not less than one standpipe shall be provided for use during construction. Such standpipes shall be installed prior to construction exceeding 40 feet (12 192 mm) in height above the lowest level of fire department vehicle access. Such standpipe shall be provided with fire department hose connections at accessible locations adjacent to usable stairways. Such standpipes shall be extended as construction progresses to within one floor of the highest point of construction having secured decking or flooring.

3313.2 Buildings being demolished. Where a building is being demolished and a standpipe is existing within such a building, such standpipe shall be maintained in an operable condition so as to be available for use by the fire department. Such standpipe shall be demolished with the building but shall not be demolished more than one floor below the floor being demolished.

3313.3 Detailed requirements. Standpipes shall be installed in accordance with the provisions of Section 905.

> **Exception:** Standpipes shall be either temporary or permanent in nature, and with or without a water supply, provided that such standpipes comply with the requirements of Section 905 as to capacity, outlets and materials.

SECTION 3314
AUTOMATIC SPRINKLER SYSTEM

3314.1 Completion before occupancy. In buildings where an *automatic sprinkler system* is required by this code or the *International Building Code*, it shall be unlawful to occupy any portion of a building or structure until the *automatic sprinkler system* installation has been tested and *approved*, except as provided in Section 105.3.4.

3314.2 Operation of valves. Operation of sprinkler control valves shall be allowed only by properly authorized personnel and shall be accompanied by notification of duly designated parties. Where the sprinkler protection is being regularly turned off and on to facilitate connection of newly completed segments, the sprinkler control valves shall be checked at the end of each work period to ascertain that protection is in service.

SECTION 3315
PORTABLE FIRE EXTINGUISHERS

3315.1 Where required. Structures under construction, *alteration* or demolition shall be provided with not less than one *approved* portable fire extinguisher in accordance with Section 906 and sized for not less than ordinary hazard as follows:

1. At each *stairway* on all floor levels where combustible materials have accumulated.
2. In every storage and construction shed.
3. Additional portable fire extinguishers shall be provided where special hazards exist including, but not limited to, the storage and use of flammable and *combustible liquids*.

SECTION 3316
MOTORIZED CONSTRUCTION EQUIPMENT

3316.1 Conditions of use. Internal-combustion-powered construction equipment shall be used in accordance with all of the following conditions:

1. Equipment shall be located so that exhausts do not discharge against combustible material.
2. Exhausts shall be piped to the outside of the building.

3. Equipment shall not be refueled while in operation.
4. Fuel for equipment shall be stored in an *approved* area outside of the building.

SECTION 3317 SAFEGUARDING ROOFING OPERATIONS

3317.1 General. Roofing operations utilizing heat-producing systems or other ignition sources shall be conducted in accordance with Sections 3317.2 and 3317.3 and Chapter 35.

3317.2 Asphalt and tar kettles. Asphalt and tar kettles shall be operated in accordance with Section 303.

3317.3 Fire extinguishers for roofing operations. Fire extinguishers shall comply with Section 906. There shall be not less than one multipurpose portable fire extinguisher with a minimum 3-A 40-B:C rating on the roof being covered or repaired.

CHAPTER 34

TIRE REBUILDING AND TIRE STORAGE

SECTION 3401
GENERAL

3401.1 Scope. Tire rebuilding plants, tire storage and tire byproduct facilities shall comply with this chapter, other applicable requirements of this code and NFPA 13. Tire storage in buildings shall also comply with Chapter 32.

3401.2 Permit required. Permits shall be required as set forth in Section 105.6.

SECTION 3402
DEFINITIONS

3402.1 Terms defined in Chapter 2. Words and terms used in this chapter and defined in Chapter 2 shall have the meanings ascribed to them as defined therein.

SECTION 3403
TIRE REBUILDING

3403.1 Construction. Tire rebuilding plants shall comply with the requirements of the *International Building Code*, as to construction, separation from other buildings or other portions of the same building, and protection.

3403.2 Location. Buffing operations shall be located in a room separated from the remainder of the building housing the tire rebuilding or tire recapping operations by a 1-hour *fire barrier*.

Exception: Buffing operations are not required to be separated where all of the following conditions are met:

1. Buffing operations are equipped with an *approved* continuous automatic water-spray system directed at the point of cutting action.
2. Buffing machines are connected to particle-collecting systems providing a minimum air movement of 1,500 cubic feet per minute (cfm) (0.71 m^3/s) in volume and 4,500 feet per minute (fpm) (23 m/s) in-line velocity.
3. The collecting system shall discharge the rubber particles to an *approved* outdoor noncombustible or fire-resistant container that is emptied at frequent intervals to prevent overflow.

3403.3 Cleaning. The buffing area shall be cleaned at frequent intervals to prevent the accumulation of rubber particles.

3403.4 Spray rooms and booths. Each spray room or spray booth where flammable or combustible solvents are applied, shall comply with Chapter 24.

SECTION 3404
PRECAUTIONS AGAINST FIRE

3404.1 Open burning. *Open burning* is prohibited in tire storage yards.

3404.2 Sources of heat. Cutting, welding or heating devices shall not be operated in tire storage yards.

3404.3 Smoking prohibited. Smoking is prohibited in tire storage yards, except in designated areas.

3404.4 Power lines. Tire storage piles shall not be located beneath electrical power lines having a voltage in excess of 750 volts or that supply power to fire emergency systems.

3404.5 Fire safety plan. The *owner* or individual in charge of the tire storage yard shall be required to prepare and submit to the *fire code official* a fire safety plan for review and approval. The fire safety plan shall include provisions for fire department vehicle access. Not less than one copy of the fire safety plan shall be prominently posted and maintained at the storage yard.

3404.6 Telephone number. The telephone number of the fire department and location of the nearest telephone shall be posted conspicuously in attended locations.

SECTION 3405
OUTDOOR STORAGE

3405.1 Individual piles. Tire storage shall be restricted to individual piles not exceeding 5,000 square feet (464.5 m^2) of continuous area. Piles shall not exceed 50,000 cubic feet (1416 m^3) in volume or 10 feet (3048 mm) in height.

3405.2 Separation of piles. Individual tire storage piles shall be separated from other piles by a clear space of not less than 40 feet (12 192 mm).

3405.3 Distance between piles of other stored products. Tire storage piles shall be separated by a clear space of not less than 40 feet (12 192 mm) from piles of other stored product.

3405.4 Distance from lot lines and buildings. Tire storage piles shall be located not less than 50 feet (15 240 mm) from *lot lines* and buildings.

3405.5 Fire breaks. Storage yards shall be maintained free from combustible ground vegetation for a distance of 40 feet (12 192 mm) from the stored material to grass and weeds; and for a distance of 100 feet (30 480 mm) from the stored product to brush and forested areas.

3405.6 Volume more than 150,000 cubic feet. Where the bulk volume of stored product is more than 150,000 cubic feet (4248 m^3), storage arrangement shall be in accordance with the following:

1. Individual storage piles shall comply with size and separation requirements in Sections 3405.1 through 3405.5.
2. Adjacent storage piles shall be considered a group, and the aggregate volume of storage piles in a group shall not exceed 150,000 cubic feet (4248 m^3).

Separation between groups shall be not less than 75 feet (22 860 mm) wide.

3405.7 Location of storage. Outdoor waste tire storage shall not be located under bridges, elevated trestles, elevated roadways or elevated railroads.

SECTION 3406
FIRE DEPARTMENT ACCESS

3406.1 Required access. New tire storage yards shall be provided with fire apparatus access roads in accordance with Section 503 and Section 3406.2. Existing tire storage yards shall be provided with fire apparatus access roads where required in Chapter 11.

3406.2 Location. Fire apparatus access roads shall be located within all pile clearances identified in Section 3405.4 and within all fire breaks required in Section 3405.5. Access roadways shall be within 150 feet (45 720 mm) of any point in the storage yard where storage piles are located, not less than 20 feet (6096 mm) from any storage pile.

SECTION 3407
FENCING

3407.1 Where required. Where the bulk volume of stored material is more than 20,000 cubic feet (566 m^3), a firmly anchored fence or other *approved* method of security that controls unauthorized access to the storage yard shall surround the storage yard.

3407.2 Construction. The fence shall be constructed of *approved* materials and shall be not less than 6 feet (1829 mm) high and provided with gates at least 20 feet (6096 mm) wide.

3407.3 Locking. Gates to the storage yard shall be locked when the storage yard is not staffed.

3407.4 Unobstructed. Gateways shall be kept clear of obstructions and be fully openable at all times.

SECTION 3408
FIRE PROTECTION

3408.1 Water supply. A public or private fire protection water supply shall be provided in accordance with Section 508. The water supply shall be arranged such that any part of the storage yard can be reached by using not more than 500 feet (152 m) of hose.

3408.2 Fire extinguishers. Buildings or structures shall be provided with portable fire extinguishers in accordance with Section 906. Fuel-fired vehicles operating in the storage yard shall be equipped with a minimum 2-A:20-B:C-rated portable fire extinguisher.

SECTION 3409
INDOOR STORAGE ARRANGEMENT

3409.1 Pile dimensions. Where tires are stored on-tread, the dimension of the pile in the direction of the wheel hole shall be not more than 50 feet (15 240 mm). Tires stored adjacent to or along one wall shall not extend more than 25 feet (7620 mm) from that wall. Other piles shall be not more than 50 feet (15 240 mm) in width.

CHAPTER 35

WELDING AND OTHER HOT WORK

SECTION 3501 GENERAL

3501.1 Scope. Welding, cutting, open torches and other hot work operations and equipment shall comply with this chapter.

3501.2 Permits. Permits shall be required as set forth in Section 105.6.

3501.3 Restricted areas. Hot work shall only be conducted in areas designed or authorized for that purpose by the personnel responsible for a Hot Work Program. Hot work shall not be conducted in the following areas unless approval has been obtained from the *fire code official*:

1. Areas where the sprinkler system is impaired.
2. Areas where there exists the potential of an explosive atmosphere, such as locations where flammable gases, liquids or vapors are present.
3. Areas with readily ignitable materials, such as storage of large quantities of bulk sulfur, baled paper, cotton, lint, dust or loose combustible materials.
4. On board ships at dock or ships under construction or repair.
5. At other locations as specified by the *fire code official.*

3501.4 Cylinders and containers. *Compressed gas* cylinders and fuel containers shall comply with this chapter and Chapter 53.

3501.5 Design and installation of oxygen-fuel gas systems. An oxygen-fuel gas system with two or more manifolded cylinders of oxygen shall be in accordance with NFPA 51.

SECTION 3502 DEFINITIONS

3502.1 Definitions. The following terms are defined in Chapter 2:

HOT WORK.

HOT WORK AREA.

HOT WORK EQUIPMENT.

HOT WORK PERMITS.

HOT WORK PROGRAM.

RESPONSIBLE PERSON.

TORCH-APPLIED ROOF SYSTEM.

SECTION 3503 GENERAL REQUIREMENTS

3503.1 General. Hot work conditions and operations shall comply with this chapter.

3503.2 Temporary and fixed hot work areas. Temporary and fixed hot work areas shall comply with this section.

3503.3 Hot work program permit. Hot work permits, issued by an *approved* responsible person under a hot work program, shall be available for review by the *fire code official* at the time the work is conducted and for 48 hours after work is complete.

3503.4 Qualifications of operators. A permit for hot work operations shall not be issued unless the individuals in charge of performing such operations are capable of performing such operations safely. Demonstration of a working knowledge of the provisions of this chapter shall constitute acceptable evidence of compliance with this requirement.

3503.5 Records. The individual responsible for the hot work area shall maintain "prework check" reports in accordance with Section 3504.3.1. Such reports shall be maintained on the premises for not less than 48 hours after work is complete.

3503.6 Signage. Visible hazard identification signs shall be provided where required by Chapter 50. Where the hot work area is accessible to persons other than the operator of the hot work equipment, conspicuous signs shall be posted to warn others before they enter the hot work area. Such signs shall display the following warning:

CAUTION

HOT WORK IN PROGRESS

STAY CLEAR

SECTION 3504 FIRE SAFETY REQUIREMENTS

3504.1 Protection of combustibles. Protection of combustibles shall be in accordance with Sections 3504.1.1 through 3504.1.9.

3504.1.1 Combustibles. Hot work areas shall not contain combustibles or shall be provided with appropriate shielding to prevent sparks, slag or heat from igniting exposed combustibles.

3504.1.2 Openings. Openings or cracks in walls, floors, ducts or shafts within the hot work area shall be tightly covered to prevent the passage of sparks to adjacent combustible areas, or shielded by metal fire-resistant guards, or curtains shall be provided to prevent passage of sparks or slag.

3504.1.3 Housekeeping. Floors shall be kept clean within the hot work area.

3504.1.4 Conveyor systems. Conveyor systems that are capable of carrying sparks to distant combustibles shall be shielded or shut down.

3504.1.5 Partitions. Partitions segregating hot work areas from other areas of the building shall be noncombustible.

In fixed hot work areas, the partitions shall be securely connected to the floor such that no gap exists between the floor and the partition. Partitions shall prevent the passage of sparks, slag, and heat from the hot work area.

3504.1.6 Floors. Fixed hot work areas shall have floors with noncombustible surfaces.

3504.1.7 Precautions in hot work. Hot work shall not be performed on containers or equipment that contain or have contained flammable liquids, gases or solids until the containers and equipment have been thoroughly cleaned, inerted or purged; except that "hot tapping" shall be allowed on tanks and pipe lines where such work is to be conducted by *approved* personnel. Hot work on flammable and *combustible liquid* storage tanks shall be conducted in accordance with Section 3510.

3504.1.8 Sprinkler protection. Automatic sprinkler protection shall not be shut off while hot work is performed. Where hot work is performed close to automatic sprinklers, noncombustible barriers or damp cloth guards shall shield the individual sprinkler heads and shall be removed when the work is completed. If the work extends over several days, the shields shall be removed at the end of each workday. The *fire code official* shall approve hot work where sprinkler protection is impaired.

3504.1.9 Fire detection systems. *Approved* special precautions shall be taken to avoid accidental operation of automatic fire detection systems.

3504.2 Fire watch. Fire watches shall be established and conducted in accordance with Sections 3504.2.1 through 3504.2.6.

3504.2.1 When required. A fire watch shall be provided during hot work activities and shall continue for not less than 30 minutes after the conclusion of the work. The *fire code official*, or the responsible manager under a hot work program, is authorized to extend the fire watch based on the hazards or work being performed.

Exception: Where the hot work area has no fire hazards or combustible exposures.

3504.2.2 Location. The fire watch shall include the entire hot work area. Hot work conducted in areas with vertical or horizontal fire exposures that are not observable by a single individual shall have additional personnel assigned to fire watches to ensure that exposed areas are monitored.

3504.2.3 Duties. Individuals designated to fire watch duty shall have fire-extinguishing equipment readily available and shall be trained in the use of such equipment. Individuals assigned to fire watch duty shall be responsible for extinguishing spot fires and communicating an alarm.

3504.2.4 Fire training. The individuals responsible for performing the hot work and individuals responsible for providing the fire watch shall be trained in the use of portable fire extinguishers.

3504.2.5 Fire hoses. Where hoselines are required, they shall be connected, charged and ready for operation.

3504.2.6 Fire extinguisher. Not less than one portable fire extinguisher complying with Section 906 and with a minimum 2-A:20-B:C rating shall be readily accessible within 30 feet (9144 mm) of the location where hot work is performed.

3504.3 Area reviews. Before hot work is permitted and not less than once per day while the permit is in effect, the area shall be inspected by the individual responsible for authorizing hot work operations to ensure that it is a fire safe area. Information shown on the permit shall be verified prior to signing the permit in accordance with Section 105.6.

3504.3.1 Pre-hot-work check. A pre-hot-work check shall be conducted prior to work to ensure that all equipment is safe and hazards are recognized and protected. A report of the check shall be kept at the work site during the work and available upon request. The pre-hot-work check shall determine all of the following:

1. Hot work equipment to be used shall be in satisfactory operating condition and in good repair.
2. Hot work site is clear of combustibles or combustibles are protected.
3. Exposed construction is of noncombustible materials or, if combustible, then protected.
4. Openings are protected.
5. Floors are kept clean.
6. No exposed combustibles are located on the opposite side of partitions, walls, ceilings or floors.
7. Fire watches, where required, are assigned.
8. *Approved* actions have been taken to prevent accidental activation of suppression and detection equipment in accordance with Sections 3504.1.8 and 3504.1.9.
9. Fire extinguishers and fire hoses (where provided) are operable and available.

SECTION 3505 GAS WELDING AND CUTTING

3505.1 General. Devices or attachments mixing air or oxygen with combustible gases prior to consumption, except at the burner or in a standard torch or blow pipe, shall not be allowed unless *approved*.

3505.2 Cylinder and container storage, handling and use. Storage, handling and use of *compressed gas* cylinders, containers and tanks shall be in accordance with this section and Chapter 53.

3505.2.1 Cylinders connected for use. The storage or use of a single cylinder of oxygen and a single cylinder of fuel gas located on a cart shall be allowed without requiring the cylinders to be separated in accordance with Section 5003.9.8 or 5003.10.3.6 when the cylinders are connected to regulators, ready for service, equipped with apparatus designed for cutting or welding and all of the following:

1. Carts shall be kept away from the cutting or welding operation in accordance with Section 3505.5 or fire-resistant shields shall be provided.

2. Cylinders shall be secured to the cart to resist movement.
3. Carts shall be in accordance with Section 5003.10.3.
4. Cylinder valves not having fixed hand wheels shall have keys, handles or nonadjustable wrenches on valve stems while the cylinders are in service.
5. Cylinder valve outlet connections shall conform to the requirements of CGA V-1.
6. Cylinder valves shall be closed when work is finished.
7. Cylinder valves shall be closed before moving the cart.

3505.2.1.1 Individual cart separation. Individual carts shall be separated from each other in accordance with Section 5003.9.8.

3505.3 Precautions. Cylinders, valves, regulators, hose and other apparatus and fittings for oxygen shall be kept free from oil or grease. Oxygen cylinders, apparatus and fittings shall not be handled with oily hands, oily gloves, or greasy tools or equipment.

3505.4 Acetylene gas. Acetylene gas shall not be piped except in *approved* cylinder manifolds and cylinder manifold connections, or utilized at a pressure exceeding 15 pounds per square inch gauge (psig) (103 kPa) unless dissolved in a suitable solvent in cylinders manufactured in accordance with DOTn 49 CFR Part 178. Acetylene gas shall not be brought in contact with unalloyed copper, except in a blowpipe or torch.

3505.5 Remote locations. Oxygen and fuel-gas cylinders and acetylene generators shall be located away from the hot work area to prevent such cylinders or generators from being heated by radiation from heated materials, sparks or slag, or misdirection of the torch flame.

3505.6 Cylinders shutoff. The torch valve shall be closed and the gas supply to the torch completely shut off when gas welding or cutting operations are discontinued for a period of 1 hour or more.

3505.7 Prohibited operation. Welding or cutting work shall not be held or supported on *compressed gas* cylinders or containers.

3505.8 Tests. Tests for leaks in piping systems and equipment shall be made with soapy water. The use of flames shall be prohibited for leak testing.

SECTION 3506 ELECTRIC ARC HOT WORK

3506.1 General. The frame or case of electric hot work machines, except internal-combustion-engine-driven machines, shall be grounded. Ground connections shall be mechanically strong and electrically adequate for the required current.

3506.2 Return circuits. Welding current return circuits from the work to the machine shall have proper electrical contact at joints. The electrical contact shall be periodically inspected.

3506.3 Disconnecting. Electrodes shall be removed from the holders when electric arc welding or cutting is discontinued for any period of 1 hour or more. The holders shall be located to prevent accidental contact and the machines shall be disconnected from the power source.

3506.4 Emergency disconnect. A switch or circuit breaker shall be provided so that fixed electric welders and control equipment can be disconnected from the supply circuit. The disconnect shall be installed in accordance with NFPA 70.

3506.5 Damaged cable. Damaged cable shall be removed from service until properly repaired or replaced.

SECTION 3507 CALCIUM CARBIDE SYSTEMS

3507.1 Calcium carbide storage. Storage and handling of calcium carbide shall comply with Chapter 50 of this code and Chapter 9 of NFPA 51.

SECTION 3508 ACETYLENE GENERATORS

3508.1 Use of acetylene generators. The use of acetylene generators shall comply with this section and Chapter 6 of NFPA 51A.

3508.2 Portable generators. The minimum volume of rooms containing portable generators shall be 35 times the total gas-generating capacity per charge of all generators in the room. The gas-generating capacity in cubic feet per charge shall be assumed to be 4.5 times the weight of carbide per charge in pounds. The minimum ceiling height of rooms containing generators shall be 10 feet (3048 mm). An acetylene generator shall not be moved by derrick, crane or hoist while charged.

3508.3 Protection against freezing. Generators shall be located where water will not freeze. Common salt such as sodium chloride or other corrosive chemicals shall not be utilized for protection against freezing.

SECTION 3509 PIPING MANIFOLDS AND HOSE SYSTEMS FOR FUEL GASES AND OXYGEN

3509.1 General. The use of piping manifolds and hose systems shall be in accordance with Section 3509.2 through 3509.7, Chapter 53 and Chapter 5 of NFPA 51.

3509.2 Protection. Piping shall be protected against physical damage.

3509.3 Signage. Signage shall be provided for piping and hose systems as follows:

1. Above-ground piping systems shall be marked in accordance with ASME A13.1.
2. Station outlets shall be marked to indicate their intended usage.
3. Signs shall be posted, indicating clearly the location and identity of section shutoff valves.

3509.4 Manifolding of cylinders. Oxygen manifolds shall not be located in an acetylene generator room. Oxygen manifolds shall be located not less than 20 feet (6096 mm) away from combustible material such as oil or grease, and gas cylinders containing flammable gases, unless the gas cylinders are separated by a *fire partition*.

3509.5 Identification of manifolds. Signs shall be posted for oxygen manifolds with service pressures not exceeding 200 psig (1379 kPa). Such signs shall include the words:

LOW-PRESSURE MANIFOLD

DO NOT CONNECT HIGH-PRESSURE CYLINDERS

MAXIMUM PRESSURE 250 PSIG

3509.6 Clamps. Hose connections shall be clamped or otherwise securely fastened.

3509.7 Inspection. Hoses shall be inspected frequently for leaks, burns, wear, loose connections or other defects rendering the hose unfit for service.

SECTION 3510 HOT WORK ON FLAMMABLE AND COMBUSTIBLE LIQUID STORAGE TANKS

3510.1 General. Hot work performed on the interior or exterior of tanks that hold or have held flammable or *combustible liquids* shall be in accordance with Section 3510.2 and Chapters 4, 5, 6, 7 and 10 of NFPA 326.

3510.2 Prevention. The following steps shall be taken to minimize hazards where hot work must be performed on a flammable or *combustible liquid* storage container:

1. Use alternative methods to avoid hot work where possible.
2. Analyze the hazards prior to performing hot work, identify the potential hazards and the methods of hazard control.
3. Hot work shall conform to the requirements of the code or standard to which the container was originally fabricated.
4. Test the immediate and surrounding work area with a combustible gas detector and provide for a means of continuing monitoring while conducting the hot work.
5. Qualified employees and contractors performing hot work shall use an industry-approved hot work permit system to control the work.
6. Personnel shall be properly trained on hot work policies and procedures regarding equipment, safety, hazard controls and job-specific requirements.
7. On-site safety supervision shall be present where hot work is in progress to protect the personnel conducting the hot work and provide additional overview of site-specific hazards.

CHAPTER 36
MARINAS

SECTION 3601 SCOPE

3601.1 Scope. Marina facilities shall be in accordance with this chapter.

3601.2 Plans and approvals. Plans for marina fire protection facilities shall be *approved* prior to installation. The work shall be subject to final inspection and approval after installation.

SECTION 3602 DEFINITIONS

3602.1 Definitions. The following terms are defined in Chapter 2:

FLOAT.

MARINA.

PIER.

VESSEL.

WHARF.

SECTION 3603 GENERAL PRECAUTIONS

3603.1 Combustible debris. Combustible debris and rubbish shall not be deposited or accumulated on land beneath marina structures, piers or wharves.

3603.2 Sources of ignition. Open-flame devices used for lighting or decoration on the exterior of a vessel, float, pier or wharf shall be *approved*.

3603.3 Flammable or combustible liquid spills. Spills of flammable or *combustible liquids* at or upon the water shall be reported immediately to the fire department or jurisdictional authorities.

3603.4 Rubbish containers. Containers with tight-fitting or self-closing lids shall be provided for temporary storage of combustible debris, rubbish and waste material. The rubbish containers shall be constructed entirely of materials that comply with any one of the following:

1. Noncombustible materials.
2. Materials that meet a peak rate of heat release not exceeding 300 kW/m^2 where tested in accordance with ASTM E 1354 at an incident heat flux of 50 kW/m^2 in the horizontal orientation.

3603.5 Electrical equipment. Electrical equipment shall be installed and used in accordance with its listing, Section 605 of this code and Chapter 5 of NFPA 303 as required for wet, damp and hazardous locations.

3603.6 Berthing and storage. Berthing and storage shall be in accordance with Chapter 7 of NFPA 303.

3603.7 Slip identification. Slips and mooring spaces shall be individually identified by an *approved* numeric or alphabetic designator. Space designators shall be posted at the space. Signs indicating the space designators located on finger piers and floats shall be posted at the base of all piers, finger piers, floats and finger floats.

SECTION 3604 FIRE PROTECTION EQUIPMENT

3604.1 General. Piers, marinas and wharves with facilities for mooring or servicing five or more vessels, and marine motor fuel-dispensing facilities shall be equipped with fire protection equipment in accordance with Sections 3604.2 through 3604.6.

3604.2 Standpipes. Marinas and boatyards shall be equipped throughout with standpipe systems in accordance with NFPA 303. Systems shall be provided with hose connections located such that no point on the marina pier or float system exceeds 150 feet (15 240 mm) from a standpipe hose connection.

3604.2.1 Identification of standpipe outlets. Standpipe hose connection locations shall be clearly identified by a flag or other *approved* means designed to be readily visible from the pier accessing the float system.

3604.3 Access and water supply. Piers and wharves shall be provided with fire apparatus access roads and water-supply systems with on-site fire hydrants where required by the *fire code official*. Such roads and water systems shall be provided and maintained in accordance with Sections 503 and 507.

3604.4 Portable fire extinguishers. One portable fire extinguisher of the ordinary (moderate) hazard type shall be provided at each required standpipe hose connection. Additional portable fire extinguishers, suitable for the hazards involved, shall be provided and maintained in accordance with Section 906.

3604.5 Communications. A telephone not requiring a coin to operate or other *approved*, clearly identified means to notify the fire department shall be provided on the site in a location *approved* by the *fire code official.*

3604.6 Emergency operations staging areas. Space shall be provided on all float systems for the staging of emergency equipment. Emergency operation staging areas shall provide a minimum of 4 feet wide by 10 feet long (1219 mm by 3048 mm) clear area exclusive of walkways and shall be located at each standpipe hose connection. Emergency operation staging areas shall be provided with a curb or barrier having a minimum height of 4 inches (102 mm) and maximum space between the bottom edge and the surface of the staging area of 2 inches (51 mm) on the outboard sides of the staging area.

An *approved* sign reading FIRE EQUIPMENT STAGING AREA—KEEP CLEAR shall be provided at each staging area.

SECTION 3605
MARINE MOTOR FUEL-DISPENSING FACILITIES

3605.1 Fuel dispensing. Marine motor fuel-dispensing facilities shall be in accordance with Chapter 23.

CHAPTER 37

COMBUSTIBLE FIBERS

SECTION 3701 GENERAL

3701.1 Scope. The equipment, processes and operations involving *combustible fibers* shall comply with this chapter.

3701.2 Applicability. Storage of *combustible fibers* in any quantity shall comply with this section.

3701.3 Permits. Permits shall be required as set forth in Section 105.6.

SECTION 3702 DEFINITIONS

3702.1 Definitions. The following terms are defined in Chapter 2:

BALED COTTON.

BALED COTTON, DENSELY PACKED.

COMBUSTIBLE FIBERS.

SEED COTTON.

SECTION 3703 GENERAL PRECAUTIONS

3703.1 Use of combustible receptacles. Ashes, waste, rubbish or sweepings shall not be placed in wood or other combustible receptacles and shall be removed daily from the structure.

3703.2 Vegetation. Grass or weeds shall not be allowed to accumulate at any point on the premises.

3703.3 Clearances. A minimum clearance of 3 feet (914 mm) shall be maintained between automatic sprinklers and the top of piles.

3703.4 Agricultural products. Hay, straw, seed cotton or similar agricultural products shall not be stored adjacent to structures or combustible materials unless a clear horizontal distance equal to the height of a pile is maintained between such storage and structures or combustible materials. Storage shall be limited to stacks of 100 tons (91 metric tons) each. Stacks shall be separated by not less than 20 feet (6096 mm) of clear space. Quantities of hay, straw, seed cotton and other agricultural products shall not be limited where stored in or near farm structures located outside closely built areas. A permit shall not be required for agricultural storage.

3703.5 Dust collection. Where located within a building, equipment or machinery that generates or emits *combustible fibers* shall be provided with an *approved* dust-collecting and exhaust system. Such systems shall comply with Chapter 22 of this code and Section 511 of the *International Mechanical Code.*

3703.6 Portable fire extinguishers. Portable fire extinguishers shall be provided in accordance with Section 906 as required for extra-hazard occupancy protection as indicated in Table 906.3(1).

3703.7 Sources of ignition. Sources of ignition shall comply with Sections 3703.7.1 and 3703.7.2.

3703.7.1 Smoking. Smoking shall be prohibited and "No Smoking" signs provided as follows:

1. In rooms or areas where materials are stored or dispensed or used in open systems.
2. Within 25 feet (7620 mm) of outdoor storage or open use areas.
3. Facilities or areas within facilities that have been designated as totally "no smoking" shall have "No Smoking" signs placed at all entrances to the facility or area. Designated areas within such facilities where smoking is permitted either permanently or temporarily shall be identified with signs designating that smoking is permitted in these areas only.

Signs required by this section shall be in English as a primary language or in symbols allowed by this code and shall comply with Section 310.

3703.7.2 Open flames. Open flames and high-temperature devices shall not be used in a manner that creates a hazardous condition and shall be listed for use with the materials stored or used. High-temperature devices and those devices utilizing an open flame shall be listed for use with the materials stored or used.

SECTION 3704 LOOSE FIBER STORAGE

3704.1 General. Loose combustible fibers, not in suitable bales or packages and stored outdoors in the open, shall comply with Section 2808 of this code. Occupancies involving the indoor storage of loose combustible fibers in amounts exceeding the *maximum allowable quantity per control area* as set forth in Section 5003.1 shall comply with Sections 3704.2 through 3704.6.

3704.2 Storage of 100 cubic feet or less. Loose *combustible fibers* in quantities of not more than 100 cubic feet (3 m^3) located in a structure shall be stored in a metal or metal-lined bin equipped with a self-closing cover.

3704.3 Storage of more than 100 cubic feet to 500 cubic feet. Loose *combustible fibers* in quantities exceeding 100 cubic feet (3 m^3) but not exceeding 500 cubic feet (14 m^3) shall be stored in rooms enclosed with 1-hour *fire barriers* constructed in accordance with Section 707 of the *International Building Code* or *horizontal assemblies* constructed in accordance with Section 711 of the *International Building Code*, or both, with openings protected by an *approved* opening protective assembly having a *fire protection rating* of $^3/_4$ hour in accordance with the *International Building Code.*

3704.4 Storage of more than 500 cubic feet to 1,000 cubic feet. Loose *combustible fibers* in quantities exceeding 500 cubic feet (14 m^3) but not exceeding 1,000 cubic feet (28 m^3) shall be stored in rooms enclosed with 2-hour *fire barriers* constructed in accordance with Section 707 of the *International Building Code* or *horizontal assemblies* constructed in accordance with Section 711 of the *International Building Code*, or both, with openings protected by an *approved* opening protective assembly having a *fire protection rating* of $1^1/_2$ hours in accordance with the *International Building Code*.

3704.5 Storage of more than 1,000 cubic feet. Loose *combustible fibers* in quantities exceeding 1,000 cubic feet (28 m^3) shall be stored in rooms enclosed with 2-hour *fire barriers* constructed in accordance with Section 707 of the *International Building Code* or *horizontal assemblies* constructed in accordance with Section 711 of the *International Building Code*, or both, with openings protected by an *approved* opening protective assembly having a *fire protection rating* of $1^1/_2$ hours in accordance with the *International Building Code*. The storage room shall be protected by an *automatic sprinkler system* installed in accordance with Section 903.3.1.1.

3704.6 Detached storage structure. Not more than 2,500 cubic feet (70 m^3) of loose *combustible fibers* shall be stored in a detached structure suitably located, with openings protected against entrance of sparks. The structure shall not be occupied for any other purpose.

SECTION 3705
BALED STORAGE

3705.1 Bale size and separation. Baled *combustible fibers* shall be limited to single blocks or piles not more than 25,000 cubic feet (700 m^3) in volume, not including aisles or clearances. Blocks or piles of baled fiber shall be separated from adjacent storage by aisles not less than 5 feet (1524 mm) wide, or by flash-fire barriers constructed of continuous sheets of noncombustible material extending from the floor to a minimum height of 1 foot (305 mm) above the highest point of the piles and projecting not less than 1 foot (305 mm) beyond the sides of the piles.

3705.2 Special baling conditions. Sisal and other fibers in bales bound with combustible tie ropes, jute and other fibers that swell when wet, shall be stored to allow for expansion in any direction without affecting building walls, ceilings or columns. A minimum clearance of 3 feet (914 mm) shall be required between walls and sides of piles, except that where the storage compartment is not more than 30 feet (9144 mm) wide, the minimum clearance at side walls shall be 1 foot (305 mm), provided that a center aisle not less than 5 feet (1524 mm) wide is maintained.

CHAPTERS 38 through 49

RESERVED

Part V—Hazardous Materials

CHAPTER 50

HAZARDOUS MATERIALS—GENERAL PROVISIONS

SECTION 5001
GENERAL

5001.1 Scope. Prevention, control and mitigation of dangerous conditions related to storage, dispensing, use and handling of hazardous materials shall be in accordance with this chapter.

This chapter shall apply to all hazardous materials, including those materials regulated elsewhere in this code, except that where specific requirements are provided in other chapters, those specific requirements shall apply in accordance with the applicable chapter. Where a material has multiple hazards, all hazards shall be addressed.

Exceptions:

1. In retail or wholesale sales occupancies, the quantities of medicines, foodstuff or consumer products and cosmetics containing not more than 50 percent by volume of water-miscible liquids and with the remainder of the solutions not being flammable shall not be limited, provided such materials are packaged in individual containers not exceeding 1.3 gallons (5 L).
2. Quantities of alcoholic beverages in retail or wholesale sales occupancies shall not be limited providing the liquids are packaged in individual containers not exceeding 1.3 gallons (5 L).
3. Application and release of pesticide and agricultural products and materials intended for use in weed abatement, erosion control, soil amendment or similar applications where applied in accordance with the manufacturers' instructions and label directions.
4. The off-site transportation of hazardous materials where in accordance with Department of Transportation (DOTn) regulations.
5. Building materials not otherwise regulated by this code.
6. Refrigeration systems (see Section 606).
7. Stationary storage battery systems regulated by Section 608.
8. The display, storage, sale or use of fireworks and *explosives* in accordance with Chapter 56.
9. *Corrosives* utilized in personal and household products in the manufacturers' original consumer packaging in Group M occupancies.
10. The storage of distilled spirits and wines in wooden barrels and casks.
11. The use of wall-mounted dispensers containing alcohol-based hand rubs classified as Class I or II liquids where in accordance with Section 5705.5.

5001.1.1 Waiver. The provisions of this chapter are waived where the *fire code official* determines that such enforcement is preempted by other codes, statutes or ordinances. The details of any action granting such a waiver shall be recorded and entered in the files of the code enforcement agency.

5001.2 Material classification. Hazardous materials are those chemicals or substances defined as such in this code. Definitions of hazardous materials shall apply to all hazardous materials, including those materials regulated elsewhere in this code.

5001.2.1 Mixtures. Mixtures shall be classified in accordance with hazards of the mixture as a whole. Mixtures of hazardous materials shall be classified in accordance with nationally recognized reference standards; by an *approved* qualified organization, individual, or Material Safety Data Sheet (MSDS); or by other *approved* methods.

5001.2.2 Hazard categories. Hazardous materials shall be classified according to hazard categories. The categories include materials regulated by this chapter and materials regulated elsewhere in this code.

5001.2.2.1 Physical hazards. The material categories listed in this section are classified as *physical hazards*. A material with a primary classification as a *physical hazard* can also pose a *health hazard*.

1. *Explosives* and blasting agents.
2. *Combustible liquids.*
3. Flammable solids, liquids and gases.
4. Organic peroxide solids or liquids.
5. Oxidizer, solids or liquids.
6. Oxidizing gases.
7. Pyrophoric solids, liquids or gases.
8. Unstable (reactive) solids, liquids or gases.
9. Water-reactive materials solids or liquids.
10. *Cryogenic fluids.*

5001.2.2.2 Health hazards. The material categories listed in this section are classified as *health hazards*. A material with a primary classification as a *health hazard* can also pose a *physical hazard*.

1. Highly toxic and toxic materials.
2. *Corrosive* materials.

5001.3 Performance-based design alternative. Where *approved* by the *fire code official*, buildings and facilities where hazardous materials are stored, used or handled shall be permitted to comply with this section as an alternative to compliance with the other requirements set forth in this chapter and Chapters 51 through 67.

5001.3.1 Objective. The objective of Section 5001.3 is to protect people and property from the consequences of unauthorized discharge, fires or explosions involving hazardous materials.

5001.3.2 Functional statements. Performance-based design alternatives are based on the following functional statements:

1. Provide safeguards to minimize the risk of unwanted releases, fires or explosions involving hazardous materials.
2. Provide safeguards to minimize the consequences of an unsafe condition involving hazardous materials during normal operations and in the event of an abnormal condition.

5001.3.3 Performance requirements. Where safeguards, systems, documentation, written plans or procedures, audits, process hazards analysis, mitigation measures, engineering controls or construction features are required by Sections 5001.3.3.1 through 5001.3.3.18, the details of the design alternative shall be subject to approval by the *fire code official*. The details of actions granting the use of the design alternatives shall be recorded and entered in the files of the jurisdiction.

5001.3.3.1 Properties of hazardous materials. The physical- and health-hazard properties of hazardous materials on site shall be known and shall be made readily available to employees, neighbors and the *fire code official*.

5001.3.3.2 Reliability of equipment and operations. Equipment and operations involving hazardous materials shall be designed, installed and maintained to ensure that they reliably operate as intended.

5001.3.3.3 Prevention of unintentional reaction or release. Safeguards shall be provided to minimize the risk of an unintentional reaction or release that could endanger people or property.

5001.3.3.4 Spill mitigation. Spill containment systems or means to render a spill harmless to people or property shall be provided where a spill is determined to be a plausible event and where such an event would endanger people or property.

5001.3.3.5 Ignition hazards. Safeguards shall be provided to minimize the risk of exposing combustible hazardous materials to unintended sources of ignition.

5001.3.3.6 Protection of hazardous materials. Safeguards shall be provided to minimize the risk of exposing hazardous materials to a fire or physical damage whereby such exposure could endanger or lead to the endangerment of people or property.

5001.3.3.7 Exposure hazards. Safeguards shall be provided to minimize the risk of and limit damage from a fire or explosion involving explosive hazardous materials whereby such fire or explosion could endanger or lead to the endangerment of people or property.

5001.3.3.8 Detection of gas or vapor release. Where a release of hazardous materials gas or vapor would cause immediate harm to persons or property, means of mitigating the dangerous effects of a release shall be provided.

5001.3.3.9 Reliable power source. Where a power supply is relied upon to prevent or control an emergency condition that could endanger people or property, the power supply shall be from a reliable source.

5001.3.3.10 Ventilation. Where ventilation is necessary to limit the risk of creating an emergency condition resulting from normal or abnormal operations, means of ventilation shall be provided.

5001.3.3.11 Process hazard analyses. Process hazard analyses shall be conducted to ensure reasonably the protection of people and property from dangerous conditions involving hazardous materials.

5001.3.3.12 Pre-startup safety review. Written documentation of pre-startup safety review procedures shall be developed and enforced to ensure that operations are initiated in a safe manner. The process of developing and updating such procedures shall involve the participation of affected employees.

5001.3.3.13 Operating and emergency procedures. Written documentation of operating procedures and procedures for emergency shut down shall be developed and enforced to ensure that operations are conducted in a safe manner. The process of developing and updating such procedures shall involve the participation of affected employees.

5001.3.3.14 Management of change. A written plan for management of change shall be developed and enforced. The process of developing and updating the plan shall involve the participation of affected employees.

5001.3.3.15 Emergency plan. A written emergency plan shall be developed to ensure that proper actions are taken in the event of an emergency, and the plan shall be followed if an emergency condition occurs. The process of developing and updating the plan shall involve the participation of affected employees.

5001.3.3.16 Accident procedures. Written procedures for investigation and documentation of accidents shall be developed, and accidents shall be investigated and documented in accordance with these procedures.

5001.3.3.17 Consequence analysis. Where an accidental release of hazardous materials could endanger people or property, either on or off-site, an analysis of the expected consequences of a plausible release shall be performed and utilized in the analysis and selection of active and passive hazard mitigation controls.

5001.3.3.18 Safety audits. Safety audits shall be conducted on a periodic basis to verify compliance with the requirements of this section.

5001.4 Retail and wholesale storage and display. For retail and wholesale storage and display of nonflammable solid and nonflammable or noncombustible liquid hazardous materials in Group M occupancies and storage in Group S occupancies, see Section 5003.11.

5001.5 Permits. Permits shall be required as set forth in Sections 105.6 and 105.7.

When required by the *fire code official*, permittees shall apply for approval to permanently close a storage, use or handling facility. Such application shall be submitted not less than 30 days prior to the termination of the storage, use or handling of hazardous materials. The *fire code official* is authorized to require that the application be accompanied by an *approved* facility closure plan in accordance with Section 5001.6.3.

5001.5.1 Hazardous Materials Management Plan. Where required by the *fire code official*, an application for a permit shall include a Hazardous Materials Management Plan (HMMP). The HMMP shall include a facility site plan designating the following:

1. Access to each storage and use area.
2. Location of emergency equipment.
3. Location where liaison will meet emergency responders.
4. Facility evacuation meeting point locations.
5. The general purpose of other areas within the building.
6. Location of all above-ground and underground tanks and their appurtenances including, but not limited to, sumps, vaults, below-grade treatment systems and piping.
7. The hazard classes in each area.
8. Locations of all control areas and Group H occupancies.
9. Emergency *exits*.

5001.5.2 Hazardous Materials Inventory Statement (HMIS). Where required by the *fire code official*, an application for a permit shall include an HMIS, such as Superfund Amendments and Reauthorization Act of 1986 (SARA) Title III, Tier II Report or other *approved* statement. The HMIS shall include the following information:

1. Product name.
2. Component.
3. Chemical Abstract Service (CAS) number.
4. Location where stored or used.
5. Container size.
6. Hazard classification.
7. Amount in storage.
8. Amount in use-*closed systems*.
9. Amount in use-*open systems*.

5001.6 Facility closure. Facilities shall be placed out of service in accordance with Sections 5001.6.1 through 5001.6.3.

5001.6.1 Temporarily out-of-service facilities. Facilities that are temporarily out of service shall continue to maintain a permit and be monitored and inspected.

5001.6.2 Permanently out-of-service facilities. Facilities for which a permit is not kept current or is not monitored and inspected on a regular basis shall be deemed to be permanently out of service and shall be closed in an *approved* manner. Where required by the *fire code official*, permittees shall apply for approval to close permanently storage, use or handling facilities. The *fire code official* is authorized to require that such application be accompanied by an *approved* facility closure plan in accordance with Section 5001.6.3.

5001.6.3 Facility closure plan. Where a facility closure plan is required in accordance with Section 5001.5 to terminate storage, dispensing, handling or use of hazardous materials, it shall be submitted to the *fire code official* not less than 30 days prior to facility closure. The plan shall demonstrate that hazardous materials that are stored, dispensed, handled or used in the facility will be transported, disposed of or reused in a manner that eliminates the need for further maintenance and any threat to public health and safety.

SECTION 5002 DEFINITIONS

5002.1 Definitions. The following terms are defined in Chapter 2:

BOILING POINT.

CEILING LIMIT.

CHEMICAL.

CHEMICAL NAME.

CLOSED CONTAINER.

CONTAINER.

CONTROL AREA.

CYLINDER.

DAY BOX.

DEFLAGRATION.

DESIGN PRESSURE.

DETACHED BUILDING.

DISPENSING.

EXCESS FLOW CONTROL.

EXHAUSTED ENCLOSURE.

EXPLOSION.

FLAMMABLE VAPORS OR FUMES.

GAS CABINET.

GAS ROOM.

HANDLING.

HAZARDOUS MATERIALS.

HEALTH HAZARD.

IMMEDIATELY DANGEROUS TO LIFE AND HEALTH (IDLH).

INCOMPATIBLE MATERIALS.

LIQUID.

LOWER EXPLOSIVE LIMIT (LEL).

LOWER FLAMMABLE LIMIT (LFL).

MATERIAL SAFETY DATA SHEET (MSDS).

MAXIMUM ALLOWABLE QUANTITY PER CONTROL AREA.

NORMAL TEMPERATURE AND PRESSURE (NTP).

OUTDOOR CONTROL AREA.

PERMISSIBLE EXPOSURE LIMIT (PEL).

PESTICIDE.

PHYSICAL HAZARD.

PRESSURE VESSEL.

SAFETY CAN.

SECONDARY CONTAINMENT.

SEGREGATED.

SOLID.

STORAGE, HAZARDOUS MATERIALS.

SYSTEM.

TANK, ATMOSPHERIC.

TANK, PORTABLE.

TANK, STATIONARY.

TANK VEHICLE.

UNAUTHORIZED DISCHARGE.

USE (MATERIAL).

VAPOR PRESSURE.

SECTION 5003 GENERAL REQUIREMENTS

5003.1 Scope. The storage, use and handling of all hazardous materials shall be in accordance with this section.

5003.1.1 Maximum allowable quantity per control area. The *maximum allowable quantity per control area* shall be as specified in Tables 5003.1.1(1) through 5003.1.1(4).

For retail and wholesale storage and display in Group M occupancies and Group S storage, see Section 5003.11.

5003.1.2 Conversion. Where quantities are indicated in pounds and where the weight per gallon of the liquid is not provided to the *fire code official*, a conversion factor of 10 pounds per gallon (1.2 kg/L) shall be used.

5003.1.3 Quantities not exceeding the maximum allowable quantity per control area. The storage, use and handling of hazardous materials in quantities not exceeding the *maximum allowable quantity per control area* indicated in Tables 5003.1.1(1) through 5003.1.1(4) shall be in accordance with Sections 5001 and 5003.

5003.1.4 Quantities exceeding the maximum allowable quantity per control area. The storage and use of hazardous materials in quantities exceeding the *maximum allowable quantity per control area* indicated in Tables 5003.1.1(1) through 5003.1.1(4) shall be in accordance with this chapter.

5003.2 Systems, equipment and processes. Systems, equipment and processes utilized for storage, dispensing, use or handling of hazardous materials shall be in accordance with Sections 5003.2.1 through 5003.2.8.

5003.2.1 Design and construction of containers, cylinders and tanks. Containers, cylinders and tanks shall be designed and constructed in accordance with *approved* standards. Containers, cylinders, tanks and other means used for containment of hazardous materials shall be of an *approved* type. Pressure vessels not meeting DOTn requirements for transportation shall comply with the ASME *Boiler and Pressure Vessel Code.*

5003.2.2 Piping, tubing, valves and fittings. Piping, tubing, valves, and fittings conveying hazardous materials shall be designed and installed in accordance with ASME B31 or other approved standards, and shall be in accordance with Sections 5003.2.2.1 and 5003.2.2.2.

5003.2.2.1 Design and construction. Piping, tubing, valves, fittings and related components used for hazardous materials shall be in accordance with the following:

1. Piping, tubing, valves, fittings and related components shall be designed and fabricated from materials that are compatible with the material to be contained and shall be of adequate strength and durability to withstand the pressure, structural and seismic stress and exposure to which they are subject.
2. Piping and tubing shall be identified in accordance with ASME A13.1 to indicate the material conveyed.
3. Readily accessible manual valves or automatic remotely activated fail-safe emergency shutoff valves shall be installed on supply piping and tubing at the following locations:
   3.1. The point of use.
   3.2. The tank, cylinder or bulk source.
4. Manual emergency shutoff valves and controls for remotely activated emergency shutoff valves shall be identified and the location shall be clearly visible, accessible and indicated by means of a sign.
5. Backflow prevention or check valves shall be provided where the backflow of hazardous materials could create a hazardous condition or cause

the unauthorized discharge of hazardous materials.

6. Where gases or liquids having a hazard ranking of:

   Health Class 3 or 4
   Flammability Class 4
   Instability Class 3 or 4

   in accordance with NFPA 704 are carried in pressurized piping above 15 pounds per square inch gauge (psig) (103 kPa), an *approved* means of leak detection and emergency shutoff or excess flow control shall be provided. Where the piping originates from within a hazardous material storage room or area, the excess flow control shall be located within the storage room or area. Where the piping originates from a bulk source, the excess flow control shall be located as close to the bulk source as practical.

Exceptions:

1. Piping for inlet connections designed to prevent backflow.
2. Piping for pressure relief devices.

5003.2.2.2 Additional regulations for supply piping for health-hazard materials. Supply piping and tubing for gases and liquids having a health-hazard ranking of 3 or 4 in accordance with NFPA 704 shall be in accordance with ASME B31.3 and the following:

1. Piping and tubing utilized for the transmission of highly toxic, toxic or highly volatile *corrosive* liquids and gases shall have welded, threaded or flanged connections throughout except for connections located within a ventilated enclosure if the material is a gas, or an *approved* method of drainage or containment is provided for connections if the material is a liquid.
2. Piping and tubing shall not be located within *corridors*, within any portion of a *means of egress* required to be enclosed in fire-resistance-rated construction or in concealed spaces in areas not classified as Group H occupancies.

Exception: Piping and tubing within the space defined by the walls of *corridors* and the floor or roof above or in concealed spaces above other occupancies where installed in accordance with Section 415.11.6.4 of the *International Building Code* for Group H-5 occupancies.

5003.2.3 Equipment, machinery and alarms. Equipment, machinery and required detection and alarm systems associated with the use, storage or handling of hazardous materials shall be listed or *approved*.

5003.2.4 Installation of tanks. Installation of tanks shall be in accordance with Sections 5003.2.4.1 through 5003.2.4.2.1.

5003.2.4.1 Underground tanks. Underground tanks used for the storage of liquid hazardous materials shall be provided with secondary containment. In lieu of providing secondary containment for an underground tank, an above-ground tank in an underground vault complying with Section 5704.2.8 shall be permitted.

5003.2.4.2 Above-ground tanks. Above-ground stationary tanks used for the storage of hazardous materials shall be located and protected in accordance with the requirements for outdoor storage of the particular material involved.

Exception: Above-ground tanks that are installed in vaults complying with Section 5303.16 or 5704.2.8 shall not be required to comply with location and protection requirements for outdoor storage.

5003.2.4.2.1 Marking. Above-ground stationary tanks shall be marked as required by Section 5003.5.

5003.2.5 Empty containers and tanks. Empty containers and tanks previously used for the storage of hazardous materials shall be free from residual material and vapor as defined by DOTn, the Resource Conservation and Recovery Act (RCRA) or other regulating authority or maintained as specified for the storage of hazardous material.

5003.2.6 Maintenance. In addition to the requirements of Section 5003.2.3, equipment, machinery and required detection and alarm systems associated with hazardous materials shall be maintained in an operable condition. Defective containers, cylinders and tanks shall be removed from service, repaired or disposed of in an *approved* manner. Defective equipment or machinery shall be removed from service and repaired or replaced. Required detection and alarm systems shall be replaced or repaired where defective.

5003.2.6.1 Tanks out of service for 90 days. Stationary tanks not used for a period of 90 days shall be properly safeguarded or removed in an *approved* manner. Such tanks shall have the fill line, gauge opening and pump connection secured against tampering. Vent lines shall be properly maintained.

5003.2.6.1.1 Return to service. Tanks that are to be placed back in service shall be tested in an *approved* manner.

5003.2.6.2 Defective containers and tanks. Defective containers and tanks shall be removed from service, repaired in accordance with approved standards or disposed of in an *approved* manner.

5003.2.7 Liquid-level limit control. Atmospheric tanks having a capacity greater than 500 gallons (1893 L) and that contain hazardous material liquids shall be equipped with a liquid-level limit control or other *approved* means to prevent overfilling of the tank.

5003.2.8 Seismic protection. Machinery and equipment utilizing hazardous materials shall be braced and anchored in accordance with the seismic design requirements of the *International Building Code* for the seismic design category in which the machinery or equipment is classified.

TABLE 5003.1.1(1)
MAXIMUM ALLOWABLE QUANTITY PER CONTROL AREA OF HAZARDOUS MATERIALS POSING A PHYSICAL HAZARD[a, j, m, n, p]

MATERIAL	CLASS	GROUP WHEN THE MAXIMUM ALLOWABLE QUANTITY IS EXCEEDED	STORAGE[b]			USE-CLOSED SYSTEMS[b]			USE-OPEN SYSTEMS[b]	
			Solid pounds (cubic feet)	Liquid gallons (pounds)	Gas (cubic feet at NTP)	Solid pounds (cubic feet)	Liquid gallons (pounds)	Gas (cubic feet at NTP)	Solid pounds (cubic feet)	Liquid gallons (pounds)
Combustible dust	NA	H-2	See Note q	NA	NA	See Note q	NA	NA	See Note q	NA
Combustible fibers[g]	Loose Baled[o]	H-3	(100) (1,000)	NA	NA	(100) (1,000)	NA	NA	(20) (200)	NA
Combustible liquid[c, i]	II IIIA IIIB	H-2 or H-3 H-2 or H-3 NA	NA	120[d,e] 330[d, e] 13,200[e, f]	NA	NA	120[d] 330[d] 13,200[f]	NA	NA	30[d] 80[d] 3,300[f]
Consumer fireworks	1.4G	H-3	125[e, l]	NA	NA	NA	NA	NA	NA	NA
Cryogenic Flammable	NA	H-2	NA	45[d]	NA	NA	45[d]	NA	NA	10[d]
Cryogenic Inert	NA	NA	NA	NA	NL	NA	NA	NL	NA	NA
Cryogenic Oxidizing	NA	H-3	NA	45[d]	NA	NA	45[d]	NA	NA	10[d]
Explosives	Division 1.1 Division 1.2 Division 1.3 Division 1.4 Division 1.4G Division 1.5 Division 1.6	H-1 H-1 H-1 or H-2 H-3 H-3 H-1 H-1	1[e, g] 1[e, g] 5[e, g] 50[e, g] 125[d, e, l] 1[e, g] 1[e, g]	(1)[e, g] (1)[e, g] (5)[e, g] (50)[e, g] NA (1)[e,g] NA	NA	0.25[g] 0.25[g] 1[g] 50[g] NA 0.25[g] NA	(0.25)[g] (0.25)[g] (1)[g] (50)[g] NA (0.25)[g] NA	NA	0.25[g] 0.25[g] 1[g] NA NA 0.25[g] NA	(0.25)[g] (0.25)[g] (1)[g] NA NA (0.25)[g] NA
Flammable gas	Gaseous Liquefied	H-2	NA	NA (150)[d, e]	1,000[d, e] NA	NA	NA (150)[d, e]	1,000[d, e] NA	NA	NA
Flammable liquid[c]	IA IB and IC	H-2 or H-3	NA	30[d, e] 120[d, e]	NA	NA	30[d] 120[d]	NA	NA	10[d] 30[d]
Flammable liquid, combination (IA, IB, IC)	NA	H-2 or H-3	NA	120[d, e, h]	NA	NA	120[d, h]	NA	NA	30[d, h]
Flammable solid	NA	H-3	125[d, e]	NA	NA	125[d]	NA	NA	25[d]	NA

(continued)

TABLE 5003.1.1(1)—continued
MAXIMUM ALLOWABLE QUANTITY PER CONTROL AREA OF HAZARDOUS MATERIALS POSING A PHYSICAL HAZARD[a, j, m, n, p]

MATERIAL	CLASS	GROUP WHEN THE MAXIMUM ALLOWABLE QUANTITY IS EXCEEDED	STORAGE[b]			USE-CLOSED SYSTEMS[b]			USE-OPEN SYSTEMS[b]	
			Solid pounds (cubic feet)	Liquid gallons (pounds)	Gas (cubic feet at NTP)	Solid pounds (cubic feet)	Liquid gallons (pounds)	Gas (cubic feet at NTP)	Solid pounds (cubic feet)	Liquid gallons (pounds)
Inert Gas	Gaseous	NA	NA	NA	NL	NA	NA	NL	NA	NA
	Liquefied	NA	NA	NA	NL	NA	NA	NL	NA	NA
Organic peroxide	UD	H-1	1[e, g]	(1)[e, g]	NA	0.25[g]	(0.25)[g]	NA	0.25[g]	(0.25)[g]
	I	H-2	5[d, e]	(5)[d, e]		1[d]	(1)[d]		1[d]	(1)[d]
	II	H-3	50[d, e]	(50)[d, e]		50[d]	(50)[d]		10[d]	(10)[d]
	III	H-3	125[d, e]	(125)[d, e]		125[d]	(125)[d]		25[d]	(25)[d]
	IV	NA	NL	NL		NL	NL		NL	NL
	V	NA	NL	NL		NL	NL		NL	NL
Oxidizer	4	H-1	1[g]	(1)[e, g]	NA	0.25[g]	(0.25)[g]	NA	0.25[g]	(0.25)[g]
	3[k]	H-2 or H-3	10[d, e]	(10)[d, e]		2[d]	(2)[d]		2[d]	(2)[d]
	2	H-3	250[d, e]	(250)[d, e]		250[d]	(250)[d]		50[d]	(50)[d]
	1	NA	4,000[e,f]	(4,000)[e,f]		4,000[f]	(4,000)[f]		1,000[f]	(1,000)[f]
Oxidizing gas	Gaseous	H-3	NA	NA	1,500[d, e]	NA	NA	1,500[d, e]	NA	NA
	Liquefied			(150)[d, e]	NA		(150)[d, e]	NA		
Pyrophoric	NA	H-2	4[e, g]	(4)[e, g]	50[e, g]	1[g]	(1)[g]	10[e, g]	0	0
Unstable (reactive)	4	H-1	1[e, g]	(1)[e, g]	10[e, g]	0.25[g]	(0.25)[g]	2[e, g]	0.25[g]	(0.25)[g]
	3	H-1 or H-2	5[d, e]	(5)[d, e]	50[d, e]	1[d]	(1)[d]	10[d, e]	1[d]	(1)[d]
	2	H-3	50[d, e]	(50)[d, e]	750[d, e]	50[d]	(50)[d]	750[d, e]	10[d]	(10)[d]
	1	NA	NL	NL	NL	NL	NL	NL	NL	NL
Water reactive	3	H-2	5[d,e]	(5)[d, e]	NA	5[d]	(5)[d]	NA	1[d]	(1)[d]
	2	H-3	50[d, e]	(50)[d, e]		50[d]	(50)[d]		10[d]	(10)[d]
	1	NA	NL	NL		NL	NL		NL	NL

For SI: 1 cubic foot = 0.02832 m^3, 1 pound = 0.454 kg, 1 gallon = 3.785 L.

NA = Not Applicable, NL = Not Limited, UD = Unclassified Detonable.

a. For use of control areas, see Section 5003.8.3.

b. The aggregate quantity in use and storage shall not exceed the quantity listed for storage.

c. The quantities of alcoholic beverages in retail and wholesale sales occupancies shall not be limited providing the liquids are packaged in individual containers not exceeding 1.3 gallons. In retail and wholesale sales occupancies, the quantities of medicines, foodstuff or consumer products and cosmetics containing not more than 50 percent by volume of water-miscible liquids with the remainder of the solutions not being flammable shall not be limited, provided that such materials are packaged in individual containers not exceeding 1.3 gallons.

d. Maximum allowable quantities shall be increased 100 percent in buildings equipped throughout with an approved automatic sprinkler system in accordance with Section 903.3.1.1. Where Note e also applies, the increase for both notes shall be applied accumulatively.

(continued)

TABLE 5003.1.1(1)—continued MAXIMUM ALLOWABLE QUANTITY PER CONTROL AREA OF HAZARDOUS MATERIALS POSING A PHYSICAL HAZARD[a, j, m, n, p]

e. Maximum allowable quantities shall be increased 100 percent where stored in approved storage cabinets, day boxes, gas cabinets, gas rooms, exhausted enclosures or in listed safety cans in accordance with Section 5003.9.10. Where Note d also applies, the increase for both notes shall be applied accumulatively.

f. Quantities shall not be limited in a building equipped throughout with an approved automatic sprinkler system in accordance with Section 903.3.1.1.

g. Allowed only in buildings equipped throughout with an approved automatic sprinkler system.

h. Containing not more than the maximum allowable quantity per control area of Class IA, Class IB or Class IC flammable liquids.

i. The maximum allowable quantity shall not apply to fuel oil storage complying with Section 603.3.2.

j. Quantities in parenthesis indicate quantity units in parenthesis at the head of each column.

k. A maximum quantity of 200 pounds of solid or 20 gallons of liquid Class 3 oxidizers is allowed where such materials are necessary for maintenance purposes, operation or sanitation of equipment where the storage containers and the manner of storage are approved.

l. Net weight of pyrotechnic composition of the fireworks. Where the net weight of the pyrotechnic composition of the fireworks is not known, 25 percent of the gross weight of the fireworks including packaging shall be used.

m. For gallons of liquids, divide the amount in pounds by 10 in accordance with Section 5003.1.2.

n. For storage and display quantities in Group M and storage quantities in Group S occupancies complying with Section 5003.11, see Table 5003.11.1.

o. Densely-packed baled cotton that complies with the packing requirements of ISO 8115 shall not be included in this material class.

p. The following shall not be included in determining the maximum allowable quantities:

1. Liquid or gaseous fuel in fuel tanks on vehicles.
2. Liquid or gaseous fuel in fuel tanks on motorized equipment operated in accordance with this code.
3. Gaseous fuels in piping systems and fixed appliances regulated by the *International Fuel Gas Code*.
4. Liquid fuels in piping systems and fixed appliances, regulated by the *International Mechanical Code*.
5. Alcohol-based hand rubs classified as Class I or II liquids in dispensers that are installed in accordance with Sections 5705.5 and 5705.5.1. The location of the alcohol-based hand rub (ABHR) dispensers shall be provided in the construction documents.

q. Where manufactured, generated or used in such a manner that the concentration and conditions create a fire or explosion hazard based on information prepared in accordance with Section 104.7.2.

TABLE 5003.1.1(2)
MAXIMUM ALLOWABLE QUANTITY PER CONTROL AREA OF HAZARDOUS MATERIAL POSING A HEALTH HAZARD[a, c, f, h, i]

MATERIAL	STORAGE[b]			USE-CLOSED SYSTEMS[b]			USE-OPEN SYSTEMS[b]	
	Solid pounds[d, e]	Liquid gallons (pounds)[d, e]	Gas cubic feet at NTP (pounds)[d]	Solid pounds[d]	Liquid gallons (pounds)[d]	Gas cubic feet at NTP (pounds)[d]	Solid pounds[d]	Liquid gallons (pounds)[d]
Corrosives	5,000	500	Gaseous 810[e] Liquefied (150)	5,000	500	Gaseous 810[e] Liquefied (150)	1,000	100
Highly Toxics	10	(10)	Gaseous 20[g] Liquefied (4)[g]	10	(10)	Gaseous 20[g] Liquified (4)[g]	3	(3)
Toxics	500	(500)	Gaseous 810[e] Liquefied (150)[e]	500	(500)	Gaseous 810[e] Liquefied (150)[e]	125	(125)

For SI: 1 cubic foot = 0.02832 m^3, 1 pound = 0.454 kg, 1 gallon = 3.785 L.

a. For use of control areas, see Section 5003.8.3.

b. The aggregate quantity in use and storage shall not exceed the quantity listed for storage.

c. In retail and wholesale sales occupancies, the quantities of medicines, foodstuff or consumer products and cosmetics, containing not more than 50 percent by volume of water-miscible liquids and with the remainder of the solutions not being flammable, shall not be limited, provided that such materials are packaged in individual containers not exceeding 1.3 gallons.

d. Maximum allowable quantities shall be increased 100 percent in buildings equipped throughout with an approved automatic sprinkler system in accordance with Section 903.3.1.1. Where Note e also applies, the increase for both notes shall be applied accumulatively.

e. Maximum allowable quantities shall be increased 100 percent where stored in approved storage cabinets, gas cabinets or exhausted enclosures. Where Note d also applies, the increase for both notes shall be applied accumulatively.

f. For storage and display quantities in Group M and storage quantities in Group S occupancies complying with Section 5003.11, see Table 5003.11.1.

g. Allowed only where stored in approved exhausted gas cabinets or exhausted enclosures.

h. Quantities in parentheses indicate quantity units in parentheses at the head of each column.

i. For gallons of liquids, divide the amount in pounds by 10 in accordance with Section 5003.1.2.

TABLE 5003.1.1(3)
MAXIMUM ALLOWABLE QUANTITY PER CONTROL AREA OF HAZARDOUS MATERIALS POSING A PHYSICAL HAZARD IN AN OUTDOOR CONTROL AREA[a, b, c, d]

MATERIAL	CLASS	STORAGE[b]			USE-CLOSED SYSTEMS[b]			USE-OPEN SYSTEMS[b]	
		Solid pounds (cubic feet)	Liquid gallons (pounds)[d]	Gas cubic feet at NTP	Solid pounds (cubic feet)	Liquid gallons (pounds)[d]	Gas cubic feet at NTP	Solid pounds (cubic feet)	Liquid gallons (pounds)[d]
Flammable gas	Gaseous Liquefied	Not Applicable	Not Applicable (300)	3,000 Not Applicable	Not Applicable	Not Applicable (150)	1,500 Not Applicable	Not Applicable	Not Applicable
Flammable solid	Not Applicable	500	Not Applicable	Not Applicable	250	Not Applicable	Not Applicable	50	Not Applicable
Inert Gas Cryogenic inert	Gaseous Liquefied Not Applicable	Not Applicable Not Applicable Not Applicable	Not Applicable Not Applicable Not Applicable	Not Limited Not Limited Not Limited	Not Applicable Not Applicable Not Applicable	Not Applicable Not Applicable Not Applicable	Not Limited Not Limited Not Limited	Not Applicable Not Applicable Not Applicable	Not Applicable Not Applicable Not Applicable
Organic peroxide	Unclassified Detonable	1	(1)	Not Applicable	0.25	(0.25)	Not Applicable	0.25	(0.25)
Organic peroxide	I II III IV V	20 200 500 1,000 Not Limited	(20) (200) (500) (1,000) Not Limited	Not Applicable	10 100 250 500 Not Limited	(10) (100) (250) (500) Not Limited	Not Applicable	2 20 50 100 Not Limited	(2) (20) (50) (100) Not Limited
Oxidizer	4 3 2 1	2 40 1,000 Not Limited	(2) (40) (1,000) Not Limited	Not Applicable	1 20 500 Not Limited	(1) (20) (500) Not Limited	Not Applicable	0.25 4 100 Not Limited	(0.25) (4) (100) Not Limited
Oxidizing gas	Gaseous Liquefied	Not Applicable	Not Applicable (600)	6,000 Not Applicable	Not Applicable	Not Applicable (300)	1,500 Not Applicable	Not Applicable	Not Applicable
Pyrophoric materials	Not Applicable	8	(8)	100	4	(4)	10	0	0
Unstable (reactive)	4 3 2 1	2 20 200 Not Limited	(2) (20) (200) Not Limited	20 200 1,000 1,500	1 10 100 Not Limited	(1) (10) (100) Not Limited	2 10 250 Not Limited	0.25 1 10 Not Limited	(0.25) (1) (10) Not Limited
Water reactive	3 2 1	20 200 Not Limited	(20) (200) Not Limited	Not Applicable	10 100 Not Limited	(10) (100) Not Limited	Not Applicable	1 10 Not Limited	(1) (10) Not Limited

For SI: 1 pound = 0.454 kg, 1 gallon = 3.785 L, 1 cubic foot = 0.02832 m^3.

a. For gallons of liquids, divide the amount in pounds by 10 in accordance with Section 5003.1.2.

b. The aggregate quantities in storage and use shall not exceed the quantity listed for storage.

c. The aggregate quantity of nonflammable solid and nonflammable or noncombustible liquid hazardous materials allowed in outdoor storage per single property under the same ownership or control used for retail or wholesale sales is allowed to exceed the maximum allowable quantity per control area where such storage is in accordance with Section 5003.11.

d. Quantities in parentheses indicate quantity units in parentheses at the head of each column.

TABLE 5003.1.1(4)
MAXIMUM ALLOWABLE QUANTITY PER CONTROL AREA OF HAZARDOUS MATERIALS POSING A HEALTH HAZARD IN AN OUTDOOR CONTROL AREA[a, b, c, f]

MATERIAL	STORAGE			USE-CLOSED SYSTEMS			USE-OPEN SYSTEMS	
	Solid pounds	Liquid gallons (pounds)	Gas cubic feet at NTP (pounds)	Solid pounds	Liquid gallons (pounds)	Gas cubic feet at NTP (pounds)	Solid pounds	Liquid gallons (pounds)
Corrosives	20,000	2,000	Gaseous 1,620 Liquefied (300)	10,000	1,000	Gaseous 810 Liquefied (150)	1,000	100
Highly toxics	20	(20)	Gaseous 40[d] Liquefied (8)[d]	10	(10)	Gaseous 20[d] Liquefied (4)[d]	3	(3)
Toxics	1,000	(1,000)[e]	Gaseous 1,620 Liquefied (300)	500	50[e]	Gaseous 810 Liquefied (150)	125	(125)[e]

For SI: 1 cubic foot = 0.02832 m^3, 1 pound = 0.454 kg, 1 gallon = 3.785 L, 1 pound per square inch absolute = 6.895 kPa,°C = [(°F)-32/1.8].

a. For gallons of liquids, divide the amount in pounds by 10 in accordance with Section 5003.1.2.

b. The aggregate quantities in storage and use shall not exceed the quantity listed for storage.

c. The aggregate quantity of nonflammable solid and nonflammable or noncombustible liquid hazardous materials allowed in outdoor storage per single property under the same ownership or control used for retail or wholesale sales is allowed to exceed the maximum allowable quantity per control area where such storage is in accordance with Section 5003.11.

d. Allowed only where used in approved exhausted gas cabinets, exhausted enclosures or under fume hoods.

e. The maximum allowable quantity per control area for toxic liquids with vapor pressures in excess of 1 psia at 77°F shall be the maximum allowable quantity per control area listed for highly toxic liquids.

f. Quantities in parentheses indicate quantity units in parentheses at the head of each column.

5003.2.9 Testing. The equipment, devices and systems listed in Section 5003.2.9.1 shall be tested at the time of installation and at one of the intervals listed in Section 5003.2.9.2. Records of the tests conducted or maintenance performed shall be maintained in accordance with the provisions of Section 107.2.1.

Exceptions:

1. Periodic testing shall not be required where *approved* written documentation is provided stating that testing will damage the equipment, device or system and the equipment, device or system is maintained as specified by the manufacturer.
2. Periodic testing shall not be required for equipment, devices and systems that fail in a fail-safe manner.
3. Periodic testing shall not be required for equipment, devices and systems that self-diagnose and report trouble. Records of the self-diagnosis and trouble reporting shall be made available to the *fire code official.*
4. Periodic testing shall not be required if system activation occurs during the required test cycle for the components activated during the test cycle.
5. *Approved* maintenance in accordance with Section 5003.2.6 that is performed not less than annually or in accordance with an *approved* schedule shall be allowed to meet the testing requirements set forth in Sections 5003.2.9.1 and 5003.2.9.2.

5003.2.9.1 Equipment, devices and systems requiring testing. The following equipment, systems and devices shall be tested in accordance with Sections 5003.2.9 and 5003.2.9.2.

1. Gas detection systems, alarms and automatic emergency shutoff valves required by Section 6004.2.2.10 for highly toxic and toxic gases.
2. Limit control systems for liquid level, temperature and pressure required by Sections 5003.2.7, 5004.8 and 5005.1.4.
3. Emergency alarm systems and supervision required by Sections 5004.9 and 5005.4.4.
4. Monitoring and supervisory systems required by Sections 5004.10 and 5005.1.6.
5. Manually activated shutdown controls required by Section 6403.1.1.1 for *compressed gas* systems conveying pyrophoric gases.

5003.2.9.2 Testing frequency. The equipment, systems and devices listed in Section 5003.2.9.1 shall be tested at one of the frequencies listed below:

1. Not less than annually.
2. In accordance with the *approved* manufacturer's requirements.
3. In accordance with *approved* recognized industry standards.
4. In accordance with an *approved* schedule.

5003.3 Release of hazardous materials. Hazardous materials in any quantity shall not be released into a sewer, storm drain, ditch, drainage canal, creek, stream, river, lake or tidal waterway or on the ground, sidewalk, street, highway or into the atmosphere.

Exceptions:

1. The release or emission of hazardous materials is allowed where in compliance with federal, state or local governmental agencies, regulations or permits.
2. The release of pesticides is allowed where used in accordance with registered label directions.
3. The release of fertilizer and soil amendments is allowed where used in accordance with manufacturer's specifications.

5003.3.1 Unauthorized discharges. Where hazardous materials are released in quantities reportable under state, federal or local regulations, the *fire code official* shall be notified and the following procedures required in accordance with Sections 5003.3.1.1 through 5003.3.1.4.

5003.3.1.1 Records. Records of the unauthorized discharge of hazardous materials by the permittee shall be maintained.

5003.3.1.2 Preparation. Provisions shall be made for controlling and mitigating unauthorized discharges.

5003.3.1.3 Control. Where an unauthorized discharge caused by primary container failure is discovered, the involved primary container shall be repaired or removed from service.

5003.3.1.4 Responsibility for cleanup. The person, firm or corporation responsible for an unauthorized discharge shall institute and complete all actions necessary to remedy the effects of such unauthorized discharge, whether sudden or gradual, without cost to the jurisdiction. Where deemed necessary by the *fire code official*, cleanup can be initiated by the fire department or by an authorized individual or firm. Costs associated with such cleanup shall be borne by the *owner*, operator or other person responsible for the unauthorized discharge.

5003.4 Material Safety Data Sheets. Material Safety Data Sheets (MSDS) shall be readily available on the premises for hazardous materials regulated by this chapter. Where a hazardous substance is developed in a laboratory, available information shall be documented.

Exception: Designated hazardous waste.

5003.5 Hazard identification signs. Unless otherwise exempted by the *fire code official*, visible hazard identification signs as specified in NFPA 704 for the specific material contained shall be placed on stationary containers and aboveground tanks and at entrances to locations where hazardous materials are stored, dispensed, used or handled in quantities requiring a permit and at specific entrances and locations designated by the *fire code official*.

5003.5.1 Markings. Individual containers, cartons or packages shall be conspicuously marked or labeled in an approved manner. Rooms or cabinets containing *compressed gases* shall be conspicuously labeled: COMPRESSED GAS.

5003.6 Signs. Signs and markings required by Sections 5003.5 and 5003.5.1 shall not be obscured or removed, shall be in English as a primary language or in symbols allowed by this code, shall be durable, and the size, color and lettering shall be *approved*.

5003.7 Sources of ignition. Sources of ignition shall comply with Sections 5003.7.1 through 5003.7.3.

5003.7.1 Smoking. Smoking shall be prohibited and "No Smoking" signs provided as follows:

1. In rooms or areas where hazardous materials are stored or dispensed or used in *open systems* in amounts requiring a permit in accordance with Section 5001.5.
2. Within 25 feet (7620 mm) of outdoor storage, dispensing or open use areas.
3. Facilities or areas within facilities that have been designated as totally "no smoking" shall have "No Smoking" signs placed at all entrances to the facility or area. Designated areas within such facilities where smoking is permitted either permanently or temporarily, shall be identified with signs designating that smoking is permitted in these areas only.
4. In rooms or areas where flammable or combustible hazardous materials are stored, dispensed or used.

Signs required by this section shall be in English as a primary language or in symbols allowed by this code and shall comply with Section 310.

5003.7.2 Open flames. Open flames and high-temperature devices shall not be used in a manner that creates a hazardous condition and shall be *listed* for use with the hazardous materials stored or used.

5003.7.3 Industrial trucks. Powered industrial trucks used in areas designated as hazardous (classified) locations in accordance with NFPA 70 shall be *listed* and *labeled* for use in the environment intended in accordance with NFPA 505.

5003.8 Construction requirements. Buildings, *control areas*, enclosures and cabinets for hazardous materials shall be in accordance with Sections 5003.8.1 through 5003.8.6.3.

5003.8.1 Buildings. Buildings, or portions thereof, in which hazardous materials are stored, handled or used shall be constructed in accordance with the *International Building Code*.

5003.8.2 Required detached buildings. Group H occupancies containing quantities of hazardous materials in excess of those set forth in Table 5003.8.2 shall be in detached buildings.

TABLE 5003.8.2
DETACHED BUILDING REQUIRED

A DETACHED BUILDING IS REQUIRED WHERE THE QUANTITY OF MATERIAL EXCEEDS THAT LISTED HEREIN			
Material	**Class**	**Solids and liquids (tons)[a, b]**	**Gases (cubic feet)[a, b]**
Explosives	Division 1.1 Division 1.2 Division 1.3 Division 1.4 Division 1.4[c] Division 1.5 Division 1.6	Maximum Allowable Quantity Maximum Allowable Quantity Maximum Allowable Quantity Maximum Allowable Quantity 1 Maximum Allowable Quantity Maximum Allowable Quantity	Not Applicable
Oxidizers	Class 4	Maximum Allowable Quantity	Maximum Allowable Quantity
Unstable (reactives) detonable	Class 3 or 4	Maximum Allowable Quantity	Maximum Allowable Quantity
Oxidizer, liquids and solids	Class 3 Class 2	1,200 2,000	Not Applicable
Organic peroxides	Detonable Class I Class II Class III	Maximum Allowable Quantity Maximum Allowable Quantity 25 50	Not Applicable
Unstable (reactives) nondetonable	Class 3 Class 2	1 25	2,000 10,000
Water reactives	Class 3 Class 2	1 25	Not Applicable
Pyrophoric gases	Not Applicable	Not Applicable	2,000

For SI: 1 pound = 0.454 kg, 1 cubic foot = 0.02832 m^3, 1 ton = 2000 lbs. = 907.2 kg.

a. For materials that are detonable, the distance to other buildings or lot lines shall be as specified in the *International Building Code*. For materials classified as explosives, the required separation distances shall be as specified in Chapter 56.

b. "Maximum Allowable Quantity" means the maximum allowable quantity per control area set forth in Table 5003.1.1(1).

c. Limited to Division 1.4 materials and articles, including articles packaged for shipment, that are not regulated as an explosive under Bureau of Alcohol, Tobacco, Firearms and Explosives regulations, or unpackaged articles used in process operations that do not propagate a detonation or deflagration between articles, providing the net explosive weight of individual articles does not exceed 1 pound.

5003.8.3 Control areas. *Control areas* shall comply with Sections 5003.8.3.1 through 5003.8.3.5.

5003.8.3.1 Construction requirements. *Control areas* shall be separated from each other by *fire barriers* constructed in accordance with Section 707 of the *International Building Code* or *horizontal assemblies* constructed in accordance with Section 711 of the *International Building Code*, or both.

5003.8.3.2 Percentage of maximum allowable quantities. The percentage of maximum allowable quantities of hazardous materials per *control area* allowed at each floor level within a building shall be in accordance with Table 5003.8.3.2.

5003.8.3.3 Number. The maximum number of *control areas* per floor within a building shall be in accordance with Table 5003.8.3.2.

5003.8.3.4 Fire-resistance-rating requirements. The required *fire-resistance rating* for *fire barriers* shall be in accordance with Table 5003.8.3.2. The floor assembly of the *control area* and the construction supporting the floor of the *control area* shall have a *fire-resistance rating* of not less than 2 hours.

Exception: The floor assembly of the *control area* and the construction supporting the floor of the *control area* is allowed to be 1-hour *fire-resistance* rated in buildings of Type IIA, IIIA and VA construction, provided that both of the following conditions exist:

1. The building is equipped throughout with an *automatic sprinkler system* in accordance with Section 903.3.1.1.
2. The building is three stories or less above grade plane.

5003.8.3.5 Hazardous material in Group M display and storage areas and in Group S storage areas. The aggregate quantity of nonflammable solid and nonflammable or noncombustible liquid hazardous materials allowed within a single *control area* of a Group M display and storage area or a Group S storage area is allowed to exceed the *maximum allowable quantities per control area* specified in Tables 5003.1.1(1) and 5003.1.1(2) without classifying the building or use as a Group H occupancy, provided that the materials are displayed and stored in accordance with Section 5003.11.

5003.8.4 Gas rooms. Where a gas room is used to increase the *maximum allowable quantity per control area* or provided to comply with the provisions of Chapter 60, the gas room shall be in accordance with Sections 5003.8.4.1 and 5003.8.4.2.

5003.8.4.1 Construction. Gas rooms shall be protected with an *automatic sprinkler system.* Gas rooms shall be separated from the remainder of the building in accordance with the requirements of the *International Building Code* based on the occupancy group into which it has been classified.

5003.8.4.2 Ventilation system. The ventilation system for gas rooms shall be designed to operate at a negative pressure in relation to the surrounding area. Highly toxic and toxic gases shall also comply with Section 6004.2.2.6. The ventilation system shall be installed in accordance with the *International Mechanical Code.*

5003.8.5 Exhausted enclosures. Where an exhausted enclosure is used to increase *maximum allowable quantity per control area* or where the location of hazardous materials in exhausted enclosures is provided to comply with the provisions of Chapter 60, the exhausted enclosure shall be in accordance with Sections 5003.8.5.1 through 5003.8.5.3.

5003.8.5.1 Construction. Exhausted enclosures shall be of noncombustible construction.

5003.8.5.2 Ventilation. Exhausted enclosures shall be provided with an exhaust ventilation system. The ventilation system for exhausted enclosures shall be designed to operate at a negative pressure in relation to the surrounding area. Ventilation systems used for highly toxic and toxic gases shall also comply with Items 1, 2 and 3 of Section 6004.1.2. The ventilation

TABLE 5003.8.3.2
DESIGN AND NUMBER OF CONTROL AREAS

FLOOR LEVEL		PERCENTAGE OF THE MAXIMUM ALLOWABLE QUANTITY PER CONTROL AREA[a]	NUMBER OF CONTROL AREAS PER FLOOR	FIRE-RESISTANCE RATING FOR FIRE BARRIERS IN HOURS[b]
Above grade plane	Higher than 9	5	1	2
	7-9	5	2	2
	6	12.5	2	2
	5	12.5	2	2
	4	12.5	2	2
	3	50	2	1
	2	75	3	1
	1	100	4	1
Below grade plane	1	75	3	1
	2	50	2	1
	Lower than 2	Not Allowed	Not Allowed	Not Allowed

a. Percentages shall be of the maximum allowable quantity per control area shown in Tables 5003.1.1(1) and 5003.1.1(2), with all increases allowed in the footnotes to those tables.

b. Separation shall include fire barriers and horizontal assemblies as necessary to provide separation from other portions of the building.

system shall be installed in accordance with the *International Mechanical Code*.

5003.8.5.3 Fire-extinguishing system. Exhausted enclosures where flammable materials are used shall be protected by an approved automatic fire-extinguishing system in accordance with Chapter 9.

5003.8.6 Gas cabinets. Where a gas cabinet is used to increase the *maximum allowable quantity per control area* or where the location of *compressed gases* in gas cabinets is provided to comply with the provisions of Chapter 60, the gas cabinet shall be in accordance with Sections 5003.8.6.1 through 5003.8.6.3.

5003.8.6.1 Construction. Gas cabinets shall be constructed in accordance with the following:

1. Constructed of not less than 0.097-inch (2.5 mm) (No. 12 gage) steel.
2. Be provided with self-closing limited access ports or noncombustible windows to give access to equipment controls.
3. Be provided with self-closing doors.
4. Gas cabinet interiors shall be treated, coated or constructed of materials that are compatible with the hazardous materials stored. Such treatment, coating or construction shall include the entire interior of the cabinet.

5003.8.6.2 Ventilation. Gas cabinets shall be provided with an exhaust ventilation system. The ventilation system for gas cabinets shall be designed to operate at a negative pressure in relation to the surrounding area. Ventilation systems used for highly toxic and toxic gases shall also comply with Items 1, 2 and 3 of Section 6004.1.2. The ventilation system shall be installed in accordance with the *International Mechanical Code*.

5003.8.6.3 Maximum number of cylinders per gas cabinet. The number of cylinders contained in a single gas cabinet shall not exceed three.

5003.8.7 Hazardous materials storage cabinets. Where storage cabinets are used to increase *maximum allowable quantity per control area* or to comply with this chapter, such cabinets shall be in accordance with Sections 5003.8.7.1 and 5003.8.7.2.

5003.8.7.1 Construction. The interior of cabinets shall be treated, coated or constructed of materials that are nonreactive with the hazardous material stored. Such treatment, coating or construction shall include the entire interior of the cabinet. Cabinets shall either be *listed* in accordance with UL 1275 as suitable for the intended storage or constructed in accordance with the following:

1. Cabinets shall be of steel having a thickness of not less than 0.0478 inch (1.2 mm) (No. 18 gage). The cabinet, including the door, shall be double walled with a $1^1/_2$-inch (38 mm) airspace between the walls. Joints shall be riveted or welded and shall be tight fitting. Doors shall be well fitted, self-closing and equipped with a self-latching device.
2. The bottoms of cabinets utilized for the storage of liquids shall be liquid tight to a minimum height of 2 inches (51 mm).

Electrical equipment and devices within cabinets used for the storage of hazardous gases or liquids shall be in accordance with NFPA 70.

5003.8.7.2 Warning markings. Cabinets shall be clearly identified in an approved manner with red letters on a contrasting background to read:

HAZARDOUS—KEEP FIRE AWAY.

5003.9 General safety precautions. General precautions for the safe storage, handling or care of hazardous materials shall be in accordance with Sections 5003.9.1 through 5003.9.10.

5003.9.1 Personnel training and written procedures. Persons responsible for the operation of areas in which hazardous materials are stored, dispensed, handled or used shall be familiar with the chemical nature of the materials and the appropriate mitigating actions necessary in the event of fire, leak or spill.

5003.9.1.1 Fire department liaison. Responsible persons shall be designated and trained to be liaison personnel to the fire department. These persons shall aid the fire department in preplanning emergency responses and identifying the locations where hazardous materials are located, and shall have access to Material Safety Data Sheets and be knowledgeable in the site's emergency response procedures.

5003.9.2 Security. Storage, dispensing, use and handling areas shall be secured against unauthorized entry and safeguarded in a manner *approved* by the *fire code official*.

5003.9.3 Protection from vehicles. Guard posts or other *approved* means shall be provided to protect storage tanks and connected piping, valves and fittings; dispensing areas; and use areas subject to vehicular damage in accordance with Section 312.

5003.9.4 Electrical wiring and equipment. Electrical wiring and equipment shall be installed and maintained in accordance with NFPA 70.

5003.9.5 Static accumulation. Where processes or conditions exist where a flammable mixture could be ignited by static electricity, means shall be provided to prevent the accumulation of a static charge.

5003.9.6 Protection from light. Materials that are sensitive to light shall be stored in containers designed to protect them from such exposure.

5003.9.7 Shock padding. Materials that are shock sensitive shall be padded, suspended or otherwise protected against accidental dislodgement and dislodgement during seismic activity.

5003.9.8 Separation of incompatible materials. *Incompatible materials* in storage and storage of materials that are incompatible with materials in use shall be separated where the stored materials are in containers having a capacity of more than 5 pounds (2 kg) or 0.5 gallon (2 L). Separation shall be accomplished by:

1. Segregating *incompatible materials* in storage by a distance of not less than 20 feet (6096 mm).
2. Isolating *incompatible materials* in storage by a noncombustible partition extending not less than 18 inches (457 mm) above and to the sides of the stored material.
3. Storing liquid and solid materials in hazardous material storage cabinets.
4. Storing *compressed gases* in gas cabinets or exhausted enclosures in accordance with Sections 5003.8.5 and 5003.8.6. Materials that are incompatible shall not be stored within the same cabinet or exhausted enclosure.

5003.9.9 Shelf storage. Shelving shall be of substantial construction, and shall be braced and anchored in accordance with the seismic design requirements of the *International Building Code* for the seismic zone in which the material is located. Shelving shall be treated, coated or constructed of materials that are compatible with the hazardous materials stored. Shelves shall be provided with a lip or guard where used for the storage of individual containers.

Exceptions:

1. Storage in hazardous material storage cabinets or laboratory furniture specifically designed for such use.
2. Storage of hazardous materials in amounts not requiring a permit in accordance with Section 5001.5.

Shelf storage of hazardous materials shall be maintained in an orderly manner.

5003.9.10 Safety cans. Safety cans shall be *listed* in accordance with UL 30 where used to increase the *maximum allowable quantities per control area* of flammable or *combustible liquids* in accordance with Table 5003.1.1(1). Safety cans *listed* in accordance with UL 1313 are allowed for flammable and *combustible liquids* where not used to increase the *maximum allowable quantities per control area* and for other hazardous material liquids in accordance with the listing.

5003.10 Handling and transportation. In addition to the requirements of Section 5003.2, the handling and transportation of hazardous materials in *corridors* or enclosures for stairways and ramps shall be in accordance with Sections 5003.10.1 through 5003.10.3.6.

5003.10.1 Valve protection. Hazardous material gas containers, cylinders and tanks in transit shall have their protective caps in place. Containers, cylinders and tanks of highly toxic or toxic *compressed gases* shall have their valve outlets capped or plugged with an *approved* closure device in accordance with Chapter 53.

5003.10.2 Carts and trucks required. Liquids in containers exceeding 5 gallons (19 L) in a *corridor* or enclosure for a stairway or ramp shall be transported on a cart or truck. Containers of hazardous materials having a hazard ranking of 3 or 4 in accordance with NFPA 704 and transported within *corridors* or interior exit stairways and ramps, shall be on a cart or truck. Where carts and trucks are required for transporting hazardous materials, they shall be in accordance with Section 5003.10.3.

Exceptions:

1. Two hazardous material liquid containers that are hand carried in acceptable safety carriers.
2. Not more than four drums not exceeding 55 gallons (208 L) each that are transported by suitable drum trucks.
3. Containers and cylinders of *compressed gases* that are transported by *approved* hand trucks, and containers and cylinders not exceeding 25 pounds (11 kg) that are hand carried.
4. Solid hazardous materials not exceeding 100 pounds (45 kg) that are transported by *approved* hand trucks, and a single container not exceeding 50 pounds (23 kg) that is hand carried.

5003.10.3 Carts and trucks. Carts and trucks required by Section 5003.10.2 to be used to transport hazardous materials shall be in accordance with Sections 5003.10.3.1 through 5003.10.3.6.

5003.10.3.1 Design. Carts and trucks used to transport hazardous materials shall be designed to provide a stable base for the commodities to be transported and shall have a means of restraining containers to prevent accidental dislodgement. *Compressed gas* cylinders placed on carts and trucks shall be individually restrained.

5003.10.3.2 Speed-control devices. Carts and trucks shall be provided with a device that will enable the operator to control safely movement by providing stops or speed-reduction devices.

5003.10.3.3 Construction. Construction materials for hazardous material carts or trucks shall be compatible with the material transported. The cart or truck shall be of substantial construction.

5003.10.3.4 Spill control. Carts and trucks transporting liquids shall be capable of containing a spill from the largest single container transported.

5003.10.3.5 Attendance. Carts and trucks used to transport materials shall not obstruct or be left unattended within any part of a *means of egress*.

5003.10.3.6 Incompatible materials. *Incompatible materials* shall not be transported on the same cart or truck.

5003.11 Group M storage and display and Group S storage. The aggregate quantity of nonflammable solid and nonflammable or noncombustible liquid hazardous materials stored and displayed within a single *control area* of a Group M occupancy, or an outdoor *control area*, or stored in a single *control area* of a Group S occupancy, is allowed to exceed the *maximum allowable quantity per control area* indicated in Section 5003.1 where in accordance with Sections 5003.11.1 through 5003.11.3.11.

5003.11.1 Maximum allowable quantity per control area in Group M or S occupancies. The aggregate amount of nonflammable solid and nonflammable or noncombustible liquid hazardous materials stored and displayed within a single *control area* of a Group M occupancy or stored in a single *control area* of a Group S occupancy shall not exceed the amounts set forth in Table 5003.11.1.

5003.11.2 Maximum allowable quantity per outdoor control area in Group M or S occupancies. The aggregate amount of nonflammable solid and nonflammable or noncombustible liquid hazardous materials stored and displayed within a single outdoor *control area* of a Group M occupancy shall not exceed the amounts set forth in Table 5003.11.1.

5003.11.3 Storage and display. Storage and display shall be in accordance with Sections 5003.11.3.1 through 5003.11.3.11.

5003.11.3.1 Density. Storage and display of solids shall not exceed 200 pounds per square foot (976 kg/m^2) of floor area actually occupied by solid merchandise. Storage and display of liquids shall not exceed 20 gallons per square foot (0.50 L/m^2) of floor area actually occupied by liquid merchandise.

5003.11.3.2 Storage and display height. Display height shall not exceed 6 feet (1829 mm) above the finished floor in display areas of Group M occupancies. Storage height shall not exceed 8 feet (2438 mm) above the finished floor in storage areas of Group M and Group S occupancies.

5003.11.3.3 Container location. Individual containers less than 5 gallons (19 L) or less than 25 pounds (11 kg) shall be stored or displayed on pallets, racks or shelves.

5003.11.3.4 Racks and shelves. Racks and shelves used for storage or display shall be in accordance with Section 5003.9.9.

TABLE 5003.11.1
MAXIMUM ALLOWABLE QUANTITY PER INDOOR AND OUTDOOR CONTROL AREA IN GROUP M AND S OCCUPANCIES—NONFLAMMABLE SOLIDS, NONFLAMMABLE AND NONCOMBUSTIBLE LIQUIDS[d, e, f]

CONDITION		MAXIMUM ALLOWABLE QUANTITY PER CONTROL AREA	
Material[a]	Class	Solids pounds	Liquids gallons
A. HEALTH-HAZARD MATERIALS—NONFLAMMABLE AND NONCOMBUSTIBLE SOLIDS AND LIQUIDS			
1. Corrosives[b, c]	Not Applicable	9,750	975
2. Highly Toxics	Not Applicable	20[b, c]	2[b, c]
3. Toxics[b, c]	Not Applicable	1,000	100
B. PHYSICAL-HAZARD MATERIALS—NONFLAMMABLE AND NONCOMBUSTIBLE SOLIDS AND LIQUIDS			
1. Oxidizers[b, c]	4	Not Allowed	Not Allowed
	3	1,150[g]	115
	2	2,250[h]	225
	1	18,000[i, j]	1,800[i, j]
2. Unstable (Reactives)[b, c]	4	Not Allowed	Not Allowed
	3	550	55
	2	1,150	115
	1	Not Limited	Not Limited
3. Water Reactives	3[b, c]	550	55
	2[b, c]	1,150	115
	1	Not Limited	Not Limited

For SI: 1 pound = 0.454 kg, 1 gallon = 3.785 L, 1 cubic foot = 0.02832 m^3.

a. Hazard categories are as specified in Section 5001.2.2.

b. Maximum allowable quantities shall be increased 100 percent in buildings equipped throughout with an approved automatic sprinkler system in accordance with Section 903.3.1.1. Where Note c also applies, the increase for both notes shall be applied accumulatively.

c. Maximum allowable quantities shall be increased 100 percent where stored in approved storage cabinets in accordance with Section 5003.8. Where Note b also applies, the increase for both notes shall be applied accumulatively.

d. See Table 5003.8.3.2 for design and number of control areas.

e. Maximum allowable quantities for other hazardous material categories shall be in accordance with Section 5003.1.

f. Maximum allowable quantities shall be increased 100 percent in outdoor control areas.

g. Maximum allowable quantities shall be increased to 2,250 pounds where individual packages are in the original sealed containers from the manufacturer or packager and do not exceed 10 pounds each.

h. Maximum allowable quantities shall be increased to 4,500 pounds where individual packages are in the original sealed containers from the manufacturer or packager and do not exceed 10 pounds each.

i. Quantities are unlimited where protected by an automatic sprinkler system.

j. Quantities are unlimited in an outdoor control area.

5003.11.3.5 Container type. Containers shall be *approved* for the intended use and identified as to their content.

5003.11.3.6 Container size. Individual containers shall not exceed 100 pounds (45 kg) for solids or 10 gallons (38 L) for liquids in storage and display areas.

5003.11.3.7 Incompatible materials. *Incompatible materials* shall be separated in accordance with Section 5003.9.8.

5003.11.3.8 Floors. Floors shall be in accordance with Section 5004.12.

5003.11.3.9 Aisles. Aisles 4 feet (1219 mm) in width shall be maintained on three sides of the storage or display area.

5003.11.3.10 Signs. Hazard identification signs shall be provided in accordance with Section 5003.5.

5003.11.3.11 Storage plan. A storage plan illustrating the intended storage arrangement, including the location and dimensions of aisles, and storage racks shall be provided.

5003.12 Outdoor control areas. Outdoor *control areas* for hazardous materials in amounts not exceeding the maximum allowable quantity per outdoor *control area* shall be in accordance with the following:

1. Outdoor *control areas* shall be kept free from weeds, debris and common combustible materials not necessary to the storage. The area surrounding an outdoor *control area* shall be kept clear of such materials for not less than 15 feet (4572 mm).
2. Outdoor control areas shall be located not closer than 20 feet (6096 mm) from a lot line that can be built upon, public street, public alley or public way.

   Exceptions:

   1. For solid and liquid hazardous materials, a 2-hour fire-resistance-rated wall without openings extending not less than 30 inches (762 mm) above and to the sides of the storage area shall be allowed in lieu of such distance.
   2. For compressed gas hazardous materials, unless otherwise specified, the minimum required distances shall not apply where *fire barriers* without openings or penetrations having a minimum *fire-resistance rating* of 2 hours interrupt the line of sight between the storage and the exposure. The configuration of the *fire barrier* shall be designed to allow natural ventilation to prevent the accumulation of hazardous gas concentrations.
3. Where a property exceeds 10,000 square feet (929 m^2), a group of two outdoor *control areas* is allowed where *approved* and where each *control area* is separated by a minimum distance of 50 feet (15 240 mm).
4. Where a property exceeds 35,000 square feet (3252 m^2), additional groups of outdoor *control areas* are allowed where approved and where each group is separated by a minimum distance of 300 feet (91 440 mm).

SECTION 5004
STORAGE

5004.1 Scope. Storage of hazardous materials in amounts exceeding the *maximum allowable quantity per control area* as set forth in Section 5003.1 shall be in accordance with Sections 5001, 5003 and 5004. Storage of hazardous materials in amounts not exceeding the *maximum allowable quantity per control area* as set forth in Section 5003.1 shall be in accordance with Sections 5001 and 5003. Retail and wholesale storage and display of nonflammable solid and nonflammable and noncombustible liquid hazardous materials in Group M occupancies and Group S storage shall be in accordance with Section 5003.11.

5004.2 Spill control and secondary containment for liquid and solid hazardous materials. Rooms, buildings or areas used for the storage of liquid or solid hazardous materials shall be provided with spill control and secondary containment in accordance with Sections 5004.2.1 through 5004.2.3.

Exception: Outdoor storage of containers on approved containment pallets in accordance with Section 5004.2.3.

5004.2.1 Spill control for hazardous material liquids. Rooms, buildings or areas used for the storage of hazardous material liquids in individual vessels having a capacity of more than 55 gallons (208 L), or in which the aggregate capacity of multiple vessels exceeds 1,000 gallons (3785 L), shall be provided with spill control to prevent the flow of liquids to adjoining areas. Floors in indoor locations and similar surfaces in outdoor locations shall be constructed to contain a spill from the largest single vessel by one of the following methods:

1. Liquid-tight sloped or recessed floors in indoor locations or similar areas in outdoor locations.
2. Liquid-tight floors in indoor locations or similar areas in outdoor locations provided with liquid-tight raised or recessed sills or dikes.
3. Sumps and collection systems.
4. Other *approved* engineered systems.

Except for surfacing, the floors, sills, dikes, sumps and collection systems shall be constructed of noncombustible material, and the liquid-tight seal shall be compatible with the material stored. Where liquid-tight sills or dikes are provided, they are not required at perimeter openings having an open-grate trench across the opening that connects to an approved collection system.

5004.2.2 Secondary containment for hazardous material liquids and solids. Where required by Table 5004.2.2 buildings, rooms or areas used for the storage of hazardous materials liquids or solids shall be provided with secondary containment in accordance with this section where the capacity of an individual vessel or the aggregate capacity of multiple vessels exceeds both of the following:

1. Liquids: Capacity of an individual vessel exceeds 55 gallons (208 L) or the aggregate capacity of multiple vessels exceeds 1,000 gallons (3785 L).
2. Solids: Capacity of an individual vessel exceeds 550 pounds (250 kg) or the aggregate capacity of multiple vessels exceeds 10,000 pounds (4540 kg).

5004.2.2.1 Containment and drainage methods. The building, room or area shall contain or drain the hazardous materials and fire protection water through the use of one of the following methods:

1. Liquid-tight sloped or recessed floors in indoor locations or similar areas in outdoor locations.
2. Liquid-tight floors in indoor locations or similar areas in outdoor locations provided with liquid-tight raised or recessed sills or dikes.
3. Sumps and collection systems.
4. Drainage systems leading to an *approved* location.
5. Other *approved* engineered systems.

TABLE 5004.2.2
REQUIRED SECONDARY CONTAINMENT—HAZARDOUS MATERIAL SOLIDS AND LIQUIDS STORAGE

<table>
<tr><th colspan="2" rowspan="2">MATERIAL</th><th colspan="2">INDOOR STORAGE</th><th colspan="2">OUTDOOR STORAGE</th></tr>
<tr><th>Solids</th><th>Liquids</th><th>Solids</th><th>Liquids</th></tr>
<tr><td colspan="6">**1. Physical-hazard materials**</td></tr>
<tr><td rowspan="3">Combustible liquids</td><td>Class II</td><td rowspan="4">Not Applicable</td><td>See Chapter 57</td><td rowspan="4">Not Applicable</td><td>See Chapter 57</td></tr>
<tr><td>Class IIIA</td><td>See Chapter 57</td><td>See Chapter 57</td></tr>
<tr><td>Class IIIB</td><td>See Chapter 57</td><td>See Chapter 57</td></tr>
<tr><td colspan="2">Cryogenic fluids</td><td>See Chapter 55</td><td>See Chapter 55</td></tr>
<tr><td colspan="2">Explosives</td><td colspan="2">See Chapter 56</td><td colspan="2">See Chapter 56</td></tr>
<tr><td rowspan="3">Flammable liquids</td><td>Class IA</td><td rowspan="3">Not Applicable</td><td>See Chapter 57</td><td rowspan="3">Not Applicable</td><td>See Chapter 57</td></tr>
<tr><td>Class IB</td><td>See Chapter 57</td><td>See Chapter 57</td></tr>
<tr><td>Class IC</td><td>See Chapter 57</td><td>See Chapter 57</td></tr>
<tr><td colspan="2">Flammable solids</td><td>Not Required</td><td>Not Applicable</td><td>Not Required</td><td>Not Applicable</td></tr>
<tr><td rowspan="6">Organic peroxides</td><td>Unclassified Detonable</td><td rowspan="5">Required</td><td rowspan="5">Required</td><td rowspan="5">Not Required</td><td rowspan="5">Not Required</td></tr>
<tr><td>Class I</td></tr>
<tr><td>Class II</td></tr>
<tr><td>Class III</td></tr>
<tr><td>Class IV</td></tr>
<tr><td>Class V</td><td>Not Required</td><td>Not Required</td><td>Not Required</td><td>Not Required</td></tr>
<tr><td rowspan="4">Oxidizers</td><td>Class 4</td><td rowspan="3">Required</td><td rowspan="3">Required</td><td rowspan="3">Not Required</td><td rowspan="3">Not Required</td></tr>
<tr><td>Class 3</td></tr>
<tr><td>Class 2</td></tr>
<tr><td>Class 1</td><td>Not Required</td><td>Not Required</td><td>Not Required</td><td>Not Required</td></tr>
<tr><td colspan="2">Pyrophorics</td><td>Not Required</td><td>Required</td><td>Not Required</td><td>Required</td></tr>
<tr><td rowspan="4">Unstable (reactives)</td><td>Class 4</td><td rowspan="3">Required</td><td rowspan="3">Required</td><td rowspan="3">Required</td><td rowspan="3">Required</td></tr>
<tr><td>Class 3</td></tr>
<tr><td>Class 2</td></tr>
<tr><td>Class 1</td><td>Not Required</td><td>Not Required</td><td>Not Required</td><td>Not Required</td></tr>
<tr><td rowspan="3">Water reactives</td><td>Class 3</td><td rowspan="2">Required</td><td rowspan="2">Required</td><td rowspan="2">Required</td><td rowspan="2">Required</td></tr>
<tr><td>Class 2</td></tr>
<tr><td>Class 1</td><td>Not Required</td><td>Not Required</td><td>Not Required</td><td>Not Required</td></tr>
<tr><td colspan="6">**2. Health-hazard materials**</td></tr>
<tr><td colspan="2">Corrosives</td><td>Not Required</td><td>Required</td><td>Not Required</td><td>Required</td></tr>
<tr><td colspan="2">Highly toxics</td><td rowspan="2">Required</td><td rowspan="2">Required</td><td rowspan="2">Required</td><td rowspan="2">Required</td></tr>
<tr><td colspan="2">Toxics</td></tr>
</table>

5004.2.2.2 Incompatible materials. *Incompatible materials* used in *open systems* shall be separated from each other in the secondary containment system.

5004.2.2.3 Indoor design. Secondary containment for indoor storage areas shall be designed to contain a spill from the largest vessel plus the design flow volume of fire protection water calculated to discharge from the fire-extinguishing system over the minimum required system design area or area of the room or area in which the storage is located, whichever is smaller. The containment capacity shall be designed to contain the flow for a period of 20 minutes.

5004.2.2.4 Outdoor design. Secondary containment for outdoor storage areas shall be designed to contain a spill from the largest individual vessel. If the area is open to rainfall, secondary containment shall be designed to include the volume of a 24-hour rainfall as determined by a 25-year storm and provisions shall be made to drain accumulations of groundwater and rainwater.

5004.2.2.5 Monitoring. An *approved* monitoring method shall be provided to detect hazardous materials in the secondary containment system. The monitoring method is allowed to be visual inspection of the primary or secondary containment, or other *approved* means. Where secondary containment is subject to the intrusion of water, a monitoring method for detecting water shall be provided. Where monitoring devices are provided, they shall be connected to *approved* visual or audible alarms.

5004.2.2.6 Drainage system design. Drainage systems shall be in accordance with the *International Plumbing Code* and all of the following:

1. The slope of floors to drains in indoor locations, or similar areas in outdoor locations shall be not less than 1 percent.
2. Drains from indoor storage areas shall be sized to carry the volume of the fire protection water as determined by the design density discharged from the automatic fire-extinguishing system over the minimum required system design area or area of the room or area in which the storage is located, whichever is smaller.
3. Drains from outdoor storage areas shall be sized to carry the volume of the fire flow and the volume of a 24-hour rainfall as determined by a 25-year storm.
4. Materials of construction for drainage systems shall be compatible with the materials stored.
5. *Incompatible materials* used in *open systems* shall be separated from each other in the drainage system.
6. Drains shall terminate in an *approved* location away from buildings, valves, *means of egress*, fire access roadways, adjoining property and storm drains.

5004.2.3 Containment pallets. Where used as an alternative to spill control and secondary containment for outdoor storage in accordance with the exception in Section 5004.2, containment pallets shall comply with all of the following:

1. A liquid-tight sump accessible for visual inspection shall be provided.
2. The sump shall be designed to contain not less than 66 gallons (250 L).
3. Exposed surfaces shall be compatible with material stored.
4. Containment pallets shall be protected to prevent collection of rainwater within the sump.

5004.3 Ventilation. Indoor storage areas and storage buildings shall be provided with mechanical exhaust ventilation or natural ventilation where natural ventilation can be shown to be acceptable for the materials as stored.

Exception: Storage areas for flammable solids complying with Chapter 59.

5004.3.1 System requirements. Exhaust ventilation systems shall comply with all of the following:

1. Installation shall be in accordance with the *International Mechanical Code*.
2. Mechanical ventilation shall be at a rate of not less than 1 cubic foot per minute per square foot [0.00508 $m^3/(s \bullet m^2)$] of floor area over the storage area.
3. Systems shall operate continuously unless alternative designs are *approved*.
4. A manual shutoff control shall be provided outside of the room in a position adjacent to the access door to the room or in an *approved* location. The switch shall be a break-glass or other *approved* type and shall be *labeled*: VENTILATION SYSTEM EMERGENCY SHUTOFF.
5. Exhaust ventilation shall be designed to consider the density of the potential fumes or vapors released. For fumes or vapors that are heavier than air, exhaust shall be taken from a point within 12 inches (305 mm) of the floor. For fumes or vapors that are lighter than air, exhaust shall be taken from a point within 12 inches (305 mm) of the highest point of the room.
6. The location of both the exhaust and inlet air openings shall be designed to provide air movement across all portions of the floor or room to prevent the accumulation of vapors.
7. Exhaust air shall not be recirculated to occupied areas if the materials stored are capable of emitting hazardous vapors and contaminants have not been removed. Air contaminated with explosive or flammable vapors, fumes or dusts; flammable, highly toxic or toxic gases; or radioactive materials shall not be recirculated.

5004.4 Separation of incompatible hazardous materials. *Incompatible materials* shall be separated in accordance with Section 5003.9.8.

5004.5 Automatic sprinkler systems. Indoor storage areas and storage buildings shall be equipped throughout with an *approved automatic sprinkler system* in accordance with Section 903.3.1.1. The design of the sprinkler system shall be not less than that required for Ordinary Hazard Group 2 with a minimum design area of 3,000 square feet (279 m^2). Where the materials or storage arrangement are required by other regulations to be provided with a higher level of sprinkler system protection, the higher level of sprinkler system protection shall be provided.

5004.6 Explosion control. Indoor storage rooms, areas and buildings shall be provided with explosion control in accordance with Section 911.

5004.7 Standby or emergency power. Where mechanical ventilation, treatment systems, temperature control, alarm, detection or other electrically operated systems are required, such systems shall be provided with an emergency or standby power in accordance with Section 604.

** For storage areas for highly toxic or toxic materials, see Sections 6004.2.2.8 and 6004.3.4.2.

5004.7.1 Exempt applications. Standby or emergency power is not required for mechanical ventilation systems for any of the following:

1. Storage of Class IB and Class IC flammable and *combustible liquids* in closed containers not exceeding $6^1/_2$ gallons (25 L) capacity.
2. Storage of Class 1 and 2 oxidizers.
3. Storage of Class II, III, IV and V organic peroxides.
4. Storage of asphyxiant, irritant and radioactive gases.

5004.7.2 Fail-safe engineered systems. Standby power for mechanical ventilation, treatment systems and temperature control systems shall not be required where an *approved* fail-safe engineered system is installed..

5004.8 Limit controls. Limit controls shall be provided in accordance with Sections 5004.8.1 and 5004.8.2.

5004.8.1 Temperature control. Materials that must be kept at temperatures other than normal ambient temperatures to prevent a hazardous reaction shall be provided with an approved means to maintain the temperature within a safe range. Redundant temperature control equipment that will operate on failure of the primary temperature control system shall be provided. Where *approved*, alternative means that prevent a hazardous reaction are allowed.

5004.8.2 Pressure control. Stationary tanks and equipment containing hazardous material liquids that can generate pressures exceeding design limits because of exposure fires or internal reaction shall have some form of construction or other *approved* means that will relieve excessive internal pressure. The means of pressure relief shall vent to an *approved* location or to an exhaust scrubber or treatment system where required by Chapter 60.

5004.9 Emergency alarm. An *approved* manual emergency alarm system shall be provided in buildings, rooms or areas used for storage of hazardous materials. Emergency alarm-initiating devices shall be installed outside of each interior *exit* or *exit access* door of storage buildings, rooms or areas. Activation of an emergency alarm-initiating device shall sound a local alarm to alert occupants of an emergency situation involving hazardous materials.

5004.10 Supervision and monitoring. Emergency alarm, detection and automatic fire-extinguishing systems required by Section 5004 shall be electrically supervised and monitored by an *approved* supervising station or, where approved, shall initiate an audible and visual signal at a constantly attended on-site location.

5004.11 Clearance from combustibles. The area surrounding an outdoor storage area or tank shall be kept clear of combustible materials and vegetation for a minimum distance of 25 feet (7620 mm).

5004.12 Noncombustible floor. Except for surfacing, floors of storage areas shall be of noncombustible construction.

5004.13 Weather protection. Where overhead noncombustible construction is provided for sheltering outdoor hazardous material storage areas, such storage shall not be considered indoor storage where the area is constructed in accordance with the requirements for weather protection as required by the *International Building Code.*

Exception: Storage of *explosive* materials shall be considered as indoor storage.

SECTION 5005
USE, DISPENSING AND HANDLING

5005.1 General. Use, dispensing and handling of hazardous materials in amounts exceeding the *maximum allowable quantity per control area* set forth in Section 5003.1 shall be in accordance with Sections 5001, 5003 and 5005. Use, dispensing and handling of hazardous materials in amounts not exceeding the *maximum allowable quantity per control area* set forth in Section 5003.1 shall be in accordance with Sections 5001 and 5003.

5005.1.1 Separation of incompatible materials. Separation of *incompatible materials* shall be in accordance with Section 5003.9.8.

5005.1.2 Noncombustible floor. Except for surfacing, floors of areas where liquid or solid hazardous materials are dispensed or used in *open systems* shall be of noncombustible, liquid-tight construction.

5005.1.3 Spill control and secondary containment for hazardous material liquids. Where required by other provisions of Section 5005, spill control and secondary containment shall be provided for hazardous material liquids in accordance with Section 5004.2.

5005.1.4 Limit controls. Limit controls shall be provided in accordance with Sections 5005.1.4.1 through 5005.1.4.4.

5005.1.4.1 High-liquid-level control. Open tanks in which liquid hazardous materials are used shall be equipped with a liquid-level limit control or other means to prevent overfilling of the tank.

5005.1.4.2 Low-liquid-level control. *Approved* safeguards shall be provided to prevent a low-liquid level in a tank from creating a hazardous condition, including but not limited to, overheating of a tank or its contents.

5005.1.4.3 Temperature control. Temperature control shall be provided in accordance with Section 5004.8.1.

5005.1.4.4 Pressure control. Pressure control shall be provided in accordance with Section 5004.8.2.

5005.1.5 Standby or emergency power. Where mechanical ventilation, treatment systems, temperature control, manual alarm, detection or other electrically operated systems are required by this code, such systems shall be provided with emergency or standby power in accordance with Section 604.

5005.1.5.1 Exempt applications. Standby power for mechanical ventilation, treatment systems and temperature control systems shall not be required where an *approved* fail-safe engineered system is installed.

5005.1.6 Supervision and monitoring. Manual alarm, detection and automatic fire-extinguishing systems required by other provisions of Section 5005 shall be electrically supervised and monitored by an *approved* supervisory service or, where *approved*, shall initiate an audible and visual signal at a constantly attended on-site location.

5005.1.7 Lighting. Adequate lighting by natural or artificial means shall be provided.

5005.1.8 Fire-extinguishing systems. Indoor rooms or areas in which hazardous materials are dispensed or used shall be protected by an automatic fire-extinguishing system in accordance with Chapter 9. Sprinkler system design shall be not less than that required for Ordinary Hazard, Group 2, with a minimum design area of 3,000 square feet (279 m^2). Where the materials or storage arrangement are required by other regulations to be provided with a higher level of sprinkler system protection, the higher level of sprinkler system protection shall be provided.

5005.1.9 Ventilation. Indoor dispensing and use areas shall be provided with exhaust ventilation in accordance with Section 5004.3.

Exception: Ventilation is not required for dispensing and use of flammable solids other than finely divided particles.

5005.1.10 Liquid transfer. Liquids having a hazard ranking of 3 or 4 in accordance with NFPA 704 shall be transferred by one of the following methods:

1. From safety cans complying with UL 30.
2. Through an *approved* closed piping system.
3. From containers or tanks by an approved pump taking suction through an opening in the top of the container or tank.
4. From containers or tanks by gravity through an approved self-closing or automatic-closing valve where the container or tank and dispensing operations are provided with spill control and secondary containment in accordance with Section 5004.2. Highly toxic liquids shall not be dispensed by gravity from tanks.
5. *Approved* engineered liquid transfer systems.

Exceptions:

1. Liquids having a hazard ranking of 4 where dispensed from approved containers not exceeding 1.3 gallons (5 L).
2. Liquids having a hazard ranking of 3 where dispensed from approved containers not exceeding 5.3 gallons (20 L).

5005.1.11 Design. Systems shall be suitable for the use intended and shall be designed by persons competent in such design. Controls shall be designed to prevent materials from entering or leaving the process or reaction system at other than the intended time, rate or path. Where failure of an automatic control could result in a dangerous condition or reaction, the automatic control shall be fail-safe.

5005.2 Indoor dispensing and use. Indoor dispensing and use of hazardous materials shall be in buildings complying with the *International Building Code* and in accordance with Section 5005.1 and Sections 5005.2.1 through 5005.2.2.4.

5005.2.1 Open systems. Dispensing and use of hazardous materials in open containers or systems shall be in accordance with Sections 5005.2.1.1 through 5005.2.1.4.

5005.2.1.1 Ventilation. Where gases, liquids or solids having a hazard ranking of 3 or 4 in accordance with NFPA 704 are dispensed or used, mechanical exhaust ventilation shall be provided to capture gases, fumes, mists or vapors at the point of generation.

Exception: Gases, liquids or solids that can be demonstrated not to create harmful gases, fumes, mists or vapors.

5005.2.1.2 Explosion control. Explosion control shall be provided in accordance with Section 5004.6 where an explosive environment can occur because of the characteristics or nature of the hazardous materials dispensed or used, or as a result of the dispensing or use process.

5005.2.1.3 Spill control for hazardous material liquids. Buildings, rooms or areas where hazardous material liquids are dispensed into vessels exceeding a 1.3-gallon (5 L) capacity or used in *open systems* exceeding a 5.3-gallon (20 L) capacity shall be provided with spill control in accordance with Section 5004.2.1.

5005.2.1.4 Secondary containment for hazardous material liquids. Where required by Table 5005.2.1.4, buildings, rooms or areas where hazardous material liquids are dispensed or used in *open systems* shall be provided with secondary containment in accordance with Section 5004.2.2 where the capacity of an individual vessel or system or the capacity of multiple vessels or systems exceeds the following:

1. Individual vessel or system: greater than 1.3 gallons (5 L).
2. Multiple vessels or systems: greater than 5.3 gallons (20 L).

5005.2.2 Closed systems. Use of hazardous materials in closed containers or systems shall be in accordance with Sections 5005.2.2.1 through 5005.2.2.4.

5005.2.2.1 Ventilation. Where *closed systems* are designed to be opened as part of normal operations, ventilation shall be provided in accordance with Section 5005.2.1.1.

5005.2.2.2 Explosion control. Explosion control shall be provided in accordance with Section 5004.6 where an explosive environment exists because of the hazardous materials dispensed or used, or as a result of the dispensing or use process.

Exception: Where process vessels are designed to contain fully the worst-case explosion anticipated within the vessel under process conditions based on the most likely failure.

5005.2.2.3 Spill control for hazardous material liquids. Buildings, rooms or areas where hazardous material liquids are used in individual vessels exceeding a 55-gallon (208 L) capacity shall be provided with spill control in accordance with Section 5004.2.1.

TABLE 5005.2.1.4
REQUIRED SECONDARY CONTAINMENT—HAZARDOUS MATERIAL LIQUIDS USE

MATERIAL		INDOOR LIQUIDS USE	OUTDOOR LIQUIDS USE
1. Physical-hazard materials			
Combustible liquids	Class II	See Chapter 57	See Chapter 57
	Class IIIA	See Chapter 57	See Chapter 57
	Class IIIB	See Chapter 57	See Chapter 57
Cryogenic fluids		See Chapter 55	See Chapter 55
Explosives		See Chapter 56	See Chapter 56
Flammable liquids	Class IA	See Chapter 57	See Chapter 57
	Class IB	See Chapter 57	See Chapter 57
	Class IC	See Chapter 57	See Chapter 57
Flammable solids		Not Applicable	Not Applicable
Organic peroxides	Unclassified Detonable	Required	Required
	Class I	Required	Required
	Class II		
	Class III		
	Class IV		
	Class V	Not Required	Not Required
Oxidizers	Class 4	Required	Required
	Class 3		
	Class 2		
	Class 1		
Pyrophorics		Required	Required
Unstable (reactives)	Class 4	Required	Required
	Class 3		
	Class 2		
	Class 1	Not Required	Required
Water reactives	Class 3	Required	Required
	Class 2		
	Class 1	Not Required	Required
2. Health-hazard materials			
Corrosives		Required	Required
Highly toxics			
Toxics			

5005.2.2.4 Secondary containment for hazardous material liquids. Where required by Table 5005.2.1.4, buildings, rooms or areas where hazardous material liquids are used in vessels or systems shall be provided with secondary containment in accordance with Section 5004.2.2 where the capacity of an individual vessel or system or the capacity of multiple vessels or systems exceeds the following:

1. Individual vessel or system: greater than 55 gallons (208 L).
2. Multiple vessels or systems: greater than 1,000 gallons (3785 L).

5005.3 Outdoor dispensing and use. Dispensing and use of hazardous materials outdoors shall be in accordance with Sections 5005.3.1 through 5005.3.9.

5005.3.1 Quantities exceeding the maximum allowable quantity per control area. Outdoor dispensing or use of hazardous materials, in either closed or open containers or systems, in amounts exceeding the *maximum allowable quantity per control area* indicated in Tables 5003.1.1(3) and 5003.1.1(4) shall be in accordance with Sections 5001, 5003, 5005.1 and 5005.3.

5005.3.2 Quantities not exceeding the maximum allowable quantity per control area. Outdoor dispensing or use of hazardous materials, in either closed or open containers or systems, in amounts not exceeding the *maximum allowable quantity per control area* indicated in Tables 5003.1.1(3) and 5003.1.1(4) shall be in accordance with Sections 5001 and 5003.

5005.3.3 Location. Outdoor dispensing and use areas for hazardous materials shall be located as required for outdoor storage in accordance with Section 5004.

5005.3.4 Spill control for hazardous material liquids in open systems. Outdoor areas where hazardous material liquids are dispensed in vessels exceeding a 1.3-gallon (5 L) capacity or used in *open systems* exceeding a 5.3-gallon (20 L) capacity shall be provided with spill control in accordance with Section 5004.2.1.

5005.3.5 Secondary containment for hazardous material liquids in open systems. Where required by Table 5005.2.1.4, outdoor areas where hazardous material liquids are dispensed or used in *open systems* shall be provided with secondary containment in accordance with Section 5004.2.2 where the capacity of an individual vessel or system or the capacity of multiple vessels or systems exceeds the following:

1. Individual vessel or system: greater than 1.3 gallons (5 L).
2. Multiple vessels or systems: greater than 5.3 gallons (20 L).

5005.3.6 Spill control for hazardous material liquids in closed systems. Outdoor areas where hazardous material liquids are used in *closed systems* exceeding 55 gallons (208 L) shall be provided with spill control in accordance with Section 5004.2.1.

5005.3.7 Secondary containment for hazardous material liquids in closed systems. Where required by Table 5005.2.1.4, outdoor areas where hazardous material liquids are dispensed or used in *closed systems* shall be provided with secondary containment in accordance with Section 5004.2.2 where the capacity of an individual vessel or system or the capacity of multiple vessels or systems exceeds the following:

1. Individual vessel or system: greater than 55 gallons (208 L).
2. Multiple vessels or systems: greater than 1,000 gallons (3785 L).

5005.3.8 Clearance from combustibles. The area surrounding an outdoor dispensing or use area shall be kept clear of combustible materials and vegetation for a minimum distance of 30 feet (9144 mm).

5005.3.9 Weather protection. Where overhead noncombustible construction is provided for sheltering outdoor hazardous material use areas, such use shall not be considered indoor use where the area is constructed in accordance with the requirements for weather protection as required in the *International Building Code.*

Exception: Use of *explosive* materials shall be considered as indoor use.

5005.4 Handling. Handling of hazardous materials shall be in accordance with Sections 5005.4.1 through 5005.4.4.

5005.4.1 Quantities exceeding the maximum allowable quantity per control area. Handling of hazardous materials in indoor and outdoor locations in amounts exceeding the *maximum allowable quantity per control area* indicated in Tables 5003.1.1(1) through 5003.1.1(4) shall be in accordance with Sections 5001, 5003, 5005.1 and 5005.4.

5005.4.2 Quantities not exceeding the maximum allowable quantity per control area. Handling of hazardous materials in indoor locations in amounts not exceeding the *maximum allowable quantity per control area* indicated in Tables 5003.1.1(1) and 5003.1.1(2) shall be in accordance with Sections 5001, 5003 and 5005.1. Handling of hazardous materials in outdoor locations in amounts not exceeding the *maximum allowable quantity per control area* indicated in Tables 5003.1.1(3) and 5003.1.1(4) shall be in accordance with Sections 5001 and 5003.

5005.4.3 Location. Outdoor handling areas for hazardous materials shall be located as required for outdoor storage in accordance with Section 5004.

5005.4.4 Dispensing, use and handling. Where hazardous materials having a hazard ranking of 3 or 4 in accordance with NFPA 704 are transported through *corridors*, interior *exit stairways* or *ramps* or *exit passageways*, there shall be an emergency telephone system, a local manual alarm station or an approved alarm-initiating device at not more than 150-foot (45 720 mm) intervals and at each *exit* and *exit access* doorway throughout the transport route. The signal shall be relayed to an approved central, proprietary or remote station service or constantly attended onsite location and shall also initiate a local audible alarm.

CHAPTER 51

AEROSOLS

SECTION 5101 GENERAL

5101.1 Scope. The provisions of this chapter, the *International Building Code* and NFPA 30B shall apply to the manufacturing, storage and display of aerosol products. Manufacturing of aerosol products using hazardous materials shall also comply with Chapter 50.

5101.2 Permit required. Permits shall be required as set forth in Section 105.6.

5101.3 Material Safety Data Sheets. Material Safety Data Sheet (MSDS) information for aerosol products displayed shall be kept on the premises at an *approved* location.

5101.4 Containers. Metal aerosol containers shall be limited to a maximum size of 33.8 fluid ounces (1000 ml). Plastic aerosol containers shall be limited to a maximum 4 fluid ounces (118 ml) except as provided in Section 5104.1.1. Glass aerosol containers shall be limited to a maximum 4 fluid ounces (118 ml).

SECTION 5102 DEFINITIONS

5102.1 Definitions. The following terms are defined in Chapter 2:

AEROSOL.

Level 1 aerosol products.

Level 2 aerosol products.

Level 3 aerosol products.

AEROSOL CONTAINER.

AEROSOL WAREHOUSE.

PROPELLANT.

RETAIL DISPLAY AREA.

SECTION 5103 CLASSIFICATION OF AEROSOL PRODUCTS

5103.1 Classification levels. Aerosol products shall be classified as Level 1, 2 or 3 in accordance with Table 5103.1 and NFPA 30B. Aerosol products in cartons that are not identified in accordance with this section shall be classified as Level 3.

TABLE 5103.1
CLASSIFICATION OF AEROSOL PRODUCTS

CHEMICAL HEAT OF COMBUSTION		AEROSOL CLASSIFICATION
Greater than (Btu/lb)	Less than or equal to (Btu/lb)	
0	8,600	1
8,600	13,000	2
13,000	—	3

For SI: 1 British thermal unit per pound = 0.002326 kJ/g.

5103.2 Identification. Cartons shall be identified on not less than one side with the classification level of the aerosol products contained within the carton as follows:

LEVEL ________ AEROSOLS

SECTION 5104 INSIDE STORAGE OF AEROSOL PRODUCTS

5104.1 General. The inside storage of Level 2 and 3 aerosol products shall comply with Sections 5104.2 through 5104.7 and NFPA 30B. Level 1 aerosol products and those aerosol products covered by Section 5104.1.1 shall be considered equivalent to a Class III commodity and shall comply with the requirements for palletized or rack storage in NFPA 13.

5104.1.1 Plastic containers. Aerosol products in plastic containers larger than 4 fluid ounces (118 ml), but not to exceed 33.8 fluid ounces (1000 ml), shall be allowed only where in accordance with this section. The commodity classification shall be Class III commodities, as defined in NFPA 13 where any of the following conditions are met:

1. Base product has no fire point where tested in accordance with ASTM D 92, and nonflammable propellant.
2. Base product has no sustained combustion as tested in accordance with Appendix H, "Method of Testing for Sustained Combustibility," in DOTn 49 CFR Part 173, and nonflammable propellant.
3. Base product contains up to 20 percent by volume (15.8 percent by weight) of ethanol and/or isopropyl alcohol in an aqueous mix, and nonflammable propellant.
4. Base product contains 4 percent by weight or less of an emulsified flammable liquefied gas propellant within an aqueous base. The propellant shall remain emulsified for the life of the product. Where such propellant is not permanently emulsified, the propellant shall be nonflammable.

5104.2 Storage in Groups A, B, E, F, I and R. Storage of Level 2 and 3 aerosol products in occupancies in Groups A, B, E, F, I and R shall be limited to the following maximum quantities:

1. A net weight of 1,000 pounds (454 kg) of Level 2 aerosol products.
2. A net weight of 500 pounds (227 kg) of Level 3 aerosol products.
3. A combined net weight of 1,000 pounds (454 kg) of Level 2 and 3 aerosol products.

The maximum quantity shall be increased 100 percent where the excess quantity is stored in storage cabinets in accordance with Section 5704.3.2.

5104.2.1 Excess storage. Storage of quantities exceeding the maximum quantities indicated in Section 5104.2 shall be stored in separate inside flammable liquid storage rooms in accordance with Section 5104.5.

5104.3 Storage in general purpose warehouses. Aerosol storage in general purpose warehouses utilized only for warehousing-type operations involving mixed commodities shall comply with Section 5104.3.1 or 5104.3.2.

5104.3.1 Nonsegregated storage. Storage consisting of solid pile, palletized or rack storage of Level 2 and 3 aerosol products not segregated into areas utilized exclusively for the storage of aerosols shall comply with Table 5104.3.1.

TABLE 5104.3.1
NONSEGREGATED STORAGE OF LEVEL 2 AND 3 AEROSOL PRODUCTS IN GENERAL PURPOSE WAREHOUSES[b]

AEROSOL LEVEL	MAXIMUM NET WEIGHT PER FLOOR (pounds)[b]			
	Palletized or solid-pile storage		Rack storage	
	Unprotected	Protected[a]	Unprotected	Protected[a]
2	2,500	12,000	2,500	24,000
3	1,000	12,000	1,000	24,000
Combination 2 and 3	2,500	12,000	2,500	24,000

For SI: 1 foot = 304.8 mm, 1 pound = 0.454 kg, 1 square foot = 0.0929 m^2.

a. Approved automatic sprinkler system protection and storage arrangements shall comply with NFPA 30B. Sprinkler system protection shall extend 20 feet beyond the storage area containing the aerosol products.

b. Storage quantities indicated are the maximum permitted in any 50,000-square-foot area.

5104.3.2 Segregated storage. Storage of Level 2 and 3 aerosol products segregated into areas utilized exclusively for the storage of aerosols shall comply with Table 5104.3.2 and Sections 5104.3.2.1 and 5104.3.2.2.

TABLE 5104.3.2
SEGREGATED STORAGE OF LEVEL 2 AND 3 AEROSOL PRODUCTS IN GENERAL PURPOSE WAREHOUSES

STORAGE SEPARATION	MAXIMUM SEGREGATED STORAGE AREA[a]		SPRINKLER REQUIREMENTS
	Percentage of building area (percent)	Area limitation (square feet)	
Separation area[e, f]	15	20,000	Notes b, c
Chain-link fence enclosure[d]	20	20,000	Notes b, c
1-hour fire-resistance-rated interior walls	20	30,000	Note b
2-hour fire-resistance-rated interior walls	25	40,000	Note b
3-hour fire-resistance-rated interior walls	30	50,000	Note b

For SI: 1 foot = 304.8 mm, 1 square foot = 0.0929 m^2.

a. The maximum segregated storage area shall be limited to the smaller of the two areas resulting from the percentage of building area limitation and the area limitation.

b. Automatic sprinkler system protection in aerosol product storage areas shall comply with NFPA 30B and be approved. Building areas not containing aerosol product storage shall be equipped throughout with an approved automatic sprinkler system in accordance with Section 903.3.1.1.

c. Automatic sprinkler system protection in aerosol product storage areas shall comply with NFPA 30B and be approved. Sprinkler system protection shall extend a minimum 20 feet beyond the aerosol storage area.

d. Chain-link fence enclosures shall comply with Section 5104.3.2.1.

e. A separation area shall be defined as an area extending outward from the periphery of the segregated aerosol product storage area as follows.

1. The limits of the aerosol product storage shall be clearly marked on the floor.
2. The separation distance shall be not less than 25 feet and maintained clear of all materials with a commodity classification greater than Class III in accordance with Section 903.3.1.1.

f. Separation areas shall only be permitted where approved.

5104.3.2.1 Chain-link fence enclosures. Chain-link fence enclosures required by Table 5104.3.2 shall comply with the following:

1. The fence shall not be less than No. 9 gage steel wire, woven into a maximum 2-inch (51 mm) diamond mesh.
2. The fence shall be installed from the floor to the underside of the roof or ceiling above.
3. Class IV and high-hazard commodities shall be stored outside of the aerosol storage area and not less than 8 feet (2438 mm) from the fence.
4. Access openings in the fence shall be provided with either self- or automatic-closing devices or a labyrinth opening arrangement preventing aerosol containers from rocketing through the access openings.
5. Not less than two *means of egress* shall be provided from the fenced enclosure.

5104.3.2.2 Aisles. The minimum aisle requirements for segregated storage in general purpose warehouses shall comply with Table 5104.3.2.2.

TABLE 5104.3.2.2
SEGREGATED STORAGE AISLE WIDTHS AND DISTANCE TO AISLES IN GENERAL PURPOSE WAREHOUSES

STORAGE CONDITION	MINIMUM AISLE WIDTH (feet)	MAXIMUM DISTANCE FROM STORAGE TO AISLE (feet)
Solid pile or palletized[a]	4 feet between piles	25
Racks with ESFR sprinklers[a]	4 feet between racks and adjacent Level 2 and 3 aerosol product storage	25
Racks without ESFR sprinklers[a]	8 feet between racks and adjacent Level 2 and 3 aerosol product storage	25

For SI: 1 foot = 304.8 mm.

a. Sprinklers shall comply with NFPA 30B.

5104.4 Storage in aerosol warehouses. The total quantity of Level 2 and 3 aerosol products in a warehouse utilized for the storage, shipping and receiving of aerosol products shall not be restricted in structures complying with Sections 5104.4.1 through 5104.4.4.

5104.4.1 Automatic sprinkler system. Aerosol warehouses shall be protected by an *approved* wet-pipe *automatic sprinkler system* in accordance with NFPA 30B. Sprinkler protection shall be designed based on the highest classification level of aerosol product present.

5104.4.2 Pile and palletized storage aisles. Solid pile and palletized storage shall be arranged so the maximum travel distance to an aisle is 25 feet (7620 mm). Aisles shall have a minimum width of 4 feet (1219 mm).

5104.4.3 Rack storage aisles. Rack storage shall be arranged with a minimum aisle width of 8 feet (2438 mm) between rows of racks and 8 feet (2438 mm) between racks and adjacent solid pile or palletized storage. Where early suppression fast-response (ESFR) sprinklers provide automatic sprinkler protection, the minimum aisle width shall be 4 feet (1219 mm).

5104.4.4 Combustible commodities. Combustible commodities other than flammable and *combustible liquids* shall be permitted to be stored in an aerosol warehouse.

Exception: Flammable and *combustible liquids* in 1-quart (946 mL) metal containers and smaller shall be permitted to be stored in an aerosol warehouse.

5104.5 Storage in inside flammable liquid storage rooms. Inside flammable liquid storage rooms shall comply with Section 5704.3.7. The maximum quantities of aerosol products shall comply with Section 5104.5.1 or 5104.5.2.

5104.5.1 Storage rooms of 500 square feet or less. The storage of aerosol products in flammable liquid storage rooms less than or equal to 500 square feet (46 m^2) in area shall not exceed the following quantities:

1. A net weight of 1,000 pounds (454 kg) of Level 2 aerosol products.
2. A net weight of 500 pounds (227 kg) of Level 3 aerosol products.
3. A combined net weight of 1,000 pounds (454 kg) of Level 2 and 3 aerosol products.

5104.5.2 Storage rooms greater than 500 square feet. The storage of aerosol products in flammable liquid storage rooms greater than 500 square feet (46 m^2) in area shall not exceed the following quantities:

1. A net weight of 2,500 pounds (1135 kg) of Level 2 aerosol products.
2. A net weight of 1,000 pounds (454 kg) of Level 3 aerosol products.
3. A combined net weight of 2,500 pounds (1135 kg) of Level 2 and 3 aerosol products.

The maximum aggregate storage quantity of Level 2 and 3 aerosol products permitted in separate inside storage rooms protected by an *approved automatic sprinkler system* in accordance with NFPA 30B shall be 5,000 pounds (2270 kg).

5104.6 Storage in liquid warehouses. The storage of Level 2 and 3 aerosol products in liquid warehouses shall comply with NFPA 30B. The storage shall be located within segregated storage areas in accordance with Section 5104.3.2 and Sections 5104.6.1 through 5104.6.3.

5104.6.1 Containment. Spill control or drainage shall be provided to prevent the flow of liquid to within 8 feet (2438 mm) of the segregated storage area.

5104.6.2 Sprinkler design. Sprinkler protection shall be designed based on the highest level of aerosol product present.

5104.6.3 Opening protection into segregated storage areas. Fire doors or gates opening into the segregated storage area shall either be self-closing or provided with automatic-closing devices activated by sprinkler water flow or an *approved* fire detection system.

5104.7 Storage in Group M occupancies. Storage of Level 2 and 3 aerosol products in occupancies in Group M shall comply with Table 5104.7. Retail display shall comply with Section 5106.

TABLE 5104.7
MAXIMUM QUANTITIES OF LEVEL 2 AND 3 AEROSOL PRODUCTS IN RETAIL STORAGE AREAS

MAXIMUM NET WEIGHT PER FLOOR (pounds)			
Floor	Nonsegregated storage[a, b]	Segregated storage	
		Storage cabinets[b]	Separated from retail area[c]
Basement	Not Permitted	Not Permitted	Not Permitted
Ground floor	2,500	5,000	Note d
Upper floors	500	1,000	Note d

For SI: 1 pound = 0.454 kg, 1 square foot = 0.0929 m^2.

a. The total aggregate quantity on display and in storage shall not exceed the maximum retail display quantity indicated in Section 5106.3.

b. Storage quantities indicated are the maximum permitted in any 50,000-square-foot area.

c. The storage area shall be separated from the retail area with a 1-hour fire-resistance-rated assembly.

d. See Table 5104.3.2.

SECTION 5105
OUTSIDE STORAGE

5105.1 General. The outside storage of Level 2 and 3 aerosol products, including storage in temporary storage trailers, shall be separated from exposures in accordance with Table 5105.1.

TABLE 5105.1
DISTANCE TO EXPOSURES FOR OUTSIDE STORAGE OF LEVEL 2 AND 3 AEROSOL PRODUCTS

EXPOSURE	MINIMUM DISTANCE FROM AEROSOL STORAGE (feet)[a]
Public alleys, public ways, public streets	20
Buildings	50
Exit discharge to a public way	50
Lot lines	20
Other outside storage	50

For SI: 1 inch = 25.4 mm, 1 foot = 304.8 mm.

a. The minimum separation distance indicated is not required where exterior walls having a 2-hour fire-resistance rating without penetrations separate the storage from the exposure. The walls shall extend not less than 30 inches above and to the sides of Level 2 and 3 aerosol products.

SECTION 5106
RETAIL DISPLAY

5106.1 General. This section shall apply to the retail display of 500 pounds (227 kg) or more of Level 2 and 3 aerosol products.

5106.2 Aerosol display and normal merchandising not exceeding 8 feet (2438 mm) high. Aerosol display and normal merchandising not exceeding 8 feet (2438 mm) in height shall be in accordance with Sections 5106.2.1 through 5106.2.4.

5106.2.1 Maximum quantities in retail display areas. Aerosol products in retail display areas shall not exceed quantities needed for display and normal merchandising and shall not exceed the quantities in Table 5106.2.1.

TABLE 5106.2.1
MAXIMUM QUANTITIES OF LEVEL 2 AND 3 AEROSOL PRODUCTS IN RETAIL DISPLAY AREAS

MAXIMUM NET WEIGHT PER FLOOR (pounds)[b]			
Floor	Unprotected[a]	Protected in accordance with Section 5106.2[a, c]	Protected in accordance with Section 5106.3[c]
Basement	Not allowed	500	500
Ground	2,500	10,000	10,000
Upper	500	2,000	Not allowed

For SI: 1 pound = 0.454 kg, 1 square foot = 0.0929 m^2.

a. The total quantity shall not exceed 1,000 pounds net weight in any one 100-square-foot retail display area.

b. Per 25,000-square-foot retail display area.

c. Minimum Ordinary Hazard Group 2 wet-pipe automatic sprinkler system throughout the retail sales occupancy.

5106.2.2 Display of containers. Level 2 and 3 aerosol containers shall not be stacked more than 6 feet (1829 mm) high from the base of the aerosol array to the top of the aerosol array unless the containers are placed on fixed shelving or otherwise secured in an *approved* manner. Where storage or retail display is on shelves, the height of such storage or retail display to the top of aerosol containers shall not exceed 8 feet (2438 mm).

5106.2.3 Combustible cartons. Aerosol products located in retail display areas shall be removed from combustible cartons.

Exceptions:

1. Display areas that use a portion of combustible cartons that consist of only the bottom panel and not more than 2 inches (51 mm) of the side panel are allowed.
2. When the display area is protected in accordance with Tables 6.3.2.7(a) through 6.3.2.7(l) of NFPA 30B, storage of aerosol products in combustible cartons is allowed.

5106.2.4 Retail display automatic sprinkler system. Where an *automatic sprinkler system* is required for the protected retail display of aerosol products, the wet-pipe *automatic sprinkler system* shall be in accordance with Section 903.3.1.1. The minimum system design shall be for an Ordinary Hazard Group 2 occupancy. The system shall be provided throughout the retail display area.

5106.3 Aerosol display and normal merchandising exceeding 8 feet (2438 mm) high. Aerosol display and merchandising exceeding 8 feet in height shall be in accordance with Sections 5106.3.1 through 5106.3.3.

5106.3.1 Maximum quantities in retail display areas. Aerosol products in retail display areas shall not exceed quantities needed for display and normal merchandising and shall not exceed the quantities in Table 5106.2.1, with fire protection in accordance with Section 5106.3.2.

5106.3.2 Automatic sprinkler protection. Aerosol display and merchandising areas shall be protected by an *automatic sprinkler system* based on the requirements set forth in Tables 6.3.2.7(a) through 6.3.2.7(l) of NFPA 30B and the following:

1. Protection shall be based on the highest level of aerosol product in the array and the packaging method of the storage located more than 6 feet (1829 mm) above the finished floor.
2. Where using the cartoned aerosol tables of NFPA 30B, uncartoned or display-cut Level 2 and 3 aerosols shall be permitted not more than 6 feet (1829 mm) above the finished floor.
3. The design area for Level 2 and 3 aerosols shall extend not less than 20 feet (6096 mm) beyond the Level 2 and 3 aerosol display and merchandising areas.
4. Where ordinary and high-temperature ceiling sprinkler systems are adjacent to each other, noncombustible draft curtains shall be installed at the interface.

5106.3.3 Separation of Level 2 and 3 aerosol areas. Separation of Level 2 and 3 aerosol areas shall comply with the following:

1. Level 2 and 3 aerosol display and merchandising areas shall be separated from each other by not less than 25 feet (7620 mm). See Table 5106.2.1.
2. Level 2 and 3 aerosol display and merchandising areas shall be separated from flammable and *combustible liquids* storage and display areas by one or a combination of the following:
   2.1. Segregating areas from each other by horizontal distance of not less than 25 feet (7620 mm).
   2.2. Isolating areas from each other by a noncombustible partition extending not less than 18 inches (457 mm) above the merchandise.
   2.3. In accordance with Section 5106.5.
3. Where Item 2.2 is used to separate Level 2 or 3 aerosols from flammable or *combustible liquids*, and the aerosol products are located within 25 feet (7620 mm) of flammable or *combustible liquids*, the area below the noncombustible partition shall be liquid tight at the floor to prevent spilled liquids from flowing beneath the aerosol products.

5106.4 Maximum quantities in storage areas. Aerosol products in storage areas adjacent to retail display areas shall not exceed the quantities in Table 5106.4.

5106.5 Special protection design for Level 2 and 3 aerosols adjacent to flammable and combustible liquids in double-row racks. The display and merchandising of Level 2 and 3 aerosols adjacent to flammable and *combustible liquids* in double-row racks shall be in accordance with Sections 5106.5.1 through 5106.5.8 or Section 5106.3.3.

5106.5.1 Fire protection. Fire protection for the display and merchandising of Level 2 and 3 aerosols in double-row racks shall be in accordance with Table 7.4.1 and Figure 7.4.1 of NFPA 30B.

5106.5.2 Cartoned products. Level 2 and 3 aerosols displayed or merchandised more than 8 feet (2438 mm) above the finished floor shall be in cartons.

5106.5.3 Shelving. Shelving in racks shall be limited to wire mesh shelving having uniform openings not more than 6 inches (152 mm) apart, with the openings comprising not less than 50 percent of the overall shelf area.

5106.5.4 Aisles. Racks shall be arranged so that aisles not less than $7^1/_2$ feet (2286 mm) wide are maintained between rows of racks and adjacent solid-piled or palletized merchandise.

5106.5.5 Flue spaces. Flue spaces in racks shall comply with the following:

1. Transverse flue spaces—Nominal 3-inch (76 mm) transverse flue spaces shall be maintained between merchandise and rack uprights.
2. Longitudinal flue spaces—Nominal 6-inch (152 mm) longitudinal flue spaces shall be maintained.

5106.5.6 Horizontal barriers. Horizontal barriers constructed of minimum $^3/_8$-inch-thick (10 mm) plywood or minimum 0.034-inch (0.086 mm) (No. 22 gage) sheet metal shall be provided and located in accordance with Table 7.4.1 and Figure 7.4.1 of NFPA 30B where in-rack sprinklers are installed.

5106.5.7 Class I, II, III, IV and plastic commodities. Class I, II, III, IV and plastic commodities located adjacent to Level 2 and 3 aerosols shall be protected in accordance with NFPA 13.

5106.5.8 Flammable and combustible liquids. Class I, II, III A and III B Liquids shall be allowed to be located adjacent to Level 2 and 3 aerosol products where both of the following conditions are met:

1. Class I, II, IIIA and IIIB liquid containers: Containers for Class I, II, IIIA and IIIB liquids shall be limited to 1.06-gallon (4 L) metal-relieving and nonrelieving style containers and 5.3-gallon (20 L) metal-relieving style containers.
2. Fire protection for Class I, II, IIIA and IIIB Liquids: Automatic sprinkler protection for Class I, II, IIIA and IIIB liquids shall be in accordance with Chapter 57.

SECTION 5107 MANUFACTURING FACILITIES

5107.1 General. Manufacturing facilities shall be in accordance with NFPA 30B.

TABLE 5106.4
MAXIMUM STORAGE QUANTITIES FOR STORAGE AREAS ADJACENT TO RETAIL DISPLAY OF LEVEL 2 AND 3 AEROSOLS

MAXIMUM NET WEIGHT PER FLOOR (pounds)			
Floor	Unseparated[a,b]	Separated	
		Storage Cabinets[b]	1-hour Occupancy Separation
Basement	Not Allowed	Not Allowed	Not Allowed
Ground	2,500	5,000	In accordance with Sections 6.3.4.3 and 6.3.4.4 of NFPA 30B
Upper	500	1,000	In accordance with Sections 6.3.4.3 and 6.3.4.4 of NFPA 30B

For SI: 1 pound = 0.454 kg, 1 square foot = 0.0929 m^2.

a. The aggregate quantity in storage and retail display shall not exceed the quantity limits for retail display.

b. In any 50,000-square-foot area.

CHAPTER 52

RESERVED

CHAPTER 53
COMPRESSED GASES

SECTION 5301
GENERAL

5301.1 Scope. Storage, use and handling of *compressed gases* in *compressed gas* containers, cylinders, tanks and systems shall comply with this chapter and NFPA 55, including those gases regulated elsewhere in this code. Partially full *compressed gas* containers, cylinders or tanks containing residual gases shall be considered as full for the purposes of the controls required.

Liquefied natural gas for use as a vehicular fuel shall also comply with NFPA 52 and NFPA 59A.

Compressed gases classified as hazardous materials shall also comply with Chapter 50 for general requirements and chapters addressing specific hazards, including Chapters 58 (Flammable Gases), 60 (Highly Toxic and Toxic Materials), 63 (Oxidizers, Oxidizing Gases and Oxidizing Cryogenic Fluids) and 64 (Pyrophoric Materials).

Compressed hydrogen (CH_2) for use as a vehicular fuel shall also comply with Chapters 23 and 58 of this code, the *International Fuel Gas Code* and NFPA 2.

Cutting and welding gases shall also comply with Chapter 35.

LP-gas shall also comply with Chapter 61 and the *International Fuel Gas Code.*

Exceptions:

1. Gases used as refrigerants in refrigeration systems (see Section 606).
2. Compressed natural gas (CNG) for use as a vehicular fuel shall comply with Chapter 23, NFPA 52 and the *International Fuel Gas Code.*
3. *Cryogenic fluids* shall comply with Chapter 55.

5301.2 Permits. Permits shall be required as set forth in Section 105.6.

SECTION 5302
DEFINITIONS

5302.1 Definitions. The following terms are defined in Chapter 2:

COMPRESSED GAS.

COMPRESSED GAS CONTAINER.

COMPRESSED GAS SYSTEM.

NESTING.

TUBE TRAILER.

SECTION 5303
GENERAL REQUIREMENTS

5303.1 Containers, cylinders and tanks. *Compressed gas* containers, cylinders and tanks shall comply with this section. *Compressed gas* containers, cylinders or tanks that are not designed for refillable use shall not be refilled after use of the original contents.

5303.2 Design and construction. *Compressed gas* containers, cylinders and tanks shall be designed, fabricated, tested, marked with the specifications of manufacture and maintained in accordance with the regulations of DOTn 49 CFR Parts 100-185 or the ASME *Boiler and Pressure Vessel Code*, Section VIII.

5303.3 Pressure relief devices. Pressure relief devices shall be in accordance with Sections 5303.3.1 through 5303.3.5.

5303.3.1 Where required. Pressure relief devices shall be provided to protect containers, cylinders and tanks containing *compressed gases* from rupture in the event of overpressure.

Exception: Cylinders, containers and tanks where exempt from the requirements for pressure relief devices specified by the standards of design *listed* in Section 5303.3.2.

5303.3.2 Design. Pressure relief devices to protect containers shall be designed and provided in accordance with CGA S-1.1, CGA S-1.2, CGA S-1.3 or the ASME *Boiler and Pressure Vessel Code*, Section VIII, as applicable.

5303.3.3 Sizing. Pressure relief devices shall be sized in accordance with the specifications to which the container was fabricated and to material-specific requirements as applicable.

5303.3.4 Arrangement. Pressure relief devices shall be arranged to discharge upward and unobstructed to the open air in such a manner as to prevent any impingement of escaping gas upon the container, adjacent structures or personnel.

Exception: DOTn specification containers having an internal volume of 30 cubic feet (0.855 m^3) or less.

5303.3.5 Freeze protection. Pressure relief devices or vent piping shall be designed or located so that moisture cannot collect and freeze in a manner that would interfere with the operation of the device.

5303.4 Marking. Stationary and portable *compressed gas* containers, cylinders, tanks and systems shall be marked in accordance with Sections 5303.4.1 through 5303.4.3.

5303.4.1 Stationary compressed gas containers, cylinders and tanks. Stationary *compressed gas* containers, cylinders and tanks shall be marked with the name of the gas and in accordance with Sections 5003.5 and 5003.6. Markings shall be visible from any direction of approach.

5303.4.2 Portable containers, cylinders and tanks. Portable *compressed gas* containers, cylinders and tanks shall be marked in accordance with CGA C-7.

5303.4.3 Piping systems. Piping systems shall be marked in accordance with ASME A13.1. Markings used for piping systems shall consist of the content's name and include a direction-of-flow arrow. Markings shall be provided at each valve; at wall, floor or ceiling penetrations; at each change of direction; and at not less than every 20 feet (6096 mm) or fraction thereof throughout the piping run.

Exceptions:

1. Piping that is designed or intended to carry more than one gas at various times shall have appropriate signs or markings posted at the manifold, along the piping and at each point of use to provide clear identification and warning.
2. Piping within gas manufacturing plants, gas processing plants, refineries and similar occupancies shall be marked in an *approved* manner.

5303.5 Security. *Compressed gas* containers, cylinders, tanks and systems shall be secured against accidental dislodgement and against access by unauthorized personnel in accordance with Sections 5303.5.1 through 5303.5.3.

5303.5.1 Security of areas. Areas used for the storage, use and handling of *compressed gas* containers, cylinders, tanks and systems shall be secured against unauthorized entry and safeguarded in an *approved* manner.

5303.5.2 Physical protection. *Compressed gas* containers, cylinders, tanks and systems that could be exposed to physical damage shall be protected. Guard posts or other *approved* means shall be provided to protect *compressed gas* containers, cylinders, tanks and systems indoors and outdoors from vehicular damage and shall comply with Section 312.

5303.5.3 Securing compressed gas containers, cylinders and tanks. *Compressed gas* containers, cylinders and tanks shall be secured to prevent falling caused by contact, vibration or seismic activity. Securing of *compressed gas* containers, cylinders and tanks shall be by one of the following methods:

1. Securing containers, cylinders and tanks to a fixed object with one or more restraints.
2. Securing containers, cylinders and tanks on a cart or other mobile device designed for the movement of *compressed gas* containers, cylinders or tanks.
3. Nesting of *compressed gas* containers, cylinders and tanks at container filling or servicing facilities or in sellers' warehouses not accessible to the public. Nesting shall be allowed provided the nested containers, cylinders or tanks, if dislodged, do not obstruct the required *means of egress*.
4. Securing of *compressed gas* containers, cylinders and tanks to or within a rack, framework, cabinet or similar assembly designed for such use.

Exception: *Compressed gas* containers, cylinders and tanks in the process of examination, filling, transport or servicing.

5303.6 Valve protection. *Compressed gas* container, cylinder and tank valves shall be protected from physical damage by means of protective caps, collars or similar devices in accordance with Sections 5303.6.1 and 5303.6.2.

5303.6.1 Compressed gas container, cylinder or tank protective caps or collars. *Compressed gas* containers, cylinders and tanks designed for protective caps, collars or other protective devices shall have the caps or devices in place except when the containers, cylinders or tanks are in use or are being serviced or filled.

5303.6.2 Caps and plugs. *Compressed gas* containers, cylinders and tanks designed for valve protection caps or other protective devices shall have the caps or devices in place. When outlet caps or plugs are installed, they shall be in place.

Exception: *Compressed gas* containers, cylinders or tanks in use, being serviced or being filled.

5303.7 Separation from hazardous conditions. *Compressed gas* containers, cylinders and tanks and systems in storage or use shall be separated from materials and conditions that pose exposure hazards to or from each other. *Compressed gas* containers, cylinders, tanks and systems in storage or use shall be separated in accordance with Sections 5303.7.1 through 5303.7.11.2.

5303.7.1 Incompatible materials. *Compressed gas* containers, cylinders and tanks shall be separated from each other based on the hazard class of their contents. *Compressed gas* containers, cylinders and tanks shall be separated from *incompatible materials* in accordance with Section 5003.9.8.

5303.7.2 Combustible waste, vegetation and similar materials. Combustible waste, vegetation and similar materials shall be kept not less than 10 feet (3048 mm) from *compressed gas* containers, cylinders, tanks and systems. A noncombustible partition, without openings or penetrations and extending not less than 18 inches (457 mm) above and to the sides of the storage area is allowed in lieu of such distance. The wall shall either be an independent structure, or the exterior wall of the building adjacent to the storage area.

5303.7.3 Ledges, platforms and elevators. *Compressed gas* containers, cylinders and tanks shall not be placed near elevators, unprotected platform ledges or other areas where falling would result in *compressed gas* containers, cylinders or tanks being allowed to drop distances exceeding one-half the height of the container, cylinder or tank.

5303.7.4 Temperature extremes. *Compressed gas* containers, cylinders and tanks, whether full or partially full, shall not be exposed to artificially created high temperatures exceeding 125°F (52°C) or subambient (low) temperatures unless designed for use under the exposed conditions.

5303.7.5 Falling objects. *Compressed gas* containers, cylinders, tanks and systems shall not be placed in areas where they are capable of being damaged by falling objects.

5303.7.6 Heating. *Compressed gas* containers, cylinders and tanks, whether full or partially full, shall not be heated by devices that could raise the surface temperature of the container, cylinder or tank to above 125°F (52°C). Heating devices shall comply with the *International Mechanical Code* and NFPA 70. *Approved* heating methods involving temperatures of less than 125°F (52°C) are allowed to be used by trained personnel. Devices designed to maintain individual *compressed gas* containers, cylinders or tanks at constant temperature shall be *approved* and shall be designed to be fail-safe.

5303.7.7 Sources of ignition. Open flames and high-temperature devices shall not be used in a manner that creates a hazardous condition.

5303.7.8 Exposure to chemicals. *Compressed gas* containers, cylinders, tanks and systems shall not be exposed to *corrosive* chemicals or fumes that could damage containers, cylinders, tanks, valves or valve-protective caps.

5303.7.9 Exhausted enclosures. Where exhausted enclosures are provided as a means to segregate *compressed gas* containers, cylinders and tanks from exposure hazards, such enclosures shall comply with the requirements of Section 5003.8.5.

5303.7.10 Gas cabinets. Where gas cabinets are provided as a means to separate *compressed gas* containers, cylinders and tanks from exposure hazards, such gas cabinets shall comply with the requirements of Section 5003.8.6.

5303.7.11 Tube trailers. Tube trailers, including those containing compatible *compressed gases,* shall be surrounded by a clear space of not less than 3 feet (914 mm) to allow for maintenance, access and inspection.

5303.7.11.1 Individual tube trailers containing incompatible materials. Increased separation distances between individual tube trailers containing incompatible gases shall be provided where required by Section 5303.7.1.

5303.7.11.2 Connections. Piping systems used to connect tube trailers to a user piping system shall not be viewed as an encroachment into the 3-foot (914 mm) clear space.

5303.8 Wiring and equipment. Electrical wiring and equipment shall comply with NFPA 70. *Compressed gas* containers, cylinders, tanks and systems shall not be located where they could become part of an electrical circuit. *Compressed gas* containers, cylinders, tanks and systems shall not be used for electrical grounding.

5303.9 Service and repair. Service, repair, modification or removal of valves, pressure-relief devices or other *compressed gas* container, cylinder or tank appurtenances shall be performed by trained personnel.

5303.10 Unauthorized use. *Compressed gas* containers, cylinders, tanks and systems shall not be used for any purpose other than to serve as a vessel for containing the product that it is designed to contain.

5303.11 Exposure to fire. *Compressed gas* containers, cylinders and tanks that have been exposed to fire shall be removed from service. Containers, cylinders and tanks so removed shall be handled by *approved*, qualified persons.

5303.12 Leaks, damage or corrosion. Leaking, damaged or corroded *compressed gas* containers, cylinders and tanks shall be removed from service. Leaking, damaged or corroded *compressed gas* systems shall be replaced or repaired in accordance with the following:

1. *Compressed gas* containers, cylinders and tanks that have been removed from service shall be handled in an *approved* manner.
2. *Compressed gas* systems that are determined to be leaking, damaged or corroded shall be repaired to a serviceable condition or removed from service.

5303.13 Surface of unprotected storage or use areas. Unless otherwise specified in Section 5303.14, *compressed gas* containers, cylinders and tanks are allowed to be stored or used without being placed under overhead cover. To prevent bottom corrosion, containers, cylinders and tanks shall be protected from direct contact with soil or unimproved surfaces. The surface of the area on which the containers are placed shall be graded to prevent accumulation of water.

5303.14 Overhead cover. *Compressed gas* containers, cylinders and tanks are allowed to be stored or used in the sun except in locations where extreme temperatures prevail. Where extreme temperatures prevail, overhead covers shall be provided.

5303.15 Lighting. *Approved* lighting by natural or artificial means shall be provided.

5303.16 Vaults. Generation, compression, storage and dispensing equipment for *compressed gases* shall be allowed to be located in either above- or below-grade vaults complying with Sections 5303.16.1 through 5303.16.14.

5303.16.1 Listing required. Vaults shall be *listed* by a nationally recognized testing laboratory.

Exception: Where *approved* by the *fire code official*, below-grade vaults are allowed to be constructed on site, provided that the design is in accordance with the *International Building Code* and that special inspections are conducted to verify structural strength and compliance of the installation with the *approved* design in accordance with Section 1707 of the *International Building Code*. Installation plans for below-grade vaults that are constructed on site shall be prepared by, and the design shall bear the stamp of, a professional engineer. Consideration shall be given to soil and hydrostatic loading on the floors, walls and lid; anticipated seismic forces; uplifting by ground water or flooding; and to loads imposed from above, such as traffic and equipment loading on the vault lid.

5303.16.2 Design and construction. The vault shall completely enclose generation, compression, storage or dispensing equipment located in the vault. There shall not be openings in the vault enclosure except those necessary for vault ventilation and access, inspection, filling, emptying or venting of equipment in the vault. The walls and floor of the vault shall be constructed of reinforced concrete not less than 6 inches (152 mm) thick. The top of an above-grade vault shall be constructed of noncombustible material and shall be designed to be weaker than the walls of the vault to ensure that the thrust of any explosion occurring inside the vault is directed upward.

The top of an at- or below-grade vault shall be designed to relieve safely or contain the force of an explosion occurring inside the vault. The top and floor of the vault and the tank foundation shall be designed to withstand the anticipated loading, including loading from vehicular traffic, where applicable. The walls and floor of a vault installed below grade shall be designed to withstand anticipated soil and hydrostatic loading. Vaults shall be designed to be wind and earthquake resistant, in accordance with the *International Building Code*.

5303.16.3 Secondary containment. Vaults shall be substantially liquid-tight and there shall not be backfill within the vault. The vault floor shall drain to a sump. For premanufactured vaults, liquid tightness shall be certified as part of the listing provided by a nationally recognized testing laboratory. For field-erected vaults, liquid tightness shall be certified in an *approved* manner.

5303.16.4 Internal clearance. There shall be sufficient clearance within the vault to allow for visual inspection and maintenance of equipment in the vault.

5303.16.5 Anchoring. Vaults and equipment contained therein shall be suitably anchored to withstand uplifting by groundwater or flooding. The design shall verify that uplifting is prevented even where equipment within the vault is empty.

5303.16.6 Vehicle impact protection. Vaults shall be resistant to damage from the impact of a motor vehicle, or vehicle impact protection shall be provided in accordance with Section 312.

5303.16.7 Arrangement. Equipment in vaults shall be *listed* or *approved* for above-ground use. Where multiple vaults are provided, adjacent vaults shall be allowed to share a common wall. The common wall shall be liquid and vapor tight and shall be designed to withstand the load imposed when the vault on either side of the wall is filled with water.

5303.16.8 Connections. Connections shall be provided to permit the venting of each vault to dilute, disperse and remove vapors prior to personnel entering the vault.

5303.16.9 Ventilation. Vaults shall be provided with an exhaust ventilation system installed in accordance with Section 5004.3. The ventilation system shall operate continuously or be designed to operate upon activation of the vapor or liquid detection system. The system shall provide ventilation at a rate of not less than 1 cubic foot per minute (cfm) per square foot [0.00508 $m^3/(s \cdot m^2)$] of floor area, but not less than 150 cfm (4 m^3/min). The exhaust system shall be designed to provide air movement across all parts of the vault floor for gases having a density greater than air and across all parts of the vault ceiling for gases having a density less than air. Supply ducts shall extend to within 3 inches (76 mm), but not more than 12 inches (305 mm), of the floor. Exhaust ducts shall extend to within 3 inches (76 mm), but not more than 12 inches (305 mm) of the floor or ceiling, for heavier-than-air or lighter-than-air gases, respectively. The exhaust system shall be installed in accordance with the *International Mechanical Code*.

5303.16.10 Monitoring and detection. Vaults shall be provided with *approved* vapor and liquid detection systems and equipped with on-site audible and visual warning devices with battery backup. Vapor detection systems shall sound an alarm when the system detects vapors that reach or exceed 25 percent of the lower explosive limit (LEL) or one-half the immediately dangerous to life and health (IDLH) concentration for the gas in the vault. Vapor detectors shall be located not higher than 12 inches (305 mm) above the lowest point in the vault for heavier-than-air gases and not lower than 12 inches (305 mm) below the highest point in the vault for lighter-than-air gases. Liquid detection systems shall sound an alarm upon detection of any liquid, including water. Liquid detectors shall be located in accordance with the manufacturers' instructions. Activation of either vapor or liquid detection systems shall cause a signal to be sounded at an *approved*, constantly attended location within the facility served by the tanks or at an *approved* location. Activation of vapor detection systems shall also shut off gas-handling equipment in the vault and dispensers.

5303.16.11 Liquid removal. Means shall be provided to recover liquid from the vault. Where a pump is used to meet this requirement, it shall not be permanently installed in the vault. Electric-powered portable pumps shall be suitable for use in Class I, Division 1 locations, as defined in NFPA 70.

5303.16.12 Relief vents. Vent pipes for equipment in the vault shall terminate not less than 12 feet (3658 mm) above ground level.

5303.16.13 Accessway. Vaults shall be provided with an *approved* personnel accessway with a minimum dimension of 30 inches (762 mm) and with a permanently affixed, nonferrous ladder. Accessways shall be designed to be nonsparking. Travel distance from any point inside a vault to an accessway shall not exceed 20 feet (6096 mm). At each entry point, a warning sign indicating the need for procedures for safe entry into confined spaces shall be posted. Entry points shall be secured against unauthorized entry and vandalism.

5303.16.14 Classified area. The interior of a vault containing a flammable gas shall be designated a Class I, Division 1 location, as defined in NFPA 70.

SECTION 5304
STORAGE OF COMPRESSED GASES

5304.1 Upright storage. *Compressed gas* containers, cylinders and tanks, except those designed for use in a horizontal position, and all *compressed gas* containers, cylinders and tanks containing nonliquefied gases, shall be stored in an upright position with the valve end up. An upright position shall include conditions where the container, cylinder or tank axis is inclined as much as 45 degrees (0.80 rad) from the vertical.

Exceptions:

1. *Compressed gas* containers with a water volume less than 1.3 gallons (5 L) are allowed to be stored in a horizontal position.
2. Cylinders, containers and tanks containing nonflammable gases, or cylinders, containers and tanks containing nonliquefied flammable gases that have been secured to a pallet for transportation purposes.

5304.2 Material-specific regulations. In addition to the requirements of this section, indoor and outdoor storage of *compressed gases* shall comply with the material-specific provisions of Chapters 54, 58 and 60 through 67.

SECTION 5305
USE AND HANDLING OF COMPRESSED GASES

5305.1 Compressed gas systems. *Compressed gas* systems shall be suitable for the use intended and shall be designed by persons competent in such design. *Compressed gas* equipment, machinery and processes shall be *listed* or *approved*.

5305.2 Controls. *Compressed gas* system controls shall be designed to prevent materials from entering or leaving process or reaction systems at other than the intended time, rate or path. Automatic controls shall be designed to be fail safe.

5305.3 Piping systems. Piping, including tubing, valves, fittings and pressure regulators, shall comply with this section and Chapter 50. Piping, tubing, pressure regulators, valves and other apparatus shall be kept gas tight to prevent leakage.

5305.4 Valves. Valves utilized on *compressed gas* systems shall be suitable for the use intended and shall be accessible. Valve handles or operators for required shutoff valves shall not be removed or otherwise altered to prevent access.

5305.5 Venting. Venting of gases shall be directed to an *approved* location. Venting shall comply with the *International Mechanical Code*.

5305.6 Upright use. *Compressed gas* containers, cylinders and tanks, except those designed for use in a horizontal position, and all *compressed gas* containers, cylinders and tanks containing nonliquefied gases, shall be used in an upright position with the valve end up. An upright position shall include conditions where the container, cylinder or tank axis is inclined as much as 45 degrees (0.80 rad) from the vertical. Use of nonflammable liquefied gases in the inverted position where the liquid phase is used shall not be prohibited provided that the container, cylinder or tank is properly secured and the dispensing apparatus is designed for liquefied gas use.

Exception: *Compressed gas* containers, cylinders and tanks with a water volume less than 1.3 gallons (5 L) are allowed to be used in a horizontal position.

5305.7 Transfer. Transfer of gases between containers, cylinders and tanks shall be performed by qualified personnel using equipment and operating procedures in accordance with CGA P-1.

Exception: The fueling of vehicles with CNG or CH_2, conducted in accordance with Chapter 23.

5305.8 Use of compressed gas for inflation. Inflatable equipment, devices or balloons shall only be pressurized or filled with compressed air or inert gases.

5305.9 Material-specific regulations. In addition to the requirements of this section, indoor and outdoor use of *compressed gases* shall comply with the material-specific provisions of Chapters 54, 58 and 60 through 67.

5305.10 Handling. The handling of *compressed gas* containers, cylinders and tanks shall comply with Sections 5305.10.1 and 5305.10.2.

5305.10.1 Carts and trucks. Containers, cylinders and tanks shall be moved using an *approved* method. Where containers, cylinders or tanks are moved by hand cart, hand truck or other mobile device, such carts, trucks or devices shall be designed for the secure movement of containers, cylinders or tanks. Carts and trucks utilized for transport of *compressed gas* containers, cylinders and tanks within buildings shall comply with Section 5003.10. Carts and trucks utilized for transport of *compressed gas* containers, cylinders and tanks exterior to buildings shall be designed so that the containers, cylinders and tanks will be secured against dropping or otherwise striking against each other or other surfaces.

5305.10.2 Lifting devices. Ropes, chains or slings shall not be used to suspend *compressed gas* containers, cylinders and tanks unless provisions at time of manufacture have been made on the container, cylinder or tank for appropriate lifting attachments, such as lugs.

SECTION 5306
MEDICAL GASES

5306.1 General. Medical gases at health care-related facilities intended for patient care, inhalation or sedation including, but not limited to, analgesia systems for dentistry, podiatry, veterinary and similar uses shall comply with Sections 5306.2 through 5306.4 in addition to other requirements of this chapter.

5306.2 Interior supply location. Medical gases shall be stored in areas dedicated to the storage of such gases without other storage or uses. Where containers of medical gases in quantities greater than the permit amount are located inside buildings, they shall be in a 1-hour exterior room, a 1-hour interior room or a gas cabinet in accordance with Section

5306.2.1, 5306.2.2 or 5306.2.3, respectively. Rooms or areas where medical gases are stored or used in quantities exceeding the *maximum allowable quantity per control area* as set forth in Section 5003.1 shall be in accordance with the *International Building Code* for high-hazard Group H occupancies.

5306.2.1 One-hour exterior rooms. A 1-hour exterior room shall be a room or enclosure separated from the remainder of the building by *fire barriers* constructed in accordance with Section 707 of the *International Building Code* or horizontal assemblies constructed in accordance with Section 711 of the *International Building Code*, or both, with a *fire-resistance rating* of not less than 1 hour. Openings between the room or enclosure and interior spaces shall be self-closing smoke- and draft-control assemblies having a *fire protection rating* of not less than 1 hour. Rooms shall have not less than one exterior wall that is provided with not less than two nonclosable louvered vents. Each vent shall have a minimum free opening area of 24 square inches (155 cm^2) for each 1,000 cubic feet (28 m^3) at normal temperature and pressure (NTP) of gas stored in the room and shall be not less than 72 square inches (465 cm^2) in aggregate free opening area. One vent shall be within 6 inches (152 mm) of the floor and one shall be within 6 inches (152 mm) of the ceiling. Rooms shall be provided with not less than one automatic sprinkler to provide container cooling in case of fire.

5306.2.2 One-hour interior room. Where an exterior wall cannot be provided for the room, automatic sprinklers shall be installed within the room. The room shall be exhausted through a duct to the exterior. Supply and exhaust ducts shall be enclosed in a 1-hour-rated shaft enclosure from the room to the exterior. *Approved* mechanical ventilation shall comply with the *International Mechanical Code* and be provided at a minimum rate of 1 cubic foot per minute per square foot [0.00508 $m^3/(s \cdot m^2)$] of the area of the room.

5306.2.3 Gas cabinets. Gas cabinets shall be constructed in accordance with Section 5003.8.6 and the following:

1. The average velocity of ventilation at the face of access ports or windows shall be not less than 200 feet per minute (1.02 m/s) with not less than 150 feet per minute (0.76 m/s) at any point of the access port or window.
2. They shall be connected to an exhaust system.
3. They shall be internally sprinklered.

5306.3 Exterior supply locations. Oxidizer medical gas systems located on the exterior of a building with quantities greater than the permit amount shall be located in accordance with Section 6304.2.1.

5306.4 Transfilling. Transfilling areas and operations including, but not limited to, ventilation and separation, shall comply with NFPA 99.

5306.5 Medical gas systems. Medical gas systems including, but not limited to, distribution piping, supply manifolds, connections, pressure regulators and relief devices and valves, shall be installed in accordance with NFPA 99 and the general provisions of this chapter. Existing medical gas systems shall be maintained in accordance with the maintenance, inspection and testing provisions of NFPA 99 for medical gas systems.

SECTION 5307 CARBON DIOXIDE (CO_2) SYSTEMS USED IN BEVERAGE DISPENSING APPLICATIONS

5307.1 General. Carbon dioxide systems with more than 100 pounds (45.4 kg) of carbon dioxide used in beverage dispensing applications shall comply with Sections 5307.2 through 5307.5.2.

5307.2 Permits. Permits shall be required as set forth in Section 105.6.

5307.3 Equipment. The storage, use, and handling of liquid carbon dioxide shall be in accordance with Chapter 53 and the applicable requirements of NFPA 55, Chapter 13. Insulated liquid carbon dioxide systems shall have pressure relief devices vented in accordance with NFPA 55.

5307.4 Protection from damage. Carbon dioxide systems shall be installed so the storage tanks, cylinders, piping and fittings are protected from damage by occupants or equipment during normal facility operations.

5307.5 Required protection. Where carbon dioxide storage tanks, cylinders, piping and equipment are located indoors, rooms or areas containing carbon dioxide storage tanks, cylinders, piping and fittings and other areas where a leak of carbon dioxide can collect shall be provided with either ventilation in accordance with Section 5307.5.1 or an emergency alarm system in accordance with Section 5307.5.2.

5307.5.1 Ventilation. Mechanical ventilation shall be in accordance with the *International Mechanical Code* and shall comply with all of the following:

1. Mechanical ventilation in the room or area shall be at a rate of not less than 1 cubic foot per minute per square foot [0.00508 $m^3/(s \bullet m^2)$].
2. Exhaust shall be taken from a point within 12 inches (305 mm) of the floor.
3. The ventilation system shall be designed to operate at a negative pressure in relation to the surrounding area.

5307.5.2 Emergency alarm system. An emergency alarm system shall comply with all of the following:

1. Continuous gas detection shall be provided to monitor areas where carbon dioxide can accumulate.
2. The threshold for activation of an alarm shall not exceed 5,000 parts per million (9,000 mg/m^3).
3. Activation of the emergency alarm system shall initiate a local alarm within the room or area in which the system is installed.

SECTION 5308 COMPRESSED GASES NOT OTHERWISE REGULATED

5308.1 General. *Compressed gases* in storage or use not regulated by the material-specific provisions of Chapters 6, 54, 55 and 60 through 67, including asphyxiant, irritant and radioactive gases, shall comply with this section in addition to other requirements of this chapter.

5308.2 Ventilation. Indoor storage and use areas and storage buildings shall be provided with mechanical exhaust ventilation or natural ventilation in accordance with the requirements of Section 5004.3 or 5005.1.9. Where mechanical ventilation is provided, the systems shall be operational during such time as the building or space is occupied

CHAPTER 54
CORROSIVE MATERIALS

SECTION 5401
GENERAL

5401.1 Scope. The storage and use of *corrosive* materials shall be in accordance with this chapter. *Compressed gases* shall also comply with Chapter 53.

Exceptions:

1. Display and storage in Group M and storage in Group S occupancies complying with Section 5003.11.
2. Stationary storage battery systems in accordance with Section 608.
3. This chapter shall not apply to R-717 (ammonia) where used as a refrigerant in a refrigeration system (see Section 606).

5401.2 Permits. Permits shall be required as set forth in Section 105.6.

SECTION 5402
DEFINITION

5402.1 Definition. The following term is defined in Chapter 2:

CORROSIVE.

SECTION 5403
GENERAL REQUIREMENTS

5403.1 Quantities not exceeding the maximum allowable quantity per control area. The storage and use of *corrosive* materials in amounts not exceeding the *maximum allowable quantity per control area* indicated in Section 5003.1 shall be in accordance with Sections 5001, 5003 and 5401.

5403.2 Quantities exceeding the maximum allowable quantity per control area. The storage and use of *corrosive* materials in amounts exceeding the *maximum allowable quantity per control area* indicated in Section 5003.1 shall be in accordance with this chapter and Chapter 50.

SECTION 5404
STORAGE

5404.1 Indoor storage. Indoor storage of *corrosive* materials in amounts exceeding the *maximum allowable quantity per control area* indicated in Table 5003.1.1(2), shall be in accordance with Sections 5001, 5003 and 5004 and this chapter.

5404.1.1 Liquid-tight floor. In addition to the provisions of Section 5004.12, floors in storage areas for *corrosive* liquids shall be of liquid-tight construction.

5404.2 Outdoor storage. Outdoor storage of *corrosive* materials in amounts exceeding the *maximum allowable quantity per control area* indicated in Table 5003.1.1(4) shall be in accordance with Sections 5001, 5003 and 5004 and this chapter.

5404.2.1 Above-ground outside storage tanks. Above-ground outside storage tanks exceeding an aggregate quantity of 1,000 gallons (3785 L) of *corrosive* liquids shall be provided with secondary containment in accordance with Section 5004.2.2.

5404.2.2 Distance from storage to exposures. Outdoor storage of *corrosive* materials shall not be within 20 feet (6096 mm) of buildings not associated with the manufacturing or distribution of such materials, *lot lines*, public streets, public alleys, *public ways* or *means of egress*. A 2-hour *fire barrier* without openings or penetrations, and extending not less than 30 inches (762 mm) above and to the sides of the storage area, is allowed in lieu of such distance. The wall shall either be an independent structure, or the *exterior wall* of the building adjacent to the storage area.

SECTION 5405
USE

5405.1 Indoor use. The indoor use of *corrosive* materials in amounts exceeding the *maximum allowable quantity per control area* indicated in Table 5003.1.1(2) shall be in accordance with Sections 5001, 5003 and 5005 and this chapter.

5405.1.1 Liquid transfer. *Corrosive* liquids shall be transferred in accordance with Section 5005.1.10.

5405.1.2 Ventilation. Where *corrosive* materials are dispensed or used, mechanical exhaust ventilation in accordance with Section 5005.2.1.1 shall be provided.

5405.2 Outdoor use. The outdoor use of *corrosive* materials in amounts exceeding the *maximum allowable quantity per control area* indicated in Table 5003.1.1(4) shall be in accordance with Sections 5001, 5003 and 5005 and this chapter.

5405.2.1 Distance from use to exposures. Outdoor use of *corrosive* materials shall be located in accordance with Section 5404.2.2.

CHAPTER 55

CRYOGENIC FLUIDS

SECTION 5501
GENERAL

5501.1 Scope. Storage, use and handling of *cryogenic fluids* shall comply with this chapter and NFPA 55. *Cryogenic fluids* classified as hazardous materials shall also comply with the general requirements of Chapter 50. Partially full containers containing residual *cryogenic fluids* shall be considered as full for the purposes of the controls required.

Exceptions:

1. Fluids used as refrigerants in refrigeration systems (see Section 606).
2. Liquefied natural gas (LNG), which shall comply with NFPA 59A.

Oxidizing *cryogenic fluids*, including oxygen, shall comply with Chapter 63, as applicable.

Flammable *cryogenic fluids*, including hydrogen, methane and carbon monoxide, shall comply with Chapters 23 and 58, as applicable.

Inert *cryogenic fluids*, including argon, helium and nitrogen, shall comply with ANSI/CGA P-18.

5501.2 Permits. Permits shall be required as set forth in Section 105.6.

SECTION 5502
DEFINITIONS

5502.1 Definitions. The following terms are defined in Chapter 2.

CRYOGENIC CONTAINER.

CRYOGENIC FLUID.

CRYOGENIC VESSEL.

FLAMMABLE CRYOGENIC FLUID.

LOW-PRESSURE TANK.

SECTION 5503
GENERAL REQUIREMENTS

5503.1 Containers. Containers employed for storage or use of *cryogenic fluids* shall comply with Sections 5503.1.1 through 5503.1.3.2 and Chapter 50.

5503.1.1 Nonstandard containers. Containers, equipment and devices that are not in compliance with recognized standards for design and construction shall be *approved* upon presentation of satisfactory evidence that they are designed and constructed for safe operation.

5503.1.1.1 Data submitted for approval. The following data shall be submitted to the *fire code official* with reference to the deviation from the recognized standard with the application for approval.

1. Type and use of container, equipment or device.
2. Material to be stored, used or transported.
3. Description showing dimensions and materials used in construction.
4. Design pressure, maximum operating pressure and test pressure.
5. Type, size and setting of pressure relief devices.
6. Other data requested by the *fire code official*.

5503.1.2 Concrete containers. Concrete containers shall be built in accordance with the *International Building Code*. Barrier materials and membranes used in connection with concrete, but not functioning structurally, shall be compatible with the materials contained.

5503.1.3 Foundations and supports. Containers shall be provided with substantial concrete or masonry foundations, or structural steel supports on firm concrete or masonry foundations. Containers shall be supported to prevent the concentration of excessive loads on the supporting portion of the shell. Foundations for horizontal containers shall be constructed to accommodate expansion and contraction of the container. Foundations shall be provided to support the weight of vaporizers or heat exchangers.

5503.1.3.1 Temperature effects. Where container foundations or supports are subject to exposure to temperatures below -130°F (-90°C), the foundations or supports shall be constructed of materials to withstand the low-temperature effects of *cryogenic fluid* spillage.

5503.1.3.2 Corrosion protection. Portions of containers in contact with foundations or saddles shall be painted to protect against corrosion.

5503.2 Pressure relief devices. Pressure relief devices shall be provided in accordance with Sections 5503.2.1 through 5503.2.7 to protect containers and systems containing *cryogenic fluids* from rupture in the event of overpressure. Pressure relief devices shall be designed in accordance with CGA S-1.1, CGA S-1.2 and CGA S-1.3.

5503.2.1 Containers. Containers shall be provided with pressure relief devices.

5503.2.2 Vessels or equipment other than containers. Heat exchangers, vaporizers, insulation casings surrounding containers, vessels and coaxial piping systems in which liquefied *cryogenic fluids* could be trapped because of leakage from the primary container shall be provided with a pressure relief device.

5503.2.3 Sizing. Pressure relief devices shall be sized in accordance with the specifications to which the container was fabricated. The relief device shall have sufficient capacity to prevent the maximum design pressure of the container or system from being exceeded.

5503.2.4 Accessibility. Pressure relief devices shall be located such that they are provided with ready access for inspection and repair.

5503.2.5 Arrangement. Pressure relief devices shall be arranged to discharge unobstructed to the open air in such a manner as to prevent impingement of escaping gas on personnel, containers, equipment and adjacent structures or to enter enclosed spaces.

Exception: DOTn-specified containers with an internal volume of 2 cubic feet (0.057 m^3) or less.

5503.2.6 Shutoffs between pressure relief devices and containers. Shutoff valves shall not be installed between pressure relief devices and containers.

Exceptions:

1. A shutoff valve is allowed on containers equipped with multiple pressure relief device installations where the arrangement of the valves provides the full required flow through the minimum number of required relief devices at all times.
2. A locking-type shutoff valve is allowed to be used upstream of the pressure relief device for service-related work performed by the supplier when in accordance with the requirements of the ASME *Boiler and Pressure Vessel Code*.

5503.2.7 Temperature limits. Pressure relief devices shall not be subjected to *cryogenic fluid* temperatures except when operating.

5503.3 Pressure relief vent piping. Pressure relief vent-piping systems shall be constructed and arranged so as to remain functional and direct the flow of gas to a safe location in accordance with Sections 5503.3.1 and 5503.3.2.

5503.3.1 Sizing. Pressure relief device vent piping shall have a cross-sectional area not less than that of the pressure relief device vent opening and shall be arranged so as not to restrict the flow of escaping gas.

5503.3.2 Arrangement. Pressure relief device vent piping and drains in vent lines shall be arranged so that escaping gas will discharge unobstructed to the open air and not impinge on personnel, containers, equipment and adjacent structures or enter enclosed spaces. Pressure relief device vent lines shall be installed in such a manner to exclude or remove moisture and condensation and prevent malfunction of the pressure relief device because of freezing or ice accumulation.

5503.4 Marking. Cryogenic containers and systems shall be marked in accordance with Sections 5503.4.1 through 5503.4.6.

5503.4.1 Identification signs. Visible hazard identification signs in accordance with NFPA 704 shall be provided at entrances to buildings or areas in which *cryogenic fluids* are stored, handled or used.

5503.4.2 Identification of contents. Stationary and portable containers shall be marked with the name of the gas contained. Stationary above-ground containers shall be placarded in accordance with Sections 5003.5 and 5003.6. Portable containers shall be identified in accordance with CGA C-7.

5503.4.3 Identification of containers. Stationary containers shall be identified with the manufacturing specification and maximum allowable working pressure with a permanent nameplate. The nameplate shall be installed on the container in an accessible location. The nameplate shall be marked in accordance with the ASME *Boiler and Pressure Vessel Code* or DOTn 49 CFR Parts 100-185.

5503.4.4 Identification of container connections. Container inlet and outlet connections, liquid-level limit controls, valves and pressure gauges shall be identified in accordance with one of the following:

1. Marked with a permanent tag or label identifying the function.
2. Identified by a schematic drawing that portrays the function and designates whether connected to the vapor or liquid space of the container.

Where a schematic drawing is provided, it shall be attached to the container and maintained in a legible condition.

5503.4.5 Identification of piping systems. Piping systems shall be identified in accordance with ASME A13.1.

5503.4.6 Identification of emergency shutoff valves. Emergency shutoff valves shall be identified and the location shall be clearly visible and indicated by means of a sign.

5503.5 Security. Cryogenic containers and systems shall be secured against accidental dislodgement and against access by unauthorized personnel in accordance with Sections 5503.5.1 through 5503.5.4.

5503.5.1 Security of areas. Containers and systems shall be secured against unauthorized entry and safeguarded in an *approved* manner.

5503.5.2 Securing of containers. Stationary containers shall be secured to foundations in accordance with the *International Building Code*. Portable containers subject to shifting or upset shall be secured. Nesting shall be an acceptable means of securing containers.

5503.5.3 Securing of vaporizers. Vaporizers, heat exchangers and similar equipment shall be anchored to a suitable foundation and its connecting piping shall be sufficiently flexible to provide for the effects of expansion and contraction due to temperature changes.

5503.5.4 Physical protection. Containers, piping, valves, pressure relief devices, regulating equipment and other appurtenances shall be protected against physical damage and tampering.

5503.6 Electrical wiring and equipment. Electrical wiring and equipment shall comply with NFPA 70 and Sections 5503.6.1 and 5503.6.2.

5503.6.1 Location. Containers and systems shall not be located where they could become part of an electrical circuit.

5503.6.2 Electrical grounding and bonding. Containers and systems shall not be used for electrical grounding. Where electrical grounding and bonding is required, the system shall comply with NFPA 70. The grounding system shall be protected against corrosion, including corrosion caused by stray electric currents.

5503.7 Service and repair. Service, repair, modification or removal of valves, pressure relief devices or other container appurtenances shall comply with Sections 5503.7.1 and 5503.7.2 and the ASME *Boiler and Pressure Vessel Code*, Section VIII or DOTn 49 CFR Parts 100-185.

5503.7.1 Containers. Containers that have been removed from service shall be handled in an *approved* manner.

5503.7.2 Systems. Service and repair of systems shall be performed by trained personnel.

5503.8 Unauthorized use. Containers shall not be used for any purpose other than to serve as a vessel for containing the product that it is designed to contain.

5503.9 Leaks, damage and corrosion. Leaking, damaged or corroded containers shall be removed from service. Leaking, damaged or corroded systems shall be replaced, repaired or removed in accordance with Section 5503.7.

5503.10 Lighting. Where required, lighting, including emergency lighting, shall be provided for fire appliances and operating facilities such as walkways, control valves and gates ancillary to stationary containers.

SECTION 5504 STORAGE

5504.1 General. Storage of containers shall comply with this section.

5504.2 Indoor storage. Indoor storage of containers shall be in accordance with Sections 5504.2.1 through 5504.2.2.3.

5504.2.1 Stationary containers. Stationary containers shall be installed in accordance with the provisions applicable to the type of fluid stored and this section.

5504.2.1.1 Containers. Stationary containers shall comply with Section 5503.1.

5504.2.1.2 Construction of indoor areas. *Cryogenic fluids* in stationary containers stored indoors shall be located in buildings, rooms or areas constructed in accordance with the *International Building Code*.

5504.2.1.3 Ventilation. Storage areas for stationary containers shall be ventilated in accordance with the *International Mechanical Code*.

5504.2.2 Portable containers. Indoor storage of portable containers shall comply with the provisions applicable to the type of fluid stored and Sections 5504.2.2.1 through 5504.2.2.3.

5504.2.2.1 Containers. Portable containers shall comply with Section 5503.1.

5504.2.2.2 Construction of indoor areas. *Cryogenic fluids* in portable containers stored indoors shall be stored in buildings, rooms or areas constructed in accordance with the *International Building Code*.

5504.2.2.3 Ventilation. Storage areas shall be ventilated in accordance with the *International Mechanical Code*.

5504.3 Outdoor storage. Outdoor storage of containers shall be in accordance with Sections 5504.3.1 through 5504.3.1.2.3.

5504.3.1 Separation from hazardous conditions. Cryogenic containers and systems in outdoor storage shall be separated from materials and conditions that pose exposure hazards to or from each other in accordance with Sections 5504.3.1.1 through 5504.3.1.1.5.

5504.3.1.1 Stationary containers. Stationary containers shall be separated from exposure hazards in accordance with the provisions applicable to the type of fluid contained and the minimum separation distances indicated in Table 5504.3.1.1.

**TABLE 5504.3.1.1
SEPARATION OF STATIONARY CONTAINERS FROM EXPOSURE HAZARDS**

EXPOSURE	MINIMUM DISTANCE (feet)
Buildings, regardless of construction type	1
Building exits	10
Wall openings	1
Air intakes	10
Lot lines	5
Places of public assembly	50
Nonambulatory patient areas	50
Combustible materials such as paper, leaves, weeds, dry grass or debris	15
Other hazardous materials	In accordance with Chapter 50

For SI: 1 foot = 304.8 mm.

5504.3.1.1.1 Point-of-fill connections. Remote transfer points and fill connection points shall not be positioned closer to exposures than the minimum distances required for stationary containers.

5504.3.1.1.2 Surfaces beneath containers. Containers shall be placed on surfaces that are compatible with the fluid in the container.

5504.3.1.1.3 Location. Containers of *cryogenic fluids* shall not be located within diked areas containing other hazardous materials.

5504.3.1.1.4 Areas subject to flooding. Stationary containers located in areas subject to flooding shall be securely anchored or elevated to prevent the containers from separating from foundations or supports.

5504.3.1.1.5 Drainage. The area surrounding stationary containers shall be provided with a means to prevent accidental discharge of fluids from endangering personnel, containers, equipment and adjacent structures or to enter enclosed spaces. The stationary container shall not be placed where spilled or discharged fluids will be retained around the container.

Exception: These provisions shall not apply where it is determined by the *fire code official* that the container does not constitute a hazard, after consideration of special features such as crushed rock utilized as a heat sink, topographical conditions, nature of occupancy, proximity to structures on the same or adjacent property, and the capacity and construction of containers and character of fluids to be stored.

5504.3.1.2 Outdoor storage of portable containers. Outdoor storage of portable containers shall comply with Section 5503 and Sections 5504.3.1.2.1 through 5504.3.1.2.3.

5504.3.1.2.1 Exposure hazard separation. Portable containers in outdoor storage shall be separated from exposure hazards in accordance with Table 5504.3.1.2.1.

**TABLE 5504.3.1.2.1
SEPARATION OF PORTABLE CONTAINERS FROM EXPOSURE HAZARDS**

EXPOSURE	MINIMUM DISTANCE (feet)
Building exits	10
Wall openings	1
Air intakes	10
Lot lines	5
Combustible materials such as paper, leaves, weeds, dry grass or debris	15
Other hazardous materials	In accordance with Chapter 50

For SI: 1 foot = 304.8 mm.

5504.3.1.2.2 Surfaces beneath containers. The surface of the area on which stationary containers are placed, including the surface of the area located below the point where connections are made for the purpose of filling such containers, shall be compatible with the fluid in the container.

5504.3.1.2.3 Drainage. The area surrounding portable containers shall be provided with a means to prevent accidental discharge of fluids from endangering adjacent containers, buildings, equipment or adjoining property.

Exception: These provisions shall not apply where it is determined by the *fire code official* that the container does not constitute a hazard.

SECTION 5505
USE AND HANDLING

5505.1 General. Use and handling of *cryogenic fluid* containers and systems shall comply with Sections 5505.1.1 through 5505.5.2.

5505.1.1 Cryogenic fluid systems. *Cryogenic fluid* systems shall be suitable for the use intended and designed by persons competent in such design. Equipment, machinery and processes shall be *listed* or *approved.*

5505.1.2 Piping systems. Piping, tubing, valves and joints and fittings conveying *cryogenic fluids* shall be installed in accordance with the material-specific provisions of Section 5501.1 and Sections 5505.1.2.1 through 5505.1.2.6.

5505.1.2.1 Design and construction. Piping systems shall be suitable for the use intended through the full range of pressure and temperature to which they will be subjected. Piping systems shall be designed and constructed to provide adequate allowance for expansion, contraction, vibration, settlement and fire exposure.

5505.1.2.2 Joints. Joints on container piping and tubing shall be threaded, welded, silver brazed or flanged.

5505.1.2.3 Valves and accessory equipment. Valves and accessory equipment shall be suitable for the intended use at the temperatures of the application and shall be designed and constructed to withstand the maximum pressure at the minimum temperature to which they will be subjected.

5505.1.2.3.1 Shutoff valves on containers. Shutoff valves shall be provided on all container connections except for pressure relief devices. Shutoff valves shall be provided with access thereto and located as close as practical to the container.

5505.1.2.3.2 Shutoff valves on piping. Shutoff valves shall be installed in piping containing *cryogenic fluids* where needed to limit the volume of liquid discharged in the event of piping or equipment failure. Pressure relief valves shall be installed where liquid is capable of being trapped between shutoff-valves in the piping system (see Section 5503.2).

5505.1.2.4 Physical protection and support. Piping systems shall be supported and protected from physical damage. Piping passing through walls shall be protected from mechanical damage.

5505.1.2.5 Corrosion protection. Above-ground piping that is subject to corrosion because of exposure to corrosive atmospheres, shall be constructed of materials to resist the corrosive environment or otherwise protected against corrosion. Below-ground piping shall be protected against corrosion.

5505.1.2.6 Testing. Piping systems shall be tested and proven free of leaks after installation as required by the standards to which they were designed and constructed. Test pressures shall be not less than 150 percent of the maximum allowable working pressure where hydraulic testing is conducted or 110 percent where testing is conducted pneumatically.

5505.2 Indoor use. Indoor use of *cryogenic fluids* shall comply with the material-specific provisions of Section 5501.1.

5505.3 Outdoor use. Outdoor use of *cryogenic fluids* shall comply with the material specific provisions of Sections 5501.1, 5505.3.1 and 5505.3.2.

5505.3.1 Separation. Distances from lot lines, buildings and exposure hazards shall comply with Section 5504.3 and the material-specific provisions of Section 5501.1.

5505.3.2 Emergency shutoff valves. Manual or automatic emergency shutoff valves shall be provided to shut off the *cryogenic fluid* supply in case of emergency. An emergency shutoff valve shall be located at the source of supply and at the point where the system enters the building.

5505.4 Filling and dispensing. Filling and dispensing of *cryogenic fluids* shall comply with Sections 5505.4.1 through 5505.4.3.

5505.4.1 Dispensing areas. Dispensing of *cryogenic fluids* with physical or *health hazards* shall be conducted in *approved* locations. Dispensing indoors shall be conducted in areas constructed in accordance with the *International Building Code*.

5505.4.1.1 Ventilation. Indoor areas where *cryogenic fluids* are dispensed shall be ventilated in accordance with the requirements of the *International Mechanical Code* in a manner that captures any vapor at the point of generation.

Exception: *Cryogenic fluids* that can be demonstrated not to create harmful vapors.

5505.4.1.2 Piping systems. Piping systems utilized for filling or dispensing of *cryogenic fluids* shall be designed and constructed in accordance with Section 5505.1.2.

5505.4.2 Vehicle loading and unloading areas. Loading or unloading areas shall be conducted in an *approved* manner in accordance with the standards referenced in Section 5501.1.

5505.4.3 Limit controls. Limit controls shall be provided to prevent overfilling of stationary containers during filling operations.

5505.5 Handling. Handling of cryogenic containers shall comply with Sections 5505.5.1 and 5505.5.2.

5505.5.1 Carts and trucks. Cryogenic containers shall be moved using an *approved* method. Where cryogenic containers are moved by hand cart, hand truck or other mobile device, such carts, trucks or devices shall be designed for the secure movement of the container.

Carts and trucks used to transport cryogenic containers shall be designed to provide a stable base for the commodities to be transported and shall have a means of restraining containers to prevent accidental dislodgement.

5505.5.2 Closed containers. Pressurized containers shall be transported in a closed condition. Containers designed for use at atmospheric conditions shall be transported with appropriate loose-fitting covers in place to prevent spillage.

CHAPTER 56

EXPLOSIVES AND FIREWORKS

SECTION 5601 GENERAL

5601.1 Scope. The provisions of this chapter shall govern the possession, manufacture, storage, handling, sale and use of *explosives*, *explosive materials*, fireworks and small arms ammunition.

Exceptions:

1. The Armed Forces of the United States, Coast Guard or National Guard.
2. *Explosives* in forms prescribed by the official United States Pharmacopoeia.
3. The possession, storage and use of small arms ammunition where packaged in accordance with DOTn packaging requirements.
4. The possession, storage and use of not more than 1 pound (0.454 kg) of commercially manufactured sporting black powder, 20 pounds (9 kg) of smokeless powder and 10,000 small arms primers for hand loading of small arms ammunition for personal consumption.
5. The use of *explosive materials* by federal, state and local regulatory, law enforcement and fire agencies acting in their official capacities.
6. Special industrial *explosive* devices that in the aggregate contain less than 50 pounds (23 kg) of *explosive materials*.
7. The possession, storage and use of blank industrial-power load cartridges where packaged in accordance with DOTn packaging regulations.
8. Transportation in accordance with DOTn 49 CFR Parts 100–185.
9. Items preempted by federal regulations.

5601.1.1 Explosive material standard. In addition to the requirements of this chapter, NFPA 495 shall govern the manufacture, transportation, storage, sale, handling and use of *explosive* materials.

5601.1.2 Explosive material terminals. In addition to the requirements of this chapter, the operation of *explosive material* terminals shall conform to the provisions of NFPA 498.

5601.1.3 Fireworks. The possession, manufacture, storage, sale, handling and use of fireworks are prohibited.

Exceptions:

1. Storage and handling of fireworks as allowed in Section 5604.
2. Manufacture, assembly and testing of fireworks as allowed in Section 5605.
3. The use of fireworks for fireworks displays as allowed in Section 5608.
4. The possession, storage, sale, handling and use of specific types of Division 1.4G fireworks where allowed by applicable laws, ordinances and regulations, provided such fireworks and facilities comply with NFPA 1124, CPSC 16 CFR Parts 1500 and 1507, and DOTn 49 CFR Parts 100–185, for consumer fireworks.

5601.1.4 Rocketry. The storage, handling and use of model and high-power rockets shall comply with the requirements of NFPA 1122, NFPA 1125 and NFPA 1127.

5601.1.5 Ammonium nitrate. The storage and handling of ammonium nitrate shall comply with the requirements of NFPA 400 and Chapter 63.

Exception: Storage of ammonium nitrate in magazines with blasting agents shall comply with the requirements of NFPA 495.

5601.2 Permit required. Permits shall be required as set forth in Section 105.6 and regulated in accordance with this section.

5601.2.1 Residential uses. Persons shall not keep or store, nor shall any permit be issued to keep or store, any *explosives* at any place of habitation, or within 100 feet (30 480 mm) thereof.

Exception: Storage of smokeless propellant, black powder and small arms primers for personal use and not for resale in accordance with Section 5606.

5601.2.2 Sale and retail display. Persons shall not construct a retail display nor offer for sale *explosives*, *explosive materials* or fireworks upon highways, sidewalks, public property or in Group A or E occupancies.

5601.2.3 Permit restrictions. The *fire code official* is authorized to limit the quantity of *explosives*, *explosive materials* or fireworks permitted at a given location. Persons possessing a permit for storage of *explosives* at any place, shall not keep or store an amount greater than authorized in such permit. Only the kind of *explosive* specified in such a permit shall be kept or stored.

5601.2.4 Financial responsibility. Before a permit is issued, as required by Section 5601.2, the applicant shall file with the jurisdiction a corporate surety bond in the principal sum of $100,000 or a public liability insurance policy for the same amount, for the purpose of the payment of all damages to persons or property that arise from, or are caused by, the conduct of any act authorized by the permit upon which any judicial judgment results. The *fire code official* is authorized to specify a greater or lesser amount when, in his or her opinion, conditions at the location of use indicate a greater or lesser amount is required.

Government entities shall be exempt from this bond requirement.

5601.2.4.1 Blasting. Before approval to do blasting is issued, the applicant for approval shall file a bond or submit a certificate of insurance in such form, amount and coverage as determined by the legal department of the jurisdiction to be adequate in each case to indemnify the jurisdiction against any and all damages arising from permitted blasting.

5601.2.4.2 Fireworks display. The permit holder shall furnish a bond or certificate of insurance in an amount deemed adequate by the *fire code official* for the payment of all potential damages to a person or persons or to property by reason of the permitted display, and arising from any acts of the permit holder, the agent, employees or subcontractors.

5601.3 Prohibited explosives. Permits shall not be issued or renewed for possession, manufacture, storage, handling, sale or use of the following materials and such materials currently in storage or use shall be disposed of in an *approved* manner.

1. Liquid nitroglycerin.
2. Dynamite containing more than 60-percent liquid *explosive* ingredient.
3. Dynamite having an unsatisfactory absorbent or one that permits leakage of a liquid *explosive* ingredient under any conditions liable to exist during storage.
4. Nitrocellulose in a dry and uncompressed condition in a quantity greater than 10 pounds (4.54 kg) of net weight in one package.
5. Fulminate of mercury in a dry condition and fulminate of all other metals in any condition except as a component of manufactured articles not hereinafter forbidden.
6. *Explosive* compositions that ignite spontaneously or undergo marked decomposition, rendering the products of their use more hazardous, when subjected for 48 consecutive hours or less to a temperature of 167°F (75°C).
7. New *explosive materials* until *approved* by DOTn, except that permits are allowed to be issued to educational, governmental or industrial laboratories for instructional or research purposes.
8. *Explosive materials* forbidden for transport by DOTn.
9. *Explosive materials* containing an ammonium salt and a chlorate.
10. *Explosives* not packed or marked as required by DOTn 49 CFR Parts 100–185.

Exception: Gelatin dynamite.

5601.4 Qualifications. Persons in charge of magazines, blasting, fireworks display or pyrotechnic special effect operations shall not be under the influence of alcohol or drugs that impair sensory or motor skills, shall be not less than 21 years of age and shall demonstrate knowledge of all safety precautions related to the storage, handling or use of *explosives, explosive materials* or fireworks.

5601.5 Supervision. The *fire code official* is authorized to require operations permitted under the provisions of Section 5601.2 to be supervised at any time by the *fire code official* in order to determine compliance with all safety and fire regulations.

5601.6 Notification. Whenever a new *explosive material* storage or manufacturing site is established, including a temporary job site, the local law enforcement agency, fire department and local emergency planning committee shall be notified 48 hours in advance, not including Saturdays, Sundays and holidays, of the type, quantity and location of *explosive materials* at the site.

5601.7 Seizure. The *fire code official* is authorized to remove or cause to be removed or disposed of in an *approved* manner, at the expense of the *owner, explosives, explosive materials* or fireworks offered or exposed for sale, stored, possessed or used in violation of this chapter.

5601.8 Establishment of quantity of explosives and distances. The quantity of *explosives* and distances shall be in accordance with Sections 5601.8.1 and 5601.8.1.1.

5601.8.1 Quantity of explosives. The quantity-distance (Q-D) tables in Sections 5604.5 and 5605.3 shall be used to provide the minimum separation distances from potential explosion sites as set forth in Tables 5601.8.1(1) through 5601.8.1(3). The classification and the weight of the *explosives* are primary characteristics governing the use of these tables. The net *explosive* weight shall be determined in accordance with Sections 5601.8.1.1 through 5601.8.1.4.

5601.8.1.1 Mass-detonating explosives (Division 1.1, 1.2 or 1.5). The total net *explosive* weight of mass-detonating explosives (Division 1.1, 1.2 or 1.5) shall be used. See Table 5604.5.2(1) or Table 5605.3 as appropriate.

Exception: Where the TNT equivalence of the *explosive material* has been determined, the equivalence is allowed to be used to establish the net *explosive* weight.

5601.8.1.2 Nonmass-detonating explosives (excluding Division 1.4). Nonmass-detonating *explosives* (excluding Division 1.4) shall be as follows:

1. Division 1.3 propellants. The total weight of the propellants alone shall be the net *explosive* weight. The net weight of propellant shall be used. See Table 5604.5.2(2).
2. Combinations of bulk metal powder and pyrotechnic compositions. The sum of the net weights of metal powders and pyrotechnic compositions in the containers shall be the net *explosive* weight. See Table 5604.5.2(2).

TABLE 5601.8.1(1)
APPLICATION OF SEPARATION DISTANCE (Q-D) TABLES—DIVISION 1.1, 1.2 AND 1.5 EXPLOSIVES[a, b, c]

ITEM	MAGAZINE	Q-D	OPERATING BUILDING	Q-D	INHABITED BUILDING	Q-D	PUBLIC TRAFFIC ROUTE	Q-D
Magazine	Table 5604.5.2(1)	IMD	Table 5605.3	ILD or IPD	Table 5604.5.2(1)	IBD	Table 5604.5.2(1)	PTR
Operating building	Table 5604.5.2(1)	ILD or IPD	Table 5605.3	ILD or IPD	Table 5604.5.2(1)	IBD	Table 5604.5.2(1)	PTR
Inhabited building	Table 5604.5.2(1)	IBD	Table 5604.5.2(1)	IBD	Not Applicable	Not Applicable	Not Applicable	Not Applicable
Public traffic route	Table 5604.5.2(1)	PTR	Table 5604.5.2(1)	PTR	Not Applicable	Not Applicable	Not Applicable	Not Applicable

For SI: 1 foot = 304.8 mm.

a. The minimum separation distance (D_o) shall be 60 feet. Where a building or magazine containing explosives is barricaded, the minimum distance shall be 30 feet.

b. Linear interpolation between tabular values in the referenced Q-D tables shall not be allowed. Nonlinear interpolation of the values shall be allowed subject to an approved technical opinion and report prepared in accordance with Section 104.7.2.

c. For definitions of Quantity-Distance abbreviations IBD, ILD, IMD, IPD and PTR, see Chapter 2.

TABLE 5601.8.1(2)
APPLICATION OF SEPARATION DISTANCE (Q-D) TABLES—DIVISION 1.3 EXPLOSIVES[a, b, c]

ITEM	MAGAZINE	Q-D	OPERATING BUILDING	Q-D	INHABITED BUILDING	Q-D	PUBLIC TRAFFIC ROUTE	Q-D
Magazine	Table 5604.5.2(2)	IMD	Table 5604.5.2(2)	ILD or IPD	Table 5604.5.2(2)	IBD	Table 5604.5.2(2)	PTR
Operating building	Table 5604.5.2(2)	ILD or IPD	Table 5604.5.2(2)	ILD or IPD	Table 5604.5.2(2)	IBD	Table 5604.5.2(2)	PTR
Inhabited building	Table 5604.5.2(2)	IBD	Table 5604.5.2(2)	IBD	Not Applicable	Not Applicable	Not Applicable	Not Applicable
Public traffic route	Table 5604.5.2(2)	PTR	Table 5604.5.2(2)	PTR	Not Applicable	Not Applicable	Not Applicable	Not Applicable

For SI: 1 foot = 304.8 mm.

a. The minimum separation distance (D_o) shall be not less than 50 feet.

b. Linear interpolation between tabular values in the referenced Q-D table shall be allowed.

c. For definitions of Quantity-Distance abbreviations IBD, ILD, IMD, IPD and PTR, see Chapter 2.

TABLE 5601.8.1(3)
APPLICATION OF SEPARATION DISTANCE (Q-D) TABLES—DIVISION 1.4 EXPLOSIVES[a, b, c, d]

ITEM	MAGAZINE	Q-D	OPERATING BUILDING	Q-D	INHABITED BUILDING	Q-D	PUBLIC TRAFFIC ROUTE	Q-D
Magazine	Table 5604.5.2(3)	IMD	Table 5604.5.2(3)	ILD or IPD	Table 5604.5.2(3)	IBD	Table 5604.5.2(3)	PTR
Operating building	Table 5604.5.2(3)	ILD or IPD	Table 5604.5.2(3)	ILD or IPD	Table 5604.5.2(3)	IBD	Table 5604.5.2(3)	PTR
Inhabited building	Table 5604.5.2(3)	IBD	Table 5604.5.2(3)	IBD	Not Applicable	Not Applicable	Not Applicable	Not Applicable
Public traffic route	Table 5604.5.2(3)	PTR	Table 5604.5.2(3)	PTR	Not Applicable	Not Applicable	Not Applicable	Not Applicable

For SI: 1 foot = 304.8 mm.

a. The minimum separation distance (D_o) shall be not less than 50 feet.

b. Linear interpolation between tabular values in the referenced Q-D table shall not be allowed.

c. For definitions of Quantity-Distance abbreviations IBD, ILD, IMD, IPD and PTR, see Chapter 2.

d. This table shall not apply to consumer fireworks, 1.4G.

5601.8.1.3 Combinations of mass-detonating and nonmass-detonating explosives (excluding Division 1.4). Combination of mass-detonating and nonmass-detonating *explosives* (excluding Division 1.4) shall be as follows:

1. Where Division 1.1 and 1.2 *explosives* are located in the same site, determine the distance for the total quantity considered first as 1.1 and then as 1.2. The required distance is the greater of the two. Where the Division 1.1 requirements are controlling and the TNT equivalence of the 1.2 is known, the TNT equivalent weight of the 1.2 items shall be allowed to be added to the total *explosive* weight of Division 1.1 items to determine the net *explosive* weight for Division 1.1 distance determination. See Table 5604.5.2(2) or Table 5605.3 as appropriate.
2. Where Division 1.1 and 1.3 *explosives* are located in the same site, determine the distances for the total quantity considered first as 1.1 and then as 1.3. The required distance is the greater of the two. Where the Division 1.1 requirements are controlling and the TNT equivalence of the 1.3 is known, the TNT equivalent weight of the 1.3 items shall be allowed to be added to the total *explosive* weight of Division 1.1 items to determine the net *explosive* weight for Division 1.1 distance determination. See Table 5604.5.2(1), 5604.5.2(2) or 5605.3, as appropriate.
3. Where Division 1.1, 1.2 and 1.3 *explosives* are located in the same site, determine the distances for the total quantity considered first as 1.1, next as 1.2 and finally as 1.3. The required distance is the greatest of the three. As allowed by paragraphs 1 and 2 above, TNT equivalent weights for 1.2 and 1.3 items are allowed to be used to determine the net weight of *explosives* for Division 1.1 distance determination. Table 5604.5.2(1) or 5605.3 shall be used where TNT equivalency is used to establish the net *explosive* weight.
4. For composite pyrotechnic items Division 1.1 and Division 1.3, the sum of the net weights of the pyrotechnic composition and the *explosives* involved shall be used. See Tables 5604.5.2(1) and 5604.5.2(2).

5601.8.1.4 Moderate fire—no blast hazards (Division 1.4). For Division 1.4 explosives, the total weight of the explosive material alone is the net weight. The net weight of the explosive material shall be used.

SECTION 5602 DEFINITIONS

5602.1 Definitions. The following terms are defined in Chapter 2:

AMMONIUM NITRATE.

BARRICADE.

Artificial barricade.

Natural barricade.

BARRICADED.

BLAST AREA.

BLAST SITE.

BLASTER.

BLASTING AGENT.

BULLET RESISTANT.

DETONATING CORD.

DETONATION.

DETONATOR.

DISCHARGE SITE.

DISPLAY SITE.

EXPLOSIVE.
- **High explosive.**
- **Low explosive.**
- **Mass-detonating explosives.**
- **UN/DOTn Class 1 explosives.**
  - **Division 1.1.**
  - **Division 1.2.**
  - **Division 1.3.**
  - **Division 1.4.**
  - **Division 1.5.**
  - **Division 1.6.**

EXPLOSIVE MATERIAL.

FALLOUT AREA.

FIREWORKS.
- **Fireworks, 1.4G.**
- **Fireworks, 1.3G.**

FIREWORKS DISPLAY.

HIGHWAY.

INHABITED BUILDING.

MAGAZINE.
- **Indoor.**
- **Type 1.**
- **Type 2.**
- **Type 3.**
- **Type 4.**
- **Type 5.**

MORTAR.

NET EXPLOSIVE WEIGHT (net weight).

OPERATING BUILDING.

OPERATING LINE.

PLOSOPHORIC MATERIAL.

PROXIMATE AUDIENCE.

PUBLIC TRAFFIC ROUTE (PTR).

PYROTECHNIC ARTICLE.

PYROTECHNIC COMPOSITION.

PYROTECHNIC SPECIAL EFFECT.

PYROTECHNIC SPECIAL-EFFECT MATERIAL.

PYROTECHNICS.

QUANTITY-DISTANCE (Q-D).

Inhabited building distance (IBD).

Intermagazine distance (IMD).

Intraline distance (ILD) or Intraplant distance (IPD).

Minimum separation distance (D_0).

RAILWAY.

READY BOX.

SMALL ARMS AMMUNITION.

SMALL ARMS PRIMERS.

SMOKELESS PROPELLANTS.

SPECIAL INDUSTRIAL EXPLOSIVE DEVICE.

THEFT RESISTANT.

SECTION 5603 RECORD KEEPING AND REPORTING

5603.1 General. Records of the receipt, handling, use or disposal of *explosive materials*, and reports of any accidents, thefts or unauthorized activities involving *explosive materials* shall conform to the requirements of this section.

5603.2 Transaction record. The permittee shall maintain a record of all transactions involving receipt, removal, use or disposal of *explosive materials*. Such records shall be maintained for a period of 5 years.

Exception: Where only Division 1.4G (consumer fireworks) are handled, records need only be maintained for a period of 3 years.

5603.3 Loss, theft or unauthorized removal. The loss, theft or unauthorized removal of *explosive materials* from a magazine or permitted facility shall be reported to the *fire code official*, local law enforcement authorities and the U.S. Department of Treasury, Bureau of Alcohol, Tobacco, Firearms and Explosives within 24 hours.

Exception: Loss of Division 1.4G (consumer fireworks) need not be reported to the Bureau of Alcohol, Tobacco, Firearms and Explosives.

5603.4 Accidents. Accidents involving the use of *explosives*, *explosive materials* and fireworks that result in injuries or property damage shall be reported to the *fire code official* immediately.

5603.5 Misfires. The pyrotechnic display operator or blaster in charge shall keep a record of all aerial shells that fail to fire or charges that fail to detonate.

5603.6 Hazard communication. Manufacturers of *explosive materials* and fireworks shall maintain records of chemicals, chemical compounds and mixtures required by DOL 29 CFR Part 1910.1200, and Section 407.

5603.7 Safety rules. Current safety rules covering the operation of magazines, as described in Section 5604.7, shall be posted on the interior of the magazine in a visible location.

SECTION 5604 EXPLOSIVE MATERIALS STORAGE AND HANDLING

5604.1 General. Storage of *explosives* and *explosive materials*, small arms ammunition, small arms primers, propellant-actuated cartridges and smokeless propellants in magazines shall comply with the provisions of this section.

5604.2 Magazine required. *Explosives* and *explosive materials*, and Division 1.3G fireworks shall be stored in magazines constructed, located, operated and maintained in accordance with the provisions of Section 5604 and NFPA 495 or NFPA 1124.

Exceptions:

1. Storage of fireworks at display sites in accordance with Section 5608.5 and NFPA 1123 or NFPA 1126.
2. Portable or mobile magazines not exceeding 120 square feet (11 m^2) in area shall not be required to comply with the requirements of the *International Building Code*.

5604.3 Magazines. The storage of *explosives* and *explosive materials* in magazines shall comply with Table 5604.3.

5604.3.1 High explosives. *Explosive materials* classified as Division 1.1 or 1.2 or formerly classified as Class A by the U.S. Department of Transportation shall be stored in Type 1, 2 or 3 magazines.

Exceptions:

1. Black powder shall be stored in a Type 1, 2, 3 or 4 magazine.
2. Cap-sensitive *explosive material* that is demonstrated not to be bullet sensitive shall be stored in a Type 1, 2, 3, 4 or 5 magazine.

5604.3.2 Low explosives. *Explosive materials* that are not cap sensitive shall be stored in a Type 1, 2, 3, 4 or 5 magazine.

5604.3.3 Detonating cord. For quantity and distance purposes, detonating cord of 50 grains per foot shall be calculated as equivalent to 8 pounds (4 kg) of high *explosives* per 1,000 feet (305 m). Heavier or lighter core loads shall be rated proportionally.

5604.4 Prohibited storage. Detonators shall be stored in a separate magazine for blasting supplies and shall not be stored in a magazine with other *explosive materials*.

5604.5 Location. The use of magazines for storage of *explosives* and *explosive materials* shall comply with Sections 5604.5.1 through 5604.5.3.3.

5604.5.1 Indoor magazines. The use of indoor magazines for storage of *explosives* and *explosive materials* shall comply with the requirements of Sections 5604.5.1.1 through 5604.5.1.7.

5604.5.1.1 Use. The use of indoor magazines for storage of *explosives* and *explosive materials* shall be limited to occupancies of Group F, H, M or S, and research and development laboratories.

TABLE 5604.3
STORAGE AMOUNTS AND MAGAZINE REQUIREMENTS FOR EXPLOSIVES, EXPLOSIVE MATERIALS AND FIREWORKS, 1.3G MAXIMUM ALLOWABLE QUANTITY PER CONTROL AREA

NEW UN/ DOTn DIVISION	OLD DOTn CLASS	ATF/OSHA CLASS	INDOOR[a] (pounds)				OUTDOOR (pounds)	MAGAZINE TYPE REQUIRED				
			Unprotected	Cabinet	Sprinklers	Sprinklers & cabinet		1	2	3	4	5
1.1[b]	A	High	0	0	1	2	1	X	X	X	—	—
1.2	A	High	0	0	1	2	1	X	X	X	—	—
1.2	B	Low	0	0	1	1	1	X	X	X	X	—
1.3	B	Low	0	0	5	10	1	X	X	X	X	—
1.4[c]	B	Low	0	0	50	100	1	X	X	X	X	—
1.5	C	Low	0	0	1	2	1	X	X	X	X	—
1.5	Blasting Agent	Blasting Agent	0	0	1	2	1	X	X	X	X	X
1.6	Not Applicable	Not Applicable	0	0	1	2	1	X	X	X	X	X

For SI: 1 pound = 0.454 kg, 1 pound per gallon = 0.12 kg per liter, 1 ounce = 28.35 g.

a. A factor of 10 pounds per gallon shall be used for converting pounds (solid) to gallons (liquid) in accordance with Section 5003.1.2.

b. Black powder shall be stored in a Type 1, 2, 3 or 4 magazine as provided for in Section 5604.3.1.

c. This table shall not apply to consumer fireworks, 1.4G.

5604.5.1.2 Construction. Indoor magazines shall comply with the following construction requirements:

1. Construction shall be fire resistant and theft resistant.
2. Exterior shall be painted red.
3. Base shall be fitted with wheels, casters or rollers to facilitate removal from the building in an emergency.
4. Lid or door shall be marked with conspicuous white lettering not less than 3 inches (76 mm) high and minimum $^1/_2$ inch (12.7 mm) stroke, reading EXPLOSIVES—KEEP FIRE AWAY.
5. The least horizontal dimension shall not exceed the clear width of the entrance door.

5604.5.1.3 Quantity limit. Not more than 50 pounds (23 kg) of *explosives* or *explosive materials* shall be stored within an indoor magazine.

Exception: Day boxes used for the storage of in-process material in accordance with Section 5605.6.4.1.

5604.5.1.4 Prohibited use. Indoor magazines shall not be used within buildings containing Group R occupancies.

5604.5.1.5 Location. Indoor magazines shall be located within 10 feet (3048 mm) of an entrance and only on floors at or having ramp access to the exterior grade level.

5604.5.1.6 Number. Not more than two indoor magazines shall be located in the same building. Where two such magazines are located in the same building, one magazine shall be used solely for the storage of not more than 5,000 detonators.

5604.5.1.7 Separation distance. Where two magazines are located in the same building, they shall be separated by a distance of not less than 10 feet (3048 mm).

5604.5.2 Outdoor magazines. Outdoor magazines other than Type 3 shall be located so as to comply with Table 5604.5.2(2) or 5604.5.2(3) as set forth in Tables 5601.8.1(1) through 5601.8.1(3). Where a magazine or group of magazines, as described in Section 5604.5.2.2, contains different classes of *explosive materials*, and Division 1.1 materials are present, the required separations for the magazine or magazine group as a whole shall comply with Table 5604.5.2(2).

5604.5.2.1 Separation. Where two or more storage magazines are located on the same property, each magazine shall comply with the minimum distances specified from inhabited buildings, public transportation routes and operating buildings. Magazines shall be separated from each other by not less than the intermagazine distances (IMD) shown for the separation of magazines.

5604.5.2.2 Grouped magazines. Where two or more magazines are separated from each other by less than the intermagazine distances (IMD), such magazines as a group shall be considered as one magazine and the total quantity of *explosive materials* stored in the group shall be treated as if stored in a single magazine. The location of the group of magazines shall comply with the intermagazine distances (IMD) specified from other magazines or magazine groups, inhabited buildings (IBD), public transportation routes (PTR) and operating buildings (ILD or IPD) as required.

5604.5.3 Special requirements for Type 3 magazines. Type 3 magazines shall comply with Sections 5604.5.3.1 through 5604.5.3.3.

TABLE 5604.5.2(1)
AMERICAN TABLE OF DISTANCES FOR STORAGE OF EXPLOSIVES AS APPROVED BY THE INSTITUTE OF MAKERS OF EXPLOSIVES AND REVISED JUNE 1991[a]

QUANTITY OF EXPLOSIVE MATERIALS[c]		DISTANCES IN FEET							
		Inhabited buildings		Public highways with traffic volume less than 3,000 vehicles per day		Public highways with traffic volume greater than 3,000 vehicles per day and passenger railways		Separation of magazines[d]	
Pounds over	Pounds not over	Barricaded	Unbarricaded	Barricaded	Unbarricaded	Barricaded	Unbarricaded	Barricaded	Unbarricaded
0	5	70	140	30	60	51	102	6	12
5	10	90	180	35	70	64	128	8	16
10	20	110	220	45	90	81	162	10	20
20	30	125	250	50	100	93	186	11	22
30	40	140	280	55	110	103	206	12	24
40	50	150	300	60	120	110	220	14	28
50	75	170	340	70	140	127	254	15	30
75	100	190	380	75	150	139	278	16	32
100	125	200	400	80	160	150	300	18	36
125	150	215	430	85	170	159	318	19	38
150	200	235	470	95	190	175	350	21	42
200	250	255	510	105	210	189	378	23	46
250	300	270	540	110	220	201	402	24	48
300	400	295	590	120	240	221	442	27	54
400	500	320	640	130	260	238	476	29	58
500	600	340	680	135	270	253	506	31	62
600	700	355	710	145	290	266	532	32	64
700	800	375	750	150	300	278	556	33	66
800	900	390	780	155	310	289	578	35	70
900	1,000	400	800	160	320	300	600	36	72
1,000	1,200	425	850	165	330	318	636	39	78
1,200	1,400	450	900	170	340	336	672	41	82
1,400	1,600	470	940	175	350	351	702	43	86
1,600	1,800	490	980	180	360	366	732	44	88
1,800	2,000	505	1,010	185	370	378	756	45	90
2,000	2,500	545	1,090	190	380	408	816	49	98
2,500	3,000	580	1,160	195	390	432	864	52	104
3,000	4,000	635	1,270	210	420	474	948	58	116
4,000	5,000	685	1,370	225	450	513	1,026	61	122
5,000	6,000	730	1,460	235	470	546	1,092	65	130
6,000	7,000	770	1,540	245	490	573	1,146	68	136
7,000	8,000	800	1,600	250	500	600	1,200	72	144
8,000	9,000	835	1,670	255	510	624	1,248	75	150
9,000	10,000	865	1,730	260	520	645	1,290	78	156
10,000	12,000	875	1,750	270	540	687	1,374	82	164
12,000	14,000	885	1,770	275	550	723	1,446	87	174
14,000	16,000	900	1,800	280	560	756	1,512	90	180
16,000	18,000	940	1,880	285	570	786	1,572	94	188
18,000	20,000	975	1,950	290	580	813	1,626	98	196
20,000	25,000	1,055	2,000	315	630	876	1,752	105	210

(continued)

TABLE 5604.5.2(1)-continued
AMERICAN TABLE OF DISTANCES FOR STORAGE OF EXPLOSIVES AS APPROVED BY THE INSTITUTE OF MAKERS OF EXPLOSIVES AND REVISED JUNE 1991[a]

QUANTITY OF EXPLOSIVE MATERIALS[c]		DISTANCES IN FEET							
		Inhabited buildings		Public highways with traffic volume less than 3,000 vehicles per day		Public highways with traffic volume greater than 3,000 vehicles per day and passenger railways		Separation of magazines[d]	
Pounds over	Pounds not over	Barricaded	Unbarricaded	Barricaded	Unbarricaded	Barricaded	Unbarricaded	Barricaded	Unbarricaded
25,000	30,000	1,130	2,000	340	680	933	1,866	112	224
30,000	35,000	1,205	2,000	360	720	981	1,962	119	238
35,000	40,000	1,275	2,000	380	760	1,026	2,000	124	248
40,000	45,000	1,340	2,000	400	800	1,068	2,000	129	258
45,000	50,000	1,400	2,000	420	840	1,104	2,000	135	270
50,000	55,000	1,460	2,000	440	880	1,140	2,000	140	280
55,000	60,000	1,515	2,000	455	910	1,173	2,000	145	290
60,000	65,000	1,565	2,000	470	940	1,206	2,000	150	300
65,000	70,000	1,610	2,000	485	970	1,236	2,000	155	310
70,000	75,000	1,655	2,000	500	1,000	1,263	2,000	160	320
75,000	80,000	1,695	2,000	510	1,020	1,293	2,000	165	330
80,000	85,000	1,730	2,000	520	1,040	1,317	2,000	170	340
85,000	90,000	1,760	2,000	530	1,060	1,344	2,000	175	350
90,000	95,000	1,790	2,000	540	1,080	1,368	2,000	180	360
95,000	100,000	1,815	2,000	545	1,090	1,392	2,000	185	370
100,000	110,000	1,835	2,000	550	1,100	1,437	2,000	195	390
110,000	120,000	1,855	2,000	555	1,110	1,479	2,000	205	410
120,000	130,000	1,875	2,000	560	1,120	1,521	2,000	215	430
130,000	140,000	1,890	2,000	565	1,130	1,557	2,000	225	450
140,000	150,000	1,900	2,000	570	1,140	1,593	2,000	235	470
150,000	160,000	1,935	2,000	580	1,160	1,629	2,000	245	490
160,000	170,000	1,965	2,000	590	1,180	1,662	2,000	255	510
170,000	180,000	1,990	2,000	600	1,200	1,695	2,000	265	530
180,000	190,000	2,010	2,010	605	1,210	1,725	2,000	275	550
190,000	200,000	2,030	2,030	610	1,220	1,755	2,000	285	570
200,000	210,000	2,055	2,055	620	1,240	1,782	2,000	295	590
210,000	230,000	2,100	2,100	635	1,270	1,836	2,000	315	630
230,000	250,000	2,155	2,155	650	1,300	1,890	2,000	335	670
250,000	275,000	2,215	2,215	670	1,340	1,950	2,000	360	720
275,000	300,000[b]	2,275	2,275	690	1,380	2,000	2,000	385	770

For SI: 1 foot = 304.8 mm, 1 pound = 0.454 kg.

a. This table applies only to the manufacture and permanent storage of commercial explosive materials. It is not applicable to transportation of explosives or any handling or temporary storage necessary or incident thereto. It is not intended to apply to bombs, projectiles or other heavily encased explosives.

b. Storage in excess of 300,000 pounds of explosive materials in one magazine is not allowed.

c. Where a manufacturing building on an explosive materials plant site is designed to contain explosive materials, such building shall be located with respect to its proximity to inhabited buildings, public highways and passenger railways based on the maximum quantity of explosive materials permitted to be in the building at one time.

d. Where two or more storage magazines are located on the same property, each magazine shall comply with the minimum distances specified from inhabited buildings, railways and highways, and, in addition, they should be separated from each other by not less than the distances shown for separation of magazines, except that the quantity of explosives in detonator magazines shall govern in regard to the spacing of said detonator magazines from magazines containing other explosive materials. Where any two or more magazines are separated from each other by less than the specified separation of magazines distances, then two or more such magazines, as a group, shall be considered as one magazine, and the total quantity of explosive materials stored in such group shall be treated as if stored in a single magazine located on the site of any magazine in the group and shall comply with the minimum distances specified from other magazines, inhabited buildings, railways and highways.

TABLE 5604.5.2(2)
TABLE OF DISTANCES (Q-D) FOR BUILDINGS AND MAGAZINES CONTAINING EXPLOSIVES—DIVISION 1.3 MASS-FIRE HAZARD[a, b, c]

QUANTITY OF DIVISION 1.3 EXPLOSIVES (NET EXPLOSIVES WEIGHT)		DISTANCES IN FEET			
Pounds over	Pounds not over	Inhabited Building Distance (IBD)	Distance to Public Traffic Route (PTR)	Intermagazine Distance (IMD)	Intraline Distance (ILD) or Intraplant Distance (IPD)
0	1,000	75	75	50	50
1,000	5,000	115	115	75	75
5,000	10,000	150	150	100	100
10,000	20,000	190	190	125	125
20,000	30,000	215	215	145	145
30,000	40,000	235	235	155	155
40,000	50,000	250	250	165	165
50,000	60,000	260	260	175	175
60,000	70,000	270	270	185	185
70,000	80,000	280	280	190	190
80,000	90,000	295	295	195	195
90,000	100,000	300	300	200	200
100,000	200,000	375	375	250	250
200,000	300,000	450	450	300	300

For SI: 1 foot = 304.8 mm, 1 pound = 0.454 kg

a. Black powder, where stored in magazines, is defined as low explosive by the Bureau of Alcohol, Tobacco, Firearms and Explosives (BATF).

b. For quantities less than 1,000 pounds, the required distances are those specified for 1,000 pounds. The use of lesser distances is allowed where supported by approved test data and/or analysis.

c. Linear interpolation of explosive quantities between table entries is allowed.

TABLE 5604.5.2(3)
TABLE OF DISTANCES (Q-D) FOR BUILDINGS AND MAGAZINES CONTAINING EXPLOSIVES—DIVISION 1.4[c]

QUANTITY OF DIVISION 1.4 EXPLOSIVES (NET EXPLOSIVES WEIGHT)		DISTANCES IN FEET			
Pounds over	Pounds not over	Inhabited Building Distance (IBD)	Distance to Public Traffic Route (PTR)	Intermagazine Distance[a, b] (IMD)	Intraline Distance (ILD) or Intraplant Distance[a] (IPD)
50	Not Limited	100	100	50	50

For SI: 1 foot = 304.8 mm, 1 pound = 0.454 kg.

a. A separation distance of 100 feet is required for buildings of other than Type I or Type II construction as defined in the *International Building Code.*

b. For earth-covered magazines, specific separation is not required.

1. Earth cover material used for magazines shall be relatively cohesive. Solid or wet clay and similar types of soil are too cohesive and shall not be used. Soil shall be free from unsanitary organic matter, trash, debris and stones heavier than 10 pounds or larger than 6 inches in diameter. Compaction and surface preparation shall be provided, as necessary, to maintain structural integrity and avoid erosion. Where cohesive material cannot be used, as in sandy soil, the earth cover over magazines shall be finished with a suitable material to ensure structural integrity.
2. The earth fill or earth cover between earth-covered magazines shall be either solid or sloped, in accordance with the requirements of other construction features, but not less than 2 feet of earth cover shall be maintained over the top of each magazines. To reduce erosion and facilitate maintenance operations, the cover shall have a slope of 2 horizontal to 1 vertical.

c. Restricted to articles, including articles packaged for shipment, that are not regulated as an explosive under Bureau of Alcohol, Tobacco, Firearms and Explosives regulations, or unpacked articles used in process operations that do not propagate a detonation or deflagration between articles. This table shall not apply to consumer fireworks, 1.4G.

5604.5.3.1 Location. Wherever practicable, Type 3 magazines shall be located away from neighboring inhabited buildings, railways, highways and other magazines in accordance with Table 5604.5.2(2) or 5604.5.2(3) as applicable.

5604.5.3.2 Supervision. Type 3 magazines shall be attended when *explosive materials* are stored within. *Explosive materials* shall be removed to appropriate storage magazines for unattended storage at the end of the work day.

5604.5.3.3 Use. Not more than two Type 3 magazines shall be located at the same blasting site. Where two Type 3 magazines are located at the same blasting site, one magazine shall used solely for the storage of detonators.

5604.6 Construction. Magazines shall be constructed in accordance with Sections 5604.6.1 through 5604.6.5.2.

5604.6.1 Drainage. The ground around a magazine shall be graded so that water drains away from the magazine.

5604.6.2 Heating. Magazines requiring heat shall be heated as prescribed in NFPA 495 by either hot water radiant heating within the magazine or by indirect warm air heating.

5604.6.3 Lighting. Where lighting is necessary within a magazine, electric safety flashlights or electric safety lanterns shall be used, except as provided in NFPA 495.

5604.6.4 Nonsparking materials. In other than Type 5 magazines, there shall not be exposed ferrous metal on the interior of a magazine containing packages of *explosives*.

5604.6.5 Signs and placards. Property upon which Type 1 magazines and outdoor magazines of Types 2, 4 and 5 are located shall be posted with signs stating: EXPLOSIVES—KEEP OFF. These signs shall be of contrasting colors with a minimum letter height of 3 inches (76 mm) with a minimum brush stroke of $^1/_2$ inch (12.7 mm). The signs shall be located to minimize the possibility of a bullet shot at the sign hitting the magazine.

5604.6.5.1 Access road signs. At the entrance to *explosive* material manufacturing and storage sites, all access roads shall be posted with the following warning sign or other *approved* sign:

DANGER!
NEVER FIGHT EXPLOSIVE FIRES.
EXPLOSIVES ARE STORED ON THIS SITE
CALL _______.

The sign shall be weather-resistant with a reflective surface and have lettering not less than 2 inches (51 mm) high.

5604.6.5.2 Placards. Type 5 magazines containing Division 1.5 blasting agents shall be prominently placarded as required during transportation by DOTn 49 CFR Part 172 and DOTy 27 CFR Part 55.

5604.7 Operation. Magazines shall be operated in accordance with Sections 5604.7.1 through 5604.7.9.

5604.7.1 Security. Magazines shall be kept locked in the manner prescribed in NFPA 495 at all times except during placement or removal of *explosives* or inspection.

5604.7.2 Open flames and lights. Smoking, matches, flame-producing devices, open flames, firearms and firearms cartridges shall not be allowed inside of or within 50 feet (15 240 mm) of magazines.

5604.7.3 Brush. The area located around a magazine shall be kept clear of brush, dried grass, leaves, trash, debris and similar combustible materials for a distance of 25 feet (7620 mm).

5604.7.4 Combustible storage. Combustible materials shall not be stored within 50 feet (15 240 mm) of magazines.

5604.7.5 Unpacking and repacking explosive materials. Containers of *explosive materials*, except fiberboard containers, and packages of damaged or deteriorated *explosive materials* or fireworks shall not be unpacked or repacked inside or within 50 feet (15 240 mm) of a magazine or in close proximity to other *explosive materials*.

5604.7.5.1 Storage of opened packages. Packages of *explosive materials* that have been opened shall be closed before being placed in a magazine.

5604.7.5.2 Nonsparking tools. Tools used for the opening and closing of packages of *explosive materials*, other than metal slitters for opening paper, plastic or fiberboard containers, shall be made of nonsparking materials.

5604.7.5.3 Disposal of packaging. Empty containers and paper and fiber packaging materials that previously contained *explosive materials* shall be disposed of or reused in a *approved* manner.

5604.7.6 Tools and equipment. Metal tools, other than nonferrous transfer conveyors and ferrous metal conveyor stands protected by a coat of paint, shall not be stored in a magazine containing *explosive materials* or detonators.

5604.7.7 Contents. Magazines shall be used exclusively for the storage of *explosive materials*, blasting materials and blasting accessories.

5604.7.8 Compatibility. Corresponding grades and brands of *explosive materials* shall be stored together and in such a manner that the grade and brand marks are visible. Stocks shall be stored so as to be easily counted and checked. Packages of *explosive materials* shall be stacked in a stable manner not exceeding 8 feet (2438 mm) in height.

5604.7.9 Stock rotation. When *explosive material* is removed from a magazine for use, the oldest usable stocks shall be removed first.

5604.8 Maintenance. Maintenance of magazines shall comply with Sections 5604.8.1 through 5604.8.3.

5604.8.1 Housekeeping. Magazine floors shall be regularly swept and be kept clean, dry and free of grit, paper, empty packages and rubbish. Brooms and other cleaning utensils shall not have any spark-producing metal parts. Sweepings from magazine floors shall be disposed of in accordance with the manufacturers' *approved* instructions.

5604.8.2 Repairs. *Explosive materials* shall be removed from the magazine before making repairs to the interior of a magazine. *Explosive materials* shall be removed from the magazine before making repairs to the exterior of the magazine where there is a possibility of causing a fire. *Explosive materials* removed from a magazine under repair shall either be placed in another magazine or placed a safe distance from the magazine, where they shall be properly guarded and protected until repairs have been completed. Upon completion of repairs, the *explosive materials* shall be promptly returned to the magazine. Floors shall be cleaned before and after repairs.

5604.8.3 Floors. Magazine floors stained with liquid shall be dealt with in accordance with instructions obtained from the manufacturer of the *explosive material* stored in the magazine.

5604.9 Inspection. Magazines containing *explosive materials* shall be opened and inspected at maximum seven-day intervals. The inspection shall determine whether there has been an unauthorized or attempted entry into a magazine or an unauthorized removal of a magazine or its contents.

5604.10 Disposal of explosive materials. *Explosive materials* shall be disposed of in accordance with Sections 5604.10.1 through 5604.10.7.

5604.10.1 Notification. The *fire code official* shall be notified immediately where deteriorated or leaking *explosive materials* are determined to be dangerous or unstable and in need of disposal.

5604.10.2 Deteriorated materials. Where an *explosive material* has deteriorated to an extent that it is in an unstable or dangerous condition, or when a liquid has leaked from an *explosive material*, the person in possession of such material shall immediately contact the material's manufacturer to obtain disposal and handling instructions.

5604.10.3 Qualified person. The work of destroying *explosive materials* shall be directed by persons experienced in the destruction of *explosive* materials.

5604.10.4 Storage of misfires. *Explosive materials* and fireworks recovered from blasting or display misfires shall be placed in a magazine until an experienced person has determined the proper method for disposal.

5604.10.5 Disposal sites. Sites for the destruction of *explosive materials* and fireworks shall be *approved* and located at the maximum practicable safe distance from inhabited buildings, public highways, operating buildings and all other exposures to ensure keeping air blast and ground vibration to a minimum. The location of disposal sites shall not be closer to magazines, inhabited buildings, railways, highways and other rights-of-way than is allowed by Tables 5604.5.2(1), 5604.5.2(2) and 5604.5.2(3). Where possible, *barricades* shall be utilized between the destruction site and inhabited buildings. Areas where *explosives* are detonated or burned shall be posted with adequate warning signs.

5604.10.6 Reuse of site. Unless an *approved* burning site has been thoroughly saturated with water and has passed a safety inspection, 48 hours shall elapse between the completion of a burn and the placement of scrap explosive materials for a subsequent burn.

5604.10.7 Personnel safeguards. Once an *explosive* burn operation has been started, personnel shall relocate to a safe location where adequate protection from air blast and flying debris is provided. Personnel shall not return to the burn area until the person in charge has inspected the burn site and determined that it is safe for personnel to return.

SECTION 5605 MANUFACTURE, ASSEMBLY AND TESTING OF EXPLOSIVES, EXPLOSIVE MATERIALS AND FIREWORKS

5605.1 General. The manufacture, assembly and testing of *explosives*, ammunition, blasting agents and fireworks shall comply with the requirements of this section and NFPA 495 or NFPA 1124.

Exceptions:

1. The hand loading of small arms ammunition prepared for personal use and not offered for resale.
2. The mixing and loading of blasting agents at blasting sites in accordance with NFPA 495.
3. The use of binary *explosives* or plosophoric materials in blasting or pyrotechnic special effects applications in accordance with NFPA 495 or NFPA 1126.

5605.2 Emergency planning and preparedness. Emergency plans, emergency drills, employee training and hazard communication shall conform to the provisions of this section and Sections 404, 405, 406 and 407.

5605.2.1 Hazardous Materials Management Plans and Inventory Statements required. Detailed Hazardous Materials Management Plans (HMMP) and Hazardous Materials Inventory Statements (HMIS) complying with the requirements of Section 407 shall be prepared and submitted to the local emergency planning committee, the *fire code official* and the local fire department.

5605.2.2 Maintenance of plans. A copy of the required HMMP and HMIS shall be maintained on site and furnished to the *fire code official* on request.

5605.2.3 Employee training. Workers who handle *explosives* or *explosive* charges or dispose of *explosives* shall be trained in the hazards of the materials and processes in which they are to be engaged and with the safety rules governing such materials and processes.

5605.2.4 Emergency procedures. *Approved* emergency procedures shall be formulated for each plant and shall include personal instruction in any anticipated emergency. Personnel shall be made aware of an emergency warning signal.

5605.3 Intraplant separation of operating buildings. *Explosives* manufacturing buildings and fireworks manufacturing buildings, including those where *explosive* charges are assembled, manufactured, prepared or loaded utilizing Division 1.1, 1.2, 1.3, 1.4 or 1.5 *explosives*, shall be separated from all other buildings, including magazines, within the confines of the manufacturing plant, at a distance not less than those shown in Table 5605.3 or 5604.5.2(3), as appropriate.

Exception: Fireworks manufacturing buildings separated in accordance with NFPA 1124.

The quantity of *explosives* in an operating building shall be the net weight of all *explosives* contained therein. Distances shall be based on the hazard division requiring the greatest separation, unless the aggregate *explosive* weight is divided by *approved* walls or shields designed for that purpose. Where dividing a quantity of *explosives* into smaller stacks, a suitable barrier or adequate separation distance shall be provided to prevent propagation from one stack to another.

Where distance is used as the sole means of separation within a building, such distance shall be established by testing. Testing shall demonstrate that propagation between stacks will not result. Barriers provided to protect against *explosive* effects shall be designed and installed in accordance with *approved* standards.

5605.4 Separation of manufacturing operating buildings from inhabited buildings, public traffic routes and magazines. Where an operating building on an *explosive* materials plant site is designed to contain *explosive* materials, such a building shall be located away from inhabited buildings, public traffic routes and magazines in accordance with Table 5604.5.2(2) or 5604.5.2(3) as appropriate, based on the maximum quantity of *explosive* materials permitted to be in the building at one time (see Section 5601.8).

Exception: Fireworks manufacturing buildings constructed and operated in accordance with NFPA 1124.

5605.4.1 Determination of net explosive weight for operating buildings. In addition to the requirements of Section 5601.8 to determine the net *explosive* weight for materials stored or used in operating buildings, quantities of *explosive materials* stored in magazines located at distances less than intraline distances from the operating building shall be added to the contents of the operating building to determine the net *explosive* weight for the operating building.

5605.4.1.1 Indoor magazines. The storage of *explosive* materials located in indoor magazines in operating buildings shall be limited to a net *explosive* weight not to exceed 50 pounds (23 kg).

5605.4.1.2 Outdoor magazines with a net explosive weight less than 50 pounds. The storage of *explosive*

TABLE 5605.3
MINIMUM INTRALINE (INTRAPLANT) SEPARATION DISTANCES (ILD OR IPD) BETWEEN BARRICADED OPERATING BUILDINGS CONTAINING EXPLOSIVES—DIVISION 1.1, 1.2 OR 1.5 MASS-EXPLOSION HAZARD[a]

NET EXPLOSIVE WEIGHT			NET EXPLOSIVE WEIGHT		
Pounds over	Pounds not over	Intraline Distance (ILD) or Intraplant Distance (IPD) (feet)	Pounds over	Pounds not over	Intraline Distance (ILD) or Intraplant Distance (IPD) (feet)
0	50	30	20,000	25,000	265
50	100	40	25,000	30,000	280
100	200	50	30,000	35,000	295
200	300	60	35,000	40,000	310
300	400	65	40,000	45,000	320
400	500	70	45,000	50,000	330
500	600	75	50,000	55,000	340
600	700	80	55,000	60,000	350
700	800	85	60,000	65,000	360
800	900	90	65,000	70,000	370
900	1,000	95	70,000	75,000	385
1,000	1,500	105	75,000	80,000	390
1,500	2,000	115	80,000	85,000	395
2,000	3,000	130	85,000	90,000	400
3,000	4,000	140	90,000	95,000	410
4,000	5,000	150	95,000	100,000	415
5,000	6,000	160	100,000	125,000	450
6,000	7,000	170	125,000	150,000	475
7,000	8,000	180	150,000	175,000	500
8,000	9,000	190	175,000	200,000	525
9,000	10,000	200	200,000	225,000	550
10,000	15,000	225	225,000	250,000	575
15,000	20,000	245	250,000	275,000	600
—	—	—	275,000	300,000	635

For SI: 1 foot = 304.8 mm, 1 pound = 0.454 kg.

a. Where a building or magazine containing explosives is not barricaded, the intraline distances shown in this table shall be doubled.

materials in outdoor magazines located at less than intraline distances from operating buildings shall be limited to a net *explosive* weight not to exceed 50 pounds (23 kg).

5605.4.1.3 Outdoor magazines with a net explosive weight greater than 50 pounds. The storage of *explosive materials* in outdoor magazines in quantities exceeding 50 pounds (23 kg) net *explosive* weight shall be limited to storage in outdoor magazines located not less than intraline distances from the operating building in accordance with Section 5604.5.2.

5605.4.1.4 Net explosive weight of materials stored in combination indoor and outdoor magazines. The aggregate quantity of *explosive materials* stored in any combination of indoor magazines or outdoor magazines located at less than the intraline distances from an operating building shall not exceed 50 pounds (23 kg).

5605.5 Buildings and equipment. Buildings or rooms that exceed the *maximum allowable quantity per control area* of *explosive materials* shall be operated in accordance with this section and constructed in accordance with the requirements of the *International Building Code* for Group H occupancies.

Exception: Fireworks manufacturing buildings constructed and operated in accordance with NFPA 1124.

5605.5.1 Explosives dust. *Explosives* dust shall not be exhausted to the atmosphere.

5605.5.1.1 Wet collector. When collecting *explosives* dust, a wet collector system shall be used. Wetting agents shall be compatible with the *explosives*. Collector systems shall be interlocked with process power supplies so that the process cannot continue without the collector systems also operating.

5605.5.1.2 Waste disposal and maintenance. *Explosives* dust shall be removed from the collection chamber as often as necessary to prevent overloading. The entire system shall be cleaned at a frequency that will eliminate hazardous concentrations of *explosives* dust in pipes, tubing and ducts.

5605.5.2 Exhaust fans. Squirrel cage blowers shall not be used for exhausting hazardous fumes, vapors or gases. Only nonferrous fan blades shall be used for fans located within the ductwork and through which hazardous materials are exhausted. Motors shall be located outside the duct.

5605.5.3 Work stations. Work stations shall be separated by distance, barrier or other *approved* alternatives so that fire in one station will not ignite material in another work station. Where necessary, the operator shall be protected by a personnel shield located between the operator and the *explosive* device or *explosive material* being processed. This shield and its support shall be capable of withstanding a blast from the maximum amount of *explosives* allowed behind it.

5605.6 Operations. Operations involving *explosives* shall comply with Sections 5605.6.1 through 5605.6.10.

5605.6.1 Isolation of operations. Where the type of material and processing warrants, mechanical operations involving *explosives* in excess of 1 pound (0.454 kg) shall be carried on at isolated stations or at intraplant distances, and machinery shall be controlled from remote locations behind *barricades* or at separations so that workers will be at a safe distance while machinery is operating.

5605.6.2 Static controls. The work area where the screening, grinding, blending and other processing of static-sensitive *explosives* or pyrotechnic materials is done shall be provided with *approved* static controls.

5605.6.3 Approved containers. Bulk *explosives* shall be kept in *approved*, nonsparking containers when not being used or processed. *Explosives* shall not be stored or transported in open containers.

5605.6.4 Quantity limits. The quantity of *explosives* at any particular work station shall be limited to that posted on the load limit signs for the individual work station. The total quantity of *explosives* for multiple workstations shall not exceed that established by the intraplant distances in Table 5605.3 or 5604.5.2(3), as appropriate.

5605.6.4.1 Magazines. Magazines used for storage in processing areas shall be in accordance with the requirements of Section 5604.5.1. *Explosive materials* shall be removed to appropriate storage magazines for unattended storage at the end of the work day. The contents of indoor magazines shall be added to the quantity of *explosives* contained at individual workstations and the total quantity of material stored, processed or used shall be utilized to establish the intraplant separation distances indicated by Table 5605.3 or 5604.5.2(3), as appropriate.

5605.6.5 Waste disposal. *Approved* receptacles with covers shall be provided for each location for disposing of waste material and debris. These waste receptacles shall be emptied and cleaned as often as necessary but not less than once each day or at the end of each shift.

5605.6.6 Safety rules. General safety rules and operating instructions governing the particular operation or process conducted at that location shall be available at each location.

5605.6.7 Personnel limits. The number of occupants in each process building and in each magazine shall not exceed the number necessary for proper conduct of production operations.

5605.6.8 Pyrotechnic and explosive composition quantity limits. Not more than 500 pounds (227 kg) of pyrotechnic or *explosive* composition, including not more than 10 pounds (5 kg) of salute powder shall be allowed at one time in any process building or area. Compositions not in current use shall be kept in covered nonferrous containers.

Exception: Composition that has been loaded or pressed into tubes or other containers as consumer fireworks.

5605.6.9 Posting limits. The maximum number of occupants and maximum weight of pyrotechnic and *explosive* composition permitted in each process building shall be posted in a conspicuous location in each process building or magazine.

5605.6.10 Heat sources. Fireworks, *explosives* or *explosive* charges in *explosive materials* manufacturing, assembly or testing shall not be stored near any source of heat.

Exception: *Approved* drying or curing operations.

5605.7 Maintenance. Maintenance and repair of *explosives*-manufacturing facilities and areas shall comply with Section 5604.8.

5605.8 Explosive materials testing sites. *Detonation* of *explosive* materials or ignition of fireworks for testing purposes shall be done only in isolated areas at sites where distance, protection from missiles, shrapnel or flyrock, and other safeguards provides protection against injury to personnel or damage to property.

5605.8.1 Protective clothing and equipment. Protective clothing and equipment shall be provided to protect persons engaged in the testing, ignition or *detonation* of *explosive materials.*

5605.8.2 Site security. Where tests are being conducted or *explosives* are being detonated, only authorized persons shall be present. Areas where *explosives* are regularly or frequently detonated or burned shall be *approved* and posted with adequate warning signs. Warning devices shall be activated before burning or detonating *explosives* to alert persons approaching from any direction that they are approaching a danger zone.

5605.9 Waste disposal. Disposal of *explosive materials* waste from manufacturing, assembly or testing operations shall be in accordance with Section 5604.10.

SECTION 5606
SMALL ARMS AMMUNITION AND SMALL ARMS AMMUNITION COMPONENTS

5606.1 General. Indoor storage and display of black powder, smokeless propellants, small arms primers and small arms ammunition shall comply with this section and NFPA 495.

5606.2 Prohibited storage. Small arms ammunition shall not be stored together with Division 1.1, Division 1.2 or Division 1.3 *explosives* unless the storage facility is suitable for the storage of *explosive materials.*

5606.3 Packages. Smokeless propellants shall be stored in *approved* shipping containers conforming to DOTn 49 CFR Part 173.

5606.3.1 Repackaging. The bulk repackaging of smokeless propellants, black powder and small arms primers shall not be performed in retail establishments.

5606.3.2 Damaged packages. Damaged containers shall not be repackaged.

Exception: *Approved* repackaging of damaged containers of smokeless propellant into containers of the same type and size as the original container.

5606.4 Storage in Group R occupancies. The storage of small arms ammunition components in Group R occupancies shall comply with Sections 5606.4.1 through 5606.4.3.

5606.4.1 Black powder. Black powder for personal use in quantities not exceeding 20 pounds (9 kg) shall be stored in original containers in occupancies limited to Group R-3. Quantities exceeding 20 pounds (9 kg) shall not be stored in any Group R occupancy.

5606.4.2 Smokeless propellants. Smokeless propellants for personal use in quantities not exceeding 20 pounds (9 kg) shall be stored in original containers in occupancies limited to Group R-3. Smokeless propellants in quantities exceeding 20 pounds (9 kg) but not exceeding 50 pounds (23 kg) and kept in a wooden box or cabinet having walls of not less than 1 inch (25 mm) nominal thickness shall be allowed to be stored in occupancies limited to Group R-3. Quantities exceeding these amounts shall not be stored in any Group R occupancy.

5606.4.3 Small arms primers. Not more than 10,000 small arms primers shall be stored in occupancies limited to Group R-3.

5606.5 Display and storage in Group M occupancies. The display and storage of small arms ammunition components in Group M occupancies shall comply with Sections 5606.5.1 through 5606.5.2.3.

5606.5.1 Display. Display of small arms ammunition components in Group M occupancies shall comply with Sections 5606.5.1.1 through 5606.5.1.3.

5606.5.1.1 Smokeless propellant. Not more than 20 pounds (9 kg) of smokeless propellants, in containers of 1 pound (0.454 kg) or less capacity each, shall be displayed in Group M occupancies.

5606.5.1.2 Black powder. Not more than 1 pound (0.454 kg) of black powder shall be displayed in Group M occupancies.

5606.5.1.3 Small arms primers. Not more than 10,000 small arms primers shall be displayed in Group M occupancies.

5606.5.2 Storage. Storage of small arms ammunition components shall comply with Sections 5606.5.2.1 through 5606.5.2.3.

5606.5.2.1 Smokeless propellant. Commercial stocks of smokeless propellants shall be stored as follows:

1. Quantities exceeding 20 pounds (9 kg), but not exceeding 100 pounds (45 kg) shall be stored in portable wooden boxes having walls of not less than 1 inch (25 mm) nominal thickness.
2. Quantities exceeding 100 pounds (45 kg), but not exceeding 800 pounds (363 kg), shall be stored in nonportable storage cabinets having walls not less than 1 inch (25 mm) nominal thickness. Not more than 400 pounds (182 kg) shall be stored in any one cabinet, and cabinets shall be separated

by a distance of not less than 25 feet (7620 mm) or by a *fire partition* having a *fire-resistance rating* of not less than 1 hour.

3. Storage of quantities exceeding 800 pounds (363 kg), but not exceeding 5,000 pounds (2270 kg) in a building shall comply with all of the following:
   3.1. The warehouse or storage room is unaccessible to unauthorized personnel.
   3.2. Smokeless propellant shall be stored in nonportable storage cabinets having wood walls not less than 1 inch (25 mm) nominal thickness and having shelves with not more than 3 feet (914 mm) of separation between shelves.
   3.3. Not more than 400 pounds (182 kg) is stored in any one cabinet.
   3.4. Cabinets shall be located against walls of the storage room or warehouse with not less than 40 feet (12 192 mm) between cabinets.
   3.5. The minimum required separation between cabinets shall be 20 feet (6096 mm) provided that *barricades* twice the height of the cabinets are attached to the wall, midway between each cabinet. The *barricades* must extend not less than 10 feet (3048 mm) outward, be firmly attached to the wall and be constructed of steel not less than $^{1}/_{4}$ inch thick (6.4 mm), 2-inch (51 mm) nominal thickness wood, brick or concrete block.
   3.6. Smokeless propellant shall be separated from materials classified as *combustible liquids*, flammable liquids, flammable solids or oxidizing materials by a distance of 25 feet (7620 mm) or by a *fire partition* having a *fire-resistance rating* of 1 hour.
   3.7. The building shall be equipped throughout with an *automatic sprinkler system* installed in accordance with Section 903.3.1.1.
4. Smokeless propellants not stored in accordance with Item 1, 2, or 3 above shall be stored in a Type 2 or 4 magazine in accordance with Section 5604 and NFPA 495.

5606.5.2.2 Black powder. Commercial stocks of black powder in quantities less than 50 pounds (23 kg) shall be allowed to be stored in Type 2 or 4 indoor or outdoor magazines. Quantities greater than 50 pounds (23 kg) shall be stored in outdoor Type 2 or 4 magazines. Where black powder and smokeless propellants are stored together in the same magazine, the total quantity shall not exceed that permitted for black powder.

5606.5.2.3 Small arms primers. Commercial stocks of small arms primers shall be stored as follows:

1. Quantities not to exceed 750,000 small arms primers stored in a building shall be arranged such that not more than 100,000 small arms primers are stored in any one pile and piles are not less than 15 feet (4572 mm) apart.
2. Quantities exceeding 750,000 small arms primers stored in a building shall comply with all of the following:
   2.1. The warehouse or storage building shall not be accessible to unauthorized personnel.
   2.2. Small arms primers shall be stored in cabinets. Not more than 200,000 small arms primers shall be stored in any one cabinet.
   2.3. Shelves in cabinets shall have vertical separation of not less than 2 feet (610 mm).
   2.4. Cabinets shall be located against walls of the warehouse or storage room with not less than 40 feet (12 192 mm) between cabinets. The minimum required separation between cabinets shall be allowed to be reduced to 20 feet (6096 mm) provided that *barricades* twice the height of the cabinets are attached to the wall, midway between each cabinet. The *barricades* shall be firmly attached to the wall and shall be constructed of steel not less than $^{1}/_{4}$ inch thick (6.4 mm), 2-inch (51 mm) nominal thickness wood, brick or concrete block.
   2.5. Small arms primers shall be separated from materials classified as *combustible liquids*, flammable liquids, flammable solids or oxidizing materials by a distance of 25 feet (7620 mm) by a *fire partition* having a *fire-resistance rating* of 1 hour.
   2.6. The building shall be protected throughout with an *automatic sprinkler system* installed in accordance with Section 903.3.1.1.
3. Small arms primers not stored in accordance with Item 1 or 2 of this section shall be stored in a magazine meeting the requirements of Section 5604 and NFPA 495.

SECTION 5607
BLASTING

5607.1 General. Blasting operations shall be conducted only by *approved*, competent operators familiar with the required safety precautions and the hazards involved and in accordance with the provisions of NFPA 495.

5607.2 Manufacturer's instructions. Blasting operations shall be performed in accordance with the instructions of the manufacturer of the *explosive materials* being used.

5607.3 Blasting in congested areas. Where blasting is done in a congested area or in close proximity to a structure, railway or highway, or any other installation, precautions shall be taken to minimize earth vibrations and air blast effects. Blasting mats or other protective means shall be used to prevent fragments from being thrown.

5607.4 Restricted hours. Surface-blasting operations shall only be conducted during daylight hours between sunrise and sunset. Other blasting shall be performed during daylight hours unless otherwise *approved* by the *fire code official*.

5607.5 Utility notification. Where blasting is being conducted in the vicinity of utility lines or rights-of-way, the blaster shall notify the appropriate representatives of the utilities not less than 24 hours in advance of blasting, specifying the location and intended time of such blasting. Verbal notices shall be confirmed with written notice.

Exception: In an emergency situation, the time limit shall not apply where *approved*.

5607.6 Electric detonator precautions. Precautions shall be taken to prevent accidental discharge of electric detonators from currents induced by radar and radio transmitters, lightning, adjacent power lines, dust and snow storms, or other sources of extraneous electricity.

5607.7 Nonelectric detonator precautions. Precautions shall be taken to prevent accidental initiation of nonelectric detonators from stray currents induced by lightning or static electricity.

5607.8 Blasting area security. During the time that holes are being loaded or are loaded with *explosive materials*, blasting agents or detonators, only authorized persons engaged in drilling and loading operations or otherwise authorized to enter the site shall be allowed at the blast site. The blast site shall be guarded or barricaded and posted. Blast site security shall be maintained until after the post-blast inspection has been completed.

5607.9 Drill holes. Holes drilled for the loading of *explosive* charges shall be made and loaded in accordance with NFPA 495.

5607.10 Removal of excess explosive materials. After loading for a blast is completed and before firing, excess *explosive materials* shall be removed from the area and returned to the proper storage facilities.

5607.11 Initiation means. The initiation of blasts shall be by means conforming to the provisions of NFPA 495.

5607.12 Connections. The blaster shall supervise the connecting of the blastholes and the connection of the loadline to the power source or initiation point. Connections shall be made progressively from the blasthole back to the initiation point.

Blasting lead lines shall remain shunted (shorted) and shall not be connected to the blasting machine or other source of current until the blast is to be fired.

5607.13 Firing control. A blast shall not be fired until the blaster has made certain that all surplus *explosive materials* are in a safe place in accordance with Section 5607.10, all persons and equipment are at a safe distance or under sufficient cover and that an adequate warning signal has been given.

5607.14 Post-blast procedures. After the blast, the following procedures shall be observed.

1. No person shall return to the blast area until allowed to do so by the blaster in charge.
2. The blaster shall allow sufficient time for smoke and fumes to dissipate and for dust to settle before returning to or approaching the blast area.
3. The blaster shall inspect the entire blast site for misfires before allowing other personnel to return to the blast area.

5607.15 Misfires. Where a misfire is suspected, all initiating circuits shall be traced and a search made for unexploded charges. Where a misfire is found, the blaster shall provide proper safeguards for excluding all personnel from the blast area. Misfires shall be reported to the blasting supervisor immediately. Misfires shall be handled under the direction of the person in charge of the blasting operation in accordance with NFPA 495.

SECTION 5608
FIREWORKS DISPLAY

5608.1 General. Outdoor fireworks displays, use of pyrotechnics before a *proximate audience* and pyrotechnic special effects in motion picture, television, theatrical and group entertainment productions shall comply with Sections 5608.2 through 5608.10 and NFPA 1123 or NFPA 1126.

5608.2 Permit application. Prior to issuing permits for a fireworks display, plans for the fireworks display, inspections of the display site and demonstrations of the display operations shall be *approved*. A plan establishing procedures to follow and actions to be taken in the event that a shell fails to ignite in, or discharge from, a mortar or fails to function over the fallout area or other malfunctions shall be provided to the *fire code official*.

5608.2.1 Outdoor fireworks displays. In addition to the requirements of Section 403, permit applications for outdoor fireworks displays using Division 1.3G fireworks shall include a diagram of the location at which the fireworks display will be conducted, including the site from which fireworks will be discharged; the location of buildings, highways, overhead obstructions and utilities; and the lines behind which the audience will be restrained.

5608.2.2 Use of pyrotechnics before a proximate audience. Where the separation distances required in Section 5608.4 and NFPA 1123 are unavailable or cannot be secured, fireworks displays shall be conducted in accordance with NFPA 1126 for *proximate audiences*. Applications for use of pyrotechnics before a *proximate audience* shall include plans indicating the required clearances for spectators and combustibles, crowd control measures, smoke control measures and requirements for standby personnel and equipment where provision of such personnel or equipment is required by the *fire code official*.

5608.3 Approved fireworks displays. *Approved* fireworks displays shall include only the *approved* fireworks 1.3G, fireworks 1.4G, fireworks 1.4S and pyrotechnic articles, 1.4G, which shall be handled by an *approved*, competent operator. The *approved* fireworks shall be arranged, located, discharged and fired in a manner that will not pose a hazard to property or endanger any person.

5608.4 Clearance. Spectators, spectator parking areas, and *dwellings*, buildings or structures shall not be located within the display site.

Exceptions:

1. This provision shall not apply to pyrotechnic special effects and fireworks displays using Division 1.4G materials before a *proximate audience* in accordance with NFPA 1126.
2. This provision shall not apply to unoccupied *dwellings*, buildings and structures with the approval of the building *owner* and the *fire code official*.

5608.5 Storage of fireworks at display site. The storage of fireworks at the display site shall comply with the requirements of this section and NFPA 1123 or NFPA 1126.

5608.5.1 Supervision and weather protection. Beginning as soon as fireworks have been delivered to the display site, they shall not be left unattended.

5608.5.2 Weather protection. Fireworks shall be kept dry after delivery to the display site.

5608.5.3 Inspection. Shells shall be inspected by the operator or assistants after delivery to the display site. Shells having tears, leaks, broken fuses or signs of having been wet shall be set aside and shall not be fired. Aerial shells shall be checked for proper fit in mortars prior to discharge. Aerial shells that do not fit properly shall not be fired. After the fireworks display, damaged, deteriorated or dud shells shall either be returned to the supplier or destroyed in accordance with the supplier's instructions and Section 5604.10.

Exception: Minor repairs to fuses shall be allowed. For electrically ignited displays, attachment of electric matches and similar tasks shall be allowed.

5608.5.4 Sorting and separation. After delivery to the display site and prior to the fireworks display, all shells shall be separated according to their size and their designation as salutes.

Exception: For electrically fired displays, or displays where all shells are loaded into mortars prior to the show, there is no requirement for separation of shells according to their size or their designation as salutes.

5608.5.5 Ready boxes. Display fireworks, 1.3G, that will be temporarily stored at the site during the fireworks display shall be stored in ready boxes located upwind and not less than 25 feet (7620 mm) from the mortar placement and separated according to their size and their designation as salutes.

Exception: For electrically fired fireworks displays, or fireworks displays where all shells are loaded into mortars prior to the show, there is no requirement for separation of shells according to their size, their designation as salutes or for the use of ready boxes.

5608.6 Installation of mortars. Mortars for firing fireworks shells shall be installed in accordance with NFPA 1123 and shall be positioned so that shells are propelled away from spectators and over the fallout area. Under no circumstances shall mortars be angled toward the spectator viewing area. Prior to placement, mortars shall be inspected for defects, such as dents, bent ends, damaged interiors and damaged plugs. Defective mortars shall not be used.

5608.7 Handling. Aerial shells shall be carried to mortars by the shell body. For the purpose of loading mortars, aerial shells shall be held by the thick portion of the fuse and carefully loaded into mortars.

5608.8 Fireworks display supervision. Whenever in the opinion of the *fire code official* or the operator a hazardous condition exists, the fireworks display shall be discontinued immediately until such time as the dangerous situation is corrected.

5608.9 Post-fireworks display inspection. After the fireworks display, the firing crew shall conduct an inspection of the fallout area for the purpose of locating unexploded aerial shells or live components. This inspection shall be conducted before public access to the site shall be allowed. Where fireworks are displayed at night and it is not possible to inspect the site thoroughly, the operator or designated assistant shall inspect the entire site at first light.

A report identifying any shells that fail to ignite in, or discharge from, a mortar or fail to function over the fallout area or otherwise malfunction, shall be filed with the *fire code official*.

5608.10 Disposal. Any shells found during the inspection required in Section 5608.9 shall not be handled until not less than 15 minutes have elapsed from the time the shells were fired. The fireworks shall then be doused with water and allowed to remain for not less than 5 additional minutes before being placed in a plastic bucket or fiberboard box. The disposal instructions of the manufacturer as provided by the fireworks supplier shall then be followed in disposing of the fireworks in accordance with Section 5604.10.

SECTION 5609 TEMPORARY STORAGE OF CONSUMER FIREWORKS

5609.1 General. Where the temporary storage of consumer fireworks, 1.4G is allowed by Section 5601.1.3, Exception 4, such storage shall comply with the applicable requirements of NFPA 1124.

CHAPTER 57

FLAMMABLE AND COMBUSTIBLE LIQUIDS

SECTION 5701 GENERAL

5701.1 Scope and application. Prevention, control and mitigation of dangerous conditions related to storage, use, dispensing, mixing and handling of flammable and *combustible liquids* shall be in accordance with Chapter 50 and this chapter.

5701.2 Nonapplicability. This chapter shall not apply to liquids as otherwise provided in other laws or regulations or chapters of this code, including:

1. Specific provisions for flammable liquids in motor fuel-dispensing facilities, repair garages, airports and marinas in Chapter 23.
2. Medicines, foodstuffs, cosmetics and commercial or institutional products containing not more than 50 percent by volume of water-miscible liquids and with the remainder of the solution not being flammable, provided that such materials are packaged in individual containers not exceeding 1.3 gallons (5 L).
3. Quantities of alcoholic beverages in retail or wholesale sales or storage occupancies, providied that the liquids are packaged in individual containers not exceeding 1.3 gallons (5 L).
4. Storage and use of fuel oil in tanks and containers connected to oil-burning equipment. Such storage and use shall be in accordance with Section 603. For abandonment of fuel oil tanks, this chapter applies.
5. Refrigerant liquids and oils in refrigeration systems (see Section 606).
6. Storage and display of aerosol products complying with Chapter 51.
7. Storage and use of liquids that do not have a fire point when tested in accordance with ASTM D 92.
8. Liquids with a *flash point* greater than 95°F (35°C) in a water-miscible solution or dispersion with a water and inert (noncombustible) solids content of more than 80 percent by weight, which do not sustain combustion.
9. Liquids without *flash points* that can be flammable under some conditions, such as certain halogenated hydrocarbons and mixtures containing halogenated hydrocarbons.
10. The storage of distilled spirits and wines in wooden barrels and casks.
11. Commercial cooking oil storage tank systems located within a building and designed and installed in accordance with Section 610 and NFPA 30.

5701.3 Referenced documents. The applicable requirements of Chapter 50, other chapters of this code, the *International Building Code* and the *International Mechanical Code* pertaining to flammable liquids shall apply.

5701.4 Permits. Permits shall be required as set forth in Sections 105.6 and 105.7.

5701.5 Material classification. Flammable and *combustible liquids* shall be classified in accordance with the definitions in Chapter 2.

When mixed with lower flash-point liquids, Class II or III liquids are capable of assuming the characteristics of the lower flash-point liquids. Under such conditions, the appropriate provisions of this chapter for the actual *flash point* of the mixed liquid shall apply.

When heated above their *flash points*, Class II and III liquids assume the characteristics of Class I liquids. Under such conditions, the appropriate provisions of this chapter for flammable liquids shall apply.

SECTION 5702 DEFINITIONS

5702.1 Definitions. The following terms are defined in Chapter 2:

ALCOHOL-BASED HAND RUB.

BULK PLANT OR TERMINAL.

BULK TRANSFER.

COMBUSTIBLE LIQUID.
Class II.
Class IIIA.
Class IIIB.

FIRE POINT.

FLAMMABLE LIQUID.
Class IA.
Class IB.
Class IC.

FLASH POINT.

FUEL LIMIT SWITCH.

LIQUID STORAGE ROOM.

LIQUID STORAGE WAREHOUSE.

MOBILE FUELING.

PROCESS TRANSFER.

REFINERY.

REMOTE EMERGENCY SHUTOFF DEVICE.

REMOTE SOLVENT RESERVOIR.

SOLVENT DISTILLATION UNIT.

TANK, PRIMARY.

SECTION 5703
GENERAL REQUIREMENTS

5703.1 Electrical. Electrical wiring and equipment shall be installed and maintained in accordance with Section 605 and NFPA 70.

5703.1.1 Classified locations for flammable liquids. Areas where flammable liquids are stored, handled, dispensed or mixed shall be in accordance with Table 5703.1.1. A classified area shall not extend beyond an unpierced floor, roof or other solid partition.

The extent of the classified area is allowed to be reduced, or eliminated, where sufficient technical justification is provided to the *fire code official* that a concentration in the area in excess of 25 percent of the lower flammable limit (LFL) cannot be generated.

5703.1.2 Classified locations for combustible liquids. Areas where Class II or III liquids are heated above their *flash points* shall have electrical installations in accordance with Section 5703.1.1.

Exception: Solvent distillation units in accordance with Section 5705.4.

5703.1.3 Other applications. The *fire code official* is authorized to determine the extent of the Class I electrical equipment and wiring location where a condition is not specifically covered by these requirements or NFPA 70.

5703.2 Fire protection. Fire protection for the storage, use, dispensing, mixing, handling and on-site transportation of flammable and *combustible liquids* shall be in accordance with this chapter and applicable sections of Chapter 9.

5703.2.1 Portable fire extinguishers and hose lines. Portable fire extinguishers shall be provided in accordance with Section 906. Hose lines shall be provided in accordance with Section 905.

5703.3 Site assessment. In the event of a spill, leak or discharge from a tank system, a site assessment shall be completed by the *owner* or operator of such tank system if the *fire code official* determines that a potential fire or explosion hazard exists. Such site assessments shall be conducted to ascertain potential fire hazards and shall be completed and submitted to the fire department within a time period established by the *fire code official*, not to exceed 60 days.

5703.4 Spill control and secondary containment. Where the *maximum allowable quantity per control area* is exceeded, and where required by Section 5004.2, rooms, buildings or areas used for storage, dispensing, use, mixing or handling of Class I, II and IIIA liquids shall be provided with spill control and secondary containment in accordance with Section 5004.2.

5703.5 Labeling and signage. The *fire code official* is authorized to require warning signs for the purpose of identifying the hazards of storing or using flammable liquids. Signage for identification and warning such as for the inherent hazard of flammable liquids or smoking shall be provided in accordance with this chapter and Sections 5003.5 and 5003.6.

5703.5.1 Style. Warning signs shall be of a durable material. Signs warning of the hazard of flammable liquids shall have white lettering on a red background and shall read: DANGER—FLAMMABLE LIQUIDS. Letters shall be not less than 3 inches (76 mm) in height and $^1/_2$ inch (12.7 mm) in stroke.

5703.5.2 Location. Signs shall be posted in locations as required by the *fire code official*. Piping containing flammable liquids shall be identified in accordance with ASME A13.1.

5703.5.3 Warning labels. Individual containers, packages and cartons shall be identified, marked, labeled and placarded in accordance with federal regulations and applicable state laws.

5703.5.4 Identification. Color coding or other *approved* identification means shall be provided on each loading and unloading riser for flammable or *combustible liquids* to identify the contents of the tank served by the riser.

5703.6 Piping systems. Piping systems, and their component parts, for flammable and *combustible liquids* shall be in accordance with Sections 5703.6.1 through 5703.6.11.

5703.6.1 Nonapplicability. The provisions of Section 5703.6 shall not apply to gas or oil well installations; piping that is integral to stationary or portable engines, including aircraft, watercraft and motor vehicles; and piping in connection with boilers and pressure vessels regulated by the *International Mechanical Code*.

5703.6.2 Design and fabrication of piping systems and components. Piping system components shall be designed and fabricated in accordance with the applicable standard listed in Table 5703.6.2 and Chapter 27 of NFPA 30, except as modified by Section 5703.6.2.1.

TABLE 5703.6.2
PIPING STANDARDS

PIPING USE	STANDARD
Power Piping	ASME B31.1
Process Piping	ASME B31.3
Pipeline Transportation Systems for Liquid Hydrocarbons and Other Liquids	ASME B31.4
Building Services Piping	ASME B31.9

5703.6.2.1 Special materials. Low-melting-point materials (such as aluminum, copper or brass), materials that soften on fire exposure (such as nonmetallic materials) and nonductile material (such as cast iron) shall be acceptable for use underground in accordance with the applicable standard listed in Table 5703.6.2. Where such materials are used outdoors in above-ground piping systems or within buildings, they shall be in accordance with the applicable standard listed in Table 5703.6.2 and one of the following:

1. Suitably protected against fire exposure.
2. Located where leakage from failure would not unduly expose people or structures.
3. Located where leakage can be readily controlled by operation of accessible remotely located valves.

In all cases, nonmetallic piping shall be used in accordance with Section 27.4.6 of NFPA 30.

TABLE 5703.1.1
CLASS I ELECTRICAL EQUIPMENT LOCATIONS[a]

LOCATION	GROUP D DIVISION	EXTENT OF CLASSIFIED AREA
Underground tank fill opening	1	Pits, boxes or spaces below grade level, any part of which is within the Division 1 or 2 classified area.
	2	Up to 18 inches above grade level within a horizontal radius of 10 feet from a loose-fill connection and within a horizontal radius of 5 feet from a tight-fill connection.
Vent—Discharging upward	1	Within 3 feet of open end of vent, extending in all directions.
	2	Area between 3 feet and 5 feet of open end of vent, extending in all directions.
Drum and container filling		
Outdoor or indoor with adequate ventilation	1	Within 3 feet of vent and fill opening, extending in all directions.
	2	Area between 3 feet and 5 feet from vent of fill opening, extending in all directions. Also up to 18 inches above floor or grade level within a horizontal radius of 10 feet from vent or fill opening.
Pumps, bleeders, withdrawal fittings, meters and similar devices		
Indoor	2	Within 5 feet of any edge of such devices, extending in all directions, and up to 3 feet above floor or grade level within 25 feet horizontally from any edge of such devices.
Outdoor	2	Within 3 feet of any edge of such devices, extending in all directions, and up to 18 inches above floor or grade level within 10 feet horizontally from an edge of such devices.
Pits		
Without mechanical ventilation	1	Entire area within pit if any part is within a Division 1 or 2 classified area.
With mechanical ventilation	2	Entire area within pit if any part is within a Division 1 or 2 classified area.
Containing valves, fittings or piping, and not within a Division 1 or 2 classified area	2	Entire pit.
Drainage ditches, separators, impounding basins		
Indoor	1 or 2	Same as pits.
Outdoor	2	Area up to 18 inches above ditch, separator or basin, and up to 18 inches above grade within 15 feet horizontal from any edge.
Tank vehicle and tank car[b]		
Loading through open dome	1	Within 3 feet of edge of dome, extending in all directions.
	2	Area between 3 feet and 15 feet from edge of dome, extending in all directions.
Loading through bottom connections with atmospheric venting	1	Within 3 feet of point of venting to atmosphere, extending in all directions.
	2	Area between 3 feet and 15 feet from point of venting to atmosphere, extending in all directions. Also up to 18 inches above grade within a horizontal radius of 10 feet from point of loading connection.

(continued)

TABLE 5703.1.1—continued
CLASS I ELECTRICAL EQUIPMENT LOCATIONS[a]

LOCATION	GROUP D DIVISION	EXTENT OF CLASSIFIED AREA
Tank vehicle and tank car[b]—continued		
Loading through closed dome with atmospheric venting	1	Within 3 feet of open end of vent, extending in all directions.
	2	Area between 3 feet and 15 feet from open end of vent, extending in all directions, and within 3 feet of edge of dome, extending in all directions.
Loading through closed dome with vapor control	2	Within 3 feet of point of connection of both fill and vapor lines, extending in all directions.
Bottom loading with vapor control or any bottom unloading	2	Within 3 feet of point of connection, extending in all directions, and up to 18 inches above grade within a horizontal radius of 10 feet from point of connection.
Storage and repair garage for tank vehicles	1	Pits or spaces below floor level.
	2	Area up to 18 inches above floor or grade level for entire storage or repair garage.
Garages for other than tank vehicles	Ordinary	Where there is an opening to these rooms within the extent of an outdoor classified area, the entire room shall be classified the same as the area classification at the point of the opening.
Outdoor drum storage	Ordinary	—
Indoor warehousing where there is no flammable liquid transfer	Ordinary	Where there is an opening to these rooms within the extent of an indoor classified area, the room shall be classified the same as if the wall, curb or partition did not exist.
Indoor equipment where flammable vapor/air mixtures could exist under normal operations	1	Area within 5 feet of any edge of such equipment, extending in all directions.
	2	Area between 5 feet and 8 feet of any edge of such equipment, extending in all directions, and the area up to 3 feet above floor or grade level within 5 feet to 25 feet horizontally from any edge of such equipment.[c]
Outdoor equipment where flammable vapor/air mixtures could exist under normal operations	1	Area within 3 feet of any edge of such equipment, extending in all directions.
	2	Area between 3 feet and 8 feet of any edge of such equipment extending in all directions, and the area up to 3 feet above floor or grade level within 3 feet to 10 feet horizontally from any edge of such equipment.
Tank—Above ground		
Shell, ends or roof and dike area	1	Area inside dike where dike height is greater than the distance from the tank to the dike for more than 50 percent of the tank circumference.
	2	Area within 10 feet from shell, ends or roof of tank. Area inside dikes to level of top of dike.
Vent	1	Area within 5 feet of open end of vent, extending in all directions.
	2	Area between 5 feet and 10 feet from open end of vent, extending in all directions.
Floating roof	1	Area above the roof and within the shell.
Office and restrooms	Ordinary	Where there is an opening to these rooms within the extent of an indoor classified location, the room shall be classified the same as if the wall, curb or partition did not exist.

For SI: 1 inch = 25.4 mm, 1 foot = 304.8 mm.

a. Locations as classified in NFPA 70.

b. When classifying extent of area, consideration shall be given to the fact that tank cars or tank vehicles can be spotted at varying points. Therefore, the extremities of the loading or unloading positions shall be used.

c. The release of Class I liquids can generate vapors to the extent that the entire building, and possibly a zone surrounding it, are considered a Class I, Division 2 location.

5703.6.3 Testing. Unless tested in accordance with the applicable section of ASME B31.9, piping, before being covered, enclosed or placed in use, shall be hydrostatically tested to 150 percent of the maximum anticipated pressure of the system, or pneumatically tested to 110 percent of the maximum anticipated pressure of the system, but not less than 5 pounds per square inch gauge (psig) (34.47 kPa) at the highest point of the system. This test shall be maintained for a sufficient time period to complete visual inspection of joints and connections. For not less than 10 minutes, there shall be no leakage or permanent distortion. Care shall be exercised to ensure that these pressures are not applied to vented storage tanks. Such storage tanks shall be tested independently from the piping.

5703.6.3.1 Existing piping. Existing piping shall be tested in accordance with this section where the *fire code official* has reasonable cause to believe that a leak exists. Piping that could contain flammable or *combustible liquids* shall not be tested pneumatically. Such tests shall be at the expense of the *owner* or operator.

Exception: Vapor-recovery piping is allowed to be tested using an inert gas.

5703.6.4 Protection from vehicles. Guard posts or other *approved* means shall be provided to protect piping, valves or fittings subject to vehicular damage in accordance with Section 312.

5703.6.5 Protection from external corrosion and galvanic action. Where subject to external corrosion, piping, related fluid-handling components and supports for both underground and above-ground applications shall be fabricated from noncorrosive materials, and coated or provided with corrosion protection. Dissimilar metallic parts that promote galvanic action shall not be joined.

5703.6.6 Valves. Piping systems shall contain a sufficient number of manual control valves and check valves to operate the system properly and to protect the plant under both normal and emergency conditions. Piping systems in connection with pumps shall contain a sufficient number of such valves to control properly the flow of liquids in normal operation and in the event of physical damage or fire exposure.

5703.6.6.1 Backflow protections. Connections to pipelines or piping by which equipment (such as tank cars, tank vehicles or marine vessels) discharges liquids into storage tanks shall be provided with check valves or block valves for automatic protection against backflow where the piping arrangement is such that backflow from the system is possible. Where loading and unloading is done through a common pipe system, a check valve is not required. However, a block valve, located so as to be readily accessible or remotely operable, shall be provided.

5703.6.6.2 Manual drainage. Manual drainage-control valves shall be located at *approved* locations remote from the tanks, diked area, drainage system and impounding basin to ensure their operation in a fire condition.

5703.6.7 Connections. Above-ground tanks with connections located below normal liquid level shall be provided with internal or external isolation valves located as close as practical to the shell of the tank. Except for liquids whose chemical characteristics are incompatible with steel, such valves, where external, and their connections to the tank shall be of steel.

5703.6.8 Piping supports. Piping systems shall be substantially supported and protected against physical damage and excessive stresses arising from settlement, vibration, expansion, contraction or exposure to fire. The supports shall be protected against exposure to fire by one of the following:

1. Draining liquid away from the piping system at a minimum slope of not less than 1 percent.
2. Providing protection with a *fire-resistance rating* of not less than 2 hours.
3. Other *approved* methods.

5703.6.9 Flexible joints. Flexible joints shall be *listed* and *approved* and shall be installed on underground liquid, vapor and vent piping at all of the following locations:

1. Where piping connects to underground tanks.
2. Where piping ends at pump islands and vent risers.
3. At points where differential movement in the piping can occur.

5703.6.9.1 Fiberglass-reinforced plastic piping. Fiberglass-reinforced plastic (FRP) piping is not required to be provided with flexible joints in locations where both of the following conditions are present:

1. Piping does not exceed 4 inches (102 mm) in diameter.
2. Piping has a straight run of not less than 4 feet (1219 mm) on one side of the connection where such connections result in a change of direction.

In lieu of the minimum 4-foot (1219 mm) straight run length, *approved* and *listed* flexible joints are allowed to be used under dispensers and suction pumps, at submerged pumps and tanks, and where vents extend above ground.

5703.6.10 Pipe joints. Joints shall be liquid tight and shall be welded, flanged or threaded except that *listed* flexible connectors are allowed in accordance with Section 5703.6.9. Threaded or flanged joints shall fit tightly by using *approved* methods and materials for the type of joint. Joints in piping systems used for Class I liquids shall be welded where located in concealed spaces within buildings.

Nonmetallic joints shall be *approved* and shall be installed in accordance with the manufacturer's instructions.

Pipe joints that are dependent on the friction characteristics or resiliency of combustible materials for liquid tightness of piping shall not be used in buildings. Piping shall be secured to prevent disengagement at the fitting.

5703.6.11 Bends. Pipe and tubing shall be bent in accordance with ASME B31.9.

SECTION 5704
STORAGE

5704.1 General. The storage of flammable and *combustible liquids* in containers and tanks shall be in accordance with this section and the applicable sections of Chapter 50.

5704.2 Tank storage. The provisions of this section shall apply to:

1. The storage of flammable and *combustible liquids* in fixed above-ground and underground tanks.
2. The storage of flammable and *combustible liquids* in fixed above-ground tanks inside of buildings.
3. The storage of flammable and *combustible liquids* in portable tanks whose capacity exceeds 660 gallons (2498 L).
4. The installation of such tanks and portable tanks.

5704.2.1 Change of tank contents. Tanks subject to change in contents shall be in accordance with Section 5704.2.7. Prior to a change in contents, the *fire code official* is authorized to require testing of a tank.

Tanks that have previously contained Class I liquids shall not be loaded with Class II or Class III liquids until such tanks and all piping, pumps, hoses and meters connected thereto have been completely drained and flushed.

5704.2.2 Use of tank vehicles and tank cars as storage tanks. Tank cars and tank vehicles shall not be used as storage tanks.

5704.2.3 Labeling and signs. Labeling and signs for storage tanks and storage tank areas shall comply with Sections 5704.2.3.1 and 5704.2.3.2.

5704.2.3.1 Smoking and open flame. Signs shall be posted in storage areas prohibiting open flames and smoking. Signs shall comply with Section 5703.5.

5704.2.3.2 Label or placard. Tanks more than 100 gallons (379 L) in capacity, which are permanently installed or mounted and used for the storage of Class I, II or III liquids, shall bear a label and placard identifying the material therein. Placards shall be in accordance with NFPA 704.

Exceptions:

1. Tanks of 300-gallon (1136 L) capacity or less located on private property and used for heating and cooking fuels in single-family *dwellings*.
2. Tanks located underground.

5704.2.4 Sources of ignition. Smoking and open flames are prohibited in storage areas in accordance with Section 5003.7.

Exception: Areas designated as smoking and hot work areas, and areas where hot work permits have been issued in accordance with this code.

5704.2.5 Explosion control. Explosion control shall be provided in accordance with Section 911 for indoor tanks.

5704.2.6 Separation from incompatible materials. Storage of flammable and *combustible liquids* shall be separated from *incompatible materials* in accordance with Section 5003.9.8.

Grass, weeds, combustible materials and waste Class I, II or IIIA liquids shall not be accumulated in an unsafe manner at a storage site.

5704.2.7 Design, fabrication and construction requirements for tanks. The design, fabrication and construction of tanks shall comply with NFPA 30. Each tank shall bear a permanent nameplate or marking indicating the standard used as the basis of design.

5704.2.7.1 Materials used in tank construction. The materials used in tank construction shall be in accordance with NFPA 30. The materials of construction for tanks and their appurtenances shall be compatible with the liquids to be stored.

5704.2.7.2 Pressure limitations for tanks. Tanks shall be designed for the pressures to which they will be subjected in accordance with NFPA 30.

5704.2.7.3 Tank vents for normal venting. Tank vents for normal venting shall be installed and maintained in accordance with Sections 5704.2.7.3.1 through 5704.2.7.3.5.3.

5704.2.7.3.1 Vent lines. Vent lines from tanks shall not be used for purposes other than venting unless *approved*.

5704.2.7.3.2 Vent-line flame arresters and pressure-vacuum vents. *Listed* or *approved* flame arresters or pressure-vacuum (PV) vents that remain closed unless venting under pressure or vacuum conditions shall be installed in normal vents of tanks containing Class IB and IC liquids.

Exception: Where determined by the *fire code official* that the use of such devices can result in damage to the tank.

Vent-line flame arresters shall be installed in accordance with their listing or API 2000 and maintained in accordance with Section 21.8.6 of NFPA 30 or API 2000. In-line flame arresters in piping systems shall be installed and maintained in accordance with their listing or API 2028. Pressure-vacuum vents shall be installed in accordance with Section 21.4.3 of NFPA 30 or API 2000 and maintained in accordance with Section 21.8.6 of NFPA 30 or API 2000.

5704.2.7.3.3 Vent pipe outlets. Vent pipe outlets for tanks storing Class I, II or IIIA liquids shall be located such that the vapors are released at a safe point outside of buildings and not less than 12 feet (3658 mm) above the finished ground level. Vapors shall be discharged upward or horizontally away from adjacent walls to assist in vapor dispersion. Vent outlets shall be located such that flammable vapors will not be trapped by eaves or other obstructions and shall be not less than 5 feet (1524 mm) from building openings or *lot lines* of properties that can be built upon. Vent outlets on atmospheric tanks storing Class IIIB liquids are allowed to discharge inside a building where the vent is a normally closed vent.

Exception: Vent pipe outlets on tanks storing Class IIIB liquid inside buildings and connected to fuel-burning equipment shall be located such that the vapors are released to a safe location outside of buildings.

5704.2.7.3.4 Installation of vent piping. Vent piping shall be designed, sized, constructed and installed in accordance with Section 5703.6. Vent pipes shall be installed such that they will drain toward the tank without sags or traps in which liquid can collect. Vent pipes shall be installed in such a manner so as not to be subject to physical damage or vibration.

5704.2.7.3.5 Manifolding. Tank vent piping shall not be manifolded unless required for special purposes such as vapor recovery, vapor conservation or air pollution control.

5704.2.7.3.5.1 Above-ground tanks. For above-ground tanks, manifolded vent pipes shall be adequately sized to prevent system pressure limits from being exceeded where manifolded tanks are subject to the same fire exposure.

5704.2.7.3.5.2 Underground tanks. For underground tanks, manifolded vent pipes shall be sized to prevent system pressure limits from being exceeded when manifolded tanks are filled simultaneously.

5704.2.7.3.5.3 Tanks storing Class I liquids. Vent piping for tanks storing Class I liquids shall not be manifolded with vent piping for tanks storing Class II and III liquids unless positive means are provided to prevent the vapors from Class I liquids from entering tanks storing Class II and III liquids, to prevent contamination and possible change in classification of less volatile liquid.

➡ **5704.2.7.4 Emergency venting.** Stationary, above-ground tanks shall be equipped with additional venting that will relieve excessive internal pressure caused by exposure to fires. Emergency vents for Class I, II and IIIA liquids shall not discharge inside buildings. The venting shall be installed and maintained in accordance with Section 22.7 of NFPA 30.

Exceptions:

1. Tanks larger than 12,000 gallons (45 420 L) in capacity storing Class IIIB liquids that are not within the diked area or the drainage path of Class I or II liquids do not require emergency relief venting.
2. Emergency vents on protected above-ground tanks complying with UL 2085 containing Class II or IIIA liquids are allowed to discharge inside the building.

5704.2.7.5 Tank openings other than vents. Tank openings for other than vents shall comply with Sections 5704.2.7.5.1 through 5704.2.7.5.8.

5704.2.7.5.1 Connections below liquid level. Connections for tank openings below the liquid level shall be liquid tight.

5704.2.7.5.2 Filling, emptying and vapor recovery connections. Filling, emptying and vapor recovery connections to tanks containing Class I, II or IIIA liquids shall be located outside of buildings in accordance with Section 5704.2.7.5.6 at a location free from sources of ignition and not less than 5 feet (1524 mm) away from building openings or *lot lines* of property that can be built upon. Such openings shall be properly identified and provided with a liquid-tight cap that shall be closed when not in use.

Filling and emptying connections to indoor tanks containing Class IIIB liquids and connected to fuel-burning equipment shall be located at a finished ground level location outside of buildings. Such openings shall be provided with a liquid-tight cap that shall be closed when not in use. A sign in accordance with Section 5003.6 that displays the following warning shall be permanently attached at the filling location:

TRANSFERRING FUEL OTHER THAN CLASS IIIB COMBUSTIBLE LIQUID TO THIS TANK CONNECTION IS A VIOLATION OF THE FIRE CODE AND IS STRICTLY PROHIBITED

5704.2.7.5.3 Piping, connections and fittings. Piping, connections, fittings and other appurtenances shall be installed in accordance with Section 5703.6.

5704.2.7.5.4 Manual gauging. Openings for manual gauging, if independent of the fill pipe, shall be provided with a liquid-tight cap or cover. Covers shall be kept closed when not gauging. If inside a building, such openings shall be protected against liquid overflow and possible vapor release by means of a spring-loaded check valve or other *approved* device.

5704.2.7.5.5 Fill pipes and discharge lines. For top-loaded tanks, a metallic fill pipe shall be designed and installed to minimize the generation of static electricity by terminating the pipe within 6 inches (152 mm) of the bottom of the tank, and it shall be installed in a manner that avoids excessive vibration.

5704.2.7.5.5.1 Class I liquids. For Class I liquids other than crude oil, gasoline and asphalt, the fill pipe shall be designed and installed in a manner that will minimize the possibility of generating static electricity by terminating within 6 inches (152 mm) of the bottom of the tank.

5704.2.7.5.5.2 Underground tanks. For underground tanks, fill pipe and discharge lines shall enter only through the top. Fill lines shall be sloped toward the tank. Underground tanks for Class I liquids having a capacity greater than 1,000 gallons (3785 L) shall be equipped with a tight fill device for connecting the fill hose to the tank.

5704.2.7.5.6 Location of connections that are made or broken. Filling, withdrawal and vapor-recovery connections for Class I, II and IIIA liquids that are made and broken shall be located outside of buildings, not more than 5 feet (1524 mm) above the finished ground level, in an *approved* location in close proximity to the parked delivery vehicle. Such location shall be away from sources of ignition and not less than 5 feet (1524 mm) away from building openings. Such connections shall be closed and liquid tight when not in use and shall be properly identified.

5704.2.7.5.7 Protection against vapor release. Tank openings provided for purposes of vapor recovery shall be protected against possible vapor release by means of a spring-loaded check valve or dry-break connections, or other *approved* device, unless the opening is a pipe connected to a vapor processing system. Openings designed for combined fill and vapor recovery shall also be protected against vapor release unless connection of the liquid delivery line to the fill pipe simultaneously connects the vapor recovery line. Connections shall be vapor tight.

5704.2.7.5.8 Overfill prevention. An *approved* means or method in accordance with Section 5704.2.9.7.6 shall be provided to prevent the overfill of all Class I, II and IIIA liquid storage tanks. Storage tanks in refineries, bulk plants or terminals regulated by Section 5706.4 or 5706.7 shall have overfill protection in accordance with API 2350.

An *approved* means or method in accordance with Section 5704.2.9.7.6 shall be provided to prevent the overfilling of Class IIIB liquid storage tanks connected to fuel-burning equipment inside buildings.

Exception: Outside above-ground tanks with a capacity of 1,320 gallons (5000 L) or less.

5704.2.7.6 Repair, alteration or reconstruction of tanks and piping. The repair, *alteration* or reconstruction, including welding, cutting and hot tapping of storage tanks and piping that have been placed in service, shall be in accordance with NFPA 30. Hot work, as defined in Section 202, on such tanks shall be conducted in accordance with Section 3510.

5704.2.7.7 Design of supports. The design of the supporting structure for tanks shall be in accordance with the *International Building Code* and NFPA 30.

5704.2.7.8 Locations subject to flooding. Where a tank is located in an area where it is subject to buoyancy because of a rise in the water table, flooding or accumulation of water from fire suppression operations, uplift protection shall be provided in accordance with Sections 22.14 and 23.14 of NFPA 30.

5704.2.7.9 Corrosion protection. Where subject to external corrosion, tanks shall be fabricated from corrosion-resistant materials, coated or provided with corrosion protection in accordance with Section 23.3.5 of NFPA 30.

5704.2.7.10 Leak reporting. A consistent or accidental loss of liquid, or other indication of a leak from a tank system, shall be reported immediately to the fire department, the *fire code official* and other authorities having jurisdiction.

5704.2.7.10.1 Leaking tank disposition. Leaking tanks shall be promptly emptied, repaired and returned to service, abandoned or removed in accordance with Section 5704.2.13 or 5704.2.14.

5704.2.7.11 Tank lining. Steel tanks are allowed to be lined only for the purpose of protecting the interior from corrosion or providing compatibility with a material to be stored. Only those liquids tested for compatibility with the lining material are allowed to be stored in lined tanks.

5704.2.8 Vaults. Vaults shall be allowed to be either above or below grade and shall comply with Sections 5704.2.8.1 through 5704.2.8.18.

5704.2.8.1 Listing required. Vaults shall be *listed* in accordance with UL 2245.

Exception: Where *approved* by the *fire code official*, below-grade vaults are allowed to be constructed on site, provided that the design is in accordance with the *International Building Code* and that special inspections are conducted to verify structural strength and compliance of the installation with the *approved* design in accordance with Section 1707 of the *International Building Code*. Installation plans for below-grade vaults that are constructed on site shall be prepared by, and the design shall bear the stamp of, a professional engineer. Consideration shall be given to soil and hydrostatic loading on the floors, walls and lid; anticipated seismic forces; uplifting by groundwater or flooding; and to loads imposed from above such as traffic and equipment loading on the vault lid.

5704.2.8.2 Design and construction. The vault shall completely enclose each tank. There shall not be openings in the vault enclosure except those necessary for access to, inspection of, and filling, emptying and venting of the tank. The walls and floor of the vault shall be constructed of reinforced concrete not less than 6 inches (152 mm) thick. The top of an above-grade vault shall be constructed of noncombustible material and shall be designed to be weaker than the walls of the vault, to ensure that the thrust of an explosion occurring inside the vault is directed upward before significantly high pressure can develop within the vault.

The top of an at-grade or below-grade vault shall be designed to relieve safely or contain the force of an explosion occurring inside the vault. The top and floor of the vault and the tank foundation shall be designed to withstand the anticipated loading, including loading from vehicular traffic, where applicable. The walls and floor of a vault installed below grade shall be designed to withstand anticipated soil and hydrostatic loading.

Vaults shall be designed to be wind and earthquake resistant, in accordance with the *International Building Code*.

5704.2.8.3 Secondary containment. Vaults shall be substantially liquid tight and there shall not be backfill around the tank or within the vault. The vault floor shall drain to a sump. For premanufactured vaults, liquid tightness shall be certified as part of the listing provided by a nationally recognized testing laboratory. For field-erected vaults, liquid tightness shall be certified in an *approved* manner.

5704.2.8.4 Internal clearance. There shall be sufficient clearance between the tank and the vault to allow for visual inspection and maintenance of the tank and its appurtenances. Dispensing devices are allowed to be installed on tops of vaults.

5704.2.8.5 Anchoring. Vaults and their tanks shall be suitably anchored to withstand uplifting by ground water or flooding, including when the tank is empty.

5704.2.8.6 Vehicle impact protection. Vaults shall be resistant to damage from the impact of a motor vehicle, or vehicle impact protection shall be provided in accordance with Section 312.

5704.2.8.7 Arrangement. Tanks shall be *listed* for above-ground use, and each tank shall be in its own vault. Compartmentalized tanks shall be allowed and shall be considered as a single tank. Adjacent vaults shall be allowed to share a common wall. The common wall shall be liquid and vapor tight and shall be designed to withstand the load imposed when the vault on either side of the wall is filled with water.

5704.2.8.8 Connections. Connections shall be provided to permit venting of each vault to dilute, disperse and remove vapors prior to personnel entering the vault.

5704.2.8.9 Ventilation. Vaults that contain tanks of Class I liquids shall be provided with an exhaust ventilation system installed in accordance with Section 5004.3. The ventilation system shall operate continuously or be designed to operate upon activation of the vapor or liquid detection system. The system shall provide ventilation at a rate of not less than 1 cubic foot per minute (cfm) per square foot of floor area [0.00508 $m^3/(s \cdot m^2)$], but not less than 150 cfm (4 m^3/min). The exhaust system shall be designed to provide air movement across all parts of the vault floor. Supply and exhaust ducts shall extend to within 3 inches (76 mm), but not more than 12 inches (305 mm), of the floor. The exhaust system shall be installed in accordance with the *International Mechanical Code*.

5704.2.8.10 Liquid detection. Vaults shall be equipped with a detection system capable of detecting liquids, including water, and activating an alarm.

5704.2.8.11 Monitoring and detection. Vaults shall be provided with *approved* vapor and liquid detection systems and equipped with on-site audible and visual warning devices with battery backup. Vapor detection systems shall sound an alarm when the system detects vapors that reach or exceed 25 percent of the lower explosive limit (LEL) of the liquid stored. Vapor detectors shall be located not higher than 12 inches (305 mm) above the lowest point in the vault. Liquid detection systems shall sound an alarm upon detection of any liquid, including water. Liquid detectors shall be located in accordance with the manufacturer's instructions. Activation of either vapor or liquid detection systems shall cause a signal to be sounded at an *approved*, constantly attended location within the facility serving the tanks or at an *approved* location. Activation of vapor detection systems shall also shut off dispenser pumps.

5704.2.8.12 Liquid removal. Means shall be provided to recover liquid from the vault. Where a pump is used to meet this requirement, the pump shall not be permanently installed in the vault. Electric-powered portable pumps shall be suitable for use in Class I, Division 1, or Zone 0 locations, as defined in NFPA 70.

5704.2.8.13 Normal vents. Vent pipes that are provided for normal tank venting shall terminate not less than 12 feet (3658 mm) above ground level.

5704.2.8.14 Emergency vents. Emergency vents shall be vapor tight and shall be allowed to discharge inside the vault. Long-bolt manhole covers shall not be allowed for this purpose.

5704.2.8.15 Accessway. Vaults shall be provided with an *approved* personnel accessway with a minimum dimension of 30 inches (762 mm) and with a permanently affixed, nonferrous ladder. Accessways shall be designed to be nonsparking. Travel distance from any point inside a vault to an accessway shall not exceed 20 feet (6096 mm). At each entry point, a warning sign indicating the need for procedures for safe entry into confined spaces shall be posted. Entry points shall be secured against unauthorized entry and vandalism.

5704.2.8.16 Fire protection. Vaults shall be provided with a suitable means to admit a fire suppression agent.

5704.2.8.17 Classified area. The interior of a vault containing a tank that stores a Class I liquid shall be designated a Class I, Division 1, or Zone 0 location, as defined in NFPA 70.

5704.2.8.18 Overfill protection. Overfill protection shall be provided in accordance with Section

5704.2.9.7.6. The use of a float vent valve shall be prohibited.

5704.2.9 Above-ground tanks. Above-ground storage of flammable and *combustible liquids* in tanks shall comply with Section 5704.2 and Sections 5704.2.9.1 through 5704.2.9.7.9.

5704.2.9.1 Existing noncompliant installations. Existing above-ground tanks shall be maintained in accordance with the code requirements that were applicable at the time of installation. Above-ground tanks that were installed in violation of code requirements applicable at the time of installation shall be made code compliant or shall be removed in accordance with Section 5704.2.14, regardless of whether such tank has been previously inspected (see Section 106.4).

5704.2.9.2 Fire protection. Fire protection for above-ground tanks shall comply with Sections 5704.2.9.2.1 through 5704.2.9.2.4.

5704.2.9.2.1 Required foam fire protection systems. Where required by the *fire code official*, foam fire protection shall be provided for above-ground tanks, other than pressure tanks operating at or above 1 pound per square inch gauge (psig) (6.89 kPa) where such tank, or group of tanks spaced less than 50 feet (15 240 mm) apart measured shell to shell, has a liquid surface area in excess of 1,500 square feet (139 m^2), and is in accordance with one of the following:

1. Used for the storage of Class I or II liquids.
2. Used for the storage of crude oil.
3. Used for in-process products and is located within 100 feet (30 480 mm) of a fired still, heater, related fractioning or processing apparatus or similar device at a processing plant or petroleum refinery as herein defined.
4. Considered by the *fire code official* as posing an unusual exposure hazard because of topographical conditions; nature of occupancy, proximity on the same or adjoining property, and height and character of liquids to be stored; degree of private fire protection to be provided; and facilities of the fire department to cope with flammable liquid fires.

5704.2.9.2.2 Foam fire protection system installation. Where foam fire protection is required, it shall be installed in accordance with NFPA 11.

5704.2.9.2.2.1 Foam storage. Where foam fire protection is required, foam-producing materials shall be stored on the premises.

Exception: Storage of foam-producing materials off the premises is allowed as follows:

1. Such materials stored off the premises shall be of the proper type suitable for use with the equipment at the installation where required.
2. Such materials shall be readily available at the storage location at all times.
3. Adequate loading and transportation facilities shall be provided.
4. The time required to deliver such materials to the required location in the event of fire shall be consistent with the hazards and fire scenarios for which the foam supply is intended.
5. At the time of a fire, these off-premises supplies shall be accumulated in sufficient quantities before placing the equipment in operation to ensure foam production at an adequate rate without interruption until extinguishment is accomplished.

5704.2.9.2.3 Fire protection of supports. Supports or pilings for above-ground tanks storing Class I, II or IIIA liquids elevated more than 12 inches (305 mm) above grade shall have a *fire-resistance rating* of not less than 2 hours in accordance with the fire exposure criteria specified in ASTM E 1529.

Exceptions:

1. Structural supports tested as part of a protected above-ground tank in accordance with UL 2085.
2. Stationary tanks located outside of buildings where protected by an *approved* water-spray system designed in accordance with Chapter 9 and NFPA 15.
3. Stationary tanks located inside of buildings equipped throughout with an *approved* automatic sprinkler system designed in accordance with Section 903.3.1.1.

5704.2.9.2.4 Inerting of tanks storing boilover liquids. Liquids with boilover characteristics shall not be stored in fixed roof tanks larger than 150 feet (45 720 mm) in diameter unless an *approved* gas enrichment or inerting system is provided on the tank.

Exception: Crude oil storage tanks in production fields with no other exposures adjacent to the storage tank.

5704.2.9.3 Supports, foundations and anchorage. Supports, foundations and anchorages for above-ground tanks shall be designed and constructed in accordance with NFPA 30 and the *International Building Code*.

5704.2.9.4 Stairways, platforms and walkways. *Stairways*, platforms and walkways shall be of noncombustible construction and shall be designed and constructed in accordance with NFPA 30 and the *International Building Code*.

5704.2.9.5 Above-ground tanks inside of buildings. Above-ground tanks inside of buildings shall comply with Sections 5704.2.9.5.1 and 5704.2.9.5.2.

5704.2.9.5.1 Overfill prevention. Above-ground tanks storing Class I, II and IIIA liquids inside buildings shall be equipped with a device or other means to prevent overflow into the building including, but not limited to: a float valve; a preset meter on the fill line; a valve actuated by the weight of the tank's contents; a low-head pump that is incapable of producing overflow; or a liquid-tight overflow pipe not less than one pipe size larger than the fill pipe and discharging by gravity back to the outside source of liquid or to an *approved* location. Tanks containing Class IIIB liquids and connected to fuel-burning equipment shall be provided with a means to prevent overflow into buildings in accordance with Section 5704.2.7.5.8.

5704.2.9.5.2 Fill pipe connections. Fill pipe connections for tanks storing Class I, II and IIIA liquids and Class IIIB liquids connected to fuel-burning equipment shall be in accordance with Section 5704.2.9.7.7.

5704.2.9.6 Above-ground tanks outside of buildings. Above-ground tanks outside of buildings shall comply with Sections 5704.2.9.6.1 through 5704.2.9.6.3.

5704.2.9.6.1 Locations where above-ground tanks are prohibited. Storage of Class I and II liquids in above-ground tanks outside of buildings is prohibited within the limits established by law as the limits of districts in which such storage is prohibited (see Section 3 of the Sample Legislation for Adoption of the *International Fire Code* on page xxi).

5704.2.9.6.1.1 Location of tanks with pressures 2.5 psig or less. Above-ground tanks operating at pressures not exceeding 2.5 psig (17.2 kPa) for storage of Class I, II or IIIA liquids, which are designed with a floating roof, a weak roof-to-shell seam or equipped with emergency venting devices limiting pressure to 2.5 psig (17.2 kPa), shall be located in accordance with Table 22.4.1.1(a) of NFPA 30.

Exceptions:

1. Vertical tanks having a weak roof-to-shell seam and storing Class IIIA liquids are allowed to be located at one-half the distances specified in Table 22.4.1.1(a) of NFPA 30, provided the tanks are not within a diked area or drainage path for a tank storing Class I or II liquids.
2. Liquids with boilover characteristics and unstable liquids in accordance with Sections 5704.2.9.6.1.3 and 5704.2.9.6.1.4.
3. For protected above-ground tanks in accordance with Section 5704.2.9.7 and tanks in at-grade or above-grade vaults in accordance with Section 5704.2.8, the distances in Table 22.4.1.1(b) of NFPA 30 shall apply and shall be reduced by one-half, but not to less than 5 feet (1524 mm).

5704.2.9.6.1.2 Location of tanks with pressures exceeding 2.5 psig. Above-ground tanks for the storage of Class I, II or IIIA liquids operating at pressures exceeding 2.5 psig (17.2 kPa) or equipped with emergency venting allowing pressures to exceed 2.5 psig (17.2 kPa) shall be located in accordance with Table 22.4.1.3 of NFPA 30.

Exception: Liquids with boilover characteristics and unstable liquids in accordance with Sections 5704.2.9.6.1.4 and 5704.2.9.6.1.5.

5704.2.9.6.1.3 Location of tanks storing boilover liquids. Above-ground tanks for storage of liquids with boilover characteristics shall be located in accordance with Table 22.4.1.4 of NFPA 30.

5704.2.9.6.1.4 Location of tanks storing unstable liquids. Above-ground tanks for the storage of unstable liquids shall be located in accordance with Table 22.4.1.5 of NFPA 30.

5704.2.9.6.1.5 Location of tanks storing Class IIIB liquids. Above-ground tanks for the storage of Class IIIB liquids, excluding unstable liquids, shall be located in accordance with Table 22.4.1.6 of NFPA 30, except where located within a diked area or drainage path for a tank or tanks storing Class I or II liquids. Where a Class IIIB liquid storage tank is within the diked area or drainage path for a Class I or II liquid, distances required by Section 5704.2.9.6.1.1 shall apply.

5704.2.9.6.1.6 Reduction of separation distances to adjacent property. Where two tank properties of diverse ownership have a common boundary, the *fire code official* is authorized to, with the written consent of the *owners* of the two properties, apply the distances in Sections 5704.2.9.6.1.2 through 5704.2.9.6.1.5 assuming a single property.

5704.2.9.6.2 Separation between adjacent stable or unstable liquid tanks. The separation between tanks containing stable liquids shall be in accordance with Table 22.4.2.1 of NFPA 30. Where tanks are in a diked area containing Class I or II liquids, or in the drainage path of Class I or II liquids, and are compacted in three or more rows or in an irregular pattern, the *fire code official* is authorized to require greater separation than specified in Table 22.4.2.1 of NFPA 30 or other means to make tanks in the interior of the pattern accessible for fire-fighting purposes.

Exception: Tanks used for storing Class IIIB liquids are allowed to be spaced 3 feet (914 mm) apart unless within a diked area or drainage path for a tank storing Class I or II liquids.

The separation between tanks containing unstable liquids shall be not less than one-half the sum of their diameters.

5704.2.9.6.3 Separation between adjacent tanks containing flammable or combustible liquids and LP-gas. The minimum horizontal separation between an LP-gas container and a Class I, II or IIIA liquid storage tank shall be 20 feet (6096 mm) except in the case of Class I, II or IIIA liquid tanks operating at pressures exceeding 2.5 psig (17.2 kPa) or equipped with emergency venting allowing pressures to exceed 2.5 psig (17.2 kPa), in which case the provisions of Section 5704.2.9.6.2 shall apply.

An *approved* means shall be provided to prevent the accumulation of Class I, II or IIIA liquids under adjacent LP-gas containers such as by dikes, diversion curbs or grading. Where flammable or *combustible liquid* storage tanks are within a diked area, the LP-gas containers shall be outside the diked area and not less than 10 feet (3048 mm) away from the centerline of the wall of the diked area.

Exceptions:

1. Liquefied petroleum gas containers of 125 gallons (473 L) or less in capacity installed adjacent to fuel-oil supply tanks of 660 gallons (2498 L) or less in capacity.
2. Horizontal separation is not required between above-ground LP-gas containers and underground flammable and *combustible liquid* tanks.

5704.2.9.7 Additional requirements for protected above-ground tanks. In addition to the requirements of this chapter for above-ground tanks, the installation of protected above-ground tanks shall be in accordance with Sections 5704.2.9.7.1 through 5704.2.9.7.9.

5704.2.9.7.1 Tank construction. The construction of a protected above-ground tank and its primary tank shall be in accordance with Section 5704.2.7.

5704.2.9.7.2 Normal and emergency venting. Normal and emergency venting for protected above-ground tanks shall be provided in accordance with Sections 5704.2.7.3 and 5704.2.7.4. The vent capacity reduction factor shall not be allowed.

5704.2.9.7.3 Secondary containment. Protected above-ground tanks shall be provided with secondary containment, drainage control or diking in accordance with Section 5004.2. A means shall be provided to establish the integrity of the secondary containment in accordance with NFPA 30.

5704.2.9.7.4 Vehicle impact protection. Where protected above-ground tanks, piping, electrical conduit or dispensers are subject to vehicular impact, they shall be protected therefrom, either by having the impact protection incorporated into the system design in compliance with the impact test protocol of UL 2085, or by meeting the provisions of Section 312, or where necessary, a combination of both. Where guard posts or other *approved* barriers are provided, they shall be independent of each above-ground tank.

5704.2.9.7.5 Overfill prevention. Protected above-ground tanks shall not be filled in excess of 95 percent of their capacity. An overfill prevention system shall be provided for each tank. During tank-filling operations, the system shall comply with one of the following:

1. The system shall:
   - 1.1. Provide an independent means of notifying the person filling the tank that the fluid level has reached 90 percent of tank capacity by providing an audible or visual alarm signal, providing a tank level gauge marked at 90 percent of tank capacity, or other *approved* means; and
   - 1.2. Automatically shut off the flow of fuel to the tank when the quantity of liquid in the tank reaches 95 percent of tank capacity. For rigid hose fuel-delivery systems, an *approved* means shall be provided to empty the fill hose into the tank after the automatic shutoff device is activated.
2. The system shall reduce the flow rate to not more than 15 gallons per minute (0.95 L/s) so that at the reduced flow rate, the tank will not overfill for 30 minutes, and automatically shut off flow into the tank so that none of the fittings on the top of the tank are exposed to product because of overfilling.

5704.2.9.7.5.1 Information signs. A permanent sign shall be provided at the fill point for the tank, documenting the filling procedure and the tank calibration chart.

Exception: Where climatic conditions are such that the sign may be obscured by ice or snow, or weathered beyond readability or otherwise impaired, said procedures and chart shall be located in the office window, lock box or other area accessible to the person filling the tank.

5704.2.9.7.5.2 Determination of available tank capacity. The filling procedure shall require the person filling the tank to determine the gallonage (literage) required to fill it to 90 percent of capacity before commencing the fill operation.

5704.2.9.7.6 Fill pipe connections. The fill pipe shall be provided with a means for making a direct connection to the tank vehicle's fuel delivery hose so that the delivery of fuel is not exposed to the open air during the filling operation. Where any portion of the fill pipe exterior to the tank extends below the level of the top of the tank, a check valve shall be installed in the fill pipe not more than 12 inches (305 mm) from the fill hose connection.

5704.2.9.7.7 Spill containers. A spill container having a capacity of not less than 5 gallons (19 L) shall be provided for each fill connection. For tanks with a top fill connection, spill containers shall be noncombustible and shall be fixed to the tank and equipped with a manual drain valve that drains into the primary tank. For tanks with a remote fill connection, a portable spill container shall be allowed.

5704.2.9.7.8 Tank openings. Tank openings in protected above-ground tanks shall be through the top only.

5704.2.9.7.9 Antisiphon devices. *Approved* antisiphon devices shall be installed in each external pipe connected to the protected above-ground tank where the pipe extends below the level of the top of the tank.

5704.2.10 Drainage and diking. The area surrounding a tank or group of tanks shall be provided with drainage control or shall be diked to prevent accidental discharge of liquid from endangering adjacent tanks, adjoining property or reaching waterways.

Exceptions:

1. The *fire code official* is authorized to alter or waive these requirements based on a technical report that demonstrates that such tank or group of tanks does not constitute a hazard to other tanks, waterways or adjoining property, after consideration of special features such as topographical conditions, nature of occupancy and proximity to buildings on the same or adjacent property, capacity, and construction of proposed tanks and character of liquids to be stored, and nature and quantity of private and public fire protection provided.
2. Drainage control and diking is not required for *listed* secondary containment tanks.

5704.2.10.1 Volumetric capacity. The volumetric capacity of the diked area shall be not less than the greatest amount of liquid that can be released from the largest tank within the diked area. The capacity of the diked area enclosing more than one tank shall be calculated by deducting the volume of the tanks other than the largest tank below the height of the dike.

5704.2.10.2 Diked areas containing two or more tanks. Diked areas containing two or more tanks shall be subdivided in accordance with NFPA 30.

5704.2.10.3 Protection of piping from exposure fires. Piping shall not pass through adjacent diked areas or impounding basins, unless provided with a sealed sleeve or otherwise protected from exposure to fire.

5704.2.10.4 Combustible materials in diked areas. Diked areas shall be kept free from combustible materials, drums and barrels.

5704.2.10.5 Equipment, controls and piping in diked areas. Pumps, manifolds and fire protection equipment or controls shall not be located within diked areas or drainage basins or in a location where such equipment and controls would be endangered by fire in the diked area or drainage basin. Piping above ground shall be minimized and located as close as practical to the shell of the tank in diked areas or drainage basins.

Exceptions:

1. Pumps, manifolds and piping integral to the tanks or equipment being served, which is protected by intermediate diking, berms, drainage or fire protection such as water spray, monitors or resistive coating.
2. Fire protection equipment or controls that are appurtenances to the tanks or equipment being protected, such as foam chambers or foam piping and water or foam monitors and hydrants, or hand and wheeled extinguishers.

5704.2.11 Underground tanks. Underground storage of flammable and *combustible liquids* in tanks shall comply with Section 5704.2 and Sections 5704.2.11.1 through 5704.2.11.4.2.

5704.2.11.1 Location. Flammable and *combustible liquid* storage tanks located underground, either outside or under buildings, shall be in accordance with all of the following:

1. Tanks shall be located with respect to existing foundations and supports such that the loads carried by the latter cannot be transmitted to the tank.
2. The distance from any part of a tank storing liquids to the nearest wall of a *basement*, pit, cellar or *lot line* shall be not less than 3 feet (914 mm).
3. A minimum distance of 1 foot (305 mm), shell to shell, shall be maintained between underground tanks.

5704.2.11.2 Depth and cover. Excavation for underground storage tanks shall be made with due care to avoid undermining of foundations of existing structures. Underground tanks shall be set on firm foundations and surrounded with not less than 6 inches (152 mm) of noncorrosive inert material, such as clean sand.

5704.2.11.3 Overfill protection and prevention systems. Fill pipes shall be equipped with a spill container and an overfill prevention system in accordance with NFPA 30.

5704.2.11.4 Leak prevention. Leak prevention for underground tanks shall comply with Sections 5704.2.11.4.1 and 5704.2.11.4.2.

5704.2.11.4.1 Inventory control. Daily inventory records for underground storage tank systems shall be maintained.

5704.2.11.4.2 Leak detection. Underground storage tank systems shall be provided with an *approved* method of leak detection from any component of the system that is designed and installed in accordance with NFPA 30.

5704.2.12 Testing. Tank testing shall comply with Sections 5704.2.12.1 and 5704.2.12.2.

5704.2.12.1 Acceptance testing. Prior to being placed into service, tanks shall be tested in accordance with Section 21.5 of NFPA 30.

5704.2.12.2 Testing of underground tanks. Before being covered or placed in use, tanks and piping connected to underground tanks shall be tested for tightness in the presence of the *fire code official*. Piping shall be tested in accordance with Section 5703.6.3. The system shall not be covered until it has been *approved*.

5704.2.13 Abandonment and status of tanks. Tanks taken out of service shall be removed in accordance with Section 5704.2.14, or safeguarded in accordance with Sections 5704.2.13.1 through 5704.2.13.2.3 and API 1604.

5704.2.13.1 Underground tanks. Underground tanks taken out of service shall comply with Sections 5704.2.13.1.1 through 5704.2.13.1.5.

5704.2.13.1.1 Temporarily out of service. Underground tanks temporarily out of service shall have the fill line, gauge opening, vapor return and pump connection secure against tampering. Vent lines shall remain open and be maintained in accordance with Sections 5704.2.7.3 and 5704.2.7.4.

5704.2.13.1.2 Out of service for 90 days. Underground tanks not used for a period of 90 days shall be safeguarded in accordance with all the following or be removed in accordance with Section 5704.2.14:

1. Flammable or *combustible liquids* shall be removed from the tank.
2. All piping, including fill line, gauge opening, vapor return and pump connection, shall be capped or plugged and secured from tampering.
3. Vent lines shall remain open and be maintained in accordance with Sections 5704.2.7.3 and 5704.2.7.4.

5704.2.13.1.3 Out of service for one year. Underground tanks that have been out of service for a period of one year shall be removed from the ground in accordance with Section 5704.2.14 or abandoned in place in accordance with Section 5704.2.13.1.4.

5704.2.13.1.4 Tanks abandoned in place. Tanks abandoned in place shall be as follows:

1. Flammable and *combustible liquids* shall be removed from the tank and connected piping.
2. The suction, inlet, gauge, vapor return and vapor lines shall be disconnected.
3. The tank shall be filled completely with an *approved* inert solid material.
4. Remaining underground piping shall be capped or plugged.
5. A record of tank size, location and date of abandonment shall be retained.
6. All exterior above-grade fill piping shall be permanently removed when tanks are abandoned or removed.

5704.2.13.1.5 Reinstallation of underground tanks. Tanks that are to be reinstalled for flammable or *combustible liquid* service shall be in accordance with this chapter, ASME *Boiler and Pressure Vessel Code* (Section VIII), API 12-P, API 1615, UL 58 and UL 1316.

5704.2.13.2 Above-ground tanks. Above-ground tanks taken out of service shall comply with Sections 5704.2.13.2.1 through 5704.2.13.2.3.

5704.2.13.2.1 Temporarily out of service. Above-ground tanks temporarily out of service shall have all connecting lines isolated from the tank and be secured against tampering.

Exception: In-place fire protection (foam) system lines.

5704.2.13.2.2 Out of service for 90 days. Above-ground tanks not used for a period of 90 days shall be safeguarded in accordance with Section 5704.2.13.1.2 or removed in accordance with Section 5704.2.14.

Exceptions:

1. Tanks and containers connected to oil burners that are not in use during the warm season of the year or are used as a backup heating system to gas.
2. In-place, active fire protection (foam) system lines.

5704.2.13.2.3 Out of service for one year. Above-ground tanks that have been out of service for a period of one year shall be removed in accordance with Section 5704.2.14.

Exception: Tanks within operating facilities.

5704.2.14 Removal and disposal of tanks. Removal and disposal of tanks shall comply with Sections 5704.2.14.1 and 5704.2.14.2.

5704.2.14.1 Removal. Removal of above-ground and underground tanks shall be in accordance with all of the following:

1. Flammable and *combustible liquids* shall be removed from the tank and connected piping.
2. Piping at tank openings that is not to be used further shall be disconnected.
3. Piping shall be removed from the ground.

   Exception: Piping is allowed to be abandoned in place where the *fire code official* determines that removal is not practical. Abandoned piping shall be capped and safeguarded as required by the *fire code official.*
4. Tank openings shall be capped or plugged, leaving a $^1/_8$-inch to $^1/_4$-inch-diameter (3.2 mm to 6.4 mm) opening for pressure equalization.
5. Tanks shall be purged of vapor and inerted prior to removal.
6. All exterior above-grade fill and vent piping shall be permanently removed.

   Exception: Piping associated with bulk plants, terminal facilities and refineries.

5704.2.14.2 Disposal. Tanks shall be disposed of in accordance with federal, state and local regulations.

5704.2.15 Maintenance. Above-ground tanks, connected piping and ancillary equipment shall be maintained in a safe operating condition. Tanks shall be maintained in accordance with their listings. Damage to above-ground tanks, connected piping or ancillary equipment shall be repaired using materials having equal or greater strength and *fire resistance* or the equipment shall be replaced or taken out of service.

5704.3 Container and portable tank storage. Storage of flammable and *combustible liquids* in closed containers that do not exceed 60 gallons (227 L) in individual capacity and portable tanks that do not exceed 660 gallons (2498 L) in individual capacity, and limited transfers incidental thereto, shall comply with Sections 5704.3.1 through 5704.3.8.5.

5704.3.1 Design, construction and capacity of containers and portable tanks. The design, construction and capacity of containers for the storage of Class I, II and IIIA liquids shall be in accordance with this section and Section 9.4 of NFPA 30.

5704.3.1.1 Approved containers. Only *approved* containers and portable tanks shall be used.

5704.3.2 Liquid storage cabinets. Where other sections of this code require that liquid containers be stored in storage cabinets, such cabinets and storage shall be in accordance with Sections 5704.3.2.1 through 5704.3.2.2.

5704.3.2.1 Design and construction of storage cabinets. Design and construction of liquid storage cabinets shall be in accordance with Sections 5704.3.2.1.1 through 5704.3.2.1.4.

5704.3.2.1.1 Materials. Cabinets shall be *listed* in accordance with UL 1275, or constructed of *approved* wood or metal in accordance with the following:

1. Unlisted metal cabinets shall be constructed of steel having a thickness of not less than 0.044 inch (1.12 mm) (18 gage). The cabinet, including the door, shall be double walled with $1^1/_2$-inch (38 mm) airspace between the walls. Joints shall be riveted or welded and shall be tight fitting.
2. Unlisted wooden cabinets, including doors, shall be constructed of not less than 1-inch (25 mm) exterior grade plywood. Joints shall be rabbeted and shall be fastened in two directions with wood screws. Door hinges shall be of steel or brass. Cabinets shall be painted with an intumescent-type paint.

5704.3.2.1.2 Labeling. Cabinets shall be provided with a conspicuous label in red letters on contrasting background that reads: FLAMMABLE—KEEP FIRE AWAY.

5704.3.2.1.3 Doors. Doors shall be well fitted, self-closing and equipped with a three-point latch.

5704.3.2.1.4 Bottom. The bottom of the cabinet shall be liquid tight to a height of not less than 2 inches (51 mm).

5704.3.2.2 Capacity. The combined total quantity of liquids in a cabinet shall not exceed 120 gallons (454 L).

5704.3.3 Indoor storage. Storage of flammable and *combustible liquids* inside buildings in containers and portable tanks shall be in accordance with Sections 5704.3.3.1 through 5704.3.3.10.

Exceptions:

1. Liquids in the fuel tanks of motor vehicles, aircraft, boats or portable or stationary engines.
2. The storage of distilled spirits and wines in wooden barrels or casks.

5704.3.3.1 Portable fire extinguishers. *Approved* portable fire extinguishers shall be provided in accordance with specific sections of this chapter and Section 906.

5704.3.3.2 Incompatible materials. Materials that will react with water or other liquids to produce a hazard shall not be stored in the same room with flammable and combustible liquids except where stored in accordance with Section 5003.9.8.

5704.3.3.3 Clear means of egress. Storage of any liquids, including stock for sale, shall not be stored near or be allowed to obstruct physically the route of egress.

5704.3.3.4 Empty containers or portable tank storage. The storage of empty tanks and containers previously used for the storage of flammable or *combustible liquids*, unless free from explosive vapors, shall be stored as required for filled containers and portable tanks. Portable tanks and containers, when emptied, shall have the covers or plugs immediately replaced in openings.

5704.3.3.5 Shelf storage. Shelving shall be of *approved* construction, adequately braced and anchored. Seismic requirements shall be in accordance with the *International Building Code*.

5704.3.3.5.1 Use of wood. Wood of not less than 1 inch (25 mm) nominal thickness is allowed to be used as shelving, racks, dunnage, scuffboards, floor overlay and similar installations.

5704.3.3.5.2 Displacement protection. Shelves shall be of sufficient depth and provided with a lip or guard to prevent individual containers from being displaced.

Exception: Shelves in storage cabinets or on laboratory furniture specifically designed for such use.

5704.3.3.5.3 Orderly storage. Shelf storage of flammable and *combustible liquids* shall be maintained in an orderly manner.

5704.3.3.6 Rack storage. Where storage on racks is allowed elsewhere in this code, a minimum 4-foot-wide (1219 mm) aisle shall be provided between adjacent rack sections and any adjacent storage of liquids. Main aisles shall be not less than 8 feet (2438 mm) wide.

5704.3.3.7 Pile or palletized storage. Solid pile and palletized storage in liquid warehouses shall be arranged so that piles are separated from each other by not less than 4 feet (1219 mm). Aisles shall be provided and arranged so that no container or portable tank is more than 20 feet (6096 mm) from an aisle. Main *aisles* shall be not less than 8 feet (2438 mm) wide.

5704.3.3.8 Limited combustible storage. Limited quantities of combustible commodities are allowed to be stored in liquid storage areas where the ordinary combustibles, other than those used for packaging the liquids, are separated from the liquids in storage by not less than 8 feet (2438 mm) horizontally, either by open aisles or by open racks, and where protection is provided in accordance with Chapter 9.

5704.3.3.9 Idle combustible pallets. Storage of empty or idle combustible pallets inside an unprotected liquid storage area shall be limited to a maximum pile size of 2,500 square feet (232 m^2) and to a maximum storage height of 6 feet (1829 mm). Storage of empty or idle combustible pallets inside a protected liquid storage area shall comply with NFPA 13. Pallet storage shall be separated from liquid storage by aisles that are not less than 8 feet (2438 mm) wide.

5704.3.3.10 Containers in piles. Containers in piles shall be stacked in such a manner as to provide stability and to prevent excessive stress on container walls. Portable tanks stored more than one tier high shall be designed to nest securely, without dunnage. Material-handling equipment shall be suitable to handle containers and tanks safely at the upper tier level.

5704.3.4 Quantity limits for storage. Liquid storage quantity limitations shall comply with Sections 5704.3.4.1 through 5704.3.4.4.

5704.3.4.1 Maximum allowable quantity per control area. For occupancies other than Group M wholesale and retail sales uses, indoor storage of flammable and *combustible liquids* shall not exceed the *maximum allowable quantities per control area* indicated in Table 5003.1.1(1) and shall not exceed the additional limitations set forth in this section.

For Group M occupancy wholesale and retail sales uses, indoor storage of flammable and *combustible liquids* shall not exceed the *maximum allowable quantities per control area* indicated in Table 5704.3.4.1.

TABLE 5704.3.4.1
MAXIMUM ALLOWABLE QUANTITY PER CONTROL AREA OF FLAMMABLE AND COMBUSTIBLE LIQUIDS IN WHOLESALE AND RETAIL SALES OCCUPANCIES[a]

TYPE OF LIQUID	MAXIMUM ALLOWABLE QUANTITY PER CONTROL AREA (gallons)		
	Sprinklered[b] in accordance with footnote densities and arrangements	Sprinklered in accordance with Tables 5704.3.6.3(4) through 5704.3.6.3(8) and Table 5704.3.7.5.1	Nonsprinklered
Class IA	60	60	30
Class IB, IC, II and IIIA	7,500[c]	15,000[c]	1,600
Class IIIB	Unlimited	Unlimited	13,200

For SI: 1 foot = 304.8 mm, 1 square foot = 0.0929 m^2, 1 gallon = 3.785 L, 1 gallon per minute per square foot = 40.75 L/min/m^2.

a. Control areas shall be separated from each other by not less than a 1-hour *fire barrier*.

b. To be considered as sprinklered, a building shall be equipped throughout with an approved automatic sprinkler system with a design providing minimum densities as follows:

1. For uncartoned commodities on shelves 6 feet or less in height where the ceiling height does not exceed 18 feet, quantities are those allowed with a minimum sprinkler design density of Ordinary Hazard Group 2.
2. For cartoned, palletized or racked commodities where storage is 4 feet 6 inches or less in height and where the ceiling height does not exceed 18 feet, quantities are those allowed with a minimum sprinkler design density of 0.21 gallon per minute per square foot over the most remote 1,500-square-foot area.

c. Where wholesale and retail sales or storage areas exceed 50,000 square feet in area, the maximum allowable quantities are allowed to be increased by 2 percent for each 1,000 square feet of area in excess of 50,000 square feet, up to not more than 100 percent of the table amounts. A control area separation is not required. The cumulative amounts, including amounts attained by having an additional control area, shall not exceed 30,000 gallons.

Storage of hazardous production material flammable and *combustible liquids* in Group H-5 occupancies shall be in accordance with Chapter 27.

5704.3.4.2 Occupancy quantity limits. The following limits for quantities of stored flammable or *combustible liquids* shall not be exceeded:

1. Group A occupancies: Quantities in Group A occupancies shall not exceed that necessary for demonstration, treatment, laboratory work, maintenance purposes and operation of equipment, and shall not exceed quantities set forth in Table 5003.1.1(1).
2. Group B occupancies: Quantities in drinking, dining, office and school uses within Group B occupancies shall not exceed that necessary for demonstration, treatment, laboratory work, maintenance purposes and operation of equipment, and shall not exceed quantities set forth in Table 5003.1.1(1).
3. Group E occupancies: Quantities in Group E occupancies shall not exceed that necessary for demonstration, treatment, laboratory work, maintenance purposes and operation of equipment, and shall not exceed quantities set forth in Table 5003.1.1(1).
4. Group F occupancies: Quantities in dining, office, and school uses within Group F occupancies shall not exceed that necessary for demonstration, laboratory work, maintenance purposes and operation of equipment, and shall not exceed quantities set forth in Table 5003.1.1(1).
5. Group I occupancies: Quantities in Group I occupancies shall not exceed that necessary for demonstration, laboratory work, maintenance purposes and operation of equipment, and shall not exceed quantities set forth in Table 5003.1.1(1).
6. Group M occupancies: Quantities in dining, office, and school uses within Group M occupancies shall not exceed that necessary for demonstration, laboratory work, maintenance purposes and operation of equipment, and shall not exceed quantities set forth in Table 5003.1.1(1). The maximum allowable quantities for storage in wholesale and retail sales areas shall be in accordance with Section 5704.3.4.1.
7. Group R occupancies: Quantities in Group R occupancies shall not exceed that necessary for maintenance purposes and operation of equipment, and shall not exceed quantities set forth in Table 5003.1.1(1).
8. Group S occupancies: Quantities in dining and office uses within Group S occupancies shall not exceed that necessary for demonstration, laboratory work, maintenance purposes and operation of equipment, and shall not exceed quantities set forth in Table 5003.1.1(1).

5704.3.4.3 Quantities exceeding limits for control areas. Quantities exceeding those allowed in *control areas* set forth in Section 5704.3.4.1 shall be in liquid storage rooms or liquid storage warehouses in accordance with Sections 5704.3.7 and 5704.3.8.

5704.3.4.4 Liquids for maintenance and operation of equipment. In all occupancies, quantities of flammable and *combustible liquids* in excess of 10 gallons (38 L) used for maintenance purposes and the operation of equipment shall be stored in liquid storage cabinets in accordance with Section 5704.3.2. Quantities not exceeding 10 gallons (38 L) are allowed to be stored outside of a cabinet where in *approved* containers located in private garages or other *approved* locations.

5704.3.5 Storage in control areas. Storage of flammable and *combustible liquids* in *control areas* shall be in accordance with Sections 5704.3.5.1 through 5704.3.5.4.

5704.3.5.1 Basement storage. Class I liquids shall be allowed to be stored in *basements* in amounts not exceeding the *maximum allowable quantity per control area* for use-*open systems* in Table 5003.1.1(1), provided that automatic suppression and other fire protection are provided in accordance with Chapter 9. Class II and IIIA liquids shall also be allowed to be stored in *basements*, provided that automatic suppression and other fire protection are provided in accordance with Chapter 9.

5704.3.5.2 Storage pile heights. Containers having less than a 30-gallon (114 L) capacity that contain Class I or II liquids shall not be stacked more than 3 feet (914.4 mm) or two containers high, whichever is greater, unless stacked on fixed shelving or otherwise satisfactorily secured. Containers of Class I or II liquids having a capacity of 30 gallons (114 L) or more shall not be stored more than one container high. Containers shall be stored in an upright position.

5704.3.5.3 Storage distance from ceilings and roofs. Piles of containers or portable tanks shall not be stored closer than 3 feet (914 mm) to the nearest beam, chord, girder or other obstruction, and shall be 3 feet (914 mm) below sprinkler deflectors or discharge orifices of water spray or other overhead *fire protection system*.

5704.3.5.4 Combustible materials. In areas that are inaccessible to the public, Class I, II and IIIA liquids shall not be stored in the same pile or rack section as ordinary combustible commodities unless such materials are packaged together as kits.

5704.3.6 Wholesale and retail sales uses. Flammable and combustible liquids in Group M occupancy wholesale and retail sales uses shall be in accordance with Sections 5704.3.6.1 through 5704.3.6.5, or Sections 10.10.2, 12.3.8, 16.4.1 through 16.4.3, 16.5.1 through 16.5.2.12, Tables 16.5.2.1 through 16.5.2.12, and Figures 16.4.1(a) through 16.14.1(c) of NFPA 30.

5704.3.6.1 Container type. Containers for Class I liquids shall be metal.

Exception: In sprinklered buildings, an aggregate quantity of 120 gallons (454 L) of water-miscible Class IB and Class IC liquids is allowed in nonmetallic containers, each having a capacity of 16 ounces (0.473 L) or less.

5704.3.6.2 Container capacity. Containers for Class I liquids shall not exceed a capacity of 5 gallons (19 L).

Exception: Metal containers not exceeding 55 gallons (208 L) are allowed to store up to 240 gallons (908 L) of the *maximum allowable quantity per control area* of Class IB and IC liquids in a control area. The building shall be equipped throughout with an *approved* automatic sprinkler system in accordance with Table 5704.3.4.1. The containers shall be provided with plastic caps without cap seals and shall be stored upright. Containers shall not be stacked or stored in racks and shall not be located in areas accessible to the public.

5704.3.6.3 Fire protection and storage arrangements. Fire protection and container storage arrangements shall be in accordance with Table 5704.3.6.3(1) or the following:

1. Storage on shelves shall not exceed 6 feet (1829 mm) in height, and shelving shall be metal.
2. Storage on pallets or in piles greater than 4 feet 6 inches (1372 mm) in height, or where the ceiling exceeds 18 feet (5486 mm) in height, shall be protected in accordance with Table 5704.3.6.3(4), and the storage heights and arrangements shall be limited to those specified in Table 5704.3.6.3(2).
3. Storage on racks greater than 4 feet 6 inches (1372 mm) in height, or where the ceiling exceeds 18 feet (5486 mm) in height shall be protected in accordance with Tables 5704.3.6.3(5), 5704.3.6.3(6), and 5704.3.6.3(7) as appropriate, and the storage heights and arrangements shall be limited to those specified in Table 5704.3.6.3(3).

Combustible commodities shall not be stored above flammable and *combustible liquids*.

5704.3.6.4 Warning for containers. Cans, containers and vessels containing flammable liquids or flammable liquid compounds or mixtures offered for sale shall be provided with a warning indicator, painted or printed on the container and stating that the liquid is flammable, and shall be kept away from heat and an open flame.

5704.3.6.5 Storage plan. Where required by fire the code official, *aisle* and storage plans shall be submitted in accordance with Chapter 50.

5704.3.7 Liquid storage rooms. Liquid storage rooms shall comply with Sections 5704.3.7.1 through 5704.3.7.5.2.

5704.3.7.1 General. Quantities of liquids exceeding those set forth in Section 5704.3.4.1 for storage in *control areas* shall be stored in a liquid storage room complying with this section and constructed and separated as required by the *International Building Code*.

5704.3.7.2 Quantities and arrangement of storage. The quantity limits and storage arrangements in liquid storage rooms shall be in accordance with Tables 5704.3.6.3(2) and 5704.3.6.3(3) and Sections 5704.3.7.2.1 through 5704.3.7.2.3.

5704.3.7.2.1 Mixed storage. Where two or more classes of liquids are stored in a pile or rack section, both of the following shall apply:

1. The quantity in that pile or rack shall not exceed the smallest of the maximum quantities for the classes of liquids stored in accordance with Table 5704.3.6.3(2) or 5704.3.6.3(3).
2. The height of storage in that pile or rack shall not exceed the smallest of the maximum heights for the classes of liquids stored in accordance with Table 5704.3.6.3(2) or 5704.3.6.3(3).

TABLE 5704.3.6.3(1)
MAXIMUM STORAGE HEIGHT IN CONTROL AREA

TYPE OF LIQUID	NONSPRINKLERED AREA (feet)	SPRINKLERED AREA[a] (feet)	SPRINKLERED WITH IN-RACK PROTECTION[a, b] (feet)
Flammable liquids:			
Class IA	4	4	4
Class IB	4	8	12
Class IC	4	8	12
Combustible liquids:			
Class II	6	8	12
Class IIIA	8	12	16
Class IIIB	8	12	20

For SI: 1 foot = 304.8 mm.

a In buildings protected by an automatic sprinkler system, the storage height for containers and portable tanks shall not exceed the maximum storage height permitted for the fire protection scheme set forth in NFPA 30 or the maximum storage height demonstrated in a full-scale fire test, whichever is greater. NFPA 30 criteria and fire test results for metallic containers and portable tanks shall not be applied to nonmetallic containers and portable tanks.

b. In-rack protection shall be in accordance with Table 5704.3.6.3(5), 5704.3.6.3(6) or 5704.3.6.3(7).

TABLE 5704.3.6.3(2)
STORAGE ARRANGEMENTS FOR PALLETIZED OR SOLID-PILE STORAGE IN LIQUID STORAGE ROOMS AND WAREHOUSES

CLASS	STORAGE LEVEL	MAXIMUM STORAGE HEIGHT			MAXIMUM QUANTITY PER PILE (gallons)		MAXIMUM QUANTITY PER ROOM[a] (gallons)	
		Drums	Containers[b] (feet)	Portable tanks[b] (feet)	Containers	Portable tanks	Containers	Portable tanks
IA	Ground floor Upper floors Basements	1 1 0	5 5 Not Allowed	Not Allowed Not Allowed Not Allowed	3,000 2,000 Not Allowed	Not Allowed Not Allowed Not Allowed	12,000 8,000 Not Allowed	Not Allowed Not Allowed Not Allowed
IB	Ground floor Upper floors Basements	1 1 0	6.5 6.5 Not Allowed	7 7 Not Allowed	5,000 3,000 Not Allowed	20,000 10,000 Not Allowed	15,000 12,000 Not Allowed	40,000 20,000 Not Allowed
IC	Ground floor[d] Upper floors Basements	1 1 0	6.5[c] 6.5[c] Not Allowed	7 7 Not Allowed	5,000 3,000 Not Allowed	20,000 10,000 Not Allowed	15,000 12,000 Not Allowed	40,000 20,000 Not Allowed
II	Ground floor[d] Upper floors Basements	3 3 1	10 10 5	14 14 7	10,000 10,000 7,500	40,000 40,000 20,000	25,000 25,000 7,500	80,000 80,000 20,000
III	Ground floor Upper floors Basements	5 5 3	20 20 10	14 14 7	15,000 15,000 10,000	60,000 60,000 20,000	50,000 50,000 25,000	100,000 100,000 40,000

For SI: 1 foot = 304.8 mm, 1 gallon = 3.785 L.

a. See Section 5704.3.8.1 for unlimited quantities in liquid storage warehouses.

b. In buildings protected by an automatic sprinkler system, the storage height for containers and portable tanks shall not exceed the maximum storage height permitted for the fire protection scheme set forth in NFPA 30 or the maximum storage height demonstrated in a full-scale fire test, whichever is greater. NFPA 30 criteria and fire test results for metallic containers and portable tanks shall not be applied to nonmetallic containers and portable tanks.

c. These height limitations are allowed to be increased to 10 feet for containers having a capacity of 5 gallons or less.

d. For palletized storage of unsaturated polyester resins (UPR) in relieving-style metal containers with 50 percent or less by weight Class IC or II liquid and no Class IA or IB liquid, height and pile quantity limits shall be permitted to be 10 feet and 15,000 gallons, respectively, provided that such storage is protected by sprinklers in accordance with NFPA 30 and that the UPR storage area is not located in the same containment area or drainage path for other Class I or II liquids.

TABLE 5704.3.6.3(3)
STORAGE ARRANGEMENTS FOR RACK STORAGE IN LIQUID STORAGE ROOMS AND WAREHOUSES

CLASS	TYPE RACK	STORAGE LEVEL	MAXIMUM STORAGE HEIGHT[b] (feet)	MAXIMUM QUANTITY PER ROOM[a] (gallons)
			Containers	Containers
IA	Double row or Single row	Ground floor Upper floors Basements	25 15 Not Allowed	7,500 4,500 Not Allowed
IB IC	Double row or Single row	Ground floor Upper floors Basements	25 15 Not Allowed	15,000 9,000 Not Allowed
II	Double row or Single row	Ground floor Upper floors Basements	25 25 15	24,000 24,000 9,000
III	Multirow Double row Single row	Ground floor Upper floors Basements	40 20 20	48,000 48,000 24,000

For SI: 1 foot = 304.8 mm, 1 gallon = 3.785 L.

a. See Section 5704.3.8.1 for unlimited quantities in liquid storage warehouses.

b. In buildings protected by an automatic sprinkler system, the storage height for containers and portable tanks shall not exceed the maximum storage height permitted for the fire protection scheme set forth in NFPA 30 or the maximum storage height demonstrated in a full-scale fire test, whichever is greater. NFPA 30 criteria and fire test results for metallic containers and portable tanks shall not be applied to nonmetallic containers and portable tanks.

TABLE 5704.3.6.3(4)
AUTOMATIC SPRINKLER PROTECTION FOR SOLID-PILE AND PALLETIZED STORAGE OF LIQUIDS IN METAL CONTAINERS AND PORTABLE TANKS[a]

STORAGE CONDITIONS		CEILING SPRINKLER DESIGN AND DEMAND				MINIMUM HOSE STREAM DEMAND (gpm)	MINIMUM DURATION SPRINKLERS AND HOSE STREAMS (hours)
Class liquid	Container size and arrangement	Density (gpm/ft^2)	Area (square feet)		Maximum spacing (square feet)		
			High temperature sprinklers	Ordinary temperature sprinklers			
IA	5 gallons or less, with or without cartons, palletized or solid pile[b]	0.30	3,000	5,000	100	750	2
	Containers greater than 5 gallons, on end or side, palletized or solid pile	0.60	5,000	8,000	80	750	
IB, IC and II	5 gallons or less, with or without cartons, palletized or solid pile[b]	0.30	3,000	5,000	100	500	2
	Containers greater than 5 gallons on pallets or solid pile, one high	0.25	5,000	8,000	100		
II	Containers greater than 5 gallons on pallets or solid pile, more than one high, on end or side	0.60	5,000	8,000	80	750	2
IB, IC and II	Portable tanks, one high	0.30	3,000	5,000	100	500	2
II	Portable tanks, two high	0.60	5,000	8,000	80	750	2
III	5 gallons or less, with or without cartons, palletized or solid pile	0.25	3,000	5,000	120	500	1
	Containers greater than 5 gallons on pallets or solid pile, on end or sides, up to three high	0.25	3,000	5,000	120	500	1
	Containers greater than 5 gallons, on pallets or solid pile, on end or sides, up to 18 feet high	0.35	3,000	5,000	100	750	2
	Portable tanks, one high	0.25	3,000	5,000	120	500	1
	Portable tanks, two high	0.50	3,000	5,000	80	750	2

For SI: 1 foot = 304.8 mm, 1 gallon = 3.785 L, 1 square foot = 0.0929 m^2, 1 gallon per minute = 3.785 L/m, 1 gallon per minute per square foot = 40.75 $L/min/m^2$.

a. The design area contemplates the use of Class II standpipe systems. Where Class I standpipe systems are used, the area of application shall be increased by 30 percent without revising density.

b. For storage heights above 4 feet or ceiling heights greater than 18 feet, an approved engineering design shall be provided in accordance with Section 104.7.2.

TABLE 5704.3.6.3(5)
AUTOMATIC SPRINKLER PROTECTION REQUIREMENTS FOR RACK STORAGE OF LIQUIDS IN METAL CONTAINERS OF 5-GALLON CAPACITY OR LESS WITH OR WITHOUT CARTONS ON CONVENTIONAL WOOD PALLETS[a]

CLASS LIQUID	CEILING SPRINKLER DESIGN AND DEMAND				IN-RACK SPRINKLER ARRANGEMENT AND DEMAND				MINIMUM HOSE STREAM DEMAND (gpm)	MINIMUM DURATION SPRINKLER AND HOSE STREAM (hours)
	Density (gpm/ft²)	Area (square feet)		Maximum spacing	Racks up to 9 feet deep	Racks more than 9 feet to 12 feet deep	30 psi (standard orifice)	Number of sprinklers operating		
		High-temperature sprinklers	Ordinary temperature sprinklers				14 psi (large orifice)			
I (maximum 25-foot height) Option 1	0.40	3,000	5,000	80 ft²/head	1. Ordinary temperature, quick-response sprinklers, maximum 8 feet 3 inches horizontal spacing 2. One line sprinklers above each level of storage 3. Locate in longitudinal flue space, staggered vertical 4. Shields required where multilevel	1. Ordinary temperature, quick-response sprinklers, maximum 8 feet 3 inches horizontal spacing 2. One line sprinklers above each level of storage 3. Locate in transverse flue spaces, staggered vertical and within 20 inches of aisle 4. Shields required where multilevel	30 psi (0.5- inch orifice)	1. Eight sprinklers if only one level 2. Six sprinklers each on two levels if only two levels 3. Six sprinklers each on top three levels, if three or more levels 4. Hydraulically most remote	750	2
I (maximum 25-foot height) Option 2	0.55	2,000[b]	Not Applicable	100 ft²/ head	1. Ordinary temperature, quick-response sprinklers, maximum 8 feet 3 inches horizontal spacing 2. See 2 above 3. See 3 above 4. See 4 above	1. Ordinary temperature, quick-response sprinklers, maximum 8 feet 3 inches horizontal spacing 2. See 2 above 3. See 3 above 4. See 4 above	14 psi (0.53-inch orifice)	See 1 through 4 above	500	2
I and II (maximum 14-foot storage height) (maximum three tiers)	0.55[c]	2,000[d]	Not Applicable	100 ft²/ head	Not Applicable None for maximum 6-foot-deep racks	Not Applicable	Not Applicable	Not Applicable	500	2
II (maximum 25-foot height)	0.30	3,000	5,000	100 ft²/ head	1. Ordinary temperature sprinklers 8 feet apart horizontally 2. One line sprinklers between levels at nearest 10-foot vertical intervals 3. Locate in longitudinal flue space, staggered vertical 4. Shields required where multilevel	1. Ordinary temperature sprinklers 8 feet apart horizontally 2. Two lines between levels at nearest 10-foot vertical intervals 3. Locate in transverse flue spaces, staggered vertical and within 20 inches of aisle 4. Shields required where multilevel	30 psi	Hydraulically most remote—six sprinklers at each level, up to a maximum of three levels	750	2
III (40-foot height)	0.25	3,000	5,000	120 ft²/ head	Same as for Class II liquids	Same as for Class II liquids	30 psi	Same as for Class II liquids	500	2

For SI: 1 inch = 25.4 mm, 1 foot = 304.8 mm, 1 square foot = 0.0929 m², 1 pound per square inch = 6.895 kPa, 1 gallon = 3.785 L, 1 gallon per minute = 3.785 L/m, 1 gallon per minute per square foot = 40.75 L/min/m².

a. The design area contemplates the use of Class II standpipe systems. Where Class I standpipe systems are used, the area of application shall be increased by 30 percent without revising density.

b. Using listed or approved extra-large orifices, high-temperature quick-response or standard element sprinklers under a maximum 30-foot ceiling with minimum 7.5-foot aisles.

c. For friction lid cans and other metal containers equipped with plastic nozzles or caps, the density shall be increased to 0.65 gpm per square foot using listed or approved extra-large orifice, high-temperature quick-response sprinklers.

d. Using listed or approved extra-large orifice, high-temperature quick-response or standard element sprinklers under a maximum 18-foot ceiling with minimum 7.5-foot aisles and metal containers.

TABLE 5704.3.6.3(6) AUTOMATIC SPRINKLER PROTECTION REQUIREMENTS OR RACK STORAGE OF LIQUIDS IN METAL CONTAINERS GREATER THAN 5-GALLON CAPACITY[a]

CLASS LIQUID	CEILING SPRINKLER DESIGN AND DEMAND			IN-RACK SPRINKLER ARRANGEMENT AND DEMAND					MINIMUM HOSE STREAM DEMAND (gpm)	MINIMUM DURATION SPRINKLER AND HOSE STREAM (hours)
	Density (gpm/ ft^2)	Area (square feet)		Maximum spacing	On-side storage racks up to 9-foot-deep racks	On-end storage (on pallets) up to 9-foot-deep racks	Minimum nozzle pressure	Number of sprinklers operating		
		High-temperature sprinklers	Ordinary temperature sprinklers							
IA (maximum 25-foot height)	0.60	3,000	5,000	80 ft^2/head	1. Ordinary temperature sprinklers 8 feet apart horizontally 2. One line sprinklers above each tier of storage 3. Locate in longitudinal flue space, staggered vertical 4. Shields required where multilevel	1. Ordinary temperature sprinklers 8 feet apart horizontally 2. One line sprinklers above each tier of storage 3. Locate in longitudinal flue space, staggered vertical 4. Shields required where multilevel	30 psi	Hydraulically most remote—six sprinklers at each level	1,000	2
IB, IC and II (maximum 25-foot height)	0.60	3,000	5,000	100 ft^2/head	1. See 1 above 2. One line sprinklers every three tiers of storage 3. See 3 above 4. See 4 above	1. See 1 above 2. See 2 above 3. See 3 above 4. See 4 above	30 psi	Hydraulically most remote—six sprinklers at each level	750	2
III (maximum 40-foot height)	0.25	3,000	5,000	120 ft^2/head	1. See 1 above 2. One line sprinklers every sixth level (maximum) 3. See 3 above 4. See 4 above	1. See 1 above 2. One line sprinklers every third level (maximum) 3. See 3 above 4. See 4 above	15 psi	Hydraulically most remote—six sprinklers at each level	500	1

For SI: 1 foot = 304.8 mm, 1 square foot = 0.0929 m^2, 1 pound per square inch = 6.895 kPa, 1 gallon = 3.785 L, 1 gallon per minute = 3.785 L/m, 1 gallon per minute per square foot = 40.75 L/min/m^2.

a. The design assumes the use of Class II standpipe systems. Where a Class I standpipe system is used, the area of application shall be increased by 30 percent without revising density.

TABLE 5704.3.6.3(7)
AUTOMATIC AFFF WATER PROTECTION REQUIREMENTS FOR RACK STORAGE OF LIQUIDS IN METAL CONTAINERS GREATER THAN 5-GALLON CAPACITY[a,b]

CLASS LIQUID	CEILING SPRINKLER DESIGN AND DEMAND			IN-RACK SPRINKLER ARRANGEMENT AND DEMAND[c]				DURATION AFFF SUPPLY (minimum)	DURATION WATER SUPPLY (hours)
	Density (gpm/ft^2)	Area (square feet)		On-end storage of drums on pallets, up to 25 feet	Minimum nozzle pressure (psi)	Number of sprinklers operating	Hose stream demand[d] (gpm)		
		High-temperature sprinklers	Ordinary temperature sprinklers						
IA, IB, IC and II	0.30	1,500	2,500	1. Ordinary temperature sprinkler up to 10 feet apart horizontally 2. One line sprinklers above each level of storage 3. Locate in longitudinal flue space, staggered vertically 4. Shields required for multilevel	30	Three sprinklers per level	500	15	2

For SI: 1 inch = 25.4 mm, 1 foot = 304.8 mm, 1 square foot = 0.0929 m^2, 1 pound per square inch = 6.895 kPa, 1 gallon = 3.785 L, 1 gallon per minute = 3.785 L/m, 1 gallon per minute per square foot = 40.75 L/min/m^2.

a. System shall be a closed-head wet system with *approved* devices for proportioning aqueous film-forming foam.

b. Except as modified herein, in-rack sprinklers shall be installed in accordance with NFPA 13.

c. The height of storage shall not exceed 25 feet.

d. Hose stream demand includes $1^1/_2$-inch inside hose connections, where required.

TABLE 5704.3.6.3(8)
AUTOMATIC SPRINKLER PROTECTION REQUIREMENTS FOR CLASS I LIQUID STORAGE IN METAL CONTAINERS OF 1-GALLON CAPACITY OR LESS WITH UNCARTONED OR CASE-CUT SHELF DISPLAY UP TO 6.5 FEET, AND PALLETIZED STORAGE ABOVE IN A DOUBLE-ROW RACK ARRAY[a]

STORAGE HEIGHT	CEILING SPRINKLER DESIGN AND DEMAND				IN-RACK SPRINKLER ARRANGEMENT AND DEMAND				MINIMUM HOSE STREAM DEMAND (gpm)	MINIMUM DURATION SPRINKLERS AND HOSE STREAM (hours)
	Density (gpm/ft^2)	Area (square feet)		Maximum spacing	Racks up to 9 feet deep	Racks 9 to 12 feet	Minimum nozzle pressure	Number of sprinklers operating		
		High temperature	Ordinary temperature							
Maximum 20-foot storage height	0.60	2,000[b]	Not Applicable	100 ft^2/head	1. Ordinary temperature, quick-response sprinklers, maximum 8 feet 3 inches horizontal spacing 2. One line of sprinklers at the 6-foot level and the 11.5-foot level of storage 3. Locate in longitudinal flue space, staggered vertical 4. Shields required where multilevel	Not Applicable	30 psi (standard orifice) or 14 psi (large orifice)	1. Six sprinklers each on two levels 2. Hydraulically most remote 12 sprinklers	500	2

For SI: 1 inch = 25.4 mm, 1 foot = 304.8 mm, 1 square foot = 0.0929 m^2, 1 pound per square inch = 6.895 kPa, 1 gallon = 3.785 L, 1 gallon per minute = 3.785 L/m, 1 gallon per minute per square foot = 40.75 L/min/m^2.

a. This table shall not apply to racks with solid shelves.

b. Using extra-large orifice sprinklers under a ceiling 30 feet or less in height. Minimum aisle width is 7.5 feet.

5704.3.7.2.2 Separation and aisles. Piles shall be separated from each other by not less than 4-foot (1219 mm) aisles. Aisles shall be provided so that all containers are 20 feet (6096 mm) or less from an aisle. Where the storage of liquids is on racks, a minimum 4-foot-wide (1219 mm) aisle shall be provided between adjacent rows of racks and adjacent storage of liquids. Main aisles shall be not less than 8 feet (2438 mm) wide.

Additional aisles shall be provided for access to doors, required windows and ventilation openings, standpipe connections, mechanical equipment and switches. Such aisles shall be not less than 3 feet (914 mm) in width, unless greater widths are required for separation of piles or racks, in which case the greater width shall be provided.

5704.3.7.2.3 Stabilizing and supports. Containers and piles shall be separated by pallets or dunnage to provide stability and to prevent excessive stress to container walls. Portable tanks stored over one tier shall be designed to nest securely without dunnage.

Requirements for portable tank design shall be in accordance with Chapters 9 and 12 of NFPA 30. Shelving, racks, dunnage, scuffboards, floor overlay and similar installations shall be of noncombustible construction or of wood not less than a 1-inch (25 mm) nominal thickness. Adequate material-handling equipment shall be available to handle tanks safely at upper tier levels.

5704.3.7.3 Spill control and secondary containment. Liquid storage rooms shall be provided with spill control and secondary containment in accordance with Section 5004.2.

5704.3.7.4 Ventilation. Liquid storage rooms shall be ventilated in accordance with Section 5004.3.

5704.3.7.5 Fire protection. Fire protection for liquid storage rooms shall comply with Sections 5704.3.7.5.1 and 5704.3.7.5.2.

5704.3.7.5.1 Fire-extinguishing systems. Liquid storage rooms shall be protected by *automatic sprinkler systems* installed in accordance with Chapter 9 and Tables 5704.3.6.3(4) through 5704.3.6.3(7) and Table 5704.3.7.5.1. In-rack sprinklers shall also comply with NFPA 13.

Automatic foam-water systems and automatic aqueous film-forming foam (AFFF) water sprinkler systems shall not be used except where *approved.*

Protection criteria developed from fire modeling or full-scale fire testing conducted at an *approved* testing laboratory are allowed in lieu of the protection as shown in Tables 5704.3.6.3(2) through 5704.3.6.3(7) and Table 5704.3.7.5.1 where *approved.*

5704.3.7.5.2 Portable fire extinguishers. Not less than one *approved* portable fire extinguisher complying with Section 906 and having a rating of not less than 20-B shall be located not less than 10 feet (3048 mm) or more than 50 feet (15 240 mm) from any Class I or II liquid storage area located outside of a liquid storage room.

Not less than one portable fire extinguisher having a rating of not less than 20-B shall be located outside of, but not more than 10 feet (3048 mm) from, the door opening into a liquid storage room.

5704.3.8 Liquid storage warehouses. Buildings used for storage of flammable or *combustible liquids* in quantities exceeding those set forth in Section 5704.3.4 for *control areas* and Section 5704.3.7 for liquid storage rooms shall comply with Sections 5704.3.8.1 through 5704.3.8.5 and shall be constructed and separated as required by the *International Building Code.*

5704.3.8.1 Quantities and storage arrangement. The total quantities of liquids in a liquid storage warehouse shall not be limited. The arrangement of storage shall be in accordance with Table 5704.3.6.3(2) or 5704.3.6.3(3).

5704.3.8.1.1 Mixed storage. Mixed storage shall be in accordance with Section 5704.3.7.2.1.

5704.3.8.1.2 Separation and aisles. Separation and *aisles* shall be in accordance with Section 5704.3.7.2.2.

5704.3.8.2 Spill control and secondary containment. Liquid storage warehouses shall be provided with spill

TABLE 5704.3.7.5.1
AUTOMATIC AFFF-WATER PROTECTION REQUIREMENTS FOR SOLID-PILE AND PALLETIZED STORAGE OF LIQUIDS IN METAL CONTAINERS OF 5-GALLON CAPACITY OR LESS[a, b]

PACKAGE TYPE	CLASS LIQUID	CEILING SPRINKLER DESIGN AND DEMAND					STORAGE HEIGHT (feet)	HOSE DEMAND (gpm)[c]	DURATION AFFF SUPPLY (minimum)	DURATION WATER SUPPLY (hours)
		Density (gpm/ft^2)	Area (square feet)	Temperature rating	Maximum spacing	Orifice size (inch)				
Cartoned	IB, IC, II and III	0.40	2,000	286°F	100 ft^2/ head	0.531	11	500	15	2
Uncartoned	IB, IC, II and III	0.30	2,000	286°F	100 ft^2/ head	0.5 or 0.531	12	500	15	2

For SI: 1 inch = 25.4 mm, 1 foot = 304.8 mm, 1 square foot = 0.0929 m^2, 1 gallon per minute = 3.785 L/m, 1 gallon per minute per square foot = 40.75 L/min/m^2, °C = [(°F)-32]/1.8.

a. System shall be a closed-head wet system with approved devices for proportioning aqueous film-forming foam.

b. Maximum ceiling height of 30 feet.

c. Hose stream demand includes $1^1/_2$-inch inside hose connections, where required.

control and secondary containment as set forth in Section 5004.2.

5704.3.8.3 Ventilation. Liquid storage warehouses storing containers greater than 5 gallons (19 L) in capacity shall be ventilated at a rate of not less than 0.25 cfm per square foot (0.075 $m^3/s \cdot m^2$) of floor area over the storage area.

5704.3.8.4 Automatic sprinkler systems. Liquid storage warehouses shall be protected by *automatic sprinkler systems* installed in accordance with Chapter 9 and Tables 5704.3.6.3(4) through 5704.3.6.3(7) and Table 5704.3.7.5.1, or Sections 16.4.1 through 16.4.3, 16.5.1 through 16.5.2.12, and Tables 16.5.2.1 through 16.5.2.12 and Figures 16.4.1(a) through 16.4.1(c) of NFPA 30. In-rack sprinklers shall also comply with NFPA 13.

Automatic foam-water systems and automatic AFFF water sprinkler systems shall not be used except where *approved*.

Protection criteria developed from fire modeling or full-scale fire testing conducted at an *approved* testing laboratory are allowed in lieu of the protection as shown in Tables 5704.3.6.3(2) through 5704.3.6.3(7) and Table 5704.3.7.5.1 where *approved*.

5704.3.8.5 Warehouse hose lines. In liquid storage warehouses, either $1^1/_2$-inch (38 mm) lined or 1-inch (25 mm) hard rubber hose lines shall be provided in sufficient number to reach all liquid storage areas and shall be in accordance with Section 903 or 905.

5704.4 Outdoor storage of containers and portable tanks. Storage of flammable and *combustible liquids* in closed containers and portable tanks outside of buildings shall be in accordance with Section 5703 and Sections 5704.4.1 through 5704.4.8. Capacity limits for containers and portable tanks shall be in accordance with Section 5704.3.

5704.4.1 Plans. Storage shall be in accordance with *approved* plans.

5704.4.2 Location on property. Outdoor storage of liquids in containers and portable tanks shall be in accordance with Table 5704.4.2. Storage of liquids near buildings located on the same lot shall be in accordance with this section.

5704.4.2.1 Mixed liquid piles. Where two or more classes of liquids are stored in a single pile, the quantity in the pile shall not exceed the smallest of maximum quantities for the classes of material stored.

5704.4.2.2 Access. Storage of containers or portable tanks shall be provided with fire apparatus access roads in accordance with Chapter 5.

5704.4.2.3 Security. The storage area shall be protected against tampering or trespassers where necessary and shall be kept free from weeds, debris and other combustible materials not necessary to the storage.

5704.4.2.4 Storage adjacent to buildings. Not more than 1,100 gallons (4163 L) of liquids stored in closed containers and portable tanks is allowed adjacent to a building located on the same premises and under the same management, provided that:

1. The building does not exceed one story in height. Such building shall be of fire-resistance-rated construction with noncombustible exterior surfaces or noncombustible construction and shall be used principally for the storage of liquids; or
2. The exterior building wall adjacent to the storage area shall have a *fire-resistance rating* of not less than 2 hours, having no openings to above-grade areas within 10 feet (3048 mm) horizontally of such storage and no openings to below-grade areas within 50 feet (15 240 mm) horizontally of such storage.

The quantity of liquids stored adjacent to a building protected in accordance with Item 2 is allowed to exceed 1,100 gallons (4163 L), provided that the maximum quantity per pile does not exceed 1,100 gallons (4163 L) and each pile is separated by a 10-foot-minimum (3048 mm) clear space along the common wall.

Where the quantity stored exceeds 1,100 gallons (4163 L) adjacent to a building complying with Item 1, or the provisions of Item 1 cannot be met, a minimum distance in accordance with Table 5704.4.2, column 7

TABLE 5704.4.2
OUTDOOR LIQUID STORAGE IN CONTAINERS AND PORTABLE TANKS

CLASS OF LIQUID	CONTAINER STORAGE—MAXIMUM PER PILE		PORTABLE TANK STORAGE—MAXIMUM PER PILE		MINIMUM DISTANCE BETWEEN PILES OR RACKS (feet)	MINIMUM DISTANCE TO LOT LINE OF PROPERTY THAT CAN BE BUILT UPON[c, d] (feet)	MINIMUM DISTANCE TO PUBLIC STREET, PUBLIC ALLEY OR PUBLIC WAY[d] (feet)
	Quantity[a, b] (gallons)	Height (feet)	Quantity[a, b] (gallons)	Height (feet)			
IA	1,100	10	2,200	7	5	50	10
IB	2,200	12	4,400	14	5	50	10
IC	4,400	12	8,800	14	5	50	10
II	8,800	12	17,600	14	5	25	5
III	22,000	18	44,000	14	5	10	5

For SI: 1 foot = 304.8 mm, 1 gallon 3.785 L.

a. For mixed class storage, see Section 5704.4.2.

b. For storage in racks, the quantity limits per pile do not apply, but the rack arrangement shall be limited to not more than 50 feet in length and two rows or 9 feet in depth.

c. If protection by a public fire department or private fire brigade capable of providing cooling water streams is not available, the distance shall be doubled.

d. When the total quantity stored does not exceed 50 percent of the maximum allowed per pile, the distances are allowed to be reduced 50 percent, but not less than 3 feet.

("Minimum Distance to Lot Line of Property That Can Be Built Upon") shall be maintained between buildings and the nearest container or portable tank.

5704.4.3 Spill control and secondary containment. Storage areas shall be provided with spill control and secondary containment in accordance with Section 5703.4.

Exception: Containers stored on *approved* containment pallets in accordance with Section 5004.2.3 and containers stored in cabinets and lockers with integral spill containment.

5704.4.4 Security. Storage areas shall be protected against tampering or trespassers by fencing or other *approved* control measures.

5704.4.5 Protection from vehicles. Guard posts or other means shall be provided to protect exterior storage tanks from vehicular damage. Where guard posts are installed, the posts shall be installed in accordance with Section 312.

5704.4.6 Clearance from combustibles. The storage area shall be kept free from weeds, debris and combustible materials not necessary to the storage. The area surrounding an exterior storage area shall be kept clear of such materials for a minimum distance of 15 feet (4572 mm).

5704.4.7 Weather protection. Weather protection for outdoor storage shall be in accordance with Section 5004.13.

5704.4.8 Empty containers and tank storage. The storage of empty tanks and containers previously used for the storage of flammable or *combustible liquids*, unless free from explosive vapors, shall be stored as required for filled containers and tanks. Tanks and containers when emptied shall have the covers or plugs immediately replaced in openings.

SECTION 5705 DISPENSING, USE, MIXING AND HANDLING

5705.1 Scope. Dispensing, use, mixing and handling of flammable liquids shall be in accordance with Section 5703 and this section. Tank vehicle and tank car loading and unloading and other special operations shall be in accordance with Section 5706.

Exception: Containers of organic coatings having no fire point and which are opened for pigmentation are not required to comply with this section.

5705.2 Liquid transfer. Liquid transfer equipment and methods for transfer of Class I, II and IIIA liquids shall be *approved* and be in accordance with Sections 5705.2.1 through 5705.2.6.

5705.2.1 Pumps. Positive-displacement pumps shall be provided with pressure relief discharging back to the tank, pump suction or other *approved* location, or shall be provided with interlocks to prevent over-pressure.

5705.2.2 Pressured systems. Where gases are introduced to provide for transfer of Class I liquids, or Class II and III liquids transferred at temperatures at or above their *flash points* by pressure, only inert gases shall be used. Controls, including pressure relief devices, shall be provided to limit the pressure so that the maximum working pressure of tanks, containers and piping systems cannot be exceeded. Where devices operating through pressure within a tank or container are used, the tank or container shall be a pressure vessel *approved* for the intended use. Air or oxygen shall not be used for pressurization.

Exception: Air transfer of Class II and III liquids at temperatures below their *flash points*.

5705.2.3 Piping, hoses and valves. Piping, hoses and valves used in liquid transfer operations shall be *approved* or *listed* for the intended use.

5705.2.4 Class I, II and III liquids. Class I liquids or, when heated to or above their flash points, Class II and Class III liquids, shall be transferred by one of the following methods:

1. From safety cans complying with UL 30.
2. Through an *approved* closed piping system.
3. From containers or tanks by an *approved* pump taking suction through an opening in the top of the container or tank.
4. For Class IB, IC, II and III liquids, from containers or tanks by gravity through an *approved* self-closing or automatic-closing valve where the container or tank and dispensing operations are provided with spill control and secondary containment in accordance with Section 5703.4. Class IA liquids shall not be dispensed by gravity from tanks.
5. *Approved* engineered liquid transfer systems.

Exception: Liquids in original shipping containers not exceeding a 5.3-gallon (20 L) capacity.

5705.2.5 Manual container filling operations. Class I liquids or Class II and Class III liquids that are heated up to or above their *flash points* shall not be transferred into containers unless the nozzle and containers are electrically interconnected. Acceptable methods of electrical interconnection include either of the following:

1. Metallic floor plates on which containers stand while filling, where such floor plates are electrically connected to the fill stem.
2. Where the fill stem is bonded to the container during filling by means of a bond wire.

5705.2.6 Automatic container-filling operations for Class I liquids. Container-filling operations for Class I liquids involving conveyor belts or other automatic-feeding operations shall be designed to prevent static accumulations.

5705.3 Use, dispensing and mixing inside of buildings. Indoor use, dispensing and mixing of flammable and *combustible liquids* shall be in accordance with Section 5705.2 and Sections 5705.3.1 through 5705.3.5.3.

5705.3.1 Closure of mixing or blending vessels. Vessels used for mixing or blending of Class I liquids and Class II or III liquids heated up to or above their *flash points* shall be provided with self-closing, tight-fitting, noncombustible lids that will control a fire within such vessel.

Exception: Where such devices are impractical, *approved* automatic or manually controlled fire-extinguishing devices shall be provided.

5705.3.2 Bonding of vessels. Where differences of potential could be created, vessels containing Class I liquids or liquids handled at or above their *flash points* shall be electrically connected by bond wires, ground cables, piping or similar means to a static grounding system to maintain equipment at the same electrical potential to prevent sparking.

5705.3.3 Heating, lighting and cooking appliances. Heating, lighting and cooking appliances that utilize Class I liquids shall not be operated within a building or structure.

Exception: Operation in single-family *dwellings*.

5705.3.4 Location of processing vessels. Processing vessels shall be located with respect to distances to *lot lines* of adjoining property that can be built on, in accordance with Tables 5705.3.4(1) and 5705.3.4(2).

Exception: Where the exterior wall facing the adjoining *lot line* is a blank wall having a *fire-resistance rating* of not less than 4 hours, the *fire code official* is authorized to modify the distances. The distance shall be not less than that set forth in the *International Building Code*, and where Class IA or unstable liquids are involved, explosion control shall be provided in accordance with Section 911.

5705.3.5 Quantity limits for use. Liquid use quantity limitations shall comply with Sections 5705.3.5.1 through 5705.3.5.3.

5705.3.5.1 Maximum allowable quantity per control area. Indoor use, dispensing and mixing of flammable and *combustible liquids* shall not exceed the *maximum allowable quantity per control area* indicated in Table 5003.1.1(1) and shall not exceed the additional limitations set forth in Section 5705.3.5.

Exception: Cleaning with Class I, II and IIIA liquids shall be in accordance with Section 5705.3.6.

Use of hazardous production material flammable and *combustible liquids* in Group H-5 occupancies shall be in accordance with Chapter 27.

5705.3.5.2 Occupancy quantity limits. The following limits for quantities of flammable and *combustible liquids* used, dispensed or mixed based on occupancy classification shall not be exceeded:

Exception: Cleaning with Class I, II, or IIIA liquids shall be in accordance with Section 5705.3.6.

1. Group A occupancies: Quantities in Group A occupancies shall not exceed that necessary for demonstration, treatment, laboratory work,

**TABLE 5705.3.4(1)
SEPARATION OF PROCESSING VESSELS FROM LOT LINES**

PROCESSING VESSELS WITH EMERGENCY RELIEF VENTING	LOCATION[a]	
	Stable liquids	Unstable liquids
Not in excess of 2.5 psig	Table 5705.3.4(2)	2.5 times Table 5705.3.4(2)
Over 2.5 psig	1.5 times Table 5705.3.4(2)	4 times Table 5705.3.4(2)

For SI: 1 pound per square inch gauge = 6.895 kPa.

a. Where protection of exposures by a public fire department or private fire brigade capable of providing cooling water streams on structures is not provided, distances shall be doubled.

**TABLE 5705.3.4(2)
REFERENCE TABLE FOR USE WITH TABLE 5705.3.4(1)**

TANK CAPACITY (gallons)	MINIMUM DISTANCE FROM LOT LINE OF A LOT WHICH IS OR CAN BE BUILT UPON, INCLUDING THE OPPOSITE SIDE OF A PUBLIC WAY (feet)	MINIMUM DISTANCE FROM NEAREST SIDE OF ANY PUBLIC WAY OR FROM NEAREST IMPORTANT BUILDING ON THE SAME PROPERTY (feet)
275 or less	5	5
276 to 750	10	5
751 to 12,000	15	5
12,001 to 30,000	20	5
30,001 to 50,000	30	10
50,001 to 100,000	50	15
100,001 to 500,000	80	25
500,001 to 1,000,000	100	35
1,000,001 to 2,000,000	135	45
2,000,001 to 3,000,000	165	55
3,000,001 or more	175	60

For SI: 1 foot = 304.8 mm, 1 gallon = 3.785 L.

maintenance purposes and operation of equipment, and shall not exceed quantities set forth in Table 5003.1.1(1).

2. Group B occupancies: Quantities in drinking, dining, office and school uses within Group B occupancies shall not exceed that necessary for demonstration, treatment, laboratory work, maintenance purposes and operation of equipment, and shall not exceed quantities set forth in Table 5003.1.1(1).

3. Group E occupancies: Quantities in Group E occupancies shall not exceed that necessary for demonstration, treatment, laboratory work, maintenance purposes and operation of equipment and shall not exceed quantities set forth in Table 5003.1.1(1).

4. Group F occupancies: Quantities in dining, office and school uses within Group F occupancies shall not exceed that necessary for demonstration, laboratory work, maintenance purposes and operation of equipment, and shall not exceed quantities set forth in Table 5003.1.1(1).

5. Group I occupancies: Quantities in Group I occupancies shall not exceed that necessary for demonstration, laboratory work, maintenance purposes and operation of equipment, and shall not exceed quantities set forth in Table 5003.1.1(1).

6. Group M occupancies: Quantities in dining, office and school uses within Group M occupancies shall not exceed that necessary for demonstration, laboratory work, maintenance purposes and operation of equipment, and shall not exceed quantities set forth in Table 5003.1.1(1).

7. Group R occupancies: Quantities in Group R occupancies shall not exceed that necessary for maintenance purposes and operation of equipment, and shall not exceed quantities set forth in Table 5003.1.1(1).

8. Group S occupancies: Quantities in dining and office uses within Group S occupancies shall not exceed that necessary for demonstration, laboratory work, maintenance purposes and operation of equipment and shall not exceed quantities set forth in Table 5003.1.1(1).

5705.3.5.3 Quantities exceeding limits for control areas. Quantities exceeding the *maximum allowable quantity per control area* indicated in Sections 5705.3.5.1 and 5705.3.5.2 shall be in accordance with the following:

1. For *open systems,* indoor use, dispensing and mixing of flammable and *combustible liquids* shall be within a room or building complying with the *International Building Code* and Sections 5705.3.7.1 through 5705.3.7.5.

2. For *closed systems,* indoor use, dispensing and mixing of flammable and *combustible liquids* shall be within a room or building complying with the *International Building Code* and Sections 5705.3.7 through 5705.3.7.4 and Section 5705.3.7.6.

5705.3.6 Cleaning with flammable and combustible liquids. Cleaning with Class I, II and IIIA liquids shall be in accordance with Sections 5705.3.6.1 through 5705.3.6.2.7.

Exceptions:

1. Dry cleaning shall be in accordance with Chapter 21.

2. Spray-nozzle cleaning shall be in accordance with Section 2403.3.5.

5705.3.6.1 Cleaning operations. Class IA liquids shall not be used for cleaning. Cleaning with Class IB, IC or II liquids shall be conducted as follows:

1. In a room or building in accordance with Section 5705.3.7; or

2. In a parts cleaner *listed, labeled* and approved for the purpose in accordance with Section 5705.3.6.2.

Exception: Materials used in commercial and industrial process-related cleaning operations in accordance with other provisions of this code and not involving facilities maintenance cleaning operations.

5705.3.6.2 Listed and approved machines. Parts cleaning and degreasing conducted in *listed* and *approved* machines in accordance with Section 5705.3.6.1 shall be in accordance with Sections 5705.3.6.2.1 through 5705.3.6.2.7.

5705.3.6.2.1 Solvents. Solvents shall be classified and shall be compatible with the machines within which they are used.

5705.3.6.2.2 Machine capacities. The quantity of solvent shall not exceed the *listed* design capacity of the machine for the solvent being used with the machine.

5705.3.6.2.3 Solvent quantity limits. Solvent quantities shall be limited as follows:

1. Machines without remote solvent reservoirs shall be limited to quantities set forth in Section 5705.3.5.

2. Machines with remote solvent reservoirs using Class I liquids shall be limited to quantities set forth in Section 5705.3.5.

3. Machines with remote solvent reservoirs using Class II liquids shall be limited to 35 gallons (132 L) per machine. The total quantities shall not exceed an aggregate of 240 gallons (908 L) per *control area* in buildings not equipped throughout with an *approved* automatic sprinkler system and an aggregate of 480 gallons (1817 L) per *control area* in buildings

equipped throughout with an *approved* automatic sprinkler system in accordance with Section 903.3.1.1.

4. Machines with remote solvent reservoirs using Class IIIA liquids shall be limited to 80 gallons (303 L) per machine.

5705.3.6.2.4 Immersion soaking of parts. Work areas of machines with remote solvent reservoirs shall not be used for immersion soaking of parts.

5705.3.6.2.5 Separation. Multiple machines shall be separated from each other by a distance of not less than 30 feet (9144 mm) or by a *fire barrier* with a minimum 1-hour *fire-resistance rating*.

5705.3.6.2.6 Ventilation. Machines shall be located in areas adequately ventilated to prevent accumulation of vapors.

5705.3.6.2.7 Installation. Machines shall be installed in accordance with their listings.

5705.3.7 Rooms or buildings for quantities exceeding the maximum allowable quantity per control area. Where required by Section 5705.3.5.3 or 5705.3.6.1, rooms or buildings used for the use, dispensing or mixing of flammable and *combustible liquids* in quantities exceeding the maximum allowable quantity per control area shall be in accordance with Sections 5705.3.7.1 through 5705.3.7.6.3.

5705.3.7.1 Construction, location and fire protection. Rooms or buildings classified in accordance with the *International Building Code* as Group H-2 or H-3 occupancies based on use, dispensing or mixing of flammable or *combustible liquids* shall be constructed in accordance with the *International Building Code*.

5705.3.7.2 Basements. In rooms or buildings classified in accordance with the *International Building Code* as Group H-2 or H-3, dispensing or mixing of flammable or *combustible liquids* shall not be conducted in *basements*.

5705.3.7.3 Fire protection. Rooms or buildings classified in accordance with the *International Building Code* as Group H-2 or H-3 occupancies shall be equipped with an *approved* automatic fire-extinguishing system in accordance with Chapter 9.

5705.3.7.4 Doors. Interior doors to rooms or portions of such buildings shall be self-closing fire doors in accordance with the *International Building Code*.

5705.3.7.5 Open systems. Use, dispensing and mixing of flammable and *combustible liquids* in *open systems* shall be in accordance with Sections 5705.3.7.5.1 through 5705.3.7.5.3.

5705.3.7.5.1 Ventilation. Continuous mechanical ventilation shall be provided at a rate of not less than 1 cfm per square foot [0.00508 $m^3/(s \cdot m^2)$] of floor area over the design area. Provisions shall be made for introduction of makeup air in such a manner to include all floor areas or pits where vapors can collect. Local or spot ventilation shall be provided where needed to prevent the accumulation of hazardous vapors. Ventilation system design shall comply with the *International Building Code* and *International Mechanical Code*.

Exception: Where natural ventilation can be shown to be effective for the materials used, dispensed or mixed.

5705.3.7.5.2 Explosion control. Explosion control shall be provided in accordance with Section 911.

5705.3.7.5.3 Spill control and secondary containment. Spill control shall be provided in accordance with Section 5703.4 where Class I, II or IIIA liquids are dispensed into containers exceeding a 1.3-gallon (5 L) capacity or mixed or used in open containers or systems exceeding a 5.3-gallon (20 L) capacity. Spill control and secondary containment shall be provided in accordance with Section 5703.4 where the capacity of an individual container exceeds 55 gallons (208 L) or the aggregate capacity of multiple containers or tanks exceeds 100 gallons (378.5 L).

5705.3.7.6 Closed systems. Use or mixing of flammable or *combustible liquids* in *closed systems* shall be in accordance with Sections 5705.3.7.6.1 through 5705.3.7.6.3.

5705.3.7.6.1 Ventilation. *Closed systems* designed to be opened as part of normal operations shall be provided with ventilation in accordance with Section 5705.3.7.5.1.

5705.3.7.6.2 Explosion control. Explosion control shall be provided where an explosive environment can occur as a result of the mixing or use process. Explosion control shall be designed in accordance with Section 911.

Exception: Where process vessels are designed to contain fully the worst-case explosion anticipated within the vessel under process conditions considering the most likely failure.

5705.3.7.6.3 Spill control and secondary containment. Spill control shall be provided in accordance with Section 5703.4 where flammable or *combustible liquids* are dispensed into containers exceeding a 1.3-gallon (5 L) capacity or mixed or used in open containers or systems exceeding a 5.3-gallon (20 L) capacity. Spill control and secondary containment shall be provided in accordance with Section 5703.4 where the capacity of an individual container exceeds 55 gallons (208 L) or the aggregate capacity of multiple containers or tanks exceeds 1,000 gallons (3785 L).

5705.3.8 Use, dispensing and handling outside of buildings. Outside use, dispensing and handling shall be in accordance with Sections 5705.3.8.1 through 5705.3.8.4.

Dispensing of liquids into motor vehicle fuel tanks at motor fuel-dispensing facilities shall be in accordance with Chapter 23.

5705.3.8.1 Spill control. Outside use, dispensing and handling areas shall be provided with spill control as set forth in Section 5703.4.

5705.3.8.2 Location on property. Dispensing activities that exceed the quantities set forth in Table 5705.3.8.2 shall not be conducted within 15 feet (4572 mm) of buildings or combustible materials or within 25 feet (7620 mm) of building openings, *lot lines*, public streets, public alleys or *public ways*. Dispensing activities that exceed the quantities set forth in Table 5705.3.8.2 shall not be conducted within 15 feet (4572 mm) of storage of Class I, II or III liquids unless such liquids are stored in tanks that are *listed* and *labeled* as 2-hour protected tank assemblies in accordance with UL 2085.

Exceptions:

1. The requirements shall not apply to areas where only the following are dispensed: Class III liquids; liquids that are heavier than water; water-miscible liquids; and liquids with viscosities greater than 10,000 centipoise (cp) (10 Pa • s).
2. Flammable and *combustible liquid* dispensing in refineries, chemical plants, process facilities, gas and crude oil production facilities and oil-blending and packaging facilities, terminals and bulk plants.

**TABLE 5705.3.8.2
MAXIMUM ALLOWABLE QUANTITIES FOR DISPENSING OF FLAMMABLE AND COMBUSTIBLE LIQUIDS IN OUTDOOR CONTROL AREAS[a, b]**

CLASS OF LIQUID	QUANTITY (gallons)
Flammable	
Class IA	10
Class IB	15
Class IC	20
Combination Class IA, IB and IC	30[c]
Combustible	
Class II	30
Class IIIA	80
Class IIIB	3,300

For SI: 1 gallon = 3.785 L.

a. For definition of "Outdoor Control Area," see Section 5002.1.

b. The fire code official is authorized to impose special conditions regarding locations, types of containers, dispensing units, fire control measures and other factors involving fire safety.

c. Containing not more than the maximum allowable quantity per control area of each individual class.

5705.3.8.3 Location of processing vessels. Processing vessels shall be located with respect to distances to *lot lines* that can be built on in accordance with Table 5705.3.4(1).

Exception: In refineries and distilleries.

5705.3.8.4 Weather protection. Weather protection for outdoor use shall be in accordance with Section 5005.3.9.

5705.4 Solvent distillation units. Solvent distillation units shall comply with Sections 5705.4.1 through 5705.4.9.

5705.4.1 Unit with a capacity of 60 gallons or less. Solvent distillation units used to recycle Class I, II or IIIA liquids having a distillation chamber capacity of 60 gallons (227 L) or less shall be *listed*, *labeled* and installed in accordance with Section 5705.4 and UL 2208.

Exceptions:

1. Solvent distillation units used in continuous through-put industrial processes where the source of heat is remotely supplied using steam, hot water, oil or other heat transfer fluids, the temperature of which is below the auto-ignition point of the solvent.
2. *Approved* research, testing and experimental processes.

5705.4.2 Units with a capacity exceeding 60 gallons. Solvent distillation units used to recycle Class I, II or IIIA liquids, having a distillation chamber capacity exceeding 60 gallons (227 L) shall be used in locations that comply with the use and mixing requirements of Section 5705 and other applicable provisions in this chapter.

5705.4.3 Prohibited processing. Class I, II and IIIA liquids that are also classified as unstable (reactive) shall not be processed in solvent distillation units.

Exception: Appliances *listed* for the distillation of unstable (reactive) solvents.

5705.4.4 Labeling. A permanent label shall be affixed to the unit by the manufacturer. The label shall indicate the capacity of the distillation chamber, and the distance the unit shall be placed away from sources of ignition. The label shall indicate the products for which the unit has been *listed* for use or refer to the instruction manual for a list of the products.

5705.4.5 Manufacturer's instruction manual. An instruction manual shall be provided. The manual shall be readily available for the user and the *fire code official*. The manual shall include installation, use and servicing instructions. It shall identify the liquids for which the unit has been *listed* for distillation purposes along with each liquid's *flash point* and auto-ignition temperature. For units with adjustable controls, the manual shall include directions for setting the heater temperature for each liquid to be instilled.

5705.4.6 Location. Solvent distillation units shall be used in locations in accordance with the listing. Solvent distillation units shall not be used in *basements*.

5705.4.7 Storage of liquids. Distilled liquids and liquids awaiting distillation shall be stored in accordance with Section 5704.

5705.4.8 Storage of residues. Hazardous residue from the distillation process shall be stored in accordance with Section 5704 and Chapter 50.

5705.4.9 Portable fire extinguishers. *Approved* portable fire extinguishers shall be provided in accordance with Section 906. Not less than one portable fire extinguisher having a rating of not less than 40-B shall be located not less than 10 feet (3048 mm) or more than 30 feet (9144 mm) from any solvent distillation unit.

5705.5 Alcohol-based hand rubs classified as Class I or II liquids. The use of wall-mounted dispensers containing alcohol-based hand rubs classified as Class I or II liquids shall be in accordance with all of the following:

1. The maximum capacity of each dispenser shall be 68 ounces (2 L).
2. The minimum separation between dispensers shall be 48 inches (1219 mm).
3. The dispensers shall not be installed above, below, or closer than 1 inch (25 mm) to an electrical receptacle, switch, appliance, device or other ignition source. The wall space between the dispenser and the floor or intervening counter top shall be free of electrical receptacles, switches, appliances, devices or other ignition sources.
4. Dispensers shall be mounted so that the bottom of the dispenser is not less than 42 inches (1067 mm) and not more than 48 inches (1219 mm) above the finished floor.
5. Dispensers shall not release their contents except when the dispenser is manually activated. Facilities shall be permitted to install and use automatically activated "touch free" alcohol-based hand-rub dispensing devices with the following requirements:
   5.1. The facility or persons responsible for the dispensers shall test the dispensers each time a new refill is installed in accordance with the manufacturer's care and use instructions.
   5.2. Dispensers shall be designed and must operate in a manner that ensures accidental or malicious activations of the dispensing device are minimized. At a minimum, all devices subject to or used in accordance with this section shall have the following safety features:
      5.2.1. Any activations of the dispenser shall only occur when an object is placed within 4 inches (98 mm) of the sensing device.
      5.2.2. The dispenser shall not dispense more than the amount required for hand hygiene consistent with label instructions as regulated by the United States Food and Drug Administration (USFDA).
      5.2.3. An object placed within the activation zone and left in place will cause only one activation.
6. Storage and use of alcohol-based hand rubs shall be in accordance with the applicable provisions of Sections 5704 and 5705.
7. Dispensers installed in occupancies with carpeted floors shall only be allowed in smoke compartments or *fire areas* equipped throughout with an *approved* automatic sprinkler system in accordance with Section 903.3.1.1 or 903.3.1.2.

5705.5.1 Corridor installations. In addition to the provisions of Section 5705.5, where wall-mounted dispensers containing alcohol-based hand rubs are installed in *corridors* or rooms and areas open to the corridor, they shall be in accordance with all of the following:

1. Level 2 and 3 aerosol containers shall not be allowed in *corridors*.
2. The maximum capacity of each Class I or II liquid dispenser shall be 41 ounces (1.21 L) and the maximum capacity of each Level 1 aerosol dispenser shall be 18 ounces (0.51 kg).
3. The maximum quantity allowed in a *corridor* within a *control area* shall be 10 gallons (37.85 L) of Class I or II liquids or 1135 ounces (32.2 kg) of Level 1 aerosols, or a combination of Class I or II liquids and Level 1 aerosols not to exceed, in total, the equivalent of 10 gallons (37.85 L) or 1,135 ounces (32.2 kg) such that the sum of the ratios of the liquid and aerosol quantities divided by the allowable quantity of liquids and aerosols, respectively, shall not exceed one.
4. The minimum *corridor* width shall be 72 inches (1829 mm).
5. Projections into a *corridor* shall be in accordance with Section 1003.3.3.

SECTION 5706
SPECIAL OPERATIONS

5706.1 General. This section shall cover the provisions for special operations that include, but are not limited to, storage, use, dispensing, mixing or handling of flammable and *combustible liquids*. The following special operations shall be in accordance with Sections 5701, 5703, 5704 and 5705, except as provided in Section 5706.

1. Storage and dispensing of flammable and *combustible liquids* on farms and construction sites.
2. Well drilling and operating.
3. Bulk plants or terminals.
4. Bulk transfer and process transfer operations utilizing tank vehicles and tank cars.
5. Tank vehicles and tank vehicle operation.
6. Refineries.
7. Vapor recovery and vapor-processing systems.

5706.2 Storage and dispensing of flammable and combustible liquids on farms and construction sites. Permanent and temporary storage and dispensing of Class I and II liquids for private use on farms and rural areas and at construction sites, earth-moving projects, gravel pits or borrow pits shall be in accordance with Sections 5706.2.1 through 5706.2.8.1.

> **Exception:** Storage and use of fuel oil and containers connected with oil-burning equipment regulated by Section 603 and the *International Mechanical Code*.

5706.2.1 Combustibles and open flames near tanks. Storage areas shall be kept free from weeds and extraneous combustible material. Open flames and smoking are prohibited in flammable or *combustible liquid* storage areas.

5706.2.2 Marking of tanks and containers. Tanks and containers for the storage of liquids above ground shall be conspicuously marked with the name of the product that they contain and the words: FLAMMABLE—KEEP FIRE AND FLAME AWAY. Tanks shall bear the additional marking: KEEP 50 FEET FROM BUILDINGS.

5706.2.3 Containers for storage and use. Metal containers used for storage of Class I or II liquids shall be in accordance with DOTn requirements or shall be of an *approved* design.

Discharge devices shall be of a type that do not develop an internal pressure on the container. Pumping devices or *approved* self-closing faucets used for dispensing liquids shall not leak and shall be well-maintained. Individual containers shall not be interconnected and shall be kept closed when not in use.

Containers stored outside of buildings shall be in accordance with Section 5704 and the *International Building Code*.

5706.2.4 Permanent and temporary tanks. The capacity of permanent above-ground tanks containing Class I or II liquids shall not exceed 1,100 gallons (4164 L). The capacity of temporary above-ground tanks containing Class I or II liquids shall not exceed 10,000 gallons (37 854 L). Tanks shall be of the single-compartment design.

Exception: Permanent above-ground tanks of greater capacity that meet the requirements of Section 5704.2.

5706.2.4.1 Fill-opening security. Fill openings shall be equipped with a locking closure device. Fill openings shall be separate from vent openings.

5706.2.4.2 Vents. Tanks shall be provided with a method of normal and emergency venting. Normal vents shall be in accordance with Section 5704.2.7.3.

Emergency vents shall be in accordance with Section 5704.2.7.4. Emergency vents shall be arranged to discharge in a manner that prevents localized overheating or flame impingement on any part of the tank in the event that vapors from such vents are ignited.

5706.2.4.3 Location. Tanks containing Class I or II liquids shall be kept outside and not less than 50 feet (15 240 mm) from buildings and combustible storage. Additional distance shall be provided where necessary to ensure that vehicles, equipment and containers being filled directly from such tanks will not be less than 50 feet (15 240 mm) from structures, haystacks or other combustible storage.

5706.2.4.4 Locations where above-ground tanks are prohibited. The storage of Class I and II liquids in above-ground tanks is prohibited within the limits established by law as the limits of districts in which such storage is prohibited (see Section 3 of the Sample Legislation for Adoption of the *International Fire Code* on page xxi).

5706.2.5 Type of tank. Tanks shall be provided with top openings only or shall be elevated for gravity discharge.

5706.2.5.1 Tanks with top openings only. Tanks with top openings shall be mounted in accordance with either of the following:

1. On well-constructed metal legs connected to shoes or runners designed so that the tank is stabilized and the entire tank and its supports can be moved as a unit.
2. For stationary tanks, on a stable base of timbers or blocks approximately 6 inches (152 mm) in height that prevents the tank from contacting the ground.

5706.2.5.1.1 Pumps and fittings. Tanks with top openings only shall be equipped with a tightly and permanently attached, *approved* pumping device having an *approved* hose of sufficient length for filling vehicles, equipment or containers to be served from the tank. Either the pump or the hose shall be equipped with a padlock to its hanger to prevent tampering. An effective antisiphoning device shall be included in the pump discharge unless a self-closing nozzle is provided. Siphons or internal pressure discharge devices shall not be used.

5706.2.5.2 Tanks for gravity discharge. Tanks with a connection in the bottom or the end for gravity-dispensing liquids shall be mounted and equipped as follows:

1. Supports to elevate the tank for gravity discharge shall be designed to carry all required loads and provide stability.
2. Bottom or end openings for gravity discharge shall be equipped with a valve located adjacent to the tank shell that will close automatically in the event of fire through the operation of an effective heat-activated releasing device. Where this valve cannot be operated manually, it shall be supplemented by a second, manually operated valve.

The gravity discharge outlet shall be provided with an *approved* hose equipped with a self-closing valve at the discharge end of a type that can be padlocked to its hanger.

5706.2.6 Spill control drainage control and diking. Indoor storage and dispensing areas shall be provided with spill control and drainage control as set forth in Section 5703.4. Outdoor storage areas shall be provided with drainage control or diking as set forth in Section 5704.2.10.

5706.2.7 Portable fire extinguishers. Portable fire extinguishers with a minimum rating of 20-B:C and complying with Section 906 shall be provided where required by the *fire code official*.

5706.2.8 Dispensing from tank vehicles. Where *approved*, liquids used as fuels are allowed to be transferred from tank vehicles into the tanks of motor vehicles or special equipment, provided:

1. The tank vehicle's specific function is that of supplying fuel to motor vehicle fuel tanks.
2. The dispensing hose does not exceed 100 feet (30 480 mm) in length.
3. The dispensing nozzle is an *approved* type.
4. The dispensing hose is properly placed on an *approved* reel or in a compartment provided before the tank vehicle is moved.
5. Signs prohibiting smoking or open flames within 25 feet (7620 mm) of the vehicle or the point of refueling are prominently posted on the tank vehicle.
6. Electrical devices and wiring in areas where fuel dispensing is conducted are in accordance with NFPA 70.
7. Tank vehicle-dispensing equipment is operated only by designated personnel who are trained to handle and dispense motor fuels.
8. Provisions are made for controlling and mitigating unauthorized discharges.

5706.2.8.1 Location. Dispensing from tank vehicles shall be conducted not less than 50 feet (15 240 mm) from structures or combustible storage.

5706.3 Well drilling and operating. Wells for oil and natural gas shall be drilled and operated in accordance with Sections 5706.3.1 through 5706.3.8.

5706.3.1 Location. The location of wells shall comply with Sections 5706.3.1.1 through 5706.3.1.3.2.

5706.3.1.1 Storage tanks and sources of ignition. Storage tanks or boilers, fired heaters, open-flame devices or other sources of ignition shall not be located within 25 feet (7620 mm) of well heads. Smoking is prohibited at wells or tank locations except as designated and in *approved* posted areas.

Exception: Engines used in the drilling, production and serving of wells.

5706.3.1.2 Streets and railways. Wells shall not be drilled within 75 feet (22 860 mm) of any dedicated public street, highway or nearest rail of an operating railway.

5706.3.1.3 Buildings. Wells shall not be drilled within 100 feet (30 480 mm) of buildings not necessary to the operation of the well.

5706.3.1.3.1 Group A, E or I buildings. Wells shall not be drilled within 300 feet (91 440 mm) of buildings with an occupancy in Group A, E or I.

5706.3.1.3.2 Existing wells. Where wells are existing, buildings shall not be constructed within the distances set forth in Section 5706.3.1 for separation of wells or buildings.

5706.3.2 Waste control. Control of waste materials associated with wells shall comply with Sections 5706.3.2.1 and 5706.3.2.2.

5706.3.2.1 Discharge on a street or water channel. Liquids containing crude petroleum or its products shall not be discharged into or on streets, highways, drainage canals or ditches, storm drains or flood control channels.

5706.3.2.2 Discharge and combustible materials on ground. The surface of the ground under, around or near wells, pumps, boilers, oil storage tanks or buildings shall be kept free from oil, waste oil, refuse or waste material.

5706.3.3 Sumps. Sumps associated with wells shall comply with Sections 5706.3.3.1 through 5706.3.3.3.

5706.3.3.1 Maximum width. Sumps or other basins for the retention of oil or petroleum products shall not exceed 12 feet (3658 mm) in width.

5706.3.3.2 Backfilling. Sumps or other basins for the retention of oil or petroleum products larger than 6 feet by 6 feet by 6 feet (1829 mm by 1829 mm by 1829 mm) shall not be maintained longer than 60 days after the cessation of drilling operations.

5706.3.3.3 Security. Sumps, diversion ditches and depressions used as sumps shall be securely fenced or covered.

5706.3.4 Prevention of blowouts. Protection shall be provided to control and prevent the blowout of a well. Protection equipment shall meet federal, state and other applicable jurisdiction requirements.

5706.3.5 Storage tanks. Storage of flammable or *combustible liquids* in tanks shall be in accordance with Section 5704. Oil storage tanks or groups of tanks shall have posted in a conspicuous place, on or near such tank or tanks, an *approved* sign with the name of the *owner* or operator, or the lease number and the telephone number where a responsible person can be reached at any time.

5706.3.6 Soundproofing. Where soundproofing material is required during oil field operations, such material shall be noncombustible.

5706.3.7 Signs. Well locations shall have posted in a conspicuous place on or near such tank or tanks an *approved* sign with the name of the *owner* or operator, name of the leasee or the lease number, the well number and the telephone number where a responsible person can be reached at any time. Such signs shall be maintained on the premises from the time materials are delivered for drilling purposes until the well is abandoned.

5706.3.8 Field-loading racks. Field-loading racks shall be in accordance with Section 5706.5.

5706.4 Bulk plants or terminals. Portions of properties where flammable and *combustible liquids* are received by tank vessels, pipelines, tank cars or tank vehicles and stored or blended in bulk for the purpose of distribution by tank vessels, pipelines, tanks cars, tank vehicles or containers shall be in accordance with Sections 5706.4.1 through 5706.4.10.4.

5706.4.1 Building construction. Buildings shall be constructed in accordance with the *International Building Code*.

5706.4.2 Means of egress. Rooms in which liquids are stored, used or transferred by pumps shall have *means of egress* arranged to prevent occupants from being trapped in the event of fire.

5706.4.3 Heating. Rooms in which Class I liquids are stored or used shall be heated only by means not constituting a source of ignition, such as steam or hot water. Rooms containing heating appliances involving sources of ignition shall be located and arranged to prevent entry of flammable vapors.

5706.4.4 Ventilation. Ventilation shall be provided for rooms, buildings and enclosures in which Class I liquids are pumped, used or transferred. Design of ventilation systems shall consider the relatively high specific gravity of the vapors. Where natural ventilation is used, adequate openings in outside walls at floor level, unobstructed except by louvers or coarse screens, shall be provided. Where natural ventilation is inadequate, mechanical ventilation shall be provided in accordance with the *International Mechanical Code*.

5706.4.4.1 Basements and pits. Class I liquids shall not be stored or used within a building having a *basement* or pit into which flammable vapors can travel, unless such area is provided with ventilation designed to prevent the accumulation of flammable vapors therein.

5706.4.4.2 Dispensing of Class I liquids. Containers of Class I liquids shall not be drawn from or filled within buildings unless a provision is made to prevent the accumulation of flammable vapors in hazardous concentrations. Where mechanical ventilation is required, it shall be kept in operation while flammable vapors could be present.

5706.4.5 Storage. Storage of Class I, II and IIIA liquids in bulk plants shall be in accordance with the applicable provisions of Section 5704.

5706.4.6 Overfill protection of Class I and II liquids. Manual and automatic systems shall be provided to prevent overfill during the transfer of Class I and II liquids from mainline pipelines and marine vessels in accordance with API 2350.

5706.4.7 Wharves. This section shall apply to all wharves, piers, bulkheads and other structures over or contiguous to navigable water having a primary function of transferring liquid cargo in bulk between shore installations and tank vessels, ships, barges, lighter boats or other mobile floating craft.

Exception: Marine motor fuel-dispensing facilities in accordance with Chapter 23.

5706.4.7.1 Transferring approvals. Handling packaged cargo of liquids, including full and empty drums, bulk fuel and stores, over a wharf during cargo transfer shall be subject to the approval of the wharf supervisor and the senior deck officer on duty.

5706.4.7.2 Transferring location. Wharves at which liquid cargoes are to be transferred in bulk quantities to or from tank vessels shall be not less than 100 feet (30 480 mm) from any bridge over a navigable waterway; or from an entrance to, or superstructure of, any vehicular or railroad tunnel under a waterway. The termination of the fixed piping used for loading or unloading at a wharf shall be not less than 200 feet (60 960 mm) from a bridge or from an entrance to, or superstructures of, a tunnel.

5706.4.7.3 Superstructure and decking material. Superstructure and decking shall be designed for the intended use. Decking shall be constructed of materials that will afford the desired combination of flexibility, resistance to shock, durability, strength and *fire resistance*.

5706.4.7.4 Tanks allowed. Tanks used exclusively for ballast water or Class II or III liquids are allowed to be installed on suitably designed wharves.

5706.4.7.5 Transferring equipment. Loading pumps capable of building up pressures in excess of the safe working pressure of cargo hose or loading arms shall be provided with bypasses, relief valves or other arrangements to protect the loading facilities against excessive pressure. Relief devices shall be tested not less than annually to determine that they function satisfactorily at their set pressure.

5706.4.7.6 Piping, valves and fittings. Piping valves and fittings shall be in accordance with Section 5703.6 except as modified by the following:

1. Flexibility of piping shall be ensured by appropriate layout and arrangement of piping supports so that motion of the wharf structure resulting from wave action, currents, tides or the mooring of vessels will not subject the pipe to repeated excessive strain.
2. Pipe joints that depend on the friction characteristics of combustible materials or on the grooving of pipe ends for mechanical continuity of piping shall not be used.
3. Swivel joints are allowed in piping to which hoses are connected and for articulated, swivel-joint transfer systems, provided the design is such that the mechanical strength of the joint will not be impaired if the packing materials fail such as by exposure to fire.
4. Each line conveying Class I or II liquids leading to a wharf shall be provided with a readily accessible block valve located on shore near the approach to the wharf and outside of any diked area. Where more than one line is involved, the valves shall be grouped in one location.
5. Means shall be provided for easy access to cargo line valves located below the wharf deck.
6. Piping systems shall contain a sufficient number of valves to operate the system properly and to

control the flow of liquid in normal operation and in the event of physical damage.

7. Piping on wharves shall be bonded and grounded where Class I and II liquids are transported. Where excessive stray currents are encountered, insulating joints shall be installed. Bonding and grounding connections on piping shall be located on the wharf side of hose riser insulating flanges, where used, and shall be accessible for inspection.
8. Hose or articulated swivel-joint pipe connections used for cargo transfer shall be capable of accommodating the combined effects of change in draft and maximum tidal range, and mooring lines shall be kept adjusted to prevent surge of the vessel from placing stress on the cargo transfer system.
9. Hoses shall be supported to avoid kinking and damage from chafing.

5706.4.7.7 Loading and unloading. Loading or discharging shall not commence until the wharf superintendent and officer in charge of the tank vessel agree that the tank vessel is properly moored and connections are properly made.

5706.4.7.8 Mechanical work. Mechanical work shall not be performed on the wharf during cargo transfer, except under special authorization by the *fire code official* based on a review of the area involved, methods to be employed and precautions necessary.

5706.4.8 Sources of ignition. Class I, II or IIIA liquids shall not be used, drawn or dispensed where flammable vapors can reach a source of ignition. Smoking shall be prohibited except in designated locations. "No Smoking" signs complying with Section 310 shall be conspicuously posted where a hazard from flammable vapors is normally present.

5706.4.9 Drainage control. Loading and unloading areas shall be provided with drainage control in accordance with Section 5704.2.10.

5706.4.10 Fire protection. Fire protection shall be in accordance with Chapter 9 and Sections 5706.4.10.1 through 5706.4.10.4.

5706.4.10.1 Portable fire extinguishers. Portable fire extinguishers with a rating of not less than 20-B and complying with Section 906 shall be located within 75 feet (22 860 mm) of hose connections, pumps and separator tanks.

5706.4.10.2 Fire hoses. Where piped water is available, ready-connected fire hose in a size appropriate for the water supply shall be provided in accordance with Section 905 so that manifolds where connections are made and broken can be reached by not less than one hose stream.

5706.4.10.3 Obstruction of equipment. Material shall not be placed on wharves in such a manner that would obstruct access to fire-fighting equipment or impor pipeline control valves.

5706.4.10.4 Fire apparatus access. Where the whar accessible to vehicular traffic, an unobstructed f apparatus access road to the shore end of the wh shall be maintained in accordance with Chapter 5.

5706.5 Bulk transfer and process transfer operation Bulk transfer and process transfer operations shall b *approved* and be in accordance with Sections 5706.5. through 5706.5.4.5. Motor fuel-dispensing facilities shal comply with Chapter 23.

5706.5.1 General. The provisions of Sections 5706.5.1.1 through 5706.5.1.18 shall apply to bulk transfer and process transfer operations; Sections 5706.5.2 and 5706.5.2.1 shall apply to bulk transfer operations; Sections 5706.5.3 through 5706.5.3.3 shall apply to process transfer operations and Sections 5706.5.4 through 5706.5.4.5 shall apply to dispensing from tank vehicles and tank cars.

5706.5.1.1 Location. Bulk transfer and process transfer operations shall be conducted in *approved* locations. Tank cars shall be unloaded only on private sidings or railroad-siding facilities equipped for transferring flammable or *combustible liquids*. Tank vehicle and tank car transfer facilities shall be separated from buildings, above-ground tanks, combustible materials, *lot lines*, public streets, public alleys or *public ways* by a distance of 25 feet (7620 mm) for Class I liquids and 15 feet (4572 mm) for Class II and III liquids measured from the nearest position of any loading or unloading valve. Buildings for pumps or shelters for personnel shall be considered part of the transfer facility.

5706.5.1.2 Weather protection canopies. Where weather protection canopies are provided, they shall be constructed in accordance with Section 5004.13. Weather protection canopies shall not be located within 15 feet (4572 mm) of a building or combustible material or within 25 feet (7620 mm) of building openings, *lot lines*, public streets, public alleys or *public ways*.

5706.5.1.3 Ventilation. Ventilation shall be provided to prevent accumulation of vapors in accordance with Section 5705.3.7.5.1.

5706.5.1.4 Sources of ignition. Sources of ignition shall be controlled or eliminated in accordance with Section 5003.7.

5706.5.1.5 Spill control and secondary containment. Areas where transfer operations are located shall be provided with spill control and secondary containment in accordance with Section 5703.4. The spill control and secondary containment system shall have a design capacity capable of containing the capacity of the largest tank compartment located in the area where transfer operations are conducted. Containment of the rainfall volume specified in Section 5004.2.2.6 is not required.

5706.5.1.6 Fire protection. Fire protection shall be in accordance with Section 5703.2.

ant

is
re
rf

.7 **Static protection.** Static protection shall be l to prevent the accumulation of static charges ransfer operations. Bonding facilities shall be d during the transfer through open domes where liquids are transferred, or where Class II and III are transferred into tank vehicles or tank cars ould contain vapors from previous cargoes of l liquids.

otection shall consist of a metallic bond wire per- ently electrically connected to the fill stem. The fill assembly shall form a continuous electrically con- tive path downstream from the point of bonding. e free end of such bond wire shall be provided with a amp or equivalent device for convenient attachment a metallic part in electrical contact with the cargo ank of the tank vehicle or tank car. For tank vehicles, protection shall consist of a flexible bond wire of adequate strength for the intended service and the electrical resistance shall not exceed 1 megohm. For tank cars, bonding shall be provided where the resistance of a tank car to ground through the rails is 25 ohms or greater.

Such bonding connection shall be fastened to the vehicle, car or tank before dome covers are raised and shall remain in place until filling is complete and all dome covers have been closed and secured.

Exceptions:

1. Where vehicles and cars are loaded exclusively with products not having a static-accumulating tendency, such as asphalt, cutback asphalt, most crude oils, residual oils and water-miscible liquids.
2. Where Class I liquids are not handled at the transfer facility and the tank vehicles are used exclusively for Class II and III liquids.
3. Where vehicles and cars are loaded or unloaded through closed top or bottom connections whether the hose is conductive or nonconductive.

Filling through open domes into the tanks of tank vehicles or tank cars that contain vapor-air mixtures within the flammable range, or where the liquid being filled can form such a mixture, shall be by means of a downspout which extends to near the bottom of the tank.

5706.5.1.8 Stray current protection. Tank car loading facilities where Class I, II or IIIA liquids are transferred through open domes shall be protected against stray currents by permanently bonding the pipe to not less than one rail and to the transfer apparatus. Multiple pipes entering the transfer areas shall be permanently electrically bonded together. In areas where excessive stray currents are known to exist, all pipes entering the transfer area shall be provided with insulating sections to isolate electrically the transfer apparatus from the pipelines.

5706.5.1.9 Top loading. When top loading a tank vehicle with Class I and II liquids without vapor control, valves used for the final control of flow shall be of the self-closing type and shall be manually held open except where automatic means are provided for shutting off the flow when the tank is full. Where used, automatic shutoff systems shall be provided with a manual shutoff valve located at a safe distance from the loading nozzle to stop the flow if the automatic system fails.

When top loading a tank vehicle with vapor control, flow control shall be in accordance with Section 5706.5.1.10. Self-closing valves shall not be tied or locked in the open position.

5706.5.1.10 Bottom loading. When bottom loading a tank vehicle or tank car with or without vapor control, a positive means shall be provided for loading a predetermined quantity of liquid, together with an automatic secondary shutoff control to prevent overfill. The connecting components between the transfer equipment and the tank vehicle or tank car required to operate the secondary control shall be functionally compatible.

5706.5.1.10.1 Dry disconnect coupling. When bottom loading a tank vehicle, the coupling between the liquid loading hose or pipe and the truck piping shall be a dry disconnect coupling.

5706.5.1.10.2 Venting. When bottom loading a tank vehicle or tank car that is equipped for vapor control and vapor control is not used, the tank shall be vented to the atmosphere to prevent pressurization of the tank. Such venting shall be at a height equal to or greater than the top of the cargo tank.

5706.5.1.10.3 Vapor-tight connection. Connections to the plant vapor control system shall be designed to prevent the escape of vapor to the atmosphere when not connected to a tank vehicle or tank car.

5706.5.1.10.4 Vapor-processing equipment. Vapor-processing equipment shall be separated from above-ground tanks, warehouses, other plant buildings, transfer facilities or nearest *lot line* of adjoining property that can be built on by a distance of not less than 25 feet (7620 mm). Vapor-processing equipment shall be protected from physical damage by remote location, guard rails, curbs or fencing.

5706.5.1.11 Switch loading. Tank vehicles or tank cars that have previously contained Class I liquids shall not be loaded with Class II or III liquids until such vehicles and all piping, pumps, hoses and meters connected thereto have been completely drained and flushed.

5706.5.1.12 Loading racks. Where provided, loading racks, *stairways* or platforms shall be constructed of noncombustible materials. Buildings for pumps or for shelter of loading personnel are allowed to be part of the loading rack. Wiring and electrical equipment located within 25 feet (7620 mm) of any portion of the loading rack shall be in accordance with Section 5703.1.1.

5706.5.1.13 Transfer apparatus. Bulk and process transfer apparatus shall be of an *approved* type.

5706.5.1.14 Inside buildings. Tank vehicles and tank cars shall not be located inside a building while transferring Class I, II or IIIA liquids, unless *approved* by the *fire code official*.

Exception: Tank vehicles are allowed under weather protection canopies and canopies of automobile motor vehicle fuel-dispensing stations.

5706.5.1.15 Tank vehicle and tank car certification. Certification shall be maintained for tank vehicles and tank cars in accordance with DOTn 49 CFR Parts 100-185.

5706.5.1.16 Tank vehicle and tank car stability. Tank vehicles and tank cars shall be stabilized against movement during loading and unloading in accordance with Sections 5706.5.1.16.1 through 5706.5.1.16.3.

5706.5.1.16.1 Tank vehicles. When the vehicle is parked for loading or unloading, the cargo trailer portion of the tank vehicle shall be secured in a manner that will prevent unintentional movement.

5706.5.1.16.2 Chock blocks. Not less than two chock blocks not less than 5 inches by 5 inches by 12 inches (127 mm by 127 mm by 305 mm) in size and dished to fit the contour of the tires shall be used during transfer operations of tank vehicles.

5706.5.1.16.3 Tank cars. Brakes shall be set and the wheels shall be blocked to prevent rolling.

5706.5.1.17 Monitoring. Transfer operations shall be monitored by an *approved* monitoring system or by an attendant. Where monitoring is by an attendant, the operator or other competent person shall be present at all times.

5706.5.1.18 Security. Transfer operations shall be surrounded by a noncombustible fence not less than 5 feet (1524 mm) in height. Tank vehicles and tank cars shall not be loaded or unloaded unless such vehicles are entirely within the fenced area.

Exceptions:

1. Motor fuel-dispensing facilities complying with Chapter 23.
2. Installations where adequate public safety exists because of isolation, natural barriers or other factors as determined appropriate by the *fire code official.*
3. Facilities or properties that are entirely enclosed or protected from entry.

5706.5.2 Bulk transfer. Bulk transfer shall be in accordance with Sections 5706.5.1 and 5706.5.2.1.

5706.5.2.1 Vehicle motor. Motors of tank vehicles or tank cars shall be shut off during the making and breaking of hose connections and during the unloading operation.

Exception: Where unloading is performed with a pump deriving its power from the tank vehicle motor.

5706.5.3 Process transfer. Process transfer shall be in accordance with Section 5706.5.1 and Sections 5706.5.3.1 through 5706.5.3.3.

5706.5.3.1 Piping, valves, hoses and fittings. Piping, valves, hoses and fittings that are not a part of the tank vehicle or tank car shall be in accordance with Section 5703.6. Caps or plugs that prevent leakage or spillage shall be provided at all points of connection to transfer piping.

5706.5.3.1.1 Shutoff valves. *Approved* automatically or manually activated shutoff valves shall be provided where the transfer hose connects to the process piping, and on both sides of any exterior fire-resistance-rated wall through which the piping passes. Manual shutoff valves shall be arranged so that they are accessible from grade. Valves shall not be locked in the open position.

5706.5.3.1.2 Hydrostatic relief. Hydrostatic pressure-limiting or relief devices shall be provided where pressure buildup in trapped sections of the system could exceed the design pressure of the components of the system.

Devices shall relieve to other portions of the system or to another *approved* location.

5706.5.3.1.3 Antisiphon valves. Antisiphon valves shall be provided where the system design would allow siphonage.

5706.5.3.2 Vents. Normal and emergency vents shall be maintained operable at all times.

5706.5.3.3 Motive power. Motors of tank vehicles or tank cars shall be shut off during the making and breaking of hose connections and during the unloading operation.

Exception: When unloading is performed with a pump deriving its power from the tank vehicle motor.

5706.5.4 Dispensing from tank vehicles and tank cars. Dispensing from tank vehicles and tank cars into the fuel tanks of motor vehicles shall be prohibited unless allowed by and conducted in accordance with Sections 5706.5.4.1 through 5706.5.4.5.

5706.5.4.1 Marine craft and special equipment. Liquids intended for use as motor fuels are allowed to be transferred from tank vehicles into the fuel tanks of marine craft and special equipment where *approved* by the *fire code official*, and where:

1. The tank vehicle's specific function is that of supplying fuel to fuel tanks.

2. The operation is not performed where the public has access or where there is unusual exposure to life and property.
3. The dispensing line does not exceed 50 feet (15 240 mm) in length.
4. The dispensing nozzle is *approved*.

5706.5.4.2 Emergency refueling. Where *approved* by the *fire code official*, dispensing of motor vehicle fuel from tank vehicles into the fuel tanks of motor vehicles is allowed during emergencies. Dispensing from tank vehicles shall be in accordance with Sections 5706.2.8 and 5706.6.

5706.5.4.3 Aircraft fueling. Transfer of liquids from tank vehicles to the fuel tanks of aircraft shall be in accordance with Chapter 20.

5706.5.4.4 Fueling of vehicles at farms, construction sites and similar areas. Transfer of liquid from tank vehicles to motor vehicles for private use on farms and rural areas and at construction sites, earth-moving projects, gravel pits and borrow pits is allowed in accordance with Section 5706.2.8.

5706.5.4.5 Commercial, industrial, governmental or manufacturing. Dispensing of Class II and III motor vehicle fuel from tank vehicles into the fuel tanks of motor vehicles located at commercial, industrial, governmental or manufacturing establishments is allowed where permitted, provided such dispensing operations are conducted in accordance with the following:

1. Dispensing shall occur only at sites that have been issued a permit to conduct mobile fueling.
2. The *owner* of a mobile fueling operation shall provide to the jurisdiction a written response plan which demonstrates readiness to respond to a fuel spill and carry out appropriate mitigation measures, and describes the process to dispose properly of contaminated materials.
3. A detailed site plan shall be submitted with each application for a permit. The site plan shall indicate: all buildings, structures and appurtenances on site and their use or function; all uses adjacent to the lot lines of the site; the locations of all storm drain openings, adjacent waterways or wetlands; information regarding slope, natural drainage, curbing, impounding and how a spill will be retained upon the site property; and the scale of the site plan.

   Provisions shall be made to prevent liquids spilled during dispensing operations from flowing into buildings or off-site. Acceptable methods include, but shall not be limited to, grading driveways, raising doorsills or other *approved* means.
4. The *fire code official* is allowed to impose limits on the times and days during which mobile fueling operations is allowed to take place, and specific locations on a site where fueling is permitted.
5. Mobile fueling operations shall be conducted in areas not accessible to the public or shall be limited to times when the public is not present.
6. Mobile fueling shall not take place within 15 feet (4572 mm) of buildings, property lines, combustible storage or storm drains.

   Exceptions:
   1. The distance to storm drains shall not apply where an *approved* storm drain cover or an *approved* equivalent that will prevent any fuel from reaching the drain is in place prior to fueling or a fueling hose being placed within 15 feet (4572 mm) of the drain. Where placement of a storm drain cover will cause the accumulation of excessive water or difficulty in conducting the fueling, such cover shall not be used and the fueling shall not take place within 15 feet (4572 mm) of a drain.
   2. The distance to storm drains shall not apply for drains that direct influent to *approved* oil interceptors.
7. The tank vehicle shall comply with the requirements of NFPA 385 and local, state and federal requirements. The tank vehicle's specific functions shall include that of supplying fuel to motor vehicle fuel tanks. The vehicle and all its equipment shall be maintained in good repair.
8. Signs prohibiting smoking or open flames within 25 feet (7620 mm) of the tank vehicle or the point of fueling shall be prominently posted on three sides of the vehicle including the back and both sides.
9. A portable fire extinguisher with a minimum rating of 40:BC shall be provided on the vehicle with signage clearly indicating its location.
10. The dispensing nozzles and hoses shall be of an *approved* and *listed* type.
11. The dispensing hose shall not be extended from the reel more than 100 feet (30 480 mm) in length.
12. Absorbent materials, nonwater-absorbent pads, a 10-foot-long (3048 mm) containment boom, an *approved* container with lid and a nonmetallic shovel shall be provided to mitigate a minimum 5-gallon (19 L) fuel spill.
13. Tank vehicles shall be equipped with a "fuel limit" switch such as a count-back switch, to limit the amount of a single fueling operation to not more than 500 gallons (1893 L) before resetting the limit switch.

    Exception: Tank vehicles where the operator carries and can utilize a remote emergency shutoff device which, when activated, immediately causes flow of fuel from the tank vehicle to cease.

14. Persons responsible for dispensing operations shall be trained in the appropriate mitigating actions in the event of a fire, leak or spill. Training records shall be maintained by the dispensing company.
15. Operators of tank vehicles used for mobile fueling operations shall have in their possession at all times an emergency communications device to notify the proper authorities in the event of an emergency.
16. The tank vehicle dispensing equipment shall be constantly attended and operated only by designated personnel who are trained to handle and dispense motor fuels.
17. Fuel dispensing shall be prohibited within 25 feet (7620 mm) of any source of ignition.
18. The engines of vehicles being fueled shall be shut off during dispensing operations.
19. Nighttime fueling operations shall only take place in adequately lighted areas.
20. The tank vehicle shall be positioned with respect to vehicles being fueled to prevent traffic from driving over the delivery hose.
21. During fueling operations, tank vehicle brakes shall be set, chock blocks shall be in place and warning lights shall be in operation.
22. Motor vehicle fuel tanks shall not be topped off.
23. The dispensing hose shall be properly placed on an *approved* reel or in an *approved* compartment prior to moving the tank vehicle.
24. The *fire code official* and other appropriate authorities shall be notified when a reportable spill or unauthorized discharge occurs.
25. Operators shall place a drip pan or an absorbent pillow under each fuel fill opening prior to and during dispensing operations. Drip pans shall be liquid-tight. The pan or absorbent pillow shall have a capacity of not less than 3 gallons (11.36 L). Spills retained in the drip pan or absorbent pillow need not be reported. Operators, when fueling, shall have on their person an absorbent pad capable of capturing diesel fuel overfills. Except during fueling, the nozzle shall face upward and an absorbent pad shall be kept under the nozzle to catch drips. Contaminated absorbent pads or pillows shall be disposed of regularly in accordance with local, state and federal requirements.

5706.6 Tank vehicles and vehicle operation. Tank vehicles shall be designed, constructed, equipped and maintained in accordance with NFPA 385 and Sections 5706.6.1 through 5706.6.4.

5706.6.1 Operation of tank vehicles. Tank vehicles shall be utilized and operated in accordance with NFPA 385 and Sections 5706.6.1.1 through 5706.6.1.11.

5706.6.1.1 Vehicle maintenance. Tank vehicles shall not be operated unless they are in proper state of repair and free from accumulation of grease, oil or other flammable substance, and leaks.

5706.6.1.2 Leaving vehicle unattended. The driver, operator or attendant of a tank vehicle shall not remain in the vehicle cab and shall not leave the vehicle while it is being filled or discharged. The delivery hose, when attached to a tank vehicle, shall be considered to be a part of the tank vehicle.

5706.6.1.3 Vehicle motor shutdown. Motors of tank vehicles or tractors shall be shut down during the making or breaking of hose connections. If loading or unloading is performed without the use of a power pump, the tank vehicle or tractor motor shall be shut down throughout such operations.

5706.6.1.4 Outage. A cargo tank or compartment thereof used for the transportation of flammable or *combustible liquids* shall not be loaded to absolute capacity. The vacant space in a cargo tank or compartment thereof used in the transportation of flammable or *combustible liquids* shall be not less than 1 percent. Sufficient space shall be left vacant to prevent leakage from or distortion of such tank or compartment by expansion of the contents caused by rise in temperature in transit.

5706.6.1.5 Overfill protection. The driver, operator or attendant of a tank vehicle shall, before making delivery to a tank, determine the unfilled capacity of such tank by a suitable gauging device. To prevent overfilling, the driver, operator or attendant shall not deliver in excess of that amount.

5706.6.1.6 Securing hatches. During loading, hatch covers shall be secured on all but the receiving compartment.

5706.6.1.7 Liquid temperature. Materials shall not be loaded into or transported in a tank vehicle at a temperature above the material's ignition temperature unless safeguarded in an *approved* manner.

5706.6.1.8 Bonding to underground tanks. An external bond-wire connection or bond-wire integral with a hose shall be provided for the transferring of flammable liquids through open connections into underground tanks.

5706.6.1.9 Smoking. Smoking by tank vehicle drivers, helpers or other personnel is prohibited while they are driving, making deliveries, filling or making repairs to tank vehicles.

5706.6.1.10 Hose connections. Delivery of flammable liquids to underground tanks with a capacity of more than 1,000 gallons (3785 L) shall be made by means of *approved* liquid and vapor-tight connections between the delivery hose and tank fill pipe. Where underground tanks are equipped with any type of vapor recovery system, all connections required to be made for the safe and proper functioning of the particular vapor recovery process shall be made. Such connections shall be made liquid and vapor tight and remain connected throughout the unloading process. Vapors shall not be discharged at grade level during delivery.

5706.6.1.10.1 Simultaneous delivery. Simultaneous delivery to underground tanks of any capacity from two or more discharge hoses shall be made by means of mechanically tight connections between the hose and fill pipe.

5706.6.1.11 Hose protection. Upon arrival at a point of delivery and prior to discharging any flammable or *combustible liquids* into underground tanks, the driver, operator or attendant of the tank vehicle shall ensure that all hoses utilized for liquid delivery and vapor recovery, where required, will be protected from physical damage by motor vehicles. Such protection shall be provided by positioning the tank vehicle to prevent motor vehicles from passing through the area or areas occupied by hoses, or by other *approved* equivalent means.

5706.6.2 Parking. Parking of tank vehicles shall be in accordance with Sections 5706.6.2.1 through 5706.6.2.3.

Exception: In cases of accident, breakdown or other emergencies, tank vehicles are allowed to be parked and left unattended at any location while the operator is obtaining assistance.

5706.6.2.1 Parking near residential, educational and institutional occupancies and other high-risk areas. Tank vehicles shall not be left unattended at any time on residential streets, or within 500 feet (152 m) of a residential area, apartment or hotel complex, educational facility, hospital or care facility. Tank vehicles shall not be left unattended at any other place that would, in the opinion of the fire chief, pose an extreme life hazard.

5706.6.2.2 Parking on thoroughfares. Tank vehicles shall not be left unattended on a public street, highway, public avenue or public alley.

Exceptions:

1. The necessary absence in connection with loading or unloading the vehicle. During actual fuel transfer, Section 5706.6.1.2 shall apply. The vehicle location shall be in accordance with Section 5706.6.2.1.
2. Stops for meals during the day or night, where the street is well lighted at the point of parking. The vehicle location shall be in accordance with Section 5706.6.2.1.

5706.6.2.3 Duration exceeding 1 hour. Tank vehicles parked at one point for longer than 1 hour shall be located off of public streets, highways, public avenues or alleys, and in accordance with either of the following:

1. Inside of a bulk plant and either 25 feet (7620 mm) or more from the nearest *lot line* or within a building *approved* for such use.
2. At other *approved* locations not less than 50 feet (15 240 mm) from the buildings other than those *approved* for the storage or servicing of such vehicles.

5706.6.3 Garaging. Tank vehicles shall not be parked or garaged in buildings other than those specifically *approved* for such use by the *fire code official*.

5706.6.4 Portable fire extinguisher. Tank vehicles shall be equipped with a portable fire extinguisher complying with Section 906 and having a minimum rating of 2-A:20-B:C.

During unloading of the tank vehicle, the portable fire extinguisher shall be out of the carrying device on the vehicle and shall be 15 feet (4572 mm) or more from the unloading valves.

5706.7 Refineries. Plants and portions of plants in which flammable liquids are produced on a scale from crude petroleum, natural gasoline or other hydrocarbon sources shall be in accordance with Sections 5706.7.1 through 5706.7.3. Petroleum-processing plants and facilities or portions of plants or facilities in which flammable or *combustible liquids* are handled, treated or produced on a commercial scale from crude petroleum, natural gasoline, or other hydrocarbon sources shall also be in accordance with API 651, API 653, API 752, API 1615, API 2001, API 2003, API 2009, API 2015, API 2023, API 2201 and API 2350.

5706.7.1 Corrosion protection. Above-ground tanks and piping systems shall be protected against corrosion in accordance with API 651.

5706.7.2 Cleaning of tanks. The safe entry and cleaning of petroleum storage tanks shall be conducted in accordance with API 2015.

5706.7.3 Storage of heated petroleum products. Where petroleum-derived asphalts and residues are stored in heated tanks at refineries and bulk storage facilities or in tank vehicles, such products shall be in accordance with API 2023.

5706.8 Vapor recovery and vapor-processing systems. Vapor-processing systems in which the vapor source operates at pressures from vacuum, up to and including 1 psig (6.9 kPa) or in which a potential exists for vapor mixtures in the flammable range, shall comply with Sections 5706.8.1 through 5706.8.5.

Exceptions:

1. Marine systems complying with federal transportation waterway regulations such as DOTn 33 CFR Parts 154 through 156, and CGR 46 CFR Parts 30, 32, 35 and 39.

2. Motor fuel-dispensing facility systems complying with Chapter 23.

5706.8.1 Over-pressure/vacuum protection. Tanks and equipment shall have independent venting for over-pressure or vacuum conditions that might occur from malfunction of the vapor recovery or processing system.

Exception: For tanks, venting shall comply with Section 5704.2.7.3.

5706.8.2 Vent location. Vents on vapor-processing equipment shall be not less than 12 feet (3658 mm) from adjacent ground level, with outlets located and directed so that flammable vapors will disperse to below the lower flammable limit (LFL) before reaching locations containing potential ignition sources.

5706.8.3 Vapor collection systems and overfill protection. The design and operation of the vapor collection system and overfill protection shall be in accordance with this section and Section 19.5 of NFPA 30.

5706.8.4 Liquid-level monitoring. A liquid knock-out vessel used in the vapor collection system shall have means to verify the liquid level and a high-liquid-level sensor that activates an alarm. For unpopulated facilities, the high-liquid-level sensor shall initiate the shutdown of liquid transfer into the vessel and shutdown of vapor recovery or vapor-processing systems.

5706.8.5 Overfill protection. Storage tanks served by vapor recovery or processing systems shall be equipped with overfill protection in accordance with Section 5704.2.7.5.8.

CHAPTER 58

FLAMMABLE GASES AND FLAMMABLE CRYOGENIC FLUIDS

SECTION 5801 GENERAL

5801.1 Scope. The storage and use of flammable gases and flammable cryogenic fluids shall be in accordance with this chapter and NFPA 55. Compressed gases shall also comply with Chapter 53 and cryogenic fluids shall also comply with Chapter 55. Flammable cryogenic fluids shall comply with Section 5806. Hydrogen motor fuel-dispensing stations and repair garages and their associated above-ground hydrogen storage systems shall also be designed, constructed and maintained in accordance with Chapter 23 and NFPA 2.

Exceptions:

1. Gases used as refrigerants in refrigeration systems (see Section 606).
2. Liquefied petroleum gases and natural gases regulated by Chapter 61.
3. Fuel-gas systems and appliances regulated under the *International Fuel Gas Code* other than gaseous hydrogen systems and appliances.
4. Pyrophoric gases in accordance with Chapter 64.

5801.2 Permits. Permits shall be required as set forth in Section 105.6.

SECTION 5802 DEFINITIONS

5802.1 Definitions. The following terms are defined in Chapter 2:

FLAMMABLE GAS.

FLAMMABLE LIQUEFIED GAS.

GASEOUS HYDROGEN SYSTEM.

HYDROGEN FUEL GAS ROOM.

METAL HYDRIDE.

METAL HYDRIDE STORAGE SYSTEM.

SECTION 5803 GENERAL REQUIREMENTS

5803.1 Quantities not exceeding the maximum allowable quantity per control area. The storage and use of flammable gases in amounts not exceeding the *maximum allowable quantity per control area* indicated in Section 5003.1 shall be in accordance with Sections 5001, 5003, 5801 and 5803.

5803.1.1 Special limitations for indoor storage and use. Flammable gases shall not be stored or used in Group A, E, I or R occupancies or in offices in Group B occupancies.

Exceptions:

1. Cylinders of nonliquefied *compressed gases* not exceeding a capacity of 250 cubic feet (7.08 m^3) or liquefied gases not exceeding a capacity of 40 pounds (18 kg) each at *normal temperature and pressure (NTP)* used for maintenance purposes, patient care or operation of equipment.
2. Food service operations in accordance with Section 6103.2.1.7.
3. Hydrogen gas systems located in a hydrogen fuel gas room constructed in accordance with Section 421 of the *International Building Code.*

5803.1.1.1 Medical gases. Medical gas system supply cylinders shall be located in medical gas storage rooms or gas cabinets as set forth in Section 5306.

5803.1.1.2 Aggregate quantity. The aggregate quantities of flammable gases used for maintenance purposes and operation of equipment shall not exceed the *maximum allowable quantity per control area* indicated in Table 5003.1.1(1).

5803.1.2 Storage containers. Cylinders and pressure vessels for flammable gases shall be designed, constructed, installed, tested and maintained in accordance with Chapter 53.

5803.1.3 Emergency shutoff. *Compressed gas* systems conveying flammable gases shall be provided with *approved* manual or automatic emergency shutoff valves that can be activated at each point of use and at each source.

5803.1.3.1 Shutoff at source. A manual or automatic fail-safe emergency shutoff valve shall be installed on supply piping at the cylinder or bulk source. Manual or automatic cylinder valves are allowed to be used as the required emergency shutoff valve where the source of supply is limited to unmanifolded cylinder sources.

5803.1.3.2 Shutoff at point of use. A manual or automatic emergency shutoff valve shall be installed on the supply piping at the point of use or at a point where the equipment using the gas is connected to the supply system.

5803.1.4 Ignition source control. Ignition sources in areas containing flammable gases in storage or in use shall be controlled in accordance with Section 5003.7.

Exception: Fuel gas systems connected to building service utilities in accordance with the *International Fuel Gas Code.*

5803.1.4.1 Static-producing equipment. Static-producing equipment located in flammable gas storage areas shall be grounded.

5803.1.4.2 Signs. "No Smoking" signs shall be posted at entrances to rooms and in areas containing flammable gases in accordance with Section 5003.7.1.

5803.1.5 Electrical. Electrical wiring and equipment shall be installed and maintained in accordance with Section 605 and NFPA 70.

5803.1.5.1 Bonding of electrically conductive materials and equipment. Exposed noncurrent-carrying metal parts, including metal gas piping systems, that are part of flammable gas supply systems located in a hazardous (electrically classified) location shall be bonded to a grounded conductor in accordance with the provisions of NFPA 70.

5803.1.5.2 Static-producing equipment. Static-producing equipment located in flammable gas storage or use areas shall be grounded.

5803.1.6 Liquefied flammable gases and flammable gases in solution. Containers of liquefied flammable gases and flammable gases in solution shall be positioned in the upright position or positioned so that the pressure relief valve is in direct contact with the vapor space of the container.

Exceptions:

1. Containers of flammable gases in solution with a capacity of 1.3 gallons (5 L) or less.
2. Containers of flammable liquefied gases, with a capacity not exceeding 1.3 gallons (5 L), designed to preclude the discharge of liquid from safety relief devices.

5803.2 Quantities exceeding the maximum allowable quantity per control area. The storage and use of flammable gases in amounts exceeding the *maximum allowable quantity per control area* indicated in Section 5003.1 shall be in accordance with Chapter 50 and this chapter.

SECTION 5804 STORAGE

5804.1 Indoor storage. Indoor storage of flammable gases in amounts exceeding the *maximum allowable quantity per control area* indicated in Table 5003.1.1(1), shall be in accordance with Sections 5001, 5003 and 5004, and this chapter.

5804.1.1 Explosion control. Buildings or portions thereof containing flammable gases shall be provided with explosion control in accordance with Section 911.

5804.2 Outdoor storage. Outdoor storage of flammable gases in amounts exceeding the *maximum allowable quantity per control area* indicated in Table 5003.1.1(3) shall be in accordance with Sections 5001, 5003 and 5004, and this chapter.

SECTION 5805 USE

5805.1 General. The use of flammable gases in amounts exceeding the *maximum allowable quantity per control area* indicated in Table 5003.1.1(1) or 5003.1.1(3) shall be in accordance with Sections 5001, 5003 and 5005, and this chapter.

SECTION 5806 FLAMMABLE CRYOGENIC FLUIDS

5806.1 General. The storage and use of flammable *cryogenic fluids* shall be in accordance with Sections 5806.2 through 5806.4.8.3 and Chapter 55.

5806.2 Limitations. Storage of flammable *cryogenic fluids* in stationary containers outside of buildings is prohibited within the limits established by law as the limits of districts in which such storage is prohibited (see Section 3 of the Sample Legislation for Adoption of the *International Fire Code* on page xxi).

5806.3 Above-ground tanks for liquid hydrogen. Above-ground tanks for the storage of liquid hydrogen shall be in accordance with Sections 5806.3 through 5806.3.2.1.

5806.3.1 Construction of the inner vessel. The inner vessel of storage tanks in liquid hydrogen service shall be designed and constructed in accordance with Section VIII, Division 1, of the ASME *Boiler and Pressure Vessel Code* and shall be vacuum jacketed in accordance with Section 5806.3.2.

5806.3.2 Construction of the vacuum jacket (outer vessel). The vacuum jacket used as an outer vessel for storage tanks in liquid hydrogen service shall be of welded steel construction designed to withstand the maximum internal and external pressure to which it will be subjected under operating conditions to include conditions of emergency pressure relief of the annular space between the inner and outer vessel. The jacket shall be designed to withstand a minimum collapsing pressure differential of 30 psi (207 kPa).

5806.3.2.1 Vacuum-level monitoring. A connection shall be provided on the exterior of the vacuum jacket to allow measurement of the pressure within the annular space between the inner and outer vessel. The connection shall be fitted with a bellows-sealed or diaphragm-type valve equipped with a vacuum gauge tube that is shielded to protect against damage from impact.

5806.4 Underground tanks for liquid hydrogen. Underground tanks for the storage of liquid hydrogen shall be in accordance with Sections 5806.4.1 through 5806.4.8.3.

5806.4.1 Construction. Storage tanks for liquid hydrogen shall be designed and constructed in accordance with ASME *Boiler and Pressure Vessel Code* (Section VIII, Division 1) and shall be vacuum jacketed in accordance with Section 5806.4.8.

5806.4.2 Location. Storage tanks shall be located outside in accordance with the following:

1. Tanks and associated equipment shall be located with respect to foundations and supports of other structures such that the loads carried by the latter cannot be transmitted to the tank.
2. The distance from any part of the tank to the nearest wall of a *basement*, pit, cellar or *lot line* shall be not less than 3 feet (914 mm).
3. A minimum distance of 1 foot (305 mm), shell to shell, shall be maintained between underground tanks.

5806.4.3 Depth, cover and fill. The tank shall be buried such that the top of the vacuum jacket is covered with not less than 1 foot (305 mm) of earth and with concrete not less than 4 inches (102 mm) thick placed over the earthen cover. The concrete shall extend not less than 1 foot (305 mm) horizontally beyond the footprint of the tank in all directions. Underground tanks shall be set on firm foundations constructed in accordance with the *International Building Code* and surrounded with not less than 6 inches (152 mm) of noncorrosive inert material, such as sand.

Exception: The vertical extension of the vacuum jacket as required for service connections.

5806.4.4 Anchorage and security. Tanks and systems shall be secured against accidental dislodgement in accordance with this chapter.

5806.4.5 Venting of underground tanks. Vent pipes for underground storage tanks shall be in accordance with Section 5503.3.

5806.4.6 Underground liquid hydrogen piping. Underground liquid hydrogen piping shall be vacuum jacketed or protected by *approved* means and designed in accordance with Chapter 55.

5806.4.7 Overfill protection and prevention systems. An *approved* means or method shall be provided to prevent the overfill of all storage tanks.

5806.4.8 Vacuum jacket construction. The vacuum jacket shall be designed and constructed in accordance with Section VIII of ASME *Boiler and Pressure Vessel Code* and shall be designed to withstand the anticipated loading, including loading from vehicular traffic, where applicable. Portions of the vacuum jacket installed below grade shall be designed to withstand anticipated soil, seismic and hydrostatic loading.

5806.4.8.1 Material. The vacuum jacket shall be constructed of stainless steel or other *approved* corrosion-resistant material.

5806.4.8.2 Corrosion protection. The vacuum jacket shall be protected by *approved* or *listed* corrosion-resistant materials or an engineered cathodic protection system. Where cathodic protection is utilized, an *approved* maintenance schedule shall be established. Exposed components shall be inspected not less than twice a year. Records of maintenance and inspection events shall be maintained.

5806.4.8.3 Vacuum-level monitoring. An *approved* method shall be provided to indicate loss of vacuum within the vacuum jacket(s).

SECTION 5807
METAL HYDRIDE STORAGE SYSTEMS

5807.1 General requirements. The storage and use of metal hydride storage systems shall be in accordance with Sections 5801, 5803, 5804, 5805 and 5807. Those portions of the system that are used as a means to store or supply hydrogen shall also comply with Chapters 50 and 53 as applicable.

5807.1.1 Classification. The hazard classification of the metal hydride storage system, as required by Section 5001.2.2, shall be based on the hydrogen stored without regard to the metal hydride content.

5807.1.2 Listed or approved systems. Metal hydride storage systems shall be *listed* or *approved* for the application and designed in a manner that prevents the addition or removal of the metal hydride by other than the original equipment manufacturer.

5807.1.3 Containers, design and construction. *Compressed gas* containers, cylinders and tanks shall be designed and constructed in accordance with Section 5303.2.

5807.1.4 Service life and inspection of containers. Metal hydride storage system cylinders, containers or tanks shall be inspected, tested and requalified for service at not less than 5-year intervals.

5807.1.5 Marking and labeling. Marking and labeling of cylinders, containers, tanks and systems shall be in accordance with Section 5303.4 and Sections 5807.1.5.1 through 5807.1.5.4.

5807.1.5.1 System marking. Metal hydride storage systems shall be marked with all of the following:

1. Manufacturer's name.
2. Service life indicating the last date the system can be used.
3. A unique code or serial number specific to the unit.
4. System name or product code that identifies the system by the type of chemistry used in the system.
5. Emergency contact name, telephone number or other contact information.
6. Limitations on refilling of containers to include rated charging pressure and capacity.

5807.1.5.2 Valve marking. Metal hydride storage system valves shall be marked with all of the following:

1. Manufacturer's name.

2. Service life indicating the last date the valve can be used.
3. Metal hydride service in which the valve can be used, or a product code that is traceable to this information.

5807.1.5.3 Pressure relief device marking. Metal hydride storage system pressure relief devices shall be marked with all of the following:

1. Manufacturer's name.
2. Metal hydride service in which the device can be used, or a product code that is traceable to this information.
3. Activation parameters to include temperature, pressure or both.

5807.1.5.3.1 Pressure relief devices integral to container valves. The required markings for pressure relief devices that are integral components of valves used on cylinders, containers and tanks shall be allowed to be placed on the valve.

5807.1.5.4 Pressure vessel markings. Cylinders, containers and tanks used in metal hydride storage systems shall be marked with all of the following:

1. Manufacturer's name.
2. Design specification to which the vessel was manufactured.
3. Authorized body approving the design and initial inspection and test of the vessel.
4. Manufacturer's original test date.
5. Unique serial number for the vessel.
6. Service life identifying the last date the vessel can be used.
7. System name or product code that identifies the system by the type of chemistry used in the system.

5807.1.6 Temperature extremes. Metal hydride storage systems, whether full or partially full, shall not be exposed to artificially created high temperatures exceeding 125°F (52°C) or subambient (low) temperatures unless designed for use under the exposed conditions.

5807.1.7 Falling objects. Metal hydride storage systems shall not be placed in areas where they are capable of being damaged by falling objects.

5807.1.8 Piping systems. Piping, including tubing, valves, fittings and pressure regulators, serving metal hydride storage systems, shall be maintained gas tight to prevent leakage.

5807.1.8.1 Leaking systems. Leaking systems shall be removed from service.

5807.1.9 Refilling of containers. The refilling of *listed* or *approved* metal hydride storage systems shall be in accordance with the listing requirements and manufacturer's instructions.

5807.1.9.1 Industrial trucks. The refilling of metal hydride storage systems serving powered industrial trucks shall be in accordance with Section 309.

5807.1.9.2 Hydrogen purity. The purity of hydrogen used for the purpose of refilling containers shall be in accordance with the listing and the manufacturer's instructions.

5807.1.10 Electrical. Electrical components for metal hydride storage systems shall be designed, constructed and installed in accordance with NFPA 70.

5807.2 Portable containers or systems. Portable containers or systems shall comply with Sections 5807.2.1 through 5807.2.2.

5807.2.1 Securing containers. Containers, cylinders and tanks shall be secured in accordance with Section 5303.5.3.

5807.2.1.1 Use on mobile equipment. Where a metal hydride storage system is used on mobile equipment, the equipment shall be designed to restrain containers, cylinders or tanks from dislodgement, slipping or rotating when the equipment is in motion.

5807.2.1.2 Motorized equipment. Metal hydride storage systems used on motorized equipment, shall be installed in a manner that protects valves, pressure regulators, fittings and controls against accidental impact.

5807.2.1.2.1 Protection from damage. Metal hydride storage systems, including cylinders, containers, tanks and fittings, shall not extend beyond the platform of the mobile equipment.

5807.2.2 Valves. Valves on containers, cylinders and tanks shall remain closed except when containers are connected to *closed systems* and ready for use.

SECTION 5808
HYDROGEN FUEL GAS ROOMS

5808.1 General. Where required by this code, hydrogen fuel gas rooms shall be designed and constructed in accordance with Sections 5808.1 through 5808.7 and the *International Building Code*.

5808.2 Location. Hydrogen fuel gas rooms shall not be located below grade.

5808.3 Design and construction. Hydrogen fuel gas rooms not exceeding the *maximum allowable quantity per control area* in Table 5003.1.1(1) shall be separated from other areas of the building in accordance with Section 509.1 of the *International Building Code*.

5808.3.1 Pressure control. Hydrogen fuel gas rooms shall be provided with a ventilation system designed to maintain the room at a negative pressure in relation to surrounding rooms and spaces.

5808.3.2 Windows. Operable windows in interior walls shall not be permitted. Fixed windows shall be permitted where in accordance with Section 716 of the *International Building Code*.

5808.4 Exhaust ventilation. Hydrogen fuel gas rooms shall be provided with mechanical exhaust ventilation in accordance with the applicable provisions of Section 2311.7.1.1.

5808.5 Gas detection system. Hydrogen fuel gas rooms shall be provided with an approved flammable gas detection system in accordance with Sections 5808.5.1 through 5808.5.4.

5808.5.1 System design. The flammable gas detection system shall be *listed* for use with hydrogen and any other flammable gases used in the hydrogen fuel gas room. The gas detection system shall be designed to activate when the level of flammable gas exceeds 25 percent of the lower flammable limit (LFL) for the gas or mixtures present at their anticipated temperature and pressure.

5808.5.2 Gas detection system components. Gas detection system control units shall be *listed* and *labeled* in accordance with UL 864 or UL 2017. Gas detectors shall be *listed* and *labeled* in accordance with UL 2075 for use with the gases and vapors being detected.

5808.5.3 Operation. Activation of the gas detection system shall result in both of the following:

1. Initiation of distinct audible and visual alarm signals both inside and outside of the hydrogen fuel gas room.
2. Activation of the mechanical exhaust ventilation system.

5808.5.4 Failure of the gas detection system. Failure of the gas detection system shall result in activation of the mechanical exhaust ventilation system, cessation of hydrogen generation and the sounding of a trouble signal in an approved location.

5808.6 Explosion control. Explosion control shall be provided where required by Section 911.

5808.7 Standby power. Mechanical ventilation and gas detection systems shall be connected to a standby power system in accordance with Section 604.

CHAPTER 59

FLAMMABLE SOLIDS

SECTION 5901 GENERAL

5901.1 Scope. The storage and use of flammable solids shall be in accordance with this chapter.

5901.2 Permits. Permits shall be required as set forth in Section 105.6.

SECTION 5902 DEFINITIONS

5902.1 Definitions. The following terms are defined in Chapter 2:

FLAMMABLE SOLID.

MAGNESIUM.

SECTION 5903 GENERAL REQUIREMENTS

5903.1 Quantities not exceeding the maximum allowable quantity per control area. The storage and use of flammable solids in amounts not exceeding the *maximum allowable quantity per control area* as indicated in Section 5003.1 shall be in accordance with Sections 5001, 5003 and 5901.

5903.2 Quantities exceeding the maximum allowable quantity per control area. The storage and use of flammable solids exceeding the *maximum allowable quantity per control area* as indicated in Section 5003.1 shall be in accordance with Chapter 50 and this chapter.

SECTION 5904 STORAGE

5904.1 Indoor storage. Indoor storage of flammable solids in amounts exceeding the *maximum allowable quantity per control area* indicated in Table 5003.1.1(1) shall be in accordance with Sections 5001, 5003, 5004 and this chapter.

5904.1.1 Pile size limits and location. Flammable solids stored in quantities greater than 1,000 cubic feet (28 m^3) shall be separated into piles each not larger than 1,000 cubic feet (28 m^3).

5904.1.2 Aisles. Aisle widths between piles shall not be less than the height of the piles or 4 feet (1219 mm), whichever is greater.

5904.1.3 Basement storage. Flammable solids shall not be stored in *basements*.

5904.2 Outdoor storage. Outdoor storage of flammable solids in amounts exceeding the *maximum allowable quantities per control area* indicated in Table 5003.1.1(1) shall be in accordance with Sections 5001, 5003, 5004 and this chapter. Outdoor storage of magnesium shall be in accordance with Section 5906.

5904.2.1 Distance from storage to exposures. Outdoor storage of flammable solids shall not be located within 20 feet (6096 mm) of a building, *lot line*, public street, public alley, *public way* or *means of egress*. A 2-hour *fire barrier* without openings or penetrations and extending 30 inches (762 mm) above and to the sides of the storage area is allowed in lieu of such distance. The wall shall either be an independent structure, or the *exterior wall* of the building adjacent to the storage area.

5904.2.2 Pile size limits. Outdoor storage of flammable solids shall be separated into piles not larger than 5,000 cubic feet (141 m^3) each. Piles shall be separated by aisles with a minimum width of not less than one-half the pile height or 10 feet (3048 mm), whichever is greater.

SECTION 5905 USE

5905.1 General. The use of flammable solids in amounts exceeding the *maximum allowable quantity per control area* indicated in Table 5003.1.1(1) or 5003.1.1(3) shall be in accordance with Sections 5001, 5003, 5005 and this chapter. The use of magnesium shall be in accordance with Section 5906.

SECTION 5906 MAGNESIUM

5906.1 General. Storage, use, handling and processing of magnesium, including the pure metal and alloys of which the major part is magnesium, shall be in accordance with Chapter 50 and Sections 5906.2 through 5906.5.8.

5906.2 Storage of magnesium articles. The storage of magnesium shall comply with Sections 5906.2.1 through 5906.4.3.

5906.2.1 Storage of greater than 50 cubic feet. Magnesium storage in quantities greater than 50 cubic feet (1.4 m^3) shall be separated from storage of other materials that are either combustible or in combustible containers by aisles. Piles shall be separated by aisles with a minimum width of not less than the pile height.

5906.2.2 Storage of greater than 1,000 cubic feet. Magnesium storage in quantities greater than 1,000 cubic feet (28 m^3) shall be separated into piles not larger than 1,000 cubic feet (28 m^3) each. Piles shall be separated by aisles with a minimum width of not less than the pile height. Such storage shall not be located in nonsprinklered buildings of Type III, IV or V construction, as defined in the *International Building Code*.

5906.2.3 Storage in combustible containers or within 30 feet of other combustibles. Where in nonsprinklered buildings of Type III, IV or V construction, as defined in the *International Building Code*, magnesium shall not be stored in combustible containers or within 30 feet (9144 mm) of other combustibles.

5906.2.4 Storage in foundries and processing plants. The size of storage piles of magnesium articles in foundries and processing plants shall not exceed 1,250 cubic feet (25 m^3). Piles shall be separated by aisles with a minimum width of not less than one-half the pile height.

5906.3 Storage of pigs, ingots and billets. The storage of magnesium pigs, ingots and billets shall comply with Sections 5906.3.1 and 5906.3.2.

5906.3.1 Indoor storage. Indoor storage of pigs, ingots and billets shall only be on floors of noncombustible construction. Piles shall not be larger than 500,000 pounds (226.8 metric tons) each. Piles shall be separated by aisles with a minimum width of not less than one-half the pile height.

5906.3.2 Outdoor storage. Outdoor storage of magnesium pigs, ingots and billets shall be in piles not exceeding 1,000,000 pounds (453.6 metric tons) each. Piles shall be separated by aisles with a minimum width of not less than one-half the pile height. Piles shall be separated from combustible materials or buildings on the same or adjoining property by a distance of not less than the height of the nearest pile.

5906.4 Storage of fine magnesium scrap. The storage of scrap magnesium shall comply with Sections 5906.4.1 through 5906.4.3.

5906.4.1 Separation. Magnesium fines shall be kept separate from other combustible materials.

5906.4.2 Storage of 50 to 1,000 cubic feet. Storage of fine magnesium scrap in quantities greater than 50 cubic feet (1.4 m^3) [six 55-gallon (208 L) steel drums] shall be separated from other occupancies by an open space of not less than 50 feet (15 240 mm) or by a *fire barrier* constructed in accordance with Section 707 of the *International Building Code*.

5906.4.3 Storage of greater than 1,000 cubic feet. Storage of fine magnesium scrap in quantities greater than 1,000 cubic feet (28 m^3) shall be separated from all buildings other than those used for magnesium scrap recovery operations by a distance of not less than 100 feet (30 480 mm).

5906.5 Use of magnesium. The use of magnesium shall comply with Sections 5906.5.1 through 5906.5.8.

5906.5.1 Melting pots. Floors under and around melting pots shall be of noncombustible construction.

5906.5.2 Heat-treating ovens. *Approved* means shall be provided for control of magnesium fires in heat-treating ovens.

5906.5.3 Dust collection. Magnesium grinding, buffing and wire-brushing operations, other than rough finishing of castings, shall be provided with *approved* hoods or enclosures for dust collection that are connected to a liquid-precipitation type of separator that converts dust to sludge without contact (in a dry state) with any high-speed moving parts.

5906.5.3.1 Duct construction. Connecting ducts or suction tubes shall be completely grounded, as short as possible, and without bends. Ducts shall be fabricated and assembled with a smooth interior, with internal lap joints pointing in the direction of airflow and without unused capped side outlets, pockets or other dead-end spaces that allow an accumulation of dust.

5906.5.3.2 Independent dust separators. Each machine shall be equipped with an individual dust-separating unit.

Exceptions:

1. One separator is allowed to serve two dust-producing units on multiunit machines.
2. One separator is allowed to serve not more than four portable dust-producing units in a single enclosure or stand.

5906.5.4 Power supply interlock. Power supply to machines shall be interlocked with exhaust airflow, and liquid pressure level or flow. The interlock shall be designed to shut down the machine it serves when the dust removal or separator system is not operating properly.

5906.5.5 Electrical equipment. Electric wiring, fixtures and equipment in the immediate vicinity of and attached to dust-producing machines, including those used in connection with separator equipment, shall be of *approved* types and shall be *approved* for use in Class II, Division 1 hazardous locations in accordance with NFPA 70.

5906.5.6 Grounding. Equipment shall be securely grounded by permanent ground wires in accordance with NFPA 70.

5906.5.7 Fire-extinguishing materials. Fire-extinguishing materials shall be provided for every operator performing machining, grinding or other processing operation on magnesium in accordance with either of the folowing:

1. Within 30 feet (9144 mm), a supply of extinguishing materials in an *approved* container with a hand scoop or shovel for applying the material.
2. Within 75 feet (22 860 mm), a portable fire extinguisher complying with Section 906.

All extinguishing materials shall be *approved* for use on magnesium fires. Where extinguishing materials are stored in cabinets or other enclosed areas, the enclosures shall be openable without the use of a key or special knowledge.

5906.5.8 Collection of chips, turnings and fines. Chips, turnings and other fine magnesium scrap shall be collected from the pans or spaces under machines and from other places where they collect not less than once each working day. Such material shall be placed in a covered, vented steel container and removed to an *approved* location.

CHAPTER 60

HIGHLY TOXIC AND TOXIC MATERIALS

SECTION 6001
GENERAL

6001.1 Scope. The storage and use of highly toxic and toxic materials shall comply with this chapter. *Compressed gases* shall also comply with Chapter 53.

Exceptions:

1. Display and storage in Group M and storage in Group S occupancies complying with Section 5003.11.
2. Conditions involving pesticides or agricultural products as follows:
   - 2.1. Application and release of pesticide, agricultural products and materials intended for use in weed abatement, erosion control, soil amendment or similar applications when applied in accordance with the manufacturer's instruction and label directions.
   - 2.2. Transportation of pesticides in compliance with the Federal Hazardous Materials Transportation Act and regulations thereunder.
   - 2.3. Storage in *dwellings* or private garages of pesticides registered by the U.S. Environmental Protection Agency to be utilized in and around the home, garden, pool, spa and patio.

6001.2 Permits. Permits shall be required as set forth in Section 105.6.

SECTION 6002
DEFINITIONS

6002.1 Definitions. The following terms are defined in Chapter 2:

CONTAINMENT SYSTEM.

CONTAINMENT VESSEL.

EXCESS FLOW VALVE.

HIGHLY TOXIC.

OZONE-GAS GENERATOR.

PHYSIOLOGICAL WARNING THRESHOLD LEVEL.

REDUCED FLOW VALVE.

TOXIC.

SECTION 6003
HIGHLY TOXIC AND TOXIC SOLIDS AND LIQUIDS

6003.1 Indoor storage and use. The indoor storage and use of highly toxic and toxic materials shall comply with Sections 6003.1.1 through 6003.1.5.3.

6003.1.1 Quantities not exceeding the maximum allowable quantity per control area. The indoor storage or use of highly toxic and toxic solids or liquids in amounts not exceeding the *maximum allowable quantity per control area* indicated in Table 5003.1.1(2) shall be in accordance with Sections 5001, 5003 and 6001.

6003.1.2 Quantities exceeding the maximum allowable quantity per control area. The indoor storage or use of highly toxic and toxic solids or liquids in amounts exceeding the *maximum allowable quantity per control area* set forth in Table 5003.1.1(2) shall be in accordance with Section 6001, Sections 6003.1.3 through 6003.1.5.3 and Chapter 50.

6003.1.3 Treatment system—highly toxic liquids. Exhaust scrubbers or other systems for processing vapors of highly toxic liquids shall be provided where a spill or accidental release of such liquids can be expected to release highly toxic vapors at *normal temperature and pressure*. Treatment systems and other processing systems shall be installed in accordance with the *International Mechanical Code*.

6003.1.4 Indoor storage. Indoor storage of highly toxic and toxic solids and liquids shall comply with Sections 6003.1.4.1 and 6003.1.4.2.

6003.1.4.1 Floors. In addition to the requirements set forth in Section 5004.12, floors of storage areas where highly toxic and toxic liquids are stored shall be of liquid-tight construction.

6003.1.4.2 Separation—highly toxic solids and liquids. In addition to the requirements set forth in Section 5003.9.8, highly toxic solids and liquids in storage shall be located in *approved* hazardous material storage cabinets or isolated from other hazardous material storage by construction in accordance with the *International Building Code*.

6003.1.5 Indoor use. Indoor use of highly toxic and toxic solids and liquids shall comply with Sections 6003.1.5.1 through 6003.1.5.3.

6003.1.5.1 Liquid transfer. Highly toxic and toxic liquids shall be transferred in accordance with Section 5005.1.10.

6003.1.5.2 Exhaust ventilation for open systems. Mechanical exhaust ventilation shall be provided for highly toxic and toxic liquids used in *open systems* in accordance with Section 5005.2.1.1.

Exception: Liquids that do not generate highly toxic or toxic fumes, mists or vapors.

6003.1.5.3 Exhaust ventilation for closed systems. Mechanical exhaust ventilation shall be provided for highly toxic and toxic liquids used in *closed systems* in accordance with Section 5005.2.2.1.

Exception: Liquids that do not generate highly toxic or toxic fumes, mists or vapors.

6003.2 Outdoor storage and use. Outdoor storage and use of highly toxic and toxic materials shall comply with Sections 6003.2.1 through 6003.2.6.

6003.2.1 Quantities not exceeding the maximum allowable quantity per control area. The outdoor storage or use of highly toxic and toxic solids or liquids in amounts not exceeding the *maximum allowable quantity per control area* indicated in Table 5003.1.1(4) shall be in accordance with Sections 5001, 5003 and 6001.

6003.2.2 Quantities exceeding the maximum allowable quantity per control area. The outdoor storage or use of highly toxic and toxic solids or liquids in amounts exceeding the *maximum allowable quantity per control area* set forth in Table 5003.1.1(4) shall be in accordance with Sections 6001 and 6003.2 and Chapter 50.

6003.2.3 General outdoor requirements. The general requirements applicable to the outdoor storage of highly toxic or toxic solids and liquids shall be in accordance with Sections 6003.2.3.1 and 6003.2.3.2.

6003.2.3.1 Location. Outdoor storage or use of highly toxic or toxic solids and liquids shall not be located within 20 feet (6096 mm) of *lot lines*, public streets, public alleys, *public ways*, *exit discharges* or *exterior wall* openings. A 2-hour *fire barrier* without openings or penetrations extending not less than 30 inches (762 mm) above and to the sides of the storage is allowed in lieu of such distance. The wall shall either be an independent structure, or the exterior wall of the building adjacent to the storage area.

6003.2.3.2 Treatment system—highly toxic liquids. Exhaust scrubbers or other systems for processing vapors of highly toxic liquid shall be provided where a spill or accidental release of such liquids can be expected to release highly toxic vapors at *normal temperature and pressure (NTP)*. Treatment systems and other processing systems shall be installed in accordance with the *International Mechanical Code*.

6003.2.4 Outdoor storage piles. Outdoor storage piles of highly toxic and toxic solids and liquids shall be separated into piles not larger than 2,500 cubic feet (71 m^3). Aisle widths between piles shall be not less than one-half the height of the pile or 10 feet (3048 mm), whichever is greater.

6003.2.5 Weather protection for highly toxic liquids and solids—outdoor storage or use. Where overhead weather protection is provided for outdoor storage or use of highly toxic liquids or solids, and the weather protection is attached to a building, the storage or use area shall either be equipped throughout with an *approved automatic sprinkler system* in accordance with Section 903.3.1.1, or storage or use vessels shall be fire resistive. Weather protection shall be provided in accordance with Section 5004.13 for storage and Section 5005.3.9 for use.

6003.2.6 Outdoor liquid transfer. Highly toxic and toxic liquids shall be transferred in accordance with Section 5005.1.10.

SECTION 6004
HIGHLY TOXIC AND TOXIC COMPRESSED GASES

6004.1 General. The storage and use of highly toxic and toxic *compressed gases* shall comply with this section.

6004.1.1 Special limitations for indoor storage and use by occupancy. The indoor storage and use of highly toxic and toxic *compressed gases* in certain occupancies shall be subject to the limitations contained in Sections 6004.1.1.1 through 6004.1.1.3.

6004.1.1.1 Group A, E, I or U occupancies. Toxic and highly toxic *compressed gases* shall not be stored or used within Group A, E, I or U occupancies.

Exception: Cylinders not exceeding 20 cubic feet (0.566 m^3) at *normal temperature and pressure (NTP)* are allowed within gas cabinets or fume hoods.

6004.1.1.2 Group R occupancies. Toxic and highly toxic *compressed gases* shall not be stored or used in Group R occupancies.

6004.1.1.3 Offices, retail sales and classrooms. Toxic and highly toxic *compressed gases* shall not be stored or used in offices, retail sales or classroom portions of Group B, F, M or S occupancies.

Exception: In classrooms of Group B occupancies, cylinders with a capacity not exceeding 20 cubic feet (0.566 m^3) at *NTP* are allowed in gas cabinets or fume hoods.

6004.1.2 Gas cabinets. Gas cabinets containing highly toxic or toxic *compressed gases* shall comply with Section 5003.8.6 and the following requirements:

1. The average ventilation velocity at the face of gas cabinet access ports or windows shall be not less than 200 feet per minute (1.02 m/s) with not less than 150 feet per minute (0.76 m/s) at any point of the access port or window.
2. Gas cabinets shall be connected to an exhaust system.
3. Gas cabinets shall not be used as the sole means of exhaust for any room or area.

4. The maximum number of cylinders located in a single gas cabinet shall not exceed three, except that cabinets containing cylinders not exceeding 1 pound (0.454 kg) net contents are allowed to contain up to 100 cylinders.
5. Gas cabinets required by Section 6004.2 or 6004.3 shall be equipped with an *approved automatic sprinkler system* in accordance with Section 903.3.1.1. Alternative fire-extinguishing systems shall not be used.

6004.1.3 Exhausted enclosures. Exhausted enclosures containing highly toxic or toxic *compressed gases* shall comply with Section 5003.8.5 and the following requirements:

1. The average ventilation velocity at the face of the enclosure shall be not less than 200 feet per minute (1.02 m/s) with not less than 150 feet per minute (0.76 m/s).
2. Exhausted enclosures shall be connected to an exhaust system.
3. Exhausted enclosures shall not be used as the sole means of exhaust for any room or area.
4. Exhausted enclosures required by Section 6004.2 or 6004.3 shall be equipped with an *approved automatic sprinkler system* in accordance with Section 903.3.1.1. Alternative fire-extinguishing systems shall not be used.

6004.2 Indoor storage and use. The indoor storage and use of highly toxic or toxic *compressed gases* shall be in accordance with Sections 6004.2.1 through 6004.2.2.10.4.

6004.2.1 Applicability. The applicability of regulations governing the indoor storage and use of highly toxic and toxic *compressed gases* shall be as set forth in Sections 6004.2.1.1 through 6004.2.1.3.

6004.2.1.1 Quantities not exceeding the maximum allowable quantity per control area. The indoor storage or use of highly toxic and toxic gases in amounts not exceeding the *maximum allowable quantity per control area* set forth in Table 5003.1.1(2) shall be in accordance with Sections 5001, 5003, 6001 and 6004.1.

6004.2.1.2 Quantities exceeding the maximum allowable quantity per control area. The indoor storage or use of highly toxic and toxic gases in amounts exceeding the *maximum allowable quantity per control area* set forth in Table 5003.1.1(2) shall be in accordance with Sections 6001, 6004.1, 6004.2 and Chapter 50.

6004.2.1.3 Ozone gas generators. The indoor use of ozone gas-generating equipment shall be in accordance with Section 6005.

6004.2.2 General indoor requirements. The general requirements applicable to the indoor storage and use of highly toxic and toxic *compressed gases* shall be in accordance with Sections 6004.2.2.1 through 6004.2.2.10.4.

6004.2.2.1 Cylinder and tank location. Cylinders shall be located within gas cabinets, exhausted enclosures or gas rooms. Portable and stationary tanks shall be located within gas rooms or exhausted enclosures.

6004.2.2.2 Ventilated areas. The room or area in which gas cabinets or exhausted enclosures are located shall be provided with exhaust ventilation. Gas cabinets or exhausted enclosures shall not be used as the sole means of exhaust for any room or area.

6004.2.2.3 Leaking cylinders and tanks. One or more gas cabinets or exhausted enclosures shall be provided to handle leaking cylinders, containers or tanks.

Exceptions:

1. Where cylinders, containers or tanks are located within gas cabinets or exhausted enclosures.
2. Where *approved* containment vessels or containment systems are provided in accordance with all of the following:
   2.1. Containment vessels or containment systems shall be capable of fully containing or terminating a release.
   2.2. Trained personnel shall be available at an *approved* location.
   2.3. Containment vessels or containment systems shall be capable of being transported to the leaking cylinder, container or tank.

6004.2.2.3.1 Location. Gas cabinets and exhausted enclosures shall be located in gas rooms and connected to an exhaust system.

6004.2.2.4 Local exhaust for portable tanks. A means of local exhaust shall be provided to capture leaks from portable tanks. The local exhaust shall consist of portable ducts or collection systems designed to be applied to the site of a leak in a valve or fitting on the tank. The local exhaust system shall be located in a gas room. Exhaust shall be directed to a treatment system in accordance with Section 6004.2.2.7.

6004.2.2.5 Piping and controls—stationary tanks. In addition to the requirements of Section 5003.2.2, piping and controls on stationary tanks shall comply with the following requirements:

1. Pressure relief devices shall be vented to a treatment system designed in accordance with Section 6004.2.2.7.

   Exception: Pressure relief devices on outdoor tanks provided exclusively for relieving pressure due to fire exposure are not required to be vented to a treatment system provided that:

   1. The material in the tank is not flammable.
   2. The tank is not located in a diked area with other tanks containing combustible materials.

3. The tank is located not less than 30 feet (9144 mm) from combustible materials or structures or is shielded by a *fire barrier* complying with Section 6004.3.2.1.1.

2. Filling or dispensing connections shall be provided with a means of local exhaust. Such exhaust shall be designed to capture fumes and vapors. The exhaust shall be directed to a treatment system in accordance with Section 6004.2.2.7.
3. Stationary tanks shall be provided with a means of excess flow control on all tank inlet or outlet connections.

   Exceptions:

   1. Inlet connections designed to prevent backflow.
   2. Pressure relief devices.

6004.2.2.6 Gas rooms. Gas rooms shall comply with Section 5003.8.4 and both of the following requirements:

1. The exhaust ventilation from gas rooms shall be directed to an exhaust system.
2. Gas rooms shall be equipped with an *approved automatic sprinkler system*. Alternative fire-extinguishing systems shall not be used.

6004.2.2.7 Treatment systems. The exhaust ventilation from gas cabinets, exhausted enclosures and gas rooms, and local exhaust systems required in Sections 6004.2.2.4 and 6004.2.2.5 shall be directed to a treatment system. The treatment system shall be utilized to handle the accidental release of gas and to process exhaust ventilation. The treatment system shall be designed in accordance with Sections 6004.2.2.7.1 through 6004.2.2.7.5 and Section 510 of the *International Mechanical Code*.

Exceptions:

1. Highly toxic and toxic gases—storage. A treatment system is not required for cylinders, containers and tanks in storage where all of the following controls are provided:
   1.1. Valve outlets are equipped with gas-tight outlet plugs or caps.

   1.2. Handwheel-operated valves have handles secured to prevent movement.

   1.3. *Approved* containment vessels or containment systems are provided in accordance with Section 6004.2.2.3.
2. Toxic gases—use. Treatment systems are not required for toxic gases supplied by cylinders or portable tanks not exceeding 1,700 pounds (772 kg) water capacity where the following are provided:
   2.1. A *listed* or *approved* gas detection system with a sensing interval not exceeding 5 minutes.

   2.2. A *listed* or *approved* automatic-closing fail-safe valve located immediately adjacent to cylinder valves. The fail-safe valve shall close when gas is detected at the permissible exposure limit (PEL) by a gas detection system monitoring the exhaust system at the point of discharge from the gas cabinet, exhausted enclosure, ventilated enclosure or gas room. The gas detection system shall comply with Section 6004.2.2.10.

6004.2.2.7.1 Design. Treatment systems shall be capable of diluting, adsorbing, absorbing, containing, neutralizing, burning or otherwise processing the contents of the largest single vessel of compressed gas. Where a total containment system is used, the system shall be designed to handle the maximum anticipated pressure of release to the system when it reaches equilibrium.

6004.2.2.7.2 Performance. Treatment systems shall be designed to reduce the maximum allowable discharge concentrations of the gas to one-half immediate by dangerous to life and health (IDLH) at the point of discharge to the atmosphere. Where more than one gas is emitted to the treatment system, the treatment system shall be designed to handle the worst-case release based on the release rate, the quantity and the IDLH for all *compressed gases* stored or used.

6004.2.2.7.3 Sizing. Treatment systems shall be sized to process the maximum worst-case release of gas based on the maximum flow rate of release from the largest vessel utilized. The entire contents of the largest *compressed gas* vessel shall be considered.

6004.2.2.7.4 Stationary tanks. Stationary tanks shall be labeled with the maximum rate of release for the *compressed gas* contained based on valves or fittings that are inserted directly into the tank. Where multiple valves or fittings are provided, the maximum flow rate of release for valves or fittings with the highest flow rate shall be indicated. Where liquefied *compressed gases* are in contact with valves or fittings, the liquid flow rate shall be utilized for computation purposes. Flow rates indicated on the label shall be converted to cubic feet per minute (cfm/min) (m^3/s) of gas at *normal temperature and pressure (NTP)*.

6004.2.2.7.5 Portable tanks and cylinders. The maximum flow rate of release for portable tanks and cylinders shall be calculated based on the total release from the cylinder or tank within the time specified in Table 6004.2.2.7.5. Where portable tanks or cylinders are equipped with *approved* excess flow or reduced flow valves, the worst-case release shall be determined by the maximum achievable flow from the valve as determined by the valve manufacturer or *compressed gas* supplier. Reduced flow and excess flow valves shall be permanently marked by the valve manufacturer to indicate the maximum design flow rate. Such markings shall indicate the flow rate for air under *normal temperature and pressure.*

TABLE 6004.2.2.7.5
RATE OF RELEASE FOR CYLINDERS AND PORTABLE TANKS

VESSEL TYPE	NONLIQUEFIED (minutes)	LIQUEFIED (minutes)
Containers	5	30
Portable tanks	40	240

6004.2.2.8 Emergency power. Emergency power shall be provided for the following systems in accordance with Section 604:

1. Exhaust ventilation system.
2. Treatment system.
3. Gas detection system.
4. Smoke detection system.
5. Temperature control system.
6. Fire alarm system.
7. Emergency alarm system.

6004.2.2.8.1 Fail-safe engineered systems. Emergency power shall not be required for mechanical exhaust ventilation, treatment systems and temperature control systems where *approved* fail-safe engineered systems are installed.

6004.2.2.9 Automatic fire detection system—highly toxic compressed gases. An *approved* automatic fire detection system shall be installed in rooms or areas where highly toxic *compressed gases* are stored or used. Activation of the detection system shall sound a local alarm. The fire detection system shall comply with Section 907.

6004.2.2.10 Gas detection system. A gas detection system shall be provided to detect the presence of gas at or below the PEL or ceiling limit of the gas for which detection is provided. The system shall be capable of monitoring the discharge from the treatment system at or below one-half the IDLH limit.

Exception: A gas detection system is not required for toxic gases when the physiological warning threshold level for the gas is at a level below the accepted PEL for the gas.

6004.2.2.10.1 Gas detection system components. Gas detection system control units shall be *listed* and *labeled* in accordance with UL 864 or UL 2017, or approved. Gas detectors shall be *listed* and *labeled* in accordance with UL 2075 for use with the gases and vapors being detected, or *approved.*

6004.2.2.10.2 Alarms. The gas detection system shall initiate a local alarm and transmit a signal to a constantly attended control station when a short-term hazard condition is detected. The alarm shall be both visual and audible and shall provide warning both inside and outside the area where gas is detected. The audible alarm shall be distinct from all other alarms.

Exception: Signal transmission to a constantly attended control station is not required where not more than one cylinder of highly toxic or toxic gas is stored.

6004.2.2.10.3 Shut off of gas supply. The gas-detection system shall automatically close the shutoff valve at the source on gas supply piping and tubing related to the system being monitored for whichever gas is detected.

Exception: Automatic shutdown is not required for reactors utilized for the production of highly toxic or toxic *compressed gases* where such reactors are:

1. Operated at pressures less than 15 pounds per square inch gauge (psig) (103.4 kPa).
2. Constantly attended.
3. Provided with readily accessible emergency shutoff valves.

6004.2.2.10.4 Valve closure. Automatic closure of shutoff valves shall be in accordance with the following:

1. Where the gas-detection sampling point initiating the gas detection system alarm is within a gas cabinet or exhausted enclosure, the shutoff valve in the gas cabinet or exhausted enclosure for the specific gas detected shall automatically close.
2. Where the gas-detection sampling point initiating the gas detection system alarm is within a gas room and *compressed gas* containers are not in gas cabinets or exhausted enclosures, the shutoff valves on all gas lines for the specific gas detected shall automatically close.
3. Where the gas-detection sampling point initiating the gas detection system alarm is within a piping distribution manifold enclosure, the shutoff valve for the compressed container of specific gas detected supplying the manifold shall automatically close.

Exception: Where the gas-detection sampling point initiating the gas-detection system alarm is at a use location or within a gas valve enclosure of a branch line downstream of a piping distribution manifold, the shutoff valve in the gas valve enclosure for the branch line located in the piping distribution manifold enclosure shall automatically close.

6004.3 Outdoor storage and use. The outdoor storage and use of highly toxic and toxic *compressed gases* shall be in accordance with Sections 6004.3.1 through 6004.3.4.

6004.3.1 Applicability. The applicability of regulations governing the outdoor storage and use of highly toxic and toxic *compressed gases* shall be as set forth in Sections 6004.3.1.1 through 6004.3.1.3.

6004.3.1.1 Quantities not exceeding the maximum allowable quantity per control area. The outdoor storage or use of highly toxic and toxic gases in amounts not exceeding the *maximum allowable quantity per control area* set forth in Table 5003.1.1(4) shall be in accordance with Sections 5001, 5003 and 6001.

6004.3.1.2 Quantities exceeding the maximum allowable quantity per control area. The outdoor storage or use of highly toxic and toxic gases in amounts exceeding the *maximum allowable quantity per control area* set forth in Table 5003.1.1(4) shall be in accordance with Sections 6001 and 6004.3 and Chapter 50.

6004.3.1.3 Ozone gas generators. The outdoor use of ozone gas-generating equipment shall be in accordance with Section 6005.

6004.3.2 General outdoor requirements. The general requirements applicable to the outdoor storage and use of highly toxic and toxic *compressed gases* shall be in accordance with Sections 6004.3.2.1 through 6004.3.2.4.

6004.3.2.1 Location. Outdoor storage or use of highly toxic or toxic *compressed gases* shall be located in accordance with Sections 6004.3.2.1.1 through 6004.3.2.1.3.

Exception: *Compressed gases* located in gas cabinets complying with Sections 5003.8.6 and 6004.1.2 and located 5 feet (1524 mm) or more from buildings and 25 feet (7620 mm) or more from an *exit discharge*.

6004.3.2.1.1 Distance limitation to exposures. Outdoor storage or use of highly toxic or toxic *compressed gases* shall not be located within 75 feet (22 860 mm) of a *lot line*, public street, public alley, *public way*, *exit discharge* or building not associated with the manufacture or distribution of such gases, unless all of the following conditions are met:

1. Storage is shielded by a 2-hour *fire barrier* that interrupts the line of sight between the storage and the exposure.
2. The 2-hour *fire barrier* shall be located not less than 5 feet (1524 mm) from any exposure.
3. The 2-hour *fire barrier* shall not have more than two sides at approximately 90-degree (1.57 rad) directions, or three sides with connecting angles of approximately 135 degrees (2.36 rad).

6004.3.2.1.2 Openings in exposed buildings. Where the storage or use area is located closer than 75 feet (22 860 mm) to a building not associated with the manufacture or distribution of highly toxic or toxic *compressed gases*, openings into a building other than for piping are not allowed above the height of the top of the 2-hour *fire barrier* or within 50 feet (15 240 mm) horizontally from the storage area whether or not shielded by a *fire barrier*.

6004.3.2.1.3 Air intakes. The storage or use area shall not be located within 75 feet (22 860 mm) of air intakes.

6004.3.2.2 Leaking cylinders and tanks. The requirements of Section 6004.2.2.3 shall apply to outdoor cylinders and tanks. Gas cabinets and exhausted enclosures shall be located within or immediately adjacent to outdoor storage or use areas.

6004.3.2.3 Local exhaust for portable tanks. Local exhaust for outdoor portable tanks shall be provided in accordance with the requirements set forth in Section 6004.2.2.4.

6004.3.2.4 Piping and controls-stationary tanks. Piping and controls for outdoor stationary tanks shall be in accordance with the requirements set forth in Section 6004.2.2.5.

6004.3.3 Outdoor storage weather protection for portable tanks and cylinders. Weather protection in accordance with Section 5004.13 shall be provided for portable tanks and cylinders located outdoors and not within gas cabinets or exhausted enclosures. The storage area shall be equipped with an *approved automatic sprinkler system* in accordance with Section 903.3.1.1.

Exception: An *automatic sprinkler system* is not required when:

1. All materials under the weather protection structure, including hazardous materials and the containers in which they are stored, are noncombustible.
2. The weather protection structure is located not less than 30 feet (9144 mm) from combustible materials or structures or is separated from such materials or structures using a *fire barrier* complying with Section 6004.3.2.1.1.

6004.3.4 Outdoor use of cylinders, containers and portable tanks. Cylinders, containers and portable tanks in outdoor use shall be located in gas cabinets or exhausted enclosures and shall comply with Sections 6004.3.4.1 through 6004.3.4.3.

6004.3.4.1 Treatment systems. The treatment system requirements set forth in Section 6004.2.2.7 shall apply to highly toxic or toxic gases located outdoors.

6004.3.4.2 Emergency power. The requirements for emergency power set forth in Section 6004.2.2.8 shall apply to highly toxic or toxic gases located outdoors.

6004.3.4.3 Gas detection system. The gas detection system requirements set forth in Section 6004.2.2.10 shall apply to highly toxic or toxic gases located outdoors.

SECTION 6005
OZONE GAS GENERATORS

6005.1 Scope. Ozone gas generators having a maximum ozone-generating capacity of 0.5 pound (0.23 kg) or more over a 24-hour period shall be in accordance with Sections 6005.2 through 6005.6.

Exceptions:

1. Ozone-generating equipment used in Group R-3 occupancies.
2. Ozone-generating equipment where used in Group H-5 occupancies where in compliance with Chapters 27 and 50 and the other provisions in this chapter for highly toxic gases.

6005.2 Design. Ozone gas generators shall be designed, fabricated and tested in accordance with NEMA 250.

6005.3 Location. Ozone generators shall be located in *approved* cabinets or ozone generator rooms in accordance with Section 6005.3.1 or 6005.3.2.

Exception: An ozone gas generator within an *approved* pressure vessel where located outside of buildings.

6005.3.1 Cabinets. Ozone cabinets shall be constructed of *approved* materials and compatible with ozone. Cabinets shall display an *approved* sign stating: OZONE GAS GENERATOR—HIGHLY TOXIC—OXIDIZER.

Cabinets shall be braced for seismic activity in accordance with the *International Building Code*.

Cabinets shall be mechanically ventilated in accordance with the *International Mechanical Code* with not less than six air changes per hour.

The average velocity of ventilation at makeup air openings with cabinet doors closed shall be not less than 200 feet per minute (1.02 m/s).

6005.3.2 Ozone gas generator rooms. Ozone gas generator rooms shall be mechanically ventilated in accordance with the *International Mechanical Code* with not less than six air changes per hour. Ozone gas generator rooms shall be equipped with a continuous gas detection system that will shut off the generator and sound a local alarm when concentrations above the permissible exposure limit occur.

Ozone gas generator rooms shall not be normally occupied, and such rooms shall be kept free of combustible and hazardous material storage. Room access doors shall display an *approved* sign stating: OZONE GAS GENERATOR—HIGHLY TOXIC—OXIDIZER.

6005.4 Piping, valves and fittings. Piping, valves, fittings and related components used to convey ozone shall be in accordance with Sections 6005.4.1 through 6005.4.3.

6005.4.1 Piping. Piping shall be welded stainless steel piping or tubing.

Exceptions:

1. Double-walled piping.
2. Piping, valves, fittings and related components located in exhausted enclosures.

6005.4.2 Materials. Materials shall be compatible with ozone and shall be rated for the design operating pressures.

6005.4.3 Identification. Piping shall be identified with the following: OZONE GAS—HIGHLY TOXIC—OXIDIZER.

6005.5 Automatic shutdown. Ozone gas generators shall be designed to shut down automatically under the following conditions:

1. When the dissolved ozone concentration in the water being treated is above saturation when measured at the point where the water is exposed to the atmosphere.
2. When the process using generated ozone is shut down.
3. When the gas detection system detects ozone.
4. Failure of the ventilation system for the cabinet or ozone-generator room.
5. Failure of the gas detection system.

6005.6 Manual shutdown. Manual shutdown controls shall be provided at the generator and, where in a room, within 10 feet (3048 mm) of the main *exit* or *exit access* door.

CHAPTER 61

LIQUEFIED PETROLEUM GASES

SECTION 6101 GENERAL

6101.1 Scope. Storage, handling and transportation of liquefied petroleum gas (LP-gas) and the installation of LP-gas equipment pertinent to systems for such uses shall comply with this chapter and NFPA 58. Properties of LP-gases shall be determined in accordance with Appendix B of NFPA 58.

6101.2 Permits. Permits shall be required as set forth in Sections 105.6 and 105.7.

Distributors shall not fill an LP-gas container for which a permit is required unless a permit for installation has been issued for that location by the *fire code official.*

6101.3 Construction documents. Where a single LP-gas container is more than 2,000 gallons (7570 L) in water capacity or the aggregate water capacity of LP-gas containers is more than 4,000 gallons (15 140 L), the installer shall submit *construction documents* for such installation.

SECTION 6102 DEFINITIONS

6102.1 Definitions. The following terms are defined in Chapter 2:

LIQUEFIED PETROLEUM GAS (LP-gas).

LP-GAS CONTAINER.

SECTION 6103 INSTALLATION OF EQUIPMENT

6103.1 General. LP-gas equipment shall be installed in accordance with the *International Fuel Gas Code* and NFPA 58, except as otherwise provided in this chapter.

6103.2 Use of LP-gas containers in buildings. The use of LP-gas containers in buildings shall be in accordance with Sections 6103.2.1 and 6103.2.2.

6103.2.1 Portable containers. Portable LP-gas containers, as defined in NFPA 58, shall not be used in buildings except as specified in NFPA 58 and Sections 6103.2.1.1 through 6103.2.1.7.

6103.2.1.1 Use in basement, pit or similar location. LP-gas containers shall not be used in a basement, pit or similar location where heavier-than-air gas might collect. LP-gas containers shall not be used in an above-grade underfloor space or basement unless such location is provided with an *approved* means of ventilation.

Exception: Use with self-contained torch assemblies in accordance with Section 6103.2.1.6.

6103.2.1.2 Construction and temporary heating. Portable LP-gas containers are allowed to be used in buildings or areas of buildings undergoing construction or for temporary heating as set forth in Sections 6.19.4, 6.19.5 and 6.19.8 of NFPA 58.

6103.2.1.3 Group F occupancies. In Group F occupancies, portable LP-gas containers are allowed to be used to supply quantities necessary for processing, research or experimentation. Where manifolded, the aggregate water capacity of such containers shall not exceed 735 pounds (334 kg) per manifold. Where multiple manifolds of such containers are present in the same room, each manifold shall be separated from other manifolds by a distance of not less than 20 feet (6096 mm).

6103.2.1.4 Group E and I occupancies. In Group E and I occupancies, portable LP-gas containers are allowed to be used for research and experimentation. Such containers shall not be used in classrooms. Such containers shall not exceed a 50-pound (23 kg) water capacity in occupancies used for educational purposes and shall not exceed a 12-pound (5 kg) water capacity in occupancies used for institutional purposes. Where more than one such container is present in the same room, each container shall be separated from other containers by a distance of not less than 20 feet (6096 mm).

6103.2.1.5 Demonstration uses. Portable LP-gas containers are allowed to be used temporarily for demonstrations and public exhibitions. Such containers shall not exceed a water capacity of 12 pounds (5 kg). Where more than one such container is present in the same room, each container shall be separated from other containers by a distance of not less than 20 feet (6096 mm).

6103.2.1.6 Use with self-contained torch assemblies. Portable LP-gas containers are allowed to be used to supply *approved* self-contained torch assemblies or similar appliances. Such containers shall not exceed a water capacity of $2^1/_2$ pounds (1 kg).

6103.2.1.7 Use for food preparation. Where *approved, listed* LP-gas commercial food service appliances are allowed to be used for food-preparation within restaurants and in attended commercial food-catering operations in accordance with the *International Fuel Gas Code,* the *International Mechanical Code* and NFPA 58.

6103.2.2 Industrial vehicles and floor maintenance machines. LP-gas containers on industrial vehicles and floor maintenance machines shall comply with Sections 11.13 and 11.14 of NFPA 58.

6103.3 Location of equipment and piping. Equipment and piping shall not be installed in locations where such equipment and piping is prohibited by the *International Fuel Gas Code.*

SECTION 6104
LOCATION OF LP-GAS CONTAINERS

6104.1 General. The storage and handling of LP-gas and the installation and maintenance of related equipment shall comply with NFPA 58 and be subject to the approval of the *fire code official*, except as provided in this chapter.

6104.2 Maximum capacity within established limits. Within the limits established by law restricting the storage of liquefied petroleum gas for the protection of heavily populated or congested areas, the aggregate capacity of any one installation shall not exceed a water capacity of 2,000 gallons (7570 L) (see Section 3 of the Sample Legislation for Adoption of the *International Fire Code* on page xxi).

Exception: In particular installations, this capacity limit shall be determined by the *fire code official*, after consideration of special features such as topographical conditions, nature of occupancy, and proximity to buildings, capacity of proposed LP-gas containers, degree of fire protection to be provided and capabilities of the local fire department.

6104.3 Container location. LP-gas containers shall be located with respect to buildings, *public ways* and *lot lines* of adjoining property that can be built upon, in accordance with Table 6104.3.

6104.3.1 Installation on roof prohibited. LP-gas containers used in stationary installations shall not be located on the roofs of buildings.

6104.3.2 Special hazards. LP-gas containers shall be located with respect to special hazards including, but not limited to, above-ground flammable or *combustible liquid*

TABLE 6104.3
LOCATION OF LP-GAS CONTAINERS

LP-GAS CONTAINER CAPACITY (water gallons)	MINIMUM SEPARATION BETWEEN LP-GAS CONTAINERS AND BUILDINGS, PUBLIC WAYS OR LOT LINES OF ADJOINING PROPERTY THAT CAN BE BUILT UPON		MINIMUM SEPARATION BETWEEN LP-GAS CONTAINERS[b, c] (feet)
	Mounded or underground LP-gas containers[a] (feet)	Above-ground LP-gas containers[b] (feet)	
Less than 125[c, d]	10	5[e]	None
125 to 250	10	10	None
251 to 500	10	10	3
501 to 2,000	10	25[e, f]	3
2,001 to 30,000	50	50	5
30,001 to 70,000	50	75	(0.25 of sum of diameters of adjacent LP-gas containers)
70,001 to 90,000	50	100	
90,001 to 120,000	50	125	

For SI: 1 foot = 304.8 mm, 1 gallon = 3.785 L.

a. Minimum distance for underground LP-gas containers shall be measured from the pressure relief device and the filling or liquid-level gauge vent connection at the container, except that all parts of an underground LP-gas container shall be not less than 10 feet from a building or lot line of adjoining property that can be built upon.

b. For other than installations in which the overhanging structure is 50 feet or more above the relief-valve discharge outlet. In applying the distance between buildings and ASME LP-gas containers with a water capacity of 125 gallons or more, not less than 50 percent of this horizontal distance shall also apply to all portions of the building that project more than 5 feet from the building wall and that are higher than the relief valve discharge outlet. This horizontal distance shall be measured from a point determined by projecting the outside edge of such overhanging structure vertically downward to grade or other level upon which the LP-gas container is installed. Distances to the building wall shall be not less than those prescribed in this table.

c. Where underground multicontainer installations are composed of individual LP-gas containers having a water capacity of 125 gallons or more, such containers shall be installed so as to provide access at their ends or sides to facilitate working with cranes or hoists.

d. At a consumer site, if the aggregate water capacity of a multicontainer installation, comprised of individual LP-gas containers having a water capacity of less than 125 gallons, is 500 gallons or more, the minimum distance shall comply with the appropriate portion of Table 6104.3, applying the aggregate capacity rather than the capacity per LP-gas container. If more than one such installation is made, each installation shall be separated from other installations by not less than 25 feet. Minimum distances between LP-gas containers need not be applied.

e. The following shall apply to above-ground containers installed alongside buildings:

1. LP-gas containers of less than a 125-gallon water capacity are allowed next to the building they serve where in compliance with Items 2, 3 and 4.
2. Department of Transportation (DOTn) specification LP-gas containers shall be located and installed so that the discharge from the container pressure relief device is not less than 3 feet horizontally from building openings below the level of such discharge and shall not be beneath buildings unless the space is well ventilated to the outside and is not enclosed for more than 50 percent of its perimeter. The discharge from LP-gas container pressure relief devices shall be located not less than 5 feet from exterior sources of ignition, openings into direct-vent (sealed combustion system) appliances or mechanical ventilation air intakes.
3. ASME LP-gas containers of less than a 125-gallon water capacity shall be located and installed such that the discharge from pressure relief devices shall not terminate in or beneath buildings and shall be located not less than 5 feet horizontally from building openings below the level of such discharge and not less than 5 feet from exterior sources of ignition, openings into direct vent (sealed combustion system) appliances, or mechanical ventilation air intakes.
4. The filling connection and the vent from liquid-level gauges on either DOTn or ASME LP-gas containers filled at the point of installation shall be not less than 10 feet from exterior sources of ignition, openings into direct vent (sealed combustion system) appliances or mechanical ventilation air intakes.

f. This distance is allowed to be reduced to not less than 10 feet for a single LP-gas container of 1,200-gallon water capacity or less, provided such container is not less than 25 feet from other LP-gas containers of more than 125-gallon water capacity.

tanks, oxygen or gaseous hydrogen containers, flooding or electric power lines as specified in Section 6.4.5 of NFPA 58.

6104.4 Multiple LP-gas container installations. Multiple LP-gas container installations with a total water storage capacity of more than 180,000 gallons (681 300 L) [150,000-gallon (567 750 L) LP-gas capacity] shall be subdivided into groups containing not more than 180,000 gallons (681 300 L) in each group. Such groups shall be separated by a distance of not less than 50 feet (15 240 mm), unless the containers are protected in accordance with one of the following:

1. Mounded in an *approved* manner.
2. Protected with *approved* insulation on areas that are subject to impingement of ignited gas from pipelines or other leakage.
3. Protected by fire walls of *approved* construction.
4. Protected by an *approved* system for application of water as specified in Table 6.4.2 of NFPA 58.
5. Protected by other *approved* means.

Where one of these forms of protection is provided, the separation shall be not less than 25 feet (7620 mm) between LP-gas container groups.

SECTION 6105 PROHIBITED USE OF LP-GAS

6105.1 Nonapproved equipment. LP-gas shall not be used for the purpose of operating devices or equipment unless such device or equipment is *approved* for use with LP-gas.

6105.2 Release to the atmosphere. LP-gas shall not be released to the atmosphere, except in accordance with Section 7.3 of NFPA 58.

SECTION 6106 DISPENSING AND OVERFILLING

6106.1 Attendants. Dispensing of LP-gas shall be performed by a qualified attendant.

6106.2 Overfilling. LP-gas containers shall not be filled or maintained with LP-gas in excess of either the volume determined using the fixed liquid-level gauge installed in accordance with the manufacturer's specifications and in accordance with Section 5.7.5 of NFPA 58 or the weight determined by the required percentage of the water capacity marked on the container. Portable LP-gas containers shall not be refilled unless equipped with an overfilling prevention device (OPD) where required by Section 5.7.3 of NFPA 58.

6106.3 Dispensing locations. The point of transfer of LP-gas from one LP-gas container to another shall be separated from exposures as specified in NFPA 58.

SECTION 6107 SAFETY PRECAUTIONS AND DEVICES

6107.1 Safety devices. Safety devices on LP-gas containers, equipment and systems shall not be tampered with or made ineffective.

6107.2 Smoking and other sources of ignition. "No Smoking" signs complying with Section 310 shall be posted where required by the *fire code official.* Smoking within 25 feet (7620 mm) of a point of transfer, while filling operations are in progress at LP-gas containers or vehicles, shall be prohibited.

Control of other sources of ignition shall comply with Chapter 3 of this code and Section 6.22 of NFPA 58.

6107.3 Clearance to combustibles. Weeds, grass, brush, trash and other combustible materials shall be kept not less than 10 feet (3048 mm) from LP-gas tanks or containers.

6107.4 Protecting containers from vehicles. Where exposed to vehicular damage due to proximity to alleys, driveways or parking areas, LP-gas containers, regulators and piping shall be protected in accordance with NFPA 58.

SECTION 6108 FIRE PROTECTION

6108.1 General. Fire protection shall be provided for installations having LP-gas storage containers with a water capacity of more than 4,000 gallons (15 140 L), as required by Section 6.25 of NFPA 58.

6108.2 Portable fire extinguishers. Portable fire extinguishers complying with Section 906 shall be provided as specified in NFPA 58.

SECTION 6109 STORAGE OF PORTABLE LP-GAS CONTAINERS AWAITING USE OR RESALE

6109.1 General. Storage of portable LP-gas containers of 1,000 pounds (454 kg) or less, whether filled, partially filled or empty, at consumer sites or distribution points, and for resale by dealers or resellers shall comply with Sections 6109.2 through 6109.15.1.

Exceptions:

1. LP-gas containers that have not previously been in LP-gas service.
2. LP-gas containers at distribution plants.
3. LP-gas containers at consumer sites or distribution points, which are connected for use.

6109.2 Exposure hazards. LP-gas containers in storage shall be located in a manner that minimizes exposure to excessive temperature rise, physical damage or tampering.

6109.3 Position. LP-gas containers in storage having individual water capacity greater than $2^1/_2$ pounds (1 kg) [nominal 1-pound (0.454 kg) LP-gas capacity] shall be positioned with the pressure relief valve in direct communication with the vapor space of the container.

6109.4 Separation from means of egress. LP-gas containers stored in buildings in accordance with Sections 6109.9 and 6109.11 shall not be located near *exit access* doors, *exits, stairways* or in areas normally used, or intended to be used, as a *means of egress*.

6109.5 Quantity. Empty LP-gas containers that have been in LP-gas service shall be considered as full containers for the purpose of determining the maximum quantities of LP-gas allowed in Sections 6109.9 and 6109.11.

6109.6 Storage on roofs. LP-gas containers that are not connected for use shall not be stored on roofs.

6109.7 Storage in basement, pit or similar location. LP-gas containers shall not be stored in a basement, pit or similar location where heavier-than-air gas might collect. LP-gas containers shall not be stored in above-grade underfloor spaces or basements unless such location is provided with an *approved* means of ventilation.

Exception: Department of Transportation (DOTn) specification cylinders with a maximum water capacity of $2^1/_2$ pounds (1 kg) for use in completely self-contained hand torches and similar applications. The quantity of LP-gas shall not exceed 20 pounds (9 kg).

6109.8 Protection of valves on LP-gas containers in storage. LP-gas container valves shall be protected by screw-on-type caps or collars that shall be securely in place on all containers stored regardless of whether they are full, partially full or empty. Container outlet valves shall be closed or plugged.

6109.9 Storage within buildings accessible to the public. Department of Transportation (DOTn) specification cylinders with maximum water capacity of $2^1/_2$ pounds (1 kg) used in completely self-contained hand torches and similar applications are allowed to be stored or displayed in a building accessible to the public. The quantity of LP-gas shall not exceed 200 pounds (91 kg) except as provided in Section 6109.11.

6109.10 Storage within buildings not accessible to the public. The maximum quantity allowed in one storage location in buildings not accessible to the public, such as industrial buildings, shall not exceed a water capacity of 735 pounds (334 kg) [nominal 300 pounds (136 kg) of LP-gas]. Where additional storage locations are required on the same floor within the same building, they shall be separated by not less than 300 feet (91 440 mm). Storage beyond these limitations shall comply with Section 6109.11.

6109.10.1 Quantities on equipment and vehicles. LP-gas containers carried as part of service equipment on highway mobile vehicles need not be considered in the total storage capacity in Section 6109.10, provided such vehicles are stored in private garages and do not carry more than three LP-gas containers with a total aggregate LP-gas capacity not exceeding 100 pounds (45.4 kg) per vehicle. LP-gas container valves shall be closed.

6109.11 Storage within rooms used for gas manufacturing. Storage within buildings or rooms used for gas manufacturing, gas storage, gas-air mixing and vaporization, and compressors not associated with liquid transfer shall comply with Sections 6109.11.1 and 6109.11.2.

6109.11.1 Quantity limits. The maximum quantity of LP-gas shall be 10,000 pounds (4540 kg).

6109.11.2 Construction. The construction of such buildings and rooms shall comply with requirements for Group H occupancies in the *International Building Code*, Chapter 10 of NFPA 58 and both of the following:

1. Adequate vents shall be provided to the outside at both top and bottom, located not less than 5 feet (1524 mm) from building openings.
2. The entire area shall be classified for the purposes of ignition source control in accordance with Section 6.22 of NFPA 58.

6109.12 Location of storage outside of buildings. Storage outside of buildings of LP-gas containers awaiting use, resale or part of a cylinder exchange program shall be located in accordance with Table 6109.12.

6109.13 Protection of containers. LP-gas containers shall be stored within a suitable enclosure or otherwise protected

TABLE 6109.12
SEPARATION FROM EXPOSURES OF LP-GAS CONTAINERS AWAITING USE, RESALE OR EXCHANGE STORED OUTSIDE OF BUILDINGS

QUANTITY OF LP-GAS STORED (pounds)	MINIMUM SEPARATION DISTANCE FROM STORED LP-GAS CYLINDERS TO (feet):						
	Nearest important building or group of buildings or line of adjoining property that may be built upon	Line of adjoining property occupied by schools, places of religious worship, hospitals, athletic fields or other points of public gathering; busy thoroughfares; or sidewalks	LP-gas dispensing station	Doorway or opening to a building with two or more means of egress	Doorway or opening to a building with one means of egress	Combustible materials	Motor vehicle fuel dispenser
720 or less	0	0	5	5	10	10	20
721 – 2,500	0	10	10	5	10	10	20
2,501 – 6,000	10	10	10	10	10	10	20
6,001 – 10,000	20	20	20	20	20	10	20
Over 10,000	25	25	25	25	25	10	20

For SI: 1 foot = 304.8 mm, 1 pound = 0.454 kg.

against tampering. Vehicle impact protection shall be provided as required by Section 6107.4.

Exception: Vehicle impact protection shall not be required for protection of LP-gas containers where the containers are kept in lockable, ventilated cabinets of metal construction.

6109.14 Alternative location and protection of storage. Where the provisions of Sections 6109.12 and 6109.13 are impractical at construction sites, or at buildings or structures undergoing major renovation or repairs, the storage of containers shall be as required by the *fire code official*.

6109.15 LP-gas cylinder exchange for resale. In addition to other applicable requirements of this chapter, facilities operating LP-gas cylinder exchange stations that are accessible to the public shall comply with the following requirements.

1. Cylinders shall be secured in a lockable, ventilated metal cabinet or other *approved* enclosure.
2. Cylinders shall be accessible only by authorized personnel or by use of an automated exchange system in accordance with Section 6109.15.1.
3. A sign shall be posted on the entry door of the business operating the cylinder exchange stating "DO NOT BRING LP-GAS CYLINDERS INTO THE BUILDING" or similar *approved* wording.
4. An emergency contact information sign shall be posted within 10 feet (3048 mm) of the cylinder storage cabinet. The content, lettering, size, color and location of the required sign shall be as required by the *fire code official*.

6109.15.1 Automated cylinder exchange stations. Cylinder exchange stations that include an automated vending system for exchanging cylinders shall comply with the following additional requirements:

1. The vending system shall only permit access to a single cylinder per individual transaction.
2. Cabinets storing cylinders shall be designed such that cylinders can only be placed inside when they are oriented in the upright position.
3. Devices operating door releases for access to stored cylinders shall be permitted to be pneumatic, mechanical or electrically powered.
4. Electrical equipment inside of or within 5 feet (1524 mm) of a cabinet storing cylinders, including but not limited to electronics associated with vending operations, shall comply with the requirements for Class I, Division 2 equipment in accordance with NFPA 70.
5. A manual override control shall be permitted for use by authorized personnel. On newly installed cylinder exchange stations, the vending system shall not be capable of returning to automatic operation after a manual override until the system has been inspected and reset by authorized personnel.
6. Inspections shall be conducted by authorized personnel to verify that all cylinders are secured, access doors are closed and the station has no visible damage or obvious defects that necessitate placing the station out of service. The frequency of inspections shall be as specified by the *fire code official*.

SECTION 6110
LP-GAS CONTAINERS NOT IN SERVICE

6110.1 Temporarily out of service. LP-gas containers whose use has been temporarily discontinued shall comply with all of the following:

1. Be disconnected from appliance piping.
2. Have LP-gas container outlets, except relief valves, closed or plugged.
3. Be positioned with the relief valve in direct communication with the LP-gas container vapor space.

6110.2 Permanently out of service. LP-gas containers to be placed permanently out of service shall be removed from the site.

SECTION 6111
PARKING AND GARAGING OF LP-GAS TANK VEHICLES

6111.1 General. Parking of LP-gas tank vehicles shall comply with Sections 6111.2 and 6111.3.

Exception: In cases of accident, breakdown or other emergencies, LP-gas tank vehicles are allowed to be parked and left unattended at any location while the operator is obtaining assistance.

6111.2 Unattended parking. The unattended parking of LP-gas tank vehicle shall be in accordance with Sections 6111.2.1 and 6111.2.2.

6111.2.1 Near residential, educational and institutional occupancies and other high-risk areas. LP-gas tank vehicles shall not be left unattended at any time on residential streets or within 500 feet (152 m) of a residential area, apartment or hotel complex, educational facility, hospital or care facility. Tank vehicles shall not be left unattended at any other place that would, in the opinion of the *fire code official*, pose an extreme life hazard.

6111.2.2 Durations exceeding 1 hour. LP-gas tank vehicles parked at any one point for longer than 1 hour shall be located as follows:

1. Off public streets, highways, public avenues or public alleys.
2. Inside of a bulk plant.
3. At other *approved* locations not less than 50 feet (15 240 mm) from buildings other than those *approved* for the storage or servicing of such vehicles.

6111.3 Garaging. Garaging of LP-gas tank vehicles shall be as specified in NFPA 58. Vehicles with LP-gas fuel systems are allowed to be stored or serviced in garages as specified in Section 11.16 of NFPA 58.

CHAPTER 62

ORGANIC PEROXIDES

SECTION 6201 GENERAL

6201.1 Scope. The storage and use of organic peroxides shall be in accordance with this chapter and Chapter 50.

Unclassified detonable organic peroxides that are capable of *detonation* in their normal shipping containers under conditions of fire exposure shall be stored in accordance with Chapter 56.

6201.2 Permits. Permits shall be required for organic peroxides as set forth in Section 105.6.

SECTION 6202 DEFINITION

6202.1 Definition. The following term is defined in Chapter 2:

ORGANIC PEROXIDE.
Class I.
Class II.
Class III.
Class IV.
Class V.
Unclassified detonable.

SECTION 6203 GENERAL REQUIREMENTS

6203.1 Quantities not exceeding the maximum allowable quantity per control area. The storage and use of organic peroxides in amounts not exceeding the *maximum allowable quantity per control area* indicated in Section 5003.1 shall be in accordance with Sections 5001, 5003, 6201 and 6203.

6203.1.1 Special limitations for indoor storage and use by occupancy. The indoor storage and use of organic peroxides shall be in accordance with Sections 6203.1.1.1 through 6203.1.1.4.

6203.1.1.1 Group A, E, I or U occupancies. In Group A, E, I or U occupancies, any amount of unclassified detonable and Class I organic peroxides shall be stored in accordance with the following:

1. Unclassified detonable and Class I organic peroxides shall be stored in hazardous materials storage cabinets complying with Section 5003.8.7.
2. The hazardous materials storage cabinets shall not contain other storage.

6203.1.1.2 Group R occupancies. Unclassified detonable and Class I organic peroxides shall not be stored or used within Group R occupancies.

6203.1.1.3 Group B, F, M or S occupancies. Unclassified detonable and Class I organic peroxides shall not be stored or used in offices, or retail sales areas of Group B, F, M or S occupancies.

6203.1.1.4 Classrooms. In classrooms in Group B, F or M occupancies, any amount of unclassified detonable and Class 1 organic peroxides shall be stored in accordance with the following.

1. Unclassified detonable and Class 1 organic peroxides shall be stored in hazardous materials storage cabinets complying with Section 5003.8.7.
2. The hazardous materials storage cabinets shall not contain other storage.

6203.2 Quantities exceeding the maximum allowable quantity per control area. The storage and use of organic peroxides in amounts exceeding the *maximum allowable quantity per control area* indicated in Section 5003.1 shall be in accordance with Chapter 50 and this chapter.

SECTION 6204 STORAGE

6204.1 Indoor storage. Indoor storage of organic peroxides in amounts exceeding the *maximum allowable quantity per control area* indicated in Table 5003.1.1(1) shall be in accordance with Sections 5001, 5003, 5004 and this chapter.

Indoor storage of unclassified detonable organic peroxides that are capable of *detonation* in their normal shipping containers under conditions of fire exposure shall be stored in accordance with Chapter 56.

6204.1.1 Detached storage. Storage of organic peroxides shall be in detached buildings where required by Section 5003.8.2.

6204.1.2 Distance from detached buildings to exposures. In addition to the requirements of the *International Building Code*, detached storage buildings for Class I, II, III, IV and V organic peroxides shall be located in accordance with Table 6204.1.2. Detached buildings containing quantities of unclassified detonable organic peroxides in excess of those set forth in Table 5003.8.2 shall be located in accordance with Table 5604.5.2(1).

6204.1.3 Liquid-tight floor. In addition to the requirements of Section 5004.12, floors of storage areas shall be of liquid-tight construction.

6204.1.4 Electrical wiring and equipment. In addition to the requirements of Section 5003.9.4, electrical wiring and equipment in storage areas for Class I or II organic peroxides shall comply with the requirements for electrical Class I, Division 2 locations.

6204.1.5 Smoke detection. An *approved* supervised smoke detection system in accordance with Section 907 shall be provided in rooms or areas where Class I, II or III organic peroxides are stored. Activation of the smoke detection system shall sound a local alarm.

Exception: A smoke detection system shall not be required in detached storage buildings equipped throughout with an *approved* automatic fire-extinguishing system complying with Chapter 9.

6204.1.6 Maximum quantities. Maximum allowable quantities per building in a mixed occupancy building shall not exceed the amounts set forth in Table 5003.8.2. Maximum allowable quantities per building in a detached storage building shall not exceed the amounts specified in Table 6204.1.2.

6204.1.7 Storage arrangement. Storage arrangements for organic peroxides shall be in accordance with Table 6204.1.7 and shall comply with all of the following:

1. Containers and packages in storage areas shall be closed.
2. Bulk storage shall not be in piles or bins.
3. A minimum 2-foot (610 mm) clear space shall be maintained between storage and uninsulated metal walls.
4. Fifty-five-gallon (208 L) drums shall not be stored more than one drum high.

6204.1.8 Location in building. The storage of Class I or II organic peroxides shall be on the ground floor. Class III organic peroxides shall not be stored in basements.

6204.1.9 Contamination. Organic peroxides shall be stored in their original DOTn shipping containers. Organic peroxides shall be stored in a manner to prevent contamination.

6204.1.10 Explosion control. Indoor storage rooms, areas and buildings containing unclassified detonable and Class I organic peroxides shall be provided with explosion control in accordance with Section 911.

6204.1.11 Standby power. Standby power shall be provided in accordance with Section 604 for the following systems used to protect Class I and unclassified detonable organic peroxide:

1. Exhaust ventilation system.
2. Treatment system.
3. Gas detection system.
4. Smoke detection system.
5. Temperature control system.
6. Fire alarm system.
7. Emergency alarm system.

6204.1.11.1 Fail-safe engineered systems. Standby power shall not be required for mechanical exhaust

TABLE 6204.1.2
ORGANIC PEROXIDES—DISTANCE TO EXPOSURES FROM DETACHED STORAGE BUILDINGS OR OUTDOOR STORAGE AREAS

ORGANIC PEROXIDE CLASS	MAXIMUM STORAGE QUANTITY (POUNDS) AT MINIMUM SEPARATION DISTANCE					
	Distance to buildings, lot lines, public streets, public alleys, public ways or means of egress			Distance between individual detached storage buildings or individual outdoor storage areas		
	50 feet	100 feet	150 feet	20 feet	75 feet	100 feet
I	2,000	20,000	175,000	2,000	20,000	175,000
II	100,000	200,000	No Limit	100,000[a]	No Limit	No Limit
III	200,000	No Limit	No Limit	200,000[a]	No Limit	No Limit
IV	No Limit	No Limit	No Limit	No Limit	No Limit	No Limit
V	No Limit	No Limit	No Limit	No Limit	No Limit	No Limit

For SI: 1 foot = 304.8 mm, 1 pound = 0.454 kg.

a. Where the amount of organic peroxide stored exceeds this amount, the minimum separation shall be 50 feet.

TABLE 6204.1.7
STORAGE OF ORGANIC PEROXIDES

ORGANIC PEROXIDE CLASS	PILE CONFIGURATION				MAXIMUM QUANTITY PER BUILDING
	Maximum width (feet)	Maximum height (feet)	Minimum distance to next pile (feet)	Minimum distance to walls (feet)	
I	6	8	4[a]	4[b]	Note c
II	10	8	4[a]	4[b]	Note c
III	10	8	4[a]	4[b]	Note c
IV	16	10	3[a, d]	4[b]	No Requirement
V	No Requirement	No Requirement	No Requirement	No Requirement	No Requirement

For SI: 1 foot = 304.8 mm.

a. Not less than one main aisle with a minimum width of 8 feet shall divide the storage area.

b. Distance to noncombustible walls is allowed to be reduced to 2 feet.

c. See Table 6204.1.2 for maximum quantities.

d. The distance shall be not less than one-half the pile height.

ventilation, treatment systems and temperature control systems where *approved* fail-safe engineered systems are installed.

6204.2 Outdoor storage. Outdoor storage of organic peroxides in amounts exceeding the *maximum allowable quantities per control area* indicated in Table 5003.1.1(3) shall be in accordance with Sections 5001, 5003, 5004 and this chapter.

6204.2.1 Distance from storage to exposures. Outdoor storage areas for organic peroxides shall be located in accordance with Table 6204.1.2.

6204.2.2 Electrical wiring and equipment. In addition to the requirements of Section 5003.9.4, electrical wiring and equipment in outdoor storage areas containing unclassified detonable, Class I or II organic peroxides shall comply with the requirements for electrical Class I, Division 2 locations.

6204.2.3 Maximum quantities. Maximum quantities of organic peroxides in outdoor storage shall be in accordance with Table 6204.1.2.

6204.2.4 Storage arrangement. Storage arrangements shall be in accordance with Table 6204.1.7.

6204.2.5 Separation. In addition to the requirements of Section 5003.9.8, outdoor storage areas for organic peroxides in amounts exceeding those specified in Table 5003.8.2 shall be located a minimum distance of 50 feet (15 240 mm) from other hazardous material storage.

SECTION 6205
USE

6205.1 General. The use of organic peroxides in amounts exceeding the *maximum allowable quantity per control area* indicated in Table 5003.1.1(1) or 5003.1.1(3) shall be in accordance with Sections 5001, 5003, 5005 and this chapter.

CHAPTER 63

OXIDIZERS, OXIDIZING GASES AND OXIDIZING CRYOGENIC FLUIDS

SECTION 6301 GENERAL

6301.1 Scope. The storage and use of oxidizing materials shall be in accordance with this chapter and Chapter 50. Oxidizing gases shall also comply with Chapter 53. Oxidizing *cryogenic fluids* shall also comply with Chapter 55.

Exceptions:

1. Display and storage in Group M and storage in Group S occupancies complying with Section 5003.11.
2. Bulk oxygen systems at industrial and institutional consumer sites shall be in accordance with NFPA 55.
3. Liquid oxygen stored or used in home health care in Group I-1, I-4 and R occupancies in accordance with Section 6306.

6301.2 Permits. Permits shall be required as set forth in Section 105.6.

SECTION 6302 DEFINITIONS

6302.1 Definitions. The following terms are defined in Chapter 2:

BULK OXYGEN SYSTEM.

LIQUID OXYGEN AMBULATORY CONTAINER.

LIQUID OXYGEN HOME CARE CONTAINER.

OXIDIZER.
Class 4.
Class 3.
Class 2.
Class 1.

OXIDIZING CRYOGENIC FLUID.

OXIDIZING GAS.

SECTION 6303 GENERAL REQUIREMENTS

6303.1 Quantities not exceeding the maximum allowable quantity per control area. The storage and use of oxidizing materials in amounts not exceeding the *maximum allowable quantity per control area* indicated in Section 5003.1 shall be in accordance with Sections 5001, 5003, 6301 and 6303. Oxidizing gases shall also comply with Chapter 53.

6303.1.1 Special limitations for indoor storage and use by occupancy. The indoor storage and use of oxidizing materials shall be in accordance with Sections 6303.1.1.1 through 6303.1.1.3.

6303.1.1.1 Class 4 liquid and solid oxidizers. The storage and use of Class 4 liquid and solid oxidizers shall comply with Sections 6303.1.1.1.1 through 6303.1.1.1.4.

6303.1.1.1.1 Group A, E, I or U occupancies. In Group A, E, I or U occupancies, any amount of Class 4 liquid and solid oxidizers shall be stored in accordance with the following:

1. Class 4 liquid and solid oxidizers shall be stored in hazardous materials storage cabinets complying with Section 5003.8.7.
2. The hazardous materials storage cabinets shall not contain other storage.

6303.1.1.1.2 Group R occupancies. Class 4 liquid and solid oxidizers shall not be stored or used within Group R occupancies.

6303.1.1.1.3 Offices and retail sales areas. Class 4 liquid and solid oxidizers shall not be stored or used in offices or retail sales areas of Group B, F, M or S occupancies.

6303.1.1.1.4 Classrooms. In classrooms of Group B, F or M occupancies, any amount of Class 4 liquid and solid oxidizers shall be stored in accordance with the following:

1. Class 4 liquid and solid oxidizers shall be stored in hazardous materials storage cabinets complying with Section 5003.8.7.
2. Hazardous materials storage cabinets shall not contain other storage.

6303.1.1.2 Class 3 liquid and solid oxidizers. Not more than 200 pounds (91 kg) of solid or 20 gallons (76 L) of liquid Class 3 oxidizer is allowed in storage and use where such materials are necessary for maintenance purposes or operation of equipment. The oxidizers shall be stored in *approved* containers and in an *approved* manner.

6303.1.1.3 Oxidizing gases. Except for cylinders of nonliquefied *compressed gases* not exceeding a capacity of 250 cubic feet (7 m3) or liquefied *compressed gases* not exceeding a capacity of 46 pounds (21 kg) each used for maintenance purposes, patient care or operation of equipment, oxidizing gases shall not be stored or used in Group A, E, I or R occupancies or in offices in Group B occupancies.

The aggregate quantities of gases used for maintenance purposes and operation of equipment shall not exceed the *maximum allowable quantity per control area* listed in Table 5003.1.1(1).

Medical gas systems and medical gas supply cylinders shall also be in accordance with Section 5306.

6303.1.2 Emergency shutoff. *Compressed gas* systems conveying oxidizing gases shall be provided with *approved* manual or automatic emergency shutoff valves that can be activated at each point of use and at each source.

6303.1.2.1 Shutoff at source. A manual or automatic fail-safe emergency shutoff valve shall be installed on supply piping at the cylinder or bulk source. Manual or automatic cylinder valves are allowed to be used as the required emergency shutoff valve where the source of supply is limited to unmanifolded cylinder sources.

6303.1.2.2 Shutoff at point of use. A manual or automatic emergency shutoff valve shall be installed on the supply piping at the point of use or at a point where the equipment using the gas is connected to the supply system.

6303.1.3 Ignition source control. Ignition sources in areas containing oxidizing gases shall be controlled in accordance with Section 5003.7.

6303.2 Class 1 oxidizer storage configuration. The storage configuration of Class I liquid and solid oxidizers shall be as set forth in Table 6303.2.

SECTION 6304 STORAGE

6304.1 Indoor storage. Indoor storage of oxidizing materials in amounts exceeding the *maximum allowable quantity per control area* indicated in Table 5003.1.1(1) shall be in accordance with Sections 5001, 5003 and 5004 and this chapter.

6304.1.1 Explosion control. Indoor storage rooms, areas and buildings containing Class 4 liquid or solid oxidizers shall be provided with explosion control in accordance with Section 911.

6304.1.2 Automatic sprinkler system. The *automatic sprinkler system* for oxidizer storage shall be designed in accordance with NFPA 400.

6304.1.3 Liquid-tight floor. In addition to Section 5004.12, floors of storage areas for liquid and solid oxidizers shall be of liquid-tight construction.

6304.1.4 Smoke detection. An *approved* supervised smoke detection system in accordance with Section 907 shall be installed in liquid and solid oxidizer storage areas. Activation of the smoke detection system shall sound a local alarm.

Exception: Detached storage buildings protected by an *approved* automatic fire-extinguishing system.

6304.1.5 Storage conditions. The maximum quantity of oxidizers per building in storage buildings shall not exceed those quantities set forth in Tables 6304.1.5(1) through 6304.1.5(3).

The storage configuration for liquid and solid oxidizers shall be as set forth in Table 6303.2 and Tables 6304.1.5(1) through 6304.1.5(3).

Class 2 oxidizers shall not be stored in *basements* except where such storage is in stationary tanks.

Class 3 and 4 oxidizers in amounts exceeding the *maximum allowable quantity per control area* set forth in Section 5003.1 shall be stored on the ground floor only.

TABLE 6303.2
STORAGE OF CLASS 1 OXIDIZER LIQUIDS AND SOLIDS

STORAGE CONFIGURATION	LIMITS (feet)
Piles	
Maximum width	24
Maximum height	20
Maximum distance to aisle	12
Minimum distance to next pile[a]	4
Minimum distance to walls[b]	2
Maximum quantity per pile	200 tons
Maximum quantity per building	No Limit

For SI: 1 foot = 304.8 mm, 1 pound = 0.454 kg, 1 ton = 0.907185 metric ton.

a. The minimum aisle width shall be equal to the pile height, but not less than 4 feet and not greater than 8 feet.

b. There shall be no minimum distance from the pile to a wall for amounts less than 9,000 pounds.

TABLE 6304.1.5(1)
STORAGE OF CLASS 2 OXIDIZER LIQUIDS AND SOLIDS

STORAGE CONFIGURATION	LIMITS		
	Control area storage	Group H occupancy storage	Detached storage
Piles			
Maximum width	16 feet	25 feet	25 feet
Maximum height	Note a	Note a	Note a
Maximum distance to aisle	8 feet	12 feet	12 feet
Minimum distance to next pile	Note b	Note b	Note b
Minimum distance to walls	2 feet	2 feet[c]	2 feet[c]
Maximum quantity per pile	MAQ	100 tons	100 tons
Maximum quantity per building	MAQ	2000 tons	No Limit

For SI: 1 foot = 304.8 mm, 1 pound = 0.454 kg, 1 ton = 0.907185 metric ton.

a. Maximum storage height in nonsprinklered buidlings is limited to 6 feet. In sprinklered buildings see NFPA 400 for storage heights based on ceiling sprinkler protection.

b. The minimum aisle width shall be equal to the pile height, but not less than 4 feet and not greater than 8 feet.

c. For protection level and detached storage under 4,500 pounds, there shall be no minimum separation distance between the pile and any wall.

TABLE 6304.1.5(2)
STORAGE OF CLASS 3 OXIDIZER LIQUIDS AND SOLIDS

STORAGE CONFIGURATION	LIMITS		
	Control area storage	Group H occupancy storage	Detached storage
Piles			
Maximum width	12 feet	16 feet	20 feet
Maximum height	Note a	Note a	Note a
Maximum distance to aisle	8 feet	10 feet	10 feet
Minimum distance to next pile	Note b	Note b	Note b
Minimum distance to walls	4 feet	4 feet[c]	4 feet[c]
Maximum quantity per pile	NA	30 tons	100 tons
Maximum quantity per building	MAQ	1200 tons	No Limit

For SI: 1 foot = 304.8 mm, 1 pound = 0.454 kg, 1 ton = 0.907185 metric ton.

a. Maximum storage height in nonsprinklered buildings is limited to 6 feet. In sprinklered buildings see NFPA 400 for storage heights based on ceiling sprinkler protection.

b. The minimum aisle width shall be equal to the pile height, but not less than 4 feet and not greater than 8 feet.

c. For protection level and detached storage under 2,300 pounds, there shall be no minimum separation distance between the pile and any wall.

TABLE 6304.1.5(3)
STORAGE OF CLASS 4 OXIDIZER LIQUIDS AND SOLIDS

STORAGE CONFIGURATION	LIMITS (feet)
Piles	
Maximum length	10
Maximum width	4
Maximum height	8
Minimum distance to next pile	8
Maximum quantity per building	No Limit

For SI: 1 foot = 304.8 mm.

6304.1.6 Separation of Class 4 oxidizers from other materials. In addition to the requirements in Section 5003.9.8, Class 4 oxidizer liquids and solids shall be separated from other hazardous materials by not less than a 1-hour *fire barrier* or stored in hazardous materials storage cabinets.

6304.1.7 Contamination. Liquid and solid oxidizers shall not be stored on or against combustible surfaces. Liquid and solid oxidizers shall be stored in a manner to prevent contamination.

6304.1.8 Detached storage. Storage of liquid and solid oxidizers shall be in detached buildings where required by Section 5003.8.2.

6304.1.8.1 Separation distance. Detached storage buildings for Class 4 oxidizer liquids and solids shall be located not less than 50 feet (15 240 mm) from other hazardous materials storage.

6304.2 Outdoor storage. Outdoor storage of oxidizing materials in amounts exceeding the *maximum allowable quantities per control area* set forth in Table 5003.1.1(3) shall be in accordance with Sections 5001, 5003, 5004 and this chapter. Oxidizing gases shall also comply with Chapter 53.

6304.2.1 Distance from storage to exposures for oxidizing gases. Outdoor storage areas for oxidizing gases shall be located in accordance with Table 6304.2.2.

6304.2.1.1 Oxidizing cryogenic fluids. Outdoor storage areas for oxidizing *cryogenic fluids* shall be located in accordance with Chapter 55.

6304.2.2 Storage configuration for liquid and solid oxidizers. Storage configuration for liquid and solid oxidizers shall be in accordance with Table 6303.2 and Tables 6304.1.5(1) through 6304.1.5(3).

6304.2.3 Storage configuration for oxidizing gases. Storage configuration for oxidizing gases shall be in accordance with Table 6304.2.2.

SECTION 6305 USE

6305.1 Scope. The use of oxidizers in amounts exceeding the *maximum allowable quantity per control area* indicated in Table 5003.1.1(1) or 5003.1.1(3) shall be in accordance with Sections 5001, 5003, 5005 and this chapter. Oxidizing gases shall also comply with Chapter 53.

SECTION 6306 LIQUID OXYGEN IN HOME HEALTH CARE

6306.1 General. The storage and use of liquid oxygen (LOX) in home health care in Group I-1, I-4 and R occupancies shall comply with Sections 6306.2 through 6306.6, or shall be stored and used accordance with Chapter 50.

TABLE 6304.2.2
OXIDIZER GASES—DISTANCE FROM STORAGE TO EXPOSURES[a]

QUANTITY OF GAS STORED (cubic feet at NTP)	DISTANCE TO A BUILDING NOT ASSOCIATED WITH THE MANUFACTURE OR DISTRIBUTION OF OXIDIZING GASES OR PUBLIC WAY OR LOT LINE THAT CAN BE BUILT UPON (feet)	DISTANCE BETWEEN STORAGE AREAS (feet)
0 – 50,000	5	5
50,001 – 100,000	10	10
100,001 or greater	15	10

For SI: 1 foot = 304.8 mm, 1 cubic foot = 0.02832 m^3.

a. The minimum required distances shall not apply where fire barriers without openings or penetrations having a minimum fire-resistance rating of 2 hours interrupt the line of sight between the storage and the exposure. The configuration of the fire barrier shall be designed to allow natural ventilation to prevent the accumulation of hazardous gas concentrations.

6306.2 Information and instructions to be provided. The seller of liquid oxygen shall provide the user with information in written form that includes, but is not limited to, the following:

1. Manufacturer's instructions and labeling for safe storage and use of the containers.
2. Locating containers away from ignition sources, *exits,* electrical hazards and high-temperature devices in accordance with Section 6306.3.3.
3. Restraint of containers to prevent falling in accordance with Section 6306.3.4.
4. Requirements for handling containers in accordance with Section 6306.3.5.
5. Safeguards for refilling containers in accordance with Section 6306.3.6.
6. Signage requirements in accordance with Section 6306.6.

6306.3 Liquid oxygen home care containers. Containers of liquid oxygen in home health care shall be in accordance with Sections 6306.3.1 through 6306.3.6.

6306.3.1 Maximum individual container capacity. Liquid oxygen home care containers shall not exceed an individual capacity of 15.8 gallons (60 L) in Group I-1, I-4 and R occupancies. Liquid oxygen ambulatory containers are allowed in Group I-1, I-4 and R occupancies. Containers of liquid oxygen in home health care shall also be stored, used and filled in accordance with Section 6306 and Sections 5503.1 and 5503.2.

6306.3.2 Manufacturer's instructions and labeling. Containers shall be stored, used and operated in accordance with the manufacturer's instructions and labeling.

6306.3.3 Locating containers. Containers shall not be located in areas where any of the following conditions exist:

1. They can be overturned due to operation of a door.
2. They are in the direct path of egress.
3. They are subject to falling objects.
4. They can become part of an electrical circuit.
5. Open flames and high-temperature devices can cause a hazard.

6306.3.4 Restraining containers. Liquid oxygen home care containers shall be restrained while in storage or use to prevent falling caused by contact, vibration or seismic activity. Containers shall be restrained by one of the following methods:

1. Restraining containers to a fixed object with one or more restraints.
2. Restraining containers within a framework, stand or assembly designed to secure the container.
3. Restraining containers by locating a container against two points of contact such as the walls of a corner of a room or a wall and a secure furnishing or object such as a desk.

6306.3.5 Container handling. Containers shall be handled by use of a cart or hand truck designed for such use.

Exceptions:

1. Liquid oxygen home care containers equipped with a roller base.
2. Liquid oxygen ambulatory containers are allowed to be hand carried.

6306.3.6 Filling of containers. The filling of containers shall be in accordance with Sections 6306.3.6.1 through 6306.3.6.3.

6306.3.6.1 Filling location. Liquid oxygen home care containers and ambulatory containers shall be filled outdoors.

Exception: Liquid oxygen ambulatory containers are allowed to be filled indoors where the supply container is specifically designed for filling such containers and written instructions are provided by the container manufacturer.

6306.3.6.2 Incompatible surfaces. A drip pan compatible with liquid oxygen shall be provided under home care container fill and vent connections during the filling process in order to protect against liquid oxygen spillage from coming into contact with combustible surfaces, including asphalt.

6306.3.6.3 Open flames and high-temperature devices. The use of open flames and high-temperature devices shall be in accordance with Section 5003.7.2.

6306.4 Maximum aggregate quantity. The maximum aggregate quantity of liquid oxygen allowed in storage and in use in each *dwelling unit* shall be 31.6 gallons (120 L).

Exceptions:

1. The maximum aggregate quantity of liquid oxygen allowed in Group I-4 occupancies shall be limited by the maximum allowable quantity set forth in Table 5003.1.1(1).
2. Where individual sleeping rooms are separated from the remainder of the *dwelling unit* by *fire barriers* constructed in accordance with Section 707 of the *International Building Code*, and *horizontal assemblies* constructed in accordance with Section 711 of the *International Building Code*, or both, having a minimum *fire-resistance rating* of 1 hour, the maximum aggregate quantity per *dwelling unit* shall be increased to allow not more than 31.6 gallons (120 L) of liquid oxygen per sleeping room.

6306.5 Smoking prohibited. Smoking shall be prohibited in rooms or areas where liquid oxygen is in use.

6306.6 Signs. Warning signs for occupancies using home health care liquid oxygen shall be in accordance with Sections 6306.6.1 and 6306.6.2.

6306.6.1 No smoking sign. A sign stating "OXYGEN—NO SMOKING" shall be posted in each room or area where liquid oxygen containers are stored, used or filled.

6306.6.2 Premises signage. Where required by the *fire code official*, each *dwelling unit* or *sleeping unit* shall have an *approved* sign indicating that the unit contains liquid oxygen home care containers.

6306.7 Fire department notification. Where required by the *fire code official*, the liquid oxygen seller shall notify the fire department of the locations of liquid oxygen home care containers.

CHAPTER 64

PYROPHORIC MATERIALS

SECTION 6401 GENERAL

6401.1 Scope. The storage and use of pyrophoric materials shall be in accordance with this chapter. *Compressed gases* shall also comply with Chapter 53.

6401.2 Permits. Permits shall be required as set forth in Section 105.6.

SECTION 6402 DEFINITION

6402.1 Definition. The following term is defined in Chapter 2:

PYROPHORIC.

SECTION 6403 GENERAL REQUIREMENTS

6403.1 Quantities not exceeding the maximum allowable quantity per control area. The storage and use of pyrophoric materials in amounts not exceeding the *maximum allowable quantity per control area* indicated in Section 5003.1 shall be in accordance with Sections 5001, 5003, 6401 and 6403.

6403.1.1 Emergency shutoff. *Compressed gas* systems conveying pyrophoric gases shall be provided with *approved* manual or automatic emergency shutoff valves that can be activated at each point of use and at each source.

6403.1.1.1 Shutoff at source. An automatic emergency shutoff valve shall be installed on supply piping at the cylinder or bulk source. The shutoff valve shall be operated by a remotely located manually activated shutdown control located not less than 15 feet (4572 mm) from the source of supply. Manual or automatic cylinder valves are allowed to be used as the required emergency shutoff valve where the source of supply is limited to unmanifolded cylinder sources.

6403.1.1.2 Shutoff at point of use. A manual or automatic emergency shutoff valve shall be installed on the supply piping at the point of use or at a point where the equipment using the gas is connected to the supply system.

6403.2 Quantities exceeding the maximum allowable quantity per control area. The storage and use of pyrophoric materials in amounts exceeding the *maximum allowable quantity per control area* indicated in Section 5003.1 shall be in accordance with Chapter 50 and this chapter.

SECTION 6404 STORAGE

6404.1 Indoor storage. Indoor storage of pyrophoric materials in amounts exceeding the *maximum allowable quantity per control area* indicated in Table 5003.1.1(1), shall be in accordance with Sections 5001, 5003 and 5004 and this chapter.

The storage of silane gas, and gas mixtures with a silane concentration of 1.37 percent or more by volume, shall be in accordance with CGA G-13.

6404.1.1 Liquid-tight floor. In addition to the requirements of Section 5004.12, floors of storage areas containing pyrophoric liquids shall be of liquid-tight construction.

6404.1.2 Pyrophoric solids and liquids. Storage of pyrophoric solids and liquids shall be limited to a maximum area of 100 square feet (9.3 m^2) per pile. Storage shall not exceed 5 feet (1524 mm) in height. Individual containers shall not be stacked.

Aisles between storage piles shall be not less than 10 feet (3048 mm) in width.

Individual tanks or containers shall not exceed 500 gallons (1893 L) in capacity.

6404.1.3 Pyrophoric gases. Storage of pyrophoric gases shall be in detached buildings where required by Section 5003.8.2.

6404.1.4 Separation from incompatible materials. In addition to the requirements of Section 5003.9.8, indoor storage of pyrophoric materials shall be isolated from incompatible hazardous materials by 1-hour *fire barriers* with openings protected in accordance with the *International Building Code.*

Exception: Storage in *approved* hazardous materials storage cabinets constructed in accordance with Section 5003.8.7.

6404.2 Outdoor storage. Outdoor storage of pyrophoric materials in amounts exceeding the *maximum allowable quantity per control area* indicated in Table 5003.1.1(3) shall be in accordance with Sections 5001, 5003 and 5004, and this chapter.

The storage of silane gas, and gas mixtures with a silane concentration of 1.37 percent or more by volume, shall be in accordance with CGA G-13.

6404.2.1 Distance from storage to exposures. The separation of pyrophoric solids, liquids and gases from buildings, *lot lines*, public streets, public alleys, *public ways* or *means of egress* shall be in accordance with the following:

1. Solids and liquids. Two times the separation required by Chapter 57 for Class IB flammable liquids.
2. Gases. The location and maximum amount of pyrophoric gas per storage area shall be in accordance with Table 6404.2.1.

6404.2.2 Weather protection. Where overhead construction is provided for sheltering outdoor storage areas of pyrophoric materials, the storage areas shall be provided with *approved* automatic fire-extinguishing system protection.

SECTION 6405
USE

6405.1 General. The use of pyrophoric materials in amounts exceeding the *maximum allowable quantity per control area* indicated in Table 5003.1.1(1) or 5003.1.1(3) shall be in accordance with Sections 5001, 5003, 5005 and this chapter.

6405.2 Weather protection. Where overhead construction is provided for sheltering of outdoor use areas of pyrophoric materials, the use areas shall be provided with *approved* automatic fire-extinguishing system protection.

6405.3 Silane gas. The use of silane gas, and gas mixtures with a silane concentration of 1.37 percent or more by volume, shall be in accordance with CGA G-13.

TABLE 6404.2.1
PYROPHORIC GASES—DISTANCE FROM STORAGE TO EXPOSURES[a]

MAXIMUM AMOUNT PER STORAGE AREA (cubic feet)	MINIMUM DISTANCE BETWEEN STORAGE AREAS (feet)	MINIMUM DISTANCE TO LOT LINES OF PROPERTY THAT CAN BE BUILT UPON (feet)	MINIMUM DISTANCE TO PUBLIC STREETS, PUBLIC ALLEYS OR PUBLIC WAYS (feet)	MINIMUM DISTANCE TO BUILDINGS ON THE SAME PROPERTY		
				Nonrated construction or openings within 25 feet	Two-hour construction and no openings within 25 feet	Four-hour construction and no openings within 25 feet
250	5	25	5	5	0	0
2,500	10	50	10	10	5	0
7,500	20	100	20	20	10	0

For SI: 1 foot = 304.8 mm, 1 cubic foot = 0.02832 m^3.

a. The minimum required distances shall be reduced to 5 feet when protective structures having a minimum fire resistance of 2 hours interrupt the line of sight between the container and the exposure. The protective structure shall be at least 5 feet from the exposure. The configuration of the protective structure shall allow natural ventilation to prevent the accumulation of hazardous gas concentrations.

CHAPTER 65

PYROXYLIN (CELLULOSE NITRATE) PLASTICS

SECTION 6501 GENERAL

6501.1 Scope. This chapter shall apply to the storage and handling of plastic substances, materials or compounds with cellulose nitrate as a base, by whatever name known, in the form of blocks, sheets, tubes or fabricated shapes.

Cellulose nitrate motion picture film shall comply with the requirements of Section 306.

6501.2 Permits. Permits shall be required as set forth in Section 105.6.

SECTION 6502 DEFINITIONS

6502.1 Terms defined in Chapter 2. Words and terms used in this chapter and defined in Chapter 2 shall have the meanings ascribed to them as defined therein.

SECTION 6503 GENERAL REQUIREMENTS

6503.1 Displays. Cellulose nitrate (pyroxylin) plastic articles are allowed to be placed on tables not more than 3 feet (914 mm) wide and 10 feet (3048 mm) long. Tables shall be spaced at least 3 feet (914 mm) apart. Where articles are displayed on counters, they shall be arranged in a like manner.

6503.2 Space under tables. Spaces underneath tables shall be kept free from storage of any kind and accumulation of paper, refuse and other combustible material.

6503.3 Location. Sales or display tables shall be so located that in the event of a fire at the table, the table will not interfere with free *means of egress* from the room in not less than one direction.

6503.4 Lighting. Lighting shall not be located directly above cellulose nitrate (pyroxylin) plastic material, unless provided with a suitable guard to prevent heated particles from falling.

SECTION 6504 STORAGE AND HANDLING

6504.1 Raw material. Raw cellulose nitrate (pyroxylin) plastic material in a Group F building shall be stored and handled in accordance with Sections 6504.1.1 through 6504.1.7.

6504.1.1 Storage of incoming material. Where raw material in excess of 25 pounds (11 kg) is received in a building or *fire area*, an *approved* vented cabinet or *approved* vented vault equipped with an *approved automatic sprinkler system* shall be provided for the storage of material.

6504.1.2 Capacity limitations. Cabinets in any one workroom shall not contain more than 1,000 pounds (454 kg) of raw material. Each cabinet shall not contain more than 500 pounds (227 kg). Each compartment shall not contain more than 250 pounds (114 kg).

6504.1.3 Storage of additional material. Raw material in excess of that allowed by Section 6504.1.2 shall be kept in vented vaults not exceeding 1,500-cubic-foot capacity (43 m^3) of total vault space, and with *approved* construction, venting and sprinkler protection.

6504.1.4 Heat sources. Cellulose nitrate (pyroxylin) plastic shall not be stored within 2 feet (610 mm) of heat-producing appliances, steam pipes, radiators or chimneys.

6504.1.5 Accumulation of material. In factories manufacturing articles of cellulose nitrate (pyroxylin) plastics, *approved* sprinklered and vented cabinets, vaults or storage rooms shall be provided to prevent the accumulation in workrooms of raw stock in process or finished articles.

6504.1.6 Operators. In workrooms of cellulose nitrate (pyroxylin) plastic factories, operators shall not be stationed closer together than 3 feet (914 mm), and the amount of material per operator shall not exceed one shift's supply and shall be limited to the capacity of three tote boxes, including material awaiting removal or use.

6504.1.7 Waste material. Waste cellulose nitrate (pyroxylin) plastic materials such as shavings, chips, turnings, sawdust, edgings and trimmings shall be kept under water in metal receptacles until removed from the premises.

6504.2 Fire protection. The manufacture or storage of articles of cellulose nitrate (pyroxylin) plastic in quantities exceeding 100 pounds (45 kg) shall be located in a building or portion thereof equipped throughout with an *approved automatic sprinkler system* in accordance with Section 903.3.1.1.

6504.3 Sources of ignition. Sources of ignition shall not be located in rooms in which cellulose nitrate (pyroxylin) plastic in excess of 25 pounds (11 kg) is handled or stored.

6504.4 Heating. Rooms in which cellulose nitrate (pyroxylin) plastic is handled or stored shall be heated by low-pressure steam or hot water radiators.

CHAPTER 66

UNSTABLE (REACTIVE) MATERIALS

SECTION 6601 GENERAL

6601.1 Scope. The storage and use of unstable (reactive) materials shall be in accordance with this chapter. *Compressed gases* shall also comply with Chapter 53.

Exceptions:

1. Display and storage in Group M and storage in Group S occupancies complying with Section 5003.11.
2. Detonable unstable (reactive) materials shall be stored in accordance with Chapter 56.

6601.2 Permits. Permits shall be required as set forth in Section 105.6.

SECTION 6602 DEFINITION

6602.1 Definition. The following term is defined in Chapter 2:

UNSTABLE (REACTIVE) MATERIAL.
Class 4.
Class 3.
Class 2.
Class 1.

SECTION 6603 GENERAL REQUIREMENTS

6603.1 Quantities not exceeding the maximum allowable quantity per control area. Quantities of unstable (reactive) materials not exceeding the *maximum allowable quantity per control area* shall be in accordance with Sections 6603.1.1 through 6603.1.2.5.

6603.1.1 General. The storage and use of unstable (reactive) materials in amounts not exceeding the *maximum allowable quantity per control area* indicated in Section 5003.1 shall be in accordance with Sections 5001, 5003, 6601 and 6603.

6603.1.2 Limitations for indoor storage and use by occupancy. The indoor storage of unstable (reactive) materials shall be in accordance with Sections 6603.1.2.1 through 6603.1.2.5.

6603.1.2.1 Group A, E, I or U occupancies. In Group A, E, I or U occupancies, any amount of Class 3 and 4 unstable (reactive) materials shall be stored in accordance with the following:

1. Class 3 and 4 unstable (reactive) materials shall be stored in hazardous material storage cabinets complying with Section 5003.8.7.
2. The hazardous material storage cabinets shall not contain other storage.

6603.1.2.2 Group R occupancies. Class 3 and 4 unstable (reactive) materials shall not be stored or used within Group R occupancies.

6603.1.2.3 Group M occupancies. Class 4 unstable (reactive) materials shall not be stored or used in retail sales portions of Group M occupancies.

6603.1.2.4 Offices. Class 3 and 4 unstable (reactive) materials shall not be stored or used in offices of Group B, F, M or S occupancies.

6603.1.2.5 Classrooms. In classrooms in Group B, F or M occupancies, any amount of Class 3 and 4 unstable (reactive) materials shall be stored in accordance with the following:

1. Class 3 and 4 unstable (reactive) materials shall be stored in hazardous material storage cabinets complying with Section 5003.8.7.
2. The hazardous material storage cabinets shall not contain other storage.

6603.2 Quantities exceeding the maximum allowable quantity per control area. The storage and use of unstable (reactive) materials in amounts exceeding the *maximum allowable quantity per control area* indicated in Section 5003.1 shall be in accordance with Chapter 50 and this chapter.

SECTION 6604 STORAGE

6604.1 Indoor storage. Indoor storage of unstable (reactive) materials in amounts exceeding the *maximum allowable quantity per control area* indicated in Table 5003.1.1(1) shall be in accordance with Sections 5001, 5003, 5004 and this chapter.

In addition, Class 3 and 4 unstable (reactive) detonable materials shall be stored in accordance with the *International Building Code* requirements for *explosives*.

6604.1.1 Detached storage. Storage of unstable (reactive) materials shall be in detached buildings when required in Section 5003.8.2.

6604.1.2 Explosion control. Indoor storage rooms, areas and buildings containing Class 3 or 4 unstable (reactive) materials shall be provided with explosion control in accordance with Section 911.

6604.1.3 Liquid-tight floor. In addition to Section 5004.12, floors of storage areas for liquids and solids shall be of liquid-tight construction.

6604.1.4 Storage configuration. Unstable (reactive) materials stored in quantities greater than 500 cubic feet

(14 m^3) shall be separated into piles, each not larger than 500 cubic feet (14 m^3). Aisle width shall not be less than the height of the piles or 4 feet (1219 mm), whichever is greater.

Exception: Materials stored in tanks.

6604.1.5 Location in building. Unstable (reactive) materials shall not be stored in *basements*.

6604.2 Outdoor storage. Outdoor storage of unstable (reactive) materials in amounts exceeding the *maximum allowable quantities per control area* indicated in Table 5003.1.1(3) shall be in accordance with Sections 5001, 5003, 5004 and this chapter.

6604.2.1 Distance from storage to exposures Class 4 and 3 (detonable) materials. Outdoor storage of Class 4 or 3 (detonable) unstable (reactive) material shall be in accordance with Table 5604.5.2(2). The number of pounds of material listed in the table shall be the net weight of the material present. Alternatively, the number of pounds of material shall be based on a trinitrotoluene (TNT) equivalent weight.

6604.2.2 Distance from storage to exposures Class 3 (deflagratable) materials. Outdoor storage of deflagratable Class 3 unstable (reactive) materials shall be in accordance with Table 5604.5.2(3). The number of pounds of material listed shall be the net weight of the material present.

6604.2.3 Distance from storage to exposures Class 2 and 1 materials. Outdoor storage of Class 2 or 1 unstable (reactive) materials shall not be located within 20 feet (6096 mm) of buildings not associated with the manufacture or distribution of such materials, *lot lines*, public streets, public alleys, *public ways* or *means of egress*. The minimum required distance shall not apply when *fire barriers* without openings or penetrations having a minimum fire-resistance rating of 2 hours interrupt the line of sight between the storage and the exposure. The *fire barrier* shall either be an independent structure or the exterior wall of the building adjacent to the storage area.

6604.2.4 Storage configuration. Piles of unstable (reactive) materials shall not exceed 1,000 cubic feet (28 m^3).

6604.2.5 Aisle widths. Aisle widths between piles shall not be less than one-half the height of the pile or 10 feet (3048 mm), whichever is greater.

SECTION 6605 USE

6605.1 General. The use of unstable (reactive) materials in amounts exceeding the *maximum allowable quantity per control area* indicated in Table 5003.1.1(1) or 5003.1.1(3) shall be in accordance with Sections 5001, 5003, 5005 and this chapter.

CHAPTER 67

WATER-REACTIVE SOLIDS AND LIQUIDS

SECTION 6701
GENERAL

6701.1 Scope. The storage and use of water-reactive solids and liquids shall be in accordance with this chapter.

Exceptions:

1. Display and storage in Group M and storage in Group S occupancies complying with Section 5003.11.
2. Detonable water-reactive solids and liquids shall be stored in accordance with Chapter 56.

6701.2 Permits. Permits shall be required as set forth in Section 105.6.

SECTION 6702
DEFINITION

6702.1 Definition. The following term is defined in Chapter 2:

WATER-REACTIVE MATERIAL.
Class 3.
Class 2.
Class 1.

SECTION 6703
GENERAL REQUIREMENTS

6703.1 Quantities not exceeding the maximum allowable quantity per control area. The storage and use of water-reactive solids and liquids in amounts not exceeding the *maximum allowable quantity per control area* indicated in Section 5003.1 shall be in accordance with Sections 5001, 5003, 6701 and 6703.

6703.2 Quantities exceeding the maximum allowable quantity per control area. The storage and use of water-reactive solids and liquids in amounts exceeding the *maximum allowable quantity per control area* indicated in Section 5003.1 shall be in accordance with Chapter 50 and this chapter.

SECTION 6704
STORAGE

6704.1 Indoor storage. Indoor storage of water-reactive solids and liquids in amounts exceeding the *maximum allowable quantity per control area* indicated in Table 5003.1.1(1), shall be in accordance with Sections 5001, 5003, 5004 and this chapter.

6704.1.1 Detached storage. Storage of water-reactive solids and liquids shall be in detached buildings where required by Section 5003.8.2.

6704.1.2 Liquid-tight floor. In addition to the provisions of Section 5004.12, floors in storage areas for water-reactive solids and liquids shall be of liquid-tight construction.

6704.1.3 Waterproof room. Rooms or areas used for the storage of water-reactive solids and liquids shall be constructed in a manner which resists the penetration of water through the use of waterproof materials. Piping carrying water for other than *approved automatic sprinkler systems* shall not be within such rooms or areas.

6704.1.4 Water-tight containers. Where Class 3 water-reactive solids and liquids are stored in areas equipped with an *automatic sprinkler system*, the materials shall be stored in closed water-tight containers.

6704.1.5 Storage configuration. Water-reactive solids and liquids stored in quantities greater than 500 cubic feet (14 m^3) shall be separated into piles, each not larger than 500 cubic feet (14 m^3). Aisle widths between piles shall not be less than the height of the pile or 4 feet (1219 mm), whichever is greater.

Exception: Water-reactive solids and liquids stored in tanks.

Class 2 water-reactive solids and liquids shall not be stored in *basements* unless such materials are stored in closed water-tight containers or tanks.

Class 3 water-reactive solids and liquids shall not be stored in *basements*.

Class 2 or 3 water-reactive solids and liquids shall not be stored with flammable liquids.

6704.1.6 Explosion control. Indoor storage rooms, areas and buildings containing Class 2 or 3 water-reactive solids and liquids shall be provided with explosion control in accordance with Section 911.

6704.2 Outdoor storage. Outdoor storage of water-reactive solids and liquids in quantities exceeding the *maximum allowable quantity per control area* indicated in Table 5003.1.1(3) shall be in accordance with Sections 5001, 5003, 5004 and this chapter.

6704.2.1 General. Outdoor storage of water-reactive solids and liquids shall be within tanks or closed water-tight containers and shall be in accordance with Sections 6704.2.2 through 6704.2.5.

6704.2.2 Class 3 distance to exposures. Outdoor storage of Class 3 water-reactive solids and liquids shall not be within 75 feet (22 860 mm) of buildings, *lot lines*, public streets, public alleys, *public ways* or *means of egress*.

6704.2.3 Class 2 distance to exposures. Outdoor storage of Class 2 water-reactive solids and liquids shall not be within 20 feet (6096 mm) of buildings, *lot lines*, public streets, public alleys, *public ways* or *means of egress*. A 2-hour *fire barrier* without openings or penetrations, and extending not less than 30 inches (762 mm) above and to the sides of the storage area, is allowed in lieu of such distance. The wall shall either be an independent structure, or the exterior wall of the building adjacent to the storage area.

6704.2.4 Storage conditions. Class 3 water-reactive solids and liquids shall be limited to piles not greater than 500 cubic feet (14 m^3).

Class 2 water-reactive solids and liquids shall be limited to piles not greater than 1,000 cubic feet (28 m^3).

Aisle widths between piles shall not be less than one-half the height of the pile or 10 feet (3048 mm), whichever is greater.

6704.2.5 Containment. Secondary containment shall be provided in accordance with the provisions of Section 5004.2.2.

SECTION 6705 USE

6705.1 General. The use of water-reactive solids and liquids in amounts exceeding the *maximum allowable quantity per control area* indicated in Table 5003.1.1(1) or 5003.1.1(3) shall be in accordance with Sections 5001, 5003, 5005 and this chapter.

CHAPTERS 68 through 79

RESERVED

Part VI—Referenced Standards

CHAPTER 80

REFERENCED STANDARDS

This chapter lists the standards that are referenced in various sections of this document. The standards are listed herein by the promulgating agency of the standard, the standard identification, the effective date and title, and the section or sections of this document that reference the standard. The application of the referenced standards shall be as specified in Section 102.7.

AASHTO

American Association of State Highway and Transportation Officials
444 North Capitol Street, Northwest, #249
Washington, DC 20001

Standard reference number	Title	Referenced in code section number
HB-17—2002	Specification for Highway Bridges, 17th Edition 2002	503.2.6

AFSI

Architectural Fabric Structures Institute
c/o Industrial Fabric Association International
1801 County Road B West
Roseville, MN 55113

Standard reference number	Title	Referenced in code section number
ASI—77	Design and Standard Manual	3103.10.2

ANSI

American National Standards Institute
25 West 43rd Street, Fourth Floor
New York, NY 10036

Standard reference number	Title	Referenced in code section number
ANSI E1.21—2006	Entertainment Technology: Temporary Ground Supported Overhead Structures Used to Cover the Stage Areas and Support Equipment in the Production of Outdoor Entertainment Events	3105.1
ANSI Z21.69/ CSA 6.16—09	Connectors for Movable Gas Appliances	609.4

API

American Petroleum Institute
1220 L Street, Northwest
Washington, DC 20005

Standard reference number	Title	Referenced in code section number
Spec 12P—3rd Edition (Reaffirmed 2008)	Specification for Fiberglass Reinforced Plastic Tanks	5704.2.13.1.5
RP 651 3rd Edition (2007)	Cathodic Protection of Aboveground Petroleum Storage Tanks	5706.7, 5706.7.1
Std 653—4th Edition (2009)	Tank Inspection, Repair, Alteration and Reconstruction	5706.7

API—continued

RP 752— 3rd Edition (2009)	Management of Hazards Associated with Location of Process Plant Buildings, CMA Managers Guide	5706.7
RP 1604—3rd Edition (1996 R2010)	Closure of Underground Petroleum Storage Tanks	5704.2.13
RP 1615—(1996) 6th Edition (2011)	Installation of Underground-petroleum Storage Systems	5704.2.13.1.5, 5706.7
Std 2000—6th Edition (2009)	Venting Atmosphere and Low-pressure Storage Tanks: Nonrefrigerated and Refrigerated	5704.2.7.3.2
RP 2001—9th Edition (2012)	Fire Protection in Refineries, 8th Edition	5706.7
RP 2003— 7th Edition (2008)	Protection Against Ignitions Arising out of Static, Lightning and Stray Currents	5706.7
Publ 2009 7th Edition— (2002, R2012)	Safe Welding and Cutting Practices in Refineries, Gas Plants and Petrochemical Plants	5706.7
Std 2015—6th Edition 2001 (R2006)	Safe Entry and Clearing of Petroleum Storage Tanks	5706.7, 5706.7.2
RP 2023 3rd Edition— (2001, R2006)	Guide for Safe Storage and Handling of Heated Petroleum-derived Asphalt Products and Crude-oil Residue	5706.7, 5706.7.3
Publ 2028 3rd Edition— (2002, R2012)	Flame Arrestors in Piping Systems	5704.2.7.3.2
Publ 2201 5th Edition— (2003, R2010)	Procedures for Welding or Hot Tapping on Equipment in Service	5706.7
RP 2350— 4th Edition (2012)	Overfill Protection for Storage Tanks in Petroleum Facilities, 3rd Edition	5704.2.7.5.8, 5706.4.6, 5706.7

ASCE/SEI

American Society of Civil Engineers
Structural Engineering Institute
1801 Alexander Bell Drive
Reston, VA 20191

Standard reference number	Title	Referenced in code section number
ASCE/SEI 24—13	Flood Resistant Design and Construction	604.1.7

ASHRAE

American Society of Heating, Refrigerating and Air-Conditioning Engineers, Inc.
1791 Tullie Circle, NE
Atlanta, GA 30329

Standard reference number	Title	Referenced in code section number
15—2013	Safety Standard for Refrigeration Systems	606.12.1

ASME

The American Society of Mechanical Engineers
Two Park Avenue
New York, NY 10016-5990

Standard reference number	Title	Referenced in code section number
A13.1—2007	Scheme for the Identification of Piping Systems	3509.3, 5003.2.2.1, 5303.4.3, 5503.4.5, 5703.5.2
ASME A17.1/CSA B44—2013	Safety Code for Elevators and Escalators	508.1.6, 607.1, 907.3.3, 1009.4
A17.3—2008	Safety Code for Existing Elevators and Escalators	1103.3.1, 1103.3.2
B16.18—2012	Cast Copper-Alloy Solder Joint Pressure Fittings	909.13.1
B16.22—2001 (R2010)	Wrought Copper and Copper-Alloy Solder-joint Pressure Fittings	909.13.1
B31.1—2012	Power Piping	5003.2.2, Table 5703.6.2
B31.3—2012	Process Piping	5003.2.2.2, Table 5703.6.2

ASME—continued

B31.4—2012	Pipeline Transportation Systems for Liquid Hydrocarbons and Other Liquids	Table 5703.6.2
B31.9—2011	Building Services Piping	Table 5703.6.2, 5703.6.3, 5703.6.11
BPVC—2010/2011 addenda	ASME Boiler and Pressure Vessel Code (Sections I, II, IV, V & VI, VIII)	5003.2.1, 5303.2, 5303.3.2, 5503.2.6, 5503.4.3, 5503.7, 5704.2.13.1.5, 5806.3.1, 5806.4.1, 5806.4.8

ASSE

American Society of Safety Engineers
1800 East Oakton Street
Des Plaines, IL 60018

Standard reference number	Title	Referenced in code section number
ANSI/ASSE Z359.1-2007	Safety Requirements for Personal Fall Arrest Systems, Subsystems and Components, Part of the Fall Protection Code	1015.6, 1015.7

ASTM

ASTM International
100 Barr Harbor Drive
West Conshohocken, PA 19428-2959

Standard reference number	Title	Referenced in code section number
B 42—10	Specification for Seamless Copper Pipe, Standard Sizes	909.13.1
B 43—09	Specification for Seamless Red Brass Pipe, Standard Sizes	909.13.1
B 68—11	Specification for Seamless Copper Tube, Bright Annealed	909.13.1
B 88—09	Specification for Seamless Copper Water Tube	909.13.1
B 251—10	Specification for General Requirements for Wrought Seamless Copper and Copper-alloy Tube	909.13.1
B 280—08	Specification for Seamless Copper Tube for Air Conditioning and Refrigeration Field Service	909.13.1
D 56—05(2010)	Test Method for Flash Point by Tag Closed Tester	202
D 86—2012	Test Method for Distillation of Petroleum Products at Atmospheric Pressure	202
D 92—12b	Test Method for Flash and Fire Points by Cleveland Open Cup	202, 2401.2, 5104.1.1, 5701.2
D 93—12	Test Method for Flash Point by Pensky-Martens Closed Up Tester	202
D 323—08	Test Method for Vapor Pressure of Petroleum Products (Reid Method)	202
D 2859—06(2011)	Standard Test Method for Ignition Characteristics of Finished Textile Floor Covering Materials	804.3.3.1, 804.3.3.2
D 3278—96(2011)	Test Methods for Flash Point of Liquids by Small Scale Closed-cup Apparatus	202
E 84—2013A	Test Method for Surface Burning Characteristics of Building Materials	202, 803.1, 803.1.1, 803.1.2, 803.5.1, 803.5.2, 803.6, 803.7, 803.10, 804.1, 804.1.1, 804.2.4
E 108—2011	Test Methods for Fire Tests of Roof Coverings	317.3
E 681—2009	Test Method for Concentration Limits of Flammability of Chemicals (Vapors and Gases)	202
E 1354—2013	Standard Test Method for Heat and Visible Smoke Release Rates for Materials and Products Using an Oxygen Consumption Calorimeter	304.3.2, 304.3.4, 318.1, 808.1, 808.2, 2310.5.3, 3304.2.3, 3603.4
E 1529—13	Test Method for Determining Effects of Large Hydrocarbon Pool Fires on Structural Members and Assemblies	5704.2.9.2.3
E 1537—13	Test Method for Fire Testing of Upholstered Furniture	805.1.1.2, 805.2.1.2, 805.3.1.2, 805.4.1.2
E 1590—13	Test Method for Fire Testing of Mattresses	805.1.2.2, 805.2.2.2, 805.3.2.2.1, 805.4.2.2
E 1966—2012A	Test Method for Fire-resistant Joint Systems	202
E 2072—10	Standard Specification for Pholuminescent (Phosphorescent) Safety Markings	1025.4
E 2404—2013E1	Standard Practice for Specimen Preparation and Mounting of Textile, Paper or Vinyl Wall or Ceiling Coverings to Assess Surface Burning Characteristics	803.5.2, 803.6, 803.7
E 2573—12	Standard Practice for Specimen Preparation and Mounting of Site-fabricated Stretch Systems to Assess Surface Burning Characteristics	803.10
F 1085—10	Standard Specification for Mattress and Box Springs for Use in Berths in Marine Vessels	805.3.2.2.2
F 2006—10	Standard/Safety Specification for Window Fall Prevention Devices for Non-Emergency Escape (Egress) and Rescue (Ingress) Windows	1015.8

ASTM—continued

F 2090—10	Specification for Window Fall Prevention Devices with Emergency Escape (Egress) Release Mechanisms	1015.8, 1015.8.1
F 2200—13	Standard Specification for Automated Vehicular Gate Construction	503.5, 503.6

BHMA

Builders Hardware Manufacturers' Association
355 Lexington Avenue, 15th Floor
New York, NY 10017

Standard reference number	Title	Referenced in code section number
A156.10—2011	American National Standard for Power-operated Pedestrian Doors	1010.1.4.2
A156.19—2013	American National Standard for Power Assist and Low-energy Power-operated Doors	1010.1.4.2
A156.27—2011	Power and Manual Operated Revolving Pedestrian Doors	1010.1.4.1, 1010.1.4.3.1

CA

State of California Department of Consumer Affairs
Bureau of Electronics and Appliance Repair, Home Furnishings and Thermal Insulation
4244 South Market Court, Suite D
Sacramento, CA 95834-1243

Standard reference number	Title	Referenced in code section number
California Technical Bulletin 129—1992	Flammability Test Procedure for Mattresses for Use in Public Buildings	805.1.1.2, 805.2.2.2, 805.3.2.2.1, 805.4.2.2
California Technical Bulletin 133—1991	Flammability Test Procedure for Seating Furniture for Use in Public Occupancies	805.1.1.2, 805.2.1.2, 805.4.1.2

CGA

Compressed Gas Association
14501 George Carter Way, Suite 103
Chantilly, VA 20151

Standard reference number	Title	Referenced in code section number
C-7—(2011)	Guide to the Preparation of Precautionary Labeling and Marking of Compressed Gas Containers	5303.4.2, 5503.4.2
G-13—(2006)	Storage and Handling of Silane and Silane Mixtures (an American National Standard)	6404.1, 6404.2, 6405.3
P-1—(2000)	Safe Handling of Compressed Gases in Containers	5305.7
ANSI/P-18—(2006)	Standard for Bulk Inert Gas Systems	5501.1
S-1.1—(2011)	Relief Device Standards—Part 1—Cylinders for Compressed Gases	5303.3.2, 5503.2
S-1.2—(2005)	Pressure Relief Device Standards—Part 2—Cargo and Portable Tanks for Compressed Gases	5303.3.2, 5503.2
S-1.3—(2008)	Pressure Relief Device Standards—Part 3—Stationary Storage Containers for Compressed Gases	5303.3.2, 5503.2
V-1—(2005)	Standard for Gas Cylinder Valve Outlet and Inlet Connections	3505.2.1

CGR

Coast Guard Regulations
c/o Superintendent of Documents
U.S. Government Printing Office
Washington, DC 20402-9325

Standard reference number	Title	Referenced in code section number
46 CFR Parts 30, 32, 35 & 39—1999	Shipping	5706.8

CPSC

Consumer Product Safety Commission
4330 East West Highway
Bethesda, MD 20814

Standard reference number	Title	Referenced in code section number
16 CFR Part 1500.41—2009	Method for Testing Primary Irritant Substances	202
16 CFR Part 1500.42—2009	Test for Eye Irritants	202
16 CFR Part 1500.44—2009	Method for Testing Extremely Flammable and Flammable Solids	202
16 CFR Part 1500—2009	Hazardous Substances and Articles; Administration and Enforcement Regulations	202, 5601.1.3
16 CFR Part 1507—2002	Fireworks Devices	5601.1.3
16 CFR Part 1630—2007	Standard for the Surface Flammability of Carpets and Rugs	804.3.3.1, 804.3.3.2

DOC

U.S. Department of Commerce
1401 Constitution Avenue, NW
Washington, DC 20230

Standard reference number	Title	Referenced in code section number
16 CFR Part 1632—2009	Standard for the Flammability of Mattress and Mattress Pads (FF 4-72, Amended)	805.1.2.1, 805.2.2.1, 805.3.2.1, 805.4.2.1

DOL

U.S. Department of Labor
c/o Superintendent of Documents
U.S. Government Printing Office
Washington, DC 20402-9325

Standard reference number	Title	Referenced in code section number
29 CFR Part 1910.1000—2009	Air Contaminants	202, 2104.2.1
29 CFR Part 1910.1200—2009	Hazard Communication	202, 5603.6

DOTn

U.S. Department of Transportation
Office of Hazardous Material Safety
1200 New Jersey Avenue, SE
East Building, 2nd Floor
Washington, DC 20590

Standard reference number	Title	Referenced in code section number
33 CFR Part 154—1998	Facilities Transferring Oil or Hazardous Material in Bulk	5706.8
33 CFR Part 155—1998	Oil or Hazardous Material Pollution Prevention Regulations for Vessels	5706.8
33 CFR Part 156 —1998	Oil and Hazardous Material Transfer Operations	5706.8
49 CFR Parts 100-185—2005	Hazardous Materials Regulations	202, 3505.4, 5303.2, 5503.4.3, 5503.7, 5601.1, 5601.1.3, 5601.3, 5706.5.1.15
49 CFR Part 172—2009	Hazardous Materials Tables, Special Provisions, Hazardous Materials Communications, Emergency Response Information and Training Requirements	5604.6.5.2
49 CFR Part 173—2009	Shippers—General Requirements for Shipments and Packagings	5104.1.1,5606.3
49 CFR Part 173.137—2009	Shippers—General Requirements for Shipments and Packagings: Class 8—Assignment of Packing Group	202

DOTy

U.S. Department of Treasury
c/o Superintendent of Documents
U.S. Government Printing Office
Washington, DC 20402-9325

Standard reference number	Title	Referenced in code section number
27 CFR Part 55—1998	Commerce in Explosives, as amended through April 1, 1998	202, 5604.6.5.2

EN

European Committee for Standardization (EN)
Central Secretariat
Rue de Stassart 36
B-10 50 Brussels

Standard reference number	Title	Referenced in code section number
European Standard EN 1081	1998 Resilient Floor Coverings—Determination of the Electrical Resistance	2309.5.1.1

FCC

Federal Communications Commission
Wireless Telecommunications Bureau (WTB)
445 12th Street, SW
Washington, DC 20554

Standard reference number	Title	Referenced in code section number
47 CFR Part 90.219—2007	Private Land Mobile Radio Services—Use of Signal Boosters	510.5.4

FM

Factory Mutual Global Research
Standards Laboratories Department
1301 Atwood Avenue, P.O. Box 7500
Johnston, RI 02919

Standard reference number	Title	Referenced in code section number
4430—12	Approval Standard for Heat and Smoke Vents	910.3.1
ANSI/FM 4996—13	Approval Standard for Classification of Pallets and Other Material Handling Products as Equivalent to Wood Pallets	3206.4.1.1

ICC

International Code Council, Inc.
500 New Jersey Avenue, NW
6th Floor
Washington, DC 20001

Standard reference number	Title	Referenced in code section number
ICC A117.1—09	Accessible and Usable Buildings and Facilities	907.5.2.3.3, 1009.9, 1009.11, 1010.14.9.7, 1012.1, 1012.6.5, 1012.10, 1013.4, 1023.9
ICC 300—12	Standard on Bleachers, Folding and Telescopic Seating and Grandstands	1029.1.1, 1029.16
IBC—15	International Building Code®	102.3, 102.4, 201.3, 202, 304.1.3, 306.1, 311.1.1, 311.3, 313.1, 317.1, 403.8.2, 403.11.4, 404.2.1, 504.1, 508.1, 508.1.2, 603.2, 603.5.2, 603.6.1, 603.8, 604.1.2, 604.1.7, 604.2.2, 604.2.6, 604.2.7, 604.2.9, 604.2.13, 604.2.16, 605.11, 607.3, 607.4, 607.5, 607.6, 608.4, 608.8, 701.1, 704.1, 801.1, 803.1, Table 803.3, 803.7.2, 803.8.1, 803.8.2, 807.3, 807.5.1.2, 808.1, 808.2, 901.4.1, 901.4.2, 901.4.3, 901.8.2, 903.2, 903.2.5.2, 903.2.8.3.2, 903.2.9.1, 903.2.10, 903.3.1.1.1, 903.3.1.2, 903.3.2, 904.13, 907.1.1, 907.2.1, 907.2.1.9, 907.2.6.2, 907.2.6.3.3, 907.2.7, 907.2.13, 907.2.18, 907.2.21, 907.5.2.1, 907.5.2.2.4, 907.6.6, 909.1, 909.2, 909.3, 909.4.3, 909.5, 909.5.3, 909.5.3.1, 909.5.3.2, 909.6.3, 909.10.5, 909.11.1, 909.18.8, 909.21.1, 910.4.5, Table 911.1, 911.2, 914.1, 914.2.1, 914.3.1, 914.3.2, 914.4.1, 914.5.3, Table 914.8.3, 914.8.3.2, 914.10, 915.1.6, 1003.2, 1003.5, Table 1004.1.2, 1004.4, 1005.7.2, 1006.2.1, 1009.2, 1009.4, 1009.5, 1009.6.4, 1010.1.4.1, 1010.1.4.3, 1010.1.5, 1010.1.7, 1010.1.9.1, 1010.1.9.11, 1011.10, 1011.11, 1011.12.2, 1012.6.3, 1012.6.4, 1014.1, 1015.2, 1015.2.1, 1016.2, 1018.3, 1018.5, 1019.3, 1019.4,1020.1, Table 1020.1, 1021.4, 1023.2, 1023.3.1, 1023.4, 1023.5, 1023.6, 1023.7, 1023.11.1, 1023.11.2, 1024.3, 1024.4, 1024.5, 1024.6, 1024.7, 1026.2, 1026.3, 1027.5, 1028.1.6, 1029.1.1.1, 1101.2, 1101.3, 1103.1, 1103.3.2, 1103.4.1, 1103.4.8, 1103.4.9.1, 1103.4.9.2.1, 1103.4.9.2.2, 1103.4.9.4, 1103.4.9.5, 1103.4.10, 1104.5, 1104.17, 1104.17.1, Table 1104.17.2, 1105.3.1, 1105.3.3.1, 1105.3.3.2, 1105.4.7, 1105.6.2, 1105.6.3, 1105.6.4, 1105.6.5, 1105.6.6, 1105.6.7, 1105.7, 2004.6, 2006.17, 2007.1, 2007.4, 2103.3, 2107.1, 2301.1, 2301.4, 2303.1, 2307.4, 2308.3, 2308.3.1, 2308.3.1.2, 2309.3.1.5.1, 2309.3.2, 2309.6.1.2.3, 2310.1, 2311.1, 2311.3.1, 2311.4.1, 2404.2, 2404.3.1, 2404.3.2.6, 2404.3.3, 2405.2, 2701.1, 2701.4, 2703.2.2, 2703.3.1, 2703.3.2, 2703.3.3, 2703.3.4, 2703.3.8, 2703.14, 2703.14.1, 2703.14.2, 2703.15.1, 2704.3.1, 2705.2.3.2, 2705.3.1, 2705.3.2.1, 2705.3.3, 2803.1, 2905.1, 2909.2, 2909.4, 2909.6, 3101.1, 3103.1, 3103.8.2, 3103.8.4, 3103.9.1, 3104.1, 3105.5, 3314.1, 3403.1, 3704.3, 3704.4, 3704.5, 5003.2.2.2, 5003.2.8, 5003.8.1, Table 5003.8.2, 5003.8.3.1, 5003.8.4.1, 5003.9.9, 5004.13, 5005.2, 5005.3.9, 5101.1, 5303.16.1, 5303.16.2, 5306.2, 5306.2.1, 5503.1.2, 5503.5.2, 5504.2.1.2, 5504.2.2.2, 5505.4.1, 5604.2, Table 5604.5.2(3), 5605.5, 5701.3, 5704.2.7.7, 5704.2.8.1, 5704.2.8.2, 5704.2.9.3, 5704.2.9.4, 5704.3.3.5, 5704.3.7.1, 5704.3.8, 5705.3.4, 5705.3.5.3, 5705.3.7.1, 5705.3.7.2, 5705.3.7.3, 5705.3.7.4, 5705.3.7.5.1, 5706.2.3, 5706.4.1, 5803.1.1, 5806.4.3, 5808.1, 5808.3, 5808.3.2, 5906.2.2, 5906.2.3, 5906.4.2, 6003.1.4.2, 6005.3.1, 6109.11.2, 6204.1.2, 6306.4, 6404.1.4, 6604.1
IEBC—15	International Existing Building Code®	1011.5.2, 1103.1
IFGC—15	International Fuel Gas Code®	201.3, 603.1, 603.1.2, 603.5.2, 603.8, 2301.1, 2301.6, 2309.3.1.2, 2309.3.1.5, 2504.5, 3001.1, 3003.1, 3004.1, 3004.2, 3104.15.1, 3104.15.2, 3104.16.1, 3303.1, 3303.3, 3306.2.1, Table 5003.1.1(1), 5301.1, 5801.1, 5803.1.4, 6103.1, 6103.2.1.7, 6103.3

ICC—continued

IMC—15	International Mechanical Code®	201.3, 202, 308.3, 603.1, 603.1.2, 603.2, 603.3, 603.3.2.4, 603.5.2, 603.8, 606.1, 606.2, 606.3, 606.4, 606.7, 606.8, 606.9, 606.16, 608.6.1, 609.1, 903.2.11.4, 904.12, 907.3.1, 909.1, 909.10.2, 909.13.1, 910.4.7, 1006.2.2.3, 1011.16, 1020.5.1, 2104.2.1, 2105.3, 2301.1, 2301.6, 2309.3.1.2, 2309.3.2.3, 2311.3.1, 2311.4.3, 2311.7.1, 2404.7, 2404.7.2, 2504.5, 2703.2.2, 2703.10.4, 2703.14, 2803.2, 2803.3, 3001.1, 3003.1, 3004.2, 3104.15.1, 3104.15.2, 3303.1, 3703.5, Table 5003.1.1(1), 5003.8.4.2, 5003.8.5.2, 5003.8.6.2, 5004.3.1, 5303.7.6, 5303.16.9, 5305.5, 5306.2.2, 5307.5.1, 5504.2.1.3, 5504.2.2.3, 5505.4.1.1, 5701.3, 5703.6.1, 5704.2.8.9, 5705.3.7.5.1, 5706.2, 5706.4.4, 6003.1.3, 6003.2.3.2, 6004.2.2.7, 6005.3.1, 6005.3.2, 6103.2.1.7
IPC—15	International Plumbing Code®	201.3, 903.3.5, 904.11.1.3, 912.6, 2311.2.3, 5004.2.2.6
IPMC—15	International Property Maintenance Code®	311.1.1
IRC—15	International Residential Code®	102.5, 202, 605.11, 605.11.1.2, 1001.1
IWUIC—15	International Wildland-Urban Interface Code®	304.1.2

IIAR

International Institute of Ammonia Refrigeration
1001 N. Fairfax Street, Suite 503
Alexandria, VA 22314

Standard reference number	Title	Referenced in code section number
IIAR-2—2014	Equipment, Design and Installation of Closed-Circuit Ammonia Mechanical Refrigerating Systems	606.12.1.1
IIAR-7—2013	Developing Operating Procedures for Closed-Circuit Ammonia Mechanical Refrigerating Systems	606.12.1.1

IKECA

International Kitchen Exhaust Cleaning Association
100 North 20th Street, Suite 400
Philadelphia, PA 19103

Standard reference number	Title	Referenced in code section number
C10—2011 ANSI/IKECA	Standard for Cleaning of Commercial Kitchen Exhaust Systems	609.3.3.2

ISO

International Organization for Standardization (ISO)
ISO Central Secretariat
1 ch, de la Voie-Creuse, Case postale 56
CH-1211 Geneva 20, Switzerland

Standard reference number	Title	Referenced in code section number
ISO 8115—86	Cotton Bales—Dimensions and Density	Table 2704.2.2.1, Table 5003.1.1(1)

NEMA

National Electrical Manufacturer's Association
1300 N. 17th Street, Suite 1752
Rosslyn, VA 22209

Standard reference number	Title	Referenced in code section number
250—2003	Enclosures for Electrical Equipment (1,000 Volt Maximum)	6005.2

NFPA

National Fire Protection Association
1 Batterymarch Park
Quincy, MA 02169-7471

Standard reference number	Title	Referenced in code section number
02—11	Hydrogen Technologies Code	2309.3.1.1, 2309.3.1.2, 5301.1, 5307.3, 5801.1
10—13	Standard for Portable Fire Extinguishers	Table 901.6.1, 906.2, 906.3, Table 906.3(1), Table 906.3(2), 906.3.2, 906.3.4, 3006.3, I101.1
11—10	Standard for Low-, Medium- and High-expansion Foam	904.7, 5704.2.9.2.2
12—11	Standard on Carbon Dioxide Extinguishing Systems	Table 901.6.1, 904.8, 904.12
12A—09	Standard on Halon 1301 Fire Extinguishing Systems	Table 901.6.1, 904.9
13—13	Standard for the Installation of Sprinkler Systems	903.3.1.1, 903.3.2, 903.3.8.2 903.3.8.5, 904.12, 905.3.4, 907.6.4, 914.3.2, 1019.3, 1103.4.8, 3201.1, 3204.2, Table 3206.2, 3206.4.1, 3206.9, 3207.2, 3207.2.1, 3208.2.2, 3208.2.2.1, 3208.4, 3210.1, 3401.1, 5104.1, 5104.1.1, 5106.5.7, 5704.3.3.9, Table 5704.3.6.3(7), 5704.3.7.5.1, 5704.3.8.4
13D—13	Standard for the Installation of Sprinkler Systems in One- and Two-family Dwellings and Manufactured Homes	903.3.1.3
13R—13	Standard for the Installation of Sprinkler Systems in Low Rise Residential Occupancies	903.3.1.2, 903.3.5.2, 903.4
14—13	Standard for the Installation of Standpipe and Hose Systems	905.2, 905.3.4, 905.4.2, 905.6.2, 905.8
15—12	Standard for Water Spray Fixed Systems for Fire Protection	5704.2.9.2.3
16—15	Standard for the Installation of Foam-water Sprinkler and Foam-water Spray Systems	904.7, 904.12
17—13	Standard for Dry Chemical Extinguishing Systems	Table 901.6.1, 904.6, 904.12
17A—13	Standard for Wet Chemical Extinguishing Systems	Table 901.6.1, 904.5, 904.12
20—13	Standard for the Installation of Stationary Pumps for Fire Protection	913.1, 913.2, 913.5.1
22—13	Standard for Water Tanks for Private Fire Protection	507.2.2
24—13	Standard for Installation of Private Fire Service Mains and Their Appurtenances	507.2.1, 2809.5
25—14	Standard for the Inspection, Testing and Maintenance of Water-based Fire Protection Systems	507.5.3, Table 901.6.1, 904.7.1, 912.7, 913.5
30—12	Flammable and Combustible Liquids Code	610.1, 5701.2, 5703.6.2, 5703.6.2.1, 5704.2.7, 5704.2.7.1, 5704.2.7.2, 5704.2.7.3.2, 5704.2.7.4, 5704.2.7.6, 5704.2.7.7, 5704.2.7.8, 5704.2.7.9, 5704.2.9.3, 5704.2.9.4, 5704.2.9.6.1.1, 5704.2.9.6.1.2, 5704.2.9.6.1.3, 5704.2.9.6.1.4, 5704.2.9.6.1.5, 5704.2.9.6.2, 5704.2.9.7.3, 5704.2.10.2, 5704.2.11.3, 5704.2.11.4.2, 5704.2.12.1, 5704.3.1, 5704.3.6, Table 5704.3.6.3(1), Table 5704.3.6.3(2), Table 5704.3.6.3(3), 5704.3.7.2.3, 5704.3.8.4, 5706.8.3
30A—15	Code for Motor Fuel-dispensing Facilities and Repair Garages	2301.4, 2301.5, 2301.6, 2306.6.3, 2310.1
30B—15	Code for the Manufacture and Storage of Aerosol Products	5101.1, 5103.1, 5104.1, Table 5104.3.1, Table 5104.3.2, Table 5104.3.2.2, 5104.4.1, 5104.5.2, 5104.6, 5106.2.3 5106.3.2, Table 5106.4,5106.5.1, 5106.5.6, 5107.1
31—11	Standard for the Installation of Oil-burning Equipment	603.1.7, 603.3.1, 603.3.3
32—11	Standard for Dry Cleaning Plants	2107.1, 2107.3
33—15	Standard for Spray Application Using Flammable or Combustible Materials	2404.3.2
34—15	Standard for Dipping, Coating and Printing Processes Using Flammable or Combustible Liquids	2405.3, 2405.4.1.1
35—11	Standard for the Manufacture of Organic Coatings	2901.3, 2905.4
40—11	Standard for the Storage and Handling of Cellulose Nitrate Film	306.2
51—13	Standard for the Design and Installation of Oxygen-fuel Gas Systems for Welding, Cutting and Allied Processes	3501.5, 3507.1, 3509.1
51A—12	Standard for Acetylene Cylinder Charging Plants	3508.1
52—13	Vehicular Gaseous Fuel System Code	5301.1
55—13	Compressed Gases and Cryogenic Fluids Code	2309.2.1, 5301.1, 5307.3, 5501.1, 5801.1, 6301.1
56—12	Standard for Fire and Explosion Prevention during Cleaning and Purging of Flammable Gas Piping Systems	3306.2.1
58—14	Liquefied Petroleum Gas Code	603.4.2.1.1, 6101.1, 6103.1, 6103.2.1, 6103.2.1.2, 6103.2.1.7, 6103.2.2, 6104.1, 6104.3.2, 6104.4, 6105.2, 6106.2, 6106.3, 6107.2, 6107.4, 6108.1, 6108.2, 6109.11.2, 6111.3
59A—13	Standard for the Production, Storage and Handling of Liquefied Natural Gas (LNG)	5301.1, 5501.1

NFPA—continued

61—13	Standard for the Prevention of Fires and Dust Explosions in Agricultural and Food Processing Facilities	Table 2204.1
69—14	Standard on Explosion Prevention Systems	911.1, 911.3, Table 2204.1
70—14	National Electrical Code	603.1.3, 603.1.7, 603.5.2, 604.1.2, 605.3, 605.4, 605.9, 605.11, 606.16, 610.6, 610.7, 904.3.1, 907.6.1, 909.12.2, 909.16.3, 910.4.6, 2006.3.4, 2104.2.3, 2108.2, Table 2204.1, 2301.5, 2305.4, 2308.8.1.2.4, 2309.2.3, 2309.6.1.2.4, 2311.3.1, 2403.2.1, 2403.2.1.1, 2403.2.1.4, 2403.2.5, 2404.6.1.2.2, 2404.9.4, 2504.5, 2603.2.1, 2606.4, 2703.7.1, 2703.7.2, 2703.7.3, 2803.4, 2904.1, 3103.12.6.1, 3104.15.7, 3304.7, 3506.4, 5003.7.3, 5003.8.7.1, 5003.9.4, 5303.7.6, 5303.8, 5303.16.11, 5303.16.14, 5503.6, 5503.6.2, 5703.1, Table 5703.1.1, 5703.1.3, 5704.2.8.12, 5704.2.8.17, 5706.2.8, 5803.1.5, 5803.1.5.1, 5807.1.10, 5906.5.5, 5906.5.6, 6109.15.1
72—13	National Fire Alarm and Signaling Code	508.1.6, 604.2.4, Table 901.6.1, 903.4.1, 904.3.5, 907.2, 907.2.6, 907.2.9.3, 907.2.11, 907.2.13.2, 907.3, 907.3.3, 907.3.4, 907.5.2.1.2, 907.5.2.2, 907.5.2.2.5, 907.6, 907.6.1, 907.6.2, 907.6.6, 907.7, 907.7.1, 907.7.2, 907.8, 907.8.2, 907.8.5, 1103.3.2
80—13	Standard for Fire Doors and Other Opening Protectives	703.1.3, 1010.1.4.3
85—15	Boiler and Combustion System Hazards Code	Table 2204.1
86—15	Standard for Ovens and Furnaces	3001.1
92—15	Standard for Smoke Control Systems	909.7, 909.8
99—15	Health Care Facilities Code	611.1, 1105.5.2, 1105.10.1, 1105.10.2, 5306.4, 5306.5
101—15	Life Safety Code	1029.6.2
105—13	Standard for Smoke Door Assemblies and Other Opening Protectives	703.1.2
110—13	Standard for Emergency and Standby Power Systems	604.1.2, 604.4, 604.5, 913.5.2, 913.5.3
111—13	Standard on Stored Electrical Energy Emergency and Standby Power Systems	604.1.2, 604.4, 604.5
120—15	Standard for Fire Prevention and Control in Coal Mines	Table 2204.1
160—11	Standard for the Use of Flame Effects Before an Audience	308.3.2
170—15	Standard for Fire Safety and Emergency Symbols	1025.2.6.1
204—15	Standard for Smoke and Heat Venting	Table 901.6.1, 910.5.1
211—13	Standard for Chimneys, Fireplaces, Vents and Solid Fuel-burning Appliances	603.2
241—13	Standard for Safeguarding Construction, Alteration and Demolition Operations	3301.1
253—15	Standard Method of Test for Critical Radiant Flux of Floor Covering Systems Using a Radiant Heat Energy Source	804.3.1, 804.3.2, 804.4
260—13	Methods of Tests and Classification System for Cigarette Ignition Resistance of Components of Upholstered Furniture	805.1.1.1, 805.2.1.1, 805.3.1.1, 805.4.1.1
261—13	Standard Method of Test for Determining Resistance of Mock-up Upholstered Furniture Material Assemblies to Ignition by Smoldering Cigarettes	805.2.1.1, 805.3.1.1, 805.4.1.1
265—11	Standard Methods of Fire Tests for Evaluating Room Fire Growth Contribution of Textile Wall Coverings in Full Height Panels and Walls	803.5.1, 803.5.1.1, 803.5.1.2, 803.5.2, 803.6
286—15	Standard Methods of Fire Tests for Evaluating Contribution of Wall and Ceiling Interior Finish to Room Fire Growth	803.1, 803.1.2, 803.1.2.1, 803.5.1, 803.5.2, 803.6, 803.7
289—13	Standard Method of Fire Test for Individual Fuel Packages	806.2, 807.4, 807.5.1.1, 808.3
303—11	Fire Protection Standard for Marinas and Boatyards	905.3.7, 3603.5, 3603.6, 3604.2
318—15	Standard for the Protection of Semiconductor Fabrication Facilities	2703.16
326—10	Standard for the Safeguarding of Tanks and Containers for Entry, Cleaning, or Repair	3510.1
385—12	Standard for Tank Vehicles for Flammable and Combustible Liquids	5706.5.4.5, 5706.6, 5706.6.1
400—13	Hazardous Materials Code	5601.1.5, 6304.1.2, Table 6304.1.5(1), Table 6304.1.5(2)
407—12	Standard for Aircraft Fuel Servicing	2006.2, 2006.3
409—11	Standard for Aircraft Hangars	914.8.3, Table 914.8.3, 914.8.3.1, 914.8.6
410—10	Standard on Aircraft Maintenance	2004.7
484—15	Standard for Combustible Metals	Table 2204.1
495—13	Explosive Materials Code	202, 911.1, 911.4, 5601.1.1, 5601.1.5, 5604.2, 5604.6.2, 5604.6.3, 5604.7.1, 5605.1, 5606.1, 5606.5.2.1, 5606.5.2.3, 5607.1, 5607.9, 5607.11, 5607.15
498—13	Standard for Safe Havens and Interchange Lots for Vehicles Transporting Explosives	5601.1.2
505—13	Fire Safety Standard for Powered Industrial Trucks, Including Type Designations, Areas of Use, Maintenance and Operation	5003.7.3

NFPA—continued

654—13	Standard for Prevention of Fire and Dust Explosions from the Manufacturing, Processing and Handling of Combustible Particulate Solids	Table 2204.1
655—12	Standard for the Prevention of Sulfur Fires and Explosions	Table 2204.1
664—12	Standard for the Prevention of Fires and Explosions in Wood Processing and Woodworking Facilities	Table 2204.1, 2805.3
701—10	Standard Methods of Fire Tests for Flame-propagation of Textiles and Films	806.2, 807.4, 807.5.1.2, 2603.5, 3104.2
703—15	Standard for Fire Retardant-Wood and Fire-Retardant Coatings for Building Materials	803.4
704—12	Standard System for Identification of the Hazards of Materials for Emergency Response	606.7, 202, 3104.2, 5003.2.2.1, 5003.2.2.2, 5003.5, 5003.10.2, 5005.1.10, 5005.2.1.1, 5005.4.4, 5503.4.1, 5704.2.3.2,
720—15	Standard for the Installation of Carbon Monoxide (CO) Detection and Warning Equipment	915.5.1, 915.5.2, 915.6
750—14	Standard on Water Mist Fire Protection Systems	202, Table 901.6.1, 904.11.1.1
914—10	Code for Fire Protection of Historic Structures	1103.1.1
1122—13	Code for Model Rocketry	5601.1.4
1123—14	Code for Fireworks Display	202, 5604.2, 5608.1, 5608.2.2, 5608.5, 5608.6
1124—06	Code for the Manufacture, Transportation, Storage and Retail Sales of Fireworks and Pyrotechnic Articles	202, 5601.1.3, 5604.2, 5605.1, 5605.3, 5605.4, 5605.5, 5609.1
1125—12	Code for the Manufacture of Model Rocket and High Power Rocket Motors	5601.1.4
1126—11	Standard for the Use of Pyrotechnics Before a Proximate Audience	5604.2, 5605.1, 5608.1, 5608.2.2, 5608.4, 5608.5
1127—13	Code for High Power Rocketry	5601.1.4
2001—15	Standard on Clean Agent Fire Extinguishing Systems	Table 901.6.1, 904.10

UL

Underwriters Laboratories LLC
333 Pfingsten Road
Northbrook, IL 60062

Standard reference number	Title	Referenced in code section number
10C—09	Positive Pressure Fire Tests of Door Assemblies	1010.1.10.1
30—95	Metal Safety Cans—with revisions through July 2009	5003.9.10, 5005.1.10, 5705.2.4
58—96	Steel Underground Tanks for Flammable and Combustible Liquids—with revisions through July 1998	5704.2.13.1.5
80—07	Steel Tanks for Oil-Burner Fuels and Other Combustible Liquids—with revisions Through August 2009	610.2
87A—12	Outline of Investigation for Power-Operated Dispensing Devices for Gasoline and Gasoline/Ethanol Blends with Nominal Ethanol Concentrations up to 85 Percent	2306.8.1
142—06	Steel Aboveground Tanks for Flammable and Combustible Liquids—with revisions through February 12, 2010	610.2, 2306.2.3
199E—04	Outline of Investigation for Fire Testing of Sprinklers and Water Spray Nozzles for Protection of Deep Fat Fryers	904.12.4.1
217—06	Single and Multiple Station Smoke Alarms—with revisions through April 2012	907.2.11, 915.4.3
268—09	Smoke Detectors for Fire Alarm Systems	907.2.6.2, 907.2.11.7, 915.5.3
294—1999	Access Control System Units—with revisions through September 2010	1010.1.9.6, 1010.1.9.8, 1010.1.9.9
300—05(R2010)	Fire Testing of Fire Extinguishing Systems for Protection of Commercial Cooking Equipment—with revisions through July 16, 2010	904.12
300A—06	Outline of Investigation for Extinguishing System Units for Residential Range Top Cooking Surfaces	904.13
305—2012	Panic Hardware	1010.1.10.1
325—02	Door, Drapery, Gate, Louver and Window Operators and Systems—with revisions through June 2013	503.5, 503.6
499—05	Standard for Electrical Heating Appliances—with revisions through February 2013	610.6
710B—2011	Recirculating Systems	609.2, 904.12
723—08	Standard for Test for Surface Burning Characteristics of Building Materials—with revisions through September 2010	202, 803.5.1, 803.5.2, 803.6, 803.7, 803.10, 804.1, 804.2.4
790—04	Standard Test Methods for Fire Tests of Roof Coverings—with revisions through October 2008	317.2, 317.3
793—08	Automatically Operated Roof Vents for Smoke and Heat—with revisions through September 2011	910.3.1

UL—continued

864—03	Control Units and Accessories for Fire Alarm Systems—with revisions through August 2012	909.12, 2311.7.2.1.1, 5808.5.2, 6004.2.2.10.1
900—04	Air Filter Units—with revisions through February 2012	2404.7.8
924—06	Standard for Safety Emergency Lighting and Power Equipment—with revisions through February 2011	1013.5, 3103.12.6.1
1037—99	Antitheft Alarms and Devices—with revisions through December 2009	506.1
1275—05	Flammable Liquid Storage Cabinets—with revisions through February 2010	5003.8.7.1, 5704.3.2.1.1
1313—93	Standard for Nonmetallic Safety Cans for Petroleum Products—with revisions through November 2012	5003.9.10
1315—95	Standard for Safety for Metal Waste Paper Containers—with revisions through September 2012	808.1, 808.2
1316—94	Glass Fiber Reinforced Plastic Underground Storage Tanks for Petroleum Products, Alcohols, and Alcohol-gasoline Mixtures—with revisions through May 2006	5704.2.13.1.5
1363—07	Relocatable Power Taps—with revisions through September 2012	605.4.1
1975—06	Fire Tests for Foamed Plastics Used for Decorative Purpose	807.4.2.1, 808.3
1994—04	Standard for Luminous Egress Path Marking Systems—with revisions through November 2010	1008.2.1, 1025.2.1, 1025.2.3, 1025.2.4, 1025.4
2017—08	General-Purpose Signaling Devices and Systems—with revisions through May 2011	2311.7.2.1.1, 5808.5.2, 6004.2.2.10.1
2034—08	Single and Multiple Station Carbon Monoxide Alarms—with revisions through February 2009	915.4.2, 915.4.3, 1103.9
2075—2013	Standard for Gas and Vapor Detectors and Sensors	915.5.1, 915.5.3, 2311.7.2.1.1, 5808.5.2, 6004.2.2.10.1
2079—04	Tests for Fire Resistance of Building Joint Systems—with revisions through December 2012	202
2085—97	Protected Above-ground Tanks for Flammable and Combustible Liquids—with revisions through September 2010	202, 2306.2.2, 2306.2.3, 5704.2.7.4, 5704.2.9.2.3, 5704.2.9.7.4, 5705.3.8.2
2196—2001	Tests for Fire Resistive Cables—with revisions through March 2012	604.3, 913.2.2
2200—2012	Stationary Engine Generator Assemblies—with revisions through June 2013	604.1.1
2208—2010	Solvent Distillation Units—with revisions through March 2011	5705.4.1
2245—06	Below-grade Vaults for Flammable Liquid Storage Tanks	5704.2.8.1
2335—10	Fire Tests of Storage Pallets—with revisions through September 2012	3206.4.1.1, 3208.2.1
2360—00	Test Methods for Determining the Combustibility Characteristics of Plastics Used in Semi-Conductor Tool Construction—with revisions through May 2013	2703.10.1.2

USC

United States Code
c/o Superintendent of Documents
U.S. Government Printing Office
Washington, DC 20402-9325

Standard reference number	Title	Referenced in code section number
18 USC Part 1, Chapter 40	Importation, Manufacture, Distribution and Storage of Explosive Materials	202
21 USC Chapter 9	United States Food, Drug and Cosmetic Act	4002.1

Part VII—Appendices

APPENDIX A

BOARD OF APPEALS

The provisions contained in this appendix are not mandatory unless specifically referenced in the adopting ordinance.

SECTION A101 GENERAL

A101.1 Scope. A board of appeals shall be established within the jurisdiction for the purpose of hearing applications for modification of the requirements of the *International Fire Code* pursuant to the provisions of Section 108 of the *International Fire Code*. The board shall be established and operated in accordance with this section, and shall be authorized to hear evidence from appellants and the *fire code official* pertaining to the application and intent of this code for the purpose of issuing orders pursuant to these provisions.

A101.2 Membership. The membership of the board shall consist of five voting members having the qualifications established by this section. Members shall be nominated by the *fire code official* or the chief administrative officer of the jurisdiction, subject to confirmation by a majority vote of the governing body. Members shall serve without remuneration or compensation, and shall be removed from office prior to the end of their appointed terms only for cause.

A101.2.1 Design professional. One member shall be a practicing design professional registered in the practice of engineering or architecture in the state in which the board is established.

A101.2.2 Fire protection engineering professional. One member shall be a qualified engineer, technologist, technician or safety professional trained in fire protection engineering, fire science or fire technology. Qualified representatives in this category shall include fire protection contractors and certified technicians engaged in *fire protection system* design.

A101.2.3 Industrial safety professional. One member shall be a registered industrial or chemical engineer, certified hygienist, certified safety professional, certified hazardous materials manager or comparably qualified specialist experienced in chemical process safety or industrial safety.

A101.2.4 General contractor. One member shall be a contractor regularly engaged in the construction, *alteration*, maintenance, repair or remodeling of buildings or building services and systems regulated by the code.

A101.2.5 General industry or business representative. One member shall be a representative of business or industry not represented by a member from one of the other categories of board members described above.

A101.3 Terms of office. Members shall be appointed for terms of 4 years. No member shall be reappointed to serve more than two consecutive full terms.

A101.3.1 Initial appointments. Of the members first appointed, two shall be appointed for a term of 1 year, two for a term of 2 years, one for a term of 3 years.

A101.3.2 Vacancies. Vacancies shall be filled for an unexpired term in the manner in which original appointments are required to be made. Members appointed to fill a vacancy in an unexpired term shall be eligible for reappointment to two full terms.

A101.3.3 Removal from office. Members shall be removed from office prior to the end of their terms only for cause. Continued absence of any member from regular meetings of the board shall, at the discretion of the applicable governing body, render any such member liable to immediate removal from office.

A101.4 Quorum. Three members of the board shall constitute a quorum. In varying the application of any provisions of this code or in modifying an order of the *fire code official*, affirmative votes of the majority present, but not less than three, shall be required.

A101.5 Secretary of board. The *fire code official* shall act as secretary of the board and shall keep a detailed record of all its proceedings, which shall set forth the reasons for its decisions, the vote of each member, the absence of a member and any failure of a member to vote.

A101.6 Legal counsel. The jurisdiction shall furnish legal counsel to the board to provide members with general legal advice concerning matters before them for consideration. Members shall be represented by legal counsel at the jurisdiction's expense in all matters arising from service within the scope of their duties.

A101.7 Meetings. The board shall meet at regular intervals, to be determined by the chairman. In any event, the board shall meet within 10 days after notice of appeal has been received.

A101.8 Conflict of interest. Members with a material or financial interest in a matter before the board shall declare such interest and refrain from participating in discussions, deliberations and voting on such matters.

A101.9 Decisions. Every decision shall be promptly filed in writing in the office of the *fire code official* and shall be open to public inspection. A certified copy shall be sent by mail or otherwise to the appellant, and a copy shall be kept publicly posted in the office of the *fire code official* for 2 weeks after filing.

A101.10 Procedures. The board shall be operated in accordance with the Administrative Procedures Act of the state in which it is established or shall establish rules and regulations for its own procedure not inconsistent with the provisions of this code and applicable state law.

APPENDIX B

FIRE-FLOW REQUIREMENTS FOR BUILDINGS

The provisions contained in this appendix are not mandatory unless specifically referenced in the adopting ordinance.

SECTION B101 GENERAL

B101.1 Scope. The procedure for determining fire-flow requirements for buildings or portions of buildings hereafter constructed shall be in accordance with this appendix. This appendix does not apply to structures other than buildings.

SECTION B102 DEFINITIONS

B102.1 Definitions. For the purpose of this appendix, certain terms are defined as follows:

FIRE-FLOW. The flow rate of a water supply, measured at 20 pounds per square inch (psi) (138 kPa) residual pressure, that is available for fire fighting.

FIRE-FLOW CALCULATION AREA. The floor area, in square feet (m^2), used to determine the required fire flow.

SECTION B103 MODIFICATIONS

B103.1 Decreases. The fire chief is authorized to reduce the fire-flow requirements for isolated buildings or a group of buildings in rural areas or small communities where the development of full fire-flow requirements is impractical.

B103.2 Increases. The fire chief is authorized to increase the fire-flow requirements where conditions indicate an unusual susceptibility to group fires or conflagrations. An increase shall not be more than twice that required for the building under consideration.

B103.3 Areas without water supply systems. For information regarding water supplies for fire-fighting purposes in rural and suburban areas in which adequate and reliable water supply systems do not exist, the *fire code official* is authorized to utilize NFPA 1142 or the *International Wildland-Urban Interface Code.*

SECTION B104 FIRE-FLOW CALCULATION AREA

B104.1 General. The fire-flow calculation area shall be the total floor area of all floor levels within the *exterior walls*, and under the horizontal projections of the roof of a building, except as modified in Section B104.3.

B104.2 Area separation. Portions of buildings which are separated by *fire walls* without openings, constructed in accordance with the *International Building Code,* are allowed to be considered as separate fire-flow calculation areas.

B104.3 Type IA and Type IB construction. The fire-flow calculation area of buildings constructed of Type IA and Type IB construction shall be the area of the three largest successive floors.

> **Exception:** Fire-flow calculation area for open parking garages shall be determined by the area of the largest floor.

SECTION B105 FIRE-FLOW REQUIREMENTS FOR BUILDINGS

B105.1 One- and two-family dwellings, Group R-3 and R-4 buildings and townhouses. The minimum fire-flow and flow duration requirements for one- and two-family *dwellings*, Group R-3 and R-4 buildings and townhouses shall be as specified in Tables B105.1(1) and B105.1(2).

B105.2 Buildings other than one- and two-family dwellings, Group R-3 and R-4 buildings and townhouses. The minimum fire-flow and flow duration for buildings other than one- and two-family *dwellings*, Group R-3 and R-4 buildings and townhouses shall be as specified in Tables B105.2 and B105.1(2).

TABLE B105.1(1)
REQUIRED FIRE-FLOW FOR ONE- AND TWO-FAMILY DWELLINGS, GROUP R-3 AND R-4 BUILDINGS AND TOWNHOUSES

FIRE-FLOW CALCULATION AREA (square feet)	AUTOMATIC SPRINKLER SYSTEM (Design Standard)	MINIMUM FIRE-FLOW (gallons per minute)	FLOW DURATION (hours)
0-3,600	No automatic sprinkler system	1,000	1
3,601 and greater	No automatic sprinkler system	Value in Table B105.1(2)	Duration in Table B105.1(2) at the required fire-flow rate
0-3,600	Section 903.3.1.3 of the *International Fire Code* or Section P2904 of the *International Residential Code*	500	$^1/_2$
3,601 and greater	Section 903.3.1.3 of the *International Fire Code* or Section P2904 of the *International Residential Code*	$^1/_2$ value in Table B105.1(2)	1

For SI: 1 square foot = 0.0929 m^2, 1 gallon per minute = 3.785 L/m.

TABLE B105.1(2)
REFERENCE TABLE FOR TABLES B105.1(1) AND B105.2

FIRE-FLOW CALCULATION AREA (square feet)					FIRE-FLOW (gallons per minute)[b]	FLOW DURATION (hours)
Type IA and IB[a]	Type IIA and IIIA[a]	Type IV and V-A[a]	Type IIB and IIIB[a]	Type V-B[a]		
0-22,700	0-12,700	0-8,200	0-5,900	0-3,600	1,500	2
22,701-30,200	12,701-17,000	8,201-10,900	5,901-7,900	3,601-4,800	1,750	
30,201-38,700	17,001-21,800	10,901-12,900	7,901-9,800	4,801-6,200	2,000	
38,701-48,300	21,801-24,200	12,901-17,400	9,801-12,600	6,201-7,700	2,250	
48,301-59,000	24,201-33,200	17,401-21,300	12,601-15,400	7,701-9,400	2,500	
59,001-70,900	33,201-39,700	21,301-25,500	15,401-18,400	9,401-11,300	2,750	
70,901-83,700	39,701-47,100	25,501-30,100	18,401-21,800	11,301-13,400	3,000	3
83,701-97,700	47,101-54,900	30,101-35,200	21,801-25,900	13,401-15,600	3,250	
97,701-112,700	54,901-63,400	35,201-40,600	25,901-29,300	15,601-18,000	3,500	
112,701-128,700	63,401-72,400	40,601-46,400	29,301-33,500	18,001-20,600	3,750	
128,701-145,900	72,401-82,100	46,401-52,500	33,501-37,900	20,601-23,300	4,000	4
145,901-164,200	82,101-92,400	52,501-59,100	37,901-42,700	23,301-26,300	4,250	
164,201-183,400	92,401-103,100	59,101-66,000	42,701-47,700	26,301-29,300	4,500	
183,401-203,700	103,101-114,600	66,001-73,300	47,701-53,000	29,301-32,600	4,750	
203,701-225,200	114,601-126,700	73,301-81,100	53,001-58,600	32,601-36,000	5,000	
225,201-247,700	126,701-139,400	81,101-89,200	58,601-65,400	36,001-39,600	5,250	
247,701-271,200	139,401-152,600	89,201-97,700	65,401-70,600	39,601-43,400	5,500	
271,201-295,900	152,601-166,500	97,701-106,500	70,601-77,000	43,401-47,400	5,750	
295,901-Greater	166,501-Greater	106,501-115,800	77,001-83,700	47,401-51,500	6,000	
—	—	115,801-125,500	83,701-90,600	51,501-55,700	6,250	
—	—	125,501-135,500	90,601-97,900	55,701-60,200	6,500	
—	—	135,501-145,800	97,901-106,800	60,201-64,800	6,750	
—	—	145,801-156,700	106,801-113,200	64,801-69,600	7,000	
—	—	156,701-167,900	113,201-121,300	69,601-74,600	7,250	
—	—	167,901-179,400	121,301-129,600	74,601-79,800	7,500	
—	—	179,401-191,400	129,601-138,300	79,801-85,100	7,750	
—	—	191,401-Greater	138,301-Greater	85,101-Greater	8,000	

For SI: 1 square foot = 0.0929 m^2, 1 gallon per minute = 3.785 L/m, 1 pound per square inch = 6.895 kPa.

a. Types of construction are based on the *International Building Code*.

b. Measured at 20 psi residual pressure.

TABLE B105.2
REQUIRED FIRE-FLOW FOR BUILDINGS OTHER THAN ONE- AND TWO-FAMILY DWELLINGS, GROUP R-3 AND R-4 BUILDINGS AND TOWNHOUSES

AUTOMATIC SPRINKLER SYSTEM (Design Standard)	MINIMUM FIRE-FLOW (gallons per minute)	FLOW DURATION (hours)
No automatic sprinkler system	Value in Table B105.1(2)	Duration in Table B105.1(2)
Section 903.3.1.1 of the *International Fire Code*	25% of the value in Table B105.1(2)[a]	Duration in Table B105.1(2) at the reduced flow rate
Section 903.3.1.2 of the *International Fire Code*	25% of the value in Table B105.1(2)[b]	Duration in Table B105.1(2) at the reduced flow rate

For SI: 1 gallon per minute = 3.785 L/m.

a. The reduced fire-flow shall be not less than 1,000 gallons per minute.

b. The reduced fire-flow shall be not less than 1,500 gallons per minute.

B105.3 Water supply for buildings equipped with an automatic sprinkler system. For buildings equipped with an approved *automatic sprinkler system*, the water supply shall be capable of providing the greater of:

1. The *automatic sprinkler system* demand, including hose stream allowance.
2. The required fire-flow.

SECTION B106
REFERENCED STANDARDS

ICC	IBC—15	International Building Code	B104.2,
ICC	IFC—15	International Fire Code	Tables B105.1(1) and B105.2
ICC	IWUIC—15	International Wildland-Urban Interface Code	B103.3
ICC	IRC—15	International Residential Code	Table B105.1(1)
NFPA	1142—12	Standard on Water Supplies for Suburban and Rural Fire Fighting	B103.3

APPENDIX C

FIRE HYDRANT LOCATIONS AND DISTRIBUTION

The provisions contained in this appendix are not mandatory unless specifically referenced in the adopting ordinance.

SECTION C101 GENERAL

C101.1 Scope. In addition to the requirements of Section 507.5.1 of the *International Fire Code*, fire hydrants shall be provided in accordance with this appendix for the protection of buildings, or portions of buildings, hereafter constructed or moved into the jurisdiction.

SECTION C102 NUMBER OF FIRE HYDRANTS

C102.1 Minimum number of fire hydrants for a building. The number of fire hydrants available to a building shall be not less than the minimum specified in Table C102.1.

SECTION C103 FIRE HYDRANT SPACING

C103.1 Hydrant spacing. Fire apparatus access roads and public streets providing required access to buildings in accordance with Section 503 of the *International Fire Code* shall be provided with one or more fire hydrants, as determined by Section C102.1. Where more than one fire hydrant is required, the distance between required fire hydrants shall be in accordance with Sections C103.2 and C103.3.

C103.2 Average spacing. The average spacing between fire hydrants shall be in accordance with Table C102.1.

Exception: The average spacing shall be permitted to be increased by 10 percent where existing fire hydrants provide all or a portion of the required number of fire hydrants.

C103.3 Maximum spacing. The maximum spacing between fire hydrants shall be in accordance with Table C102.1.

SECTION C104 CONSIDERATION OF EXISTING FIRE HYDRANTS

C104.1 Existing fire hydrants. Existing fire hydrants on public streets are allowed to be considered as available to meet the requirements of Sections C102 and C103. Existing fire hydrants on adjacent properties are allowed to be considered as available to meet the requirements of Sections C102 and C103 provided that a fire apparatus access road extends between properties and that an easement is established to prevent obstruction of such roads.

TABLE C102.1
REQUIRED NUMBER AND SPACING OF FIRE HYDRANTS

FIRE-FLOW REQUIREMENT (gpm)	MINIMUM NUMBER OF HYDRANTS	AVERAGE SPACING BETWEEN HYDRANTS[a, b, c, f, g] (feet)	MAXIMUM DISTANCE FROM ANY POINT ON STREET OR ROAD FRONTAGE TO A HYDRANT[d, f, g]
1,750 or less	1	500	250
2,000-2,250	2	450	225
2,500	3	450	225
3,000	3	400	225
3,500-4,000	4	350	210
4,500-5,000	5	300	180
5,500	6	300	180
6,000	6	250	150
6,500-7,000	7	250	150
7,500 or more	8 or more[e]	200	120

For SI: 1 foot = 304.8 mm, 1 gallon per minute = 3.785 L/m.

a. Reduce by 100 feet for dead-end streets or roads.

b. Where streets are provided with median dividers that cannot be crossed by fire fighters pulling hose lines, or where arterial streets are provided with four or more traffic lanes and have a traffic count of more than 30,000 vehicles per day, hydrant spacing shall average 500 feet on each side of the street and be arranged on an alternating basis.

c. Where new water mains are extended along streets where hydrants are not needed for protection of structures or similar fire problems, fire hydrants shall be provided at spacing not to exceed 1,000 feet to provide for transportation hazards.

d. Reduce by 50 feet for dead-end streets or roads.

e. One hydrant for each 1,000 gallons per minute or fraction thereof.

f. A 50-percent spacing increase shall be permitted where the building is equipped throughout with an approved automatic sprinkler system in accordance with Section 903.3.1.1 of the *International Fire Code*.

g. A 25-percent spacing increase shall be permitted where the building is equipped throughout with an approved automatic sprinkler system in accordance with Section 903.3.1.2 or 903.3.1.3 of the *International Fire Code* or Section P2904 of the *International Residential Code*.

SECTION C105
REFERENCED STANDARDS

ICC	IFC—15	International Fire Code	C101.1, C103.1, Table C102.1
ICC	IRC—15	International Residential Code	Table C102.1

APPENDIX D

FIRE APPARATUS ACCESS ROADS

The provisions contained in this appendix are not mandatory unless specifically referenced in the adopting ordinance.

SECTION D101 GENERAL

D101.1 Scope. Fire apparatus access roads shall be in accordance with this appendix and all other applicable requirements of the *International Fire Code*.

SECTION D102 REQUIRED ACCESS

D102.1 Access and loading. Facilities, buildings or portions of buildings hereafter constructed shall be accessible to fire department apparatus by way of an *approved* fire apparatus access road with an asphalt, concrete or other *approved* driving surface capable of supporting the imposed load of fire apparatus weighing at least 75,000 pounds (34 050 kg).

SECTION D103 MINIMUM SPECIFICATIONS

D103.1 Access road width with a hydrant. Where a fire hydrant is located on a fire apparatus access road, the minimum road width shall be 26 feet (7925 mm), exclusive of shoulders (see Figure D103.1).

D103.2 Grade. Fire apparatus access roads shall not exceed 10 percent in grade.

Exception: Grades steeper than 10 percent as *approved* by the fire chief.

D103.3 Turning radius. The minimum turning radius shall be determined by the *fire code official*.

D103.4 Dead ends. Dead-end fire apparatus access roads in excess of 150 feet (45 720 mm) shall be provided with width and turnaround provisions in accordance with Table D103.4.

TABLE D103.4
REQUIREMENTS FOR DEAD-END FIRE APPARATUS ACCESS ROADS

LENGTH (feet)	WIDTH (feet)	TURNAROUNDS REQUIRED
0-150	20	None required
151-500	20	120-foot Hammerhead, 60-foot "Y" or 96-foot diameter cul-de-sac in accordance with Figure D103.1
501-750	26	120-foot Hammerhead, 60-foot "Y" or 96-foot diameter cul-de-sac in accordance with Figure D103.1
Over 750	Special approval required	

For SI: 1 foot = 304.8 mm.

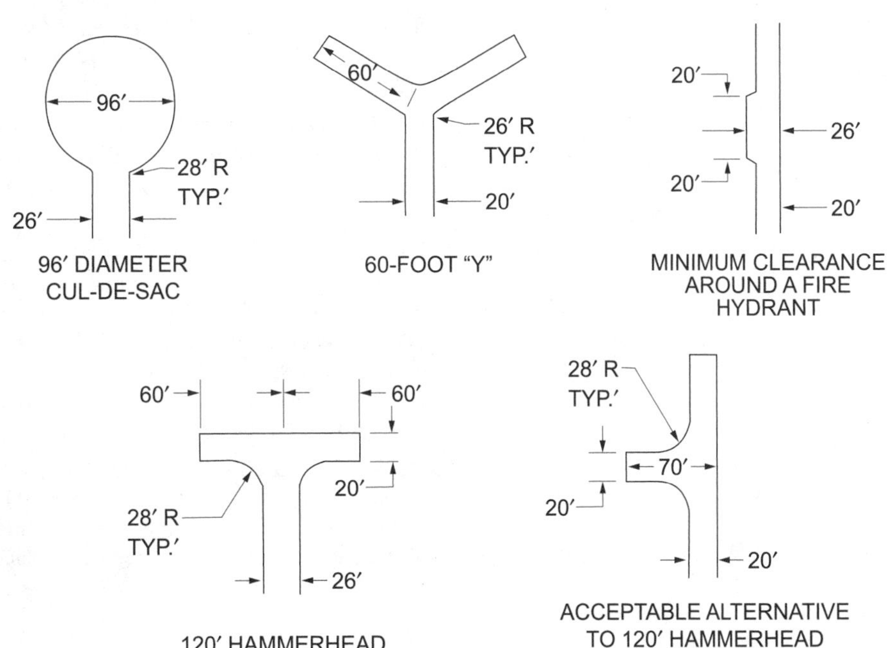

For SI: 1 foot = 304.8 mm.

FIGURE D103.1
DEAD-END FIRE APPARATUS ACCESS ROAD TURNAROUND

D103.5 Fire apparatus access road gates. Gates securing the fire apparatus access roads shall comply with all of the following criteria:

1. Where a single gate is provided, the gate width shall be not less than 20 feet (6096 mm). Where a fire apparatus road consists of a divided roadway, the gate width shall be not less than 12 feet (3658 mm).
2. Gates shall be of the swinging or sliding type.
3. Construction of gates shall be of materials that allow manual operation by one person.
4. Gate components shall be maintained in an operative condition at all times and replaced or repaired when defective.
5. Electric gates shall be equipped with a means of opening the gate by fire department personnel for emergency access. Emergency opening devices shall be *approved* by the *fire code official.*
6. Methods of locking shall be submitted for approval by the *fire code official.*
7. Electric gate operators, where provided, shall be *listed* in accordance with UL 325.
8. Gates intended for automatic operation shall be designed, constructed and installed to comply with the requirements of ASTM F 2200.

D103.6 Signs. Where required by the *fire code official*, fire apparatus access roads shall be marked with permanent NO PARKING—FIRE LANE signs complying with Figure D103.6. Signs shall have a minimum dimension of 12 inches (305 mm) wide by 18 inches (457 mm) high and have red letters on a white reflective background. Signs shall be posted on one or both sides of the fire apparatus road as required by Section D103.6.1 or D103.6.2.

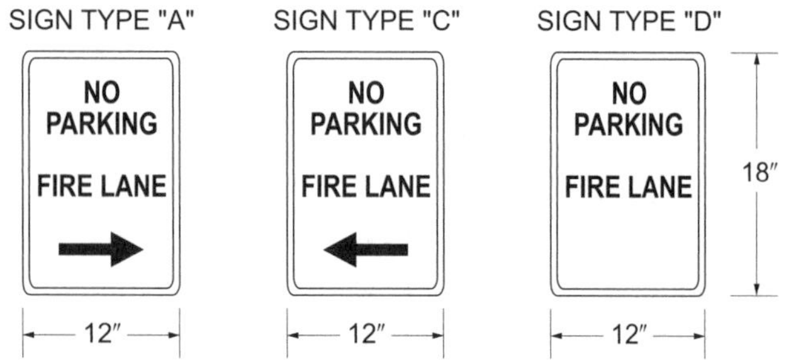

FIGURE D103.6
FIRE LANE SIGNS

D103.6.1 Roads 20 to 26 feet in width. Fire lane signs as specified in Section D103.6 shall be posted on both sides of fire apparatus access roads that are 20 to 26 feet wide (6096 to 7925 mm).

D103.6.2 Roads more than 26 feet in width. Fire lane signs as specified in Section D103.6 shall be posted on one side of fire apparatus access roads more than 26 feet wide (7925 mm) and less than 32 feet wide (9754 mm).

SECTION D104
COMMERCIAL AND INDUSTRIAL DEVELOPMENTS

D104.1 Buildings exceeding three stories or 30 feet in height. Buildings or facilities exceeding 30 feet (9144 mm) or three stories in height shall have at least two means of fire apparatus access for each structure.

D104.2 Buildings exceeding 62,000 square feet in area. Buildings or facilities having a gross *building area* of more than 62,000 square feet (5760 m^2) shall be provided with two separate and *approved* fire apparatus access roads.

Exception: Projects having a gross *building area* of up to 124,000 square feet (11 520 m^2) that have a single *approved* fire apparatus access road when all buildings are equipped throughout with *approved automatic sprinkler systems.*

D104.3 Remoteness. Where two fire apparatus access roads are required, they shall be placed a distance apart equal to not less than one half of the length of the maximum overall diagonal dimension of the lot or area to be served, measured in a straight line between accesses.

SECTION D105
AERIAL FIRE APPARATUS ACCESS ROADS

D105.1 Where required. Where the vertical distance between the grade plane and the highest roof surface exceeds 30 feet (9144 mm), approved aerial fire apparatus access roads shall be provided. For purposes of this section, the highest roof surface shall be determined by measurement to the eave of a pitched roof, the intersection of the roof to the exterior wall, or the top of parapet walls, whichever is greater.

D105.2 Width. Aerial fire apparatus access roads shall have a minimum unobstructed width of 26 feet (7925 mm), exclusive of shoulders, in the immediate vicinity of the building or portion thereof.

D105.3 Proximity to building. At least one of the required access routes meeting this condition shall be located within a minimum of 15 feet (4572 mm) and a maximum of 30 feet (9144 mm) from the building, and shall be positioned parallel to one entire side of the building. The side of the building on which the aerial fire apparatus access road is positioned shall be approved by the *fire code official.*

D105.4 Obstructions. Overhead utility and power lines shall not be located over the aerial fire apparatus access road or between the aerial fire apparatus road and the building. Other obstructions shall be permitted to be placed with the approval of the *fire code official.*

SECTION D106
MULTIPLE-FAMILY RESIDENTIAL DEVELOPMENTS

D106.1 Projects having more than 100 dwelling units. Multiple-family residential projects having more than 100 *dwelling units* shall be equipped throughout with two separate and *approved* fire apparatus access roads.

> **Exception:** Projects having up to 200 *dwelling units* may have a single *approved* fire apparatus access road when all buildings, including nonresidential occupancies, are equipped throughout with *approved automatic sprinkler systems* installed in accordance with Section 903.3.1.1 or 903.3.1.2.

D106.2 Projects having more than 200 dwelling units. Multiple-family residential projects having more than 200 *dwelling units* shall be provided with two separate and *approved* fire apparatus access roads regardless of whether they are equipped with an *approved automatic sprinkler system*.

D106.3 Remoteness. Where two fire apparatus access roads are required, they shall be placed a distance apart equal to not less than one-half of the length of the maximum overall diagonal dimension of the property or area to be served, measured in a straight line between accesses.

SECTION D107
ONE- OR TWO-FAMILY RESIDENTIAL DEVELOPMENTS

D107.1 One- or two-family dwelling residential developments. Developments of one- or two-family dwellings where the number of *dwelling units* exceeds 30 shall be provided with two separate and *approved* fire apparatus access roads.

Exceptions:

1. Where there are more than 30 *dwelling units* on a single public or private fire apparatus access road and all *dwelling units* are equipped throughout with an *approved automatic sprinkler system* in accordance with Section 903.3.1.1, 903.3.1.2 or 903.3.1.3 of the *International Fire Code*, access from two directions shall not be required.
2. The number of *dwelling units* on a single fire apparatus access road shall not be increased unless fire apparatus access roads will connect with future development, as determined by the *fire code official*.

D107.2 Remoteness. Where two fire apparatus access roads are required, they shall be placed a distance apart equal to not less than one-half of the length of the maximum overall diagonal dimension of the property or area to be served, measured in a straight line between accesses.

SECTION D108
REFERENCED STANDARDS

ASTM	F 2200—13	Standard Specification for Automated Vehicular Gate Construction	D103.5
ICC	IFC—15	International Fire Code	D101.1, D107.1
UL	325—02	Door, Drapery, Gate, Louver, and Window Operators and Systems, with Revisions through June 2013	D103.5

APPENDIX E

HAZARD CATEGORIES

This appendix is for information purposes and is not intended for adoption.

SECTION E101 GENERAL

E101.1 Scope. This appendix provides information, explanations and examples to illustrate and clarify the hazard categories contained in Chapter 50 of the *International Fire Code*. The hazard categories are based upon the DOL 29 CFR. Where numerical classifications are included, they are in accordance with nationally recognized standards.

This appendix should not be used as the sole means of hazardous materials classification.

SECTION E102 HAZARD CATEGORIES

E102.1 Physical hazards. Materials classified in this section pose a *physical hazard*.

E102.1.1 Explosives and blasting agents. The current UN/DOT classification system recognized by international authorities, the Department of Defense and others classifies all *explosives* as Class 1 materials. They are then divided into six separate divisions to indicate their relative hazard. There is not a direct correlation between the designations used by the old DOT system and those used by the current system nor is there correlation with the system (high and low) established by the Bureau of Alcohol, Tobacco, Firearms and Explosives (BATF). Table 5604.3 of the *International Fire Code* provides some guidance with regard to the current categories and their relationship to the old categories. Some items may appear in more than one division, depending on factors such as the degree of confinement or separation, by type of packaging, storage configuration or state of assembly.

In order to determine the level of hazard presented by *explosive materials*, testing to establish quantitatively their *explosive* nature is required. There are numerous test methods that have been used to establish the character of an *explosive material*. Standardized tests, required for finished goods containing *explosives* or *explosive materials* in a packaged form suitable for shipment or storage, have been established by UN/DOT and BATF. However, these tests do not consider key elements that should be examined in a manufacturing situation. In manufacturing operations, the condition and/or the state of a material may vary within the process. The in-process material classification and classification requirements for materials used in the manufacturing process may be different from the classification of the same material where found in finished goods depending on the stage of the process in which the material is found. A classification methodology must be used that recognizes the hazards commensurate with the application to the variable physical conditions as well as potential variations of physical character and type of *explosive* under consideration.

Test methods or guidelines for hazard classification of energetic materials used for in-process operations shall be *approved* by the *fire code official*. Test methods used shall be DOD, BATF, UN/DOT or other *approved* criteria. The results of such testing shall become a portion of the files of the jurisdiction and be included as an independent section of any Hazardous Materials Management Plan (HMMP) required by Section 5605.2.1 of the *International Fire Code*. Also see Section 104.7.2 of the *International Fire Code*.

Examples of materials in various Divisions are as follows:

1. Division 1.1 (High *Explosives*). Consists of *explosives* that have a mass explosion hazard. A mass explosion is one that affects almost the entire pile of material instantaneously. Includes substances that, where tested in accordance with *approved* methods, can be caused to detonate by means of a blasting cap where unconfined or will transition from *deflagration* to a *detonation* where confined or unconfined. Examples: dynamite, TNT, nitroglycerine, C-3, HMX, RDX, encased *explosives*, military ammunition.
2. Division 1.2 (Low *Explosives*). Consists of *explosives* that have a projection hazard, but not a mass explosion hazard. Examples: nondetonating encased *explosives*, military ammunition and the like.
3. Division 1.3 (Low *Explosives*). Consists of *explosives* that have a fire hazard and either a minor blast hazard or a minor projection hazard or both, but not a mass explosion hazard. The major hazard is radiant heat or violent burning, or both. Can be deflagrated where confined. Examples: smokeless powder, propellant *explosives*, display fireworks.
4. Division 1.4. Consists of *explosives* that pose a minor explosion hazard. The *explosive* effects are largely confined to the package and no projection of fragments of appreciable size or range is expected. An internal fire must not cause virtually instantaneous explosion of almost the entire contents of the package. Examples: squibs (nondetonating igniters), *explosive* actuators, *explosive* trains (low-level detonating cord).
5. Division 1.5 (Blasting Agents). Consists of very insensitive *explosives*. This division comprises substances that have a mass explosion hazard, but are so

insensitive that there is very little probability of initiation or of transition from burning to *detonation* under normal conditions of transport. Materials are not cap sensitive; however, they are mass detonating where provided with sufficient input. Examples: oxidizer and liquid fuel slurry mixtures and gels, ammonium nitrate combined with fuel oil.

6. Division 1.6. Consists of extremely insensitive articles that do not have a mass *explosive* hazard. This division comprises articles that contain only extremely insensitive detonating substances and that demonstrate a negligible probability of accidental initiation or propagation. Although this category of materials has been defined, the primary application is currently limited to military uses. Examples: Low vulnerability military weapons.

Explosives in each division are assigned a compatibility group letter by the Associate Administrator for Hazardous Materials Safety (DOT) based on criteria specified by DOTn 49 CFR. Compatibility group letters are used to specify the controls for the transportation and storage related to various materials to prevent an increase in hazard that might result if certain types of *explosives* were stored or transported together. Altogether, there are 35 possible classification codes for *explosives*, e.g., 1.1A, 1.3C, 1.4S, etc.

E102.1.2 Compressed gases. Examples include:

1. Flammable: acetylene, carbon monoxide, ethane, ethylene, hydrogen, methane. Ammonia will ignite and burn although its flammable range is too narrow for it to fit the definition of "Flammable gas."

   For binary mixtures where the hazardous component is diluted with a nonflammable gas, the mixture shall be categorized in accordance with CGA P-23.

2. Oxidizing: oxygen, ozone, oxides of nitrogen, chlorine and fluorine. Chlorine and fluorine do not contain oxygen but reaction with flammables is similar to that of oxygen.
3. *Corrosive*: ammonia, hydrogen chloride, fluorine.
4. Highly toxic: arsine, cyanogen, fluorine, germane, hydrogen cyanide, nitric oxide, phosphine, hydrogen selenide, stibine.
5. Toxic: chlorine, hydrogen fluoride, hydrogen sulfide, phosgene, silicon tetrafluoride.
6. Inert (chemically unreactive): argon, helium, krypton, neon, nitrogen, xenon.
7. Pyrophoric: diborane, dichloroborane, phosphine, silane.
8. Unstable (reactive): butadiene (unstabilized), ethylene oxide, vinyl chloride.

E102.1.3 Flammable and combustible liquids. Examples include:

1. Flammable liquids.

   Class IA liquids shall include those having *flash points* below 73°F (23°C) and having a *boiling point* at or below 100°F (38°C).

   Class IB liquids shall include those having *flash points* below 73°F (23°C) and having a *boiling point* at or above 100°F (38°C).

   Class IC liquids shall include those having *flash points* at or above 73°F (23°C) and below 100°F (38°C).

2. *Combustible liquids.*

   Class II liquids shall include those having *flash points* at or above 100°F (38°C) and below 140°F (60°C).

   Class IIIA liquids shall include those having *flash points* at or above 140°F (60°C) and below 200°F (93°C).

   Class IIIB liquids shall include those liquids having *flash points* at or above 200°F (93°C).

E102.1.4 Flammable solids. Examples include:

1. Organic solids: camphor, cellulose nitrate, naphthalene.
2. Inorganic solids: decaborane, lithium amide, phosphorous heptasulfide, phosphorous sesquisulfide, potassium sulfide, anhydrous sodium sulfide, sulfur.
3. Combustible metals (except dusts and powders): cesium, magnesium, zirconium.

E102.1.5 Combustible dusts and powders. Finely divided solids that could be dispersed in air as a dust cloud: wood sawdust, plastics, coal, flour, powdered metals (few exceptions).

E102.1.6 Combustible fibers. See Section 5202.1.

E102.1.7 Oxidizers. Examples include:

1. Gases: oxygen, ozone, oxides of nitrogen, fluorine and chlorine (reaction with flammables is similar to that of oxygen).
2. Liquids: bromine, hydrogen peroxide, nitric acid, perchloric acid, sulfuric acid.
3. Solids: chlorates, chromates, chromic acid, iodine, nitrates, nitrites, perchlorates, peroxides.

E102.1.7.1 Examples of liquid and solid oxidizers according to hazard.

Class 4: ammonium perchlorate (particle size greater than 15 microns), ammonium permanganate, guanidine nitrate, hydrogen peroxide solutions more than 91 percent by weight, perchloric acid solutions more than 72.5 percent by weight, potassium superoxide, tetranitromethane.

Class 3: ammonium dichromate, calcium hypochlorite (over 50 percent by weight), chloric acid (10 percent maximum concentration), hydrogen peroxide solutions (greater than 52 percent up to 91 percent), mono-(trichloro)-tetra-(monopotassium dichloro)-penta-s-triazinetrione, nitric acid, (fuming - more than 86 percent concentration), perchloric acid solutions (60 percent to 72 percent by weight), potassium bromate, potassium chlorate, potassium dichloro-s-triazinetrione (potassium dichloro-isocyanurate), potassium perchlorate (99 percent), potassium permanganate (greater than 97.5 percent), sodium bromate, sodium chlorate, sodium chlorite (over 40 percent by weight) and sodium dichloro-s-triazinetrione anhydrous (sodium dichloro-isocyanurate anhydrous).

Class 2: barium bromate, barium chlorate, barium hypochlorite, barium perchlorate, barium permanganate, 1-bromo-3-chloro-5, 5-dimethylhydantoin, calcium chlorate, calcium chlorite, calcium hypochlorite (50 percent or less by weight), calcium perchlorate, calcium permanganate, calcium peroxide (75 percent), chromium trioxide (chromic acid), copper chlorate, halane (1, 3-di-chloro-5, 5-dimethylhydantoin), hydrogen peroxide (greater than 27.5 percent up to 52 percent), lead perchlorate, lithium chlorate, lithium hypochlorite (more than 39 percent available chlorine), lithium perchlorate, magnesium bromate, magnesium chlorate, magnesium perchlorate, mercurous chlorate, nitric acid (more than 40 percent but less than 86 percent), perchloric acid solutions (more than 50 percent but less than 60 percent), potassium peroxide, potassium superoxide, silver peroxide, sodium chlorite (40 percent or less by weight), sodium perchlorate, sodium perchlorate monohydrate, sodium permanganate, sodium peroxide, sodium persulfate (99 percent), strontium chlorate, strontium perchlorate, thallium chlorate, urea hydrogen peroxide, zinc bromate, zinc chlorate and zinc permanganate.

Class 1: all inorganic nitrates (unless otherwise classified), all inorganic nitrites (unless otherwise classified), ammonium persulfate, barium peroxide, hydrogen peroxide solutions (greater than 8 percent up to 27.5 percent), lead dioxide, lithium hypochlorite (39 percent or less available chlorine), lithium peroxide, magnesium peroxide, manganese dioxide, nitric acid (40 percent concentration or less), perchloric acid solutions (less than 50 percent by weight), potassium dichromate, potassium monopersulfate (45 percent $KHSO_5$ or 90 percent triple salt), potassium percarbonate, potassium persulfate, sodium carbonate peroxide, sodium dichloro-s-triazinetrione dihydrate, sodium dichromate, sodium perborate (anhydrous), sodium perborate monohydrate, sodium perborate tetra-hydrate, sodium percarbonate, strontium peroxide, trichloro-s-triazinetrione (trichloroisocyanuric acid) and zinc peroxide.

E102.1.8 Organic peroxides. Organic peroxides contain the double oxygen or peroxy (-o-o) group. Some are flammable compounds and subject to explosive decomposition. They are available as:

1. Liquids.
2. Pastes.
3. Solids (usually finely divided powders).

E102.1.8.1 Classification of organic peroxides according to hazard.

Unclassified: Unclassified organic peroxides are capable of *detonation* and are regulated in accordance with Chapter 56 of the *International Fire Code*.

Class I: acetyl cyclohexane sulfonyl 60-65 percent concentration by weight, fulfonyl peroxide, benzoyl peroxide over 98 percent concentration, t-butyl hydroperoxide 90 percent, t-butyl peroxyacetate 75 percent, t-butyl peroxyisopropylcarbonate 92 percent, diisopropyl peroxydicarbonate 100 percent, di-n-propyl peroxydicarbonate 98 percent, and di-n-propyl peroxydicarbonate 85 percent.

Class II: acetyl peroxide 25 percent, t-butyl hydroperoxide 70 percent (with DTBP and t-BuOH diluents), t-butyl peroxybenzoate 98 percent, t-butyl peroxy-2-ethylhexanoate 97 percent, t-butyl peroxyisobutyrate 75 percent, t-butyl peroxyisopropyl-carbonate 75 percent, t-butyl peroxypivalate 75 percent, dybenzoyl peroxydicarbonate 85 percent, di-sec-butyl peroxydicarbonate 98 percent, di-sec-butyl peroxydicarbonate 75 percent, 1,1-di-(t-butylperoxy)-3,5,5-trimethyecyclohexane 95 percent, di-(2-ethythexyl) peroxydicarbonate 97 percent, 2,5-dymethyl-2-5 di (benzoylperoxy) hexane 92 percent, and peroxyacetic acid 43 percent.

Class III: acetyl cyclohexane sulfonal peroxide 29 percent, benzoyl peroxide 78 percent, benzoyl peroxide paste 55 percent, benzoyl peroxide paste 50 percent peroxide/50 percent butylbenzylphthalate diluent, cumene hydroperoxide 86 percent, di-(4-butylcyclohexyl) peroxydicarbonate 98 percent, t-butyl peroxy-2-ethylhexanoate 97 percent, t-butyl peroxyneodecanoate 75 percent, decanoyl peroxide 98.5 percent, di-t-butyl peroxide 99 percent, 1,1-di-(t-butylperoxy)3,5,5-trimethylcyclohexane 75 percent, 2,4-dichlorobenzoyl peroxide 50 percent, di-isopropyl peroxydicarbonate 30 percent, 2,-5-di-methyl-2,5-di-(2-ethylhexanolyperoxy)-hexane 90 percent, 2,5-dimethyl-2,5-di-(t-butylperoxy) hexane 90 percent and methyl ethyl ketone peroxide 9 percent active oxygen diluted in dimethyl phthalate.

Class IV: benzoyl peroxide 70 percent, benzoyl peroxide paste 50 percent peroxide/15 percent water/35 percent butylphthalate diluent, benzoyl peroxide slurry 40 percent, benzoyl peroxide powder 35 percent, t-butyl hydroperoxide 70 percent, (with water diluent), t-butyl peroxy-2-ethylhexanoate 50 percent, decumyl peroxide 98 percent, di-(2-ethylhexal) peroxydicarbonate 40 percent, laurel peroxide 98 percent, p-methane hydroperoxide 52.5 percent, methyl ethyl ketone peroxide 5.5 percent active oxygen and methyl ethyl ketone peroxide 9 percent active oxygen diluted in water and glycols.

Class V: benzoyl peroxide 35 percent, 1,1-di-t-butyl peroxy 3,5,5-trimethylcyclohexane 40 percent, 2,5-di-(t-butyl peroxy) hexane 47 percent and 2,4-pentanedione peroxide 4 percent active oxygen.

E102.1.9 Pyrophoric materials. Examples include:

1. Gases: diborane, phosphine, silane.
2. Liquids: diethylaluminum chloride, di-ethylberyllium, diethylphosphine, diethylzinc, dimethylarsine, triethylaluminum etherate, tri-ethylbismuthine, triethylboron, trimethylaluminum, trimethylgallium.
3. Solids: cesium, hafnium, lithium, white or yellow phosphorous, plutonium, potassium, rubidium, sodium, thorium.

E102.1.10 Unstable (reactive) materials. Examples include:

Class 4: acetyl peroxide, dibutyl peroxide, dinitrobenzene, ethyl nitrate, peroxyacetic acid and picric acid (dry) trinitrobenzene.

Class 3: hydrogen peroxide (greater than 52 percent), hydroxylamine, nitromethane, paranitroaniline, perchloric acid and tetrafluoroethylene monomer.

Class 2: acrolein, acrylic acid, hydrazine, methacrylic acid, sodium perchlorate, styrene and vinyl acetate.

Class 1: acetic acid, hydrogen peroxide 35 percent to 52 percent, paraldehyde and tetrahydrofuran.

E102.1.11 Water-reactive materials. Examples include:

Class 3: aluminum alkyls such as triethylaluminum, isobutylaluminum and trimethylaluminum; bromine pentafluoride, bromine trifluoride, chlorodiethylaluminium and diethylzinc.

Class 2: calcium carbide, calcium metal, cyanogen bromide, lithium hydride, methyldichlorosilane, potassium metal, potassium peroxide, sodium metal, sodium peroxide, sulfuric acid and trichlorosilane.

Class 1: acetic anhydride, sodium hydroxide, sulfur monochloride and titanium tetrachloride.

E102.1.12 Cryogenic fluids. The cryogenics listed will exist as *compressed gases* where they are stored at ambient temperatures.

1. Flammable: carbon monoxide, deuterium (heavy hydrogen), ethylene, hydrogen, methane.
2. Oxidizing: fluorine, nitric oxide, oxygen.
3. *Corrosive*: fluorine, nitric oxide.
4. Inert (chemically unreactive): argon, helium, krypton, neon, nitrogen, xenon.
5. Highly toxic: fluorine, nitric oxide.

E102.2 Health hazards. Materials classified in this section pose a *health hazard*.

E102.2.1 Highly toxic materials. Examples include:

1. Gases: arsine, cyanogen, diborane, fluorine, germane, hydrogen cyanide, nitric oxide, nitrogen dioxide, ozone, phosphine, hydrogen selenide, stibine.
2. Liquids: acrolein, acrylic acid, 2-chloroethanol (ethylene chlorohydrin), hydrazine, hydrocyanic acid, 2-methylaziridine (propylenimine), 2-methyl-acetonitrile (acetone cyanohydrin), methyl ester isocyanic acid (methyl isocyanate), nicotine, tetranitromethane and tetraethylstannane (tetraethyltin).
3. Solids: (aceto) phenylmercury (phenyl mercuric acetate), 4-aminopyridine, arsenic pentoxide, arsenic trioxide, calcium cyanide, 2-chloroacetophenone, aflatoxin B, decaborane(14), mercury (II) bromide (mercuric bromide), mercury (II) chloride (*corrosive* mercury chloride), pentachlorophenol, methyl parathion, phosphorus (white) and sodium azide.

E102.2.2 Toxic materials. Examples include:

1. Gases: boron trichloride, boron trifluoride, chlorine, chlorine trifluoride, hydrogen fluoride, hydrogen sulfide, phosgene, silicon tetrafluoride.
2. Liquids: acrylonitrile, allyl alcohol, alpha-chlorotoluene, aniline, 1-chloro-2,3-epoxypropane, chloroformic acid (allyl ester), 3-chloropropene (allyl chloride), o-cresol, crotonaldehyde, dibromomethane, diisopropylamine, diethyl ester sulfuric acid, dimethyl ester sulfuric acid, 2-furaldehyde (furfural), furfural alcohol, phosphorus chloride, phosphoryl chloride (phosphorus oxychloride) and thionyl chloride.
3. Solids: acrylamide, barium chloride, barium (II) nitrate, benzidine, p-benzoquinone, beryllium chloride, cadmium chloride, cadmium oxide, chloroacetic acid, chlorophenylmercury (phenyl mercuric chloride), chromium (VI) oxide (chromic acid, solid), 2,4-dinitrotoluene, hydroquinone, mercury chloride (calomel), mercury (II) sulfate (mercuric sulfate), osmium tetroxide, oxalic acid, phenol, P-phenylenediamine, phenylhydrazine, 4-phenylmorpholine, phosphorus sulfide, potassium fluoride, potassium hydroxide, selenium (IV) disulfide and sodium fluoride.

E102.2.3 Corrosives. Examples include:

1. Acids: Examples: chromic, formic, hydrochloric (muriatic) greater than 15 percent, hydrofluoric, nitric (greater than 6 percent, perchloric, sulfuric (4 percent or more).
2. Bases (alkalis): hydroxides-ammonium (greater than 10 percent), calcium, potassium (greater than 1 percent), sodium (greater than 1 percent); certain carbonates-potassium.
3. Other *corrosives*: bromine, chlorine, fluorine, iodine, ammonia.

Note: *Corrosives* that are oxidizers, e.g., nitric acid, chlorine, fluorine; or are *compressed gases*, e.g., ammonia, chlorine, fluorine; or are water-reactive, e.g., concentrated sulfuric acid, sodium hydroxide, are *physical hazards* in addition to being *health hazards*.

SECTION E103 EVALUATION OF HAZARDS

E103.1 Degree of hazard. The degree of hazard present depends on many variables that should be considered individually and in combination. Some of these variables are as shown in Sections E103.1.1 through E103.1.5.

E103.1.1 Chemical properties of the material. Chemical properties of the material determine self reactions and reactions that could occur with other materials. Generally, materials within subdivisions of hazard categories will exhibit similar chemical properties. However, materials with similar chemical properties could pose very different hazards. Each individual material should be researched to determine its hazardous properties and then considered in relation to other materials that it might contact and the surrounding environment.

E103.1.2 Physical properties of the material. Physical properties, such as whether a material is a solid, liquid or gas at ordinary temperatures and pressures, considered along with chemical properties will determine requirements for containment of the material. Specific gravity (weight of a liquid compared to water) and vapor density (weight of a gas compared to air) are both physical properties that are important in evaluating the hazards of a material.

E103.1.3 Amount and concentration of the material. The amount of material present and its concentration must be considered along with physical and chemical properties to determine the magnitude of the hazard. Hydrogen peroxide, for example, is used as an antiseptic and a hair bleach in low concentrations (approximately 8 percent in water solution). Over 8 percent, hydrogen peroxide is classed as an oxidizer and is toxic. Above 90 percent, it is a Class 4 oxidizer "that can undergo an explosive reaction when catalyzed or exposed to heat, shock or friction," a definition that incidentally also places hydrogen peroxide over 90-percent concentration in the unstable (reactive) category. Small amounts at high concentrations could present a greater hazard than large amounts at low concentrations.

E103.1.3.1 Mixtures. Gases—toxic and highly toxic gases include those gases that have an LC_{50} of 2,000 parts per million (ppm) or less when rats are exposed for a period of 1 hour or less. To maintain consistency with the definitions for these materials, exposure data for periods other than 1 hour must be normalized to 1 hour. To classify mixtures of *compressed gases* that contain one or more toxic or highly toxic components, the LC_{50} of the mixture must be determined. Mixtures that contain only two components are binary mixtures. Those that contain more than two components are multicomponent mixtures. Where two or more hazardous substances (components) having an LC_{50} below 2,000 ppm are present in a mixture, their combined effect, rather than that of the individual substance components, must be considered. In the absence of information to the contrary, the effects of the hazards present must be considered as additive. Exceptions to the above rule could be made when there is a good reason to believe that the principal effects of the different harmful substances (components) are not additive.

For binary mixtures where the hazardous component is diluted with a nontoxic gas such as an inert gas, the LC_{50} of the mixture is estimated by use of the methodology contained in CGA P-20. The hazard zones specified in CGA P-20 are applicable for DOTn purposes and shall not be used for hazard classification.

E103.1.4 Actual use, activity or process involving the material. The definition of handling, storage and use in *closed systems* refers to materials in packages or containers. Dispensing and use in open containers or systems describes situations where a material is exposed to ambient conditions or vapors are liberated to the atmosphere. Dispensing and use in *open systems*, then, are generally more hazardous situations than handling, storage or use in *closed systems*. The actual use or process could include heating, electric or other sparks, catalytic or reactive materials and many other factors that could affect the hazard and must therefore be thoroughly analyzed.

E103.1.5 Surrounding conditions. Conditions such as other materials or processes in the area, type of construction of the structure, fire protection features (e.g., *fire walls*, sprinkler systems, alarms, etc.), occupancy (use) of adjoining areas, normal temperatures, exposure to weather, etc., must be taken into account in evaluating the hazard.

E103.2 Evaluation questions. The following are sample evaluation questions:

1. What is the material? Correct identification is important; exact spelling is vital. Check labels, MSDS, ask responsible persons, etc.
2. What are the concentration and strength?
3. What is the physical form of the material? Liquids, gases and finely divided solids have differing requirements for spill and leak control and containment.
4. How much material is present? Consider in relation to permit amounts, *maximum allowable quantity per control area* (from Group H occupancy requirements), amounts that require detached storage and overall magnitude of the hazard.
5. What other materials (including furniture, equipment and building components) are close enough to interact with the material?
6. What are the likely reactions?
7. What is the activity involving the material?
8. How does the activity impact the hazardous characteristics of the material? Consider vapors released or hazards otherwise exposed.
9. What must the material be protected from? Consider other materials, temperature, shock, pressure, etc.
10. What effects of the material must people and the environment be protected from?
11. How can protection be accomplished? Consider:
    11.1. Proper containers and equipment.

11.2. Separation by distance or construction.

11.3. Enclosure in cabinets or rooms.

11.4. Spill control, drainage and containment.

11.5. Control systems-ventilation, special electrical, detection and alarm, extinguishment, explosion venting, limit controls, exhaust scrubbers and excess flow control.

11.6. Administrative (operational) controls-signs, ignition source control, security, personnel training, established procedures, storage plans and emergency plans.

Evaluation of the hazard is a strongly subjective process; therefore, the person charged with this responsibility must gather as much relevant data as possible so that the decision will be objective and within the limits prescribed in laws, policies and standards.

It could be necessary to cause the responsible persons in charge to have tests made by qualified persons or testing laboratories to support contentions that a particular material or process is or is not hazardous. See Section 104.7.2 of the *International Fire Code*.

SECTION E104
REFERENCED STANDARDS

CGA (2009)	P-20—	Standard for Classification of Toxic Mixtures	E103.1.3.1
CGA (2008)	P-23—	Standard for Categorizing Gas Mixtures Containing Flammable and Nonflammable Components	E102.1.2
ICC	IFC—15	International Fire Code	E101.1, E102.1.1, E102.1.8.1, E103.2

APPENDIX F
HAZARD RANKING

The provisions contained in this appendix are not mandatory unless specifically referenced in the adopting ordinance.

SECTION F101
GENERAL

F101.1 Scope. Assignment of levels of hazards to be applied to specific hazard classes as required by NFPA 704 shall be in accordance with this appendix. The appendix is based on application of the degrees of hazard as defined in NFPA 704 arranged by hazard class as for specific categories defined in Chapter 2 of the *International Fire Code* and used throughout.

F101.2 General. The hazard rankings shown in Table F101.2 have been established by using guidelines found within NFPA 704. As noted in Section 4.2 of NFPA 704, there could be specific reasons to alter the degree of hazard assigned to a specific material; for example, ignition temperature, flammable range or susceptibility of a container to rupture by an internal combustion explosion or to metal failure while under pressure or because of heat from external fire. As a result, the degree of hazard assigned for the same material can vary when assessed by different people of equal competence.

The hazard rankings assigned to each class represent reasonable minimum hazard levels for a given class based on the use of criteria established by NFPA 704. Specific cases of use or storage may dictate the use of higher degrees of hazard in certain cases.

SECTION F102
REFERENCED STANDARDS

ICC	IFC—15	International Fire Code	F101.1
NFPA	704—12	Identification of the Hazards of Materials for Emergency Response	F101.1, F101.2

TABLE F101.2
FIRE FIGHTER WARNING PLACARD DESIGNATIONS BASED ON HAZARD CLASSIFICATION CATEGORIES

HAZARD CATEGORY	DESIGNATION
Combustible liquid II	F2
Combustible liquid IIIA	F2
Combustible liquid IIIB	F1
Combustible dust	F4
Combustible fiber	F3
Cryogenic flammable	F4, H3
Cryogenic oxidizing	OX, H3
Explosive	R4
Flammable solid	F2
Flammable gas (gaseous)	F4
Flammable gas (liquefied)	F4
Flammable liquid IA	F4
Flammable liquid IB	F3
Flammable liquid IC	F3
Organic peroxide, UD	R4
Organic peroxide I	F4, R3
Organic peroxide II	F3, R3
Organic peroxide III	F2, R2
Organic peroxide IV	F1, R1
Organic peroxide V	None
Oxidizing gas (gaseous)	OX
Oxidizing gas (liquefied)	OX
Oxidizer 4	OX4
Oxidizer 3	OX3
Oxidizer 2	OX2
Oxidizer 1	OX1
Pyrophoric gases	F4
Pyrophoric solids, liquids	F3
Unstable reactive 4D	R4
Unstable reactive 3D	R4
Unstable reactive 3N	R2
Unstable reactive 2	R2
Unstable reactive 1	None
Water reactive 3	W3
Water reactive 2	W2
Corrosive	H3, COR
Toxic	H3
Highly toxic	H4

F—Flammable category.
R—Reactive category.
H—Health category.
W—Special hazard: water reactive.
OX—Special hazard: oxidizing properties.
COR—Corrosive.
UD—Unclassified detonable material.
4D—Class 4 detonable material.
3D—Class 3 detonable material.
3N—Class 3 nondetonable material.

APPENDIX G

CRYOGENIC FLUIDS—WEIGHT AND VOLUME EQUIVALENTS

This appendix is for information purposes and is not intended for adoption.

SECTION G101 GENERAL

G101.1 Scope. This appendix is used to convert from liquid to gas for *cryogenic fluids*.

G101.2 Conversion. Table G101.2 shall be used to determine the equivalent amounts of *cryogenic fluids* in either the liquid or gas phase.

G101.2.1 Use of the table. To use Table G101.2, read horizontally across the line of interest. For example, to determine the number of cubic feet of gas contained in 1.0 gallon (3.785 L) of liquid argon, find 1.000 in the column entitled "Volume of Liquid at Normal *Boiling Point.*" Reading across the line under the column entitled "Volume of Gas at NTP" (70°F and 1 atmosphere/14.7 psia), the value of 112.45 cubic feet (3.184 m^3) is found.

G101.2.2 Other quantities. If other quantities are of interest, the numbers obtained can be multiplied or divided to obtain the quantity of interest. For example, to determine the number of cubic feet of argon gas contained in a volume of 1,000 gallons (3785 L) of liquid argon at its normal *boiling point*, multiply 112.45 by 1,000 to obtain 112,450 cubic feet (3184 m^3).

TABLE G101.2
WEIGHT AND VOLUME EQUIVALENTS FOR COMMON CRYOGENIC FLUIDS

CRYOGENIC FLUID	WEIGHT OF LIQUID OR GAS		VOLUME OF LIQUID AT NORMAL BOILING POINT		VOLUME OF GAS AT NTP	
	Pounds	Kilograms	Liters	Gallons	Cubic feet	Cubic meters
Argon	1.000	0.454	0.326	0.086	9.67	0.274
	2.205	1.000	0.718	0.190	21.32	0.604
	3.072	1.393	1.000	0.264	29.71	0.841
	11.628	5.274	3.785	1.000	112.45	3.184
	10.340	4.690	3.366	0.889	100.00	2.832
	3.652	1.656	1.189	0.314	35.31	1.000
Helium	1.000	0.454	3.631	0.959	96.72	2.739
	2.205	1.000	8.006	2.115	213.23	6.038
	0.275	0.125	1.000	0.264	26.63	0.754
	1.042	0.473	3.785	1.000	100.82	2.855
	1.034	0.469	3.754	0.992	100.00	2.832
	0.365	0.166	1.326	0.350	35.31	1.000
Hydrogen	1.000	0.454	6.409	1.693	191.96	5.436
	2.205	1.000	14.130	3.733	423.20	11.984
	0.156	0.071	1.000	0.264	29.95	0.848
	0.591	0.268	3.785	1.000	113.37	3.210
	0.521	0.236	3.339	0.882	100.00	2.832
	0.184	0.083	1.179	0.311	35.31	1.000
Oxygen	1.000	0.454	0.397	0.105	12.00	0.342
	2.205	1.000	0.876	0.231	26.62	0.754
	2.517	1.142	1.000	0.264	30.39	0.861
	9.527	4.321	3.785	1.000	115.05	3.250
	8.281	3.756	3.290	0.869	100.00	2.832
	2.924	1.327	1.162	0.307	35.31	1.000
Nitrogen	1.000	0.454	0.561	0.148	13.80	0.391
	2.205	1.000	1.237	0.327	30.43	0.862
	1.782	0.808	1.000	0.264	24.60	0.697
	6.746	3.060	3.785	1.000	93.11	2.637
	7.245	3.286	4.065	1.074	100.00	2.832
	2.558	1.160	1.436	0.379	35.31	1.000
LNG[a]	1.000	0.454	1.052	0.278	22.968	0.650
	2.205	1.000	2.320	0.613	50.646	1.434
	0.951	0.431	1.000	0.264	21.812	0.618
	3.600	1.633	3.785	1.000	82.62	2.340
	4.356	1.976	4.580	1.210	100.00	2.832
	11.501	5.217	1.616	0.427	35.31	1.000

For SI: 1 pound = 0.454 kg, 1 gallon = 3.785 L, 1 cubic foot = 0.02832 m^3, °C = [(°F)-32]/1.8, 1 pound per square inch atmosphere = 6.895 kPa.

a. The values listed for liquefied natural gas (LNG) are "typical" values. LNG is a mixture of hydrocarbon gases, and no two LNG streams have exactly the same composition.

APPENDIX H

HAZARDOUS MATERIALS MANAGEMENT PLAN (HMMP) AND HAZARDOUS MATERIALS INVENTORY STATEMENT (HMIS) INSTRUCTIONS

The provisions contained in this appendix are not mandatory unless specifically referenced in the adopting ordinance.

SECTION H101 HMMP

H101.1 Part A (See Example Format in Figure 1).

1. Fill out items and sign the declaration.
2. Part A of this section is required to be updated and submitted annually, or within 30 days of a process or management change.

H101.2 Part B–General Facility Description/Site Plan (See Example Format in Figure 2).

1. Provide a site plan on $8^1/_2$ by 11 inch (215 mm by 279 mm) paper, showing the locations of all buildings, structures, outdoor chemical control or storage and use areas, parking lots, internal roads, storm and sanitary sewers, wells and adjacent property uses. Indicate the approximate scale, northern direction and date the drawing was completed.

H101.3 Part C–Facility Storage Map–Confidential Information (See Example Format in Figure 3).

1. Provide a floor plan of each building identified on the site plan as containing hazardous materials on $8^1/_2$-inch by 11-inch (215 mm by 279 mm) paper, identifying the northern direction, and showing the location of each storage and use area.
2. Identify storage and use areas, including hazard waste storage areas.
3. Show the following:
   - 3.1. Accesses to each storage and use area.
   - 3.2. Location of emergency equipment.
   - 3.3. Location where liaison will meet emergency responders.
   - 3.4. Facility evacuation meeting point locations.
   - 3.5. The general purpose of other areas within the building.
   - 3.6. Location of all aboveground and underground tanks to include sumps, vaults, below-grade treatment systems, piping, etc.
   - 3.7. Show hazard classes in each area.
   - 3.8. Show locations of all Group H occupancies, control areas, and exterior storage and use areas.
   - 3.9. Show emergency exits.

SECTION H102 HMIS

H102.1 Inventory statement contents.

1. HMIS Summary Report (see Example Format in Figure 4).
   - 1.1. Complete a summary report for each control area and Group H occupancy.
   - 1.2. The storage summary report includes the HMIS Inventory Report amounts in storage, use-closed and use-open conditions.
   - 1.3. Provide separate summary reports for storage, use-closed and use-open conditions.
   - 1.4. IBC/IFC Hazard Class.
   - 1.5. Inventory Amount. [Solid (lb), Liquid (gal), Gas (cu ft, gal or lbs)].
   - 1.6. IBC/IFC Maximum Allowable Quantity per control area (MAQ). (If applicable, double MAQ for sprinkler protection and/or storage in cabinets. For wholesale and retail sales occupancies, go to Tables 5003.11.1 and 5704.3.4.1 of the *International Fire Code* for MAQs.).
2. HMIS Inventory Report (see Example Format in Figure 5).
   - 2.1. Complete an inventory report by listing products by location.
   - 2.2. Product Name.
   - 2.3. Components. (For mixtures specify percentages of major components if available.)
   - 2.4. Chemical Abstract Service (CAS) Number. (For mixtures list CAS Numbers of major components if available.)
   - 2.5. Location. (Identify the control area or, if it is a Group H occupancy, provide the classification, such as H-2, H-3, etc.)
   - 2.6. Container with a capacity of greater than 55 gallons (208 L). (If product container, vessel or tank could exceed 55 gallons, indicate yes in column.)
   - 2.7. Hazard Classification. (List applicable classifications for each product.)
   - 2.8. Stored. (Amount of product in storage conditions.)

2.9. Closed. (Amount of product in use-closed systems.)

2.10. Open. (Amount of product in use-open systems.)

Facilities that have prepared, filed and submitted a Tier II Inventory Report required by the U.S. Environmental Protection Agency (USEPA) or required by a state that has secured USEPA approval for a similar form shall be deemed to have complied with this section.

SECTION H103 EMERGENCY PLAN

1. Emergency Notification. (See Example Format in Figure 6.)
2. Where OSHA or state regulations require a facility to have either an Emergency Action Plan (EAP) or an Emergency Response Plan (ERP), the EAP or ERP shall be included as part of the HMMP.

SECTION H104 REFERENCED STANDARDS

ICC	IBC—15	International Building Code	H102.1
ICC	IFC—15	International Fire Code	H102.1

FIGURE 1
HAZARDOUS MATERIALS MANAGEMENT PLAN
SECTION I: FACILITY DESCRIPTION

1. Business Name:______________________________ Phone:______________
   Address:____

2. Person Responsible for the Business
   Name: ______________ Title: ______________ Phone: ______________

3. Emergency Contacts:

Name:	Title:	Home Number:	Work Number:

4. Person Responsible for the Application/Principal Contact:
   Name: ______________ Title: ______________ Phone: ______________

5. Principal Business Activity:
   ____
   ____
   ____

6. Number of Employees:__________

7. Number of Shifts: ______________
   a. Number of Employees per Shift:
   ____
   ____
   ____

8. Hours of Operation: ____________

FIGURE 2
HAZARDOUS MATERIALS MANAGEMENT PLAN SECTION I: FACILITY DESCRIPTION

FIGURE 3
HAZARDOUS MATERIALS MANAGEMENT PLAN SECTION I: FACILITY DESCRIPTION PART C—FACILITY MAP

Business Name	Date
Address	Page of

FIGURE 4
SECTION II—HAZARDOUS MATERIALS INVENTORY STATEMENT (HMIS) HMIS SUMMARY REPORT[a] (Storage[b] Conditions)[c]

IBC/IFC HAZARD CLASS	HAZARD CLASS	INVENTORY AMOUNT			IBC/IFC MAXIMUM ALLOWABLE QUANTITY[d]		
	(Abbrev)	Solid (lb)	Liquid (gal)	Gas (cu ft, gal, lb)	Solid (lb)	Liquid (gal)	Gas (cu ft, gal, lb)
Combustible Liquid	C2		5			120	
	C3A					330	
	C3B		6			13,200	
Combustible Fiber	Loose/Baled						
Cryogenics, Flammable	Cryo-Flam					45	
Cryogenic, Oxidizing	Cryo-OX					45	
Flammable Gas	FLG						
(Gaseous)				150			1,000
(Liquefied)						30	
Flammable Liquid	F1A					30	
	F1B & F1C		5			120	
Combination (1A, 1B, 1C)			5			120	
Flammable Solid	FLS				125		
Organic Peroxide	OPU				0		
	OP1				5		
	OP2				50		
	OP3				125		
	OP4				NL		
	OP5				NL		
Oxidizer	OX4				0		
	OX3				10		
	OX2				250		
	OX1				4,000		

a. Complete a summary report for each control area and Group H occupancy.
b. Storage = storage + use-closed + use-open systems.
c. Separate reports are required for use-closed and use-open systems.
d. Include increases for sprinklefrs or storage in cabinets, if applicable.

(This is an example; add additional hazard classes as needed.)

FIGURE 5
SECTION II — HAZARDOUS MATERIALS INVENTORY STATEMENT (HMIS) HMIS INVENTORY REPORT
(Sort Products Alphabetically by Location of Product and then Alphabetically by Product Name)

Product Name (Components)[c]	CAS Number	Location[a]	Container > 55 gal[b]	Haz Class 1	Haz Class 2	Haz Class 3	Stored (lbs)	Stored (gal)	Stored (gas)[d]	Closed (lbs)	Closed (gal)	Closed gas[d]	Open (lbs)	Open (gal)
ACETYLENE (Acetylene gas)	74-86-2	Control Area 1		FLG	UR2				150					
BLACK AEROSOL SPRAY PAINT (Mixture)	Mixture	Control Area 1		A-L3			24							
GASOLINE, UNLEADED (Gasoline-Mixture) Methyl-t-Butyl-Ether-15% Diisopropyl Ether-7% Ethanol-11% Toluene-12% Xylene-11%	8006-61-9 1634-04-4 108-20-3 64-17-5 108-88-3 1330-20-7	Control Area 1		F1B				5						
MOTOR OIL-10W40 (Hydrotreated Heavy Paraffinic Distillate-85%; Additives-20%)	64742-54-7 Mixture	Control Area 1		C3B				3						
DIESEL (Diesel-99-100%; Additives)	68476-34-6 Proprietary	Control Area 2	Yes	C2				225						
TRANSMISSION FLUID (Oil-Solvent-Neutral; Performance Additives)	64742-65-0	Control Area 2		C3B				3						
OXYGEN, GAS (Oxygen)	7782-44-7	H-3		OXG					5,000					

a. Identify the control area or, if it is a Group H occupancy, provide the classification, such as H-2, H-3, etc.
b. If the product container, vessel or tank could exceed 55 gallons, indicate yes in the column.
c. Specify percentages of main components if available.
d. In cubic feet, gallons or pounds.

(This is an example; add additional hazard classes as needed.)

FIGURE 6
HAZARDOUS MATERIALS MANAGEMENT PLAN
SECTION III: EMERGENCY PLAN

1. In the event of an emergency, the following shall be notified:
   a. Facility Liaison

Name	Title	Home Number	Work Number
________	________	________	________
________	________	________	________
________	________	________	________
________	________	________	________
________	________	________	________

   b. Agency

Agency	Contact	Phone Number
Fire Department	________	________
LEPC	________	________
Other	________	________

APPENDIX I

FIRE PROTECTION SYSTEMS—NONCOMPLIANT CONDITIONS

The provisions contained in this appendix are not mandatory unless specifically referenced in the adopting ordinance.

SECTION I101 NONCOMPLIANT CONDITIONS

I101.1 General. This appendix is intended to identify conditions that can occur where fire protection systems are not properly maintained or components have been damaged. This appendix is not intended to provide comprehensive inspection, testing and maintenance requirements, which are found in NFPA 10, 25 and 72. Rather, its intent is to identify problems that are readily observable during fire inspections.

I101.2 Noncompliant conditions requiring component replacement. The following conditions shall be deemed noncompliant and shall cause the related component(s) to be replaced to comply with the provisions of this code:

1. Sprinkler heads having any of the following conditions:
    - 1.1. Signs of leakage.
    - 1.2. Paint or other ornamentation that is not factory applied.
    - 1.3. Evidence of corrosion including, but not limited to, discoloration or rust.
    - 1.4. Deformation or damage of any part.
    - 1.5. Improper orientation of sprinkler head.
    - 1.6. Empty glass bulb.
    - 1.7. Sprinkler heads manufactured prior to 1920.
    - 1.8. Replacement sprinkler heads that do not match existing sprinkler heads in orifice size, K-factor temperature rating, coating or deflector type.
    - 1.9. Sprinkler heads for the protection of cooking equipment that have not been replaced within one year.
2. Water pressure and air pressure gauges that have been installed for more than 5 years and have not been tested to within 3 percent accuracy.

I101.3 Noncompliant conditions requiring component repair or replacement. The following shall be deemed noncompliant conditions and shall cause the related component(s) to be repaired or replaced to comply with the provisions of this code:

1. Sprinkler and standpipe system piping and fittings having any of the following conditions:
    - 1.1. Signs of leakage.
    - 1.2. Evidence of corrosion.
    - 1.3. Misalignment.
    - 1.4. Mechanical damage.
2. Sprinkler piping support having any of the following conditions:
    - 2.1. Materials resting on or hung from sprinkler piping.
    - 2.2. Damaged or loose hangers or braces.
3. Class II and Class III standpipe systems having any of the following conditions:
    - 3.1. No hose or nozzle, where required.
    - 3.2. Hose threads incompatible with fire department hose threads.
    - 3.3. Hose connection cap missing.
    - 3.4. Mildew, cuts, abrasions and deterioration evident.
    - 3.5. Coupling damaged.
    - 3.6. Gaskets missing or deteriorated.
    - 3.7. Nozzle missing or obstructed.
4. Hose racks and cabinets having any of the following conditions:
    - 4.1. Difficult to operate or damaged.
    - 4.2. Hose improperly racked or rolled.
    - 4.3. Inability of rack to swing 90 degrees (1.57 rad) out of the cabinet.
    - 4.4. Cabinet locked, except as permitted by this code.
    - 4.5. Cabinet door will not fully open.
    - 4.6. Door glazing cracked or broken.
5. Portable fire extinguishers having any of the following conditions:
    - 5.1. Broken seal or tamper indicator.
    - 5.2. Expired maintenance tag.
    - 5.3. Pressure gauge indicator in "red."
    - 5.4. Signs of leakage or corrosion.
    - 5.5. Mechanical damage, denting or abrasion of tank.
    - 5.6. Presence of repairs such as welding, soldering or brazing.
    - 5.7. Damaged threads.
    - 5.8. Damaged hose assembly, couplings or swivel joints.

6. Fire alarm and detection control equipment, initiating devices and notification appliances having any of the following conditions:
    - 6.1. Corroded or leaking batteries or terminals.
    - 6.2. Smoke detectors having paint or other ornamentation that is not factory-applied.
    - 6.3. Mechanical damage to heat or smoke detectors.
    - 6.4. Tripped fuses.
7. Fire department connections having any of the following conditions:
    - 7.1. Fire department connections are not visible or accessible from the fire apparatus access road.
    - 7.2. Couplings or swivels are damaged.
    - 7.3. Plugs and caps are missing or damaged.
    - 7.4. Gaskets are deteriorated.
    - 7.5. Check valve is leaking.
    - 7.6. Identification signs are missing.
8. Fire pumps having any of the following conditions:
    - 8.1. Pump room temperature is less than 40°F (4.4°C).
    - 8.2. Ventilating louvers are not freely operable.
    - 8.3. Corroded or leaking system piping.
    - 8.4. Diesel fuel tank is less than two-thirds full.
    - 8.5. Battery readings, lubrication oil or cooling water levels are abnormal.

SECTION I102 REFERENCED STANDARDS

NFPA 10—13	Portable Fire Extinguishers	I101.1
NFPA 25—13	Inspection, Testing and Maintenance of Water-based Fire Protection Systems	I101.1
NFPA 72—13	National Fire Alarm Code	I101.1

APPENDIX J
BUILDING INFORMATION SIGN

The provisions contained in this appendix are not mandatory unless specifically referenced in the adopting ordinance.

SECTION J101
GENERAL

J101.1 Scope. New buildings shall have a building information sign(s) that shall comply with Sections J101.1.1 through J101.7. Existing buildings shall be brought into conformance with Sections J101.1 through J101.9 when one of the following occurs:

1. The fire department conducts an annual inspection intended to verify compliance with this section, or any required inspection.
2. When a change in use or occupancy has occurred.

Exceptions:

1. Group U occupancies.
2. One- and two-family dwellings.

J101.1.1 Sign location. The building information sign shall be placed at one of the following locations:

1. Upon the entry door or sidelight at a minimum height of 42 inches (1067 mm) above the walking surface on the address side of the building or structure.
2. Upon the exterior surface of the building or structure on either side of the entry door, not more than 3 feet (76 mm) from the entrance door, at a minimum height of 42 inches (1067 mm) above the walking surface on the address side of the building or structure.
3. Conspicuously placed inside an enclosed entrance lobby, on any vertical surface within 10 feet (254 mm) of the entrance door at a minimum height of 42 inches (1067 mm) above the walking surface.
4. Inside the building's fire command center.
5. On the exterior of the fire alarm control unit or on the wall immediately adjacent to the fire alarm control unit door where the alarm panel is located in the enclosed main lobby.

J101.1.2 Sign features. The building information sign shall consist of all of the following:

1. White reflective background with red letters.
2. Durable material.
3. Numerals shall be Roman or Latin numerals, as required, or alphabet letters.
4. Permanently affixed to the building or structure in an approved manner.

J101.1.3 Sign shape. The building information sign shall be a Maltese cross as shown in Figure J101.1.3.

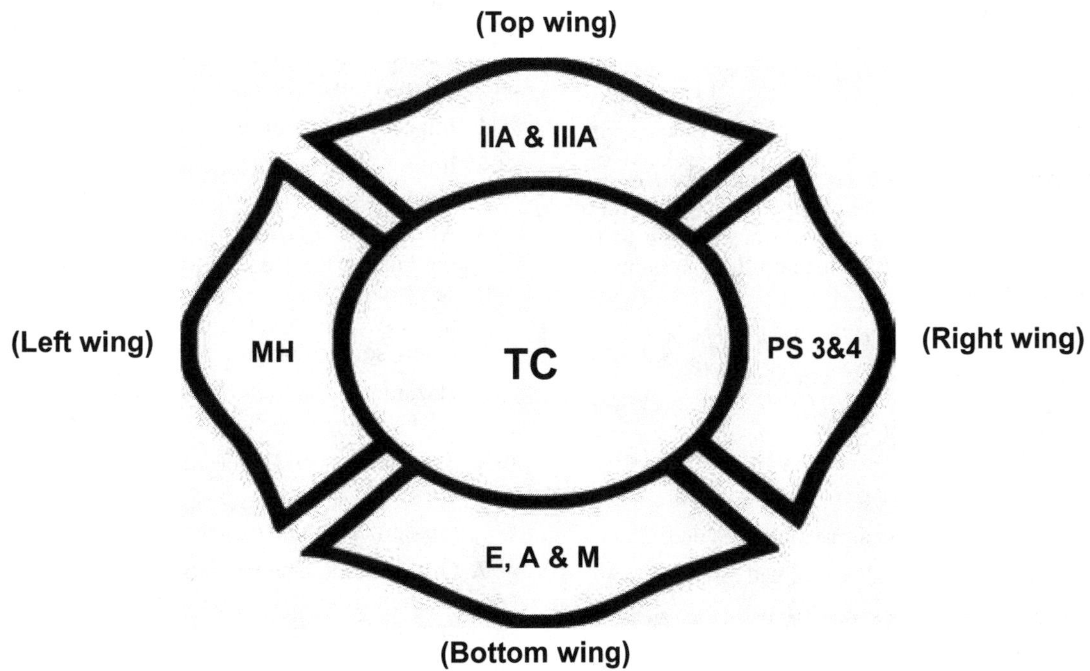

FIGURE J101.1.3
EXAMPLE OF COMPLETED BUILDING INFORMATION SIGN

J101.1.4 Sign size and lettering. The minimum size of the building information sign and lettering shall be in accordance with the following:

1. The width and height shall be 6 inches by 6 inches (152 mm by 152 mm).
2. The height or width of each Maltese cross wing area shall be $1^1/_8$ inches (29 mm) and have a stroke width of $^1/_2$ inch (13 mm).
3. The center of the Maltese cross, a circle or oval, shall be 3 inches (76 mm) in diameter and have a stroke width of $^1/_2$ inch (6 mm).
4. All Roman numerals and alphabetic designations, shall be $1^1/_4$ inch (32 mm) height and have a stroke width of $^1/_4$ inch (6 mm).

J101.2 Sign designations. Designations shall be made based upon the construction type, content, hazard, *fire protection systems*, life safety and occupancy. Where multiple designations occur within a classification category, the designation used shall be based on the greatest potential risk.

J101.3 Construction type (top wing). The construction types shall be designated by assigning the appropriate Roman numeral, and letter, placed inside the top wing of the Maltese cross. The hourly rating provided is for the structural framing in accordance with Table 601 of the *International Building Code*,

CONSTRUCTION TYPE	FIRE-RESISTANCE RATING
IA—Noncombustible	3 Hours
IB—Noncombustible	2 Hours
IIA—Noncombustible	1 Hour
IIB—Noncombustible	0 Hours
IIIA—Noncombustible/combustible	1 Hour
IIIB—Noncombustible/combustible	0 Hours
IV—Heavy timber (HT)	HT
VA—Combustible	1 Hour
VB—Combustible	0 Hours

J101.4 Fire protection systems (right wing). The *fire protection system* shall be designated by determining its level of protection and assigning the appropriate designation to the right wing of the Maltese cross. Where multiple systems are provided, all shall be listed:

AS *Automatic sprinkler system* installed throughout

DS Dry sprinkler system and designated areas

FA Fire alarm system

FP Fire pump

FW *Fire wall* and designated areas

PAS Pre-action sprinkler system and designated floor

PS Partial *automatic sprinkler system*, and designate floor

CES Chemical extinguishing system and designated area

CS Combination sprinkler and standpipe system

S Standpipe system

NS No system installed

J101.5 Occupancy type (bottom wing). The occupancy of a building or structure shall be designated in accordance with the occupancy classification found in Section 302.1 of the *International Building Code* and the corresponding designation shall be placed in the bottom wing of the Maltese cross. Where a building or structure contains a mixture of uses and occupancies; all uses and occupancies shall be identified.

A Assembly

B Business

E Educational

F Factory or Industrial

H High Hazard

I Institutional

M Mercantile

R Residential

J101.6 Hazards of content (left wing). The hazards of building contents shall be designated by one of the following classifications as defined in NFPA 13 and the appropriate designation shall be placed inside the left wing of the Maltese cross:

LH Light hazard

MH Moderate hazard

HH High hazard

J101.7 Tactical considerations (center circle). The center circle shall include the name of the local fire service and when required the letters TC for tactical considerations. Where fire fighters conduct preplan operations, a unique situation(s) for tactical considerations shall be identified and the information provided to the fire dispatch communications center to further assist fire fighters in identifying that there is special consideration(s) for this occupancy. Special consideration designations include, but are not limited to:

1. Impact-resistant drywall.
2. Impact-resistant glazing, such as blast or hurricane-type glass.
3. All types of roof and floor structural members including but not limited to post-tension concrete, bar joists, solid wood joists, rafters, trusses, cold-formed galvanized steel, I-joists and I-beams; green roof with vegetation, soil and plants.
4. Hazardous materials (explosives, chemicals, plastics, etc.).
5. Solar panels and DC electrical energy.
6. HVAC system; and smoke management system for pressurization and exhaust methods.
7. Other unique characteristic(s) within the building that are ranked according to a potential risk to occupants and fire fighters.

J101.8 Sign classification maintenance, building information. Sign maintenance shall comply with each of the following:

1. Fire departments in the jurisdiction shall define the designations to be placed within the sign.
2. Fire departments in the jurisdiction shall conduct annual inspections to verify compliance with this section of the code and shall notify the *owner*, or the *owner's* agent, of any required updates to the sign in accordance with fire department designations and the *owner*, or the *owner's* agent, shall comply within 30 days.
3. The owner of a building shall be responsible for the maintenance and updates to the sign in accordance with fire department designations.

J101.9 Training. Jurisdictions shall train fire department personnel on Sections J101.1 through J101.9.

SECTION J102 REFERENCED STANDARDS

ICC	IBC—15	International Building Code	J101.3, J101.5
NFPA	13—13	Installation of Sprinkler Systems	J101.6

APPENDIX K

CONSTRUCTION REQUIREMENTS FOR EXISTING AMBULATORY CARE FACILITIES

The provisions contained in this appendix are not mandatory unless specifically referenced in the adopting ordinance.

SECTION K101 GENERAL

K101.1 Scope. The provisions of this chapter shall apply to existing buildings containing ambulatory care facilities in addition to the requirements of Chapter 11 of the *International Fire Code*. Where the provisions of this chapter conflict with either the construction requirements in Chapter 11 of the *International Fire Code* or the construction requirements that applied at the time of construction, the most restrictive provision shall apply.

K101.2 Intent. The intent of this appendix is to provide a minimum degree of fire and life safety to persons occupying and existing buildings containing ambulatory care facilities where such buildings do not comply with the minimum requirements of the *International Building Code*.

SECTION K102 FIRE SAFETY REQUIREMENTS FOR EXISTING AMBULATORY CARE FACILITIES

K102.1 Separation. Ambulatory care facilities where the potential exists for four or more care recipients to be incapable of self-preservation at any time, whether rendered incapable by staff or staff has accepted responsibility for a care recipient already incapable, shall be separated from adjacent spaces, corridors or tenants with a fire partition installed in accordance with Section 708 of the *International Building Code*.

K102.2 Smoke compartments. Where the aggregate area of one or more ambulatory care facilities is greater than 10,000 square feet (929 m^2) on one story, the story shall be provided with a *smoke barrier* to subdivide the *story* into no fewer than two *smoke compartments*. The area of any one such *smoke compartment* shall be not greater than 22,500 square feet (2092 m^2). The travel distance from any point in a *smoke compartment* to a *smoke barrier* door shall be not greater than 200 feet (60 960 mm). The *smoke barrier* shall be installed in accordance with Section 709 of the *International Building Code* with the exception that *smoke barriers* shall be continuous from an outside wall to an outside wall, a floor to a floor, or from a *smoke barrier* to a *smoke barrier* or a combination thereof.

K102.2.1 Refuge area. Not less than 30 net square feet (2.8 m^2) for each nonambulatory care recipient shall be provided within the aggregate area of *corridors*, care recipient rooms, treatment rooms, lounge or dining areas and other low-hazard areas within each *smoke compartment*. Each occupant of an ambulatory care facility shall be provided with access to a refuge area without passing through or utilizing adjacent tenant spaces.

K102.2.2 Independent egress. A *means of egress* shall be provided from each *smoke compartment* created by smoke barriers without having to return through the *smoke compartment* from which *means of egress* originated.

K102.3 Automatic sprinkler systems. An *automatic sprinkler system* shall be provided throughout the entire floor containing an ambulatory care facility in Type IIB, IIIB and VB construction where either of the following conditions exist at any time:

1. Four or more care recipients are incapable of self-preservation, whether rendered incapable by staff or staff has accepted responsibility for care recipients already incapable.
2. One or more care recipients that are incapable of self-preservation are located at other than the *level of exit discharge* serving such a facility.

In buildings where ambulatory care is provided on levels other than the *level of exit discharge*, an *automatic sprinkler system* shall be installed throughout the entire floor where such care is provided and all floors below, and all floors between the level of ambulatory care and the nearest *level of exit discharge*, including the *level of exit discharge*.

K102.4 Automatic fire alarm system. *Fire areas* containing ambulatory care facilities shall be provided with an electronically supervised automatic smoke detection system installed within the ambulatory care facility and in public use areas outside of tenant spaces, including *public corridors* and elevator lobbies.

Exception: Buildings equipped throughout with an *automatic sprinkler system* in accordance with Section 903.3.1.1 of the *International Fire Code*, provided the occupant notification appliances will activate throughout the notification zones upon sprinkler waterflow.

SECTION K103 INCIDENTAL USES IN EXISTING AMBULATORY CARE FACILITIES

K103.1 General. Incidental uses associated with and located within existing ambulatory care facilities required to be separated by Section 422 in the *International Building Code*, and that generally pose a greater level of risk to such occupancies, shall comply with the provisions of Sections K103.2 through K103.4.2.1. Incidental uses in ambulatory care facilities required to be separated by Section 422 of the *International Building Code* are limited to those listed in Table K103.1.

K103.2 Occupancy classification. Incidental uses shall not be individually classified in accordance with Section 302.1 of the *International Building Code*. Incidental uses shall be included in the building occupancies in which they are located.

K103.3 Area limitations. Incidental uses shall not occupy more than 10 percent of the building area of the story in which they are located.

K103.4 Separation and protection. The incidental uses listed in Table K103.1 shall be separated from the remainder of the building or equipped with an *automatic sprinkler system*, or both, in accordance with the provisions of that table.

K103.4.1 Separation. Where Table K103.1 specifies a fire-resistance-rated separation, the incidental uses shall be separated from the remainder of the building in accordance with Section 509.4.1 of the *International Building Code*.

K103.4.2 Protection. Where Table K103.1 permits an *automatic sprinkler system* without a fire-resistance-rated separation, the incidental uses shall be separated from the remainder of the building by construction capable of resisting the passage of smoke in accordance with Section 509.4.2 of the *International Building Code*.

K103.4.2.1 Protection limitation. Except as otherwise specified in Table K103.1 for certain incidental uses, where an *automatic sprinkler system* is provided in accordance with Table K103.1, only the space occupied by the incidental use need be equipped with such a system.

SECTION K104 MEANS OF EGRESS REQUIREMENTS FOR EXISTING AMBULATORY CARE FACILITIES

K104.1 Size of doors. The minimum width of each door opening shall be sufficient for the *occupant load* thereof and shall provide a clear width of not less than 28 inches (711 mm). Where this section requires a minimum clear width of 28 inches (711 mm) and a door opening includes two door leaves without a mullion, one leaf shall provide a clear opening width of 28 inches (711 mm). In ambulatory care facilities, doors serving as *means of egress* from patient treatment rooms shall provide a clear width of not less than 32 inches (813 mm). The maximum width of a swinging door leaf shall be 48 inches (1219 mm) nominal. The height of doors openings shall be not less than 80 inches (2032 mm).

Exceptions:

1. Door openings to storage closets less than 10 square feet (0.93 m^2) in area shall not be limited by the minimum width.
2. Width of door leafs in revolving doors that comply with Section 1008.1.4.1 shall not be limited.
3. *Exit access* doors serving a room not larger than 70 square feet (6.5 m^2) shall be not less than 24 inches (610 mm) in door width.
4. Door closers and door stops shall be permitted to be 78 inches (1980 mm) minimum above the door.

K104.2 Corridor and aisle width. *Corridor* width shall be as determined in Section 1005.1 of the *International Fire Code* and this section. The minimum width of *corridors* and aisles that serve gurney traffic in areas where patients receive care that causes them to be incapable of self-preservation shall be not less than 72 inches (1829 mm).

K104.3 Existing elevators. Existing elevators, escalators, dumbwaiters and moving walks shall comply with the requirements of Sections K104. 3.1 and K104.3.2.

K104.3.1 Elevators, escalators, dumbwaiters and moving walks. Existing elevators, escalators, dumbwaiters and moving walks in ambulatory care facilities required to be separated by Section 422 of the *International Building Code* shall comply with ASME A17.3.

TABLE K103.1
INCIDENTAL USES IN EXISTING AMBULATORY CARE FACILITIES

ROOM OR AREA	SEPARATION AND/OR PROTECTION
Furnace room where any piece of equipment is over 400,000 Btu per hour input	1 hour or provide automatic sprinkler system
Rooms with boilers where the largest piece of equipment is over 15 psi and 10 horsepower	1 hour or provide automatic sprinkler system
Refrigerant machinery room	1 hour or provide automatic sprinkler system
Hydrogen fuel gas rooms, not classified as Group H	1 hour in ambulatory care facilities
Incinerator rooms	2 hours and provide automatic sprinkler system
Laboratories not classified as Group H	1 hour or provide automatic sprinkler system
Laundry rooms over 100 square feet	1 hour or provide automatic sprinkler system
Waste and linen collection rooms with containers with total volume of 10 cubic feet or greater	1 hour or provide automatic sprinkler system
Storage rooms greater than 100 square feet	1 hour or provide automatic sprinkler system
Stationary storage battery systems having a liquid electrolyte capacity of more than 50 gallons for flooded lead-acid, nickel cadmium or VRLA, or more than 1,000 pounds for lithium-ion and lithium metal polymer used for facility standby power, emergency power or uninterruptible power supplies	1 hour in ambulatory care facilities

For SI: 1 square foot = 0.0929 m^2, 1 pound per square inch (psi) = 6.9 kPa, 1 British thermal unit (Btu) per hour = 0.293 watts, 1 horsepower = 746 watts, 1 gallon = 3.785 L.

K104.3.2 Elevator emergency operation. Existing elevators with a travel distance of 25 feet (7620 mm) or more above or below the main floor or other level of a building and intended to serve the needs of emergency personnel for fire-fighting or rescue purposes shall be provided with emergency operation in accordance with ASME A17.3.

SECTION K105 REFERENCED STANDARDS

ICC	IBC—15	International Building Code	K101.2, K102.1, K102.2, K103.1, K103.2, K103.4.1, K103.4.2, K104.3.1,
ICC	IFC—15	International Fire Code	K101.1, K102.4, K104.2
ASME A17.3-08			K104.3.1, K104.3.2

APPENDIX L

REQUIREMENTS FOR FIRE FIGHTER AIR REPLENISHMENT SYSTEMS

The provisions contained in this appendix are not mandatory unless specifically referenced in the adopting ordinance.

SECTION L101
GENERAL

L101.1 Scope. Fire fighter air replenishment systems (FARS) shall be provided in accordance with this appendix. The adopting ordinance shall specify building characteristics or special hazards that establish thresholds triggering a requirement for the installation of a FARS. The requirement shall be based upon the fire department's capability of replenishing fire fighter breathing air during sustained emergency operations. Considerations shall include:

1. Building characteristics, such as number of stories above or below grade plane, floor area, type of construction and fire-resistance of the primary structural *frame* to allow sustained fire-fighting operations based on a rating of not less than 2 hours.
2. Special hazards, other than buildings, that require unique accommodations to allow the fire department to replenish fire fighter breathing air.
3. Fire department staffing level.
4. Availability of a fire department breathing air replenishment vehicle.

SECTION L102
DEFINITIONS

L102.1 Definitions. For the purpose of this appendix, certain terms are defined as follows:

FIRE FIGHTER AIR REPLENISHMENT SYSTEM (FARS). A permanently installed arrangement of piping, valves, fittings and equipment to facilitate the replenishment of breathing air in self contained breathing apparatus (SCBA) for fire fighters engaged in emergency operations.

SECTION L103
PERMITS

L103.1 Permits. Permits shall be required to install and maintain a FARS. Permits shall be in accordance with Sections L103.2 and L103.3.

L103.2 Construction permit. A construction permit is required for installation of or modification to a FARS. The construction permit application shall include documentation of an acceptance and testing plan as specified in Section L105.

L103.3 Operational permit. An operational permit is required to maintain a FARS.

SECTION L104
DESIGN AND INSTALLATION

L104.1 Design and installation. A FARS shall be designed and installed in accordance with Sections L104.2 through L104.15.

L104.2 Standards. Fire fighter air replenishment systems shall be in accordance with Sections L104.2.1 and L104.2.2.

L104.2.1 Pressurized system components. Pressurized system components shall be designed and installed in accordance with ASME B31.3.

L104.2.2 Air quality. The system shall be designed to convey breathing air complying with NFPA 1989.

L104.3 Design and operating pressure. The minimum design pressure shall be 110 percent of the fire department's normal SCBA fill pressure. The system design pressure shall be marked in an approved manner at the supply connections, and adjacent to pressure gauges on any fixed air supply components. Pressure shall be maintained in the system within 5 percent of the design pressure.

L104.4 Cylinder refill rate. The FARS shall be capable of refilling breathing air cylinders of a size and pressure used by the fire department at a rate of not less than two empty cylinders in 2 minutes.

L104.5 Breathing air supply. Where a fire department mobile air unit is available, the FARS shall be supplied by an external mobile air connection in accordance with Section L104.14. Where a fire department mobile air unit is not available, a stored pressure air supply shall be provided in accordance with Section L104.5.1. A stored pressure air supply shall be permitted to be added to a system supplied by an external mobile air connection provided that a means to bypass the stored pressure air supply is located at the external mobile air connection.

L104.5.1. Stored pressure air supply. A stored pressure air supply shall be designed based on Chapter 24 of NFPA 1901 except that provisions applicable only to mobile apparatus or not applicable to system design shall not apply. A stored pressure air supply shall be capable of refilling not less than 50 empty breathing air cylinders of a size and pressure used by the fire department.

L104.5.2. Retrofit of external mobile air connection. A FARS not initially provided with an external mobile air connection due to the lack of a mobile air unit shall be retrofitted with an external mobile air connection where a mobile air unit becomes available. Where an external mobile air connection is provided, a means to bypass the stored pressure air supply shall be located at the external mobile air connection. The retrofit shall be completed not more than 12 months after notification by the *fire code official*.

L104.6 Isolation valves. System isolation valves that are accessible to the fire department shall be installed on the system riser to allow piping beyond any air cylinder refill panel to be blocked.

L104.7 Pressure relief valve. Pressure relief valves shall be installed at each point of supply and at the top or end of every riser. The relief valve shall meet the requirements of CGA S-1.3 and shall not be field adjustable. Pressure relief valves shall discharge in a manner that does not endanger personnel who are in the area. Valves, plugs or caps shall not be installed in the discharge of a pressure relief valve. Where discharge piping is used the end shall not be threaded.

L104.8 Materials and equipment. Pressurized system components shall be *listed* or *approved* for their intended use and rated for the maximum allowable design pressure in the system. Piping and fittings shall be stainless steel.

L104.9 Welded connections. Piping connections that are concealed shall be welded.

L104.10 Protection of piping. System piping shall be protected from physical damage in an *approved* manner.

L104.11 Compatibility. Fittings and connections intended to be used by the fire department shall be compatible with the fire department's equipment.

L104.12 Security. Connections to a FARS shall be safeguarded from unauthorized access in an *approved* manner.

L104.13 Fill stations. Fire fighter air replenishment fill stations shall comply with Section L104.13.1 through L104.13.3.

L104.13.1 Location. Fill stations for refilling breathing air cylinders shall be located as follows:

1. Fill stations shall be provided at the fifth floor above and below the ground level floor and every third floor level thereafter.
2. On floor levels requiring fill stations, one fill station shall be provided adjacent to a required exit stair at a location designated by the *fire code official*. In buildings required to have three or more exit stairs, additional fill stations shall be provided at a ratio of one fill station for every three stairways.

L104.13.2 Design. Fill stations for breathing air cylinders shall be designed to meet the following requirements:

1. A pressure gauge and pressure-regulating devices and controls shall be provided to allow the operator to control the fill pressure and fill rate on each cylinder fill hose.
2. Valves controlling cylinder fill hoses shall be slow-operating valves.
3. A separate flow restriction device shall be provided on each fill hose.
4. A method shall be provided to bleed each cylinder fill hose.
5. The fill station shall be designed to provide a containment area that fully encloses any cylinder being filled and flexible cylinder fill hoses, and directs the energy from a failure away from personnel. Fill stations shall be designed to prohibit filling of cylinders that are not enclosed within the containment area.

Exception: Where required or *approved* by the fire chief, fill stations providing for the direct refilling of the fire fighters' breathing air cylinders using Rapid Intervention Crew/Company Universal Air Connection (RIC/UAC) fittings shall be used in lieu of cylinder fill stations that utilize containment areas.

L104.13.3 Cylinder refill rate. Fill stations shall be capable of simultaneously filling two or more empty breathing air cylinders equivalent to those used by the fire department to the cylinders' design pressure within 2 minutes.

L104.14 External mobile air connection. An external mobile air connection shall be provided for fire department mobile air apparatus where required by Section L104.5 to supply the system with breathing air.

L104.14.1 Location. The location of the external mobile air connection shall be accessible to mobile air apparatus and *approved* by the fire chief.

L104.14.2 Protection from vehicles. A means of vehicle impact protection in accordance with Section 312 shall be provided to protect mobile air connections that are subject to vehicular impact.

L104.14.3 Clear space around connections. A working space of not less than 36 inches (914 mm) in width, 36 inches (914 mm) in depth and 78 inches (1981 mm) in height shall be provided and maintained in front of and to the sides of external mobile air connections.

L104.15 Air monitoring system. An *approved* air monitoring system shall be provided. The system shall automatically monitor air quality, moisture and pressure on a continual basis. The air monitoring system shall be equipped with not less than two content analyzers capable of detecting carbon monoxide, carbon dioxide, nitrogen, oxygen, moisture and hydrocarbons.

L104.15.1 Alarm conditions. The air monitoring system shall transmit a supervisory signal when any of the following levels are detected:

1. Carbon monoxide exceeds 5 ppm.
2. Carbon dioxide exceeds 1,000 ppm.
3. An oxygen level below 19.5 percent or above 23.5 percent.
4. A nitrogen level below 75 percent or above 81 percent.
5. Hydrocarbon (condensed) content exceeds 5 milligrams per cubic meter of air.

6. The moisture concentration exceeds 24 ppm by volume.
7. The pressure falls below 90 percent of the maintenance pressure specified in Section L104.3.

L104.15.2 Alarm supervision, monitoring and notification. The air monitoring system shall be electrically supervised and monitored by an *approved* supervising station, or where *approved*, shall initiate audible and visual supervisory signals at a constantly attended location.

L104.15.3 Air quality status display. Air quality status shall be visually displayed at the external mobile air connection required by Section L104.14.

SECTION L105 ACCEPTANCE TESTS

L105.1 Acceptance tests. Upon completion of the installation, a FARS shall be acceptance tested to verify compliance with equipment manufacturers' instructions and design documents. Oversight of the acceptance tests shall be provided by a registered design professional. Acceptance testing shall include all of the following:

1. A pneumatic test in accordance with ASME B31.3 of the complete system at a minimum test pressure of 110 percent of the system design pressure using oil free dry air, nitrogen or argon shall be conducted. Test pressure shall be maintained for not less than 24 hours. During this test, all fittings, joints and system components shall be inspected for leaks. Defects in the system or leaks detected shall be documented and repaired.
2. A cylinder-filling performance test shall be conducted to verify compliance with the required breathing air cylinder refill rate from the exterior mobile air connection and, where provided, a stored air pressure supply system.
3. The air quality monitoring system shall be tested to verify both of the following conditions:
    3.1. Visual indicators required by Section L104.15.1 function properly.
    3.2. Supervisory signals are transmitted as required by Section L104.15.2 for each sensor based on a sensor function test.
4. Connections intended for fire department use shall be confirmed as compatible with the fire department's mobile air unit, SCBA cylinders and, where provided, RIC/UAC connections.
5. Air samples shall be taken from not less than two fill stations and submitted to an *approved* gas analysis laboratory to verify compliance with NFPA 1989. The FARS shall not be placed into service until a written report verifying compliance with NFPA 1989 has been provided to the *fire code official*.

SECTION L106 INSPECTION, TESTING AND MAINTENANCE

L106.1 Periodic inspection, testing and maintenance. A FARS shall be continuously maintained in an operative condition and shall be inspected not less than annually. Not less than quarterly, an air sample shall be taken from the system and tested to verify compliance with NFPA 1989. The laboratory test results shall be maintained on site and readily available for review by the *fire code official*.

SECTION L107 REFERENCED STANDARDS

ASME B31.3—2012	Process Piping	L104.2.1, L105.1
CGA S-1.3—2008	Pressure Relief Device Standards – Part 3 Stationary Storage Containers for Compressed Gases	L104.7
NFPA 1901—09	Standard for Automotive Fire Apparatus	L104.5.1,
NFPA 1989—13	Breathing Air Quality for Fire Emergency Services Respiratory Protection	L104.2.2, L105.1, L106.1

APPENDIX M

HIGH-RISE BUILDINGS—RETROACTIVE AUTOMATIC SPRINKLER REQUIREMENT

The provisions contained in this appendix are not mandatory unless specifically referenced in the adopting ordinance.

SECTION M101
SCOPE

M101.1 Scope. An *automatic sprinkler system* shall be installed in all existing high-rise buildings in accordance with the requirements and compliance schedule of this appendix.

SECTION M102
WHERE REQUIRED

M102.1 High-rise buildings. An *automatic sprinkler system* installed in accordance with Section 903.3.1.1 of the *International Fire Code* shall be provided throughout existing high-rise buildings.

Exceptions:

1. Airport traffic control towers.
2. Open parking structures.
3. Group U occupancies.
4. Occupancies in Group F-2.

SECTION M103
COMPLIANCE

M103.1 Compliance schedule. Building owners shall file a compliance schedule with the *fire code official* not later than 365 days after the first effective date of this code. The compliance schedule shall not exceed 12 years for an *automatic sprinkler system* retrofit.

SECTION M104
REFERENCED STANDARDS

ICC	IFC—15	International Fire Code	M102.1

INDEX

C

E

F

G

H

I

J

K

L

N

O

Q

R

S

T

Y

Z